D0161789

The Diversity *and* Evolution *of* Plants

Lorentz C. Pearson

CRC Press
Boca Raton New York London Tokyo

Library of Congress Cataloging-in-Publication Data

Pearson, Lorentz C.
The diversity and evolution of plants/by Lorentz C. Pearson.
p. cm.
Includes bibliographical references and index.
ISBN 0-8493-2483-1
1. Botany. 2. Botany—Classification. 3. Plants—Evolution.
I. Title.
QK47.P33 1995
581—dc20

94–37366
CIP

Direct all inquiries to CRC Press, Inc., 2000 Corporate Blvd., N.W., Boca Raton, Florida 33431.

International Standard Book Number 0–8493–2483–1
Library of Congress Card Number 94–37366
Printed in the United States of America 1 2 3 4 5 6 7 8 9 0
Printed on acid-free paper

PREFACE

This is a book about plant diversity and about evolution as the ultimate cause of diversity. Man's activities in the 20th century have been increasingly exploitative, threatening both plant and animal diversity, including the continuing existance of man himself. Hopefully, the 21st century will be one of conservation rather than exploitation as man becomes more aware of his ecological niche and attempts to become one with nature. This is possible only to the extent that our educational and political leaders of tomorrow—our students of today—understand and appreciate biological diversity and its causes.

The 20th century has also been a century of exploration and discovery, not so much in the geographical sense, but in the discovery of ideas and development of scientific theories and concepts. Progress in developing new ideas and concepts in any discipline begins with careful observation and analysis of materials; this book is about plants, the materials that a botanist uses in his search for new knowledge. The student is introduced to each class of plants and learns what its members look like, when in the past they most likely originated, what their basic ecological requirements and contributions are, how they are or may be used in world economy, and what their potential in scientific research seems to be. These are the things that appear to be of greatest value and interest to our students: those who study botany in order to expand their appreciation of the role of science in modern culture, and those who plan on using plants in research or teaching or as raw materials to be processed for human use.

It is axiomatic that some plants are better suited for certain experiments than are other plants. Mendel would never have discovered the basic laws of heredity had he chosen oaks instead of peas for his experiments, nor would we today understand as much as we do about the basic processes of photosynthesis without the help of bacteria and green algae. In this book, all classes and divisions of plants are examined. We do not know which group will be of greatest significance in future research, but we can be grateful that over the years there were always a few people who were well acquainted with some of the so-called "obscure" groups of plants. Our knowledge of genetics today is greater than it would otherwise have been because someone had studied the life cycles in the "insignificant" Sphaeriales and could see the possibilities for research possessed by *Neurospora crassa, Chlorella pyrenoidosa, Streptomyces griseus, Vinca rosea,* and *Evernia prunastri* are other "unimportant" members of obscure families or classes which today are being used in research or medicine to the benefit of mankind.

The information in this book has been organized in such a way that similarities among closely related taxa are emphasized in order to make the task of remembering them easier for the student. I have used the simplest classification system consistent with the facts of evolution that I could find. Traditionally, all living things have been classified as either plant or animal, and I have held with this tradition. I like the convenience of a two-kingdom system partly because in our teaching and research we still have botanists and zoologists, and while they often work on very similar problems, there are often sharp differences in their research interests and methods. I have classified the Plant Kingdom into three subkingdoms: in each may be found simple, unicellular plants and highly complex, multicellular plants differentiated into distinct tissues and organs. Fossil evidence indicates that since the late Precambrian, plant evolution has been going on within each of these three subkingdoms.

Those who have helped make possible the writing and publication of this book include numerous teachers, colleagues, and students. My own hypotheses of evolution among plants stem in part from the stimulus provided by four great teachers over 20 years ago at a Botanical Society of America-sponsored summer institute: John Couch, who introduced me to significant differences in plant flagellation; Harold Strain, who shared his findings on pigmentation in plants; E. G. Pringsheim, who greatly stimulated my curiosity in basic differences among the different groups of algae; and Harlan Banks, under whom I collected and studied plant fossils. Edmund Williams, my colleague in field studies at Ricks College, and Blaine Passey, Rexburg physician and enthusiastic student of paleontology, have helped me locate and date fossils. Other colleagues at Ricks College and Uppsala University—especially Delbert Lindsay, Ray Brown, Jerald Oldham, Elisabet and Lars Eric Henriksson, Nils and Lisbet Fries, Rolf Santesson, and Angelica von Hofsten—have brought to my attention numerous kinds of plants and pointed out interesting patterns of plant diversity. Hundreds of students have added interest and purpose to my field studies: a cousin, Bill Pearson, my daughter, Suzanne Benson, Tom Gallup, Bob Walpole, Demetrios Agathangelides, Monica Williams, and many others, have made significant contributions to my knowledge of plant diversity and evolution. Howard Peterson and William Bennett at Utah State University; Seville Flowers, Walter Cottam, and Robert Vickery at the University of Utah; and Donald Lawrence, Ernst Abbe, and Clyde Christensen at the University of Minnesota are among the teachers who very early stimulated my interest in plant diversity and evolution. Suzanne Young, Laura Jensen, Tara Williams, and Laura Clark typed some of the earlier drafts, and Laura Pearson typed the final draft of the manuscript. Ellen Pearson did most of the art work; Arlen Wilcock and his associates, especially Steve Klamm and Wendy Wilcock, most of the graphics. My wife, my daughters, and my sister helped out with art work and with the assemblying of illustrations, tables, and text; my wife put up with me and provided moral and spiritual support. I acknowledge the help of all of these and numerous other teachers, colleagues, and friends with sincere gratitude.

THE AUTHOR

Lorentz C. Pearson, Ph.D., Professor of Botany and Curator of the Cryptogamic Herbarium, Ricks College, Rexburg, Idaho, is a consultant with the Environmental Science and Research Foundation, Idaho Falls, and Adjunct Associate Professor, Department of Botany and Range Science, Brigham Young University, Provo, Utah.

Dr. Pearson graduated from Utah State University in 1952 with a B.S. in agronomy; he obtained his M.S. degree from the University of Utah in 1952 in genetics and cytology, and his Ph.D. from the University of Minnesota in 1958 in plant genetics with a minor in botany (emphasis on ecology). He joined the faculty at Ricks College in 1952 where he became Head of the Department of Biology from 1959 to 1970, Chairman of the Division of Life Sciences from 1964 to 1973, and Co-director of the Ricks College Natural Science Field Expedition, 1972 to 1992.

Dr. Pearson is a member of the American Association for the Advancement of Science, Botanical Society of America, American Bryological and Lichenological Society, National Association of Biology Teachers, and the Idaho Academy of Science. He is a former member of the Minnesota Academy of Science, the American Society of Agronomy, the National Association of College Teachers of Agriculture, and the Alfalfa Improvement Conference.

He is a member of the honorary societies, Phi Kappa Phi and Sigma Xi, and is a Fellow of the American Association for the Advancement of Science. He has served as President of the Idaho Academy of Science (1974–75 and 1990–91) and on the Executive Board of the National Association of College Teachers of Agriculture. He has been the recipient of grants from the National Science Foundation to conduct research on productivity of desert area ecosystems and autecology of grasses and shrubs. He received a National Science Foundation Postdoctoral Fellowship to study lichen physiology at Uppsala, Sweden, 1963–64, and was awarded an NSF Faculty Fellowship to Uppsala in 1975–76. He is a recipient of the Thomas E. Ricks research fellowship and has also participated in Desert Biome studies of the U.S. International Biology Program. He was a delegate to all International Botanical Congresses (IBC) from 1969 to 1993. In 1984, he participated in a People to People Botanical Delegation to South Africa giving talks on biomonitoring air pollution and revegetating gold mine tailings. In 1991 he was presented the outstanding faculty award at Ricks College and he presented the honors lecture "In Quest of the Elusive Degeneria" on stimulating students to learn by encouraging them to conduct and write up original research.

Dr. Pearson is the author of Principles of Agronomy, The Mushroom Manual, numerous book chapters on lichen physiology and biomonitoring of air quality, as well as over 40 articles in journals including *Science, American Journal of Botany, Oikos, Ecology, American Midland Naturalist,* and *The Bryologist.* He has developed inexpensive, environmentally friendly biomonitoring methods for early detection of very low levels of air pollution, and his current research interests include development of an inexpensive screening test for monitoring transuranic and other polluting elements. He presented his latest findings in biomonitoring air quality at the XVth IBC in Yokohama, September 1993.

CONTENTS

PART I. INTRODUCTION

PART II. THE RED LINE

PART I

INTRODUCTION

Plant Diversity and Classification Plant Diversity in Time and Space

There is good evidence, direct and indirect, that life on our planet originated about 3.5 billion years ago. The first organisms were **prokaryotic** (having cells lacking membrane-bound organelles) and **heterotrophic** (not capable of producing their own food). Life was precarious, but became more stable when plants capable of photosynthesis evolved. Three basic polymers—a nucleic acid, a respiratory enzyme, and a photosynthetic pigment—became associated with each other in a tiny, primitive cell similar to that of a modern **cyanophyte.** During the next 2 billion years, a rich diversity of life evolved: all species were prokaryotic, resembling **bacteria** and **blue-green algae** of today, both morphologically and biochemically.

About 1.3 or 1.4 billion years ago, according to the fossil record, the first **eukaryotic** species, similar to today's green algae, appeared. By late Proterozoic time, about 1 billion years ago, plants similar to **dinoflagellates, xanthophytes,** and **water molds** had evolved, and distinct beginnings of three lines of plant evolution were apparent. There is no evidence of animals, however, until near the beginning of the Paleozoic era, but when they evolved, their impact was great. Many plants became extinct because of heavy grazing by early herbivores.

Macroscopic multicellular plants appear in the fossil record early in the Paleozoic era: first, in the ocean where lime-encrusted green algae or Siphonophycopsida appeared, followed 50 million years later by the Rhodophycopsida; later, on land when the first Tracheophyta or vascular plants evolved. Flowering plants first appear in the fossil record in Cretaceous (late Mesozoic era) deposits about 100 million years old.

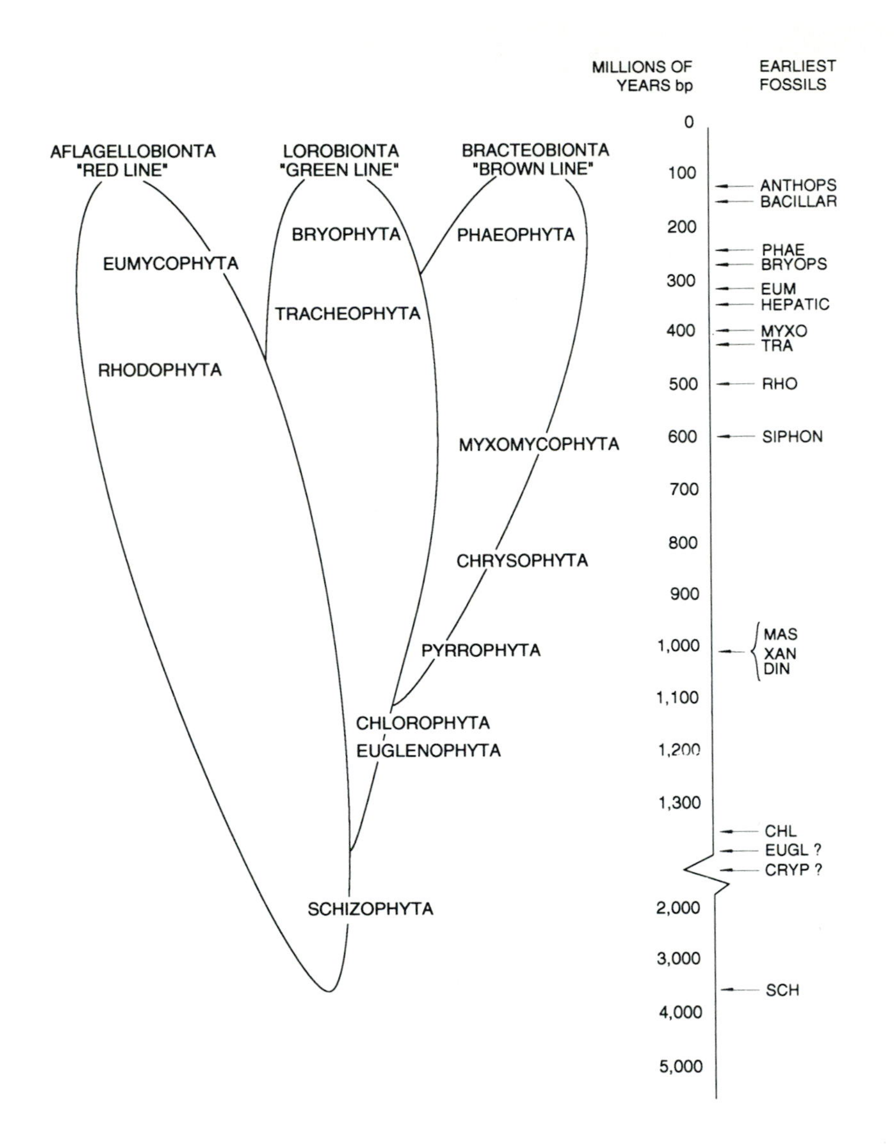

PART I. INTRODUCTION

Part I is a survey of plant diversity, of how and why plants are classified within three subkingdoms, into eleven divisions, and approximately 29 classes, and when each of the divisions and classes originated according to the geological record. This chart traces the evolution of plants from the earliest primitive bacteria and blue-green algae to the three kinds of complex and highly evolved plants of today. Abbreviations of the division (3 letters) and class (4–8 letters) names along the vertical time scale at the right edge of the chart suggest when the first representatives of each class originated.

1 PLANT DIVERSITY AND CLASSIFICATION

There are 27 different kinds of trees in my backyard, most of them **exotics** introduced from Europe or Asia, but a few natives, too, like bur oak, black walnut, blue spruce, and white fir. There are another 50 species of flowering plants in the flower and vegetable gardens: lettuce, beans, squash, carrots, and other edibles, tulips, dahlias, marigolds, snapdragons, asters, and begonias, along with 20 or 30 kinds of unwanted waifs commonly called weeds: quack grass, black nightshade, rough pigweed, matrimony vine, purslane, and mouse ear chickweed. On tree trunks and rocks there are several kinds of lichens, in damp corners there are ferns I have planted and mosses I have not planted but watch with interest, and in the springtime and summer, mushrooms (some of them delicious to eat) pop up in the lawn or on the periphery of a compost heap—all this in a city lot a quarter acre, or the tenth of a hectare, in size. As I survey my surroundings on a cool summer evening, watching bees and moths and listening to the birds, I am amazed at the diversity of life in this tiny ecosystem, and I cherish the enjoyment my own knowledge of plants, though all too meagre, gives me.

Knowledge about plant **diversity** is important to everyone, not only to an ecologist-geneticist like me or to the tropical taxonomist, daily awed by how meagre our knowledge of diversity really is, but to every soul who eats and breathes. Nevertheless, few people seem to appreciate how dependent we are on plants, even though they and their products are all about us. Human life, and indeed all life, is unthinkable without the food and oxygen which only green plants can produce.

Throughout the 20th century, man's **exploitative** activities have been a constant threat to botanical diversity. Tens of thousands of species of plants became extinct as the result of changes in the environment wrought by human abuse of the earth's natural resources. Possibly hundreds of thousands of species not yet discovered will be extinct before we find them, name them, and study them, simply because of our destruction of habitat, especially in the tropics, and our pollution of lakes, oceans, and air. On the other hand, recent polls suggest that in the minds of most Americans and Europeans, this loss of species diversity is the most critical problem we face as we approach a new century. If the public is sincere in its concern, the 21st century could be a century of **conservation** instead of one of exploitation.

Photosynthesis, the process in green plants which converts light energy into the chemical energy of food, gives off oxygen gas as a by-product. Photosynthesis is furthermore the source of all the fibers from which our clothes are made, the medicines we use, the paper on which we write, dyes, perfumes, pesticides, plastics, spices, the furniture in our homes, and much of the building material for our shelter. The world's beauty is enhanced by plants, both in natural settings and in the landscaping of cities, towns, and college campuses. Even the fuel oil and coal which heat our homes and power our industries are the products of photosynthesis in plants dead for millions of years. Unfortunately, modern industrialization and specialization have deadened our appreciation of the plants which have made modern civilization possible.

It is estimated that there are at least 300,000 different kinds of plants now living (Table 1-1), possibly two to ten times that many. Each species has its unique **ecological niche,** and the **ecosystem** is to some extent impoverished if any species (or subspecies or variety) of plant is removed. All plants contribute to the delicate balance of nature that maintains life on earth. The survival of mankind is dependent on the survival of healthy ecosystems, rich in diversity, although few people nowadays appreciate the significance of this diversity. The average person driving through a forest, a prairie, or a desert sees only the largest and most obvious plants—pine, birch, sagebrush, or cactus, for example—unaware of the scores of other species which could be found in half an hour of close observation.

Often these "hidden" plants, the ones only the observant are aware of, are the most interesting and exciting of all. Dyes and perfumes can be made from lichens; life-saving drugs are produced by many kinds of fungi; diatoms and desmids are strikingly beautiful; with the aid of a microscope, **protoplasmic streaming** can be observed in slime molds and liverworts. Frequently new species, new in the sense that no one has ever before described them, not in the sense that they have only recently evolved, are discovered as botanists delve more deeply into the secrets of nature. This has led some ecologists and taxonomists to infer that the number of species inhabiting our earth may be much greater than the estimated 300,000 suggested above, perhaps ten times more. Because of the speed with which man is destroying ecosystems today, many ecologists fear that the majority of these undiscovered species will be forever gone before any of us has even had a chance to see them (Repetto 1990).

Keeping track of 300,000 species of plants requires a good system of classification. One person cannot master all the individual characteristics of every plant species on earth, but when species are arranged into higher taxonomic categories, it becomes a simple task to learn and make use of unifying, or general, characteristics of these natural groupings, or **taxa.** Recognizing which of 29 classes a plant belongs to is not difficult for those who understand the principles involved in arranging these classes into more inclusive units: 11 divisions and 3 subkingdoms, for example (Table 1-1). With relatively little effort, a good botanist can learn the characteristics of a few dozen, or even a few hundred, of the higher taxa, such as families and orders, and thereby make better use of plant diversity in research and teaching.

The grouping of organisms into a hierarchy of categories—species, genera, families, orders, classes, divisions, and kingdoms—is called **classification.** Classification is man's basic method of coping with the otherwise chaotic multiplicity of individual objects in the world around him.

CONCEPTS OF CLASSIFICATION

Plants and animals may be classified in many ways. To be of maximum value, the classification system should reflect true genetic relationships. In other words, it should be a **natural,** as opposed to an **artificial,** system. For example, buttercups and sunflowers can be classified together on the basis of flower color, but when we have stated that both have yellow flowers, what more is there to say? On the other hand, when we classify sunflowers and asters as members of the same botanical family, we can *predict* that asters probably possess many of the characteristics known to be typical of sunflowers: exudation of sticky resins from the stem, similar internal stem structure, a taproot, flowers produced in a dense head, seeds enclosed in hard fruits called achenes. Furthermore, we would expect both plants to have similar life cycles, chromosomes, biochemical pathways and, within rather broad limits, similar ecological requirements. The more closely related two plants are to each other, the more genes they share and the more nearly alike they are in basic structure (**homologous** features), even though they may differ considerably in superficial ways. On the other hand, two plants that are not related, but are both well adapted to similar habitats, frequently resemble each other superficially even though they differ in basic structure and gene complement. These superficial similarities are called **analogous** features. To develop a natural system of plant classification, as many criteria as possible should be employed (Table 1-2) and special attention paid to homologous, as opposed to analogous, structures.

Once a good, natural system of classification has been developed for a group of plants, we can use it to predict all kinds of unknown or hidden characteristics. Thus the selective herbicides used by farmers in controlling weeds work because closely related plants function alike physiologically. Good systems of classification not only enable us to cope with nature's complexities, but are of great practical value in agriculture, medicine, forestry, geology, and other fields.

Phylogeny and Classification

The theory of evolution is the basis of modern natural systems of plant classification. According to this theory, all plants are genetically related to each other. Darwin (1859) said, "A natural arrangement must be geneological." In other words, a good system of classification should be **monophyletic,** meaning that all of the species in a given taxon are descended from a common ancestor. Evidence indicates that the Plant Kingdom is monophyletic, and that each of the three subkingdoms described in Tables 1-1 and 1-3 is monophyletic as well. This is illustrated in Figure 1-1.

Cronquist (1968) in discussing the advantages of a natural system over artificial systems of classification, wrote:

> The reasons why an evolutionary classification is preferred to one which cuts across evolutionary relationships are simple. Only if our taxa represent truly evolutionary groups will new information, from characters as yet unstudied, fall into the pattern which has been established on a relatively limited amount of information. If the system is to have predictive value . . . it must have an evolutionary foundation. Artificial classifications, using only a few arbitrarily selected characters, are easily devised, but they do not have the predictive value of a natural classification; new information will not tend to fall into line. . . . A proper taxonomic system has predictive value not only for previously studied characters on unstudied individuals, but also for characters . . . which are indeed wholly unstudied or unknown. The predictions will not always be correct, but they have a better than random chance of being so. The more closely related the taxa, the more likely they are to be similar in characters as yet unstudied.

TABLE 1-1

SYNOPSIS OF THE MAJOR PLANT TAXA

Subkingdoms	Divisions	Classes	Common Names of Classes	Number of Included Taxa			
				ORD	FAM	GEN	SPP
Aflagellobionta	Schizophyta	Schizomycopsida	Bacteria	9	48	208	1,490
		Cyanophycopsida	Blue-green algae or cyanobacteria	5	20	73	1,275
	Rhodophyta	Rhodophycopsida	Red seaweeds	7	47	184	2,500
	Eumycophyta	Ascomycopsida	Sac fungi, lichens, and most fungi imperfecti	12	72	2,900	56,000
		Zygomycopsida	Terrestrial algal fungi	2	16	77	350
		Basidiomycopsida	Club fungi, including most mushrooms	13	62	465	25,000
Bracteobionta	Pyrrophyta	Cryptophycopsida	Cryptomonads and chloromonads	5	9	19	74
		Dinophycopsida	Dinoflagellates	7	29	126	1,025
	Myxomycophyta	Acrasiomycopsida	Cellular slime molds	1	4	9	21
		Labyrinthulomycopsida	Net slime molds	3	4	5	23
		Myxomycopsida	Plasmodial slime molds	7	16	42	320
	Chrysophyta	Xanthophycopsida	Heterokonts or yellow-green algae	6	17	49	275
		Mastigomycopsida	Flagellated fungi	7	26	122	960
		Chrysophycopsida	Golden algae	7	26	72	400
		Bacillariophycopsida	Diatoms	2	20	170	5,500
	Phaeophyta	Phaeophycopsida	Brown seaweeds, kelps	11	31	170	900

Lorobionta	Euglenophyta	Euglenophycopsida	Euglenids	2	3	14	550
	Chlorophyta	Chlorophycopsida	Green algae, ditch moss	7	36	320	6,000
		Siphonophycopsida	Calcareous green algae	2	6	35	370
		Charophycopsida	Stoneworts	1	1	8	200
	Bryophyta	Hepaticopsida	Liverworts	8	24	235	8,300
		Bryopsida	Mosses	8	42	640	13,500
	Tracheophyta	Psilopsida	Whisk ferns	1	1	2	3
		Lycopsida	Club mosses and quillworts	3	3	5	850
		Sphenopsida	Horsetails or jointgrass	1	1	1	25
		Filicopsida	Ferns	5	16	185	8,600
		Pinopsida	Gymnosperms	3	9	61	700
		Gnetopsida	Gnetophytes	3	3	3	75
		Anthopsida	Angiosperms or flowering plants	53	275	11,000	220,000

Note: The last column (Number of Included Taxa) gives the number of orders (ORD), families (FAM), genera (GEN), and species (SPP) in each class.

TABLE 1–2

CRITERIA USEFUL IN THE CLASSIFICATION OF PLANTS

Criterion	Description	Comments
Genetics	Breeding behavior of plants, including fertility of hybrids, study of linkage groups	Most useful at the species and genus levels
Karyology	Study of chromosome number and morphology and other nucleus characteristics	Useful at all levels, especially genus and species
Cytology	Detailed study of cells and cellular organelles including ultrastructure of mitochondria, plastids, flagella, etc.	Especially useful in distinguishing among classes and divisions
Anatomy	Structure and arrangement of plant tissues	Has been profitably used in the Tracheophyta at all levels, especially class to family; shows promise in Phaeophyta, Rhodophyta, Eumycophyta, and Bryophyta
Gross morphology	Outward appearance of reproductive and other structures	Overemphasized in the past; most useful at species through family levels
Ontogeny	Life cycles, life history patterns, and embryonic development, especially the last	Useful in taxa having complex life cycles; at all levels, especially genus through class
Morphogenesis	Study of the forces and chemicals responsible for cell, tissue, and organ differentiation	As yet little used; should be especially useful at family and order levels
Biochemistry	Analysis of chlorophylls, carotenoids, phycobilins, fatty acids, nucleic acids, and other substances in plants	All levels, including class, division, and subkingdom
Physiology	Chemical and mechanical process involved in photosynthesis, respiration, and other metabolic processes	All levels; isozymes (enzymes having same function but produced by different genes) may be analogous only
Serology	Branch of biochemistry dealing with protein structure; antigen-antibody reactions are compared for similarities	Family and order levels; useful in identifying family limits
Ecology	Study of relationships between plants and their environment	Complex adaptations to unique ecological niches may indicate genetic relationship, but care is needed to distinguish between analogous and homologous structures
Paleobotany	Study of fossils: location, similarity to other fossils and to modern species, and observation of changes through time	All taxonomic levels; this is our best guide to the origin of orders, classes, and divisions
DNA Analysis	Sequence of nucleotide pairs ascertained by means of electrophoresis and thin layer chromatography	All levels, especially genus and species
Negative Characteristics	Absence of a structure, a chemical substance, or a physiological process	Usually of little value in studying relationships, although very widely used; may be justified when widespread and consistent

TABLE 1–3

SUMMARY OF CHARACTERISTICS OF THE THREE SUBKINGDOMS OF PLANTS

	Aflagellobionta	Bracteobionta	Lorobionta
Photosynthetic pigments	Chlorophyll *a*, sometimes *d;* phycobilins	Chlorophyll *a* with small amounts of *c* or *e;* brown carotenoids	Chlorophylls *a* and *b* in approximately 1:1 ratio
Most abundant carotenoids	β-Carotene, zeaxanthin	Diadinoxanthin, diatoxanthin, fucoxanthin	Lutein, violaxanthin, neoxanthin
Flagellation	No true flagella; proterokonts, lacking the 9 + 2 structure of true flagella; occur in the Schizomycopsida	Typically two flagella of different length: one whiplash, one tinsel with one or two rows of mastigonemes, or a modified tinsel type lacking mastigonemes	Typically two or more flagella of equal length, all of them whiplash; a modified whiplash type with a single row of mastigonemes occurs in the Euglenophyta
Food reserves	Highly branched carbohydrates with 14 to 16 glucose units, α-1,4 and 1,6 linkage	Principally oils; also slightly branched carbohydrates with 20 or 8 glucose units, β-1,3 and 1,6 linkage	Unbranched carbohydrates with 200 to 1000 glucose units plus branched carbohydrates with 20 to 25 units, α-1,4 and 1,6 linkage
Ultrastructure	Unstacked thylakoids	Thylakoids stacked in sets of 3, or sometimes in sets of 2 or 4	Thylakoids stacked in localized grana with 20 to 40 thylakoids per granum
Life cycles	Complex: often 3-phase diplohaplontic, but also haplontic; parasexual processes common	Diplontic and sporophyte dominant diplohaplontic; haplontic common in lower groups	Gametophyte dominant and sporophyte dominant diplohaplontic; haplontic common in lower groups
Tissues in the higher plants	Prosenchyma and pseudoparenchyma	Prosenchyma, pseudoparenchyma, and parenchyma; inner cortex consisting of sieve tubes; secondary tissues in perennials; growth trichothallic or from an apical meristem	Parenchyma and related types; primary and secondary xylem and phloem; primary growth from apical meristem, secondary growth from a cambium
Nucleus	Nuclear membrane intact during mitosis or else lacking; nuclei small	Chromatin granular, centrioles often present, large; or chromosomes condensed during interphase and lacking spindle fibers	Chromatin granular, centrioles seldom present; if present, associated with flagella
Other	Primary and secondary pit connections common; trichogynes present; herpokinetic motility occurs in several groups	Plastids small, disk-like, and smooth; silicification of cell walls common in some groups	Plastid morphology diverse but plastids often rough, granular, or lobed
Most common habitats	Tropical oceans, hot springs, and saprophyte niche on land; reefs	Cold oceans and damp sites on land; some groups essentially ubiquitous	Autotrophic niche on land and in freshwater ponds, lakes, and streams; reefs
Included groups	Bacteria, blue-green algae, red seaweeds, sac fungi, terrestrial algal fungi, club fungi, lichens, mushrooms, mycorrhizae	Dinoflagellates, cryptomonads, slime molds, heterokonts, flagellated fungi, diatoms, kelp, other brown seaweeds	Euglenids, green algae, stoneworts, liverworts, mosses, ferns, fern allies, gymnosperms, angiosperms

Note: The names refer to flagella characteristics typical of each subkingdom: *aflagella* = without flagella; *bracteo* = tinsel, referring to the mastigonemes on the flagella; and *loro* = thong, referring to whiplash flagella lacking mastigonemes.

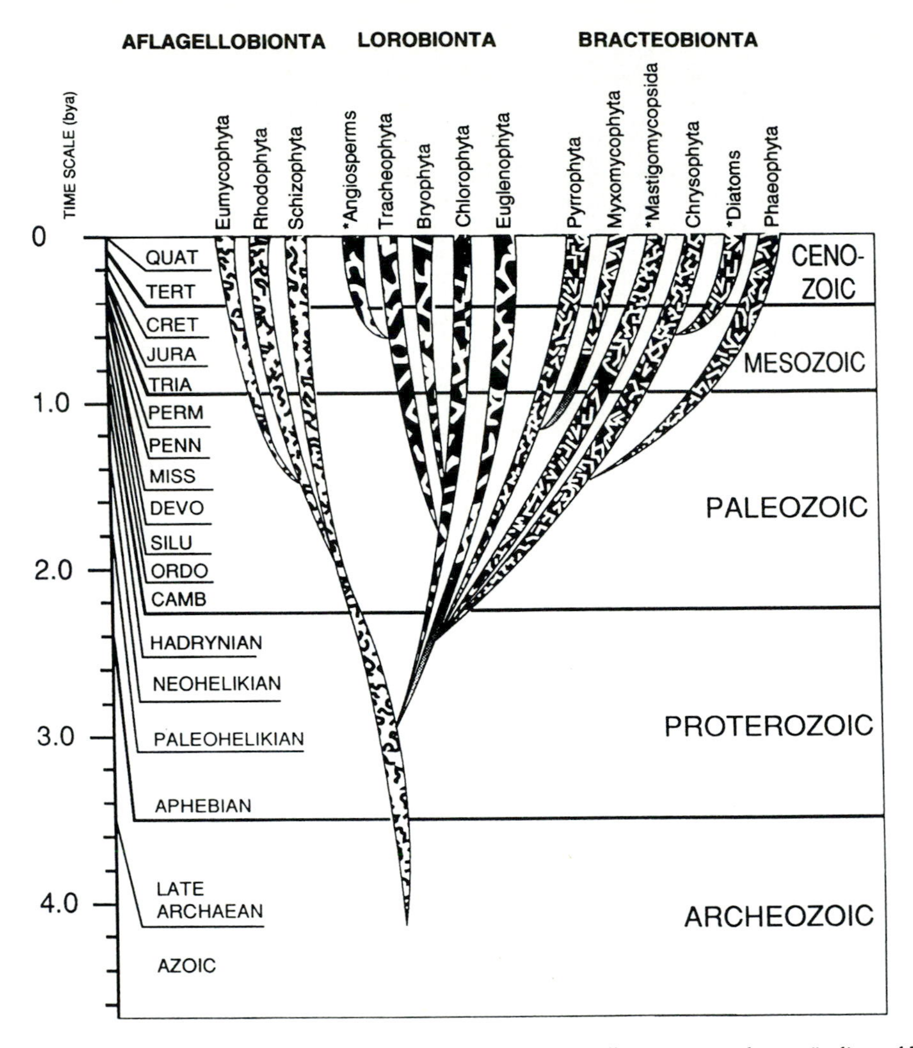

Figure 1–1 Origin of the eleven divisions of plants plus three ecologically important classes (indicated by stars; see also Table 2-1). Names of the later geological periods are abbreviated to their first four letters. (Modified from Pearson, L.C. *Am. Biol. Teach.*, 50 (8), 487, 1988. With permission.)

Origin of Life

Life probably originated in hot mineral pools, similar to those in Yellowstone Park, at a time when there was no free oxygen in the atmosphere (Oparin 1928). The environment then was a **reducing**, rather than an **oxidizing**, one. High level energy was available, making the synthesis of complex organic molecules from inorganic materials highly probable. Under such environmental conditions, sugars, amino acids, purines, pyrimidines, and other organic substances slowly accumulated until these hot pools were like soup. A similar warm broth developed in the tropical oceans. Nucleic acid molecules were eventually synthesized, becoming capable of producing duplicates of themselves using the energy released from anaerobic breakdown of the organic matter in the broth. These earliest "organisms" must have been **heterotrophs,** taking their energy from the broth which they metabolized in such a way as to carry on life's functions, simple as they were. When an oxygen-releasing form of photosynthesis came into existence, resulting in the first **autotrophs,** the environment gradually changed to its present oxidizing condition, and the origin of new forms of life from abiotic substances became less likely and eventually impossible. All species of plants and animals, therefore, are descended from life which originated 3.5 billion years ago. (See Special Interest Essay 1–1.)

Mechanics of Evolution

Four factors trigger the evolution of new species: a source of variation, natural selection, isolation, and the accumulation of chromosome aberrations or other barriers to hybridization.

The ultimate source of variation is **mutation:** a DNA molecule changes slightly, and the new enzyme produced causes a slightly different chemical reaction, resulting in a new **phenotype:** a plant which is different from others, either in morphological appearance or in physiological behavior. If the new phenotype is better adapted to the environment, an organism with this phenotype will possess an advantage in surviving and reproducing in that environment. This is **natural selection.** Often the new phenotype will have an advantage in one habitat or ecological niche, while the old phenotype will have the advantage in another habitat or niche. If the two types become isolated from each other and remain isolated for a long period of time, **chromosomal aberrations** may accumulate that make it difficult or impossible for the two populations to interbreed successfully. Under such conditions, new species result.

While the ultimate source of variation is mutation, **genetic recombination** is a more immediate source of variation, since it enables a population to recall stored mutations that have accumulated over hundreds or thousands of years. Through genetic recombination, new phenotypes are almost constantly being created. "Hidden genes," for example, frequently crop up; some of these, by themselves disadvantageous, possess the potential to be advantageous when brought together in new combinations. These hidden genes may be **recessive** genes, covered up by their dominant alleles, or **hypostatic** genes, hidden by genes other than their alleles; in either case they have the capability of remaining in the population for numerous generations even if they are disadvantageous to the individual carrying them in the present environment. Some mutations, "harmful" in their originating environment, may become "beneficial" in a new environment.

Immigration is also important for it brings together mutations stored in different populations. Usually, the introduced population is not well adapted to its new habitat and fails to survive beyond the first few generations. Nevertheless, there may be ample opportunity in that time for its genes to become well established in the existing population. Occasionally, new immigrants possess an adaptive edge; surviving well in the new habitat, they may largely or completely replace the original population.

Mountain ridges, deserts, and bodies of water are among the barriers that isolate populations from each other. However, it has often been pointed out that distance can be a very effective barrier. With isolation established, natural selection can operate independently within each habitat, and each population will be characterized by its own set of adaptations.

Figure 1–2 illustrates how **chromosome aberrations** (loss, gain, or rearrangement of segments of or total chromosomes) may be involved in the creation of new species. **Chromosome**

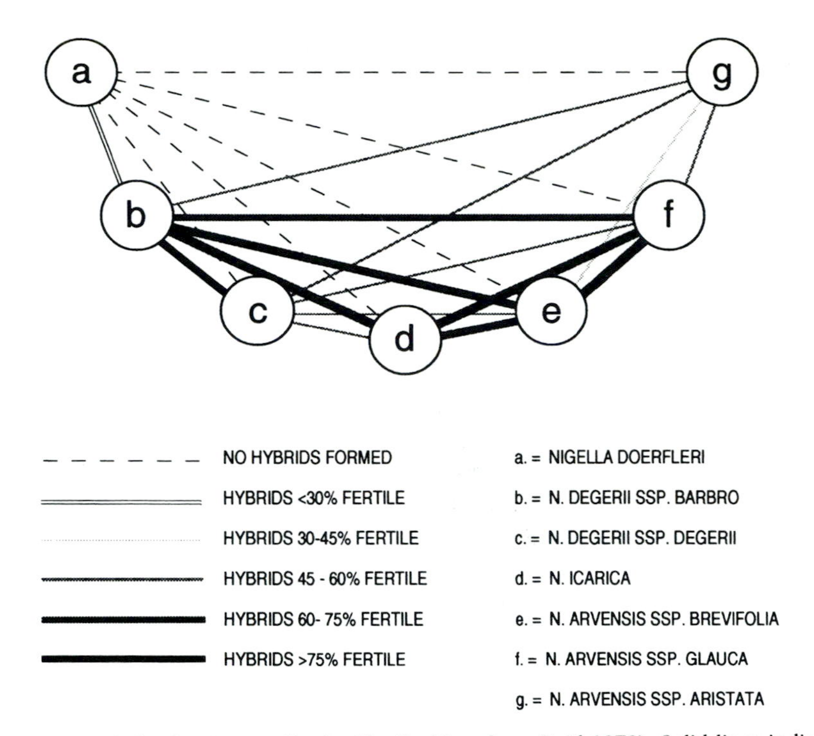

Figure 1–2 Results from hybridization studies in *Nigella* (data from Strid 1970). Solid lines indicate that hybrids were obtained, dashed lines indicate failure to obtain any hybrids; the width of the lines indicates the degree of fertility in the hybrids.

interchanges (also called translocations), **heteroploidy**, and **inversions** all result in partial or complete sterility in heterozygous individuals, although homozygous individuals are completely fertile. Frequently, a continuum of ecotypes arises in which each population differs from the adjacent one in only one interchange or other aberration, the two populations being therefore relatively cross-fertile, **e.g.,** Clausen 1949). The populations at the extremes of the continuum, however, differ from each other in so many aberrations that their hybrids are highly sterile (Figure 1-2). These extreme populations also differ from each other in their adaptations to their respective habitats. If any of these adaptations are morphological in nature, they may differ greatly in appearance as well. If the intermediate populations did not exist, the surviving populations would clearly be classified as separate species. Undoubtedly, similar continua have existed in time, as well as space, and account for evolution of modern species from populations of the past. The same forces have produced the thousands of genera and the hundreds of families that now exist or have existed.

Cladistics and Classification

In recent years, cladistic analysis, in which the taxonomic distance between two taxa is calculated by comparing the presence or absence of each of several characteristics, has been employed as an aid in developing natural systems of classification (Hennig 1966; Funk and Brooks, 1981). It is a valuable tool for some groups of organisms, especially animals. Cladistic analysis of plants is complicated by the frequent occurrence of hybridization and polyploidy, by parallel evolution, and by poorly understood family and order criteria (Funk and Brooks, 1981). Nevertheless, many botanists are finding cladistic analysis helpful, especially at the species and genus levels (Figure 1-3). Scores of research papers having cladistic analysis as their theme were presented at the Fifteenth International Botanical Congress held in Yokohama, Japan, in 1993.

Similar to cladistic analysis is **phenetic analysis.** It differs primarily by using numerical values in measuring the taxonomic distance between two taxa rather than simple presence or absence decisions. The greatest value of these two methods is in providing additional criteria for evaluating the accuracy of family trees developed by other means. If two, three, or more independent methods all agree that the species within taxonomic "Group I" are more closely related to each other than they are to the species in "Group II," we can be quite confident that we really have developed a natural classification for these plant taxa (Figure 1-3).

Botanists run into some difficulties that zoologists do not face as they attempt to use cladistic and phenetic techniques. A special problem often surfaces if **allopolyploidy** is involved. An allopolyploid originates when two species of plants cross and produce a sterile hybrid which, through a doubling of the chromosome number, becomes fertile. Chromosome doubling can be accomplished with the aid of colchicine in the botanical laboratory, but it occurs in nature as well, resulting in a plant which is fertile, which will ordinarily not hybridize with either of its parents, and which behaves like a diploid but has a polyploid number of chromosomes. When its characteristics are compared with those of its parental species, and a cladistic analysis performed, this new species may appear to be the ancestral rather than the daughter species because it has more traits in common with both parents than they will have with each other (see Figure 1-3). It is estimated that up to 95% of the ferns and between 30 and 80% of angiosperms are polyploids; polyploidy in animals, on the other hand, is extremely rare.

Parallel evolution also presents problems to botanists attempting cladistic analyses unless precautions are taken to avoid choosing, for the analysis, pairs of characteristics that are analogous rather than homologous. Two unrelated species that have evolved adaptations to identical or similar ecological niches will have many structural and physiological characteristics in common with each other. The botanist, even more so than the zoologist, must turn to cellular structure and basic biochemical structure in order to evaluate the genetic relationships of populations.

Paleobotany and Classification

Knowledge of past forms of life contributes to the study of how modern plants are related. Botanists have been remarkably more successful in discovering and interpreting plant fossils than most people realize. This is especially true of the red seaweeds, calcareous green algae, diatoms, and vascular plants (Table 2-2). Unfortunately, paleobotanical successes have never been as widely heralded as the successes of paleozoologists in their studies of the origins of the horse, dinosaurs, birds, and insects.

Paleobotanists must work closely with **geologists** in studying plants of the past and how they relate to modern species. Knowing when certain plant characteristics originated and how they have changed through the millenia helps in developing a good system of classification of the plants possessing these characteristics. As an aid to this, geologists have divided the history

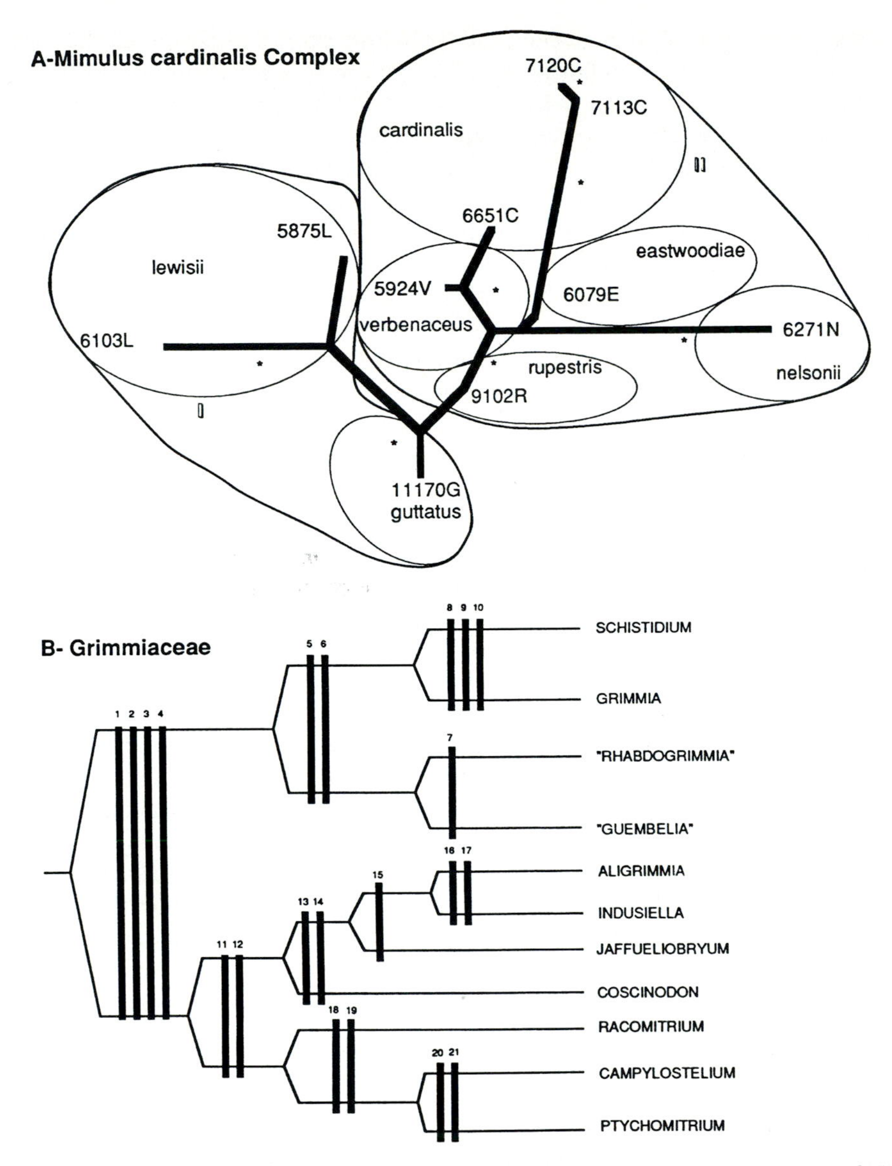

Figure 1–3 Two examples of cladistic analysis. (A) Phylogenetic distances among 10 populations of *Mimulus*, based on distribution of floral flavonoid pigments. Fine lines show the limits of the traditional species, thicker lines the limits of two major groupings. (Adapted from Vickery, R.K., Jr. and Wullstein, B.M., *Syst. Bot.*, 12, 339, 1987. With permission.) (B) Phylogenetic tree for the genera of mosses in the family Grimmiaceae. All 11 genera possess similar middrib, calyptra, peristome, and sporeling characteristics (1–4); they separate into two major groups on the basis of thickness of outer peristome layer (6 vs. 12) and thence into several minor groups (From Churchill, S.P., in *Advances in Cladistics,* Funk, A. and Brooks, D.R., Eds., New York Botanical Garden, New York, 127. With permission.)

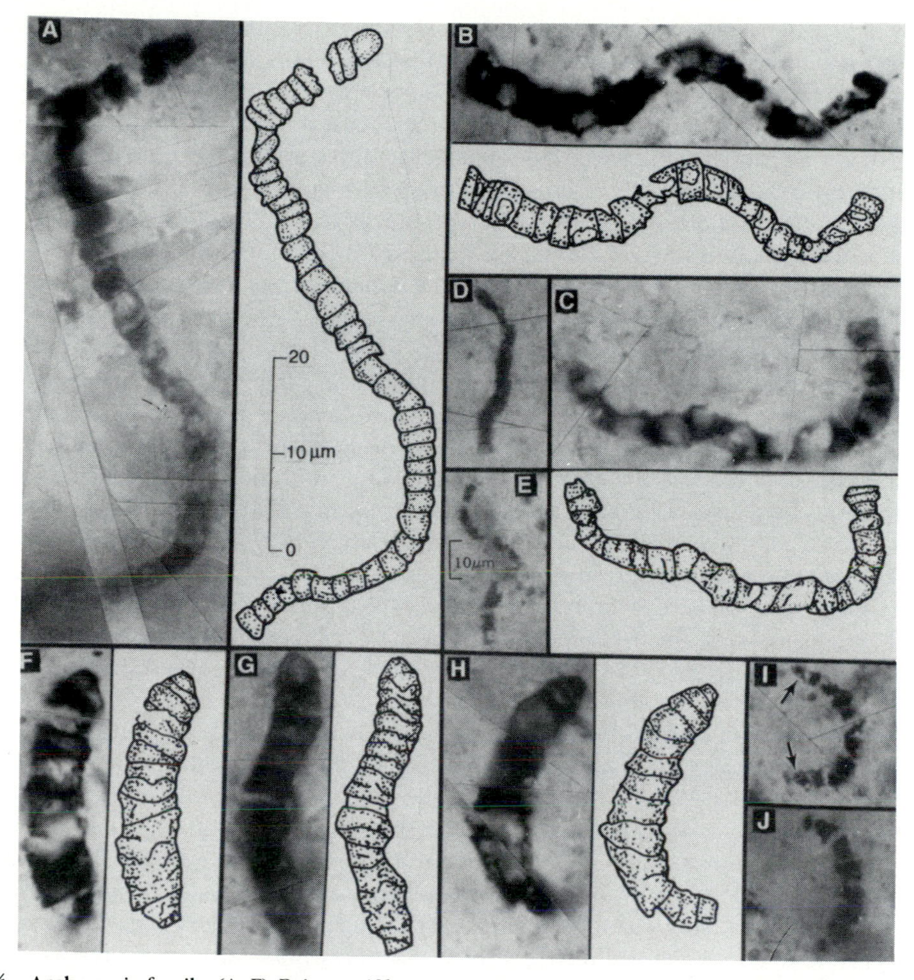

Figure 1–4 Archeozoic fossils. (A–E) *Primaevifilum amoenum*, a marine cyanophyte from the Apex chert, northwestern Western Australia, about 3.465 million years old, similar in appearance to modern *Oscillatoria* (compare with Figure 1-6d). (F–J) *Primaevifilum conicoterminatum*, from the same location and formation; arrows in I point to conical terminal cells characteristic of this species and also typical of *Oscillatoria formosa*, a common modern species of Cyanophycopsida. (From Schopf, J.W., Microfossils of the early Archean Apex Chert: New evidence of the antiquity of life, *Science*, 260, 640, 1993. Copyright 1993 by the AAAS. With permission.)

of the earth into about five **eras** and approximately 18 geological **periods** (Figure 1-1). A student must become well acquainted with names and dates of each era and period in order to study effectively the origin and evolution of living plants.

Plants have a long history of evolutionary development beginning long before the first animals came into existence 700 million years ago. *Primaevifilum*, a genus of blue-green algae found in geological formations 3.465 billion years old, is essentially identical in appearance to modern species of *Oscillatoria* (compare Figures 1-4 and 1-6d), illustrating how evolution may slow down, or even stop, in species that are well adapted to a stable environment. *Polybessurus bipartitus* is a unique marine cyanophyte that became extinct some 800 million years ago. It was very similar to some modern species of *Cyanostylon* that live in freshwater. (Figure 3-12) *Epiphyton,* a genus of marine plants that lived 500 million years ago, possessed characteristics which are intermediate between blue-green algae and some of the primitive red algae, suggesting evolution of the latter from the bluegreens. Some of the earliest angiosperm fossils, which closely resemble our modern magnolias, are also similar in appearance to various cycads which became extinct almost 100 million years ago. These and other examples illustrate how the careful study of plant fossils has often aided taxonomists in confirming or rejecting hypotheses concerning evolution of specific groups and may even reveal relationships that otherwise would have been difficult to discover.

Coevolution of plants and animals has resulted in numerous adaptations which enable organisms possessing them to fit into the environment better. The mutual dependency of many kinds of flowering plants on insects for pollination and of the insects upon the plants for food is an example of this. Frequently, two unrelated groups of plants will develop almost identical adaptations as a result of coevolution with similar animals. An example can be observed in biochemical, physiological, and morphological similarities between columbines (order Ranales) and evening primroses (order Myrtales), the former pollinated by hummingbirds, the latter by the superficially similar hawk moth. Both animals also pollinate honeysuckles (order Rubiales) and other plants with tubular corollas that produce copious amounts of fragrant, sweet organic compounds, but neither visits alfalfas, clovers, or peas, plants in which the nectaries and the stamens are covered and shielded by the corolla.

Most birds and insects cannot reach the nectar in honeysuckles, columbines, and evening primroses, but hawk moths and hummingbirds with their long, slender tongues and ability to hover over these blossoms, with pedicels too delicate to hold an insect's weight, find and feast on the nourishing food; while gathering nectar, they incidentally gather pollen which they bear on their bodies and brush on to the stigmas of the next flowers they visit. These adaptations, however, are analogous only. Without the fossil record to guide us, we might not be aware of this. We might conclude that columbines and evening primroses and honeysuckes are related. **Polyphyletic** taxa result when two or more plant taxa possessing structures which are analogous only are grouped together. An example of this is the grouping of all "fungi" into one division or kingdom, ignoring the homologous similarities of each group of fungus to different "algal" taxa.

CATEGORIES OF CLASSIFICATION

The basic unit of plant classification is the **species.** Although it is not possible to define species to the full satisfaction of everyone, it is important to understand what most biologists mean when they use the term. A widely accepted modern view is that the species, in most cases, is a real, natural unit, isolated from similar units by means of genetic barriers, and that biologists should strive to identify and describe species as they are, rather than strain to fit natural species into man-made definitions (Turesson 1925, Mayr 1992). Nevertheless, our lack of knowledge about some plants, and the nature of the plants themselves in many instances, prevents us from defining, with a high degree of certainty, the biological limits of every species. (See Figure 1-2 for an example of this type of problem.) In difficult cases, especially in taxa in which sexual reproduction is infrequent or unknown, we must accept that a species is a group of closely related individual plants, and then rely on authorities in the field to set more-or-less arbitrary limits which we can subsequently use. Nevertheless, most species have distinct biological limits that are easily recognized. To quite a degree, the same is true of genera.

Every species has its unique **habitat** and **ecological niche.** Its habitat is where it may be found, what type of ecosystem it is capable of surviving in. Its ecological niche is what its contribution to that ecosystem is; whether, for example, it is a producer organism or a consumer, or whether it blooms in the spring, the summer, or the fall. According to the "Voltera Principle" (Brower and Zar 1977), two species having exactly the same ecological niche cannot both survive in the same ecosystem.

Closely related species are included in a common genus, and closely related genera make up a family. If we think of the species as the largest unit within which the members are capable of interbreeding to produce **fertile** offspring (Turesson 1925, Mayr 1992), then the **genus** may be thought of as the largest unit within which any gene exchange is conceivably possible by means of the normal processes that occur in nature. On the other hand, we have no such natural definition for family. Many botanists define **family** as the largest unit within which similarities are more pronounced than differences.

As a rule, fewer and more distinctive species occur in families and genera that originated anciently than in those that have originated more recently. Ancient genera and families are also more likely to be **monotypic** than modern genera and families. A monotypic genus or family has only one species. Some genera contain hundreds of species, and some families have hundreds of genera.

The Higher Categories of Classification

While species, and to a large extent genera, are mostly natural units, the higher categories of classification—family, order, class, division, and kingdom—are obviously man-made conveniences. Over the years, we have found it convenient to classify every plant (and animal) in these seven essential categories. No natural law demands this, but experience indicates its advantages. However, extremely large taxa can be subdivided into intermediate units: subclasses, superorders, subfamilies, tribes, and subtribes, for example.

Probably the best way to become acquainted with the diversity of plant life on our planet is to learn first the characteristics of the divisions and classes. This book emphasizes these two categories. The classification system used divides the plant kingdom into 3 subkingdoms, 11 divisions, and 29 classes (Table 1-1). By learning first the general characteristics of the subkingdoms (Table 1-3), and then the specific deviations from these that characterize the divisions and classes, it is not difficult to learn the salient characteristics of all 29 classes of plants. Once they have fixed in mind the class characteristics, most students can examine any unfamiliar plant specimen and almost without exception recognize its class and division.

KINGDOMS OF LIVING ORGANISMS

Traditionally, the living world was classified in two kingdoms, the plant and the animal. More recently, multikingdom systems have come into wide use. These new systems generally separate the procaryotic organisms from eukaryotes, unicellular organisms from multicellular ones, and the heterotrophic fungi from the autotrophic green plants. Old systems should be discarded in favor of new ones whenever new evidence suggests an improvement, but change just for the sake of change is never justified. The popularity of the multikingdom systems seems to stem from their use of simple criteria, but the result is considerable artificiality (Bremer and Wanntorp 1981). Hence, they are of little value in predicting the nature of unknown or hidden characteristics. "The five-kingdom system is widely used today, not because it is natural but because it is convenient" (Arms et al. 1994). In this book, the "conservative" classification of all organisms as either plant or animal is followed.

Justification for retaining the traditional classification is based on ecological, physiological, and cytological evidence (Gibbs 1962, Rembert 1972; see also Roe 1961). Ecologically and physiologically, the most important dichotomy in the biotic world is between **producer organisms,** or plants, and **consumer organisms,** or animals. All living things are also **decomposers,** breaking down organic matter into carbon dioxide, water, and minerals. Although bacteria, fungi, mites, and small beetles are especially effective decomposers, all plants and animals produce inorganic compounds from the food they metabolize. The ability of most fungi to synthesize amino acids in the same way that green plants do, something animals cannot do, emphasizes their relationship to other plants. Engelmann (1961) has pointed out that in detritus-energized ecosystems, the producer organisms are fungi fed upon by a variety of soil arthropods and other consumers.

The traditional two-kingdom system encourages studying as a unit the many structures and processes common to and generally unique to plants as opposed to animals: e.g., photosynthesis, biochemistry of cell walls and of food reserves, amino acid synthesis from inorganic salts, geometry and anatomy of plant organs, and life cycles involving alternation of generations, regardless of whether these plants are protists or multicellular (Rembert 1972). Photosynthesis is essentially the same in blue-green monerans and green protists as it is in vascular plants, kelp, and Irish moss (Kok 1965, Simonis and Urbach 1973, Govindjee 1982). In many ways, we find that two-kingdom classification is more convenient than five.

All of this suggests that the plant kingdom is monophyletic and that all plants are descended from ancient Cyanophycopsida (Figure 1-1). Likewise, the animal kingdom is probably monophyletic, with all of the metazoan phyla having evolved from early Chrysophycopsida. The Protozoa, from which the other animals have evolved, originated shortly before the beginning of the Cambrian Period (Walter 1976, Klein and Cronquist 1967, Scagel et al. 1965). Table 1-4 summarizes the differences between plants and animals. Many organisms, especially those closely related to the ancient Pyrrophyta or Chrysophyta from which the earliest animals evolved, show characteristics of both kingdoms.

THE SUBKINGDOMS, DIVISIONS, AND CLASSES OF PLANTS

Based on numerous similarities and differences, including those listed in Table 1-3, the Plant Kingdom can be divided into three subkingdoms, each having three or four divisions and several classes. Each subkingdom contains morphologically simple, physiologically general, unicellular plants, and also morphologically complex, physiologically specialized, multicellular plants. Differences among the three subkingdoms are due in part to the forces of natural selection, which favors different adaptations under different environmental conditions, and in part to **genetic drift,** or differences due merely to chance (Pearson 1983, 1988).

TABLE 1–4

CHARACTERISTICS WHICH DISTINGUISH PLANTS FROM ANIMALS, USEFUL IN DISTINGUISHING ONE-CELLED, COLONIAL, AND OTHER SIMPLE PLANTS FROM ONE-CELLED SIMPLE ANIMALS

	Plants	Animals
Ecological niche	Producer organisms in either solar-energized or detritus-energized exosystems	Consumer organisms, either herbivorous, carnivorous, or both (omnivorous)
Nourishment	Mostly autotrophic; green because of the presence of chlorophyll; food intake by diffusion, rarely holozoic. Amino acids synthesized from ammonia or nitrates and sulfates, even in heterotrophs	Always heterotrophic; chlorophyll never present; food intake mostly holozoic. Amino acids obtained only by digestion of plant or animal protein, not synthesized
Organs of locomotion	Absent; motility of reproductive cells (spores and/or gametes) very common; motility of vegetative cells (also by flagella) is less common and often sluggish when it occurs. No multicellular organs of locomotion ever occur	Present in most multicellular animals, and composed of skeletal and muscular tissues. Motility of somatic as well as reproductive cells is very common and often very rapid, with quick responses to chemical stimuli
Cell covering	Cell wall composed of cellulose, nitrogenated cellulose (chitin), pectin and hemicellulose, or sometimes naked	Cells naked or enclosed in a proteinaceous periplast or pellicle inside the plasma membrane
Food storage	Starch and starch-like carbohydrates; also alcohols and soft oils; protein content usually low, about 15% in vegetative cells	Mostly fats and proteins with proteins making up about 40% of stored food; also glycogen or "animal starch"
Dictyosomes and Golgi bodies	Small dictyosomes, few and difficult to observe	Golgi bodies abundant and easily observable; typically rather large
Food translocation	Diffusion through plasma membranes from cell to cell	Translocation by means of a circulatory system and organic pump
Irritability and response to external stimuli	Generally each organ and often each cell responds individually to heat, light, and other environmental factors like humidity and soil moisture; no brain or integrated central nervous system	An integrated nervous system reacts to changes in temperature, light, and to chemicals (odors and flavors) and sends nerve impulses from a nerve center (brain) to all organs
Nucleus and mitosis	Centrosomes and centrioles, if present, not involved in mitosis but only in formation of flagella; no asters present	Centrosomes initiate mitosis; centrioles form asters which develop spindle fibers
Polyploidy	Extremely common	Extremely rare
Life cycles	Usually a distinct alternation of haploid and diploid generations; haplontic life cycle occurs in many algal and fungal groups	No alternation of generations; life cycles always diplontic
Morphogenesis and growth pattern	Growth continuous and essentially unlimited; reproductive structures not highly specialized but can arise de novo from vegetative parts	Growth limited; when maturity is reached, growth ceases; reproductive structures determined during early embryonic development
Genetic affinities	Obviously closely related to other organisms which are unequivocally plants	Obviously closely related to other organisms which are unequivocally animals

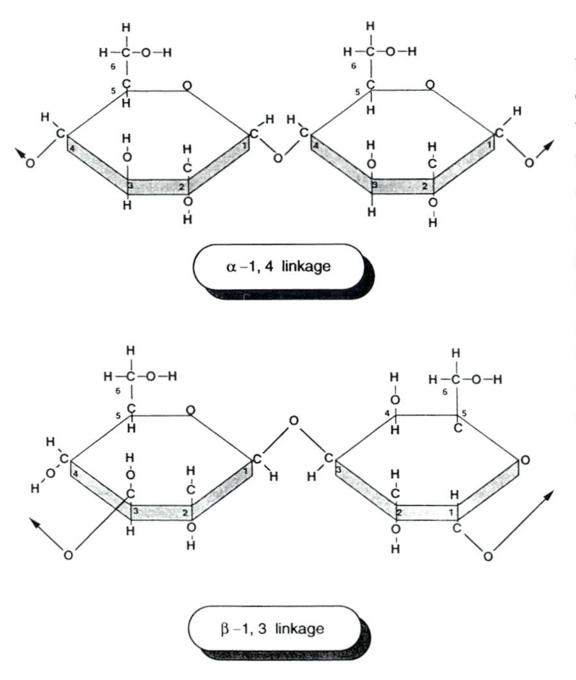

Figure 1–5 Structure of complex carbohydrates in plants. The no. 1 carbon atom of one glucose unit is connected to the no. 4 carbon of the next unit by an alpha linkage in most plants, but to the no. 3 atom by a beta linkage in the Bracteobionta.

Genetic drift occurs as large populations break up into smaller ones that become isolated from each other. Purely by chance, gene frequencies will vary widely among these isolated, small populations. As they become large populations, their gene frequencies stabilize and they differ from each other in characteristics originally determined by chance. These genes are passed on to their descendants even if they possess no adaptive advantage. For example, glucose units are linked by oxygen atoms in a beta configuration in the Bracteobionta with the no. 1 carbon atom of one unit linked to the no. 3 carbon of the next unit. In other plants, the linkage is between the no. 1 carbon and the no. 4 carbon in an alpha configuration (Figure 1-5). Since there seems to be no adaptive advantage to either type of linkage, the difference probably arose through genetic drift at the time the first bracteobionts were evolving.

Aflagellobionta

This subkingdom contains the most primitive of all organisms, yet also some highly advanced plants. Many are well adapted to warm oceans, others to relatively dry habitats. Parasites and saprophytes are common. The name refers to the total lack of true flagella in the life cycle of any species in the subkingdom. Three divisions are included: the Schizophyta, the Rhodophyta, and the Eumycophyta (Figure 1-6).

All prokaryotic organisms (organisms lacking organized nuclei and other organelles) are included in the Schizophyta, which probably originated about 3.5 billion years ago in hot pools similar to the thermal pools common today in many parts of the world. Included are the bacteria **(Schizomycopsida)** and the blue-green algae or cyanobacteria **(Cyanophycopsida).** Even young children know that bacteria are found everywhere and that they can cause diseases, such as pneumonia, tuberculosis, "lock jaw" or tetanus, "strep throat," and rheumatic fever. Fewer people are aware of how essential bacteria are in the cycling of nutrients in the soil and of how they contribute flavor to cheese, yoghurt, and other foods. The blue-green algae are also very common, occurring in ditches, ponds, and especially in hot springs where they often form firm bluish green masses on rocks. Many species of both classes still live in hot pool habitats and have changed very little, morphologically or biochemically, in the 2 or 3 billion years since they originated (Figure 1-4).

The Rhodophyta contains only one class, the **Rhodophycopsida** or red seaweeds. They are especially abundant in tropical oceans, but several species occur in cold oceans, and a few are found in freshwater streams and ponds. Biochemical and cytological characteristics suggest that they evolved directly from blue-green algae about 500 million years ago. Most of the species are much more complex structurally and biochemically than the blue-green algae, but the basic biochemical substances they produce indicate a close relationship to them.

The Eumycophyta includes the many kinds of mushrooms most people enjoy, and a variety of molds and plant disease incitants. The **lichens,** slow-growing symbiotic plants noted for their tolerance to extreme cold and drouth and their great sensitivity to air pollution, are also included in this division. The three classes are the **Ascomycopsida** or sac fungi, the **Zygomycopsida** or terrestrial algal fungi, and the **Basidiomycopsida** or club fungi. Many morphological and biochemical similarities indicate a close relationship between the Ascomycopsida, from which the other two classes evolved, and the red seaweeds, and also between the sac fungi and club fungi. The taxonomic position of the Zygomycopsida is less certain.

Morphological and paleobotanical studies suggest that the sac fungi originated on driftwood in the tropical oceans about 350 million years ago (Denison and Carroll 1965; Demoulin 1974). Like the red seaweeds, the Eumycophyta, or "higher fungi," have small nuclei with the nuclear membrane intact during mitosis. Lacking plastids, they cannot produce sugars, but they can produce amino acids just as other plants do.

Bracteobionta

As a group, these plants are better adapted to cold aquatic habitats than are any other plants. Even the terrestrial species are most commonly found in cool, wet places. The name refers to the presence in most members of this subkingdom of "tinsel type" flagella, namely, slender flagella that are blunt at the apex and usually have one or two rows of long, slender projections called **mastigonemes.** There are four divisions: Pyrrophyta, Myxomycophyta, Chrysophyta, and Phaeophyta. Only the last contains complex tissue-structured plants with distinctly differentiated organs (Figure 1-7).

The Pyrrophyta are almost entirely unicellular and contain two classes, the **Cryptophycopsida** and the **Dinophycopsida** or dinoflagellates. Most species in both classes are motile. Flagella are **pectinate;** that is, they have a single row of mastigonemes. Marine dinoflagellates are of special interest because of the impressive displays of bioluminescence put on by many species and because of the poisonous "tides" caused by some.

Slime molds can be observed on rotten logs or on the soil in damp forest habitats. They sometimes resemble slugs (shell-less snails) as they slowly creep along, but they resemble other plants in many ways and are especially similar to the Pyrrophyta, from which they probably evolved. Therefore most, though not all, biologists include the slime molds in the Plant Kingdom. There are three classes: the **Acrasiomycopsida** (or cellular slime molds), the **Labyrinthulomycopsida** (net slime molds), and the **Myxomycopsida** (plasmodial or true slime molds). Acrasiomycopsida are common in soil, especially forest soil, and are easily cultured in the laboratory on a simple medium containing lactose and peptone (Bonner 1967). Plasmodial slime molds can often be cultured by taking pieces of bark from deciduous trees or shrubs, decaying leaves, pine needles, or animal dung and placing this material on autoclaved filter paper using the moist chamber technique (Martin et al. 1969). Net slime molds are aquatic and largely marine; the most common species is a parasite on eel grass.

The Chrysophyta are often called the ubiquitous algae because they are found almost everywhere. Ecologically, they are among the most important of all plants. Three classes of aquatic autotrophs and one of fungi make up the division. Judging from present paleobotanical data,

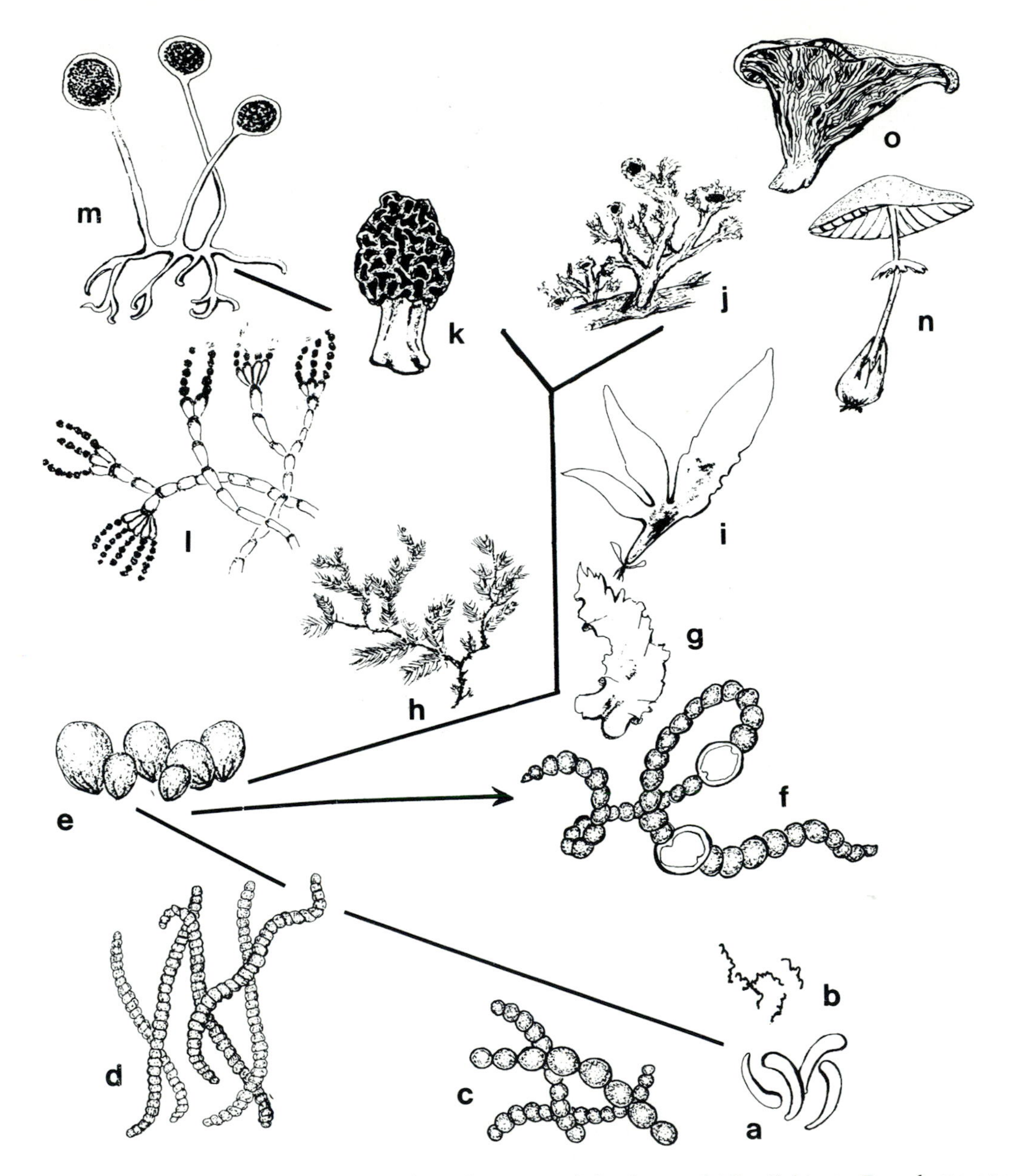

Figure 1–6 Representative species of the three divisions and six classes of Aflagellobionta. From bottom to top: Schizomycopsida (a) *Vibrio comma,* incitant of cholera; (b) *Treponema pallidum,* incitant of syphilis; (c) *Streptococcus lactis,* a lactic acid producer in silage; Cyanophycopsida (d) *Oscillatoria tenuis;* (e) colonies of *Nostoc pruniforme,* the largest 4 cm in diameter; (f) filaments of *N. pruniforme*; Rhodophycopsida (g) purple laver or nori, *Porphyra naiadum;* (h) *Pterochondria woodii,* (i) dulce or søl, *Rhodymenia palmata*; Ascomycopsida (j) wolf lichen, *Letharia columbiana;* (k) common morel, *Morchella esculenta;* (l) citrus fruit blue mold, *Penicillium notatum*); Zygomycopsida (m) common bread mold, Rhizopus nigricans; Basidiomycopsida (n) destroying angel mushroom, *Amanita virosa,* (o) golden chantarelle *Cantherellus cibarius.* (Original magnifications are a, c: ×2500; b, d, f: ×1,000; e, j, k, n, o: ×1/3; g, h, i: ×1/8; l, m: ×100.)

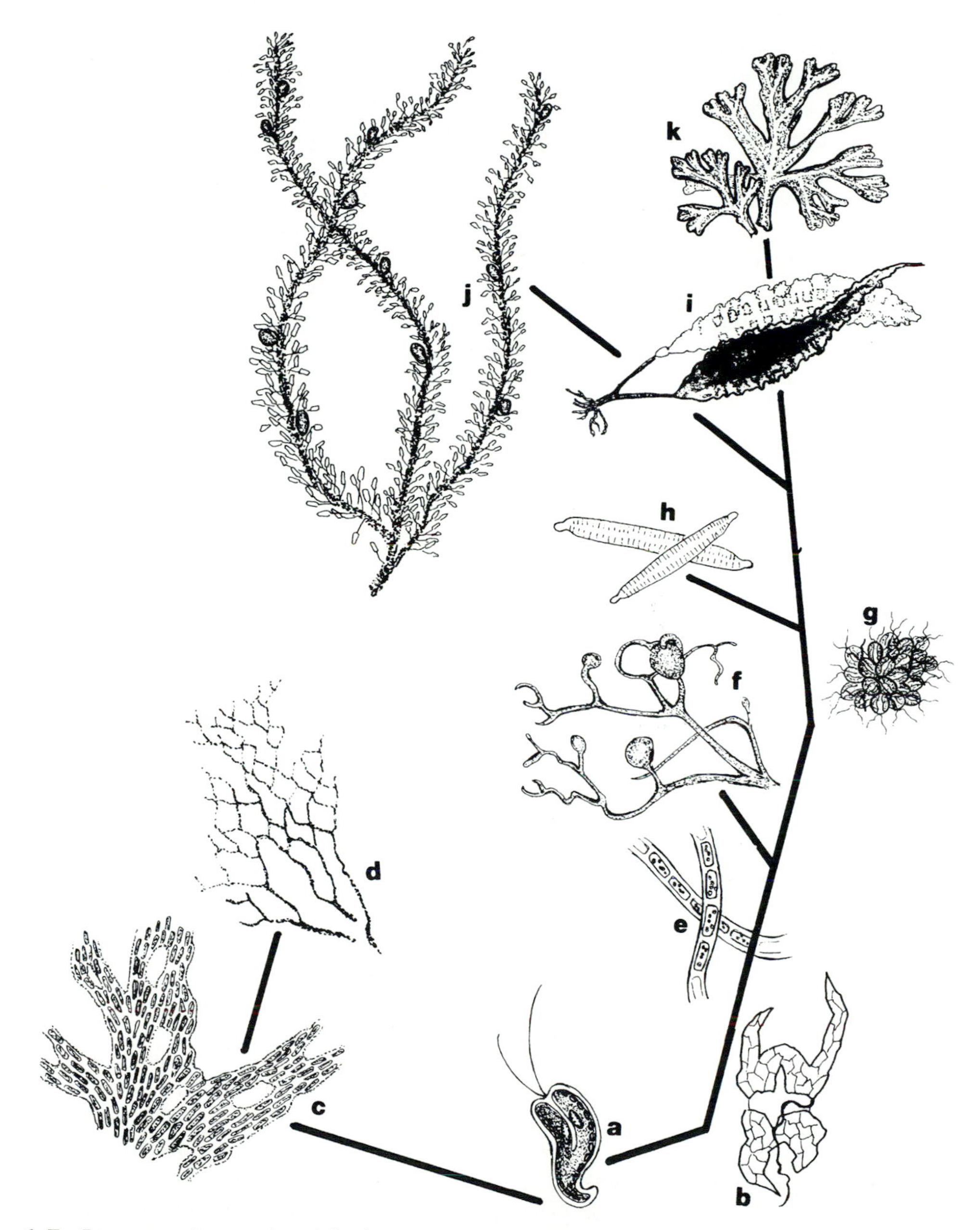

Figure 1–7 Representative species of the four divisions and nine of the ten classes of Bracteobionta. From bottom to top: Cryptophycopsida (a) *Cryptomonas marssonii;* Dinophycopsida (b) *Ceratium hirundinella;* Acrasiomycopsida (c) an Idaho grassland slime mold tentatively identified as *Dictyostelium* sp.; Myxomycopsida (d) *Physarum nutans;* Xanthophycopsida (e) *Tribonema bombycina;* Mastigomycopsida (f) *Saprolegnia ferax,* showing mode of sexual reproduction; Chrysophycopsida (g) *Synura uvella;* Bacillariophycopsida (h) *Navicula appendiculatum;* Phaeophycopsida (i) rock weed, *Fucus vesiculosus,* (j) sugar kelp, *Laminaria agardhii,* (k) jumping rope kelp or feather boa, *Egregia laevigata.* (Original magnifications are (a, e, g, h) ×250; (b, f) ×150; (c, d) ×1; (i) ×1/3; (j) ×1/6; (k) ×1/25.)

the yellow-green algae or heterokonts **(Xanthophycopsida),** and the flagellated fungi **(Mastigomycopsida)** are the most ancient of the four classes; fossils representative of both have been found in formations a billion years old (Schopf 1968, 1978). The **Chrysophycopsida,** or golden algae, are also an ancient group with fossils dating from Cambrian time. Diatoms **(Bacillariophycopsida),** on the other hand, are a recent arrival, having evolved in the ocean during the Cretaceous Period. Every student who has ever examined collections of algae from streams or ponds has probably seen a few species of diatoms, and perhaps remarked on their beauty, their symmetry, and their delicate striations.

The division Phaeophyta includes the **kelps** and other brown seaweeds. Like most Bracteobionta, the brown algae are especially common in cold oceans. Scuba divers sometimes run into trouble when they get tangled in the **stipes** (stem-like structures) of kelp. When they do, they can relate to the Old Testament poet who found himself in a similar predicament (Jonah 2:2–6). The kelps are also important commercially, being harvested for the algin they produce. Algin has many uses, including making ice cream and other desserts. The division contains only one class, the **Phaeophycopsida.** Unique among the organs characteristic of many brown seaweeds are the floaters or **pneumatocysts** which keep the leaflike fronds near the surface of the ocean where light is favorable and conditions ideal for maximum photosynthesis. Interesting physiological adaptations enable these plants, like their close relatives in the Chrysophyta, to carry on photosynthesis at very low temperatures, below the physiological zero temperature of most plants.

Lorobionta

The Lorobionta have reached their greatest development on land, but they are also found in the ocean and in freshwater streams and lakes. The name refers to the thong-like or whiplash flagella characteristic of the subkingdom. There are four divisions: Euglenophyta, Chlorophyta, Bryophyta, and Tracheophyta. The first two are mostly aquatic and are typically either unicellular or filamentous; the last two are mostly terrestrial and are always multicellular with specialized tissues and organs making up the plant body (Figure 1–8).

There is only one class **(Euglenophycopsida)** of almost exclusively unicellular organisms in the Euglenophyta. Zoologists often claim these organisms for the Animal Kingdom, but their well-developed plastids, with the same type of thylakoid arrangement as in other Lorobionta; the nature of their chlorophyll and carotenoid pigments; similarity in the carbohydrates they store; and many other characteristics they share with the Chlorophycopsida suggest that they should be included in the Plant Kingdom and in the Lorobionta. They have no fossil record that we are aware of, but they probably evolved from Cyanophycopsida about the same time and from the same parental stock as the earliest Chlorophycopsida, circa 1.4 billion years ago.

Three classes of plants are included in the Chlorophyta. The green algae **(Chlorophycopsida)** are the most primitive of these. Fossil green algae have been found in Helikian formations dated at 1.3 to 1.4 billion years ago. Some species are unicellular or colonial, others are filamentous. They are common both in the ocean and in fresh water and may form extensive blooms, especially in late summer and fall. Fishermen often get their tackle tangled in masses of these algae which they call "moss." Many Chlorophycopsida live symbiotically with sac fungi as lichens. The calcareous green algae **(Siphonophycopsida)** tend toward simple differentiation of tissues and simple organs. The cells are often very large and are usually **coenocytic,** each cell having several nuclei. They are most often found in tropical and semitropical marine

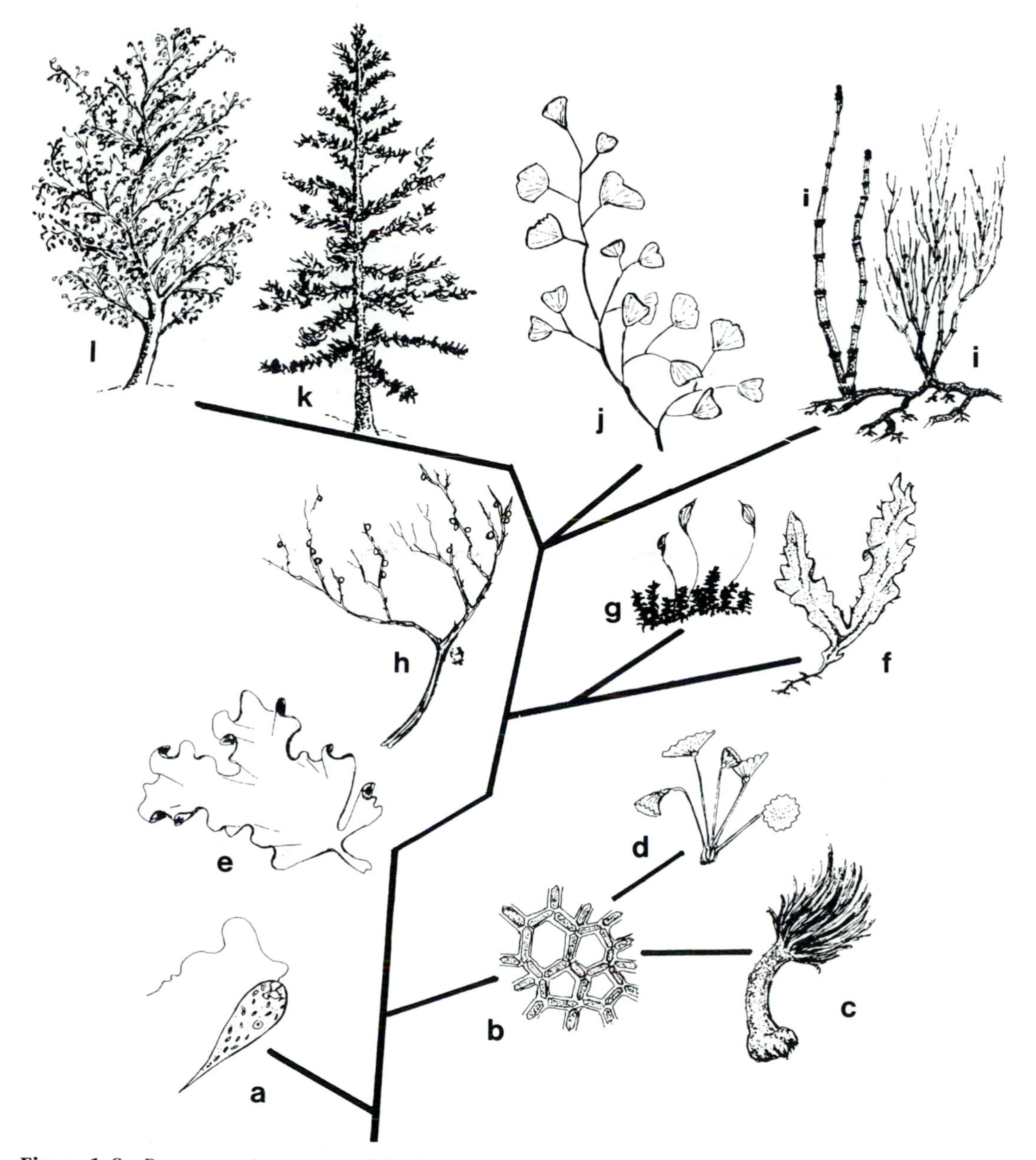

Figure 1–8 Representative species of the four divisions and ten of the thirteen classes of Lorobionta. From bottom to top: Euglenophycopsida (a) *Euglena piciformis;* Chlorophycopsida (b) water net, *Hydrodyctyon reticulatum;* (e) sea lettuce, *Ulva lactuca;* Siphonophycopsida (c) Neptune's shaving brush, *Penicillus capitatus,* (d) mermaid's wineglass, *Acetabularia crenulata;* Hepaticopsida (f) *Marchantia polymorpha;* Bryopsida (g) *Funaria hygrometrica,* growing under potted plants in greenhouse; Psilopsida (h) whisk fern, *Psilotum nudum,* growing on philodendron in greenhouse; Sphenopsida (i) horsetail, *Equisetum hyemale,* with vegetative shoots to the right and spore-bearing shoot to the left; Filicopsida (j) maidenhair fern, *Adiantum pedatum;* Pinopsida (k) tamarack or larch, *Larix occidentalis;* Anthopsida (l) paper birch, *Betula papyrifera.* (Original magnifications are (a) ×50; (b) ×150; (c, d, e, j) ×1/3; (f, g) ×1; (h, i) ×1/10; (k) ×1/300; (l) ×1/200.)

habitats where they are important reef builders. The Siphonophycopsida have left an excellent fossil record dating from early Cambrian to the present time. Stoneworts **(Charophycopsida)** have also left a good fossil record dating from early Devonian. They are characterized by an erect stem, heavily impregnated with lime and having whorls of "leaves" at each node. Each internode consists of a single cell or coenocyte. Being intolerant of excessive concentrations of phosphorus (Forsberg 1963), their absence from streams and lakes where they were formerly abundant indicates serious pollution problems.

Liverworts **(Hepaticopsida)** and mosses **(Bryopsida** or **Musci)** make up the Bryophyta, generally regarded as the most primitive of the land plants. However, paleobotanical studies do not support this hypothesis of primitiveness inasmuch as fossils of early vascular plants have been found in formations 100 million years older than those of the earliest bryophytes. Both liverworts and mosses favor damp, shady habitats, and they probably evolved as ecological subordinates in ecosystems already dominated by arborescent vascular plants. Liverworts are confined largely to tropical forests, while mosses are most common in temperate and colder climates and are not as limited to very wet sites as the liverworts are. Some species are even adapted to desert life. Many mosses thrive above timberline in the high mountains and also north of the Arctic Circle. In these cold, often dry habitats, they share dominance with lichens. The first liverworts were probably similar to modern *Anthoceros* and possessed adaptations which enabled them to carry on photosynthesis at very low light intensities, just as modern liverworts and most mosses do.

There are seven classes of vascular plants (Tracheophyta): the **Psilopsida** (whisk ferns), **Lycopsida** (club mosses and quillworts), **Sphenopsida** (horsetails), **Filicopsida** (ferns), **Pinopsida** (gymnosperms), **Gnetopsida** (*Ephedra* and its relatives), and **Anthopsida** (angiosperms or flowering plants). These, especially the Anthopsida, are the organisms most people think of when they hear the word "plant." The first vascular plants were similar in appearance to modern Psilopsida, and apparently evolved on marine mudflats during the Silurian period, 400 million years ago. Only three species of Psilopsida, all tropical, have survived. Their ancestors were probably green algae related to the common *Ulothrix.* A terrestrial alga, *Fritschiella,* has been suggested by some botanists as a proto-type of the first vascular plant ancestor, but other botanists point to ancient stoneworts (Charophycopsida) as a more likely progenitor. Lycopods, horsetails, and ferns evolved from these primitive vascular plants. Great forests of lycopods and horsetails thrived during the Carboniferous periods; their remains (coal) are used today to heat homes and run factories. The Pinopsida evolved from fernlike plants in mid-Devonian time. The other seed plants (Gnetopsida and Anthopsida) evolved much later, during the Cretaceous period. The Gnetopsida are especially well adapted to desert conditions. Angiosperms or flowering plants (Anthopsida) are the most successful, ecologically, of all land plants. At least 30%, and possibly as much as 40 to 50%, of the photosynthesis taking place in our biosphere occurs in angiosperms. They seem to have evolved from cycadlike plants approximately 100 million years ago. They are now abundant in almost all terrestrial ecosystems and in many aquatic systems as well.

BOTANICAL RESEARCH

Modern civilization is dependent upon our knowledge of all kinds of plants and how they function. We could not support a world population of 6 billion people today if we were

TABLE 1-5

TOOLS AND TECHNIQUES USED IN MODERN BOTANICAL RESEARCH

Tool/Technique	Inventor or Examples of Early Use	How Used
Botanical key	Carolus Linnaeus (or Carl Linné) 1730	Identification of unknown species or other taxa
Compound optical microscope	Zaccharias Janssen, 1590	Magnification of objects by means of refracted light
Electron microscope	German scientists, 1937	Magnification of cell components by streams of electrons which activate photographic film or TV tube
Camera	Louis Daguerre, 1839	Produce permanent record of appearance of plants or plant structures
Dendrochronology	A. E. Douglas, 1919	Estimate age of plants; analyze past patterns of climate over many centuries
Fossil peels	J. Walton, about 1930	Simplified method to study the anatomy of fossil wood
Chromatography	M. Tswett, 1903-1906	Separation of substances in mixtures; identification of substances in solution
Electrophoresis	Arne Tiselius, 1940	Separation and identification of proteins and nucleic acids
Warburg respirometer	Otto Warburg, Albert Einstein, 1910	Measure rates of respiration and photosynthesis in bits of plant tissue
Cartesian diver microrespirometer	H. Holter, 1943	Measure rates of respiration in single cells or minute bits of plant tissue
Infrared gas analyzer (IRGA)	B. Huber, 1945	Instantaneous measurement of photosynthesis and respiration under relatively "natural" conditions
Geiger counter	Hans Geiger and Ernest Rutherford, 1908	Measures rate of radiation given off by labeled plant structures; can measure steps in photosynthetic paths and other processes
Ultracentrifuge	Theodor Svedberg, 1920	Separate organelles, polymers from mixtures; rotor rides on cushion of air propelled by jets of compressed air
Plankton nets	Charles Darwin, about 1830	Collection of microscopic marine plants (plankters)
Computer	J. V. Atanasoff, 1937; Turing and Newman, 1943; Mauchly and Eckert, 1945	Make logical choices and solve mathematical problems rapidly and accurately by means of electronic connections previously programmed into it

Note: Perusal of articles in scientific journals will reveal many additional tools.

dependent only on the botanical knowledge of the early 1900s. The advances that have been made in plant physiology, genetics, medicine, forestry, horticulture, and other pure and applied branches of botany during the 20th century are the result of the work of dedicated botanists who carefully chose the species of plants that would be most useful in their work and who made use of proper tools and equipment to aid them in this research (Table 1-5; Figure 1-9). Often, as different kinds of plants or new tools were used, the results obtained surprised the

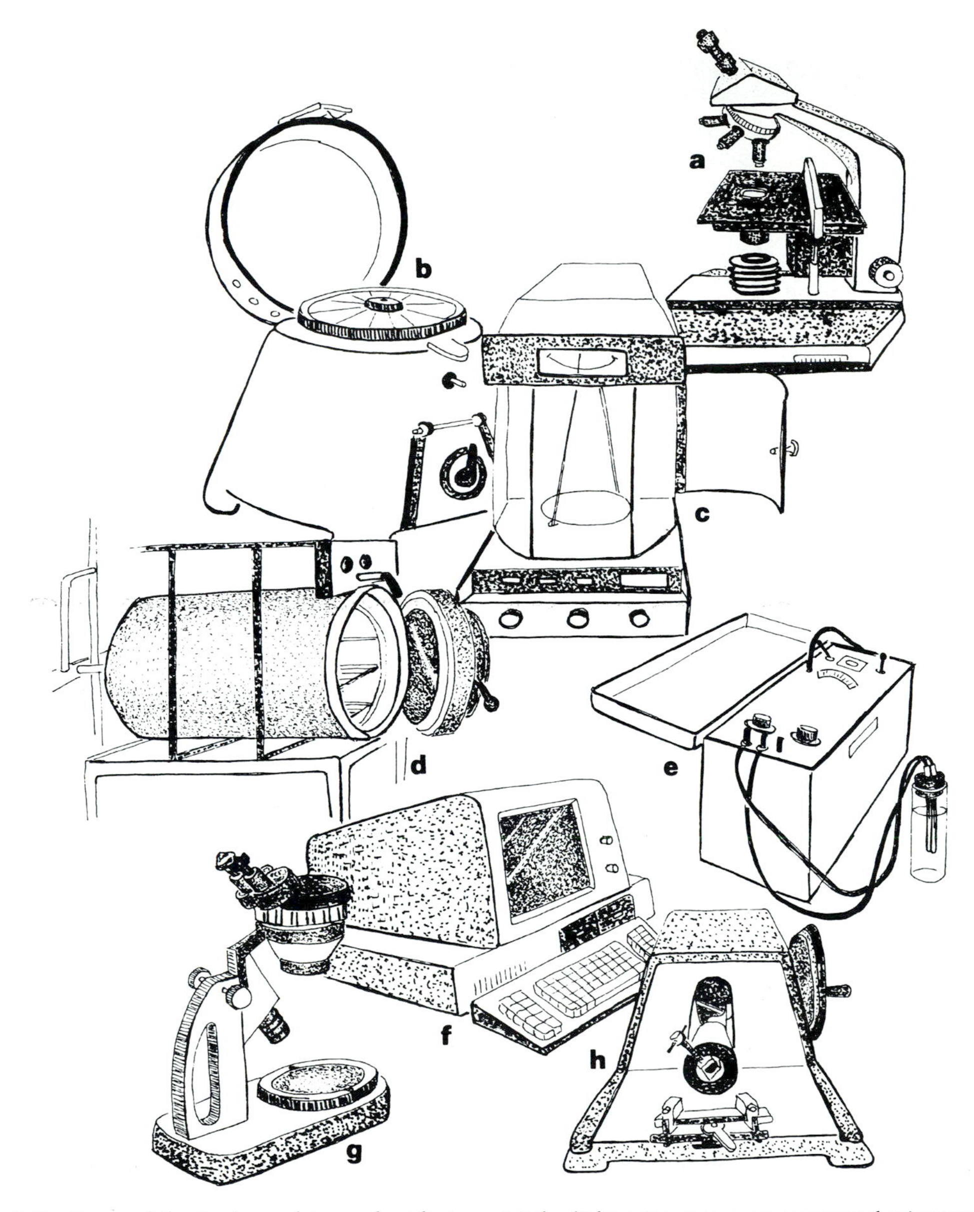

Figure 1–9 Some of the tools used in modern botany: (a) the light microscope or compound microscope; (b) a centrifuge for separating substances in solution; (c) a balance or scales for weighing materials; (d) an autoclave for sterilizing media and glassware; (e) a conductivity meter to measure the electrolytes in a solution; (f) a computer for rapid calculations; (g) a dissecting microscope; (h) a microtome for cutting thin sections of plant material.

researchers. Old hypotheses had to be discarded. That is part of the excitement of science, and for lovers of nature, perhaps no field can offer more surprises or greater excitement than botany.

The science of botany consists of basic biological principles and their application to plants: the cell theory, principles of photosynthesis and respiration, the gene theory, ecosystem theory, principles of differentiation and the effects of hormones on plant cells and tissues, the theory of evolution, and other basic concepts (Essay 1-2).

All 29 classes of plants possess great potentials for uncovering botanical mysteries. Our knowledge of photosynthesis, for example, has been gained from studies involving bacteria, blue-green algae, green algae, and angiosperms; our knowledge of genetics has come largely from studies with *Neurospora,* a sac fungus, and a few species of angiosperms; freshwater algae, especially Cyanophycopsida, Chlorophycopsida, and Bacillariophycopsida, have been of prime importance in the development of our modern models of ecosystem structure and function. Knowledge of plant anatomy has led to better understanding of the principles of food and mineral transport; most of our present knowledge has come from studies of Pinopsida and Anthopsida, but the Phaeophycopsida seem to be considerably more efficient in the transport of food than any of the Tracheophyta. In the future, kelps and other seaweeds will probably be utilized more in research in this area. Lichens, in addition to angiosperms, have provided us with significant gains in our knowledge of biochemistry. In the past, biological research has too often been centered on a very small number of species; in the future, as botanists learn more about the plants that most of us are not familiar with at the present time, more research will be conducted with algae, seaweeds, molds, mushrooms, mosses, and other plants. Some surprising, exciting, and fascinating results may be anticipated.

BOTANICAL NOMENCLATURE

Botanists make use of scientific names to keep track of the great diversity of living things. Common names are not reliable enough for scientific purposes, because the same common name is often applied to several unrelated species, and the same species may have several common names. Botanists use Latin, a universal language, in naming plants. One of the advantages of scientific names is that every species has the same name in every part of the world regardless of the language spoken.

The **scientific name** of a plant consists of a species name and the author of that name. The first time a species is referred to in a scientific paper, the writer gives the complete scientific name. Only the species name is usually given whenever the species is again mentioned. The **species name** of a plant is a **binomial,** meaning that it consists of two parts: the name of the **genus** to which it belongs, which is always a Latin or Latinized noun, and the **species epithet,** which is always a Latin or Latinized adjective. In scientific papers, the genus part of the name is sometimes indicated by its first letter if the entire name has already been used in the same paragraph and there is no danger of confusion with another genus. The genus name is always capitalized. Nowadays, the species epithet always begins with a lower-case letter, although in older literature, species epithets derived from proper nouns were usually capitalized. All plants having the same genus name as a part of the species name are genetically closely related to each other. Thus, a second major advantage of scientific names over common names is that relationships are indicated.

There are other advantages of scientific names over common names: every plant has a name. No two species have the same name. Biologists have agreed that the first scientific name given to a genus or a species is the only correct one, providing the name has not already been applied to another species and providing that the classification of that species into that genus can be logically defended. The year Carolus Linnaeus (or Carl Linné) published his *Species Plantarum,* 1753, is the year agreed upon for the establishment of which name is the oldest.

Figure 1–10 Quackgrass, *Agropyron repens* (L.) Beauv., showing the rhizomes or underground stems which function as propagules making quackgrass a troublesome weed; there are bunches of roots at each node. An herb tea brewed from its rhizomes contains a diuretic formerly used by traveling "quack doctors," hence its common name. Originally named *Triticum repens* in 1753 by Carolus Linnaeus (L.), its name was changed to *Agropyron repens* in 1812 by Ambroice Beauvois (Beauv.).

Agropyron repens (L.) Beauv., a weed commonly called either "quackgrass" or "couchgrass", is illustrated in Figure 1–10. Its scientific name consists of three parts, the genus name, the species epithet, and abbreviations of the authors' names. Named *Triticum repens* by the Swedish botanist, Carl Linné (L.), it is listed in Linné's *Species Plantarum. Triticum* is the Latin name for wheat; *repens* means to repent and refers to the creeping habit of the rhizomes. A German botanist, Josef Gaertner, created the genus *Agropyron* in 1770, based on an annual weed found growing in wheat fields in Russia, which he named *Agropyron triticeum, Agro* being Greek for wild and *puros* Greek for wheat. *A. triticeum* became the **type species** for the new genus. The French botanist, Ambroise Marie François Joseph Palisot de Beauvois (Beauv.), transferred quackgrass from *Triticum* to *Agropyron* in 1812. The scientific name indicates Linné (or Linnaeus) as the original author and Beauvois as the author of the present binomial (or species name).

Names given to the higher categories of classification also follow distinct rules. Names of families always end in -*aceae,* names of orders in *-ales,* and so on, as indicated in Table 1–6. Large classes, like the Ascomycopsida and Anthopsida, with many families and orders, may be divided into subclasses for convenience. Where the subclasses logically and naturally show some close relationships, the closely related groups have sometimes been organized into "series" as was done by Bentham and Hooker (Lawrence 1951) and others in their classifications of the Anthopsida. This has been done in Table 1–6, even though "series" is not an officially recognized category.

Some family and order names that have been widely used for many years, however, are **conserved,** meaning that they may be used even though they do not have the prescribed ending (Table 1–7). Gramineae for the grass family, also known as the Poaceae, Leguminosae for the legume or bean family (known also as both Fabaceae and Papilionaceae), and Compositae for the daisy family (Asteraceae) are well-known examples. Musci for the mosses (Bryopsida), Equisetineae for the horsetails (Sphenopsida), and Angiospermae for the flowering plants (Anthopsida) are examples of class names that have been conserved.

Choice of Scientific Names

Ideally, the name given a plant taxon should be descriptive in some way of the plants in that taxon. It may, for example, describe a **morphological characteristic.** The white-blossomed sweet clover is called *Melilotus alba: meli* means honey or sweet, *lotus* is the Latin name of a flower well known to the ancient Romans, *alba* means white. *Melilotus alba,* therefore, is a lotus-like plant which attracts honey bees and which has white blossoms. On the other hand, the name may describe an **ecological characteristic** or a **geographical area** where the taxon

TABLE 1-6

CATEGORIES OF CLASSIFICATION WITH SUFFIXES INDICATING THE CATEGORY

Category	Suffix	Examples
Kingdom	—	Animalia, Plantae
Subkingdom	-bionta	Aflagellobionta, Bracteobionta, Lorobionta
Division	-phyta	Rhodophyta, Eumycophyta, Tracheophyta
Subdivision	-phytina	Microphyllophytina, Megaphyllophytina
Class	-opsida	Rhodophycopsida, Bryopsida, Anthopsida
"Series"[a]	-ae	Monocotyledonae, Dicotyledonae
Subclass	-idae	Floridae, Liliidae, Commelinidae, Rosidae
Order	-ales	Pezizales, Bryales, Restionales, Rosales
Suborder	-ineae	Pezizineae, Pottiineae, Rosineae
Family	-aceae	Coralinaceae, Poaceae, Rosaceae, Fabaceae
Subfamily	-oideae	Panicoideae, Rosoideae, Lotoideae
Tribe	-eae	Hordeae, Cichorieae, Trifolieae, Vicieae
Subtribe	-inae	Triticinae, Hordinae, Medicaginae, Pisinae
Genus	none	*Triticum, Peziza, Rosa, Tortula, Pisum*
Subgenus	none	
Section	none	
Species	none	*Triticum aestivum* (wheat), *Pisum sativum* (pea), *Pleurotus ostreatus* (oyster mushroom), *Chondrus crispus* (Irish moss)

Note: Nonessential (convenience) categories are indented.

[a] Series is not an officially recognized category but is used here to fill the need for a category between class and subclass. See Lawrence 1951, pp. 205–206 and Stace 1980.

commonly grows. *Agropyron desertorum* is the *Agropyron* that grows well in the desert; *Prunus virginana,* the choke cherry, is a native of Virginia. Other plants may bear an **ancient Latin name** used by the Romans: *Triticum, Lotus,* and *Avena* are examples of genera with names used in ancient Rome. Frequently, the name refers to a **person,** often the man who first described the taxon. Many botanists feel this practice should be used very sparingly; nevertheless, it has been and still is used quite often: for example, *Poa sandbergii, Agropyron smithii,* and *Boletus eastwoodii.*

When a taxon is given a descriptive name, no implication is made that all individuals in that taxon fit the description. Some *Melilotus alba* plants have yellow blossoms, choke cherries grow in Idaho and Arizona as well as in Virginia, not all red seaweeds (Rhodophycopsida) are red, and some blue-green algae (Cyanophycopsida) are. Nevertheless, it is desirable, wherever possible, to have names that refer to characteristics relatively constant for the entire taxon.

Occasionally, a species is accidentally given two or more scientific names. Different botanists, each believing that he was the first to name a plant, publish descriptions of the same species in scientific journals. Each gives the species a different name. Only one of the the names is correct, but it may take years to ferret out which one. A similar problem develops when two botanists have different opinions as to which genus a certain species belongs in, each feeling that his opinion is more defensible than the other. These problems of communication and differences in interpretation make it difficult to estimate accurately the number of species and

TABLE 1-7

LIST OF CONSERVED NAMES: ORDERS AND FAMILIES OF TRACHEOPHYTA

Conserved Name	New Name (Alternative names)	Common Name
Coniferae	Pinales	
Helobiae	Najadales, Potamogetonales	
Glumiflorae	Poales, Graminales, Restionales	
Principes	Arecales, Pandanales	
Synanthae	Cyclanthales, Pandanales	
Spathiflorae	Arales, Lemnales	
Farinosae	Mayacales, Bromeliales, Commulinales	
Liliiflorae	Liliales	
Scitamineae	Zingiberales	
Verticillatae	Casuarinales	
Myrtiflorae	Myrtales	
Contortae	Gentianales	
Tubiflorae	Polemoniales	
Campanulatae	Campanulales, Asterales	
Palmae	Arecaceae	Palm Family
Gramineae	Poaceae	Grass family
Cruciferae	Brassicaceae	Mustard family
Leguminosae	Fabaceae, Papilionaceae	Pea family
Guttiferae	Clusiaceae	Garcinia family
Umbelliferae	Apiaceae, Ammiaceae	Parsley family
Labiatae	Lamiaceae	Mint family
Compositae	Asteraceae	Daisy family

genera in the plant kingdom or in any of its divisions or classes (Table 1-1). To complicate matters further, there are many examples on record of plants which have been observed by a trained botanist but once, given a name, and never seen again.

Type Species

When a species has been described and named, a representative specimen, or group of specimens, is collected, pressed, and deposited in a **herbarium.** This specimen is called the **type** and can be used by other botanists for comparison purposes to ensure correct identification of the material that they have collected.

In a herbarium, all of the species belonging to a common genus are filed together. Usually, all of the genera belonging to a common family are also filed together, making it easy for a botanist to find the specimens he is interested in. A herbarium is, in some respects, an extension of the botanical library. Here the botanist can look up the actual preserved specimens of any plant. A good herbarium contains many specimens of each species so that the range of morphological diversity can be readily observed.

The type specimen for a species will usually be the type specimen for the genus, family, and order as well. Families and orders are generally named after the type genus; for example, the Rosaceae, named after the genus *Rosa,* is included in the order Rosales along with the Leguminosae, or Fabaceae, the Crassulaceae, Saxifragaceae, and several other families. *Euglena* is the type genus for the Euglenaceae, the Euglenales, the Euglenophycopsida, and the Euglenophyta.

SUGGESTED READING

Two excellent books which present the basic principles of classification and review evolution theory as it applies to plant classification are *Plant Taxonomy and Biosystematics* by Stace (1980) and *Variation and Evolution in Plants* by Stebbins (1950). Both emphasize genetics and the study of chromosome morphology; Stace also includes a brief discussion of phenetic and cladistic analyses, while Stebbins goes into much more detail on the importance of polyploidy and chromosome aberrations in evolution.

Taxonomy of Vascular Plants by Lawrence (1951) compares different systems of classifying the vascular plants. It also contains synopses of most of the families and orders of Tracheophyta, with comments as to how they have been classified by different botanists. Excellent histories of taxonomic development are presented by Stace and by Lawrence.

Evolution and Classification of Vascular Plants by Cronquist (1968) and *Flowering Plants, Origin and Dispersal* by Takhtajan (1961) present modern systems of classifying the Anthopsida, together with the logic used in developing these systems. The systems are very similar; however, Takhtajan is a "splitter" and divides the Anthopsida into many more familes and orders than does Cronquist, who is somewhat of a "lumper"; Takhtajan believes the class Anthopsida is monophyletic, but Cronquist believes that flowering plants had a polyphyletic origin.

The modern synthetic theory of evolution as it relates to both plants and animals is discussed in considerable detail by G. Ledyard Stebbins (1966) in his small book, *Processes of Organic Evolution* and in Verne Grant's *Plant Speciation,* both of which discuss the mechanics of evolution. *The Molecular Basis of Evolution* by Anfinsen (1959) summarizes the use of biochemistry in studying the evolution of plants and animals. Evidence supporting the method of classification of plants into three subkingdoms employed in this book is published in *Speculations in Science and Technology,* Vol. 6, pp. 21–32, and in the *American Biology Teacher,* Vol. 50, pp. 487–495 (Pearson 1983, 1988). Causes and plausible solutions to problems involved in preservation of endangered species are discussed in "Deforestation in the Tropics" by Robert Repetto, published in *Scientific American,* April 1990, 262 (4): 36–42, and in *The Last Extinction* by Les Kaufman and Kenneth Mallory, MIT Press, 1987.

The Making of a Scientist by Ann Roe (1955) is an excellent little book for anyone starting out on an educational carreer to read and ponder.

STUDENT EXERCISES

1. How would you classify each of the following pairs of populations? Indicate your answers as (S) same species but separate populations, (G) same genus but separate species, or (D) different genera.
   (a) The American sycamore is obviously related to the London plane tree, yet the two are easily distinguished by leaf and other characteristics. Hybrids between the two are commonly grown in parks and are completely fertile.

(b) The sugar maple and the red maple are obviously closely related though not difficult to tell apart. They often grow together in the same ecosystems yet intermediate types are very seldom found; apparently they hybridize occasionally, but the hybrids are sterile.

(c) Siberian alfalfa has yellow blossoms, narrow leaflets, sickle pods, a shallow root system, is very winter hardy, and becomes dormant for several days after cutting or following grazing. Common alfalfa, or lucerne, has blue flowers, coiled pods, broad leaflets, a deep taproot, is seldom winter hardy, and begins growth immediately after cutting or grazing. When grown together, they hybridize readily and the hybrids are intermediate in appearance and highly fertile. The F_2 plants have blue, green, yellow, or white flowers and are highly variable in other characteristics.

(d) Mallards and pintails have very different migration, courting, and nesting habits, and seldom mate in nature. In captivity, they mate readily and the hybrids are fertile.

(e) Four populations (CR, M, F, and V) of "tidy tips" are common in Central California. Hybrids of the CR (coastal range) and V (valley) populations are almost completely sterile. Hybrids between CR and M (Munz) populations, on the other hand, are moderately fertile. So are hybrids between M and F (Fremont) populations. Hybrids between F and V are highly fertile. Examination of pachytene chromosomes in pollen mother cells of the CR × V hybrids reveals that nearly all of the chromosomes are involved in chromosome interchanges (or translocations). Chromosome interchanges involving two or three chromosomes are also present in CR × M and M × F hybrids. How will you classify the CR and V populations in respect to each other?

(f) The yellow poplar and the Lombardy poplar are tall, deciduous, broadleaf trees common in various parts of North America. They have never been successfully hybridized. The yellow poplar has large magnolia-like blossoms that are cup shaped, like tulips, with 30 to 40 stamens and numerous unicarpel pistils; the xylem consists mostly of tracheids with some poorly developed vessels and fibers; the pollen grains are similar to those of cycads and some other gymnosperms. The Lombardy poplar has small, inconspicuous flowers lacking petals and sepals and having 12 to 60 stamens and a single pistil made up of four carpels. The pollen grains have intricate patterns on the outer wall. The wood contains both well-developed vessels and fibers but no tracheids. Both have 38 chromosomes. How will you classify these two poplars relative to each other? Why?

2. From memory, write down the names and chief distinguishing characteristics of the three subkingdoms of plants.

3. From memory, list the names of the divisions of each of the three subkingdoms.

4. From memory, list the names of the classes in each of the divisions of the Plant Kingdom.

5. Several species of sugar kelp live in the coastal waters of North America. Indicate the

hierarchy of classification of one of these species by matching the list of taxa in the second column with the taxonomic categories in the first column:

Subkingdom	a. Phaeophyta
Division	b. *Laminaria*
Class	c. Bracteobionta
Order	d. Laminariaceae
Family	e. Phaeophycopsida
Genus	f. *Laminaria agardhii*
Species	g. Laminariales

6. Which of the following pairs of chararacteristics in plants would you judge as being homologous? Which are probably analogous only?
   (a) Phycocyanin-c is a very complex, water-soluble photosynthetic pigment. It differs from phycocyanin-r and phycoerythrin-c in minor details of how the carbon, hydrogen, and oxygen atoms are attached to each other. It is found in both Cyanophycopsida and Rhodophycopsida, including both freshwater and marine species. Is production of phycocyanin-c analogous or homologous in these two groups of plants?
   (b) Maples, elms, and dandelions all produce fruits that are wind disseminated. Maples and elms produce **samaras,** which are fruits with membranous wings that develop from the **pericarp** of the ovary. In elms, the wings completely surround the seed; in maples, the wings are elongated and grow out from the seed. Dandelion fruits consist of **parachutes** that have developed from the calyx of the flower. Are the samaras of maples and elms analogous or homologous to the parachutes of the dandelions? Are they analogous or homologous to each other?
   (c) Buttercups and cinquefoils have bright yellow flowers which attract pollinating insects. In buttercups, the outermost whorl of floristic structures (the calyx) is brightly colored; in cinquefoils, the inner whorl (corolla) is brightly colored. Are the colored floral structures analogous or homologous in these two plants?
   (d) In most eucaryotic plants, mitosis, or equational cell division, goes through distinct stages commonly called *prophase, metaphase, anaphase,* and *telophase,* followed by a series of stages collectively called *interphase* in which the chromosomes are completely decondensed and generally not distinguishable. The nucleus of **red algae** is enclosed within a cell membrane during metaphase and anaphase. The nuclear membrane in the common **green alga,** *Spirogyra,* disintegrates and a spindle forms during metaphase and anaphase. The chromosomes in the nucleus of **dinoflagellates** are condensed and countable during interphase; during metaphase and anaphase there is no spindle in these organisms. Is mitosis in these three kinds of plants analogous or homologous?

SPECIAL INTEREST ESSAY 1-1

Origin of Eukaryotic Organisms

The first, and only, living things for some 2 billion years after life originated, were prokaryotic organisms. The first eukaryotes appeared about 1.2 to 1.4 billion years ago. The term *-karyo* means *kernel,* or nucleus; eukaryotes have a "true" nucleus (eu = true or good), and they also have other membrane-bound organelles, but prokaryotes do not.

Prokaryotes are organized on a one-envelope system: water, minerals, DNA, and enzymes are enclosed in a single cell membrane. Eukaryotes, on the other hand, have several envelopes within a larger one, each small envelope containing enzymes or other chemicals that function together to perform a specific job. In one membrane-bound envelope, or organelle, is DNA; respiratory pigments are in a second, chlorophyll with its associated pigments in a third, and so on. Eukaryotes are like a package from grandparents, filled with gifts and goodies wrapped individually for each member of the family. Prokaryotes have unwrapped goodies all jumbled together.

Until recently, it was assumed that eukaryotes evolved from prokaryotes by invagination of the plasma membrane resulting in development of an extensive cell membrane system enclosing each organelle. Margulis (1968, 1971) and others have challenged this assumption and suggested that eukaryotes evolved through symbiosis. A heterotrophic cell containing DNA and various enzymes was invaded by an autotrophic cell similar to a modern unicellular blue-green alga. After several cell generations, the invading cell had lost most of its DNA and enzymes and had become a specialized chloroplast, providing food for the larger host cell as well as for itself. Most of the DNA from the invading cell gradually became incorporated into the nucleus of the host cell. According to this hypothesis, the host cell was again invaded, at a later time, by small prokaryotes which gradually specialized into mitochondria. A third invasion by spirochetes gave rise to flagella along some evolutionary lines.

If there is any truth to this hypothesis, one would expect to find DNA and other prokaryote substances in plastids, mitochondria, and flagella. One does indeed find DNA in the plastids of most higher plants. Furthermore, the DNA is more like prokaryote DNA than the nuclear DNA of eukaryotes. Ribosomes are also present in plastids, and they also resemble prokaryote ribosomes.

The evidence that mitochondria are also of symbiotic origin is less convincing, and the evidence favoring a symbiotic origin of flagella is very flimsy. Nevertheless, the hypothesis presents to paleobotanists an interesting challenge. Ultimately, it will stand or fall on the evidence presented by fossils. As more and more data are gathered, we should be able to decide whether the plastids and other membrane-bound organelles originated through invagination of the plasma membrane or by symbiosis. If good evidence fails to materialize, the endosymbiotic hypothesis will be discarded as a passing fad, as so many hypotheses in the past have done.

Symbiosis is a common phenomenon in the plant and animal worlds. In alfalfa and other legumes, symbiotic bacteria provide the host plant with nitrogen and the host provides the bacteria with food. A lichen is a symbiosis between a sac fungus and an alga, either a green alga or a blue-green. The symbiont is quite different in appearance from either of the bionts and is capable of producing chemical substances that neither biont can produce by itself. In coral, dinoflagellates or other unicellular plants live inside the animal tissues, where they carry on photosynthesis and provide the host with food. A similar symbiosis exists between some primitive chordates called ascidians and a green prokaryote, *Prochloron.* All in all, Margulis' endosymbiotic hypothesis of the origin of eukaryotes does not appear to be as far-fetched as it might seem to be at first glance.

SPECIAL INTEREST ESSAY 1-2

Methods and Materials in Modern Botany

The great advances that have been made in botanical science in recent years are the result of careful application of the "scientific method." Observations lead to hypotheses. The hypotheses are tested by careful experimentation involving replication and randomization and the use of representative materials and methods. When it becomes possible to **predict** accurately the occurrence of a phenomenon, we become confident that we have found an explanation for the cause of the phenomenon.

The materials of greatest importance in botanical research are the plants themselves. This book was written to better acquaint students with this most basic of materials used in botanical research. In addition to the plant material, tools have been invented over the years. Of special importance has been the light microscope. The electron microscope has added a new dimension to microscopy. Paper, column, thin layer, and gas chromatography are also very important tools used in modern botany. The computer has made it possible to fit data gathered from observations of various kinds to the hypotheses that are being tested. Some of the tools used in modern botany are listed in Table 1-5 and illustrated in Figure 1-9.

2 PLANT DIVERSITY IN TIME AND SPACE

It has been said that the only constant feature of life is change. As we travel from one part of the world to another, the patterns of vegetation change: forest gives way to savanna, the prairie changes to desert, timberline separates the forest from the meadows of the tundra, the productive waters of the estuaries and reefs open on to an unproductive open sea.

If we could travel from one geological period to another, we would also observe great changes in the patterns of plant distribution. If our time machine stopped at an early Cretaceous station, we could view forests of cycads, conifers, and ginkgoes, with flowering plants limited to a few gravel bars. If we should stop at a Mississippian station, we would find great forests of horsetails and club mosses in the swampy lowlands, but no flowering plants, and no life of any type on the drier uplands. If our station were in the Ordovician, the land would be completely bare of vegetation, and there would be so little oxygen in the air that we would not be able to breathe. Seaweeds and plankton would abound, however, in the ocean. Table 2-1 is a geological time chart summarizing many of the most significant changes in plant life during the last 3.5 billion years.

We see yet another type of change each year at the time of the vernal equinox, or about March 21 in the northern hemisphere, as the snows melt and the soil gradually warms up, a change observed in individual plants. Seeds germinate, dormant buds open, perennial herbs send up new shoots, and in a matter of a few days, plants that appeared to be dead are covered with leaves and blossoms. In some years the transition from winter to spring is very rapid; in other years, nature **vernalizes** more slowly. Cooler temperatures or a slight increase in precipitation may slow down the change from winter to spring.

The earth's surface is also constantly changing, rising or sinking or shifting horizontally. The "everlasting hills" wear away and new mountains arise; even today's continents are different from those of the Cretaceous period, 100 million years ago. As a result of these geological changes, the climate in each area also changes, having profound effects on the vegetation and complicating our study of the history of plant life through the ages as well as our study of the geographical patterns of plant diversity at any time, present or past.

The study of changes in vegetation through time and space is an important aspect of some of the basic botanical disciplines. **Paleobotany** is the study of plant and other fossils to ascertain how patterns of plant diversity have varied in the billions of years since life came into existence. **Ecology** is the study of patterns of diversity in space, best defined as the study of the structure and function of plant communities in different ecosystems. **Ontogeny** and **phenology** also treat plant changes in time, ontogeny being the study of how individual plants develop from the spore and zygote stages to mature gametophyte and sporophyte stages, respectively, and phenology the study of how ecological factors affect the growth and development of plants at all stages of their life cycles.

TABLE 2-1

GEOLOGICAL TIME TABLE

Era	Period	Some Major Epochs	Began Millions of Years Ago	Major Botanical Events: Terrestrial	Major Botanical Events: Aquatic (Marine and Freshwater)
CENOZOIC	Quaternary		1	Many species of Anthopsida cultivated by man	Aquatic ecosystems little changed since Tertiary
CENOZOIC	Tertiary	Pliocene Miocene Oligocene Eocene Paleocene	7 26 38 54 63	Anthopsida and Pinopsida dominate terrestrial ecosystems	Pennate diatoms evolve; red and brown seaweeds dominate coastal marine ecosystems, plankters the open ocean ecosystems
MESOZOIC	Cretaceous		135	Early angiosperm lines appear and expand; zenith of cycad dominance	Centric diatoms evolve and dominate the open seas; other phytoplankton abundant
MESOZOIC	Jurassic		180	Decline of ginkgos; conifers and cycads dominate on land; no desert vegetation	First "modern" red alga, *Lithothamnion* appears; golden algae dominate seas
MESOZOIC	Triassic		230	Early "fern allies" become extinct; ginkgos and cycads dominate on land	Phaeophycopsida become abundant in the cold oceans, red seaweeds common in warm seas
PALEOZOIC	Permian		280	First ginkgos evolve; earliest Bryopsida fossils appear	Solenoporaceae (Rhodophycopsida) continue to dominate tropical continent shelves
PALEOZOIC	Carboniferous: Pennsylvanian		310	Great forests of Lycopsida, Sphenopsida, and Filicopsida cover vast areas of land; Ascomycopsida and Basidiomycopsida fossils are found in middle to late Carboniferous formations	Reef-forming Rhodophycopsida and Siphonophycopsida are abundant in the oceans as are Pyrrophyta, Chrysophyta, and Cyanophyta; aquatic Sphaeriales (Ascomycopsida) are found on driftwood very early in the period
PALEOZOIC	Carboniferous: Mississippian		345		

TABLE 2-1

GEOLOGICAL TIME TABLE (continued)

Era	Period		Some Major Epochs	Began Millions of Years Ago	Major Botanical Events: Terrestrial	Major Botanical Events: Aquatic (Marine and Freshwater)
PALEOZOIC	Devonian	Upper	Famennian Frasnian	356 365	Lepidodendron and Sigillaria dominate; Hepaticopsida appear	Fossils resembling brown seaweeds occur but identification has not been verified
		Mid	Givetian Eifelian	373 380		
		Lower	Emsian Siegenian Gedinnian	384 400 405	Ancient psilophytes become extinct; Pinopsida appear	Reef-forming red seaweeds abundant
					Ancient psilophytes abundant; lycopods and horsetails appear	Freshwater habitats colonized by Charophycopsida; filamenous Chlorophycopsida abundant
	Silurian			425	First vascular plants (Psilopsida) appear	Siphonophycopsida dominate ocean; rhodophytes increase
	Ordovician			500		All life limited to marine habitats, Chlorophycopsida dominant; first Rhodophycopsida appear
	Cambrian			600		Cyanophycopsida and Chlorophycopsida continue to dominate ocean habitats; Siphonophycopsida evolve
PROTEROZOIC	Hadrynian			1,000		Cyanophytes and chlorophytes dominate all ecosystems; first Dinophycopsida, Xanthophycopsida, and Mastigomycopsida appear
	Helikian	Neohelikian		1,400		Cyanophycopsida very abundant, forming stromatolites and dominating all marine ecosystems; first eukaryotic organisms originate, evolve into a diversity of types, and increase in numbers
		Paleohelikian		1,750		
	Aphebian			2,500		Prokaryotic Schizophyta are the only forms of life on Earth; filamentous cyanophytes evolve into a diversity of forms; heterocysts present
ARCHEOZOIC	Late Archean			3,500		First organisms, various unicellular prokaryotes (Schizophyta) appear; near middle of period the existence of banded iron deposits suggest that photosynthesis was going on
	Early Archean			4,800	No evidence of cellular life, plant or animal, has yet been found in the fossil record of this period	

Figure 2–1 The Grand Canyon of the Colorado, northern Arizona, contains a record of 2½ billion years of the earth's geological and biological history. The oldest formations, found in the inner gorge, record more than 2 billion years of history in the form of stromatolites and sediments from the Proterozoic era. Above the 400-meter deep inner gorge lie 1300 meters of sedimentary rock laid down over a period of 300 million years and containing fossils from most of the periods of the Paleozoic era.

THE FOSSIL RECORD

Remains of plants embedded in rocks have left us a valuable record. Where erosion has exposed layer upon layer of sedimentary formations, as in the Grand Canyon of the Colorado in Arizona (Figure 2-1), we are able to study the kinds of vegetation once there from these fossils. We are limited, of course, in the amount of detail which we can observe: in addition to gross morphology, anatomy is readily studied in fossils, but it is more difficult to study plant physiology, life cycles, genetics, or the fine details of cytology. Nevertheless, we can infer much even in these areas. Deposition of calcium carbonates in the cells of some fossils, silica accumulation in others, reveals considerable physiological detail, and the frequent occurrence of leaves, pollen grains, stems, seedlings, and even an occasional fossilized flower allows us to predict with considerable confidence the ontogenetic development of the species. We now know how to study the biochemical characteristics, including details of DNA, in many fossils. From such studies we are learning more and more about physiology, genetics, and life cycles of plants which lived on land and in the sea millions of years ago.

Some groups of plants (Table 2-2) have left very good fossil records. But even when good fossils are abundant, our studies are often biased, as some kinds of plants fossilize more readily

TABLE 2-2

SOME GROUPS OF PLANTS HAVING HARD PARTS OR OTHER STRUCTURES THAT PRODUCE EXCELLENT FOSSILS

Taxonomic Group	Kinds of Fossils	Methods Used in Study
Vascular plants (Tracheophyta)	Petrified stems, spores, pollen grains; leaf compressions	Peels; maceration and examination of remains
Diatoms (Bacillariophycopsida)	Intact silicate-impregnated cell walls	Microscopic examination of diatomaceous earth
Calcareous green and red algae (Rhodophycopsida, Siphonophycopsida)	Limestone-impregnated remains of tissues	Examination of ancient reefs
Cyanophycopsida, Dinophycopsida, Xanthophycopsida, Chrysophycopsida, Chlorophycopsida	Stromatolites	Comparison of structures with modern stromatolites; chemical analysis of organic residues; microscopic examination of cells and filaments
Stoneworts (Charophycopsida)	Limestone-impregnated plant remains	Comparison of reproductive structures and of vegetative remains with extant species
Mosses (Bryopsida or Musci)	"Pickled" remains of plants; impressions, compressions, molds	Examination of entire structures preserved in acid bogs

than others. The interests, background, and training of the botanists who study fossils result in further bias. Nevertheless, the achievements of paleobotanists have been much greater than generally perceived.

The Geological Ages

Geologists have divided the history of the world into five **eras,** identified by unique names, about 18 or 20 named **periods,** and approximately 70 named **epochs** (Table 2-1). As methods of dating geological events have improved, experts have been able to estimate with increasing accuracy the number of years since each period or epoch began. For example, the Devonian period was once believed to have begun less than a million years ago, but its beginning is now estimated—thanks to the development of radioactive methods of dating—to be 405 million years ago (Special Interest Essay 2-1).

Because it is customary practice to indicate the age of fossils by period and epoch, using names that were originally given them by early geologists, rather than by estimates of the number of years since the formation it is in was laid down, the ages reported in scientific publications for fossil discoveries made 100 years ago or more are still useful, even though our estimates as to when those periods or epochs began have changed.

Early methods of dating geological periods and epochs depended on estimates of how rapidly sediments accumulate in the deltas of rivers and smaller streams and comparing these estimates with the thickness of sedimentary deposits. Later refinements took into account alternating rise and fall of land masses and the effects of erosion during periods when the land was above sea level. The development of radioactive dating, involving the decay of uranium to

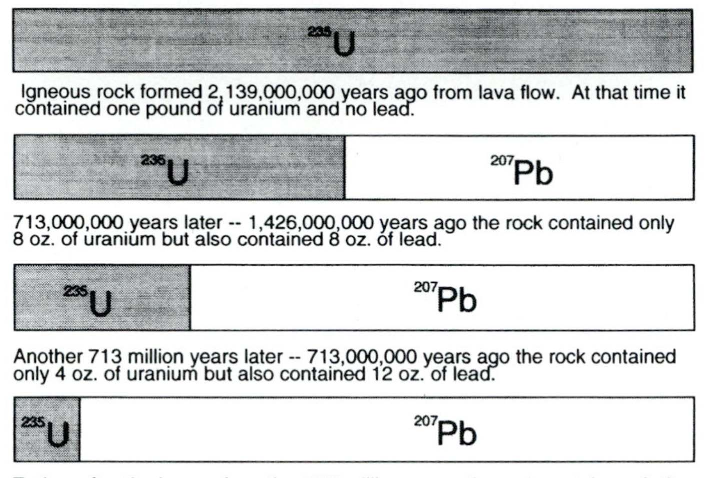

Figure 2–2 Radioactive decay of uranium 235 to lead 207. The half life of uranium 235 is 713 million years; consequently, at the end of 713 million years, half of the uranium had been changed into lead. By weighing the lead and the uranium in a rock sample, the age of the rock can be calculated. Because uranium 235 gives off radiation, its weight can be calculated indirectly by the amount of radiation emitted.

lead (Figure 2-2), made it possible to further refine the earlier estimates. Potassium-argon dating has especially improved dating accuracy and has made it possible to arrive at an independent date of the origin of the earth by estimating the amount of argon in the atmosphere. All of the argon which today makes up 1% of total atmospheric gases must have been produced by radioactive potassium since any argon in the pristine atmosphere would have been driven off by the extremely high temperatures of the primitive earth. All of these methods correlate well with each other, giving us confidence in the accuracy of our estimates of the age of the earth and of the different periods and eras as well. A recent method developed by paleozoologists involves counting the number of daily growth rings within each annual growth ring of fossil corals. We have discovered from these interesting animals that the Devonian year, during the Eifelian epoch, consisted of 400 days, and the Devonian day, therefore, was only 21.9 hours long. Comparing this with estimates by geophysicists of the rate at which tidal friction slows down the rotation of the earth on its axis each year tells us that the Eifelian epoch began about 370 million years ago (mya), which agrees with estimates derived from radiometric dating.

The classification of geological time presented in Table 2-1 is adapted from a system used by Stockwell (1982) and other Canadian geologists. Stockwell's Precambrian *eons* are called *eras* in Table 2-1; the names of the eras in the table are therefore the same as in classical biology books (e.g., Fuller and Tippo's general botany and Woodbury's general ecology textbooks from the 1950s). The names Stockwell gives to the oldest geological periods are different from those used by Fuller and Tippo and also different from the names used in most Asian and European geology and paleontology publications; for example, the Hadrynian period in Table 2-1 is roughly equivalent to the Vendian plus the Riphean periods of some of the European publications. Most American geologists lump all the periods of the first two eras, the Archeozoic and Proterozoic, into a hodgepodge called the "Precambrian." For the botanist, this is unfortunate, because many of the most important and especially interesting events of plant evolution occurred during the Proterozoic era.

Five points in time are especially significant in the evolution of plants: (1) the middle Archeozoic Era, some 3.5 billion years ago (bya), when the first prokaryotic life came into existence; (2) the middle Proterozoic era, at the end of the Paleohelikian period, 1.4 bya, when the first eukaryotic cells evolved; (3) the Hadrynian period, beginning about 1000 mya, when the oldest fossils yet discovered of dinoflagellates, heterokonts, and flagellated fungi were laid down; (4) the Devonian period, beginning about 400 mya, during which the genealogical roots of most of the vascular plants were established and the stoneworts, sac fungi, liverworts, and possibly the brown seaweeds originated; and (5) the Cretaceous period, 100 mya, when the diatoms became widespread and the flowering plants and gnetophytes originated. The first three of these five periods are of special significance in plant evolution, although they are generally ignored by geologists, as well as paleozoologists, inasmuch as animals did not come into existence until approximately the end of the Hadrynian Period, some 600 to 700 mya, and geologists traditionally work primarily with animal fossils in dating geological periods and epochs.

In recent years, paleontologists have reconstructed excellent models of plant and animal life for some periods, and fairly good records for some groups of plants for all periods (Table 2-1). For a few groups of plants, the record is complete enough that we can trace the evolution of individual species from one geological period to another. However, it is unlikely that paleobotanists will ever discover a record so complete that the entire evolutionary history of the plant world could be traced in geological formations. The Grand Canyon of Arizona displays an excellent history of much of the time since the first plants originated (Figure 2-1). But even there, reconstructing the history of plant life from the record in the rocks is much like studying U.S. history with a book from which chunks of pages have been randomly torn out.

Basic Geological Principles

Studying the development of modern ecosystems and the plants and animals in them would be simpler if we could assume that past ecosystems in any given area had the same climate, soil, and topography which now characterize that area. Frequently, however, the fossil record reveals ecosystems so different from those of today in the same sites that we know there has been an appreciable change in climate and/or other factors of the environment. For example, Bryant and Benson (1973) observed that during the Miocene epoch, the plains near what is now the Salmon River in east central Idaho were covered with a dense forest dominated by dawn redwood and containing considerable elm, birch, beech, and apple. Today, that forest is covered with 15 million years' accumulation of volcanic ash. Atop the layers of ash lies semidesert rangeland, vegetated with sagebrush and various bunchgrasses (Figure 2-3). The rise of the Cascade Mountains about a million years ago cut off the moist winds from the Pacific and turned much of Idaho, along with the eastern parts of Oregon and Washington, into a rain shadow desert.

This example illustrates the application of the first two of five basic principles geologists have discovered over the years and find useful in their study of ancient life. The **first** of these five principles is that older deposits almost always lie beneath younger ones. There are rare exceptions, however, in which folding and distortion have resulted in older formations above the younger, but these can usually be spotted through careful study of the total series of strata. The **second** principle relates to the plasticity of the earth's surface: mountains rise where there

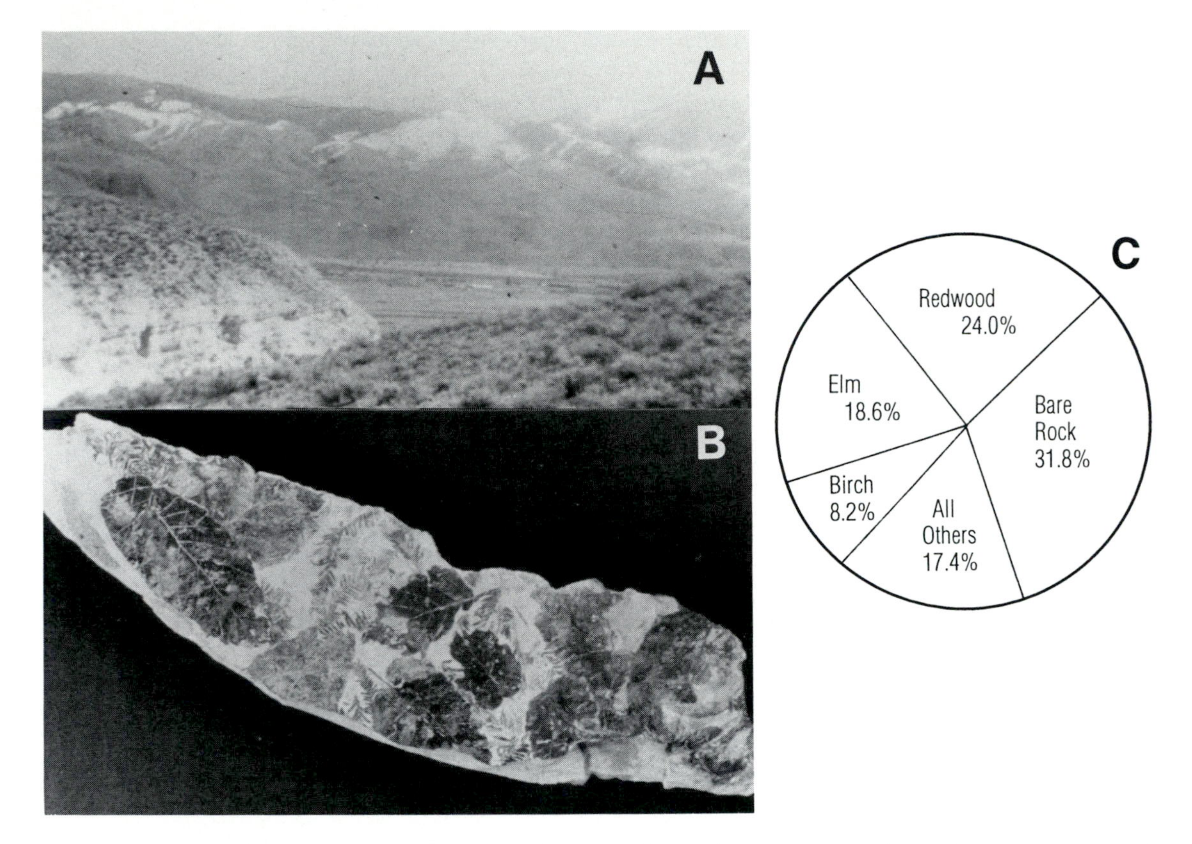

Figure 2–3 Carmen formation, east central Idaho. (A) Semiarid rangeland above the Salmon River showing one of the gullies in which Miocene fossils are exposed. (B) Fossils from the Carmen formation: m = *Malus idahoensis,* s = *Metasequoia chinensis,* u = *Ulmus* sp. (C) Relative abundance of the most common fossil leaves based on average percent of the surface of 41 rocks of average size covered by each species.

were no mountains before, high places sink and are covered by sea, and it is only when an area is physically lower than its surroundings that sedimentary rocks can be formed. During periods when an area is higher than its surroundings, no new sediments are deposited, and former sediments may be eroded away by wind or water, leaving gaps or unconformities in the strata. A **third** principle is that pebbles found in conglomerates are older than the matrix of the conglomerate; similarly, intrusions of igneous rock (which are often easily dated by radiometric methods) into a layer of sedimentary or metamorphosed rock are younger than the material into which they are intruded. A **fourth** principle involves the use of plants or animals to correlate the strata from different regions; species which existed for a very short time but were widespread during that time are the most valuable **reference species** for this purpose. Some biologists feel that previously too much emphasis was placed on animals and too little on plants as reference species. Regardless of whether plants or animals are used, dating is most reliable when it is based on several reference species rather than on only one or two and when it is confirmed by other reference species in strata immediately below or above. A **fifth**

TABLE 2–3

EVIDENCES OF AGE OF EARTH AND EARLY LIFE

Kind of Evidence	Time Involved (Half Lives)	Comments
A. Age of rock formations ascertained from decay of radioactive minerals		
1. Uranium 238 to lead 206	4.5 by	Horizontal strata are older than any igneous intrusions in them
2. Uranium 237 to lead 207	710 my	
3. Thorium 232 to lead 208	14 by	
4. Rubidium 87 to strontium 87	47 by	Igneous rocks can be dated if original and decay atoms are both present
5. Potassium 40 to argon 40	1.3 by	
6. Carbon 14 to nitrogen 14	5,600 years	
B. Age of atmosphere from decay of radioactive elements		All Ar in primordial atmosphere was lost to outer space before earth cooled
1. Potassium 40 to argon 40	1.3 by	
C. Other kinds of evidence		
1. Ratio of carbon-12 to carbon-13		
2. Daily and annual growth rings in corals		Earth's rotation is slowing down; there are fewer days per year now than anciently
3. Rates of erosion losses and sediment accumulation		Give very rough estimates of the age of formations
4. Red beds and banded iron formations		Are formed only in the presence of free oxygen (O_2)

principle involves the use of radioactive elements such as uranium 238, uranium 235, and potassium 40: because no amount of pressure or change in temperature affects the rate of decay of these isotopes, they provide an accurate clock for dating rock formations in which they occur, providing both the radioactive isotope and its product of decay can be weighed or otherwise quantitatively measured (Figure 2–2).

Origin of the Classes and Divisions of Plants

Judging from the best fossil evidence available, coupled with indirect evidence of various types (Table 2–3), the first living organisms originated about 3.5 bya during the Archeozoic era (Gurin 1980). The earth was more than a billion years old at that time. These organisms were undoubtedly preceded by noncellular, "virus-like" particles. However, we have no fossils or other direct evidence of the nature of these particles.

The first cellular organisms were small prokaryotes similar to modern Schizophyta (Figure 2–4). They probably originated in hot pools where periodic evaporation would increase the probability of polymerization and hence origin of life (Woese 1981). For over 2 billion years, this type of organism, lacking cellular organelles and apparently incapable of true sexual reproduction, was the only type in existence. During that time, a rich diversity of prokaryotic life evolved (Figure 2–4). Some of the later prokaryotes probably contained chlorophyll *b,* in addition to chlorophyll *a* and various phycobilins, and presumably gave rise to the Euglenophyta and Chlorophyta. The obligate symbiont, *Prochloron,* which lives in the cloaca of

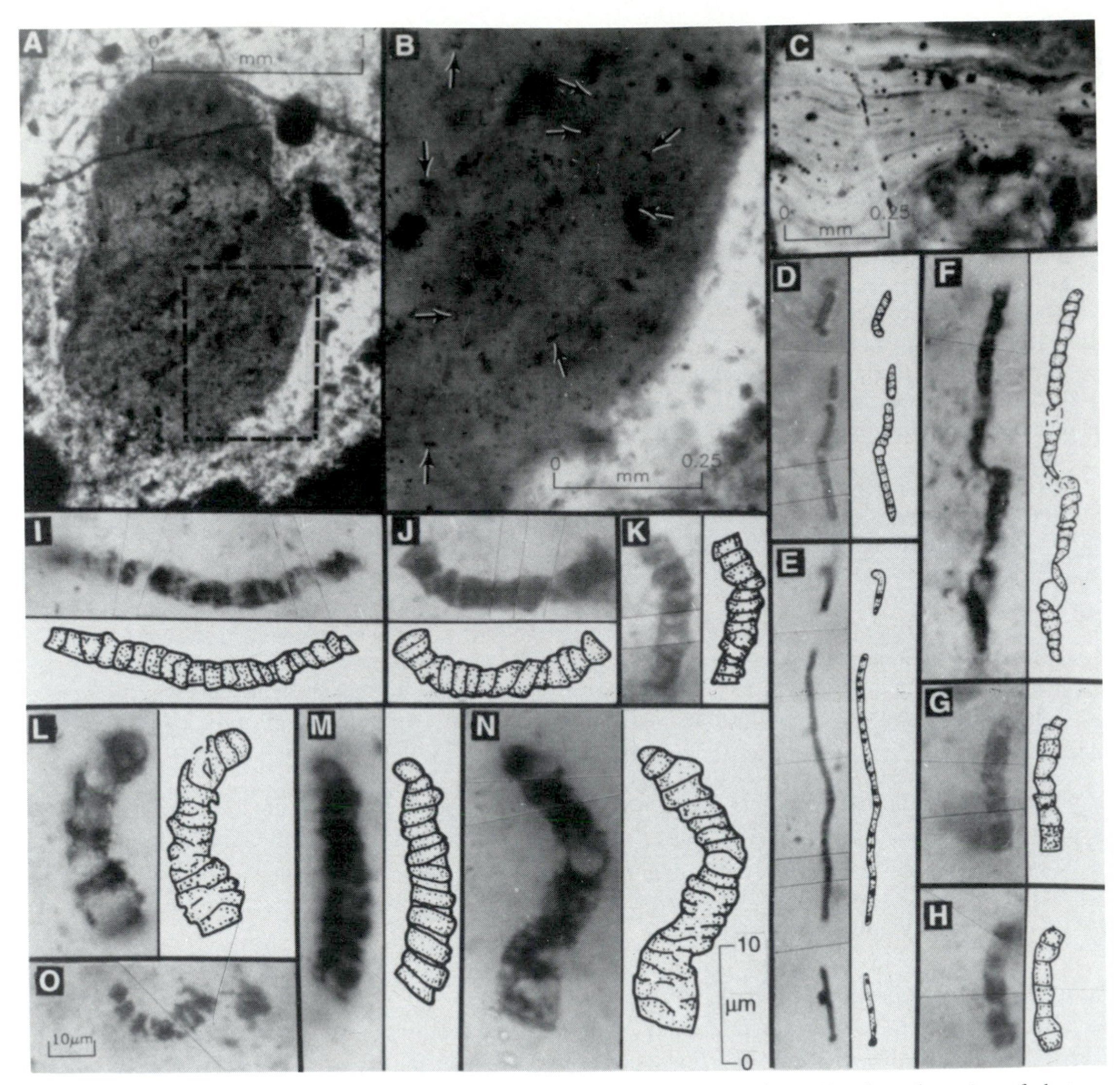

Figure 2–4 Fossils of some early plants. (A) Microfossiliferous fragment or clast; (B) enlarged portion of clast enclosed within dotted lines in A; arrows point to minute filamentous microfossils, randomly oriented; (C) portion of a clast showing lamination typical of stromatolites; (D–E) *Archaeotrichion septatum;* (F) *Eoleptonema apex;* (G–H) *Primaevifilum minutum;* (I–K) *Primaevifilum delicatulum;* (L–O) *Archaeoscillatoriopsis disciformis.* (From Schopf, J. W., Microfossils of the Early Archean Apex Chert: New evidence of the antiquity of life. *Science,* 260; 640, 1993. Copyright 1993 by AAAS. With permission.)

ascidians, is a modern prokaryote which has both chlorophyll *a* and *b* and may be a descendent of ancient free-living green prokaryotes.

The first eukaryotic species originated about 1.3 or 1.4 bya and became both diverse and abundant during the Neohelikian period. Two hypotheses attempt to explain the origin of the eukaryotic cell. According to the older or traditional hypothesis, organelles—the nucleus, plastids, mitochondria, and others—arose by invagination of the plasma membrane. The newer or symbiosis hypothesis suggests that the first organelles appeared when a colorless, heterotrophic cell was invaded by smaller autotrophic cells. These small cells then evolved into plastids. Mitochondria and some other organelles may have arisen by similar invasions. The symbiosis hypothesis has been widely acclaimed, but certainly not universally accepted (Raff and Mahler 1972).

By the beginning of the Hadrynian period, 1 billion years ago, Schizomycopsida, Cyanophycopsida, and Chlorophycopsida were abundant and widespread. Many of the species closely resembled species in existence today. In formations laid down at that time, the oldest fossils that can be classified as Dinophycopsida, Xanthophycopsida, and Mastigomycopsida have been found. The periods in which other classes probably originated are shown in Table 2-1. The dates suggested in Table 2-1 and in the following discussion are the best estimates available at the present time, but in some cases the actual origin of the group may have been tens of millions, or even hundreds of millions, of years earlier. Difficulty in distinguishing among fossils of the various eukaryotic classes, and even between eukaryotic and prokaryotic fossils, hampers the search for the exact time of origin of each of the plant divisions. The search has been further hampered by the fact that some organisms fossilize more readily than others (Table 2-2).

Throughout early and middle Hadrynian, the abundance and diversity of blue-green and green algae continued to increase as evidenced by the abundance of **stromatolites,** limestone deposits produced by algae (Figure 2-5), in formations from that time. But late Hadrynian formations show a decrease both in stromatolite abundance and in species diversity. Although there is little direct evidence of animals prior to the Cambrian period, it is believed that unicellular and other simple animals evolved during late Hadrynian. These soon grazed to extinction many species of plants, especially cyanophytes, and reduced the abundance of others.

Multicellular plants and animals evolved during the Cambrian period, which began almost 600 mya. The Siphonophycopsida were the first multicellular plants, other than filamentous species, and they were followed during the next period, the Ordovician (500 mya), by the Rhodophycopsida. All plants at this time were still marine, but late in the Silurian period, about 410 mya, vascular plants similar to *Psilotum* (Figure 15-3) evolved on mudflats of the receding sea. There are no traces of organic matter in the terrestrial habitats of the Paleozoic era prior to the appearance of these plants. Therefore, it is safe to conclude that they were the first land plants. The Emsian epoch, 25 million years later, boasted at least 42 species of vascular plants (Chaloner 1967). This number had increased to over 75 species by the end of the Devonian period an additional 30 million years later.

Freshwater habitats were invaded during early Devonian time by the stoneworts (Charophycopsida). It has been suggested that land plants may have evolved from stoneworts (Stewart and Mattox 1975); however, the current fossil record does not support this suggestion. Stoneworts fossilize very well, yet the oldest stonewort fossils are considerably younger than any of the early vascular plant fossils. Furthermore, stoneworts are freshwater species, and the early vascular plants seem to have come from the sea. More study on the origin of the Psilopsida is obviously needed.

The fossil record of the Bryophyta is poor, even though mosses (Bryopsida) have decay-resistant tissues and should fossilize very well (Table 2-2). Liverworts and mosses are well

Figure 2–5 Modern stromatolites at Shark Bay, western Australia. Similar stromatolites were produced anciently by the same forces that are producing these. (From Playford, R. E. and Cockbain, A. E., in *Stromatolites,* Walter, M. R., Ed., Elsevier Scientific Publishing Company, Amsterdam, 1976. With permission)

adapted to living at very low light intensities and probably originated as shade-tolerant subordinate species in ecosystems already dominated by large vascular plants, a niche they still occupy. Fossil liverworts first appear in late Devonian formations and fossil mosses in Carboniferous rocks.

The Devonian was a time of great evolutionary activity. In addition to the Charophycopsida and Hepaticopsida, four classes of Tracheophyta—the Lycopsida, Sphenopsida, Filicopsida, and Pinopsida—and one of terrestrial fungi, the Ascomycopsida, evolved then. Fossils resembling brown seaweeds have also been found in Devonian formations, although the first good fossils of Phaeophycopsida do not occur until the Triassic period, at which time they become quite abundant. During the Carboniferous periods, the Basidiomycopsida and Bryopsida appeared.

The Cretaceous period was also a time of great evolutionary activity. The earliest undisputed diatom fossils are found in Cretaceous formations, and diatoms became dominant in marine ecosystems during that period. They are still the most productive of all aquatic plants. Also in the Cretaceous, the first flowering plants (Anthopsida) evolved. They resembled some of the cycads in many ways. However, even the early angiosperms possessed characteristics unique to Anthopsida: closed carpels, pollen adapted for insect pollination, large flowers that were attractive to insects, and leaves with several orders of venation. Most of today's families of flowering plants had evolved by the end of the Cretaceous. The Gnetopsida evolved late in the Cretaceous period, about 70 mya. They were the first plants to evolve adaptations especially well suited for survival under extreme desert conditions. Two genera of modern gnetophytes, *Welwitschia* and *Ephedra,* are among the most **xeric** of all plants. Except for the greater abundance of cycads, and an occasional sighting of a dinosaur, a botanist visiting from the 20th century would feel at home in the late Cretaceous.

Examination of the fossil record indicates that evolutionary processes were very slow in the beginning. Although it took only a billion years from the time the earth took form as a spherical body revolving about the sun until the first **Monerans** came into existence, it took an additional 2 billion years before the first **eukaryotes** evolved. The first eukaryotes that we have record of were green algae similar to modern *Chlamydomonas* and *Protococcus;* they were followed by other chlorophyll-containing **protists** (unicellular plants) that resemble modern dinoflagellates and heterokonts. The animal-like protists that grazed the early algae almost to extinction about 700 mya did not evolve until some 700,000,000 years after the first eukaryotic green protists evolved. About 600 mya, the first multicellular **Animalia** and shortly thereafter the first terrestrial **Plantae** evolved; all animal life was limited to the ocean until about 400 mya.

Representatives of all three subkingdoms had evolved by the beginning of the Paleozoic era, 600 million years ago, and most, perhaps all, of the 11 divisions were represented by the end of the Devonian period (about 350 mya). All evolution since the late Cretaceous has occurred within each of the 29 classes. However, we have not yet found convincing fossils of euglenids, cryptophytes, or slime molds; fossils resembling terrestrial algal molds (Zygomycopsida) have been reported, but their identification is not certain. Lack of interest on the part of botanists in the morphology of these groups, together with lack of hard parts in the plants themselves, have contributed to this void.

The euglenids and the cryptophytes appear to be ancient taxa, and probably originated during the Proterozoic era about the same time as their near relatives in the Chlorophyta and Pyrrophyta. The Zygomycopsida probably evolved from sac fungi in the same time frame as the Basidiomycopsida. The Myxomycopsida and Acrasiomycopsida feed on bacteria and fungal spores, and some biologists, therefore, think they may be of ancient origin. But because the species in both classes are terrestrial and always associated with vascular plants, it is more reasonable to assume that they did not evolve until the Devonian period or later. The Labyrinthulomycopsida are so poorly known that it is difficult even to guess when they originated. Those species which are parasitic on angiosperms must be of relatively recent origin, of course, but other net slime molds, being marine and parasitic on Chlorophycopsida, could have originated prior to the Devonian period, and possibly during the Proterozoic era.

Evolution of Ecosystems

We should keep in mind that species did not evolve as independent entities but as parts of ecosystems. Margalef once said, "Evolution cannot be understood except in the frame of ecosystems." According to the Volterra concept, or "law of competitive exclusion" (Gause 1934, 1935), only one species can exist in any ecological niche at any given time, and it will be the species best adapted to that niche. The first traits to develop, giving an adaptive advantage to an ecological niche, are generally controlled by single pairs of genes. As the trait becomes well established, interactions among many pairs of genes develop, providing increased stability (Stebbins 1966). Consequently, evolution may be expected to occur in "explosive bursts," as new niches become available, followed by periods of stability in which change is much slower (Figure 2-6). Examination of the fossil record provides support to this expectation.

When a new species of plant or animal evolves, it often creates new ecological niches. There was no niche for the mosses or the mushrooms until there were vascular plants on land. There was no niche for carnivorous animals until there were herbivores. In like manner, a species migrating into an ecosystem from another area may also create new niches. On the other hand, a species may adapt to a niche occupied by another organism, and therefore compete with a population that is already there. In that case, the better adapted population should survive and the poorly adapted one should become extinct.

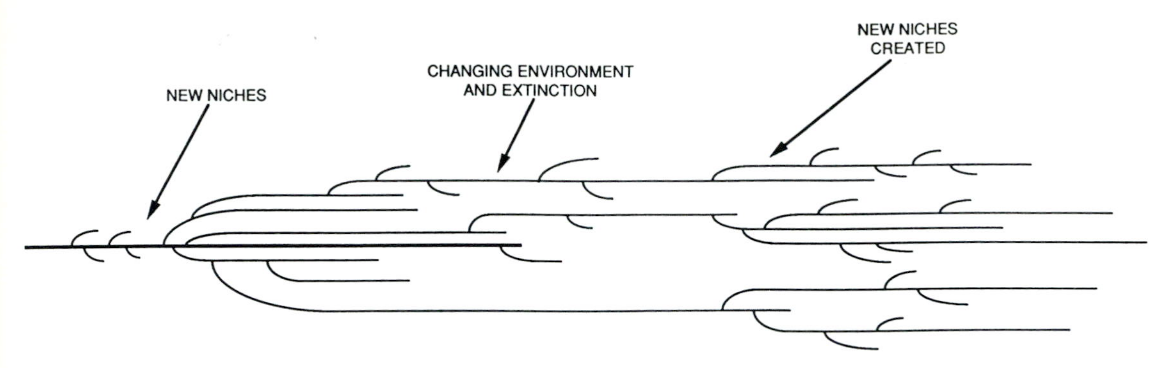

Figure 2–6 Evolution occurs in bursts as new ecological niches open up. Every time a new species evolves, it creates additional niches: for parasitic and saprophytic fungi and for herbivorous animals. At first, a newly evolved population does not need to be especially well adapted in order to survive, but as competition among populations increases, only the best adapted survive.

A basic principle of ecosystem evolution is the concept of niche succession and balance. Whenever a completely new type of habitat, vastly different from existing habitats, is first invaded by living organisms, green plants are the first to evolve adaptations making survival there possible, then herbivores, and finally carnivores. Ecosystems tend to be unstable until the carnivores arrive, for without them, overgrazing, often to the point of extinction of the most susceptible plants, threatens the survival of the producer organisms (Gebelein 1976, Walter 1976, Scott and Taylor 1983). In terrestrial ecosystems, greatly accelerated soil erosion results when the vegetation is destroyed; without soil, most terrestrial ecosystems cannot survive, illustrating again the importance of predators as protectors of ecosystems. The fossil record reveals evidence of heavy grazing by early arthropods during the early Devonian epochs, accompanied by constant fluctuation in presence and abundance of many of the plant species; ecosystems were much more stable during the Mississippian, and there is little evidence of extremely heavy grazing (Stewart and Mattox 1975). By then, carnivorous insects and amphibians had evolved.

Environmental Changes in Time

While the forces of climate and other environmental factors determine the distribution and productivity of plants, plants also determine the nature of the environment. The early atmosphere of the earth contained no free oxygen. In the absence of oxygen, and hence ozone, ultraviolet radiation was very intense. In a reducing atmosphere consisting of some combination of CH_4, CO_2, CO, H_2, and/or N_2, acted upon by high energy level radiation in the form of ultraviolet light, it was inevitable that numerous kinds of organic chemicals would be synthesized (Oparin 1928, Haldane 1929). Life originated when complex chemicals, capable of duplicating themselves, and at the same time capable of catalyzing other chemical reactions, came into existence. These were similar to deoxyribonucleic acid (DNA) in their action.

By 3.5 bya, nucleic acids and respiratory pigments had evolved and were packaged together in a primitive type of cell. These bacteria-like organisms were capable of living heterotrophically off the primordial broth postulated by Oparin and Haldane. Such life was, of course, unstable: it was dependent upon the inorganic synthesis of energy-containing food substances. But when a compound similar to chlorophyll was incorporated into these primitive

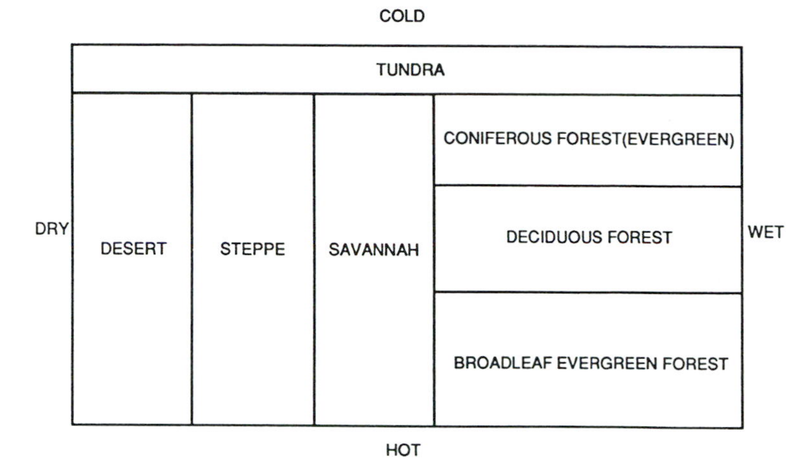

Figure 2–7 Major terrestrial biomes, showing their correlation with climate. Where moisture is limiting, similar vegetational types are found in both cold and hot habitats; where moisture is abundant, temperature becomes the limiting factor. Tundra is found in both wet and dry habitats where temperatures are too low for vegetation other than very cold-resistant species (primarily grasses, sedges, willows, mosses, and lichens).

cells, life became more stable. Before there was chlorophyll, food accumulated arithmetically while the numbers of heterotrophs, and hence their use of food, increased geometrically. After a few generations, food supplies were inevitably exhausted or at least drastically reduced. With chlorophyll present, the food supply increases or decreases at about the same rate as the biomass of the producers.

The production of oxygen by blue-green algae in the early ecosystems had a far-reaching and irreversible effect on the environment of the earth, beginning with the atmosphere. According to a model developed by Berkner and Marshall (1965), free oxygen in the atmosphere gradually increased to about 1% of the present level (or 0.2% O_2, 99.8% other gases) at the beginning of the Cambrian period and to 10% of the present level at the beginning of the Devonian. They calculated that when the concentration of O_2 in the atmosphere reached 2%, a protective ozone layer would be present and life on land would become possible.

ECOLOGY AND PLANT DIVERSITY

The basic unit of ecology is the ecosystem. An **ecosystem** consists of all the plants and animals, together with the factors and forces of their abiotic environment, that exist together in a contiguous and distinct geographical area. Each ecosystem is separated from adjacent ecosystems by **ecotones.** Examples of ecotones are the shore of a lake, timberline on a high mountain, or the border between a wheat and a potato field.

Ecosystems having similar structure and associated with each other in the same general area are grouped together in **biomes.** The distribution of terrestrial biomes in relationship to the major factors of climate is diagrammed in Figure 2-7. Our knowledge of marine biomes is more limited. The diversity of species in any biome, terrestrial or aquatic, is controlled by the number of ecosystems in the biome, the number of ecological niches in each ecosystem, the effectiveness of barriers between ecosystems, and the effectiveness of plant propagules (seeds, spores, soredia, tubers, stem or root fragments, etc.) in distribution of the species. Marine biomes appear to be more extensive in area and less effectively isolated from each other than terrestrial biomes, and the dominant species in ocean habitats appear to have more effective modes of dispersion than do the seed plants on land; therefore diversity, as measured by number of species in each class or other taxonomic unit, is less in the ocean than on land (Table 1-1).

Environment and Plant Productivity

Table 2-4 lists the major biomes and the estimated average productivity of each, with the main producers indicated. Hundreds of studies were reviewed, several of which are indicated in the endnotes, in arriving at these estimates. Productivity varies greatly from place to place within each biome as well as from year to year, month to month, and biome to biome. This is especially true in aquatic ecosystems in which species composition often changes significantly in a matter of a few days. Therefore, different investigators have arrived at markedly different estimates of productivity.

Because of significant variations in ecosystem productivity, early estimates, which were based at best on very small samples and at worst on extrapolation from limited data, were often misleading. Analysis of seawater collected in 1839 by J. D. Hooker off the coast of Antarctica, for example, led biologists for many years to believe that plant life in the open ocean was much more abundant that it now appears to be. Many ecologists for over a century believed that 90% of global production of organic matter was by marine plants, diatoms contributing at least 40% of the total. Recent estimates indicate that the ocean, even though it covers more than 70% of the total world area, contributes only about half of the food and oxygen produced. Over 90% of the ocean is essentially a desert, a desert in which mineral nutrients rather than water are limiting, and therefore is not very productive. On the other hand, Repetto (1990) has reported that the phytoplankton contains species which have the ability to concentrate nitrates, phosphates, and other minerals and carry on quite rapid photosynthesis. Like terrestrial deserts, the ocean contains many interesting kinds of plants and animals; also like terrestrial deserts, there are very productive oases around hydrothermal vents where minerals are abundant (Corliss and Ballard 1977). Passengers on an ocean liner enjoy watching the flying fish by day and flagellate-produced luminescence by night as they traverse the deepest oceans. But in the marine food chains, these may be of relatively little trophic significance. According to Marden (1978), 90% of the human food coming from the sea is produced on the continental shelves.

As late as 1955, few ecologists had conducted studies of productivity of terrestrial ecosystems and even fewer of marine ecosystems. Among the pioneers in conducting productivity research were Juday (1940), Lindeman (1942), Riley (1944, 1957), and Lawrence et al. (1957–58). Farmers and agronomists had good data on **agronomic productivity** of most crops, but essentially no data on root production, primary consumption by insects, or even agronomic productivity in pastures and rangeland. However, Odum's textbook on ecology emphasized both structure and function of ecosystems, D. B. Lawrence calculated annual energy budgets in Minnesota forests, and J. D. Ovington (1957) conducted similar studies about the same time in Great Britain, but most ecologists were more interested in describing, comparing, and naming plant communities than in measuring productivity. Part of the problem was the great complexity of ecosystems. Then, in the 1960s, with sophisticated computers a reality, a cooperative international biological program (IBP) was inaugurated involving biologists from all

TABLE 2-4

GROSS PRIMARY PRODUCTIVITY OF THE MAJOR BIOMES OF THE WORLD

Biome	**Total Area**[1] ($km^2 \times 10^3$)	**Annual Productivity** (g/m^2)	**Total Production** ($tons \times 10^9$)	**Chief Producers**
Ice snowfield rock sand	24,000	5[2]	.12	ASCO (lichens)
Arctic/alpine tundra	8,000	150[2]	1.20	ASCO BRY ANTH
Deserts (excluding oases)	18,000	120[3,9]	2.16	ANTH GNET
Steppe	9,000	400[2,3,9,11]	3.60	ANTH
Dry woodland/scrubland	8,500	600[2]	5.10	ANTH PIN
Savannah	15,000	700[2]	10.50	ANTH
Boreal forest	12,000	1,000[2]	12.00	PIN ANTH
Temperate evergreen forest	5,000	2,000[3,5]	10.00	PIN ANTH
Temperate deciduous forest	7,000	1,350[2,10]	9.45	ANTH FIL
Tropical rainforest	17,000	2,200[2,13,14]	37.40	ANTH PIN FIL
Swamp and marsh	2,000	1,800[2,3]	3.60	ANTH CHL SPH
Lake and stream	2,000	500[2,17]	1.00	CHL BAC ANG
Irrigated cropland	1,150	1,200[1,3,14]	1.38	ANTH
Other cropland	12,700	500[2,3,12]	6.35	ANTH
Total terrestrial	**148,850**		**115.11**	
Open ocean, tropical	140,900	230[2,16,17]	32.41	DIN CRYP CYAN
Open ocean, temperate	176,850	200[2,17,18]	35.37	BAC DIN CHRY
Open ocean, polar	13,200	170[17,18,19]	2.24	BAC CHRY
Reefs and atolls	2,700	6,500[4,7,17]	17.55	RHO DIN BAC
Upwelling areas	600	1,200[13,17]	.72	BAC DIN CHRY
Continental shelves				
Tropical	11,300	750[6,15,17]	8.48	RHO SIPH DIN CYAN
Temperate	14,200	700[6,15,17]	9.94	PHAE RHO BAC DIN
Polar	1,500	620[6,15,17,19]	.93	PHAE BAC CHRY
Estuaries	1,400	1,700[8,17,19]	2.38	BAC RHO PHAE ANG
Total marine	**362,650**		**110.02**	
Total global	**511,500**		**225.13**	

[1]Goode's *Atlas of the World,* Whittaker 1970, Thorne and Peterson 1954; Lieth 1970
[2]Whittaker 1970; Woodwell and Whittaker 1978
[3]Pearson 1965, 1966
[4]Summerhays 1981; Kaplan 1982
[5]Ovington 1956
[6]Marden 1978
[7]Kohn and Helfrich 1957
[8]Odum and Odum 1959; Thorn et al. 1994
[9]Walter 1973; Lange 1969
[10]Ovington and Pearsall 1957; Bray et al. 1959, 1964
[11]Penfound 1956; Singh and Yadeva 1974
[12]Pearson 1967
[13]Woodwell 1970
[14]Gentry 1993
[15]Riley 1955
[16]Riley 1957
[17]Russell-Hunter 1970
[18]Raymont 1963
[19]Bogorov 1967; Peterson et al. 1993

parts of the world and emphasizing ecosystem productivity. Models of ecosystems were developed that would take into account the effects of weather, herbivores, microorganisms, geological formations, soil, and other factors, and from these data predict changes in ecosystem composition. As data were fed into computers, predictions based on the models were compared with changes actually taking place in nature. As a result, we now have solid information on productivity of many ecosystems; however, many gaps in our knowledge still need filling.

In the past, as at the present time, productivity was correlated with temperature, moisture, and other environmental factors. Environments continually change (Figure 2-8). Year-to-year variations in temperature and precipitation result in slight but measurable changes in population density of some species of both plants and animals. Extreme changes in temperature and precipitation, like those that brought on the ice ages, have resulted in mass extinctions.

Seven major ice ages are recorded in geological formations since the beginning of the Proterozoic era, as illustrated in Figure 2-8. The largest of these was the Gondwanan, at the end of the Paleozoic era, at which time the ice cap reached to 30° latitude. Following the Gondwanan glaciation, the earth was free of ice during most of the Mesozoic era. Prior to the Aphebian period, the interior of the earth was still so hot, the crust so thin, and methane and carbon dioxide levels so high, that there were probably no glaciers at any time or place.

A cluster of three ice ages during the Hadrynian, the largest of which was the Varangian at the end of the Proterozoic era, must have reduced water temperatures in the oceans to the point where many plants could no longer survive. The last of the ice ages, the Pleistocene, also had a tremendous effect on plant life, especially in Asia and Europe. In North America, the effects were less severe: mountain ranges run north and south and so species migrated southward during the last ice age and northward as it ended, whereas in Asia and Europe, where the mountain ranges run east and west, species could not migrate. The Pleistocene ice age consisted of four stages, the latest beginning to subside only 5,000 to 10,000 years ago. At the present time, there are 16,000,000 km^3 of ice on earth, in contrast to some of the interglacial periods when there were no glaciers, even at the poles (John 1979; Imbrie and Imbrie 1979).

The most productive ecosystems are tropical reefs, some estuaries, some warm springs, and well-stratified tropical forests. Fossilized reefs bear witness of the rich abundance of life anciently in these ecosystems. Temperate-zone rain forests are also very productive, as are most of the marine ecosystems on the continental shelves and in upwelling areas. The least productive are deserts, where water is the limiting factor; the vast expanses of open ocean, where there is a shortage of available mineral nutrients; and the tundra, where temperature is not favorable for good growth. It should be kept in mind that within each biome, productivity of ecosystems may vary many fold. This is especially true of desert areas. In a study of 18 ecosystems in eastern Idaho and adjacent parts of Wyoming and Utah, the most productive ecosystem was 80 times as productive as the least (Pearson 1966).

Natural selection has resulted in plants and animals very well adapted to survival in specific ecosystems. The number of species in an ecosystem depends in part on the number of ecological niches available and in part on the productivity of the producer organisms, the green plants. Productivity, in turn, depends on a number of environmental factors: temperature, precipitation, day length, light intensity, substrate characteristics, humidity, dew, herbivory, predation, competition, and others. Each species in every ecosystem influences the survival of at least some of the other organisms that are there. However, every ecosystem contains a few **dominant** species, usually not more than five or six, which are photosynthetically more productive than all of the other species combined. Often the dominants are responsible for 80%

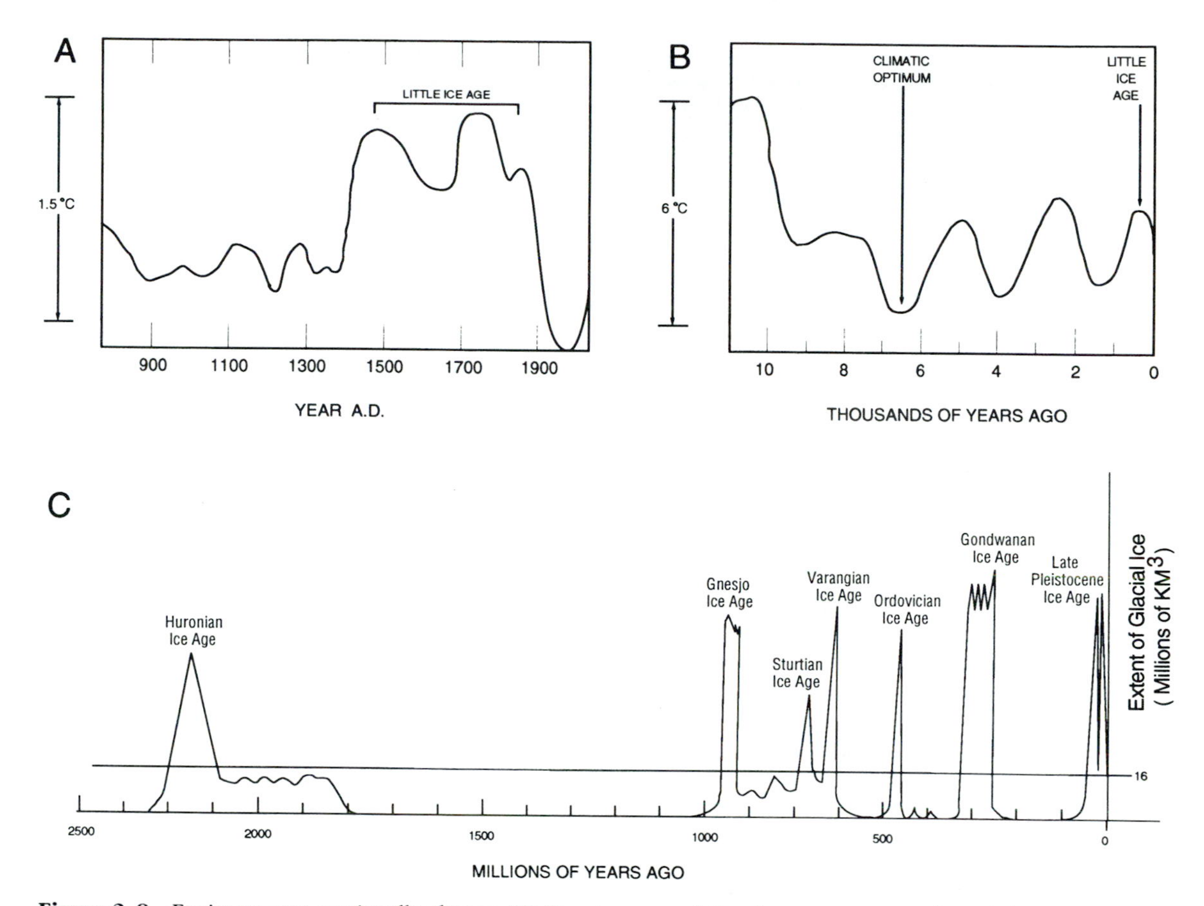

Figure 2–8 Environments continually change. (A) Year-to-year variation in temperature and precipitation result in barely measurable change in density of plant and animal populations. (B) Prolonged changes in temperature and precipitation, like those of the late Pleistocene epoch, have resulted in mass extinctions, especially in Europe and Asia. (C) We know of seven major ice ages since the beginning of the Proterozoic era. The Varangian, at the end of the Proterozoic era, and the Gondwanan, at the end of the Paleozoic era, were the greatest of these. At the present time, there are 16,000,000 km^3 of ice on earth, as indicated by the horizontal line; during some of the warm periods of the past, there have been no glaciers, even at the poles. (Data from John 1979.)

or more of the food and oxygen produced in that ecosystem; occasionally the productivity of the dominants will exceed 95% of the total (Bray et al. 1959; Pearson 1966).

Typically, the dominant species in an ecosystem all belong to the same taxonomic class. Furthermore, most of the ecosystems within each biome will be dominated by species belonging to the same class. In Table 2-5, the data from Table 2-4 are summarized by the class to which the dominants belong.

Four of the 29 classes of plants—Anthopsida, Bacillariophycopsida, Dinophycopsida, and Pinopsida—stand out in total productivity (Tables 2-4 and 2-5). Over 75% of the total global photosynthesis is contributed annually by plants in these four classes. Other classes are locally significant. In the tropics, the Rhodophycopsida and Siphonophycopsida are significant along the continental shelves and on reefs. In colder oceans, the Phaeophycopsida are significant on the continental shelves. Lichens and mosses are important in both arctic and alpine tundra, and

TABLE 2-5

ESTIMATED GLOBAL PRIMARY PRODUCTIVITY BY CLASS, SUMMARIZED FROM TABLE 2-4

Class	Production (Tons × 10^9)	Per Cent of Total
Anthopsida	82.4	36.6
Bacillariophycopsida	38.7	17.2
Dinophycopsida	26.0	11.6
Pinopsida	21.4	9.5
Rhodophycopsida	13.7	6.1
Cyanophycopsida	8.3	3.7
Cryptophycopsida	8.2	3.6
Chrysophycopsida	5.8	2.6
Phaeophycopsida	3.5	1.6
Others	17.1	7.5
	225.1	100.0

in some forested areas for mammals like the woodland caribou. Chlorophycopsida are the main photosynthesizers in lichens; they are also the most important producers in ponds and streams and significant in marshes and swamps. As more data are gathered, the estimates in Tables 2-4 and 2-5 will gradually be revised and improved.

LIFE CYCLES

The sequence of events beginning when a mature individual of one generation produces a cell or group of cells which, following several stages of development, becomes a mature individual of a new generation, is referred to as the **life cycle** of that individual or of the species. Life cycles are relatively uniform within each class and division, but vary considerably among the divisions. These variations will be discussed in each of the following chapters in considerable detail, but some of the general patterns are discussed in this chapter.

In most plants, there is a distinct alternation of haploid (n) and diploid (2n) generations. We may think of the life cycle beginning with the production of meiospores by a mature diploid plant as a result of meiosis; the meiospores germinate to produce haploid **gametophytes** which, upon maturation, produce haploid gametes (eggs and sperm, or isogametes, depending on the species). The zygotes resulting from **syngamy,** or fusion of gametes, divide mitotically to produce diploid proembryos that develop into embryos and then mature **sporophytes,** and the life cycle is complete (Figure 2-9).

Some botanists refer to the sporophytes as "asexual plants" and to the gametophytes as "sexual plants." This practice is misleading; the sporophytes, or spore-producing plants, in which meiosis takes place, are just as "sexual" as the gametophytes, or gamete-producing plants. Both meiosis and syngamy are essential to the sexual process. Only those structures directly involved in asexual reproduction can properly be referred to as asexual structures; these include such things as tubers and rhizomes in flowering plants, pycnidia and acervuli in fungi, and soredia and isidia in lichens.

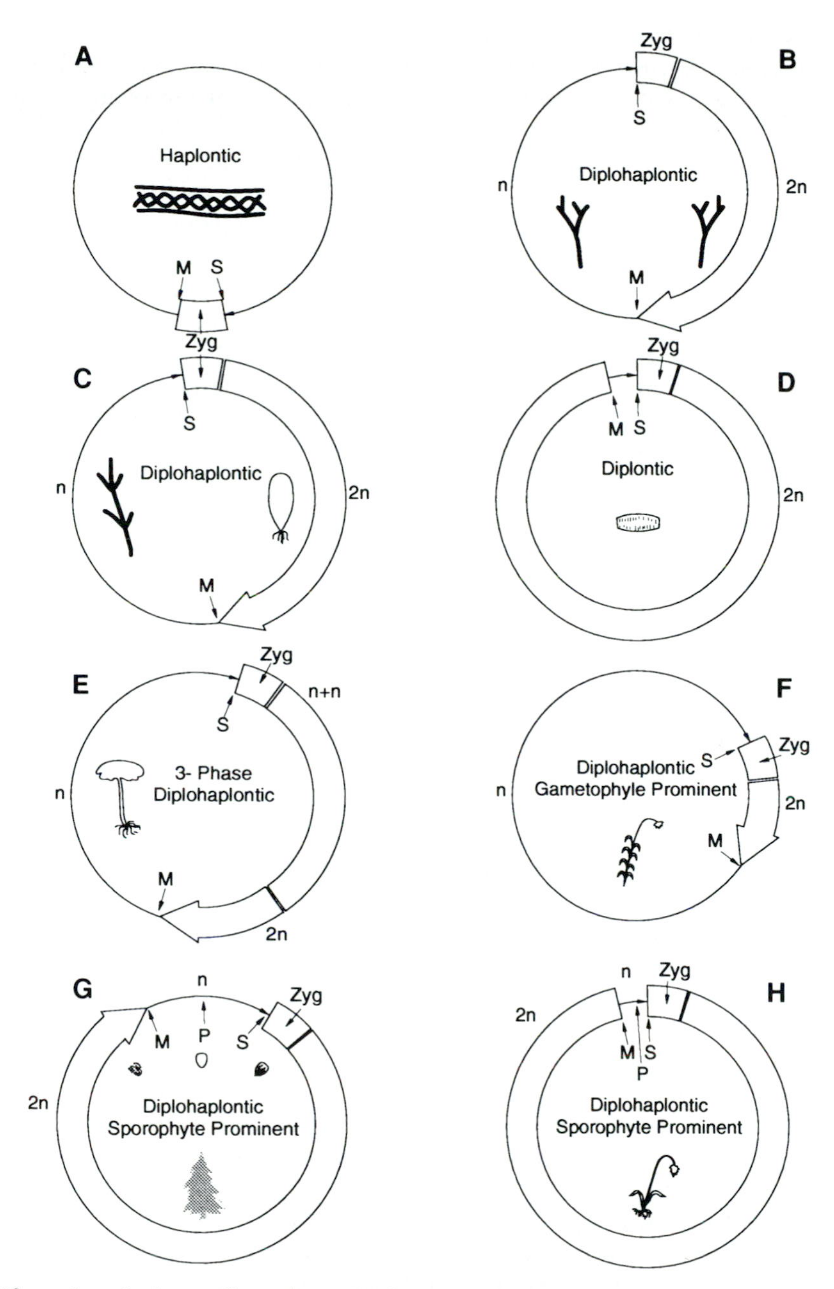

Figure 2–9 Life cycles of plants. Chorophyta: (A) *Spirogyra decima,* (B) *Cladophora subriana,* (C) *Derbesia marina;* Bacillariophycopsida: (D) *Navicula viridis;* Eumycophyta: (E) *Agaricus campestris;* Bryophyta: (F) *Polytrichum commune;* Tracheophyta: (G) *Picea excelsa,* (H) *Erythronium grandiflorum.* (M indicates the time of meiosis, S the time of syngamy; Zyg = xygote, P = nutritionally dependent ("parasitic") stage; n = haploid generation, n plus n = dikaryotic generation, 2n = diploid generation.)

Sexual Reproduction in Plants

Sexual reproduction may be described as a process in which the number of chromosomes and, hence, genes, is reduced by means of meiosis and then restored through syngamy. The advantage of sexual reproduction is that it makes possible **ecological homeostasis** or the ability of populations to survive in changing or fluctuating environments. Sexual reproduction also contributes to speed and efficiency in the evolution of new ecotypes and species.

Ecological homeostasis is greatest in sexually reproducing populations partly because there is a pool of **recessive** and **hypostatic** genes stored in the individuals making up each population, and partly because most genes are **pleiotropic** and affect two or more characteristics. The effects of recessive genes are hidden by their dominant alleles, or genes with which they can pair. Hypostatic genes may be either dominant or recessive, but in either case their effects are hidden by **epistatic** genes, or genes which are not their alleles. Some recessive and hypostatic genes are recalled from storage each generation through random mating.

An example of a pleiotropic gene is one which enables a plant to survive extremely cold weather but which also reduces the number of seeds which that plant can produce. If the environment were free of all danger of winter-kill, the cold-tolerant gene would be disadvantageous and would gradually be eliminated from the population—but environments are constantly changing (Figure 2-8) and a gene which is disadvantageous at one time may be advantageous under other circumstances.

If, as a result of pleiotropy, a plant lacking winter hardiness (or cold tolerance) should produce more seeds than those possessing winter hardiness during years with long summers, but during years with early frosts could not survive long enough for the seeds to mature, its genotypic frequency would fluctuate from year to year. If winter hardiness is determined by several pairs of genes, and we know that it is in some species (e.g., alfalfa, Pearson and Elling 1961), with some of the genes epistatic over the others, it would take many years of favorable weather (long summers) to eliminate it completely. The "little ice age" of the early 19th century did not last long enough to eliminate many species, if any at all (Figure 2-8A), because of the "buffering" effect of recessive and hypostatic genes. During the last stage of the Cenozoic (or Pleistocene) ice age, on the other hand, 90% of the plant species in northwestern Europe became extinct (Figure 2-8B), and it is estimated that during the Permian ice age (the last substage of the Gondwanan), 75 to 80% of the families of amphibians and reptiles forever disappeared (John 1979).

When heterozygous plants reproduce sexually, their progeny vary both phenotypically and genotypically. In asexual reproduction, the progeny are genetically identical to their parents and to one another; any variation is due to environment alone. Asexual reproduction is especially effective as a means of increasing numbers and dispersing progeny, but it cannot provide homeostasis. Because of its advantages in species dispersal and multiplication of progeny, it is very common in some groups of plants, especially in those species which inhabit environments that are stable or unchanging from year to year. Frequently, both sexual and asexual methods of reproduction are well developed in the same species.

Kinds of Life Cycles

There are three main types of life cycles that occur in plants: haplontic, diplontic, and diplohaplontic (Table 2-6, Figure 2-9). In the first of these, the zygote is the only stage in which the nucleus contains the diploid number of chromosomes; every cell in every vegetative plant

TABLE 2-6

DIVERSITY OF REPRODUCTIVE LIFE CYCLES IN PLANTS

	Monoploid Generation	Diploid Generation	Examples
Haplontic—			
Prokaryote type	Vegetative; does not produce gametes; meromixis in some species	None	Bacteria and blue-green algae
Eukaryote type	Vegetative; produces isogametes by mitosis or by cell differentiation	Zygote only, which germinates by meiosis to produce meiospores	*Chlamydomonas, Tribonema, Cladophora keutzginiana, Spirogyra*
Diplontic	Consists of gametes only	Vegetative; produces gametes (eggs, sperm, etc.) by meiosis	*Navicula* and many other diatoms; *Cladophora glomerata*
Diplohaplontic Gametophyte prominent	Vegetative; produces gametes (eggs and sperm) by cell differentiation	May be vegetative or not, often dependent on the gametophyte, produces meiospores	*Bryum, Riccia,* and other mosses and liverworts
Equiprominent type, isomorphic or heteromorphic	Vegetative, generally autotrophic and free living; produces gametes by cell differentiation	Vegetative, generally autotrophic and free living, produces meiospores by means of meiosis	*Derbesia (Halocystis), Cutleria, Cladophora fracta*
Sporophyte prominent	Vegetative, may be autotrophic or heterotrophic	Vegetative, usually large, prominent, and autotrophic	*Laminaria, Poa, Psilotum, Azolla, Pinus, Cycas, Pisum*
Multiphase	Vegetative; produce gametes by differentiation of prosenchymatous cells	Two or more diploid generations, one generally inconspicuous and dependent on the monoploid generation	*Polysiphonia, Gelidium, Gracilaria, Amanita, Morchella, Usnea*

has the haploid number of chromosomes. In such species, the zygote divides meiotically to produce haploid meiospores. These mature into typical haploid plants. Gametes are produced by mitosis or by simple differentiation of existing haploid cells. This type of life cycle is called **haplontic.** It is very common in the Chlorophycopsida, the Xanthophycopsida, the Mastigomycopsida, and some other groups of algae and fungi. The life cycle of *Spirogyra decimina* is illustrated in Figure 2-9A.

In a few species of plants, gametes are produced in diploid plants by means of meiosis, just as in most animals, and there is no haploid generation. Such species, an example of which is *Cladophora glomerata,* have **diplontic** life cycles. Many diatoms (Figure 2-9D) and some slime molds also have diplontic life cycles. Some botanists interpret the life cycle of *Fucus* and other Fucales (Phaeophycopsida) as diplontic.

Most plants have **diplohaplontic** life cycles with a distinct alternation of haploid and diploid generations. The generations may be **isomorphic,** as in *Cladophora subriana* (Figure 2-9B), or **heteromorphic** as in ferns, other vascular plants (Figure 2–9G, H), mosses (Figure 2–9F), kelps, and many other seaweeds and fungi. In plants with isomorphic alternation of generations the only morphological differences between the **sporophytes,** or spore-producing plants, and the **gametophytes,** or gamete-producing plants, are that the former are made up entirely of diploid cells and produce spores in sporangia by means of meiosis, while the latter are made up entirely of haploid cells and produce gametes by means of simple differentiation of existing cells or by mitosis.

In some species with heteromorphic alternation of generations, the sporophytes are small and inconspicuous, often parasitic on the much larger gametophytes. Such is the case in the Bryophyta. In mosses, the gametophyte is green and leafy; the slender sporophyte, which is typically yellow or brown and obtains all its nourishment from the gametophyte, grows out of an archegonium at the apex of the gametophyte. At the top of this slender stalk, which may be much longer than the stem of the gametophyte, there is a capsule in which meiosis takes place and meiospores are produced (Figure 2–9F).

In other species having a heteromorphic alternation of generations, the gametophytes are small, inconspicuous, and sometimes parasitic on the much larger sporophytes. Such is the case in the Tracheophyta. In the ferns, the gametophytes are free-living and autotrophic, but are much smaller than the sporophytes (Figure 16–5). In seed plants, there are two kinds of meiospores, megaspores or **ovules** and microspores or **pollen grains;** these develop into two kinds of gametophytes, female and male, respectively. The gametophytes obtain their nourishment from the parent sporophytes. Following syngamy, the embryonic sporophytes develop within the female gametophytes. In angiosperms, the mature sporophytes have roots, stems, leaves, and flowers; xylem and phloem tissues are complex and conduct food and water from one part of the plant to another. The gametophytes, on the other hand, are very small, have neither organs nor tissues, and spend their entire short life within the ovary or, if male, growing within the pistil toward the ovary. Gymnosperm gametophytes, although not as small as angiosperm gametophytes, are also dependent on the mother sporophytes with their complex tissues and well-developed organs: roots, stems, and leaves. In some gymnosperms, however, cycads and pinyon pines, for example, the gametophytes are relatively large and fleshy and are used for human food. Pinyon nuts can be torn open and the sporophyte embryo, which resembles a miniature pine seedling, readily observed.

In some plants, the two generations of the same species are so different in appearance that botanists for many years classified them in different genera, and even different families. An example is the calcareous green alga, *Derbesia marina,* in the Derbesiaceae, order Siphonales. The sporophyte is a branched, filamentous, coenocyte a few millimeters long which produces an abundance of stephanokont meiospores in ovoid sporangia. These swim around until they are attracted to a segment of either *Lithothamnion* or *Lithophyllum,* which are common genera of calcareous red algae. There they germinate and develop into egg shaped, green epiphytes, about a centimeter in diameter. These were formerly classified as *Halicystis ovalis* in the family Halicystaceae, order Siphonales. Half of the spores develop into male, half into female gametophytes. At maturity, the females produce large numbers of green, flagellated, motile eggs which are forcibly ejected from the gametangia each morning a few minutes after the sun comes up. Each egg has several disk-shaped plastids. After several minutes, the male gametophytes discharge colorless, motile sperm. Gametic union results in a zygote which settles on a sandy or muddy bottom, germinates, and grows into a typical, filamentous *Derbesia* plant (Figure 2–9C).

In the Rhodophyta and Eumycophyta, life cycles are generally complex, often consisting of three or more generations. In the common commercial mushroom, for example, a haploid **mycelium,** which has grown out of a **basidiospore** (meiospore), produces gametes; two gametes fuse in the act of syngamy to produce a **dicaryotic** mycelium having two haploid nuclei in each cell. This mycelium makes up the great bulk of the vegetative part of the plant. When conditions are favorable, the two nuclei in some of the cells fuse and diploid hyphae grow into a part of the sporocarp, or spore fruit, that we buy in the store and eat. On the gills of these spore fruits, diploid cells divide by meiosis to give rise to **meiospores.** Meiospores are called basidiospores in the Basidiomycopsida and **ascospores** in the Ascomycopsida.

Phenology

Some plants have a narrow range of tolerance for each factor, or for some of the factors, of the environment; other plants have a wide range of tolerance. Some plants grow best at relatively high temperatures, others at low temperatures; some require much water and high humidity, others do well with very little water or under arid atmospheric conditions. These ecological requirements, however, are not constant at all stages of the life cycle. In general, plants have narrower ranges of tolerance for most environmental factors during the early stages of development. Phenology is the study of how plants or animals are affected by different factors or forces of the environment during the different stages of the life cycle.

In *Oryzopsis hymenoides,* productivity of the mature plant is affected by temperature and humidity during the early weeks of spring. Additional water applied early in the season reduces final yield, probably in part because there is already sufficient moisture in the soil so that lack of water is not a limiting factor, and partly because the additional water cools the soil at a time when low temperatures are already limiting growth. On the other hand, additional water later in the season, when the soil is warm and dry, increases final productivity, unless it comes so late that the plant has already passed through the vegetative stage and does not respond to additional moisture (Pearson 1979).

Knowledge of phenology is of special importance to farmers and gardeners. In areas with cold winters, crops must be planted late enough that the danger of a late spring killing frost destroying the crop is removed, but early enough that the crop can mature before the autumnal cold season begins. Cold weather when the peach trees are in bloom can essentially eliminate a crop, resulting in high prices for consumers as well as financial ruin for many farmers.

BOTANY AND HUMAN SURVIVAL

Marston Bates (1960), in *The Forest and the Sea,* refers to two games, both ridiculous, if not downright dangerous, that people nowadays play. One he calls the ecological game: "Let's pretend that Man doesn't exist." The other is the game played by economists and engineers: "Let's pretend that Nature doesn't exist." Of course, Man and Nature both exist, and the challenge we face today is how to ensure that both continue to exist. Paleobotany teaches us that Nature can survive, and did survive for about 3½ billion years, without Man. Ecology teaches us that Man, on the other hand, cannot survive for even *one* year without Nature. In this chapter, we have shown you how Nature has changed over the years and have attempted to help you understand what the forces are that have been responsible for these changes: interaction among genes and between genes and the environment, for example. We have also discussed changes in space and how diverse different ecosystems are today or at any other point in time. To understand diversity in time and diversity in space, we need to understand sexual life cycles in plants, sexual and asexual reproduction, the concept of ecological homeostasis, and the many phenological effects of the environment. All of these have been discussed in this chapter.

In the chapters that follow, each of the 29 classes of plants will be described. Morphological and physiological characteristics will be presented first. Included in this part of the chapter will be descriptions of sexual reproduction and the kinds of life cycles that exist within the class. Paleobotanical data will be presented next along with current hypotheses as to the phylogeny and classification of the class. A brief description of each order in the class will help you gain a better understanding of the degree of diversity in the class. To help you recognize members of each class, representative species and/or genera will be described and illustrated. The ecological and economic significance of the class and its potential research significance will round out each chapter.

As a botanist or a student of botany, people will come to you with plants they have never seen before or which they have seen and been fascinated by but know nothing about. Their first question to you will be, "What is it?" If you can give them a name, they will go away happy. If you don't know the species, but can tell them what class and division it is in, this will often satisfy them. They will then usually ask you three more questions: (1) Where does it live? (2) What does it do? and (3) What good is it? You should be able to answer the first two questions, at least in a general way, after having completed your study of this book. The third question is more difficult. You may be tempted to counter with: "What good are you?" Hopefully, your knowledge of the topics presented in this chapter, and the ones which follow, will help you answer these three questions.

SUGGESTED READING

An article by R. K. Bambach in the *American Scientist,* vol. 68, pp. 26–38, "Before Pangaea: the Geography of the Paleozoic World," describes how the world was changing in form during the time period when many of the most significant events in the establishment of Plant Kingdom divisions were happening. *Historical Plant Geography* by Philip A. Stott (1981) shows some of the changes in continental form during late Mesozoic time.

Harlan Bank's *Evolution and Plants of the Past* is a small book that every student of plant diversity should read. In addition to reviewing the origin of each of the groups of plants, it describes the methods used by paleobotanists in studying fossils.

M. R. Walter's *Stromatolites,* E. Flügel's *Fossil Algae,* and M. D. Brasier's *Microfossils* are other publications containing a wealth of material for any student interested in paleobotany. Marston Bates' book, *The Forest and the Sea,* is a popular presentation of conservation and ecology that you will find interesting.

STUDENT EXERCISES

1. Assume that sediments were uniformly deposited in the northeast Pacific at a constant rate of 1 mm/1000 years from the beginning of the Archeozoic era 4.5 bya until 1½

million years ago. Then assume that since 1½ mya this region has been rising at the rate of 3 mm/year, forming a mountain range in what is now western North America.
 (a) How deep would the accumulated sediments have become by the time the uplift began 1.5 mya?
 (b) How thick would that portion of the deposit be that accumulated during the Hadrynian Period?
 (c) How high would the highest peak in this mountain range be by now?
 (d) At what level going up the mountain side would you expect to find the most ancient stromatolites?
 (e) How far from the very top of the highest peak would you expect to find the most ancient flowering plants?

2. The life cycle of *Triticum aestivum* is diplohaplontic, consisting of a prominent sporophyte generation having leaves, stems, and roots; and a microscopic gametophyte generation. Knowing that a leaf cell has 42 chromosomes, how many chromosomes would you expect to find in each of the following cells? (a) a root tip cell; (b) a pollen grain; (c) a sperm cell produced by a pollen grain; (d) a cell in the pistil; (e) a zygote; (f) a pollen mother cell; (g) a cell in the endosperm.

3. Why are most mutations harmful? Under what kind of condition would you predict a higher than average percentage of new mutations to be beneficial?

4. What are some of the factors that have contributed to the great abundance and diversity of new kinds of fossils found in Devonian formations?

5. Which of the following plants or plant structures would you most likely find fossilized in Hadrynian formations a billion years old? (a) a rose—Anthopsida; (b) a sugar kelp—Phaeophycopsida; (c) a Nostoc ball—Cyanophycopsida;(d) a lichen—Ascomycopsida; (e) a maidenhair fern—Filicopsida; (f) a pinyon nut—Pinopsida; (g) some diatomaceous earth—Bacillariophycopsida.

6. Which of the above plants or plant structures might you find in Devonian formations, 350 million years old?

7. Obtain 10 pinyon nuts, tear open the gametophyte tissue, and examine the embryos. How many cotyledons does each embryo have? What is the average number of cotyledons per embryo?

SPECIAL INTEREST ESSAY 2-1

Age of the Earth and the Oldest Fossils

The oldest dated rocks on earth are 3.8 billion years old. No conclusive evidence of life, direct or indirect, has as yet been found in these ancient formations from Greenland. Because they are of igneous origin, intruded into older material since eroded away, we are, therefore, confident that the earth is more than 3.8 billion years old.

If the entire solar system originated simultaneously, as is commonly believed, then the age of the earth should be the same as that of meteorites which enter the earth's atmosphere from time to time, or about 4.5 to 4.8 billion years.

Russian chemists have calculated the age of the earth at about 5 billion years by measuring the amount of argon in the atmosphere and the amount of potassium 40 in the earth. It is generally assumed that the earth had no atmosphere when it was first formed; therefore, all of the argon there is from radioactive decay of potassium 40. This confirms the estimates made from meteorite age.

During the first billion or more years of the earth's existence, it was apparently void of life. Much of the oldest evidence of life is indirect: the occurrence of banded iron deposits in formations about 3.3 billion years old (Table 2–3) and of some carbonate formations in which the isotopes of carbon are in the same ratio as in living cells rather than in the usual ratio of purely inorganic carbonate. Banded iron deposits are significant in that they suggest a very small amount of free oxygen in the oceans or pools where these deposits were accumulating. Recently, microfossils have been found at North Pole in western Australia in formations reliably dated at 3.5 billion years according to Gurin (1980). These appear to be prokaryotic, similar in size and form to bacteria; blue-green algae of the same age were reported by Schopf (1993). Prior to these discoveries, the oldest known fossils were of bacteria and blue-green algae from the Fig Tree formation of South Africa, dated at slightly more than 3 billion years (Schopf and Barghoorn 1967).

PART II

THE RED LINE

Schizophyta
Rhodophyta
Eumycophyta

Plants probably originated in hot mineral pools similar to those now existing in Yellowstone Park. At high temperatures, chemical conversion of organic substances, obtained from the "broth" in which these plants were growing, made energy for synthesis of nucleic acids available, even though the respiratory enzymes present were inefficient. Earliest fossils are similar to modern Schizophyta, the bacteria and blue-green algae of today.

The Schizophyta are the only plants commonly found in hot mineral pools. They and other Aflagellobionta also tend to be abundant in the somewhat similar habitats of the tropical oceans. The Rhodophycopsida are successful on the continental shelves, in estuaries, and on reefs; the Cyanophycopsida are productive in the vast reaches of the open ocean. Heterotrophic Eumycophyta, morphologically and biochemically similar to the Rhodophycopsida, are abundant in terrestrial ecosystems, where they and the Schizomycopsida are important decomposers.

The Aflagellobionta are so named because true flagella never occur in this subkingdom. They are also commonly referred to as the "red line" because most Rhodophycopsida and many Cyanophycopsida contain red, water-soluble, photosynthetic pigments called phycoerythrin. Food reserves in the Aflagellobionta are highly branched carbohydrates made up of about 14 to 16 glucose units. Cell walls are built from similar glucose and glucosamine units. Structurally, cells are relatively uniform in this subkingdom and different from those of all other plants, suggesting evolution along a path independent from that of all other organisms for at least the last half billion years.

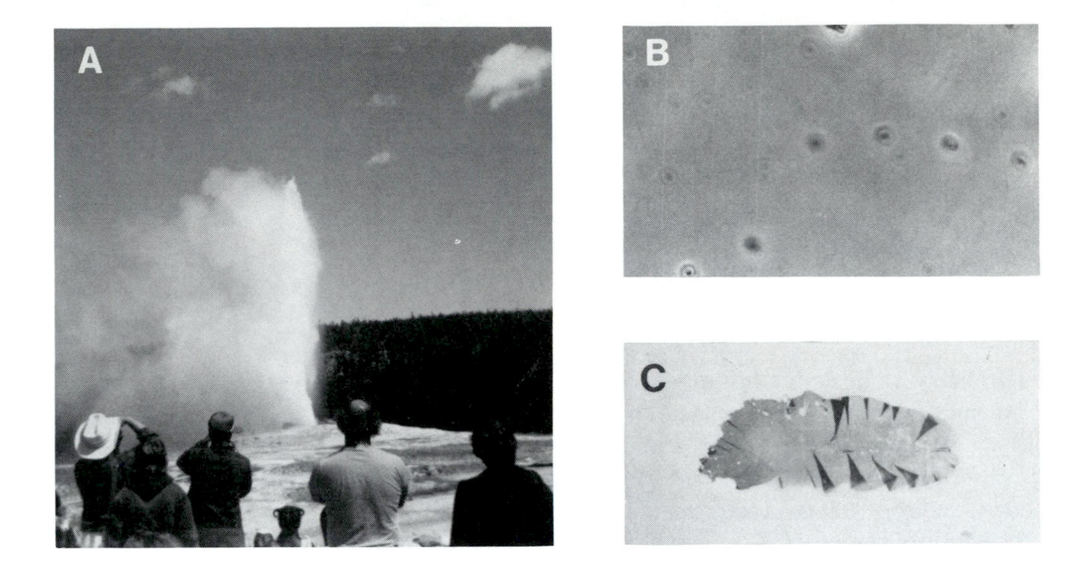

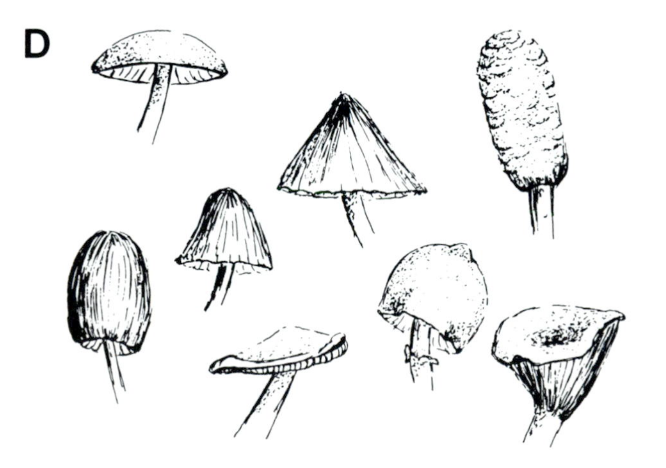

PART II. THE RED LINE

From simple prokaryotes which originated in hot pools and geysers similar to those in Yellowstone Park (A), the unicellular (B) and filamentous bluegreen algae and red seaweeds, like purple laver (C), evolved. The most complex and highly evolved of the Aflagellobionta are the chanterelles, inky caps, shaggy mane, and other mushrooms in the Division Eumycophyta (D).

3 PROKARYOTES OR FISSION PLANTS

Thermal springs throughout the world attract thousands of visitors for recreational purposes every year. A myriad of ailments such as arthritis, neuritis, high blood pressure, rheumatism, and other painful conditions are said to be relieved by bathing in hot mineral water. Centuries before the Romans built their baths at Bath in southwestern England, the sick and afflicted thronged to hot springs assured of being cured by the miraculous mineral waters.

But there is more in these springs than minerals. A visitor looking carefully near the water source will usually find slick crusts having a bluish gray, or even a deep blue, hue. Tearing a bit of the crust apart and examining it under a microscope, one sees small cells and hair-like filaments, typically 1 to 5 μm in diameter. With the aid of a botanical key, these can be identified as species of Cyanophycopsida: *Chroococcus, Aphanothece, Oscillatoria, Spirulina, Lyngbya,* and possibly *Nostoc* and *Tolypothrix,* all surviving at temperatures too high for most other forms of life; only sulfur bacteria and a few other Schizomycopsida, cousins to the Cyanophycopsida, are as heat hardy. And they too are usually evident. "How interesting," the vacationer may say before returning to important things, things that might have an influence on his life.

However, bacteria and blue-green algae do have an influence on our lives, far more than most people realize. Numerous examples may be found in agriculture of the importance of these microscopic organisms in production of food and other farm products. It was known by farmers in ancient Rome, for instance, that wheat and other crops gave higher yields when planted in a field where clover was grown the year before. Only in the last century and a half, however, have we become acquainted with the microorganisms responsible for the increased yield.

Less than 4% of the American working force is directly engaged in agriculture, each American farmer producing, on the average, enough food and fiber to feed and clothe 25 people. In many countries where over 70% of the population is so engaged, on the other hand, including some of the most densely populated areas of the earth, crop yields are low and starvation is common. What makes the difference? Largely, the availability of fertilizers. Each year in the U.S., mountains of commercial fertilizer are applied to farmland, making possible great agricultural productivity and hence a high standard of living. The involvement of bacteria and blue-green algae in the actual and potential production of fertilizers is one of the topics addressed in this chapter.

An obvious solution to the problem of low agricultural productivity in much of the world immediately comes to mind: sell more fertilizer to fertilizer-poor countries and teach the farmers how to make wise use of it. But "obvious" solutions do not always work. Even if farmers in "third world" countries had funds with which to buy fertilizers, and they do not, the merchant marine fleets of the world do not have enough ships to move it. Other solutions must be found; and they are being found. Well-trained Indian botanists, for example, have discovered that seeding rice paddies with a common **nitrogen-fixing** blue-green alga will increase rice yields significantly. Other botanists are testing species of legumes and strains of symbiotic bacteria in attempts to find the best combinations for high rates of nitrogen fixation. These combinations will then be recommended for inclusion in crop rotations. One of the most exciting areas of botanical research today involves the development of a technology for transferring nitrogen-fixation genes from the Schizophyta to the nuclei of crop plants such as wheat.

The most primitive of all plants, the bacteria and blue-green algae, make up the division Schizophyta (*schizo* = split or fission, *phyton* = plant), commonly called the fission plants and also known as the **Prokaryota** or the **Monerans.** Many of these plants reveal a remarkable degree of adaptation to environments not tolerated by other organisms. Some species, as already mentioned, survive in hot springs where temperatures very closely approach the boiling point of water; others exist under conditions of extreme salinity such as typify shallow desert lakes like the Salton Sea, the Dead Sea, and the Great Salt Lake. While the division is not large in number of species, its members are common and occur in all ecosystems of the earth. There are about 3000 species of Schizophyta classified in 15 orders and 69 families.

The two classes resemble each other and differ from all other plants and animals in several ways:

1. Both are prokaryotic (Table 3-1). The prokaryotic cell has been described as a "one envelope system," since the photosynthetic pigments, respiratory enzymes, DNA, RNA, and other chemicals responsible for the vital metabolic processes that occur in all living things are not segregated from each other in individual packages, or organelles, enclosed within a more inclusive container as they are in all **eukaryotic** organisms. Schizophyta thus have primitive **nucleoids** instead of nuclei, no plastids, no endoplasmic reticulum, and no mitochondria or other organelles, although the chemical ingredients and/or enzyme systems normally found in organelles are present.
2. The cells in both classes are very small, usually less than 5 µm in diameter and often less than 1 or 2 µm. Because of this, surface area relative to volume is large, and both bacteria and blue-green algae tend to have very high metabolic rates per unit mass.
3. The two classes seem to have originated about the same time and for 2 billion years were the only living things on the earth. There are fossils resembling bacteria and blue-green algae which are more than twice as old as fossils of any other organisms.
4. Many species of both classes live, grow, and reproduce in hot springs habitats too extreme for the survival of most organisms. It is believed that life originated in that type of environment, and if this is so it is not surprising that modern Schizophyta, which most nearly resemble the oldest fossils we have found, are the only living things today to inhabit such ecosystems.

5. Some species in both classes are capable of "fixing nitrogen," that is, of converting atmospheric nitrogen, N_2, into proteins, nitrates, and ammonia; we have no undisputed record of any other organisms possessing this ability.
6. A **parasexual** method of reproduction, **meromixis,** occurs in some bacteria, and good evidence of a similar process has been reported in the blue-green algae. True sexual reproduction involving meiosis and syngamy is not known in either class.
7. A number of unique biochemical similarities have been noted, including cell wall structure and fatty acid content. In addition, the two classes resemble each other and the other Aflagellobionta, but differ from other organisms, in the characteristics listed in Table 1-3.

Even though we refer to modern monerans as "primitive", because they are probably more like the first living things on earth than any other presently existing group of organisms, the prokaryotic cell is extremely complex, morphologically and physiologically. Studies of photosynthesis, respiration, and protein synthesis, for example, suggest that several hundred

TABLE 3-1

DIFFERENCES BETWEEN EUKARYOTIC AND PROKARYOTIC CELLS

Prokaryotes	**Eukaryotes**
No membrane-bound organelles occur	Membrane-bound nuclei, mitochondria, chloroplasts, lysosomes, etc. present
Few internal membranes occur; mesosome membrane associated with DNA and protein synthesis	Elaborate endoplasmic reticulum present and associated with protein synthesis
Cells very small, only slightly larger than eukaryotic mitochondria	Even the smallest cells are generally 5 to 10 times larger than typical procaryotic cells
Single circular DNA molecules or chromosome lacking centromere and telomeres	Usually several linear chromosomes each with a centromere, telomeres, and other chromomeres
Each gene present only once	Repetitive genes common
No nucleolus or nucleolus counterpart	One or more nucleoli that code for rRNA and consist of tandemly linked genes
Formylated methionine initiates all polypeptides	Unformylated methionine initiates all polypeptides
No microtubules or spindle fibers; cell division not affected by colchicine	Microtubules (e.g., spindle fibers) subject to degradation by colchicine present
70S ribosomes with 30S and 50S subunits	80S ribosomes with 40S and 60S subunits
Proteins loosely bound to DNA; no histones	Five kinds of histones are tightly bound to DNA
Equational cell division amitotic (no metaphase or anaphase stages occur); meiosis does not occur; no actin or myosin-like proteins; polycystronic mRNA common; only one kind of RNA polymerase; no introns, exons, or other intervening sequences in DNA molecules; no cap on RNA molecules; Pribnow box (TATAATG) near the RNA start point of promoter regions	Distinct metaphase and anaphase stages in mitosis; sexual reproduction involving meiosis and syngamy occurs; actin and myosin proteins present; only monocystronic mRNA occurs; three major forms of RNA polymerase (one each for mRNA, rRNA, and tRNA; introns and exons common; cap of guanosine nucleoside at 5′ end of mRNA molecules; Hogness box (TATA box) is the RNA polymerase II promoter

Note: In spite of the differences in DNA structure, gene splicing between prokaryotes and eukaryotes is possible; much of our knowledge of eukaryote genetics has come from studies of prokaryotic genetics.

enzymes, the production of each of which is controlled by a separate gene, are carefully correlated to make this possible. Each cell contains numerous ribosomes; unlike the ribosomes of eucaryotic organisms, they lie free in the cytoplasm and they are also smaller than ribosomes of other plants and animals and therefore sediment more slowly. The **nucleoid** seems to consist of a single circular chromosome, attached at one point to the plasma membrane, plus zero to several **plasmids.** Division of the nucleoid appears to be similar to mitosis; however, the chromosome does not migrate to the equatorial plane, as in metaphase of mitosis, and the two "chromatids" simply separate from each other without going through an anaphase-like stage (DeLamater 1951, von Hofsten and Pearson 1965). The so-called "central body," reported by early cytologists, consists of the nucleoid surrounded by numerous ribosomes. Because they could not distinguish between DNA and RNA, early botanists failed to observe the chromosome and its division, because it was hidden by the ribosomes. The respiratory enzymes, though not packaged by a special membrane, are located on infoldings of the plasma membrane adjacent to the nucleoid; this structure is called the **mesosome.**

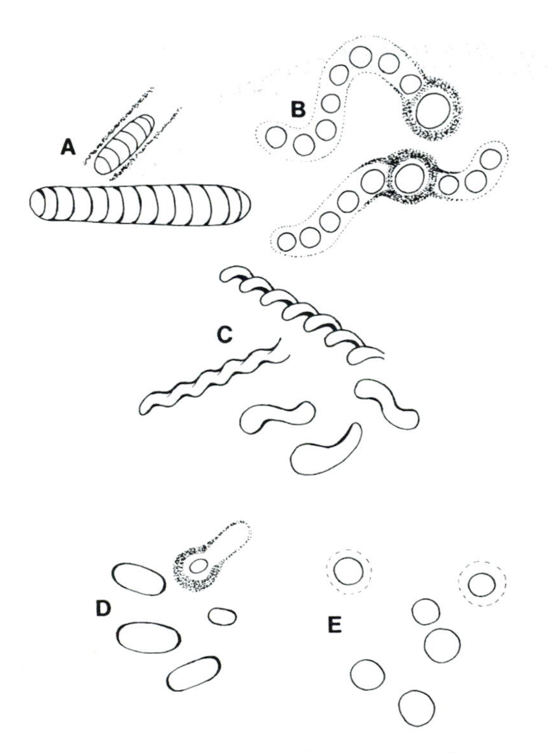

Figure 3–1 Morphological variation in cell types in the Schizophyta. (A) Filaments of short cylindrical cells, upper filament enveloped in a gelatinous sheath; (B) chains of spherical or ovoid cells with terminal and intercalary heterocysts; (C) long and short spirals; (D) rod-shaped cells (bacilli), one with endospore; (E) spherical cells (cocci) with and without gelatinous sheath surrounding individual cells or pairs of cells (diplococcus).

There is little differentiation of cells in the Schizophyta. Individual cells may be spheres, rods, or spirals (Figure 3-1). In some species of both classes, **trichomes** (or filaments) of cells are produced, but within the trichome the cells are mostly alike in appearance and function. Usually the cells are enclosed in a gelatinous sheath. In some genera, for example *Nostoc,* the sheaths fuse together to form large masses with more or less distinct macroscopic form. However, tissue and organs having distinct, specialized functions never occur.

As in other Aflagellobionta, true flagella are never present. Nevertheless, motility is relatively common and is of two types; (1) In many blue-green algae and in some bacteria—as well as in a few red seaweeds—**herpokinetic** or "snake-like" motility may be observed. This is a gliding or waving motion, usually smooth but sometimes jerky, which is most commonly observed in sheathless, filamentous species of Oscillatoriales such as *Oscillatoria* and *Spirulina.* Several hypotheses have been advanced to explain such motility, but none has proved completely satisfactory as yet. (2) In the Schizomycopsida, motility due to the lashing of **proterokonts** is common (Figure 3-2). While proterokonts are frequently referred to as flagella, they differ from true flagella structurally and chemically, being single stranded, for example, whereas true flagella, which occur in both Bracteobionta and Lorobionta, consist of eleven strands, a core of two surrounded by a ring of nine.

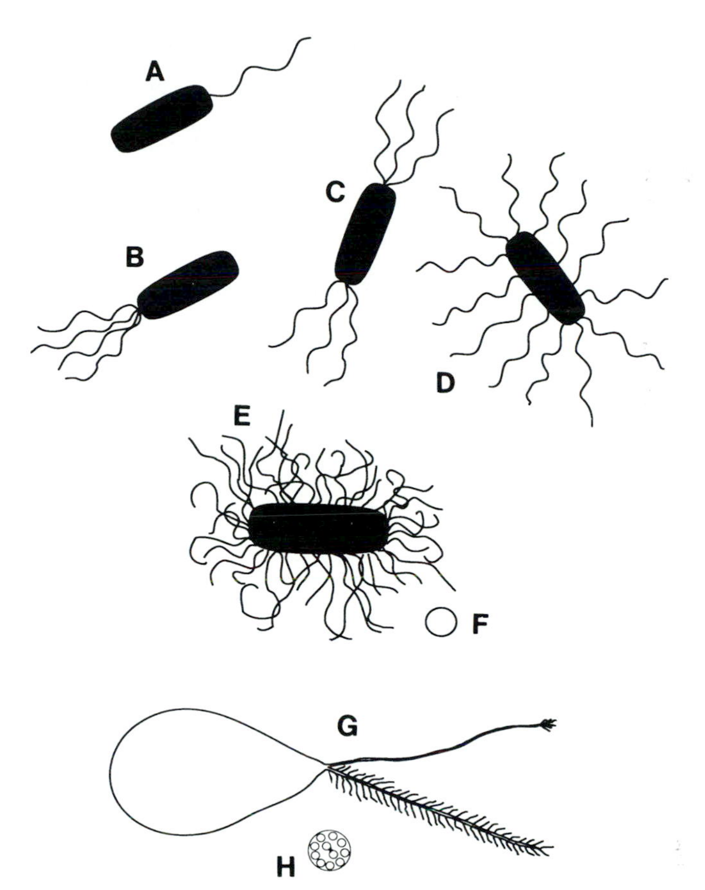

Figure 3–2 Proterokonts and flagella. (A–D) Bacterial proterokonts. Commonly called flagella, they differ from true flagella in lacking the 9+2 basic structure. Special staining techniques are used to make proterokonts visible in the light microscope: (A) monotrichous; (B) lophotrichous; (C) amphitrichous; (D–E) peritrichous; (E) *Salmonella* cell with peritrichous proterokonts, drawn from electron micrographs; (F) Cross section through one of the proterokonts in (E); (G) Sperm cell of *Saprolegnia* with two flagella, one "tinsel type" with many mastigonemes, and the other "whiplash type"; (H) Cross section through one of the flagella in (G).

Because of their distinctive characteristics, especially the small size of the cells and the absence of vesicular mitochondria, plastids, or other organelles, the Schizophyta are not often confused with other plants. Occasionally, an exceptionally small green alga or xanthophyte might be mistaken for a blue-green, but careful observation reveals the presence of a plastid or the lack of characteristic bluish coloration. Colonies of yeasts sometimes resemble bacteria, but individual cells are larger and staining reveals a distinct nucleus.

Food storage is primarily in the form of glycogen-like polysaccharides: bacterial starch in the Schizomycopsida and cyanophycean starch in the Cyanophycopsida. Proteins, fats, and other food substances are also produced. The fatty acid content of blue-green algae differs from that of bacteria; in studies conducted in the 1960s, it was found that saturated and monounsaturated 16-carbon acids were the most common fatty acids in both classes. Polyunsaturated fatty acids are common in the Cyanophycopsida but rare in the Schizomycopsida (Table 3-2).

TABLE 3-2

FATTY ACID CONTENT OF SOME REPRESENTATIVE SCHIZOMYCOPSIDA AND CYANOPHYCOPSIDA

Family and Order[a]		Number of Taxa	Taxa with Branched Chain Acids (%)	Saturated Fatty Acids				Monounsaturated Acids			Polyunsaturated Acids		
				10:0 + 12:0	15:0	17:0	14:0 + 16:0 + 18:0	16:1	17:1	14:1 + 18:1	16:2	18:2	18:3
Bacillaceae	EUBA	2	100	0	Pr	25	55	0	0	0	0	0	0
Micrococcaceae	EUBA	1	100	—	—	—	Pr	—	—	—	—	—	—
—	PSEU	4	100	0	25	5	27	15[b]	5	11	0	0	0
Chroococcaceae	CHRO	5	0	Tr	1	0	49	27	0.2	12	1	6	2
Entophysalidaceae	CHRO	2	0	0.2	0	0	43	18	1	20	0	13	6
Oscillatoriaceae-A	OSCI[c]	4	0	0.6	1	0	41	19	0	21	4	7	6
Oscillatoriaceae-B	OSCI[c]	2	0	42	2	0	31	6	0	11	0	3	13
Nostocaceae	NOST	3	0	0.7	1	0	36	17	2	14	0	13	16
Scytonemataceae	NOST	1	0	Tr	1	0	38	13	0	11	5	11	6
Stigonemataceae	STIG	1	0	0.6	0	0	58	24	0	18	0	0	0

Note: Acids are symbolized C:B, where C = number of carbon atoms in the chain and B = the number of double or triple bonds present. Data from Kanada 1963, Parker et al. 1967, and Holton et al. 1968. (Pr = present, indicating fatty acid was present but the per cent was not reported; Tr = trace, indicating less than 0.05% present; — = no data reported.)

[a] Order names are abbreviated by the first four letters; in the Schizomycopsida, EUBA = Eubacteriales and PSEU = Pseudomonadales; in the Cyanophycopsida, CHRO = Chroococcales, OSCI = Oscillatoriales, NOST = Nostocales, and STIG = Stigonematales.

[b] Marine culture G contained 1.6% 15:1 fatty acid and 0% 16:1 acid.

[c] Oscillatoriaceae group A consisted of two species of *Oscillatoria* and one each of *Lyngbya* and *Microcoleus;* Oscillatoriaceae group B consisted of two species of *Trichodesmium.*

Branched chain acids are abundant in all bacteria that have been investigated but in none of the blue-green algae.

The most obvious difference between the Schizomycopsida and the Cyanophycopsida is color. Most blue-green algae contain chlorophyll and can carry on photosynthesis; most bacteria are colorless and therefore heterotrophic. Cyanophycopsida cells also contain red and blue **phycobilin** pigments, which aid in photosynthesis; none of the bacteria do. Nevertheless, two groups of bacteria synthesize food. The **photoautotrophic** bacteria possess thylakoids similar to those of the Cyanophycopsida but containing bacteriochlorophylls instead of chlorophyll *a*. The bacteriochlorophylls differ from chlorophyll *a* not only structurally (Figure 3-3) but also physiologically, since they utilize H_2S instead of H_2O, function only anaerobically, and do not give off oxygen gas. **Chemoautotrophic** bacteria obtain energy for food synthesis from the oxidation of sulfur or other substances and can therefore live and produce food in total darkness.

Many modern biologists are of the opinion that the Schizophyta are so different from other plants and animals that they should be segregated as a separate and distinct kingdom variously designated the Monera or the Prokaryota (e.g., Whittaker 1969). This has the advantage of stressing the importance of vesicular organelles and the differences in chromosome and ribosome structure, along with other chemical differences (Table 3-1), but it has the disadvantage of ignoring the plant-like features of Schizophyta and the many similarities between the Cyanophycopsida and Rhodophycopsida. Despite their small size and other unique characteristics, the Cyanophycopsida and also the autotrophic bacteria are ecologically plants and resemble other plants in so many ways that it seems best to accept the traditional hypothesis and include them in the plant kingdom.

Classification presents special problems in the Schizophyta. In most plants and animals, sexual reproduction provides some of the chief clues to genetic relationships and phylogeny, but here true sexual reproduction is lacking; therefore, other means of establishing relationships are needed. Because nucleoids are very small, with little variation among species, karyological studies are of little or no help. The gaps that occur in the spectra of morphological variations divide groups of organisms into smaller units within which less variation occurs. These smaller units are assumed to be species. Probable relationships among the 15 orders of Schizophyta are suggested in Figure 3-4.

CLASS 1. SCHIZOMYCOPSIDA

The Schizomycopsida (from the Greek, *schizo* = to cleave, and *myceto* = mold or fungus), or fission fungi, more commonly called bacteria, are known as the Schizomycophyta by those who prefer to place the bacteria and blue-green algae in separate divisions. Most schizomycetes are heterotrophs, either saprophytes or parasites, but a few species are either photosynthetic or chemoautotrophic. In terms of total global food production, autotrophic bacteria are of little significance, but in some ecosystems, for example, sewage lagoons and caves, they play an important role. Blind beetles and other insects in some cave ecosystems are apparently dependent on the food produced by chemoautotrophic bacteria (Naseath 1974). Because heterotrophic bacteria are able to utilize a wide variety of organic substances, rapidly breaking them down to carbon dioxide, water, and minerals, they are efficient decomposers and consequently very important components of all natural ecosystems.

Bacteria are easily obtained. They can be recovered from soil by taking a sterilized petri dish of nutrient agar—containing beef extract, peptone, and granulated agar—and pouring on it some water from a suspension of garden soil. Other petri dishes can be streaked with a piece of spoiled meat. Simply breathing on some will usually result in colonies of bacteria forming. The bacterial colonies will be visible in 2 or 3 days and can be recognized by their relatively small size, smooth or slightly rough surface and edges, and symmetrical form. Some will be brightly colored. Colonies of fungi are cotton-like and soon overgrow everything else; yeast colonies are usually dull, flat, asymmetrical, and more extensive than the bacterial colonies. Microscopic examination is needed to make the final decisions as to the type of organisms present as well as the exact species.

General Morphology and Physiology

Likely more is known about the morphology and physiology of bacteria than any other class of plants excepting the Anthopsida. Many human diseases are caused by bacteria; therefore, people have become intensely interested in their causes, and a science of bacteriology has developed (Figure 3-5). Actually, only a small percentage of bacterial species are disease producing. Most bacteria are indispensable parts of the ecosystems in which we live. As decomposers, they convert complex organic compounds into simple minerals; without them neither human life nor other forms of life could continue. Our understanding of nondisease-producing bacteria, unfortunately, is still rather meager.

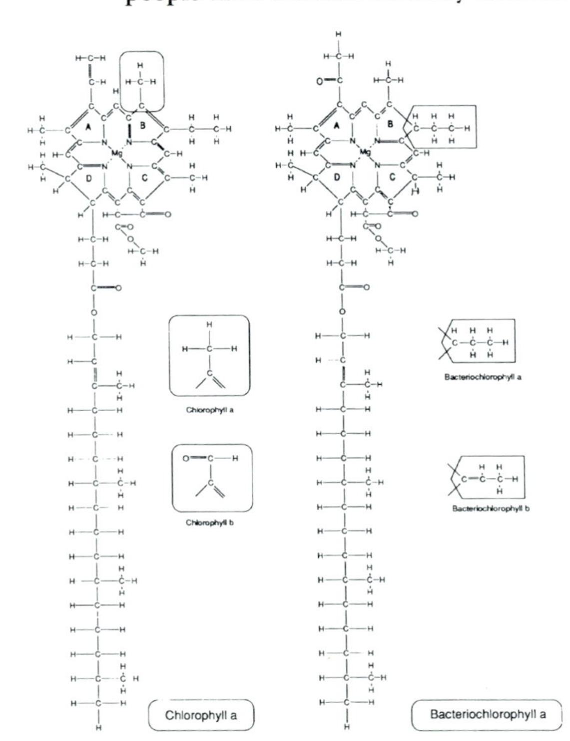

Figure 3-3 Photosynthetic pigments in Schizophyta. Chlorophyll *a* and bacteriochlorophyll *a* differ from each other in the 2-carbon group attached to the A ring. Chlorophyll *b* differs from chlorophyll *a*, and bacteriochlorophyll *b* differs from bacteriochlorophyll *a* in the groups attached to the B ring.

Because bacterial cells are extremely small, only a few morphological types can be recognized: coccoid or berry-shaped cells, bacilloid or rod-shaped, and spirilloid or curved cells are the most common morphological types. Individual cells may be grouped in chains, clusters, or pairs, or they may occur free of each other (Figure 3-1). **Proterokonts,** called "flagella" in many books and articles, are attached to the cells of several species (Figure 3-2). Some species are **peritrichous,** with proterokonts covering the entire cell surface; others are **monotrichous** with a single proterokont or **lophotrichous** with a tuft of proterokonts at one end of each rod-shaped cell. Many species, of course, lack proterokonts and are nonmotile. Similar in appearance to proterokonts, but shorter and seemingly more rigid, are **pili.** Each pilus originates from the cell membrane and is proteinaceous in nature. Pili occur almost exclusively on **Gram negative** bacteria. Vaccination with purified pili

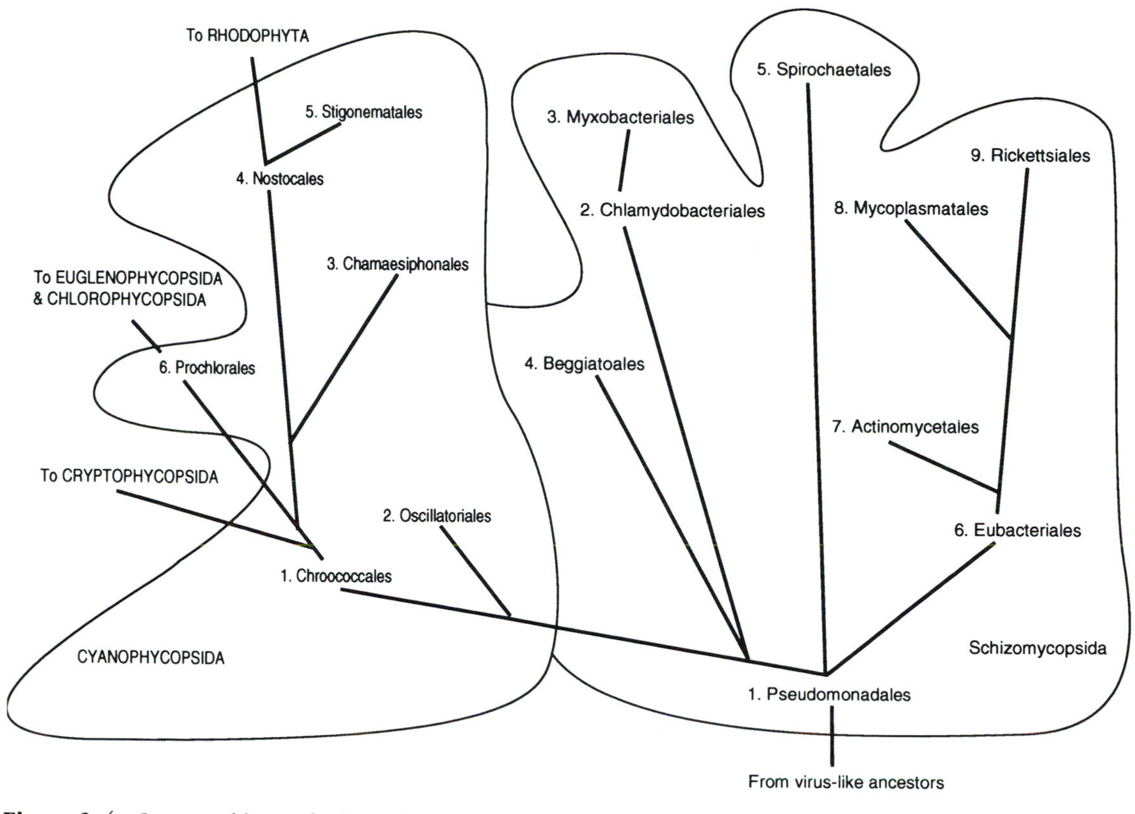

Figure 3–4 Suggested lines of relationship among the orders of the Schizophyta.

protects the host from infection. Like proterokonts, pili are visible with the light microscope only on cells that have been prepared with special staining methods. Further morphological detail is not readily discernible in such small cells, even with special staining techniques; therefore, physiology and biochemistry are more useful than morphology in identifying bacteria. Bacterial keys are based largely on biochemical tests.

One of the simplest, and certainly the most widely used, of the biochemical tests employed in studying bacteria is a differential staining technique introduced in 1884 by Christian Gram. This technique allows the student to classify all bacteria into at least four categories: Gram-positive rods, Gram-positive cocci, Gram-negative rods, and Gram-negative cocci. An initial stain and a counterstain of a different color are used in making the test. After applying the initial stain, a weak solution of iodine is added. The slide is then washed in a solution of 95% alcohol, after which the counterstain is added. When bacteria containing muramic acid or other carbohydrates and very little lipid in the cell wall are stained with the initial stain and then treated with iodine, the combination of dye, iodine, and carbohydrates produces a stable pigmented substance. This pigment is not affected by ordinary fat solvents such as alcohol and acetone. Such cells retain the pigmented (usually blue) cell wall complex and are designated Gram positive. Cells that are low in carbohydrate and high in lipids do not retain it, but show instead

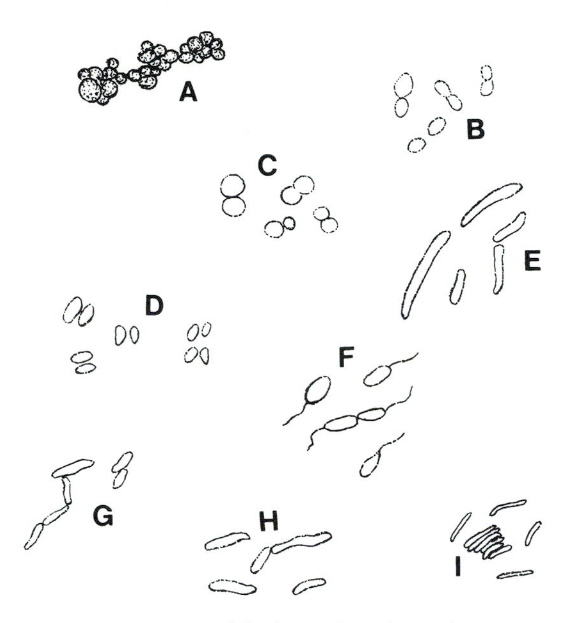

Figure 3–5 Some of the bacteria pathogenic to man, with ailments they incite in parentheses: (A) *Micrococcus (Staphylococcus) pyogenes* (boils and carbuncles); (B) *Diplococcus pneumoniae* (lobar pneumonia); (C) *Neisseria gonorrhoeae* (gonorrhoea); (D) *Neisseria meningitidis* (spinal meningitis); (E) *Proteus vulgaris* (common saprophyte, not ordinarily pathogenic to man but may cause peritonitis); (F) *Salmonella typhosa* (typhoid fever); (G) *Shigella dysenteriae* (dysentery); (H) *Corynebacterium diphtheriae*, (diphtheria); (I) *Mycobacterium leprae* (leprosy or Hansen's Disease).

the color of the counterstain (usually red), and are called Gram negative. Other differential stains operate on the same principle; the best known of these is the acid-fast technique which is valuable in the identification of the bacteria which cause tuberculosis and leprosy. Some characteristics of Gram-positive and Gram-negative bacteria are shown in Table 3-3.

The cell wall of Gram-positive bacteria is composed of *N*-acetyl muramic acid, a carbohydrate similar in basic structure to cellulose but simpler. Muramic acid consists of a molecule of glucosamine to which is attached a molecule of lactic acid. Glucosamine has exactly the same structure as *D*-glucose, the basic building block of starch and cellulose, except that the OH group attached to the no. 2 carbon atom has been replaced by an NH_2 group (Figure 5–4). Formation of glucosamine from glucose is apparently an adaptation to the saprophyte niche, an ecological niche in which amine groups (NH_2) are present in excess. In autotrophic organisms, organic compounds containing nitrogen are scarce and often become limiting factors because they are essential to photosynthesis. In heterotrophs, on the other hand, they may build up and become toxic unless bound in a biologically inert compound like muramic acid.

Probably as a result of their small cell size, bacteria are characterized by very rapid metabolic rates and a large variety of metabolic processes. Ecologically, bacteria are therefore classified as **decomposer organisms.** While all plants and animals, including humans, function as decomposers to some extent, breaking down organic substances into inorganic products such as carbon dioxide, water, and ammonia, bacteria carry on these reactions far more rapidly than do other organisms. This is accomplished by a wide variety of digestive enzymes.

Bacteria multiply rapidly: a cell generation in bacteria may be less than half an hour in some species, compared to 12 to 24 hours in many rapidly growing eukaryotic organisms. In one to a few days, therefore, a single bacterial cell may have produced a visible colony consisting of billions of cells. Since colony morphology varies greatly from one species to another, an experienced biologist is often able to recognize many species of bacteria on the basis of colony morphology alone.

Respiration in bacteria, as in many other organisms, may be either **aerobic** or **anaerobic.** In anaerobic respiration, O_2 is not involved. Many bacteria are **obligate anaerobes** and can carry on their metabolic processes only in the absence of oxygen. An example of an obligate anaerobe is *Clostridium tetani,* the organism that causes tetanus. A normal inhabitant of soil,

TABLE 3–3

REACTION OF BACTERIA TO GRAM'S STAINING TECHNIQUE

	Gram Positive	**Gram Negative**
End results	Cells retain initial stain (blue); are not affected by the counterstain	Cells lose the initial stain; show the counterstain (red)
Cell wall characteristics	Walls contain high levels of muramic acid	Walls contain little or no muramic acid; contain alcohol-soluble lipids
Reaction to iodine	Muramic acid reacts with Gram's iodine plus stain to form an alcohol-insoluble substance	Cells do not form an alcohol-insoluble substance
Reaction to alcohol and other fat solvents	Cell wall substances are not dissolved	Initial stain is removed along with alcohol-soluble cell wall substances
Reaction to penicillin and other antibiotics	Penicillin interferes with muramic acid synthesis, thus killing the cells	Penicillin has no effect on cells
Examples of genera	*Staphylococcus, Streptococcus, Treponema*	*Neisseria, Proteus, Klebsiella*

it is capable of **nitrogen fixation** and is therefore ecologically important, especially in water-logged soils. *C. tetani* is able to convert N_2 to a useful form of nitrogen because water-logged soils are depleted of free O_2. It is also able to live and grow in puncture wounds in animal tissue because deep in a healing wound no oxygen can penetrate. A toxin it produces affects the nervous system and often causes death. It is believed that *Clostridium* and other anaerobes are relicts of a bacterial flora that thrived in Archeozoic times when oxygen was a poisonous gas rather than an essential substance.

Several species of aerobic bacteria are also capable of nitrogen fixation. However, nitrogen fixation can take place only in the absence of free oxygen; therefore, these bacteria have special adaptations to ensure freedom from oxygen. *Rhizobium* bacteria form **nodules** on legumes and some other plants. Anaerobic conditions are maintained inside the nodules. Carbohydrates are provided by the legume host and organic forms of nitrogen, beneficial to both host and bacterium, are produced by the bacteria. These bacteria are therefore called **symbiotic nitrogen-fixing bacteria** because both host and bacterium profit from the relationship.

Photosynthesis occurs in several species of bacteria but is quite different from photosynthesis in other plants, including blue-green algae, in three important respects: (1) the chlorophyll of bacterial cells is chemically different from that of all other plants, (2) hydrogen sulfide rather than water is involved in the reduction reactions, and (3) there is no oxygen evolved in bacterial photosynthesis.

There are three groups of photosynthetic bacteria. (1) In the Chlorobacteriaceae, bacterioviridin is the photosynthetic pigment which captures the energy of the sun and converts hydrogen sulfide and carbon dioxide into food and elemental sulfur. (2) In the Thiorhodaceae, the photosynthetic pigment is bacteriochlorophyll (bacteriopurpurin) and the chemical reactions are the same as in the Chlorobacteriaceae, with hydrogen sulfide again providing the hydrogen for the reaction. The first light reaction is

$$H_2S \rightarrow 2\ H^+ + 2\ e^- + S \qquad (1)$$

in both of these groups. Therefore, the net photosynthetic reaction is

$$2\ H_2S + CO_2 \rightarrow (CH_2O) + 2\ S + H_2O \qquad (2)$$

(3) In the Athiorhodaceae, the pigment which traps the solar energy is bacteriochlorophyll, as in the Thiorhodaceae, but organic compounds instead of hydrogen sulfide serve as hydrogen donors by reacting with carbon dioxide in food production.

Chemosynthesis has been studied in detail in at least one bacterium, *Beggiatoa alba,* and is known in several other species. In *B. alba,* the energy needed to drive the synthesis reactions comes from oxidation of sulfur to sulfates by means of an enzyme system:

$$2\ S + 2\ H_2O + 3\ O_2 \rightarrow 2\ H_2SO_4 \qquad (3)$$

and the reactions can therefore take place in the dark as readily as in the light. The remaining reactions are apparently identical to corresponding photosynthetic reactions in the Chlorobacteriaceae and Thiorhodaceae:

$$2\ H_2S + CO_2 \rightarrow (CH_2O) + 2\ S + H_2O \qquad (4)$$

The difference between Equations 2 and 4 is that light is the source of energy for equation 2, whereas the chemical energy released from oxidation of hydrogen sulfide or other inorganic substance is the source of energy for equation 4.

The nitrifying bacteria which convert ammonia to nitrites and nitrites to nitrates in the soil and the iron bacteria of bogs and similar habitats have also been reported to be chemoautotrophic. The nitrifying bacteria obtain energy to drive chemosynthetic reactions since nitrogen is oxidized from its ammonium form to the nitrite and nitrate forms. The energy released as ferrous ions are converted to ferric is reported to be the source of energy needed to drive the synthesis reactions in bog bacteria; however, these reports have not been well substantiated and some biologists doubt that bog bacteria are true chemoautotrophs. Naseath (1974) has suggested that iron bacteria of the order Myxobacteriales convert ferrous ions in the Snake River basalt to ferric ions, releasing energy which they utilize in producing food. This food enables the bacterial colonies to grow and reproduce in the total darkness of the caves. According to this hypothesis, the blind beetles inhabiting the ice cave ecosystems of the area survive by feeding on the bacteria.

Several years ago, methods were developed which now enable us to preserve cell structures when preparing bacteria for electron microscope studies. Since then, several species of bacteria have been studied in considerable detail. A cell of *Escherichia coli* measuring 1 μm by 2 μm, for example, may contain 5000 distinguishable components. The genetic information is carried by a single DNA molecule large enough to code thousands of proteins.

There is no endoplasmic reticulum in bacteria cells, but ribosomes are present. These contain about 60% RNA and slightly less protein than the ribosomes of eukaryotic plants. They are also slightly smaller than the ribosomes of eukaryotes, sedimenting at 70s. There is no Golgi apparatus. Proterokonts, when present, are usually peritrichous, being distributed over the entire cell surface. Figure 3-6 is an electron micrograph of *Pseudomonas syringae,* the incitant of lilac wilt disease. The 40-nm thick cell wall and thin plasma membrane (<5 nm) are readily visible. There are no membrane-bound organelles, but hundreds of ribosomes, each about 15 nm in diameter, can be counted.

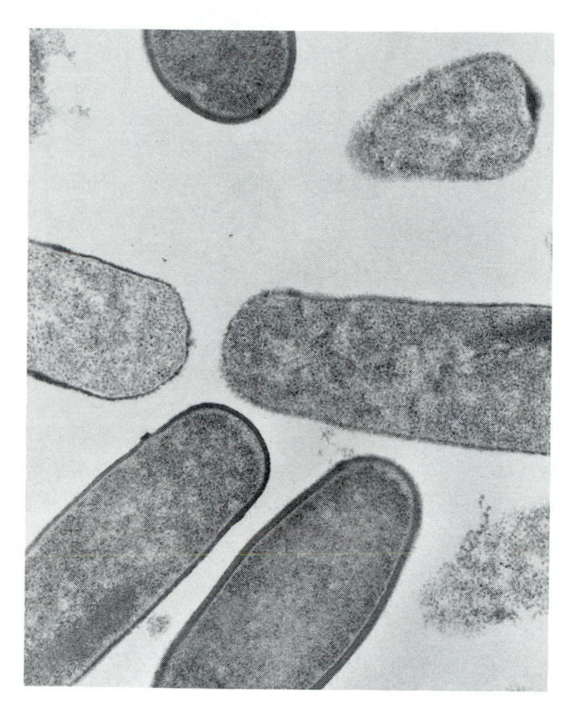

Figure 3–6 Electron micrograph of *Pseudomonas syringae,* incitant of lilac wilt. Slight desiccation in the cell at lower center has caused the plasma membrane to pull away from the cell wall. Ribosomes are numerous and appear as small, unattached, spherical bodies tending to occur in clumps; there is no endoplasmic reticulum in bacterial cells. (Photograph courtesy W. M. Hess, Brigham Young University Microscopy Laboratory, Provo, Utah.)

Phylogeny and Classification

Over the last 3.5 billion years, evolution in the Schizomycopsida has apparently proceeded along three lines, as indicated in Figure 3-4. The bacteria which today most nearly resemble the first primitive species are believed to be the Pseudomonadales; similar to them are the Beggiatoales and other rod-shaped groups which seem to form a link between the Schizomycopsida and Cyanophycopsida. The species along this line tend to be Gram-negative with a tuft of proterokonts at each pole of the cells. The second line of evolution has led to the spirochaetes, long, slender spirals common in polluted water high in organic matter. Parasitism is especially common in the orders along the third line leading to the Rickettsiales and Mycoplasmatales, proterokonts tend to be peritrichous, and the species are mostly Gram-positive rods and cocci. The degree of diversity in the Schizomycopsida is shown in Table 3-4, which lists the chief characteristics of the nine orders.

Special problems not encountered elsewhere in the Plant Kingdom complicate the taxonomy of the Schizophyta. Definitions of species as natural units within which genetic exchange of DNA may freely occur do not apply to the Schizomycopsida because of the absence of sexual reproduction. Small cell size and lack of vesicular nuclei reduce the effectiveness of cytology as a taxonomic tool. Bacterial taxonomists, therefore, base their systems on the one gene–one enzyme principle and classify in the same genera and families bacteria which have identical or similar groups of enzymes as revealed by various physiological and biochemical tests.

TABLE 3–4

GENERAL CHARACTERISTICS OF BACTERIAL ORDERS, SHOWING THE RANGE OF MORPHOLOGICAL TYPES IN THE SCHIZOMYCOPSIDA

Orders (Fam-Gen-Spp)	Cell Form and Staining Properties	Proterokont Characteristics	Representative Genera
Pseudomonadales (11–76–433)	Straight or curved Gram-negative rods, rarely in pairs or chains	Tuft on each end of cell, monotrichous or bipolar	*Thiosarcina, Chromatium, Vibrio, Nitrosomonas, Nitrobacter*
Chlamydobacteriales (3–8–26)	Gram-negative rods in long chains with ferric hydroxide sheaths; false branching common	None; motility herpokinetic	*Leptothrix, Pelonema, Sphaerotilus, Clonothrix, Crenothrix*
Myxobacteriales (5–12–71)	Naked, rod-shaped, Gram-negative cells containing abundant carotenoids; sporangia brightly colored	None; motility ameboid	*Chondromyces, Cytophaga, Archangium, Polyangium, Synangium, Chondrococcus, Myxococcus, Melittangium*
Beggiatoales (4–9–34)	Short Gram-negative rods in relatively long filaments	None; motility herpokinetic	*Beggiatoa, Leucothrix, Thiothrix, Thioploca, Thiospirillopsis*
Spirochaetales (2–6–49)	Long spiral Gram-negative cells	Probably tuft on each end of cell	*Spirochaeta, Treponema, Saprosira, Cristispira, Borrelia*
Eubacteriales (14–69–621)	Short, mostly Gram-positive rods or spheres	Peritrichous	*Azotobacter, Lactobacillus, Escherichia, Salmonella*
Actinomycetales (6–9–228)	Slender, elongated acid-fast cells having a tendency to branch	None, or rarely peritrichous	*Streptomyces, Mycobacterium, Nocardia, Actinomyces, Micromonospora*
Mycoplasmatales (1–3–15)	Sphaeroidal, naked cells resembling minute fried eggs, do not stain; form transparent colonies	None	*Mycoplasma, Acholeplasma, Ureaplasma*
Rickettsiales (4–18–67)	Difficult-to-stain spheres or rods so small they pass bacterial filters	None or rarely peritrichous	*Rickettsia, Coxiella, Migagawanella, Cowdria, Dermacentroxenus, Bartonella*

Note: The number of families (Fam), genera (Gen), and species (Spp.) in each order is indicated in the first column in parentheses.

Recent advances in biochemical technology are leading to new hypotheses concerning bacterial phylogeny (Woese 1981, 1985; Berry and Jensen 1988). For example, *Pseudomonas,* with over 100 species, is a widespread, well-known, and probably very ancient genus. Berry and Jensen, employing the latest biochemical technology, found so much variation within this one genus that they proposed dividing it into five superfamilies. This suggests that we can expect many changes in bacterial taxonomy in the next few years.

Staining characteristics are important in determining genetic relationships among bacteria. For example, cell walls in the Actinomycetales respond to the **acid-fast stain,** but most other bacteria give a negative reaction to this stain. Infections caused by Gram-positive bacteria are more likely to be controlled by antibiotics than are those caused by Gram-negative species. Motility, or lack of it, is another criterion used in classifying bacteria. The nature of the proterokonts (or bacterial "flagella") was early considered significant by Migula in 1900 (Breed 1957) and others (Figure 3-2). In the line leading to the Beggiatoales and Myxobacteriales, autotrophism is common, and so is herpokinetic motility. The presence or absence of endospores is also used in classifying species and genera into families in both major lines (Figure 3-1D).

At the species and genus level, cell size has been widely (but not wisely) employed in bacterial classification. In *Chromatium,* 12 species were described by early bacteriologists. These ranged in size from the tiny, spherical *C. minutissimum,* with cells 0.5 μm in diameter, to the large, egg-shaped *C. gobii,* with cells about 10 × 25 μm. There seemed to be no clear-cut characteristics other than size to separate the 12 species, and this aroused suspicion in the minds of many bacteriologists over the years. Finally, a single species of *Chromatium,* isolated from a California sewage lagoon, was grown in pure culture in the laboratory, and all 12 "species" were recovered simply by manipulating the environmental conditions under which the bacteria were grown. This experience has thrown doubt on the value of size as a taxonomic criterion in bacteria.

Bergey's *Manual of Determinative Bacteriology* (Breed 1957; Staley et al. 1989) is the most widely accepted authority on bacterial taxonomy. The orders shown in Figure 3-4 and described in Table 3-4 are adapted from recent editions of Bergey's *Manual,* modified to emphasize similarities. Included in the 9 orders are 48 families, 208 genera, and 1,544 species. Most of the species belong to 2 orders, Pseudomonadales and the Eubacteriales, commonly called the "true bacteria." The Actinomycetales, commonly called "ray fungi," also contains many species.

Representative Genera and Species

Table 3-5 lists several genera and species of bacteria. Included are a number of disease-producing parasites along with some ecologically important saprophytes and some species of commercial importance. Some of these are illustrated in Figures 3-5 and 3-6.

In the Pseudomonadales, *Vibrio comma* (also called *Vibrio cholerae*), the incitant of cholera, is a well-known representative. Its cells are slightly curved rods or shortened spirals. *Nitrosomonas* and *Nitrobacter* are chemosynthetic genera, utilizing the energy released by oxidizing ammonia to nitrites or nitrites to nitrates to synthesize carbohydrates. The photosynthetic bacteria are also in this primitive order. Fossil bacteria of the Archeozoic era, like *Eobacterium isolatum* from the 3 billion year old Fig Tree chert of South Africa, represent types from which all forms of modern life may have evolved (Figure 1-4).

TABLE 3–5

SOME BACTERIA OF SPECIAL ECOLOGICAL, ECONOMIC, MEDICAL, OR RESEARCH INTEREST

Species	Family and Order		Ecological or Economic Significance
Thiobacillus thiooxidans	Thiobacteraceae	PSEU	Autotrophic; oxidizes sulfur to sulfuric acid
Vibrio comma	Spirillaceae	PSEU	Incitant of cholera
Cellfascicula spp.	Spirillaceae	PSEU	Cellulose decomposer in soil
Chromatium spp.	Chlorobacteraceae	PSEU	Photosynthetic free-living nitrogen fixers
Nitrosococcus spp.	Nitrobacteraceae	PSEU	Chemoautotrophic; converts ammonia to nitrites in soil; favors alkaline soils
Nitrosocystis spp.	Nitrobacteraceae	PSEU	Converts ammonia to nitrite in forest soils; chemoautotrophic
Nitrobacter spp.	Nitrobacteraceae	PSEU	Converts nitrite to nitrate; chemoautotrophic
Cytophaga lutea	Cytophagaceae	MYXO	Cellulose decomposer in soils
Cytophaga johnsonii	Cytophagaceae	MYXO	Chitin decomposer in soils
Beggiatoa alba	Beggiatoaceae	BEGG	Chemoautotrophic; oxidizes hydrogen sulfide to sulfur in hot springs
Spirochaeta cytophaga	Spirochaetaceae	SPIR	Cellulose decomposer in soils and water; common in sewage
Treponema pallidum	Treponemataceae	SPIR	Incitant of syphilis and yaws
Rhizobium leguminosarum	Rhizobiaceae	EUBA	Symbiotic nitrogen-fixing bacterium; at least seven cross-inoculating groups occur
Azotobacter beijerincki	Azotobacteraceae	EUBA	Free-living; aerobic nitrogen fixer in soil; decomposes straw
Escherichia coli	Enterobacteraceae	EUBA	Ubiquitous colon bacterium widely used in genetic research; test species for polluted water
Salmonella typhosa	Enterobacteraceae	EUBA	Incitant of typhoid fever
Salmonella typhimurirum			Pathogen of mice; widely used in genetic research
Urobacillus pasteurii	Enterobacteraceae	EUBA	Converts urea to ammonia in soil
Streptococcus hemolyticus	Lactobacillaceae	EUBA	Incitant of several diseases including scarlet fever, rheumatic fever, etc.
Streptococcus lactis			Converts lactose to lactic acid; important in cheese making, etc.

Lactobacillus acidophilus	Lactobacillaceae	EUBA	Imparts flavor to cheeses; converts milk to yoghurt; alimentary canal symbiont
Pasteurella	Brucellaceae	EUBA	
pestis			Incitant of bubonic plague ("black death")
tularensis			Incitant of tularemia
Brucella abortis	Brucellaceae	EUBA	Incitant of Bang's disease in cattle and Malta fever (undulant fever) in man
Hemophillus	Brucellaceae	EUBA	
pertussis			Incitant of whooping cough
ducreyi			Incitant of chancroid
Neisseria	Neisseriaceae	EUBA	
meningitidis			Incitant of spinal meningitis
ghonorheae			Incitant of gonorrhea
Corynebacterium	Corynebacteriaceae	EUBA	
insidiosum			Incitant of bacterial wilt of alfalfa and other legumes
diphtheriae			Incitant of diphtheria in man
Clostridium	Bacillaceae	EUBA	
welchii			Anaerobic soil decomposer and nitrogen fixer; cause of gas gangrene
tetani			Decomposer in water-logged soils (anaerobic); cause of tetanus (lockjaw)
Mycobacterium	Mycobacteraceae	ACTI	
leprae			Incitant of Hansen's disease (leprosy)
tuberculosis			Incitant of tuberculosis
Streptomyces griseus	Streptomycetaceae	ACTI	Source of antibiotics (streptomycin)
Mycoplasma spp.	Mycoplasmatacease	MYCO	Incitants of several plant and livestock diseases
Rickettsia	Rickettsiacease	RICK	
prowazekii			Incitant of epidemic typhus
rickettsii			Incitant of Rocky Mountain spotted fever
Coxiella burnetii	Rickettsiaceae	RICK	Incitant of Q fever

Beggiatoa alba is a common white sulfur bacterium found especially in hot mineral springs, but also in other aquatic habitats, both marine and freshwater. Its presence is easily inferred from the long strings of sulfur that accumulate near its cells as it converts "rotten egg gas" (H_2S) into food and oxygen as a result of its chemoautotrophic activities.

Cytophaga lutea and *C. johnsonii* are often abundant in soil and are important decomposers of cellulose and chitin. These and other slime bacteria are often mistaken for slime molds (Myxomycophyta in the Bracteobionta) because of their brightly colored stalks, their pseudoplasmodia, sporangia, and ameboid-like motility. The prokaryotic cellular structure reveals their true relationship to other bacteria; the superficial resemblance to slime molds is due to parallel evolution.

Leptothrix and other iron bacteria occur where there is abundant dissolved iron in the water in addition to organic matter. They are common in reservoirs, wells, pipes, and bogs, where the rod-shaped cells grow in long chains enclosed either in a sheath of colloidal ferric hydroxide or in a sheath of organic matter containing granules of ferric hydroxide. Although common, they are probably the least perfectly known of all bacteria.

Collection bottles of algae which have stood in the laboratory with insufficient water in them frequently develop a whitish scum at the surface. There is usually an abundance of spirochaetes in the scum. Members of the Spirochaetales are important decomposers in water high in organic matter, such as sewage lagoons. If the water is high in hydrogen sulfide, the "giant" species *Spirochaeta eurystrepta* and *S. stenostrepta* are especially abundant. *Treponema pallidum* is the incitant of the widespread sexually transmitted disease (venereal disease), syphilis; it is also the incitant of yaws, bejel, and pinta, nonvenereal diseases transmitted primarily by utensils used for eating and drinking in areas where there is poor sanitation.

Most human diseases are caused by Eubacteriales. *Streptococcus pyogenes,* for example, is the incitant of pharyngitis, impetigo, puerperal fever, erysipelas, scarlet fever, rheumatic fever, nephritis, and other diseases, including some plant diseases. Other disease organisms are listed in Tables 3-5 and 3-6. While *Streptococcus pyogenes* and *S. viridans* cause serious diseases in humans, most species of *Streptococcus* are either harmless or beneficial, causing the desired flavors in cheese and other dairy products and aiding digestion. A closely related species, *Lactobacillus acidophilus,* which is always present in the alimentary canal of healthy breastfed infants, is used in the commercial preparation of yoghurt.

Hansen's disease, or leprosy, and tuberculosis are serious diseases caused by Actinomycetales. A common soil bacterium, *Streptomyces griseus,* produces the antibiotic streptomycin, which has been of great value in treating diseases caused by Gram-positive bacteria.

The type species of the Rickettsiales, *Rickettsia prowazeki,* was named in honor of H. T. Ricketts, who first discovered this organism in the alimentary canal of lice which had fed on typhus fever patients, and A. von Prowazek, who furthered our knowledge of human typhus. Both of these men died of typhus, contracted while studying the disease.

Vaccination, in which weakened or dead bacteria or viruses are inoculated on or into a person's body in creating **active immunity,** and inoculation with antiserum to develop **passive immunity** virtually eliminated all the dread diseases of the past during the first half of the 20th century (Table 3-6), and most Americans and Europeans believed that plagues were a thing of the past, that disease was fully and finally conquered. Nature, always full of surprises, came up with a new set of diseases beginning in the late 1970s (Table 3-7) and warfare between medical science and the Schizomycophyta has been declared anew.

TABLE 3-6

MORTALITY FROM LEADING DISEASES IN THE U.S. HAS DECLINED DURING THE 20th CENTURY. DEVELOPMENT OF VACCINES ANTISERA, THE SULFA "MIRACLE DRUGS," AND PENICILLIN ALL CONTRIBUTED TO THE DECLINE

Disease	Agent	1901	1910	1920	1930	1940	1948	1950	1980	1985	1990	1991
Pneumonia	EUBACT											
All forms		39.7	34.6	40.0	58.8	32.4	27.8	26.9	22.0	27.1	30.6	29.2
Lobar		13.9	19.2	25.2	31.7	17.8	9.7	na	na	na	na	na
Tuberculosis	MYCOB	58.7	66.4	66.6	49.8	35.6	23.7	22.5	0.8	0.7	0.7	0.6
Syphilis	SPIROCH	1.08	1.71	5.3	6.2	11.2	6.3	5.0	0.1	<0.1	<0.1	0.1
Influenza	VIRUS	9.61	6.0	6.8	13.6	11.9	2.74	4.4	1.1	0.8	0.8	0.4
Poliomyelitis	VIRUS	na	na	na	0.81	0.60	1.02	1.1	<0.1	—	—	—
Whooping cough	EUBACT	2.9	4.7	7.3	3.4	1.72	0.62	0.7	<0.1	—	—	—
Typhoid fever	EUBACT	9.6	9.7	4.5	3.3	0.81	0.11	0.1	<0.1	—	—	—
Smallpox	VIRUS	1.03	0.16	0.34	0.014	0.008	0.003	—	—	—	—	—

Note: Ranked by 1948 data. *na* = data not available; — = very low values, not reported.

Data from Culbertson and Cowan (1952) and the U.S. Department of Health, Education and Welfare.

TABLE 3-7

RECENT NEW DISEASES[a]

	Number of New Diseases Caused by			
Decade	**Viruses**	**Bacteria (including *Rickettsia*)**	**Old Agent, New Location**	**Examples**
1950	2	0	0	O'nyong-nyong; India tick fever
1960	0	0	0	
1970	0	3	0	Toxic shock syndrome; legionaire disease; lyme disease
1980	2	2	0	HIV-A; hepatitis C; Ehrlichia; cat-scratch disease
1990	2	1	2	Guanarito; hepatitis E; yellow fever; hantavirus; cholera 0139

[a] Vaccines and antibiotics were so effective in eradicating human diseases that by 1950 it was generally believed that plagues were a thing of the past. However, since 1970, new diseases have evolved at a surprising and alarmingly rapid rate.

Data from Levins et al. 1994: includes only the diseases reported by them.

Ecological Significance

The ecological significance of the Schizomycopsida stems from four factors: (1) their importance as decomposer organisms, which is related to their small size and accompanying large surface area, high rate of respiration, and rapid rate of reproduction; (2) their role in mineral cycles, especially in nitrogen fixation, nitrification, and oxidation of sulfur; (3) the tendency of many bacteria to be parasitic on plants or animals, including man, and hence incitants of disease; and (4) their role in various symbioses.

Because matter can be neither created nor destroyed, chemical substances must be used over and over again in nature. The carbon and nitrogen atoms used by green plants in synthesizing sugars, lipids, and amino acids, and then consumed by us and other animals, are released in the form of carbon dioxide, ammonia, and other inorganic materials and used again in the process of photosynthesis. This release of inorganic raw materials is accomplished by decomposer organisms. To some extent, every plant and animal functions as a decomposer, but the bacteria are especially efficient in the this regard. The cyclic use and reuse of carbon is illustrated in the upper half of Figure 3-7 and the nitrogen cycle in the lower half.

Bacteria are efficient decomposers, partly because they are able to break down many different kinds of substances and partly because of their rapid rates of metabolism. While it is the versatility of the Schizomycopsida as a group that is of greatest significance in decomposition, some individual species are, nevertheless, highly versatile. *Pseudomonas fluorescens,* for example, is able to metabolize about 200 different organic compounds which occur in nature. *Cytophaga lutea* is one of the few bacteria that can oxidize cellulose, but it is inhibited by other kinds of carbohydrates, and *C. johnsonii* is an important decomposer of chitin; in contrast, the omnivorous *Ps. fluorescens* can metabolize neither cellulose nor chitin.

Being small, bacteria are affected more by microclimate than macroclimate. The microhabitats in which bacteria occur are highly uniform both in time and space and relatively independent of macroclimate. For example, soil temperatures are relatively constant year round

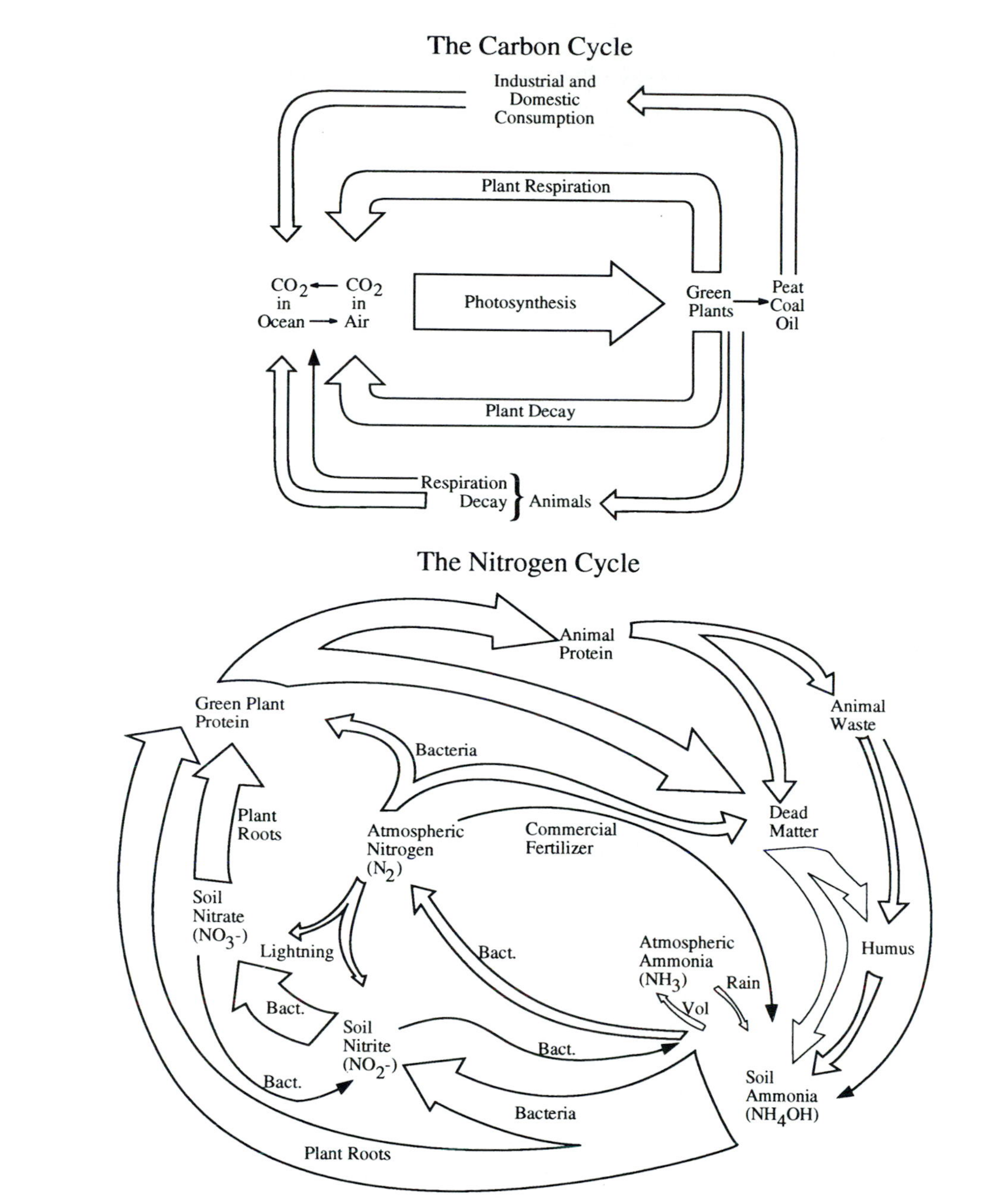

Figure 3–7 Biogeochemical cycles. Because matter can be neither created nor destroyed, every atom of carbon, nitrogen, and other minerals must be used over and over again. The width of the lines suggests the relative importance of the different processes.

compared to air temperatures, and also vary far less world wide. Consequently, each species of bacterium tends to be widespread in distribution. The rumen of a cow living in northern Scandinavia is ecologically very similar to the rumen of a cow living in the Arabian Desert, especially if both are fed alike; therefore, the same species of bacteria will likely be found in both animals. Because individual species are so widespread, the number of species found in any cubic meter of soil is generally quite large, often several hundred, even though the total number of species for the entire world is not large, only about 1500.

Many of the delicate balances in nature are dependent upon pathogenic bacteria. Epidemics have directly destroyed large human populations many times in the past; plant disease epidemics have indirectly destroyed just as many. We know less about the plant and animal diseases that have not had an effect on man, but we do know that diseases are common on native plants growing in natural ecosystems, and there is evidence that bacterial epidemics have been common for millions of years, long before man came into existence. The number of species of pathogenic bacteria is not large, yet most people immediately think of this aspect of bacteriology as soon as the word "bacteria" is mentioned.

Fewer people are aware of the ecological significance of symbiotic bacteria. Pasteur taught, in the 19th century, that some types of animals would not be able to survive if it were not for the bacteria in their intestinal tracts. However, it is only recently that the role of bacteria in the rumen of cattle and digestive tracts of other animals has been understood to any great extent. It is now known that vitamins of the B complex and vitamin K are produced in the stomach of cattle, the cecum of horses, and the large intestine of rabbits and some other animals, including man, by symbiotic bacteria living there. A person or an animal given large doses of antibiotics prior to an operation may have the microflora of the digestive tract so greatly reduced that he could experience a vitamin K deficiency and hemorrhage seriously unless this vitamin is administered artificially. Bacteria also build proteins and fats needed in animal metabolism, and they are important in water relations and in movement of materials through the digestive system. Antibiotics along with many pain-relieving and other medicines, including aspirin, can destroy the normal alimentary tract bacteria, such as the acidophilus bacterium, resulting in serious constipation and other problems. Since 1980, alimentary tract infections have become increasingly common as normal bacterial populations have been eliminated by medical drugs.

Symbiotic bacteria are as important to plants as they are to man and other animals. When the proper strain of *Rhizobium leguminosarum* is present in the soil, it will invade the young roots of legume seedlings and cause gall-like growths, or nodules, to form (Figure 3-8). These vary in size, shape, and color, depending on the species of legume and the strain of bacterium. Alfalfa has egg-shaped nodules that sometimes branch dichotomously; red clover has very small spherical nodules; the nodules on velvetbean are large and spherical, often attaining the size of walnuts. Nodules of "good" strains of *Rhizobium* are pink and healthy looking; those of "bad" strains are typically brown. The galls of nematodes and of pathogenic bacteria are sometimes confused with nitrogen-fixing nodules, but they are less distinctive in general appearance and are usually brown to black.

Species of legumes can be classified into cross-inoculating groups based on which bacterial strain will fix nitrogen in the species of that group. The six agronomically most important groups are listed in Table 3-8. Bacteria from one group are not only ineffective nitrogen fixers when inoculated into any of the other groups, but often seem to be pathogenic. Hence, a "good" strain of bacteria becomes a "bad" strain when present on the "wrong" species of legume. Some strains of bacteria cause nodules to form on a wide range of leguminous plants, whereas others are apparently capable of functioning on one or only a few species. Members of the Mimosa subfamily of legumes, such as honeylocust and Kentucky coffeetree, apparently cannot form nodules under any conditions.

In addition to the symbiotic nitrogen-fixing bacteria that occur in legumes, there are symbiotic bacteria that form nodules on the roots of birch, alder, and other vascular plants. Some Actinomycetales form galls or nodules on the stems of rabbitbrush and other desert shrubs, and there is good evidence that these galls fix nitrogen (Ray Farnsworth, Brigham Young University, personal communication). There are also several species of free-living nitrogen-fixing bacteria, such as *Azotobacter* and *Clostridium* (Table 3-9). All are important in some ecosystems, but our knowledge of most of them is very limited.

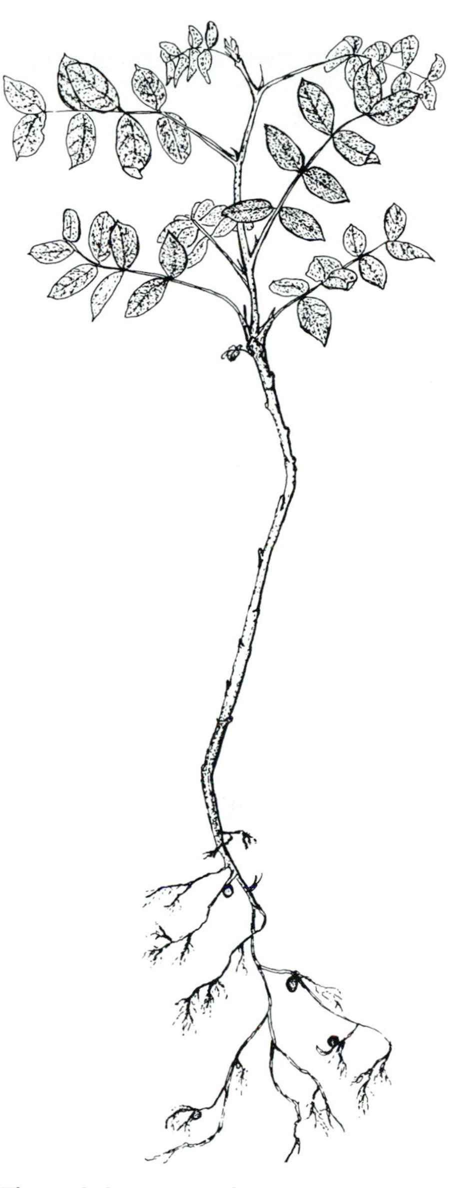

Figure 3–8 Nitrogen-fixing nodules on the roots of the Siberian pea tree, *Caragana arborescens.*

Economic Significance

Agricultural and industrial uses of bacteria have been important since ancient times. The cheese industry, for example, flourished in the days of King Saul of Israel. Butter, vinegar, and sauerkraut are examples of other food products produced by means of bacterial activity. Species of *Lactobacillus* and *Streptococcus* are largely responsible for the flavor in butter. Each type of cheese has its own specific fungus or bacterium to which it owes its flavor and texture: Swiss cheese, for example, is ripened by gas-forming bacteria. Lactic acid bacteria are responsible for the preservation and flavor of sauerkraut and silage; acetic acid bacteria aid in the production of vinegar. The **retting** of flax and hemp is accomplished by subjecting the plants to the action of bacteria. Symbiotic nitrogen-fixing bacteria living in root nodules of alfalfa, clovers, and other legumes are important in crop production (Table 3-8, Figure 3-8). Both the bacteria and the processes used in modern industry are the same as those used anciently, but the environmental conditions and the strains of bacteria used now are more carefully controlled than they were in the past.

The nodules of alfalfa, peas, beans, clovers, and other legumes can be dissected and the nodule-forming bacteria streaked on plates of nutrient agar in laboratory flasks under sterile

TABLE 3–8

CROSS-INOCULATION STRAINS OF *RHIZOBIUM*

Strain	Species Benefited
A. Alfalfa group	*Medicago sativa, Medicago* spp. (alfalfa, bur clover); *Melilotus alba, Melilotus officinalis* (sweet clovers), *Trigonella* spp. (fenugreek, etc.)
B. Clover group	*Trifolium pratense, Trifolium repens* (red and white clovers), *Trifolium* spp.
C. Pea group	*Pisum sativum* (garden pea), *Pisum* spp., *Lathyrus odorata* (sweetpea)
D. Common vetch group	*Vicia sativa* (tare or vetch)
E. Horse bean group	*Vicia faba* (horse bean or broad bean)
F. Bean group	*Phaseolus vulgaris* (garden bean), *Phaseolus* spp.
G. Soybean group	*Glycine max* (soybean), *Glycine javanica* (kudzu vine)
H. Lupine group	*Lupinus albus* (wolf bean), *Lupinus luteus* (yellow lupine), *Lupinus angustifolius* (blue bonnet), *Ulex europeus* (gorse), *Ornithopus* spp.
I. Sainfoin group	*Onobrychus sativa* (sainfoin)
J. Cowpea group	*Vigna capensis* (cow pea), *Vigna sinensis* (catjang bean), *Indigofera anil* (indigo), *Phaseolus lunatus* (lima bean), *Phaseolus aureus* (mung), *Arachis hypogaea* (peanut), *Lespedeza striata* (common lespedeza)
K. Siberian pea group	*Caragana arborescens* (Siberian pea tree), *Glycyrrhiza glabra* (licorice)

Comments: Total of approximately 20 cross-inoculation groups have been studied and reported on in botanical and bacteriological literature. Inoculation with the "wrong" strain of *Rhizobium* will sometimes give results almost as good as the "right" strain, but in other cases the "wrong" strains seem pathogenic. Groups C, D, and to a lesser extent E, for example, can be used interchangeably with moderately good results; the same is true for groups I and J. Strains C–E inoculated on any of the other groups, on the other hand, are reported to give poor and often pathogenic results.

Data from Wheeler 1950, Waksman 1952, Darlington 1945, and U.S. Department of Agriculture bulletins.

conditions. The resulting bacterial cultures are tested for effectiveness in fixing nitrogen. Pure cultures of the best strains are packaged in moist peat or similar material and sold to farmers. In farming communities where the soils are acid or otherwise not conducive for survival of natural *Rhizobium* populations or where farmers are practicing long rotations, the sale of these cultures amounts to thousands of dollars annually. Even in fields where there are abundant bacteria in the soil, farmers often obtain better yields by inoculating with a proven culture. For best results, the cultures should be applied to the legume seed with addition of water. Drying is detrimental to the bacteria and sometimes to the seed as well. Hence, prompt planting is essential. The drills that are used for sowing may need to be adjusted to accommodate the swollen moist seed (Leonard 1937).

In order to avoid contaminating our environment with toxic pesticides, agronomists and botanists have tried to develop biological methods to control weeds, insects, and even fungi. The best known of these attempts is the successful control of the prickly pear cactus, *Opuntia*

TABLE 3-9

GENERA OF BACTERIA AND BLUE-GREEN ALGAE KNOWN TO BE NITROGEN FIXERS

Genus	Known Nitrogen-Fixing Species	Rate of Nitrogen Fixation[a]	
		Laboratory (mg/g CHO)	Field (kg/Ha)
Bacteria			
Achromobacter	—	1	—
Azotobacter	1	10-20	15-35
Beijerinckia	—	10-20	—
Derxia	—	25	—
Rhizobium	—	—	30-800
Bacillus	—	12	—
Clostridium	13	2-27	25-50
Aerobacter	—	4-5	—
Klebsiella	—	5	—
Rhodospirillum complex	—	—	—
Chromatium-Chlorobium	—	—	—
Pseudomonas	—	1-4	—
Desulfovibrio complex	—	—	—
Blue-green Algae			
Anabaena	11	—	—
Anabaenopsis	1	—	—
Aulosira	1	—	—
Chloroglaea	1	—	—
Cylindrospermum	4	—	—
Nostoc	8	—	—
Collema[b]	3	—	35
Calothrix	5	—	—
Scytonema	1	—	—
Tolyptothrix	1	—	25-60
Fischerella	2	—	—
Hapalosiphon	1	—	—
Mastigocladus	1	—	—
Stigonema	1	—	—
Westelliopsis	1	—	—

[a] In laboratory tests, rate of N fixation is calculated on the basis of the milligrams of nitrogen fixed per gram sugar or other carbohydrate consumed. In the field tests, the increase over control plots in protein, converted to nitrogen, is calculated.

[b] The lichen, *Collema tenax*, on desert soil in western U.S. fixes 35 kg/Ha of nitrogen based on nitrogenase activity in laboratory tests and density of *Collema* thalli on the soil. The phycobiont of *C. tenax* is an unnamed species of *Nostoc*.

inermis, a noxious weed in Australia (Pearson 1967). In the U.S., successful control of many species of mites and insects, especially moth larvae, by a bacterium, *Bacillus thuringiensis,* (or b.t.) has been widely heralded as an example of ideal insect repression. It has been a boon to backyard gardeners who treat their cabbage, broccoli, and green beans with b.t. and never worry that their little children will be poisoned by an insecticide or that the birds in their yard will be affected. Field crops on which b.t. has been effective include potatoes, corn, and wheat.

Among the advantages of dusting or spraying crops with suspensions of b.t. in preference to chemical insecticides are its effectiveness on a wide range of insects and mites, its lack of toxicity to mammals (including man) and birds, and its specificity for pest insects—it has no harmful effects on predaceous insects and scavengers that eat the killed herbivorous pests.

Nevertheless, even ideal methods have problems. Just as insecticide resistance is a formidable complication of the use of chemical insecticides, recent studies have revealed that many insects are developing resistance to b.t. Theoretically, this ought not to be—or so it was believed at the time b.t. was released. There are many b.t. toxins and their mode of action is so complex that it was doubted that any insect would develop simultaneous mutations giving it resistance to all of the toxins. But McGaughey and Whalon (1992) have reported resistance to b.t. in the diamondback moth, and several other insects are beginning to show a low level of resistance.

In contrast to nitrogen-fixing and pest-controlling bacteria, there are other bacteria which are of economic significance in a negative way. Plant diseases cost American farmers several billion dollars per year in reduced crop yields and impaired crop quality; about 10% of this, or approximately half a billion dollars, is due to bacterial disease. Livestock diseases cost the American farmer another billion dollars per year, most of it due to bacteria. Americans spend over 40 billion dollars a year on human medicines and doctor bills; about half of this is due to bacterial infections. Bacteria are also responsible for some corrosion of materials, rotting of wooden structures, spoilage of hay and grain and of dairy products and other grocery items. The high markup of vegetables and fruits in the grocery store, as compared to the very low markup of canned goods and sugar, is necessary because a large percent of the fruit and vegetables will be spoiled by bacteria before it can be sold.

Bacteria in Modern Research

The Schizomycopsida are favorite research organisms because of the contributions they have made to biological science. Four areas of special importance are enzymology, serology, nutrition, and genetics. The first three are often referred to as **bacterial biochemistry.**

Bacteriology and biochemistry, including **enzymology,** are so closely related that some biologists tend to think of them as synonymous terms. Although numerous species of plants and animals are employed in biochemical study today, bacteria continue to make the greatest contributions, in ideas and in data. Our knowledge of the enzymes involved in the Krebs cycle of respiration, for example, has come largely from studies with bacteria.

Much of our current knowledge of protein chemistry has been gained from **serological** studies, especially the **antigen–antibody** reactions involving disease-producing bacteria. **Gene splicing** and other aspects of molecular genetics are based largely on this and other research with bacteria. Bacteria have also been useful in studying how food substances affect metabolism.

Penicillin, an antibiotic produced by ascomycetous fungi, is the miracle drug that saved millions of lives during World War II. It has virtually eliminated the fear associated with many diseases which only a few years ago had very high mortality rates (Table 3-6). Penicillin is very effective in killing Gram-positive bacteria but generally ineffective in controlling diseases caused by Gram-negative bacteria or other organisms. Although the reason for this is not fully understood, hypotheses have been developed and tested that at least partly explain how penicillin works. Cell walls of Gram-positive bacteria are composed primarily of muramic acid; the cell walls of Gram-negative bacteria are composed mostly of lipids. Muramic acid consists of a lactic acid molecule linked to a glucosamine molecule. According to bacterial biochemists, the reaction which links the two molecules is catalyzed by an enzyme on the outer surface of the cell membrane; penicillin interferes with the action of that enzyme.

Vaccination and antibiotics produced such a dramatic reduction in mortality from human disease during the first half of the 20th century that one prominant biologist wrote in 1975, "During the last 150 years the Western world has virtually eliminated death due to infectuous diseases." A change in direction, however, occurred almost immediately. Many new diseases have appeared since then, a sample of which is given in Table 3-7. Some of these have never before been recorded; others have reappeared in new places, and others have taken on new symptoms (Levins et al. 1994).

Some vitamins and other substances are needed in such minute quantity that ordinary chemical tests are not sensitive enough to detect them in the quantity needed. A bioassay can be used to ascertain the presence of substances essential in minute quantity. A medium is prepared containing all of the nutrients needed by the test bacterium except the vitamin or other nutrient being tested. Extracts from different foods are added to the basic medium and the degree of growth of the bacterium is a measure of the amount of vitamin or other growth substance present in the food. A slight variation of this method is used in assaying for poisonous substances. In the American space program, luminescent bacteria have been employed to increase safety conditions. A culture of a luminescent bacterium which is very sensitive to a chemical used in the fuel system is kept in the space capsule. If a leak in the system should develop, bacterial **luminescence** decreases, and finally stops at concentrations of the pollutant much too low to be detected by other means.

In genetics, the contributions of bacteria have been highly significant. Each cell has a single circular chromosome, consisting primarily of DNA, plus zero to four or five small plasmids which also contain DNA. In 1928, **bacterial transformations** were discovered in *Diplococcus pneumoniae,* but aroused less interest than consternation for the next 2 decades. In 1952, a similar process, **bacterial transduction,** was discovered in *Salmonella typhimurium.* Both phenomena were highly significant in leading to our present knowledge of DNA structure and function.

Frederick Griffith discovered transformations when he inoculated a mixture of heat-killed **virulent** bacteria and live nonvirulent bacteria into healthy experimental animals. The animals died and virulent bacteria were cultured from their bodies. Yet neither dead virulent bacteria nor live nonvirulent ones produced adverse effects by themselves. The "Griffith effect" was long ignored by geneticists, since it seemed to make no sense, but we now know that molecules of DNA from the dead bacteria had been taken into the cells of the live ones. The transduction phenomenon differs in that DNA is transferred from one culture to another by means of a **bacteriophage** or bacterial virus.

Discovery of **conjugation** in bacteria helps to understand these two phenomena. Conjugation is the first step in a parasexual process called **meromixis** (Figure 3–9). In *Escherichia coli,* a conjugation tube forms between two cells, and the chromosome from the **donor** cell moves through the tube and synapses with the chromosome of the **recipient** cell. These studies have furthered our knowledge of linkage patterns in bacteria; this knowledge has in turn led to a better understanding of structure of DNA. These, together with studies of rates of mutation in bacteria, have led to the work being done at the present time in gene splicing and genetic engineering.

Phenylketonuria (PKU) is a genetic disease caused by a recessive gene. Individuals homozygous for the gene lack the liver enzyme which converts phenylalanine into tyrosine. As phenylalanine builds up in the body, phenylpyruvic acid and ketones are produced from the excess and do considerable harm, including damage to the brain. During pregnancy, the mother's enzymes are able to break down any excess phenylalanine, and a perfectly normal baby is born. Within 72 hours following birth, the level of phenylalanine in the blood of a PKU baby increases from a normal 2 or 3 mg per ml blood to more than 60 mg/ml. Within 6 months, irreversible brain damage will have occurred unless a special diet, low (but not lacking) in phenylalanine, is provided.

Most states (43 of the 50 as of 1983) require that each child about 3 days after birth be tested for PKU. The test consists of putting a drop of the infant's blood on a small disk of absorbent paper which is then placed on an agar plate which has been seeded with *Bacillus subtilis.* The agar contains enough thienyladanine to inhibit the growth of bacteria at the level of phenylalanine normally present in blood. If the baby's blood contains a high level of phenylalanine, however, the bacteria thrive and a halo of dense bacterial growth develops around the disk (Kjerstin Schick, Budge Clinic, Logan, Utah, personal communication).

CLASS 2. CYANOPHYCOPSIDA

Also known as Myxophyceae (slime algae), Schizophyceae (fission algae), Cyanobacteria, and Cyanophyta, the Cyanophycopsida (from the Greek, *cyano* = dark blue, *phycus* = seaweed or alga) are so called because they are often bluish or bluish-gray in color. Like bacteria, blue-green algae have small cells organized on the one envelope system, glycogen-like food storage, ability to fix nitrogen, and primarily asexual reproduction. They differ from bacteria in many ways, the most significant of which is the possession of chlorophyll *a,* making possible the oxygen-releasing type of photosynthesis. Blue and red water-soluble pigments aid in photosynthesis in most species, and similar pigments are found in some species of Rhodophycopsida and Cryptophycopsida. They also differ from bacteria in the types of fatty acids that occur (Table 3–2). Like other photoautotrophic plants, they contain various carotenoids.

The water-soluble blue and red photosynthetic pigments are collectively called **phycobilins.** It is commonly believed that the red phycobilins have an adaptive advantage to the Cyanophycopsida and Rhodophycopsida, enabling them to carry on photosynthesis at greater depths in the ocean than would otherwise be possible. There are no experimental data available to support this assertion; however, it has been demonstrated that in species producing both red and blue phycobilins, plants growing at greater depths have relatively more phycoerythrin and relatively less phycocyanin than plants growing in shallow water.

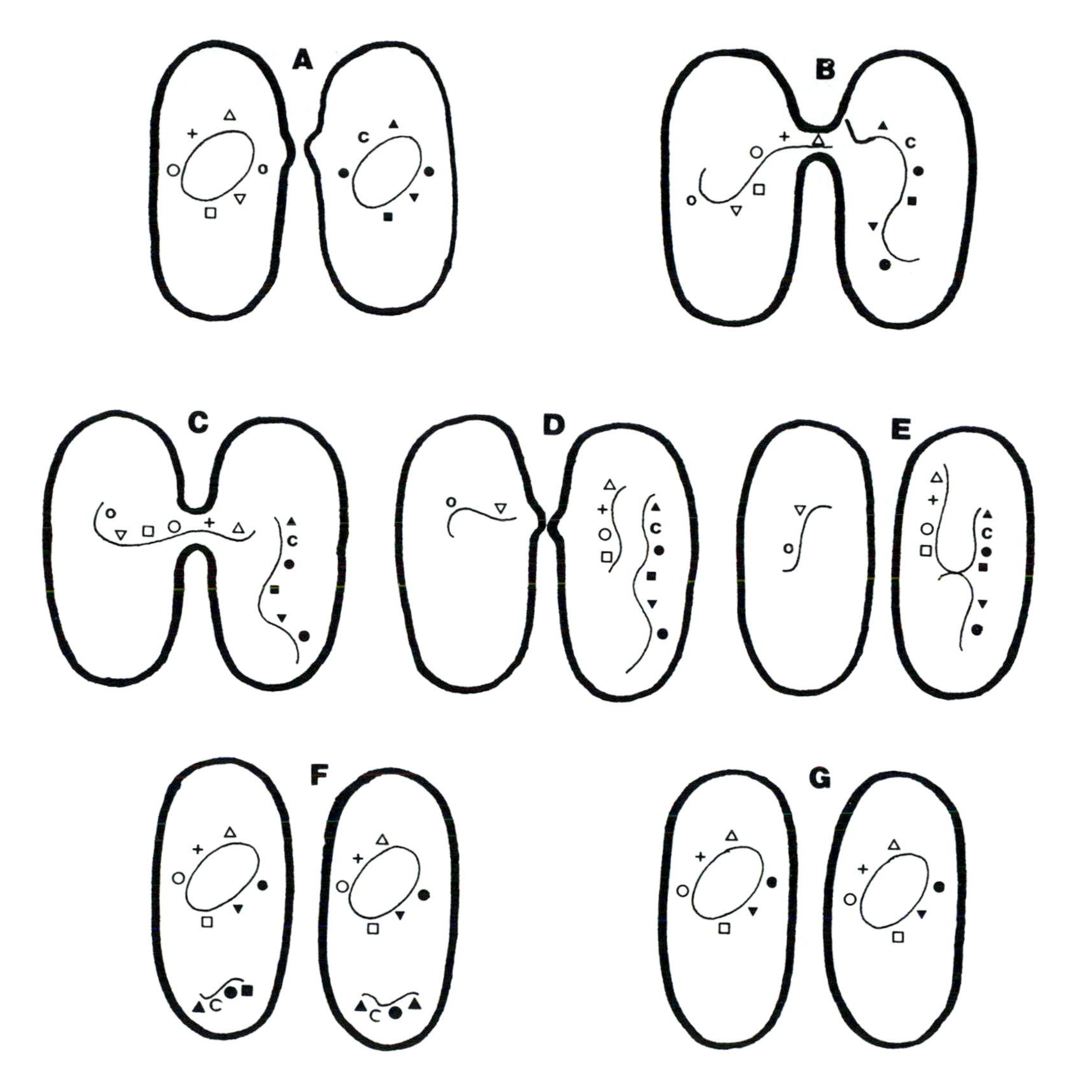

Figure 3–9 Meromixis in *Escherichia coli*: (A) Adjacent cells begin to form conjugation tube; (B) conjugation tube complete, chromosome from the donor cell, with genes represented by open symbols, migrates into the recipient cell; (C) synapsis of the two chromosomes begins at point of attachment of the recipient cell to the plasmalemma; (D) conjugation tube is broken when culture is placed in a blender and gently agitated; (E) chiasma forms between donor and recipient chromosomes resulting in a new sequence of genes in the recipient cell, whereas the donor cell begins to disintegrate; (F) recipient cell divides; (G) remnant of unattached chromosome disintegrates in the daughter cells.

Blue-green algae occur in most aquatic ecosystems, both freshwater and marine, and can be recovered for microscopic examination by (1) collecting samples of filamentous algae floating on the surface of ponds, streams, or estuaries; (2) collecting the bluish-green, slimy material that collects around hot springs on rocks, pipes, or deposits of sulfur; (3) using a plankton net with as small mesh space as possible; or (4) filtering a large sample of water through filter paper and examining the residue. The first two are the easiest, fastest, and generally most productive. Samples can be collected in any type of bottle, can, or cup and kept in the laboratory for weeks. Stored in the laboratory near a north window, the blue-greens will hold their own or even increase in density unless the ratio of algae to water is so great that they begin rotting. Associated with the blue-greens will be green algae, diatoms, and often other algal types.

General Morphology and Physiology

Except for color, the Cyanophycopsida closely resemble bacteria; however, cells tend to be somewhat larger and there is a tendency toward a filamentous structure. As in bacteria, cell walls are relatively thin (50 to 60 nm) and are composed of muramic acid. Outside the cell wall, a **sheath** of gelatinous material, which is sometimes pigmented, gives the cells their characteristic slimy feature; in "sheathless" species, this layer is only 25 to 30 nm thick and is discernable only with aid of the electron microscope (Figure 3–10), but in *Lyngbya* and many other cyanophytes, it is up to 1 μm thick and readily observed with the light microscope. There is no endoplasmic reticulum, but there are numerous small ribosomes in each cell. The thylakoids are generally unstacked and extend from one end of the cell to the other (Figure 3–10). Often oil droplets and starch grains can be observed. Polyhedral bodies, called **carboxysomes,** are apparently involved in the Calvin cycle of photosynthesis. Neither true flagella nor proterokonts have been observed in any member of the class, but herpokinetic motility is common in some groups, having been reported in 17 genera, including the common *Oscillatoria, Anabaena,* and *Chroococcus.* In masses around hot springs, Cyanophycopsida can often be recognized by the slimy nature of the mass and the bluish color. Under the microscope, their small size, color, and lack of organized plastids make recognition easy.

The color of cyanophytes is due primarily to two kinds of photosynthetic pigments, chlorophyll *a* and various **phycobilins,** the most abundant of which is ***c*-phycocyanin,** a dark blue, water-soluble pigment (Figure 3–11). Carotenoids are also usually present, especially β-carotene, the same pigment that gives carrots their orange color. Echinenone and myxoxanthophyll are deep red to red-orange pigments unique to the Cyanophycopsida; pale yellow carotenoids, especially zeaxanthin, are also usually present (Strain 1958; Goodwin 1952, 1957).

Cyanophytes often form firm, bluish mats around hot springs and in some other habitats. The pigment chiefly responsible for the color, *c*-phycocyanin, also occurs in Rhodophycopsida and Cryptophycopsida. Two other phycobilins, allophycocyanin and *c*-phycoerythrin, the former a blue pigment and the latter a red one, occur in blue-green algae and are also photosynthetically active (Table 4–2). Each phycobilin molecule consists of a bile pigment tightly bound by covalent linkage to an apoprotein (Figure 3–11). Phytochrome, the pigment which mediates photomorphogenesis in vascular plants and possibly in some algae, is also a **biliprotein** similar in structure to the phycobilins.

As in other Aflagellobionta, the thylakoids are unstacked (Figure 3–10). Molecules of chlorophyll are arranged in groups of four on the surface of the thylakoids with their hydrocarbon tails extending into the intrathylakoid space. The four molecules function as a unit in carrying on the light reactions of photosynthesis.

The phycobilins are located in small bodies called **phycobilisomes,** attached to chloroplast lamellae. The individual units come in two sizes, 5 to 7 nm and 10 to 20 nm, which aggregate to form larger bodies 35 nm in diameter. They appear as 35-nm granules when phycoerythrin predominates or as disks or tubules when phycocyanin is predominating.

The gelatinous sheath around individual cells or filaments can be discerned by adjusting the light very carefully. Many of the characteristics of the blue-greens are associated with these sheaths. In species with thick sheaths, for example, herpokinetic motility does not occur. In some species, the sheaths are fused together so that several filaments lie parallel to each other as a single unit. Gleocapsins, red or blue pigments, are sometimes present.

Pseudofilaments occur in some unicellular species, like *Heterohormogonium schizodicochomum* and *Entophysalis magnoliae,* in which the cells are held together by gelatinous sheaths (Figure 3-12). **False branching** occurs where two filaments lie side by side in a common sheath and one breaks out from it (Figure 3-13f; see also Fig. 3-16). In the Nostocaceae, the chain-like filaments are often held together in macroscopic gelatinous masses characteristic for the genus and often for the species; these masses may be hard and rubbery, almost like golf balls, or they may be soft, slimy, and indefinite in form (Figure 3-13d,e). According to Moldenke and Moldenke (*Plants of the Bible,* 1965), the grape-like, edible balls of a *Nostoc* species were eaten by the ancient Israelites and referred to in their records as manna (from the Hebrew, *man hur,* meaning "what is it?").

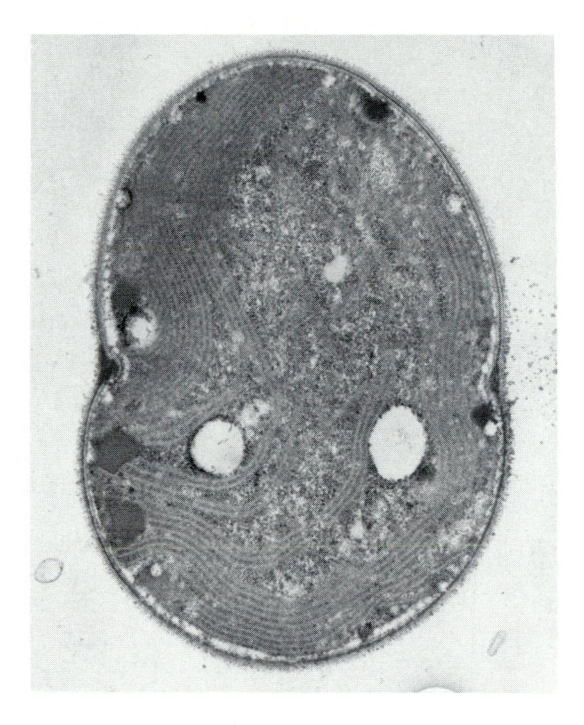

Figure 3–10 Electron micrograph of a blue-green alga cell with a very narrow band of cytoplasm between the thylakoids and the plasma membrane. Starch grains are surrounded by numerous mitochondria. The dark polyhedral bodies attached to the thylakoids and lying adjacent to the plasma membrane, e.g., lower left, are carboxysomes, believed to be involved in the Calvin cycle of photosynthesis. Ribosomes appear as small spherical bodies and are not attached to an endoplasmic reticulum. (Courtesy W. M. Hess and Brigham Young University Microscopy Laboratory, Provo, Utah.)

Species with very thin sheaths, observable only with the aid of the electron microscope, often exhibit **herpokinetic motility.** This is the gliding motility, characteristic of *Oscillatoria* and *Spirulina,* in which the filament slowly rotates or waves back and forth, not unlike the waving of a cobra as the fakir plays his flute. Sometimes the gliding is forward and backward rather than side to side. The causal mechanisms are not known although several hypotheses have been advanced. Herokinetic motility, at least in Cyanophycopsida, is correlated with protoplasmic streaming, the gliding of the filaments reversing at the same time as the protoplasmic streaming reverses. Both protoplasmic rotation and herpokinetic motility ceased when a weak salt solution was added to the medium in which the algae were growing. Herpokinetic motility, which has also been observed in a few Schizomycopsida and Rhodophycosida, is especially common in the Cyanophycopsida.

Reproduction is primarily asexual, either by means of **hormogonia,** which are filamentous fragments formed when some cells become weak or die, by means of endospores produced

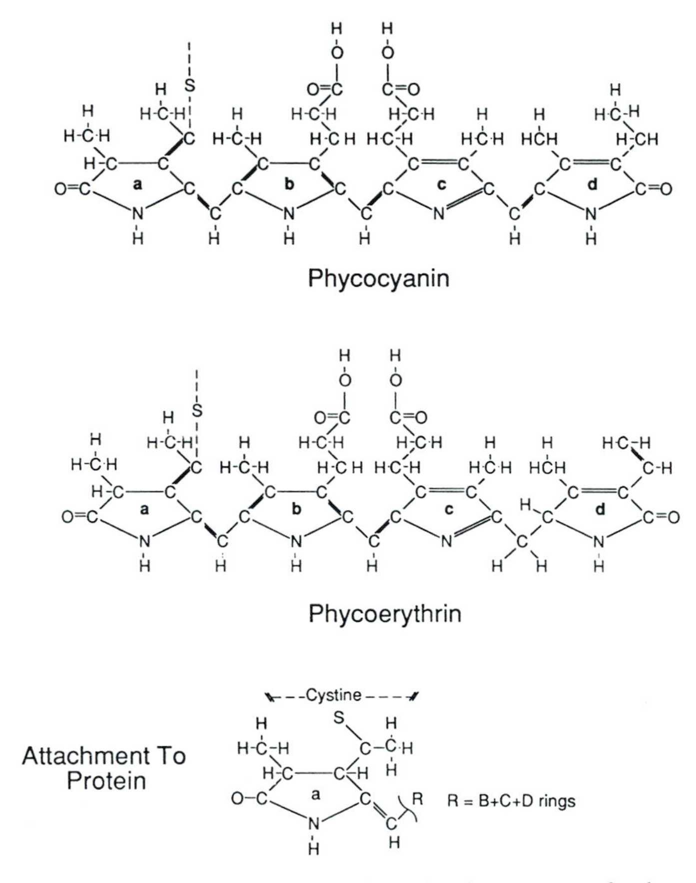

Figure 3–11 Photosynthetic pigments in Cyanophycopsida. Molecular structure of *c*-phycocyanin and *c*-phycoerythrin are shown; they are attached to protein molecules by a cysteine unit.

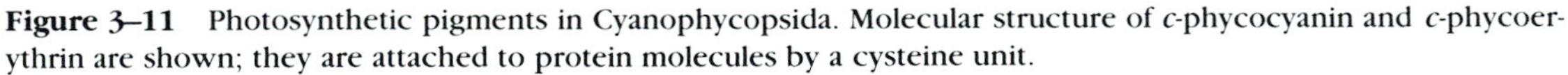

within the cell wall from the mother protoplast, by simple cell fission, or possibly by means of **heterocysts** (Figure 3-13e). Heterocysts are enlarged cells, devoid of chlorophyll, which occur only in cyanophytes, especially in the Nostocaceae. They have long been suspected of being reproductive structures, and Desikachary (1946) has reported laboratory germination of heterocysts following a short exposure to low temperature. However, his observations have not been duplicated by others. On the other hand, it has also long been known that heterocysts occur only in species in which nitrogen fixation occurs, suggesting that they play a role in this process. It is now well established that they are indeed involved in nitrogen fixation; whether they are also involved in reproduction needs to be further confirmed.

Nitrogen fixation can only take place in the absence of free oxygen. The thick cell walls of the heterocysts enable them to maintain the oxygen-free atmosphere necessary for nitrogen fixation. The fact that heterocysts have been identified in fossils from the Archeozoic era indicates that nitrogen fixation is of ancient origin and was taking place at that time (Schopf 1978).

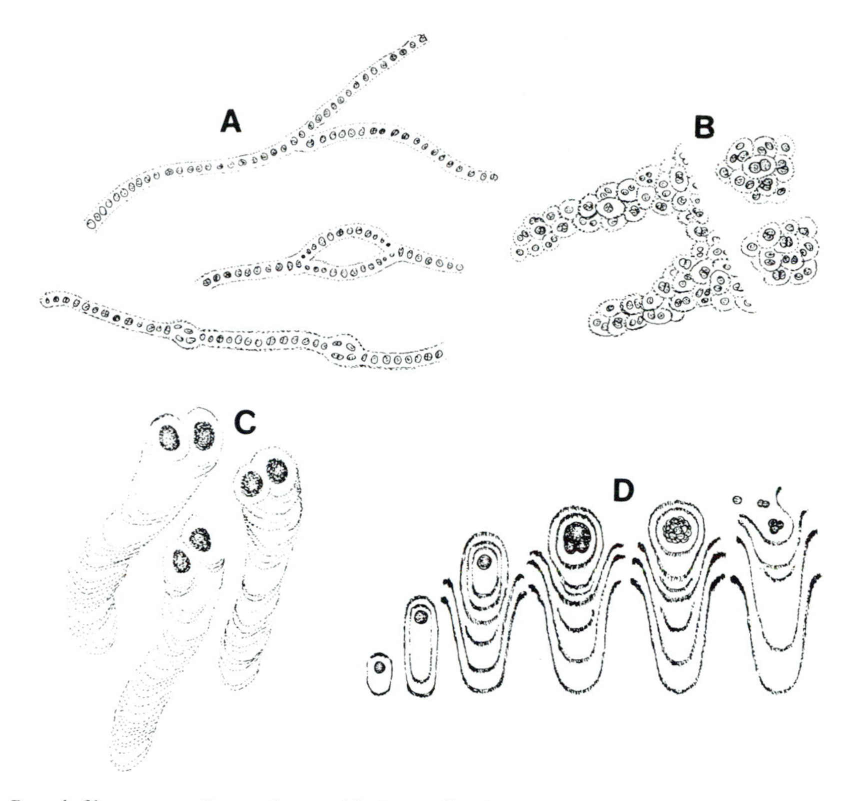

Figure 3–12 Pseudofilaments in Cyanophycopsida (magnification × 1000); (A) *Heterohormogonium schizodichotomum;* (B) *Entophysalis magnoliae;* (C) *Cyanostylon* sp.; (D) *Epiphyton* sp. from the late Hadrynian Period. As *Cyanostylon* cells divide there is pronounced unidirectional gel production resulting in gelatinous stalks with the dividing cells at the apex of the stalks. These may become calcified from the sediments in which they are growing; presumably, similar activity resulted in the layered, unidirectional stalks associated with some *Epiphyton* cells.

At the time heterocysts are being formed from regular vegetative cells of *Nostoc* and other bluegreens, the photosynthetic thylakoids become greatly distorted. It is believed that nitrogenase and other enzymes involved in nitrogen fixation are attached to these distorted thylakoids.

Cell division in Cyanophycopsida is similar to mitosis, though apparently not identical to it (Leak and Wilson 1960; von Hofsten and Pearson 1965). Stages resembling the stages of equational cell division of higher plants have been observed, but they are probably not homologous to mitosis in vascular plants. If cell division is in one plane only, as in *Oscillatoria, Heterohormogonium,* and *Nostoc,* filaments or pseudofilaments result. If cell divisions are in two planes at right angles to each other, as in *Merismopedia,* flat sheets or plates result. Three-dimensional colonies of various forms result when cell division occurs in all planes, as in *Chroococcus.*

Parasexual processes, similar to meromixis in the Schizomycopsida, may occur in Cyanophycopsida. R. N. and J. H. Singh (1962, 1974) have reported genetic recombinations in

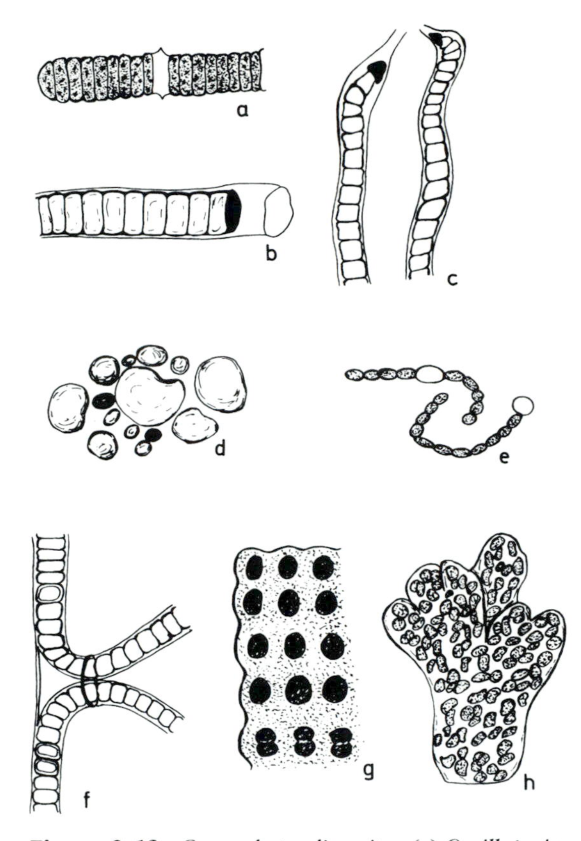

Figure 3–13 Cyanophyte diversity; (a) *Oscillatoria sancta* filament consisting of two hormogonia; (b) *Lyngbya martensiana* filament with gelatinous sheath; (c) *Phormidium uncinatum;* (d) Nostoc balls from a mountain stream; (e) *Nostoc pruniforme* filaments from (d); (f) false branching in *Scytonema tolypotrichoides;* (g) *Merismopedia elegans;* (h) *Aphanothe utahensis* from the Great Salt Lake. (Magnification: (d) natural size; all others, × 2000.)

Cylindrospermum majus, and Lazaroff and Vishniac have described anastomoses between filaments in some Cyanophycopsida, suggesting possible gene exchange between clones.

Phylogeny and Classification

Fossil evidence of the Cyanophycopsida, partly in the form of **stromatolites,** indicates they evolved in early Proterozoic time or earlier (Barghoorn and Tyler 1965, Cloud, personal communication). The fossils indicate little change in morphology in over 2 billion years (Figure 3–14). Unicellular, spherical blue-green algae from the Fig Tree formation (South Africa), very similar in appearance to *Gloeocapsa* and other Chroococcales, have been dated at 2.7 billion years. Filamentous blue-greens similar to *Oscillatoria* have been found in the Gunflint formation of Ontario, which has been dated at 2 billion years and in the Apex chert of Australia dated at 3.5 billion years (Figures 2–4 and 3–14). Many present-day Cyanophycopsida grow in environments that are highly constant year after year—for example, hot springs and shallow saline lakes—and under such conditions evolution is expected to progress very slowly since mutations would not be favored in the competition to survive.

There are about 1300 species of Cyanophycopsida classified in 6 orders, 21 families, and 74 genera. Characteristics of the orders are presented in Table 3–10 and suggested relationships among them are illustrated in Figure 3–4. Evolution seems to have proceeded from the first primitive unicellular cyanophytes along four lines, one leading to the filamentous Oscillatoriales, a second leading to the unicellular Chroococcales and Cryptophycopsida, a third leading by way of the Prochlorales to the Euglenophycopsida and Chlorophycopsida, and the fourth leading by way of the Nostocales to the Rhodophycopsida and Ascomycopsida.

The Chroococcales are morphologically and biochemically similar to some of the Pseudomonadales (Figure 3–4), whereas the Oscillatoriales resemble the Beggiatoales and Myxobacteriales in general morphology, motility, and biochemistry. Judging from the fossil record, these resemblances must be due to parallel evolution. The Chroococcales and Oscillatoriales appear to be equally ancient.

Formerly, most botanists included the Oscillatoriales and the Nostocales in a common order, the Hormogonales, but recently, more and more evidence has accumulated suggesting

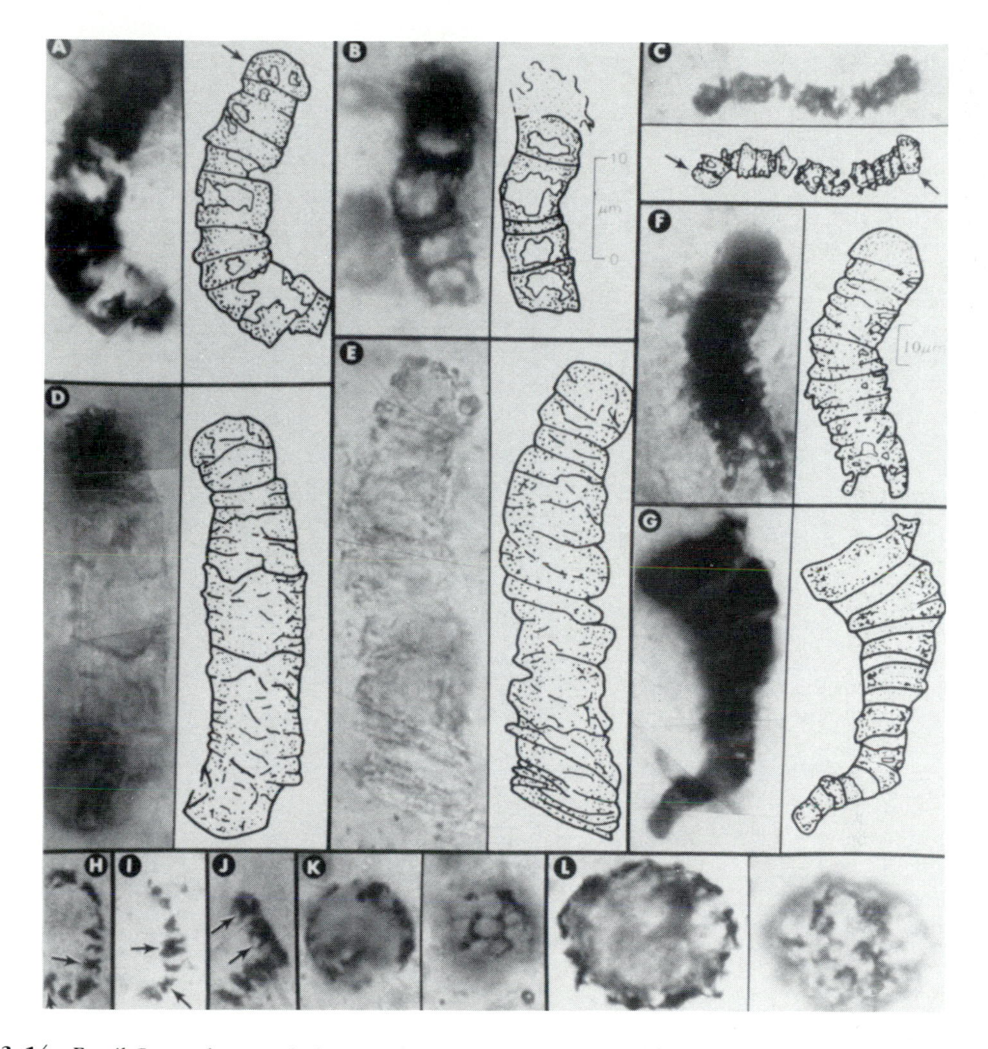

Figure 3–14 Fossil Cyanophycopsida from the Early Archean Apex chert, northwestern Western Australia, ≈3465 million years old. (A–C) *Primaevifilum laticellulosum* with "pillow-shaped" terminal cells similar to those in *Oscillatoria prolifica* indicated by arrows; (D–E) *Archaeoscillatoriopsis grandis;* (F) *Archaeoscillatoriopsis maxima;* (G) *Primaevifilum attenuatum;* (H–J) poorly preserved trichomes showing bifurcated cells and cell pairs (at arrows); (K–L) solitary *Chroococcus*-like "possible microfossils," in equatorial (left) and polar views (right). (From Schopf, J. W., Microfossils of the Early Archean Apex Chert: New evidence of the antiquity of life. *Science* 260; 640, 646, 1993. Copyright 1993 by the A.A.A.S. With permission.)

that the Oscillatoriaceae originated from a different source and at a different time from the other families of the Hormogonales, and the present tendency is to separate them (Figure 3–4).

The Stigonematales resemble both the Nostocales and Rhodophycopsida in many ways, seeming to form a morphological link between the two. Of special significance are the occurrence of primary pit connections and of multiseriate, branching filaments in both the Stigonematales and the Rhodophycopsida. Differentiation is further advanced in the Stigonematales

TABLE 3-10

CHARACTERISTICS OF THE ORDERS OF CYANOPHYCOPSIDA

Order (Fam-Gen-Spp)	General Cell Morphology	Other Characteristics	Habitat	Representative Genera
Chroococcales (2-13-221)	Cells spherical, oval, or oblong, unicellular or in loose colonies, sometimes held together by gelatinous sheaths into pseudofilaments	Herpokinetic motility reported in some genera	Mostly freshwater and in hot springs; also in nano plankton of oceans, saline lakes, and snowbanks	*Aphanothece, Gloeocapsa, Synechococcus, Microcystic, Chroococcus, Merismopedia, Heterohormogonium*
Oscillatorialse (1-9-450)	Cylindrical filaments, often held together by gelatinous sheaths; reproduction by hormogones	Herpokinetic motility common; heterocysts lacking	Mostly freshwater, hot springs, and geysers; also on soil and wet rocks	*Spirulina, Oscillatoria, Lyngbya, Phormidium, Microcoleus, Schizothrix*
Chamaesiphonales (5-13-93)	Unicellular, filamentous, or pseudofilamentous	Reproduce by means of endospores; thallus may be pseudoparenchymatous	Mostly marine in highly saline lakes	*Pleurocapsa, Hyella, Cyanidium, Xenococcus, Dermocarpon*
Nostocales (5-21-450)	Chain-like or cylindrical filaments, often tapering; heterocysts present terminal or intercalary	Herpokinetic motility common; often fix nitrogen in their heterocysts	Mostly freshwater and hot springs, also in cold mountain streams	*Nostoc, Rivularia, Tolyptothrix, Scytonema, Anabaena, Cylindrospermum*
Stigonematales (7-17-66)	Simple and branched filaments; thallus often pseudoparenchymatous with primary pits	Some species fix nitrogen	Both marine and freshwater	*Mastigocladus, Loriella, Westiella, Stigonema, Capsosira, Fisherella*
Prochlorales (1-1-4)	Cells spherical to oval; chlorophylls *a* and *b* present, no phycobilins; lipids and cell walls typical for Cyanophycopsida	Lack phycobilisomes, thylakoid structure intermediate between Cyano- and Chlorophycopsida	Obligate symbionts in cloacal cavity of sea squirts and on their tests	*Prochloron*

Note: The number of families (Fam), genera (Gen), and species (Spp.) in each order is indicated in the first column.

than in the other orders, with branch filaments much narrower and longer than those of the main filament. On the other hand, *Epiphyton,* an extinct genus of blue-greens which has sometimes been classified in the Rhodophycopsida, resembles both *Rivularia* in the Nostocales and *Solenopora* in the Rhodophycopsida, suggesting that the origin of the red seaweeds lies in the Nostocales (Figure 3-4).

A genus of prokaryotic green algae, *Prochloron,* was discovered in 1976 in grooves on the test of a small, sessile marine animal, the ascidian or sea squirt. Species of *Prochloron* have now been found in Australia, in Hawaii, and off the coast of Baja, California, living either in the cloaca or embedded in the epidermis of ascidians, and it is well established that *Prochloron* contributes significantly to the food supply of these primitive chordates (Thinh 1978). The relationship is probably an obligate symbiosis.

Prochloron resembles *Chroococcus* in many ways: both are prokaryotic and unicellular, the cell walls of both contain muramic acid derivatives, the carbohydrates are similar to those of other Schizophyta, and lipids are similar to those of the Cyanophycopsida. However, *Prochloron* cells contain chlorophyll *b;* they lack phycobilin pigments, and hence there are no phycobilisomes present. The thylakoids are arranged in a manner intermediate between the Cyanophycopsida and the Chlorophycopsida. All in all, *Prochloron* appears to be a "missing link" between the Aflagellobionta and the Lorobionta. However, it is certainly not a relict of the original population which gave rise to the Euglenophyta and/or Chlorophyta, both of which it resembles in many ways. Both of these divisions originated hundreds of millions of years before the first chordates. Rather, *Prochloron* is likely the only genus still surviving from a much larger order that linked the two subkingdoms together. Because of the differences between *Prochloron* and other Cyanophycopsida in pigmentation and thylakoid structure, some botanists place *Prochloron* in a class, or a division, by itself, the Prochlorophycopsida and/or Prochlorophyta.

Representative Genera and Species

The Chroococcales are abundant in both marine and freshwater habitats, but are often overlooked by students because of their small size and lack of distinctive cell shape. Several species of *Gloeocapsa, Aphanothece, Synechococcus,* and *Chroococcus* are especially common. Species differ from each other in cell size and habitat preferences as well as in genetic differences of cell shape and sheath characteristics (Figures 3-12 to 3-14). Specimens of the fossil species *Archaeosphaeroides barbertonensis,* a marine cyanophyte from the Fig Tree formation, South Africa (late Archaean), are not readily distinguished from specimens of *Gloeocapsa alpina,* collected in freshwater habitats today (see Figure 3-14 K,L).

Oscillatoriales and Nostocales are abundant in many habitats, primarily freshwater and hot springs, and immediately attract the attention of the student. Several species of *Oscillatoria* can be recognized in the blue-green, slimy material deposited on the rocks around hot springs; they differ from each other in cell size and in dimensions, characteristics of the apical cell, presence or absence of granular material in the cells, and coiling of the filaments. They usually exhibit herpokinetic motility. Species of *Lyngbya, Phormidium,* and *Tolypothrix* are also common and are readily differentiated from *Oscillatoria* by the presence of a thick, gelatinous sheath. Also common is *Spirulina majus,* recognized by its very fine filaments (less than 1 μm in diameter), tight spirals, and lack of septations in the filament. The Archean genera *Primaevifilum* and *Archaeooscillatoria* from Australia (Schopf 1993) and the Aphebian genus *Animikiea* from the Gunflint formation in southern Ontario (Banks 1970) are essentially identical in appearance to modern *Oscillatoria* (compare Figures 3-14f and 3-13a).

Nostoc balls (Figure 3-13d) are common both in cold streams and in many hot springs. They may be so soft that they flatten out when dropped on the floor or on a table, or they may be firm as golf balls and bounce when dropped. Many species look like green grapes, and some people say they taste like them. The cells are arranged in chain-like series (Figure 3-13e) with **heterocysts** within the chain in some species or at one end in others.

Another representative of the Nostocales is the genus *Epiphyton* (Figure 3-12D), which flourished in the Cambrian period. Originally classified as a cyanophyte by Pia (1927), it was reclassified in the Rhodophycopsida by Korde (1953; Brasier 1980) because of the level of cell differentiation it exhibits and the presence of primary pit connections. Following a very critical comparison of *Epiphyton* with *Nostoc, Stigonema,* and other Cyanophycopsida, as well as Rhodophycopsida, Chuvaskov (1963) concluded that its true taxonomic affinities are with the former.

Ecological and Economic Significance

Blue-green algae are abundant in both marine and freshwater habitats. Several species thrive in hot springs where the temperature may exceed 80° C; others are found in saline pools and lakes, in snowbanks, and in other extreme habitats. Farm ponds often develop extensive "water blooms" of blue-green algae during the summer months. The succession of dominant species in these ponds is very rapid. The blue-greens are by no means limited to warm water or summer conditions; species of *Phormidium,* for example, have been collected in Antarctic lakes.

The ecological significance of the Cyanophycopsida is due to (1) high rates of photosynthesis and importance as primary producers, (2) production of toxic substances which tend to regulate the abundance of other plants and animals in the ecosystem, and (3) importance as nitrogen fixers. In eutrophic lakes they are among the most important of all producers; their distribution and importance in marine ecosystems is less well known. Recent studies in the tropical oceans have demonstrated that the **nanoplankton,** the organisms that are so small that they are not retained by the plankton nets commonly used in oceanographic studies, may be by far the most important producers there. Cyanophycopsida always make up the largest portion of the nanoplankton, but because of their small size have generally been overlooked in the past.

Stromatolites are excellent paleoecological indicators. In most areas where algal blooms occur, grazing by small aquatic animals prevents the development of stromatolites. However, in Shark Bay in western Australia, the water is too saline for most herbivores, and stromatolites are being constructed today just as they were in the past, hundreds of millions of years ago before grazers evolved (Figure 3-15). Stromatolites are also building up in some of the hot mineral pools of Yellowstone National Park. Over 100 recognized taxa of stromatolites can be found in the early to middle Hadrynian period (McMenam and McMenam 1990), but fewer than 30 in late Hadrynian, suggesting that grazing by herbivorous animals had become a significant ecological force.

Some species of *Microcystis, Anabaena,* and *Aphanizomenon* produce substances that are toxic to fish or to warm-blooded animals, or to both (Desikachary 1959, Carmichael 1994). Occasionally livestock are poisoned from drinking water containing such toxins. At other times the water in livestock ponds and city reservoirs becomes unfit to drink because of blue-greens which produce ill-tasting, though nontoxic, substances. "Swimmers itch" is occasionally caused by contact with *Lyngbya majuscula,* a marine species, and has resulted in severe skin irritation (*not* an allergy) on parts of the body covered by swim suits of bathers on the windward side of Oahu (Schwimmer and Schwimmer 1968).

The toxins produced by species of *Nodularia, Microcystis, Anaboena, Oscillatoria,* and some other blue-greens are being studied in detail by several botanists (Carmichael 1994), and four toxins have been isolated and used in controlled experiments. Anatoxin-*a* and anatoxin-*a*(s) are neurotoxins found in relatively few, though rather widespread, species; saxitoxin and neosaxitoxin are hepatotoxins. Both groups are responsible for deaths in livestock and wild game animals but as yet have not caused human deaths. Research with these substances and with cytotoxins produced by the same species is concentrating in some laboratories on possible benefits. Since cytotoxins target individual cells and anatoxins affect the nervous system, it is hoped that some combination of these poisons can be used as a wonder drug to kill brain cells that are responsible for Alzheimer's syndrome and other forms of senility, yet allow harmless cells to live.

The ability of some Cyanophycopsida to convert atmospheric N_2 into useful nitrogen compounds was discovered by Winter in *Nostoc;* since then, several other genera, all of them heterocyst-producing, have been discovered to be nitrogen fixers. Nitrogen fixation in lichens in which the phycobiont is a cyanophyte was discovered by Henriksson (1951), who has since demonstrated the importance of this process in terrestrial ecosystems.

Recent research has indicated that pollution of lakes with phosphorous-containing detergents has hastened the eutrophication of many bodies of water. In oligotrophic lakes, low temperatures or low levels of available phosphorous or of trace minerals may be the limiting factors keeping blue-green algae scarce. Other algae are more likely limited by lack of nitrogen. When phosphorus from detergents and trace minerals from organic pollutants are supplied and the increased turbidity from the organic pollutants raises the water temperature, conditions become favorable for the Cyanophycopsida. They in turn fix large quantities of nitrogen, which then creates ideal conditions for other algae, and eutrophication is

Figure 3–15 Modern stromatolites being formed in shallow water of Shark Bay, Australia, on the Indian Ocean. (From Hoffman, P., in *Stromatolites,* Walter, M. R., Ed., 1976, 262. With permission of Elsevier Scientific Publishing Company, Amsterdam.)

on its way. Eutrophication is looked upon by many conservationists as one of the most serious problems in the U.S. at the present time; eutrophication, however, is a natural process which may be regarded as good or bad depending on one's point of view. When it is accelerated to the extent that it has been in the U.S. in recent years, it becomes a serious problem. In areas where greater fish production is desired for human consumption (not sport fishing, however) it may be regarded as a good thing and encouraged.

The potential economic significance of the Cyanophycopsida in agriculture is only now beginning to be recognized. In India, inoculation of rice paddies with *Tolypothrix tenuis* (Figure 3-16) has resulted in 15% to 20% increase in rice yields, according to Desikachary (1959). Since neither commercial fertilizers nor the capital with which to purchase them are available in developing countries, nitrogen fixation by blue-green algae is a promising source of nitrogen needed to improve crop yields. The potential value of *Spirulina platensis* in human nutrition is more direct. When first introduced to health stores, some enthusiasts referred to it as the "algal superstar of the '80's"; it is reported to have twice as much vitamin B_{12} as calf liver, the second best source, in addition to well-balanced proteins. While some of the claims of *Spirulina* enthusiasts have been substantiated by empirical studies, others have not, and concern has been expressed regarding potential dangers (Carmichael 1994). Because the botanically untrained workers who harvest *Spirulina* from ponds in Mexico and elsewhere may not be able to distinguish between *Spirulina platensis* and toxic species of *Oscillatoria,* which it resembles, or even *Anabaena,* which is quite different in appearance (to a botanist), and because federal laws do not require control checks by competent botanists prior to selling the product, this concern is understandable.

The antiviral activity of cultured blue-green algae has only recently begun to be appreciated. Health food advocates, in extolling the virtues of *Spirulina platensis,* often suggested that its merits go far beyond simple vitamin content (Challem 1981) extending to the prevention of ulcers, forestalling the onset of senility, and curing a variety of diseases. If so, not only *Spirulina platensis* but its close relatives might have health care value. Patterson et al. (1993) screened almost 700 strains of Cyanophycopsida, representing more than 300 species, for antiviral activity and found that over 10% of the strains inhibited HIV-1 retrovirus reproduction and development. The genus *Lyngbya* in the Oscillatoriales was especially effective. Various Cyanophycopsida, especially in the Chroococcales, inhibited one or both of two other viruses. It is too early to make recommendations to the medical profession as to the administration of any new medicines, but testing of these cyanophytes and their metabolic products will continue with the hope that the viral diseases concerned may be controlled.

Research Potential

The Cyanophycopsida are potentially of special value in the study of photosynthesis, respiration, and nitrogen fixation. Water-soluble phycobilins offer real possibilities in research on photosynthesis, especially where the lipid-soluble photocatalysts, the chlorophylls and carotenoids, have failed to yield the needed basic information. In respiration research, Cyanophycopsida should be of greatest value in studying the anaerobic phase. The rate of hydrogen evolution under anaerobic conditions is high enough in many blue-greens that Spruit concluded that such studies were practical. In *Synechococcus elongatum,* the enzyme hydrogenase is dormant under aerobic conditions but becomes slowly activated when the plant is subjected to anaerobic conditions. Studies with *Nostoc, Anabaena,* and other blue-greens as well as the lichen, *Peltigera,* are yielding valuable information in the physiology of nitrogen fixation and the nature of the enzyme, nitrogenase, that makes fixation possible (Figure 3-17).

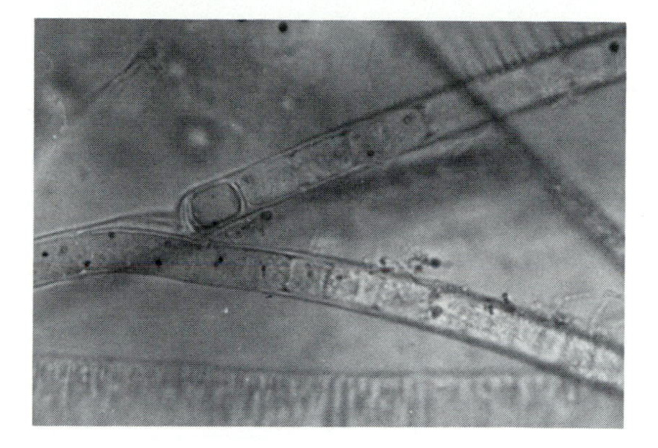

Figure 3–16 *Tolypothrix tenuis* collected in Henry's Fork of the Snake River, eastern Idaho (Photo courtesy Richard Clark, Ricks College.)

Two challenges of the 21st century, inherited from the late 20th century, are waste disposal and food production. The activity of blue-green algae in sewage lagoons (Figure 3-18) is important to both: waste disposal directly and food production through water conservation and mineral recycling. In many cases the nutrients made available through sewage treatment can be used to fertilize crops, but in other cases, the increased load of soluble nutrients in the outlet streams from the lagoons encourages development of algal blooms and eutrophication. If there were some economical way of getting these nutrients, or the wastes from which they are produced, away from the freshwater lakes and on to agricultural lands, or away from estuaries and continental shelves where they are a nuisance and out into the open sea where productivity is limited by a lack of minerals, new sources of food could be made available.

VIRUSES

Viruses appear to be at the borderline between the living and the nonliving, and some biologists, thinking of them as nonliving, ignore them completely as far as taxonomic treatment is concerned. Breed (1957) treated them as primitive plants and classified them as an order, the Virales, of the class Microtatabiotes, division Protophyta of the Plant Kingdom. He also considered the Rickettsiales to be an order in the Microtatabiotes; other bacteria and blue-green algae made up the other classes of the division. Other biologists have either placed the viruses in a kingdom separate from both plants and animals or by themselves as a division of the Plant Kingdom.

Viruses are noncellular, obligate parasites, lacking organelles of any type. The life processes, including reproduction, take place within a host organism by utilizing the energy of respiration of the host. The host may be an animal, a green plant, a fungus, or a bacterium; bacterial viruses are commonly called *Bacteriophages,* or simply *phages.*

Because of their small size, viruses were extremely difficult to study prior to the development of the transmission electron microscope (TEM). We could study the effects of different kinds of viruses on plants and animals, we even learned how to control many of the diseases caused by viruses, and we were able to analyze some viruses chemically, but we knew nothing of their morphology. Now, thanks to the electron microscope, we have succeeded in studying the structure of different viruses (Burns 1972, Nermut 1972, Weber et al. 1994). Note the pronounced structural differences between the influenza virus and the T-4 bacteriophage in Figure 3-19.

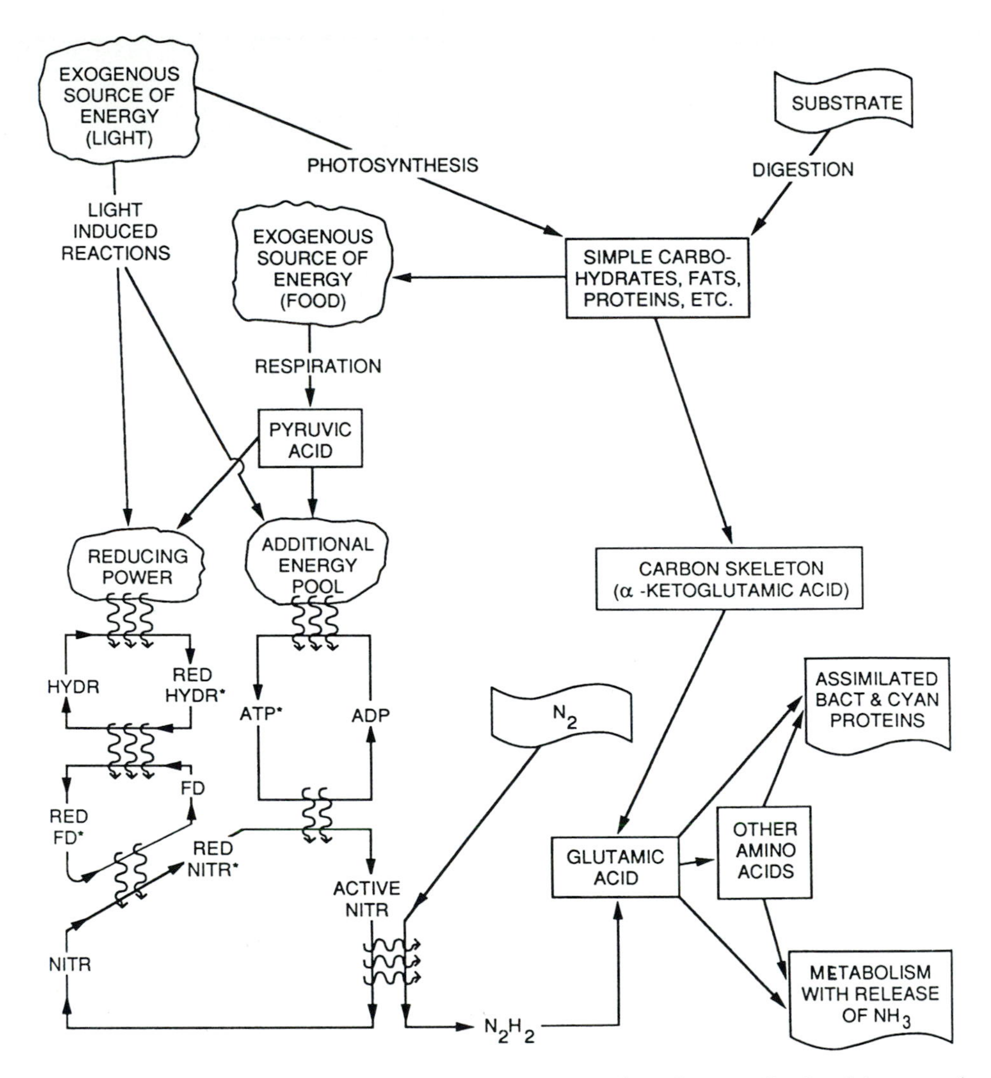

Figure 3–17 Model of nitrogen fixation in bacteria and blue-green algae (from studies involving several species). The energy needed to activate the enzyme nitrogenase can come from either an exogenous source (light) or from an endogenous source (food). Endogenous energy may be produced directly from photosynthesis or obtained from an organic substrate by nonphotosynthetic species. (HYDR = hydrogenase, FD = ferrodoxin, NITR = nitrogenase, ADP and ATP = adenosine diphosphate and adenosine triphosphate, respectively.)

Viruses are often classified by the symptoms of the diseases they produce and their immunity reactions (Table 3-11). Each virus particle consists of a DNA core enveloped by a protein sheath. The protein portion attaches to the cell of the host organism and the viral DNA enters into the cell where it reproduces. In humans and other animals, any foreign protein, such as a virus particle, is called an **antigen** and will stimulate the host organism into producing specific antibodies. In future invasions by the protein, antigen-antibody reactions occur which aid the host in rejecting or destroying the antigen. The simple chemistry of virus protein content

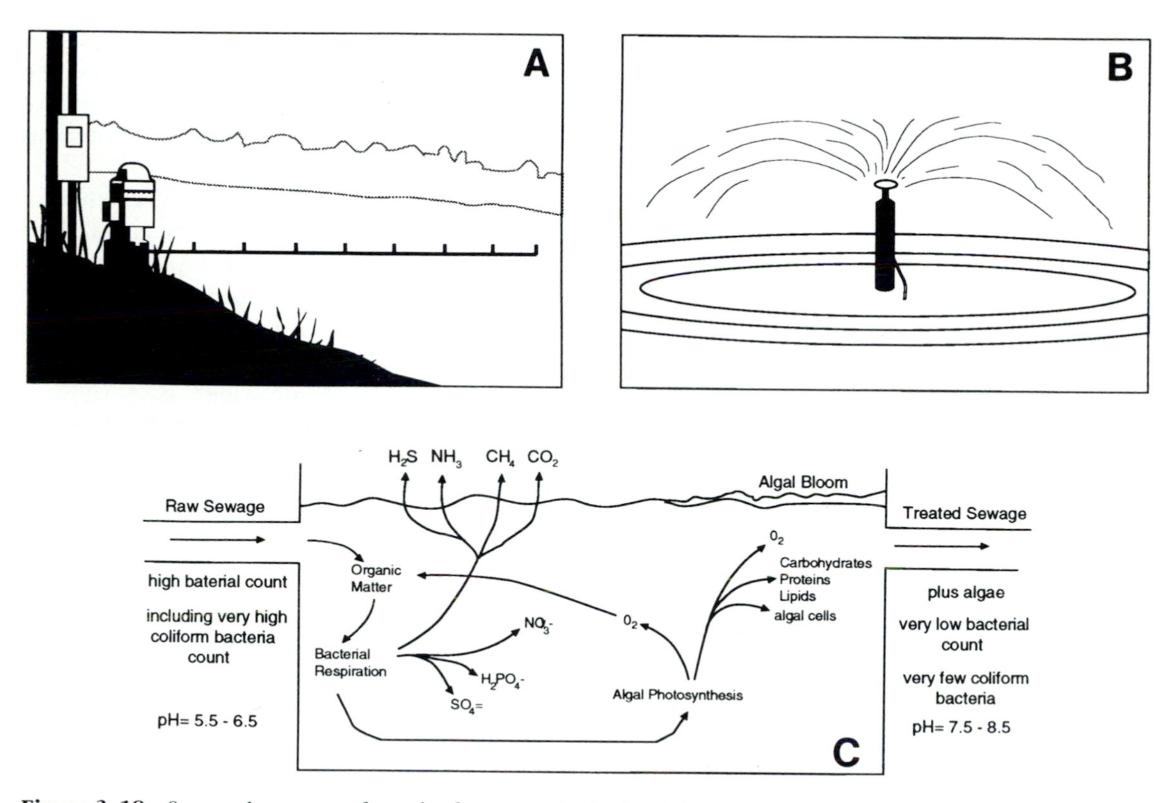

Figure 3–18 Sewage lagoons perform the first stages in the breakdown of sewage. (A) A lagoon in the springtime. In late summer, algal blooms often develop at the end of the lagoon farthest from the sewage inlet; a line of pumps for aeration is set up near inlet. (B) Pump in operation; aeration increases rate of sewage breakdown. (C) Diagram illustrating the biochemical reactions which take place in the lagoon.

has been exploited by the medical profession in developing immunity to many diseases. The viruses that cause serious diseases are killed or weakened by chemical or physical means and used as **vaccine** to provide protection against possible future attacks of the virus. The smallpox vaccine has been so successful that smallpox has been relegated to the status of a dread disease of historical significance only.

Viruses have contributed significantly to genetic research in recent years. In 1957, Heinz Fraenkel-Conrat and co-workers isolated the tobacco mosaic virus (TMV) from tobacco leaves and a similar virus (HRV) from the common plantain, a lawn weed. They separated the protein from the nucleic acid, RNA in this case, and then recombined TMV protein with HRV nucleic acid. When they infected tobacco leaves with the reconstituted hybrid virus, it formed the type of lesions on the tobacco leaves that are characteristic for the HRV disease that occurs on plantain leaves, thus proving that the genetic material is in the nucleic acid, not the protein. A. D. Hershey and M. Chase used a bacterial virus, the T2 virus, in similar studies in which protein labeled with radioactive sulfur and RNA labeled with radioactive phosphorus were introduced into bacterial cultures. As the viruses reproduced, the new particles contained

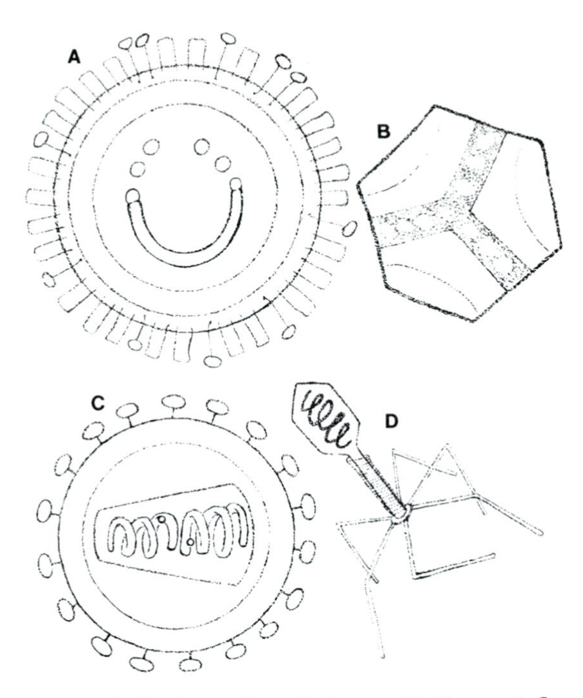

Figure 3–19 Examples of viruses: (A) Human influenza virus; in the center are nucleoprotein units enclosed in a double-layered envelope consisting of an inner protein shell and an outer lipid bilayer from which project two kinds of structures called peptomers, one containing hemagglutinin and the other neuraminidase; (B) murine adenovirus; (C) a retrovirus: the HIV virus; (D) the T-4 bacteriophage, useful in bacterial genetic studies. (A–C, interpretive drawings by L. C. and E. L. Pearson; D, from Burns, G. W., *The Science of Genetics*, Macmillan; New York, 1972. With permission.)

radioactive phosphorus but no radioactive sulfur, demonstrating again that the nucleic acid, not the protein, was the genetic material. The transfer of nucleic acid by virus vectors is known as transduction.

Tobacco mosaic virus (TMV) is an example of an "organism" in which RNA rather than DNA is the basic genetic material. In most plants, animals, and viruses, mitosis or equational cell division begins with synthesis of a new molecule of DNA which then transcribes its genetic information onto a molecule of messenger RNA (mRNA) which moves to a ribosome, organizes ribosomal RNA (rRNA), and initiates production of an enzyme or other protein. In RNA viruses, the genetic information is contained in a molecule of RNA.

Single-stranded RNA viruses which replicate by means of double-stranded DNA intermediates are called **retroviruses.** When a cell is invaded by a retrovirus, the RNA molecule operates in reverse order to alter the DNA of the host and thus reproduce. Most retroviruses multiply rapidly, often causing tumors to develop. Over 90 different cancer-producing retroviruses are known.

Since about 1980, a retrovirus found in humans, the human immunodeficiency virus (HIV), has spread rapidly throughout the world as a sexually transmitted agent responsible for the disease known as AIDS (acquired immunodeficiency syndrome). AIDS has now reached epidemic proportions in much of the world.

Three types of viruses, a DNA virus (incitant of *herpes simplex*), a single-stranded RNA retrovirus (HIV-1), and a double-stranded RNA virus (RSV, incitant of a serious infantile respiratory disease), were studied by Patterson et al. (1993) in a screening test of the effectiveness of blue-green algae as potential antiviral agents. All were inhibited to some extent by the algae. Work is now continuing on isolation of the chemical substances responsible for the inhibitory effect. The Chroococcales were especially effective in this study.

We know little about the relationships or origin of viruses; however, some hypotheses have been advanced. Many biologists regard them as the most primitive of all living things, the link between the living and the nonliving, relicts of the ancestors of all plants and animals. Other biologists look on them as degenerate bacteria. Still others consider them to be escaped genes, nuclear organelles that have managed to free themselves from their cells and retain the ability to reproduce when introduced into a living cell once more. The hypotheses are not mutually

TABLE 3-11

SOME HUMAN DISEASES CAUSED BY VIRUSES

Disease	Virus Group	Virus
Variola or smallpox	Pox group	Borreliota variolae
Varicella or chickenpox	Pox group	Briareus varicellae
Herpes zoster or shingles	Pox group	Briareus varicellae
Rubeola or measles	Pox group	Briareus morbillorum
Herpes simplex: cold sores, fever blisters, herpes VD	Pox group	Scelestus recurrens
Verrucae or warts	Pox group	Molitor verrucae
Poliomyelitis or infantile paralysis	Encephalitis group	Legio debilitans
Rabies or hydrophobia	Encephalitis group	Formido inexorabilis
Yellow fever	Yellow fever group	Charon evagatus
Influenza, Type A	Yellow fever group	Tarpeia alpha
Acute rhinitis or common cold	Yellow fever group	Tarpeia premens
Epidemic parotitis or mumps	Mumps group	Rabula inflans
Acquired immunodeficiency syndrome (or AIDS)	Retrovirus group	Human immunodeficiency virus (HIV)

exclusive; all three views could be correct. Some viruses could be relicts of the "missing link" between nonliving organic matter and primitive organisms and others could be "degenerate bacteria." Perhaps most viruses are "escaped genes." Whichever hypothesis is true, at the present we are interested in viruses because of (1) their importance as causal agents of disease and (2) their importance in genetic research.

SUGGESTED READING

M. R. Walter presents an excellent article on the occurrence and significance of stromatolites in the September–October 1977 issue of *American Scientist.* Selman A. Waksman's *Soil Microbiology* discusses in detail rates of decomposition of organic matter in soil by bacteria and fungi. W. D. P. Stewart's *Nitrogen Fixation* summarizes the current state of knowledge on this important process as it occurs in bacteria and blue-green algae.

A number of articles by C. L. V. Monty and others, dealing with stromatolites and other ancient fossils, were published in the 1970s and 1980s in a variety of scientific journals (see *Agricultural and Biological Index* for titles and sources). Some of these can be found in *Stromatolites,* edited by M. Walter and published in 1976 by Elsevier.

STUDENT EXERCISES

1. Present arguments for and against the proposal to separate the bacteria and blue-green algae from the Plant Kingdom and place them in a kingdom by themselves.

2. From time to time, reports are received of plants other than Schizophyta carrying on nitrogen fixation. If any of these reports are true, in which group of plants would you expect nitrogen fixation to occur?

3. Conjugation and probable meromixis have been reported in some Cyanophycopsida. Diagram the process of meromixis as you would expect it to occur in these plants.
4. Show the expected chemical equation for chemosynthesis in nongreen, autotrophic bacteria. What would be the source of energy for this process to take place?
5. If you were to triple the amount of thienyladenine added to the bacterial medium in running the Guthrie test for PKU, what effect, if any, would it have on the sensitivity of the test? Would this increase or decrease your ability to make an accurate diagnosis? What would happen if you cut the amount of thienyladenine to one third the usual amount?
6. What are the reasons for classifying bacteria and blue-green algae in the same division?
7. Why are bacteria more effective decomposer organisms than most plants or animals?
8. Examining phycobilisomes with the aid of an electron microscope, you observe that in cell A they are granular but in cell B they are disk-like. What color was cell A before it was fixed and stained for EM study? What color was cell B?
9. What are the characteristics of heterocysts which enable *Nostoc* species to carry on both photosynthesis and nitrogen fixation?

4 RED SEAWEEDS

Red seaweeds are among the most beautiful of all living things. Anyone who visits the seashore and observes the underwater flora and fauna is fascinated by the *Polysiphonias,* with their graceful, lacy fronds, the laurentias, sturdier but equally attractive, and the pretty purples and reds of the lavers (*porphyra* spp.). Especially beautiful are the many species that only divers ordinarily see, those that grow in deep tropical waters. Often as divers surface, after studying reefs where red seaweeds abound, they spontaneously exclaim, "Oh, what a beautiful world!"

Equally fascinating are other features of the Rhodophyta: their complex three-phase life cycles, their ancient evolutionary history revealed in fossilized reefs which they built in early Paleozoic time, the similarities they share in tissue complexity and reproductive structures with the Ascomycopsida, and their many uses as human food. Centuries-old anecdotal evidence of the value of some red seaweeds in reducing cancer and circulatory diseases in humans has been reinforced in recent years by empirical evidence involving careful experiments conducted in China, Japan, and the U.S.

The Rhodophyta are the most complex and highly evolved of the autotrophic plants classified in the Aflagellobionta. They are especially well adapted to tropical marine habitats, but species are found in all of the oceans, and a few live in freshwater. They are so different from all other autotrophic organisms that most botanists agree they should be taxonomically isolated from all other plants. Their only close relatives among the algae are the Cyanophycopsida, which they resemble in possessing blue and red phycobilin pigments, lacking flagellated cells throughout their life cycles, storing food as highly branched, short-chained carbohydrates, and with which they share a number of other morphological and cytological similarities (Table 1-3). They differ from the Cyanophycopsida in having well-developed nuclei and plastids, with the genetic material and the photosynthetic pigments enclosed in membranes, in more complex tissue-oriented plant bodies, and in several biochemical and cytological details. Upon analysis of all the known similarities and differences it seems best to place the Rhodophycopsida and Cyanophycopsida in separate divisions of a common subkingdom.

The Rhodophyta apparently evolved in tropical oceans from Cyanophycopsida during the Ordovician period some 450 million years ago. Fossils of Chlorophyta, Chrysophyta, Pyrrophyta, and Schizophyta are associated with the earliest red algae fossils (Table 2-1). By middle

Devonian, 380 million years ago, reef-forming Rhodophyta were abundant in the tropical oceans.

In 1950, it was estimated that there were approximately 2500 species of red algae. An updated estimate would possibly be well over 3500 species; a study of rhodophycean literature by Dawson (1967) suggests that on the average about 30 new species are discovered and named each year. Probable relationships among the eight orders, one of which is extinct, are suggested in Figure 4-1, with a brief description of each in Table 4-1.

CLASS 1. RHODOPHYCOPSIDA

The Rhodophycopsida (*rhodo* = a rose and *phycus* = seaweed or alga), commonly called red seaweeds or red algae, is the only class in the Rhodophyta. These are important plants on the continental shelves of all the oceans, but they are especially abundant in the tropics. Most red seaweeds are tissue-structured and easily recognized by the much-branched thallus, often leaf-like in appearance, which is frequently anchored to the substrate by a root-like **holdfast;** a few species, however, are either unicellular or filamentous. They are typically, but not always, red due to water-soluble photosynthetic pigments, the **phycoerythrins.** They occur at greater depths than any other plants, and the ones that grow deepest are usually the brightest red. A random sample of seaweeds taken along any seashore at any time of year is almost certain to include several species of Rhodophycopsida.

While most red algae are marine, a number of freshwater species are also known. These may be olive or bluish green in color, rather than red, due to the presence of **phycocyanins,** blue pigments similar in structure to phycoerythrins (Figure 3-3). Except for size and eukaryotic structure, unicellular and filamentous freshwater species often resemble Cyanophycopsida.

Because of their distinctive color and differentiated organs, the typical marine Rhodophycopsida are easily recognized. Nevertheless, the novice occasionally confuses them with other seaweeds, especially Phaeophycopsida. The latter usually have more clear-cut differentiation of organs, with stem-like **stipes** often attaching the blades to the holdfasts, and they are often characterized by pneumatocysts or floaters, which the red seaweeds never have.

Unicellular and filamentous red algae are also sometimes confused with other algae. The presence of plastids rules out confusion with Cyanophycopsida, and color is also helpful: freshwater rhodophytes vary from olive green to red, whereas Chlorophycopsida are generally grassy green, Chrysophycopsida bright gold, and Xanthophycopsida yellowish green to grassy green. Unicellular and filamentous Rhodophycopsida usually have one large, more or less star-shaped plastid, whereas the Xanthophycopsida, and often the Chlorophycopsida, have several small disk-shaped plastids. Filamentous Rhodophycopsida exhibit false branching, similar to that seen in many Cyanophycopsida (Figure 3-14), but never in other filamentous algae. The diversity of the Rhodophycopsida is suggested in Table 4-1.

General Morphological Characteristics

Most Rhodophycopsida are multicellular with their own unique macroscopic characteristics, distinctive growth form, and arrangement of tissues.

Unicellular and filamentous red algae, together with some of the multicellular species which are less complex in structure, are found in freshwater streams in many parts of the world,

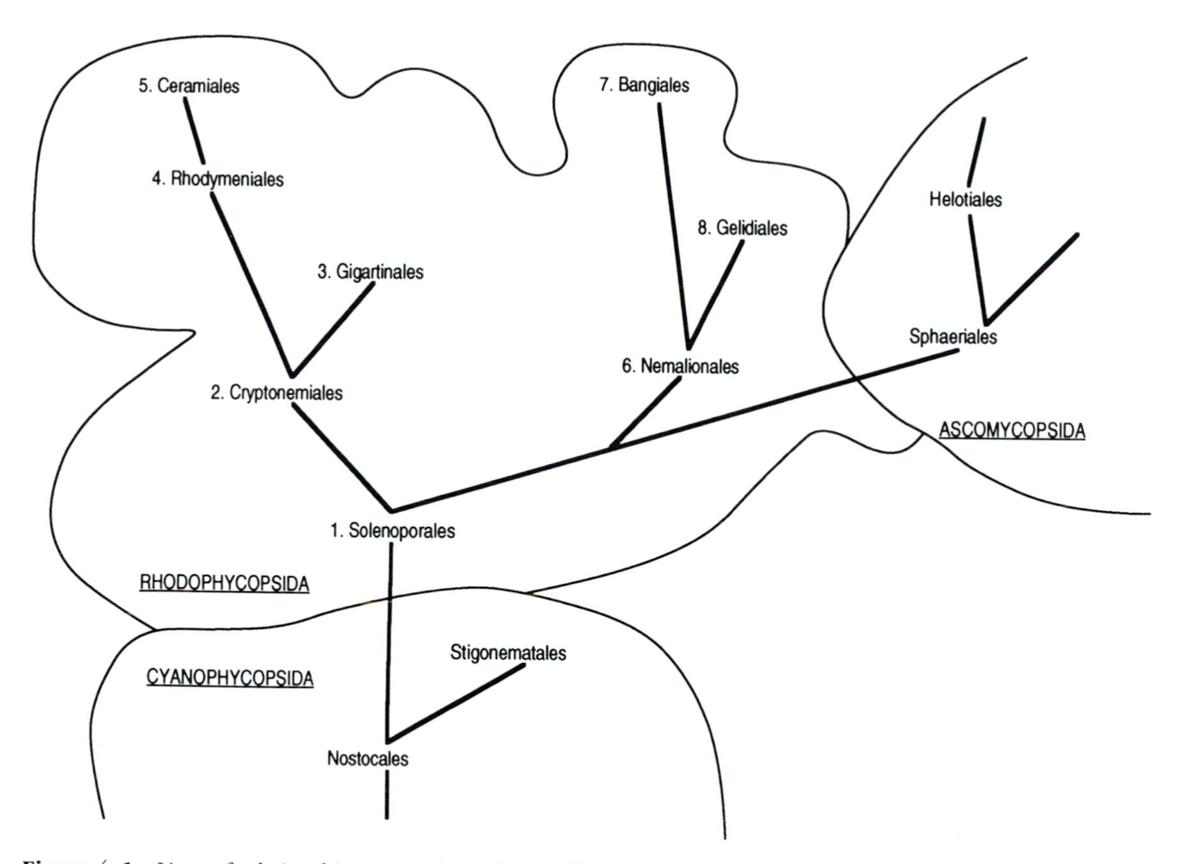

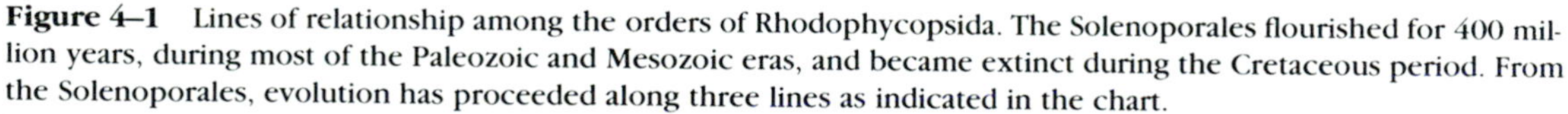
Figure 4–1 Lines of relationship among the orders of Rhodophycopsida. The Solenoporales flourished for 400 million years, during most of the Paleozoic and Mesozoic eras, and became extinct during the Cretaceous period. From the Solenoporales, evolution has proceeded along three lines as indicated in the chart.

and also in some marine habitats, but are seldom common. Like other simple algae, they are identifiable only with the aid of a microscope. The more complex multicellular red algae, on the other hand, are most common in marine habitats and are easily identified by macroscopic characteristics.

Multicellular red algae may be either **uniaxial** or **multiaxial.** Uniaxial plants have a main axis consisting of a single filament of large cells with secondary filaments made up of smaller cells branching off in whorls. They are often microscopic in size as well as simple in structure.

Multiaxial seaweeds consist of many filaments lying more or less parallel with each other; often the cells are differentiated to the extent that their filamentous nature is not apparent. Growth form may be crustose, foliose, or fruticose. **Foliose** species, often resembling long, broad, erect leaves, are especially common just below low tide level in many areas. **Crustose** species, tightly adherent to the rocks or other substrate on which they grow, are often calcified and difficult to remove from the substrate. **Fruticose,** or shrub-like species, may also be heavily calcified, but some species are fleshy and more or less round in cross section, and others are "worm-like," resembling pieces of red wire stuck in the bottom of the bay. Some common red

TABLE 4-1

CHARACTERISTICS OF THE ORDERS OF RHODOPHYCOPSIDA

Order (Fam–Gen–Spp)	General Morphology	Life Cycles/Auxiliary Cells	Representative Genera/ Species
Solenoporales (Extinct)	Cortex/medulla prosenchymatous, not differentiated from each other; cells similar to those of Cryptonemiales but larger and more angular; sporocarps raised	Unknown	*Solenopora, Parachaetetes*
Nemalionales (71-21-271)	Filamentous or uniaxial multicellular; stem-like thalli often dichotomously branched, thalli sometimes lime encrusted; cortex and medulla both prosenchymatous; cells uninucleate with single stellate, axial, or lateral plastid	Mostly haplontic; usually no auxiliary cell but nurse cells perform similar function; nurse and auxiliary cells formed prior to fertilization	*Nemalion, Liagora, Cumagloia, Scinaia, Batrachospermum, Asparagopsis, "Falkenbergia", Galaxaura*
Gelidiales (2-4-57)	Thalli cylindrical and wiry or flattened; usually pinnately branched; tissues usually pseudoparenchymatous	Three-phrase diplohaplontic life cycles; no auxiliary cells formed	*Celidium, Subria, Pterocladia,* and *Wurdemannia*
Bangiales (3-18-88)	Unicellular, simple filaments or membranous thalli; large, single axial plastid with pyrenoids in each cell	Haplontic; no auxiliary cells	*Asterocytis, Erythrotrichia, Frythrocladia, Porphyra, Boldia, Porphyridium, Compsopogon, Bangia, Goniotrichum*
Cryptonemiales (9-35-662)	Foliose, fruticose with cylindrical stipes and elicate linear "leaves," or crustose, often lime encrusted; cortex pseudoparenchymatous, medulla prosenchymatous	Three-phase diplohaplontic; auxiliary cell in special filament growing from a vegetative cell, secondary pits elongated to an ooblast forming prior to fertilization	*Corallina, Fosliella, Melobesia, Euthora, Lithothamnion, Bossea, Cryptosiphonia, Peyssonelia, Callophyllis, Choreocolax, Halymenia, Hildenbrandia, Lithothrix, Goniolithon*

Gigartinales (20–49–540)	Crustose to fruticose, foliose types with thin, broad fronds; fruticose types with flattened, usually dichotomously branched, often fleshy thalli; tissues as in Cryptonemiales	Three-phase diplohaplontic; a vegetative cell serves as auxiliary cell, developing prior to fertilization; secondary pit often elongated into an ooblast	*Iridaea, Chondrus, Agardhiella, Eucheuma, Gigartinia, Harveyella, Gracilaria, Gracilariopsis, Furcellaria, Schizymenia, Hypnea, Ahnfeltia*
Rhodymeniales (2–15–104)	Foliose or fruticose, filiform to fleshy-membranous, sometimes hollow; branching usually dichotomous; cortex and medulla both parenchymatous, outer cortex of smaller cells containing plastids, inner cortex of larger cells	Three-phase diplohaplontic; auxiliary cell formed prior to fertilization in a parallel filament cut off from the supporting cell	*Rhodymenia, Champia, Halosaccion, Gastroclonium, Lomentaria, Chrysymenia*
Ceramiales (4–42–784)	Slender, filamentous or uniaxial bush-like plants lacking clear-cut differentiation of cortex and medulla tissues; simpler anatomy accompanied by more complex reproductive structures is characteristic of this order	Three-phase diplohaplontic; auxiliary cell formed after fertilization in a parallel filament cut off from the supporting cell	*Polysiphonia, Ptilota, Delesseria, Grinnellia, Callithamnion, Odonthalia, Spermathamnion, Acrosorium, Spermothamnion, Platysiphonia, Laurencia*

Note: The number of families (Fam), genera (Gen), and species (Spp.) in each order is indicated in the first column.

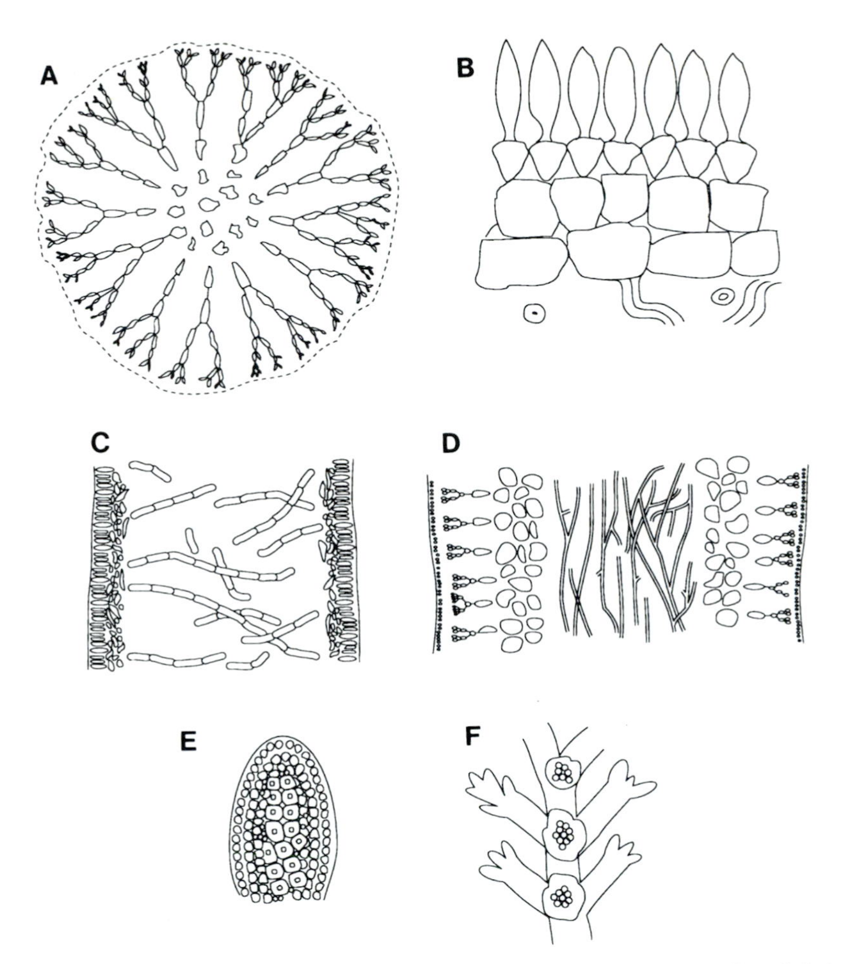

Figure 4–2 Morphology and anatomy of some common red algae. (A) *Nemalion lubricum* (Nemalionales); (B) *Galaxaura oblongata* (Nemalionales); (C) *Cryptonemia ovalifolia* (Cryptonemiales); (D) *Agardhiella tenera* (Cryptonemiales); (E) *Gelidium crinale* (Gelidiales); (F) *Melobesia farinosa* (Cryptonemiales). (Redrawn from Taylor 1957 and 1960, courtesy of University of Michigan Press, Ann Arbor.)

seaweeds are illustrated in Figure 4-2; *Nemalion* species (Figure 4-2A) are uniaxial with simple prosenchymatous tissues. A unique type of thick-walled cell, called a rhizine, occurs in the medulla of some species of *Gelidium* (Figure 4-2E).

Erect foliose and fruticose species are anchored to the substrate by means of **haptera** or **holdfasts,** root-like branched, cylindrical organs well adapted to this purpose. Holdfasts may be rope-like strands of interwoven filaments resembling the rhizines of lichens, or they may be flat disks. Prostrate and crustose species are sometimes anchored by **rhizoids,** simple filaments with swollen tips, growing out from the medulla or lower surface of the thallus (Figures 4-2B and 4-3A). Holdfasts are nourished by food produced in the **lamina** or **blades,** flat leaflike organs of photosynthesis, and transported to them either directly from the lamina or through cylindrical, stem-like **stipes.**

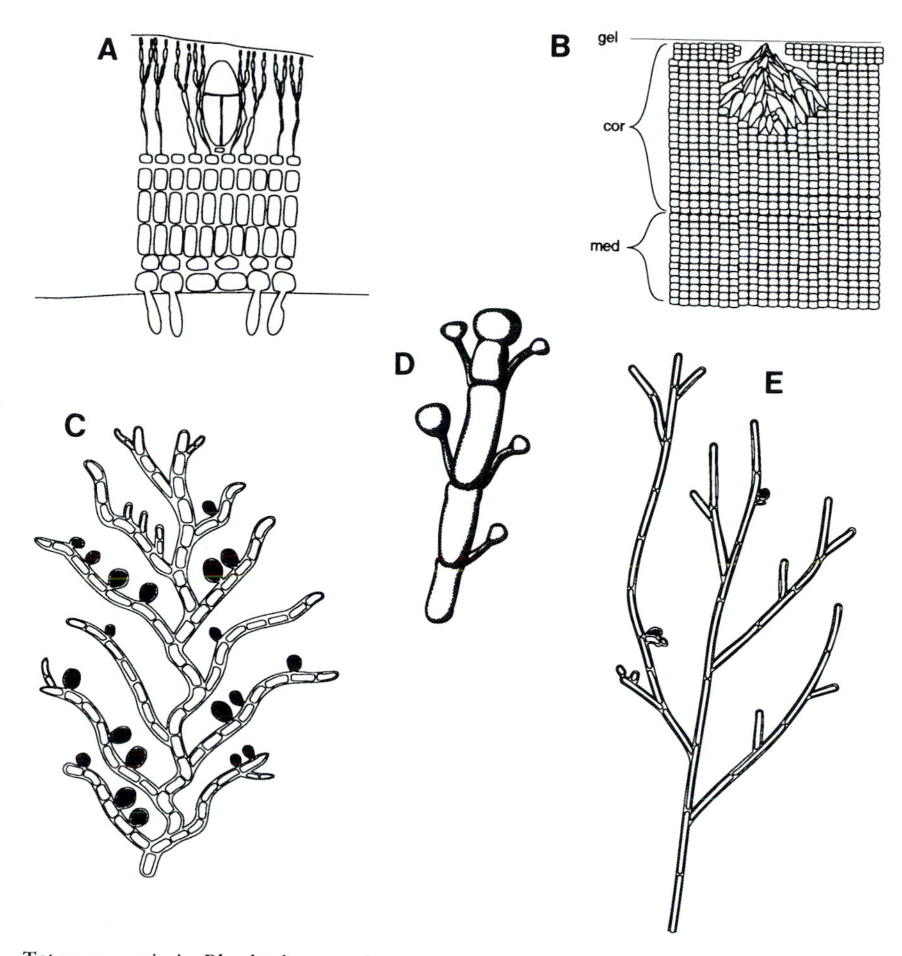

Figure 4–3 Tetrasporangia in Rhodophycopsida: (A) Laminal sporangia of *Peyssonnelia rosenvingii* (Cryptonemiales); note the abundant rhizoids which attach the thallus firmly to the rock formations; (B) laminal sporangia of *Hildenbrandia prototypus* (Cryptonemiales; both (A) and (B) are crustose species); (C) apical sporangia of *Corallina gracilis* (Cryptonemiales); (D) marginal sporangia of *Callithamnion rupicolum* (Ceramiales); (E) axial sporangia of *Spermothamnion turneri;* (Ceramiales). (Redrawn from Taylor 1957 and 1960, courtesy of University of Michigan Press, Ann Arbor.)

Two types of tissues are common in the more advanced multiaxial red seaweeds: **prosenchyma** (Figures 4-2A, C, D) and **pseudoparenchyma** (Figures 4-2B, E, and 4-3A). In both types, cell divisions are primarily in one plane, resulting in simple or branched filaments. Prosenchyma is obviously filamentous and is especially common in the medulla or innermost tissue in both lamina and stipes. Pseudoparenchyma develops from filaments, but the cells are shorter and as they grow they become rounded so that the filamentous nature is no longer obvious. In true **parenchyma** (Figure 4-2E), which is less common in red seaweeds than the other tissues, cell divisions occur in three planes; and no filamentous structure is involved in any stage of development.

In cross section, three or four distinctive tissues can be observed in the lamina of foliose reds (Figures 4-2C, E; Figure 4-3B). Uppermost is an **epidermis** of closely packed, rounded cells; next is a pseudoparenchymatous **cortex,** in which most of the photosynthesis takes place, and below the cortex a prosenchymatous or pseudoparenchymatous **medulla.** In the medulla, or sometimes the cortex, long, slender cells with very thick cell walls often occur; these are called **rhizines** (Figure 4-2E). They are not to be confused with the rhizines of lichens, which are specialized organs of anchorage similar to the haptera of some red seaweeds. In prostrate foliose species, there may be a **lower cortex (hypothallus)**, beneath the medulla, different in structure from the upper cortex.

In crustose species, the medulla is typically pseudoparenchymatous and the upper epidermis and cortex are often impregnated with lime. They generally lack a lower cortex, and the medulla is attached directly to the substrate by rhizoids. Fruticose red seaweeds have stiff, usually round, branching thalli and resemble the stems and twigs of small shrubs. They typically have a central, often pseudoparenchymatous medulla, which may be surrounded by a highly branched, prosenchymatous cortex, the outer part differentiated into an epidermis. Fruticose seaweeds often start out as crustose thalli which gradually develop into the fruticose form.

In the Solenoporales, an order of Rhodophycopsida which was very abundant during the Paleozoic and Mesozoic eras, but is now extinct, there was no distinct differentiation of tissues into epidermis, cortex, and medulla. The large, angular prosenchymatous cells were all much alike (Figure 4-4). The same is true of the Nemalionales, Gelidiales, and Bangiales.

The other four orders are characterized by more complex anatomical structure (Figure 4-5). The Cryptonemiales closely resemble the Solenoporales, but the cells are smaller and rounder, the medulla is filamentous, and the cortex is pseudoparenchymatous. In the Rhodymeniales, the tissues are all parenchymatous with distinct differentiation, involving both cell size and function, into an epidermis, a cortex, and a medulla.

In all red seaweeds, the cells are copiously covered with rather firm gelatinous matter. In the Corallinaceae (Cryptonemiales), the thalli are also so heavily encrusted with lime that they resemble corals. Many Corallinaceae are erect, bushy plants with flexible articulations between the calcified segments. Corallinaceae are difficult to identify because the lime encrustation not only effectively hides the surface characteristics but also interferes with sectioning; this is especially true of the many crustose species.

The cell walls of red algae are double layered, the inner layer being of cellulose and the outer of pectin. The rhizines, in the medulla of some red seaweeds and in the cortex of some others, have especially thick cell walls and are useful in species identification. Sheaths of colloidal polysaccharides, similar to the sheaths of cyanophycopsida, generally surround the cells or fill the intercellular spaces in the tissues. Best known of the rhodophycean colloids are the agars, gelans, and carrageenins. Colloids are less abundant in the Bangiales, where the cells lie tightly appressed to each other, than in the other orders in which cells tend to be rather widely spaced with gelatinous materials holding them together. Strands of protoplasm connect adjacent cells by means of **primary pits,** and often by **secondary pits.**

Primary pit connections result from incomplete septation as the cells undergo mitosis (Figure 4-5). Typically, the two daughter cells push away from each other, resulting in a tube-like connection between them. Gelatinous materials surround the cells and fill in the spaces around the pit connections. Secondary pits begin as protuberances on adjacent cells that may or may not be in the same filament. As the protuberances meet, an opening develops at the apex of each, resulting in a pit through which starch grains and other substances, and even organelles, may migrate. The origin of a primary pit connection is illustrated in Figure 4-5.

In the Bangiales and Nemalionales, there is usually one nucleus per cell; in other species, there are often several. Each nucleus is very small and the chromosomes are enclosed in a nuclear membrane which remains intact during metaphase and anaphase. Average size of nuclei is about 3 µm, with even smaller nuclei, sometimes less than 1 µm, occurring in many of the Cryptonemiales and Gigartinales. Nuclei in some of the Ceramiales are much larger, about 30 µm in the largest.

There is usually one relatively large chloroplast per cell in the Bangiales and Nemalionales, several smaller ones in other red seaweeds. Each plastid is enclosed in a double-layered membrane. Ultrastructure studies reveal that the thylakoids are single and unstacked, as in the Cyanophycopsida (Figure 3-10). Vesicular mitochondria definitely occur in some red algae, e.g., *Porphyridium cruentum,* but apparently are lacking in many species.

Studies with the electron microscope have revealed the presence of **phycobilisomes** within the chloroplasts. The phycobilin pigments (Figure 3-11), phycocyanin and phycoerythrin, are localized in these bodies (Table 4-2). In species like *Porphyridium aerugineum,* which are bluish green because only phycocyanin is present, the phycobilisomes are disk-shaped. In red species, like *P. cruentum,* in which both phycoerythrin and phycocyanin occur but the former is more abundant, most of the phycobilisomes are spherical (Gantt et al. 1984). Phycobilins occur in most species of Cyanophycopsida, with the exception of the Prochlorales and in a few species of Cryptophycopsida; in the Cyanophycopsida they are also localized in phycobilisomes.

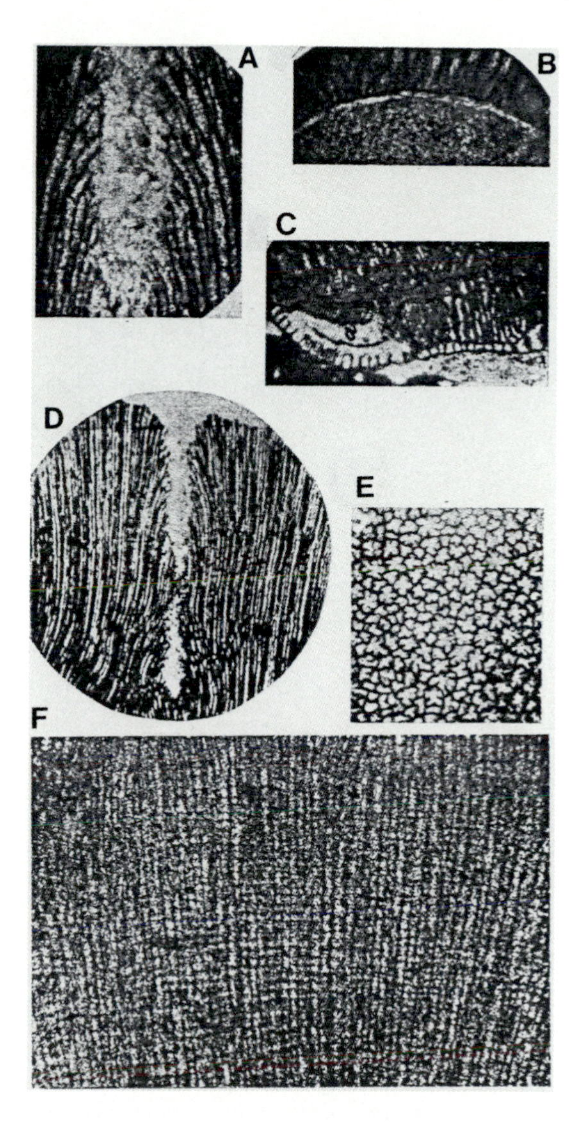

Figure 4–4 *Solenopora spongioides* and *S. nigra* from Ordovician formations in Estonia: (A) and (D) Longitudinal sections of *S. spongioides;* (E) cross section of *S. spongioides;* (B, C, F) longitudinal sections of *Solenopora* sp., tentatively identified as *S. nigra.* (D x12; others x40; from Johnson, J. H., *Studies of Ordovician algae, Q. Col. School Mines,* Courtesy Colorado School of Mines 56, 1961.)

Rhodophyta and Eumycophyta have unique life cycles: a three-phase type consisting of an independent, haploid gametophyte generation; an often dependent diploid or dicaryotic

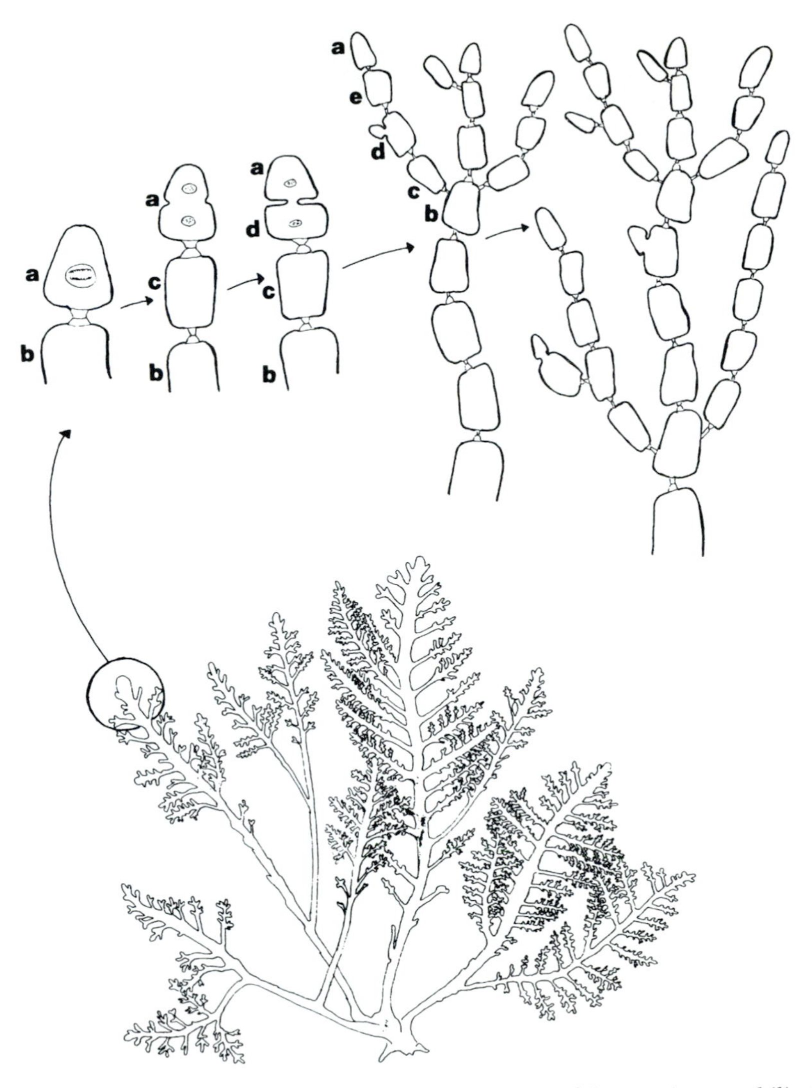

Figure 4–5 Development of primary pit connections in a growing tip of *Laurencia spectabilis,* Ceramiales. Note the position of the original basal cell (b) and apical cell (a) as the plant continues to grow. Cell (e) has resulted from intercalary growth between cells (a) and (d).

generation; and an independent or dependent, diploid meiosporophyte generation. One or more of the generations may be much reduced or lacking.

The gametophytes of red seaweeds may be **monoecious,** having male and female gametangia on the same plant, or **dioecious.** If dioecious, the male and female gametophytes are usually quite similar in appearance. They may be crustose, foliose, or fruticose, depending on species. On female plants, a carpogonial initial, called the **supporting cell,** divides mitotically to produce a **carpogonial filament** (Figure 4-6). The **carpogonium** is the terminal cell on the carpogonial filament.

TABLE 4-2

PIGMENTATION IN THE RHODOPHYTA, CYANOPHYCOPSIDA, AND CRYPTOPHYCOPSIDA

Pigment	Color	Rhodophycopsida	Cyanophycopsida	Cryptophycopsida
Chlorophyll *a*	Green	5	5	5
Chlorophyll *d*	Green	2	0	0
Allophycocyanin	Blue	1	3	0
b-Phycocyanin	Blue	1	0	0
c-Phycocyanin	Blue	2	5	2[a]
r-Phycocyanin	Blue	4	0	0
b-Phycoerythrin	Red	2	0	0
c-Phycoerythrin	Red	4	3	0
r-Phycoerythrin	Red	5	0	0
x-Phycoerythrin	Red	0	0	1

Note: Abundance indicated on scale of 0 (seldom if ever present) to 5 (present in most or all species of the class).

[a] In most Cryptophycopsida, the phycocyanin, if present, is slightly different from the phycocyanins in the Cyanophycopsida

With exception of some of the Bangiales, the carpogonium is always provided with an extension called a **trichogyne.** The trichogyne is sometimes very long and slender, sometimes even spiraled. On male plants, a **spermatangium** produces spermatia which, since they are nonmotile, are carried passively by water currents; some of them come in contact with trichogynes where enzymes dissolve the adjacent cell walls and the spermatium nucleus migrates down the trichogyne and fuses with the egg, or female nucleus, in the carpogonium. In some species of *Nemalion,* the nuclei occasionally fail to fuse; as a result, the carpogonium is **dicaryotic,** containg two haploid nuclei instead of one diploid nucleus (Smith 1938). Development of the carpogonium typically takes place in a cavity on the thallus surface, usually near the margin, called a **conceptacle** (Figure 4-3).

In most species of red seaweeds, the zygote migrates through a secondary pit connection into an **auxiliary cell** immediately after syngamy. There it undergoes mitosis, developing into a dependent diploid plant, the **carposporophyte,** which can be thought of as "parasitic" because it obtains its nourishment from the female gametophyte which produced it. It is typically minute, consisting of relatively few cells, and is part of a structure called a **sporocarp,** unique to the Rhodophyta and Eumycophyta, made up of both haploid and diploid tissues. As the carposporophyte attains maturity, which usually takes about a week, depending on species, it produces and disseminates **carpospores.**

Polysiphonia is frequently used to illustrate the more "typical" life cycles of the red algae (Figure 4-7). Like most Rhodophycopsida, *Polysiphonia* goes through three distinct phases in completing its life cycle. Gametophytes are heterothallic and the male and female plants are identical in appearance except for the sex organs of the mature plants. The carpogonia are located near the apex of hair-like branches, or female trichoblasts, and consist of a supporting cell and four almost colorless cells, the distal one of which develops a trichogyne. Spermatia are released from numerous spermatangia on male trichoblasts and lodge against the trichogynes, where they discharge their nuclei. A zygote is formed when a sperm nucleus fuses with the egg nucleus after migrating through the trichogyne. A secondary pit forms between the carpogonium, now containing a diploid zygote nucleus, and the basal cell of a short filament

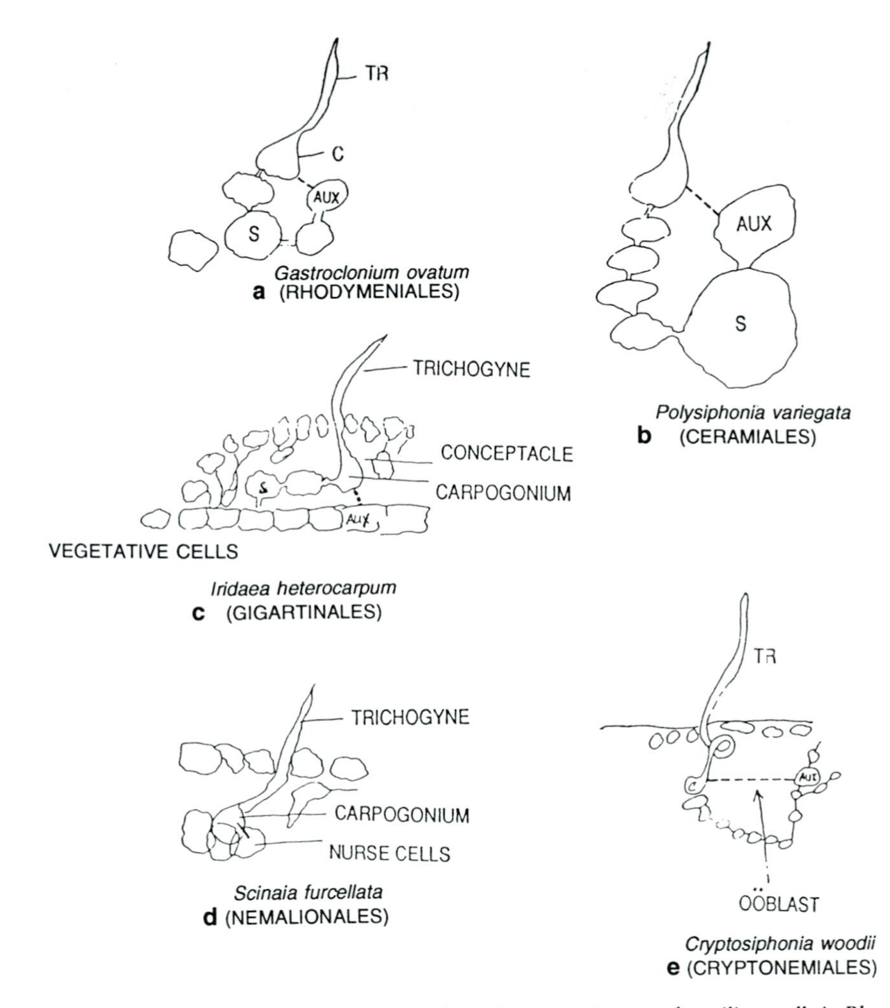

Figure 4–6 Formation of carpogonial filaments, secondary pit connections, and auxiliary cells in Rhodophycopsida. Dashed lines indicate location of secondary pits through which the zygote nucleus migrates. S = spermatia; C = carpogonium; T = tetraspore.

arising from the supporting cell at the base of the carpogonial branch. The zygote migrates through this pit connection. The haploid cell into which it migrates is the auxiliary cell. The haploid nucleus atrophies and the auxiliary cell, which now contains the zygote nucleus, increases considerably in size. The zygote nucleus undergoes mitosis and forms a large number of diploid cells. The adjacent haploid cells also enlarge and develop into a fleshy tissue called the pericarp which completely encloses the carpogonial branch and auxiliary cell except for an opening at the distal end. The nuclei developing from the zygote become carpospores and are discharged through the opening of the urn-shaped sporocarp. Upon germination, carpospores produce diploid *tetrasporophytes,* identical in appearance to the gametophytes except that they produce meiosporangia instead of gametangia. They never produce carpogonia. The diploid tetrasporophytes produce tetrasporangia on short stalks. Meiosis takes place in the tetrasporangia. The tetrasporangial wall breaks and sheds the tetraspores, which germinate to produce gametophytes, two male and two female plants from each tetrasporangium.

The alternation of haploid and diploid generations in plants provides an evolutionary advantage: natural selection operates more effectively in that deleterious genes are readily selected against and culled out of a population during the haploid stage.

The genetic cost of this is a slight loss in ecological homeostasis, which is largely made up for during the diploid stage when gametes from different gene pools come together. In seaweeds, isolation is usually achieved by distance; *Gracilaria verrucosa* (Gigartinales) populations are effectively isolated if they are separated by at least 50 meters (Richerd et al. 1993). There was no evidence of inbreeding depression in this species; nevertheless, heterosis (or hybrid vigor) was apparent in laboratory crosses between plants collected 100 meters apart.

Polysiphonia denudata has been cultured under laboratory conditions and its life cycle studied in detail by Edwards who reported in the 1968 *Journal of Phycology* that 3½ months were required to complete the full life cycle: 50 days from meiospore germination to fertilization, 7 days from fertilization to mature carposporophyte, and 44 days from carpospore germination to mature tetrasporophtes capable of producing meiospores.

Upon germination, a carpospore, which is usually diploid, develops into another spore-producing plant, the **tetrasporophyte.** This plant is diploid, free-living, and often identical in general appearance to the gametophytes. When it reaches maturity, it produces meiospores in pyramidal sets of four. These are called **tetraspores.** When they germinate, they produce gametophytes. Out of each tetrad of spores, two will become female gametophytes, two will become male (Figure 4-7). Sometimes, as in *Galaxaura,* the tetrasporophytes are slightly (or even greatly) different in appearance from the gametophytes. Formerly, the two stages were often regarded as distinct species, but because of morphological similarities were referred to as "species pairs," and recognized as belonging to the same genus (Taylor 1960).

In *Porphyra* and some other genera (Figure 4-8), the zygote reportedly divides by meiosis and the carpospores are therefore haploid. They develop into haploid **monosporophytes,** which produce **monospores,** instead of into diploid tetrasporophytes as in most Rhodophyta (Figure 4-7). Like tetraspores, monospores are haploid, but the former are produced by meiosis, the latter by mitosis. The monosporophytes of *Porphyra* species are so different in general appearance from the gametophytes that they were long thought to be distinct species and were classified in a different genus, *Conchocelis,* and often classified in a different family as well. Monospores, like the tetraspores of other rhodophytes, develop into independent, free-living gametophytes, completing the life cycle (Figure 4-8).

It has recently been suggested, however, that *Porphyra purpurea* monosporophytes, "*Conchocelis rosea*", are diploid, and meiosis occurs as the conchospores germinate (Mitman and van der Meer 1994). All four spores develop into the gametophyte thallus: the two occupying the dorsal position develop into female tissue, making up most of the frond, the other two into rhizoids, holdfast, and the ventral part of the frond.

A unique feature of red algae life cycles is the formation of the auxiliary cell, following fertilization, into which the zygote nucleus migrates (Figure 4-6). The time of formation of the auxiliary cell and its location are important clues in classification of red algae (Table 4-1). Following migration of the zygote into the auxiliary cell, mitotic divisions take place and a **gonimoblast filament** is produced. The carposporophyte begins development as a diploid

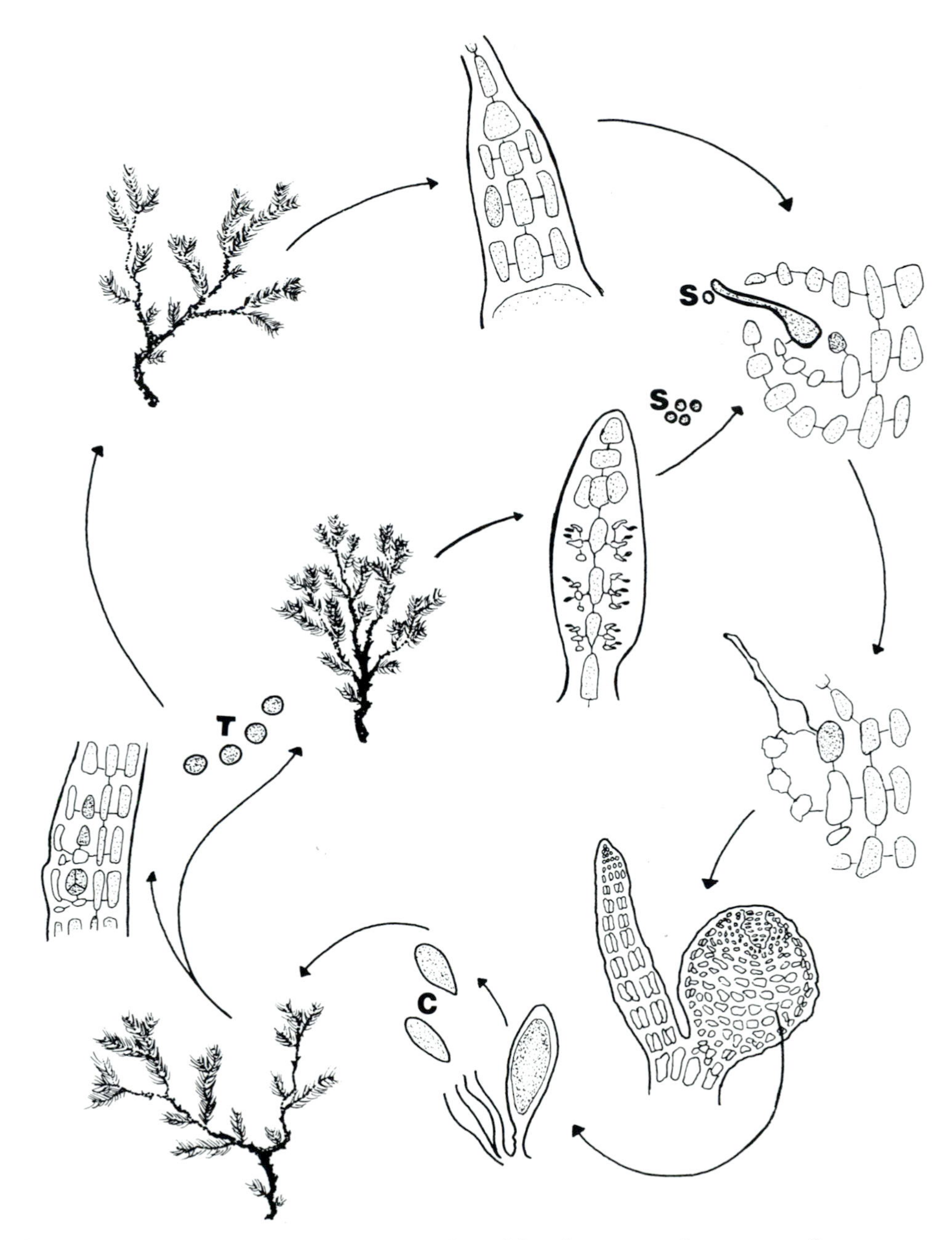

Figure 4–7 Life cycle of *Polysiphonia variegata,* Ceramiales. S = spermatia or sperm; C = carpospores; T = tetraspores.

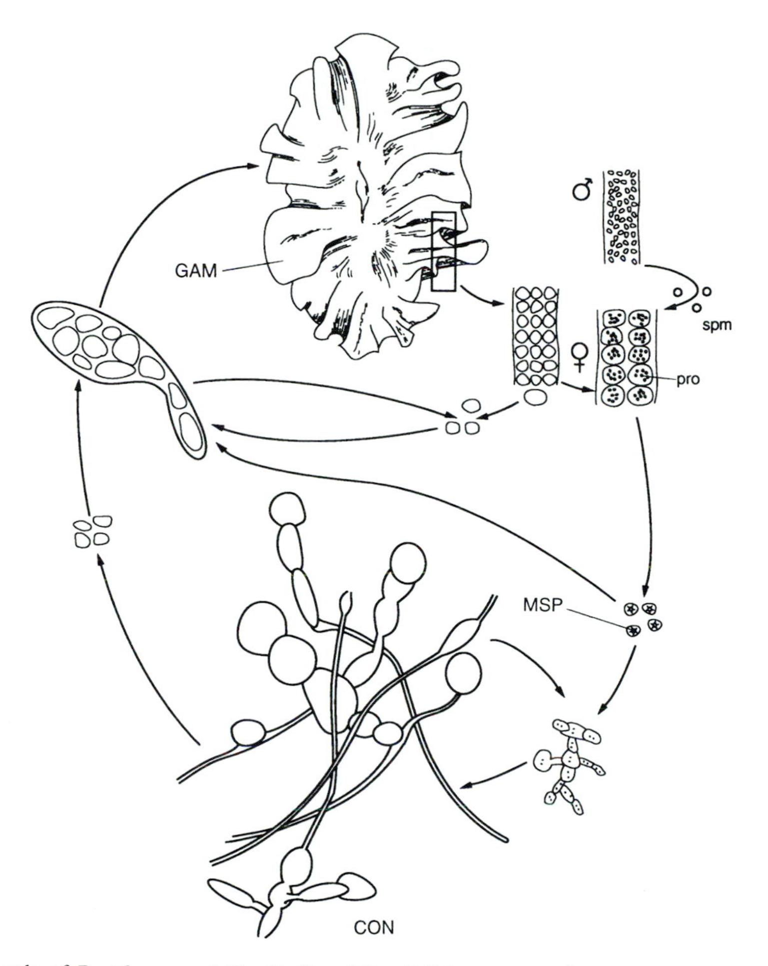

Figure 4–8 Life cycle of *Porphyra umbilicalis,* Bangiales. GAM = gametophyte or *Porphyra* stage; CON = *Conchocelis* stage. MSP = monospore, csp = carpospore, spm = spermatium (or sperm), pro = carpogonial protuberance or trichogyne. Asexual reproduction may occur in either stage, by means of aplanospores in the gametophyte stage and by means of monospores in the monosporophyte stage. Formerly, the conchocelis stage was believed to be a different species, *Conchocelis rosea,* classified in the Nemalionales.

gonimoblast filament growing out of the auxiliary cell. The carposporophyte together with associated haploid filaments makes up the sporocarp or **spore fruit.**

The development of the **auxiliary cell** varies among the orders of Rhodophycopsida (Table 4-1, Figure 4-6). In the Bangiales and Gelidiales, no auxiliary cell is formed. In some Nemalionales, **nurse cells** develop from one of the cells in the carpogonial filament and function in a way similar to that of auxiliary cells (Figure 4-6d). They are called nurse cells because they are primarily nutritive in nature. In *Scinaia* (Nemalionales), four nurse cells develop from and surround the middle cell of the carpogonial filament; following fertilization, a secondary pit forms between the carpogonium and one of the nurse cells, and the zygote nucleus migrates through the pit into the nurse cell. In the Gelidiales, the zygote remains in the carpogonium and the gonimoblast filament and carposporophyte develop there.

Typical auxiliary cells occur in all of the more advanced orders of red seaweeds. In the Gigartinales, a secondary pit forms between the carpogonium and a vegetative cell which may be at some distance from the carpogonial filament, though it is often near by. The zygote migrates through this pit connection. In the Cryptonemiales, the auxiliary cell is in a special filament, often quite distantly removed from the carpogonial filament. The zygote migrates through an elongated secondary pit connection called an oöblast. In the Rhodymeniales and Ceramiales, a secondary pit forms between the carpogonium and a cell cut off from the supporting cell of the carpogonial filament, and the zygote migrates into it; in each case, the cell into which the zygote migrates is called an auxiliary cell. Formation of the auxiliary cell takes place prior to fertilization in the Rhodymeniales, in those Nemalionales in which there is an auxiliary cell, and in the Cryptonemiales and Gigartinales, but is delayed until after fertilization in the Ceramiales. At the same time the auxiliary cells are being produced, parallel filaments are growing out of the basal cell of the carpogonial filament; these become the **pericarp** of the spore fruit.

Sporocarp formation is another unique characteristic of the Rhodophyta and Eumycophyta. Sporocarps may be primarily filamentous or they may be pseudoparenchymatous. In the latter case, they are often fleshy in texture. Large sporocarps, commonly called mushrooms, occur in some of the Eumycophyta, but in the Rhodophyta, the sporocarps are minute, urn-shaped, fleshy structures, about ½ to 1 mm in diameter. The outer tissue is the **pericarp** and is made up of haploid gametophyte cells. The **placenta** consists of diploid cells and is part of the carposporophyte. The outer layers of the placenta differentiate into carpospores.

Rhodophycean sporocarps are also called **cystocarps,** although formerly this term was often applied only to the carposporophyte instead of to the entire structure. A similar sporocarp, called a perithecium (Figure 5-4), is produced in many species of Ascomycopsida. In some red algae and in most fungi, meiosis takes place almost immediately following syngamy, and haploid meiospores, which germinate to produce gametophytes, are produced in the spore fruits. In most Rhodophycopsida, however, diploid carpospores produce independent tetrasporophytes in which meiosis takes place. The distinctive characteristics of the sporocarp makes it useful in identification of species in both the Rhodophyta and Emycophyta. Table 4-3 summarizes some of the spore and sporocarp characteristics of both divisions.

Gametophytes and tetrasporophytes are generally very similar to each other, often essentially identical, both in appearance and in ontogenetic development or morphogenesis. Upon germination, diploid carpospores and haploid tetraspores develop into embryonic meiosporophytes (tetrasporophytes) and gametophytes, respectively.

In the higher red algae, the spore typically divides two or three times by mitosis to produce a short filament 4 to 8 cells long. Each of these cells divides transversely to form an encircling layer of **pericentral cells.** The number of encircling cells varies from 4 to 24 and is fairly constant for each species. The pericentral cells send out rhizoids which help anchor the embryo. Near the apex of the embryonic filament, trichoblast initials are cut off in some of the highly advanced red seaweeds. These divide and grow, forming a dichotomously branched, tapering **trichoblast,** or hair-like filament, from which prosenchymatous tissues of the young sporelings are developed in some species; in others, they absciss quite early. Mature plants

TABLE 4-3

KINDS OF SPOROCARPS IN THE RHODOPHYTA AND EUMYCOPHYTA

Type	Occurrence	Size	Shape	Kinds of Spores	Comments
Gonimoblast	Primitive Rhodophycopsida	Small: less than 0.1 mm	Filamentous	Haploid carpospores	Composed of fertile haploid filaments; zygote undergoes meiosis
Cystocarp	Advanced Rhodophycopsida	Relatively small: 0.25–1.0 mm	Urn-shaped	Diploid carpospores	Fertile diploid cells surrounded by sterile haploid pericarp
Perithecium	Ascomycopsida: Sphaeriales to Aspergillales line	Relative small: 0.25–2.0 mm	Flask-shaped	Haploid meiospores called ascospores, 8 per ascus	Fertile diploid cells which undergo meiosis, surrounded by sterile haploid and dikaryotic tissue
Apothecium	Ascomycopsida: Helotiales and Pezizales line	Medium to large: 5 mm–15 cm or more	Disk- or cup-shaped	Haploid meiospores, 8 per ascus	As above
Basidiocarp	Basidiomycopsida	Medium to large: 5 mm–60 cm or more	Umbrella-shaped, coral-like or spheres	Haploid meiospores called basidiospores, 4 per basidium	Fertile diploid cells surrounded by sterile dikaryotic cells

developing from these sporelings may retain the prosenchymatous nature of the embryos, or the older portions of the thallus may become **corticated,** with distinct epidermis, cortex, and medulla tissues differentiated. When the thallus has reached maturity, conceptacles form and the spores or gametes are formed in these. The conceptacles may be apical, marginal, or laminal (Figure 4-3), depending on the species.

Life cycles of *Porphyra occidentalis* and *P. umbilicalis,* in the Bangiales, are examples of a modified three-phase life cycle in which the diploid phase is missing. The cycle is essentially, therefore, haplontic, but with two haploid stages: a minute, filamentous or **conchocelis** stage and a larger, foliose gametophyte stage. The mature gametophytes are membranous foliose seaweeds of the intertidal zone, deep red to purple in color, and favorite food items in Japan and China. Male *Porphyra umbilicalis* gametophytes can be recognized by the whitish margin at the edge of the lamina where large numbers of colorless sperm have been produced in sunken spermatangia, or **antheridia.** On the surface of the thallus of female plants, there are small protuberances which function as trichogynes. The sperm migrate into these surface cells, or **carpogonia,** and unite with their nuclei. Germination of the zygote is apparently by means of meiosis. Four **carpospores** are produced, each of which germinates by mitosis, after being discharged from the carpogonium, to produce an embryonic gametophyte or else a **conchocelis filament.** The conchocelis filaments may produce either new conchocelis filaments or embryonic *Porphyra* gametophytes.

Formerly, the conchocelis phase of *Porphyra umbilicalis* was known as *Conchocelis rosea* and was believed to be a distinct species. Many "species" of *Conchocelis* had been described and named over the years. They were classified in the order Nemalionales. It was known that the filaments produced **monospores** which produced new conchocelis filaments. Ongoing studies of life cycles have revealed that the various species of "*Conchocelis*" are stages in the life cycles of different species of *Porphyra.* The monospores, like the carpospores, produce not only conchocelis filaments, as formerly believed, but also embryonic gametophytes. In **homothallic** species of *Porphyra,* male and female organs are on the same plants; *P. umbilicalis* is an example of a **heterothallic** species, having male and female organs on different plants.

Phylogeny and Classification

By the beginning of the Paleozoic era, free oxygen in the atmosphere had increased to a level that was fairly effective in screening out ultraviolet rays. Marine plants were therefore becoming abundant even in relatively shallow water, creating new ecological niches which were being filled with macroscopic species. Species of Siphonophycopsida, from early Cambrian, and Rhodophycopsida, were among these.

Rhodophycopsida originated in early Paleozoic time, probably lower Ordovician, possibly late Cambrian. Fossil seaweeds similar to modern *Corallina,* but with larger, more angular cells and external sporocarps, have been found in Ordovician formations (Table 2-1); the earliest have been named *Solenopora* and classified in the Rhodophycopsida. The tissues in Solenoporales are prosenchymatous, lacking differentiation of epidermis, cortex, or medulla (Figure 4-4).

The fossil genera *Epiphyton* and *Chabakovia,* classified by Pia (1927), and Chuvaskov (1963) in the Cyanophycopsida, and by Korde (1963) in the Rhodophycopsida, were abundant during the Cambrian. They represent the group of Cyanophycopsida from which the Rhodophycopsida probably evolved. *Epiphyton* is especially common in early Paleozoic formations and is intermediate in appearance to the Nostocales, especially *Rivularia,* and the Stigonematales, on the one hand, and to the Solenoporales, Cryptonemiales, and Nemalionales, on the other. Some 80 species have been described.

It was formerly hypothesized that the first red algae probably resembled the Nemalionales and Bangiales of today, especially the former. Although several species of red seaweeds are common in late Paleozoic and Mesozoic formations (Johnson 1961a), none of them resemble either Nemalionales or Bangiales. *Nemalion*-like fossils first appear in Miocene formations. *Solenopora,* on the other hand, resembled modern Nostocales in several ways, was morphologically similar to fossil *Epiphyton,* also in the Nostocales, and was common in early Paleozoic times.

Cytological, anatomical, and paleobotanical evidence suggests that evolution has proceeded along three major lines in the Rhodophycopsida. This is illustrated in Figure 4-1.

In the line leading from the ancient Solenoporales to the Nemalionales, Gelidiales, and Bangiales, and from these groups to the Ascomycopsida, growth tends to be general (cell division is not restricted to certain portions of the thallus as in most red algae), primary pits are often lacking, the cells are uninucleate, contain only one large plastid, and have no central vacuole, and auxiliary cells are generally not formed. Several species in this line have become adapted to freshwater habitats. The Gelidiales are characterized by organ specialization and tissue formation, but in the Bangiales and some of the Nemalionales, there has been a trend toward simplification, and both filamentous and unicellular species occur in these two orders.

In the second line, leading from the Solenoporales and Cryptonemiales to the Gigartinales, the auxiliary cell is borne either on a special filament originating on a vegetative cell, sometimes located some distance from the carpogonial filament, or else is such a vegetative cell. Elongated secondary pits connect the carpogonium and the auxiliary cell. Trichogynes tend to be long and are sometimes coiled. The thalli of many species are lime-encrusted; crustose species are common. This is also the line in which the best quality carrageenins are found.

In the third line, leading to the Ceramiales by way of the Rhodymeniales, the auxiliary cell is the terminal cell of a short filament arising from the supporting cell of the carpogonial filament. The lamina tend to be rather complex with differentiation of a layer of small cells making up a thin epidermis, and with several layers of larger, closely packed cells, the cortex, surrounding a looser, large-celled medulla. Both foliose and fruticose, or shrub-like, species are common. Branches may arise from **trichocysts,** branched embryo-like filaments, crudely analogous to the buds of flowering plants, which have arisen from one of the cells of the carpogonial filament.

The 2500 to 3500 extant species of Rhodophycopsida make up 7 orders, 47 families, and about 200 genera of marine and freshwater plants. Not all botanists agree with this classification; Bold and Wynne (1985), for example, gave order status to several of the traditional families, thus dividing the Rhodophycopsida into 14 orders.

Many botanists classify the eight orders in two subclasses, the Bangiophycidae with one order and the Floridae with seven. The most consistent difference between them is the lack of pit connections in the Bangiales ("Bangiophycidae"). The discovery that *Conchocelis* species, which are classified in the Nemalionales, are really a stage in the life cycle of some species of Bangiales has thrown considerable doubt on the validity of maintaining separate subclasses, or even maintaining Bangiales as a distinct order. "*Conchocelis*" species not only have pit connections, but they resemble the Floridae in many other ways (Taylor 1957). For these reasons, it is assumed in this book that the Bangiales and Nemalionales are so closely related to each other and to other Floridae that they should not be classified in separate subclasses. The bulk of the evidence indicates that neither order is primitive but that both have evolved from more advanced species by means of simplification.

Cells are smaller in the Corallinaceae than in the Solenoporaceae and the sporocarps are embedded in tissue. *Solenopora* sporocarps were apparently completely external, like apothecia in the Ascomycopsida; therefore, they have not been well preserved. Although both families are usually included in the Cryptonemiales, the Solenoporaceae have been separated as a distinct order in Figure 4-1 and Table 4-1, stressing the pronounced differences between the two families in cell and tissue morphology and in sporocarp position. It also calls attention to the similarity of *Solenopora* sporocarps to those of the Ascomycopsida.

Representative Species and Genera

Red seaweeds are common in the intertidal and subtidal zones on rocky coasts; sandy beaches and mudflats have few seaweeds. A few species of Rhodophycopsida grow on pilings in docking areas, but the variety is poor and the specimens mostly unhealthy looking. Even vegetation, such as eelgrass and kelp, and various animals, such as barnacles and snails, have red algae growing on them. Some biological supply houses sell samples of seaweeds harvested by dredging to colleges located at a great distance from the ocean. Some of the species commonly present in these samples are included in the following examples and are listed in Table 4-1. Detailed descriptions, along with illustrations, can be found in the references listed at the end of the chapter.

Nemalion is a genus of marine summer annuals many of which at first glance look like pieces of red wire left in the water (Figure 4-2). Their morphological appearance is suggested by the generic name, which means "worm-like." Also in the Nemalionales is *Liagora*, a genus of soft and bushy tropical marine algae with compact, filamentous medulla and thin cortex. The similarities in life cycles between *L. tetrasporifera*, of the eastern hemisphere, and some pyrenomycetous sac fungi, especially *Mycosphaerella* (Pseudosphaeriales), was pointed out by Bessey (1950) as evidence that the Nemalionales, or a closely related and possibly extinct order of Rhodophycopsida, could logically be regarded as representing the type which gave rise to the Ascomycopsida.

The Corallinaceae (Cryptonemiales) is made up of plants that tend to be highly calcified and that have been important reef builders since the Jurassic period. *Corallina officinalis* is an erect, tufted "shrub" with tripinnately compound "leaves" (Figure 4-3C), found in tidal pools and on rock faces of relatively deep water from Long Island to Newfoundland. On the Pacific coast, *Lithothamnion californicum* is a common crustose species, with circular, whitish-pink thalli up to 10 cm in diameter.

Polysiphonia species (Ceramiales) are common in many parts of the world: *P. pacifica* occurs from Alaska to central California, *P. urceolata* from North Carolina to the Arctic, and *P. havanensis* from the Carolinas southward and throughout the Caribbean. All three are shallow water species, differing from each other morphologically as well as in ecological preferences. The life cycle of a *Polysiphonia* species is shown in Figure 4-7.

Several species of red algae live in freshwater habitats. Most are members of either the Nemalionales or the Bangiales. *Batrachospermum moniliforme* is especially common, growing attached to rocks in cool streams and in the outflow from springs during spring and early summer, especially where there is shade. Where culinary water is not chlorinated, it is occasionally found in drinking fountains. It is macroscopically **moniliform,** gelatinous in texture, and appears as little patches of olive green, or sometimes violet, tufts less than a millimeter thick. Its haploid carpospores produce prostrate filaments formerly thought to be a species of *Chantransia.*

Boldia erythrosiphon, found in two streams in Virginia, has saccate thalli, sometimes 20 cm long, and superficially resembles *Enteromorpha intestinalis,* a very common species of Chlorophycopsida, except for color. It is closely related to *Compsopogon coeruleus,* a weed in irrigation ditches of warm climates. *Asterocytis smaragdina* is an epiphyte on freshwater species of *Cladophora. Porphyridium aerugineum* and *P. cruentum* are unicellular species, the former found in freshwater, the latter in the ocean.

Ecological Significance

Ecologically, the red algae are the chief dominants of tropical continental shelves, reefs, and tropical estuaries, the niche they occupy being similar in many ways to that occupied by trees in land ecosystems, for there are other species that live in the shade of their fronds. Although continental shelves make up only 7.6% of the ocean area, this percentage of the tropical and subtropical oceans is approximately 15,000,000 km^2, 10% more area than is occupied by *all* cultivated crops (and almost 90% the area dominated by coniferous forests; see Table 1-5). The food produced by the vast amount of red seaweed occupying this area is of considerably less direct importance to man than that obtained from Anthopsida, but it is important to other animals, and the oxygen produced is of great significance, even to man.

The most productive of all ecosystems, terrestrial or marine, are tropical reefs (Figure 4-9), where red seaweeds are usually the most important producer organisms. Reefs and atolls are often called "coral reefs" because of the abundance and diversity of primitive animals called corals; however, ecosystem theory dictates that plants must be of greater ecological significance than corals or other animals in any stable ecosystem. A reef is made up of two parts, a skeleton or superstructure which supports it, and a vast amount of fill material which holds it all together. The superstructure often consists of the polyps of corals, but it is frequently made up of red seaweeds and sometimes calcareous green seaweeds. The fill material is primarily seaweeds and smaller algae. According to marine biologists who have conducted carefully controlled research on reef ecology, calcareous algae generally contribute more to the building of "coral" reefs, even the superstructure in many cases, than do animals. In productivity, the most important plants of reef ecosystems, in addition to Rhodophycopsida (especially the Corallinaceae of the Cryptonemiales), are the Dinophycopsida, which live symbiotically with many species of coral as zooxanthellae, and the Siphonophycopsida.

In the coralline algae of the order Cryptonemiales, deposits of calcium carbonate accumulate in the intercellular spaces; the building of ocean reefs and some "coral" islands is largely attributable to these plants. Paleozoic reefs, now often exposed in the sedimentary rocks of mountains, contain an excellent fossil record of the Rhodophycopsida.

Included in the family Solenoporaceae are the genera *Solenopora* and *Parachaetetes,* both of which date from the Ordovician, or possibly earlier. Both were widespread by late Paleozoic. During the Mesozoic, they became important reef builders. Apparently the Corallinaceae, which were first abundant in the late Mesozoic, evolved from the Solenoporaceae during the Jurassic. Occupying much the same ecological niches, the Corallinaceae were

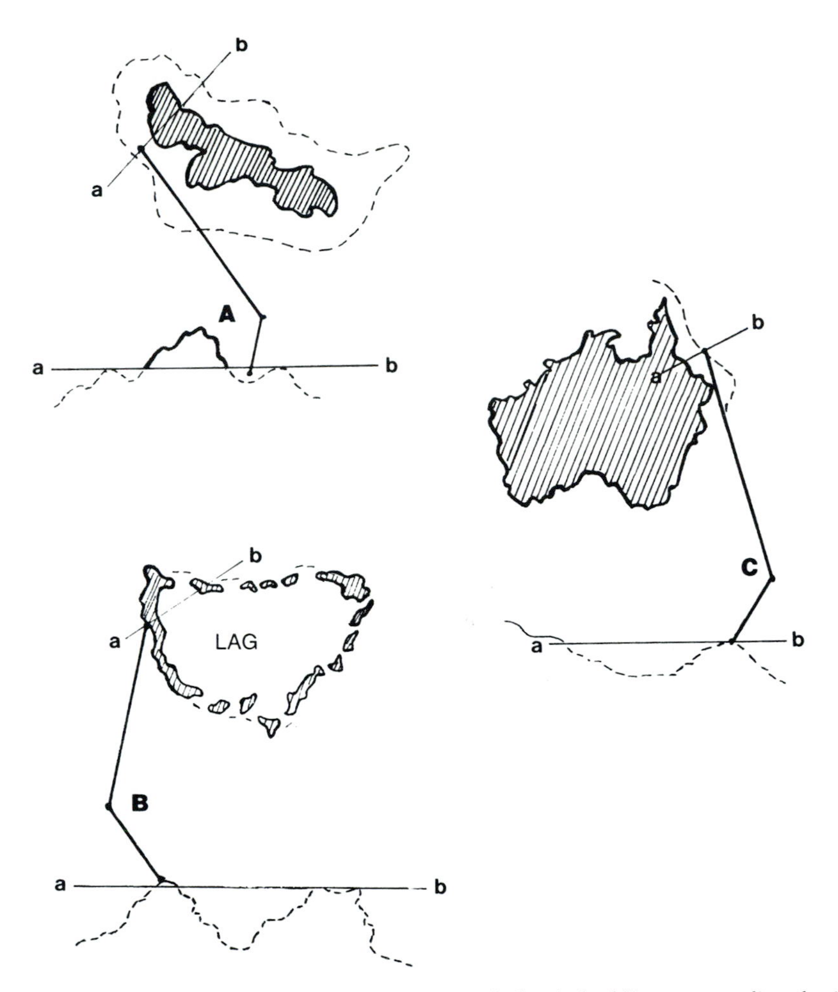

Figure 4–9 Kinds of reefs and reef formation: (A) Fringing reef, the dashed line surrounding the island indicates location of the reef; (B) atoll; (C) barrier reef. A cross section (a-b) through the reef and its surroundings is shown in the lower diagram of each pair of diagrams. LAG = lagoon.

able to compete very effectively with the Solenoporaceae and, over a period of several millions of years, replaced them in one ecosystem after another. *Lithothamnion* species were the first of these coralline competitors; *Lithothamnion* fossils from Jurassic formations are the most ancient of any of the modern Rhodophycopsida (Table 2-1). A genus of about 120 extant species, it occurs in all of the oceans. During the Cretaceous, the Solenoporaceae, after 400 million years of ecological importance, became extinct.

There are over 35 major reefs in the Atlantic, Pacific, and Indian oceans. The largest of these is the Great Barrier Reef off the east coast of Australia, which covers an area of 207,000 km^2 (Summerhays 1981). Fringing reefs and atolls are other kinds of reefs, in addition to barrier reefs, on which Rhodophycopsidea are ecologically important (Figure 4-9). It is estimated that 6% of the global productivity of organic matter and oxygen is produced by Rhodophycopsida (Table 2-5), much of it by reef-building species.

Algae are often classified as low light species, high light species, and species indifferent in light requirements. Red algae are classified as low light species. This is probably an oversimplification; nevertheless, it is true that red seaweeds grow at greater depths in the ocean than any other plants. In the clear, transparent water of the Mediterranean, for example, some red seaweeds are found in water 200 meters deep. It is believed that they are able to grow at such depths because their red pigments absorb violet and blue light. These colors penetrate deeper than red and orange light. Regardless of whether this is true or not, it is well established that Rhodophycopsida growing at great depths are brighter red than the same or closely related species growing in shallower water, and that Rhodophycopsida growing in very shallow water are often bluishgreen.

Many Rhodophycopsida live in intertidal areas where they are submersed in water half the day and exposed to sun and drying wind the other half. They are able to survive these conditions because of gelatinous materials that can absorb several times their weight in water, becoming permeable to water and gases while wet and losing this permeability as they dry out, thus protecting the tissues from desiccation.

Several red seaweeds are parasitic on other algae. The relatively high incidence of parasitism among red algae, when compared to other classes of plants made up primarily of autotrophs, may be due to their ability to form secondary pit connections (Bold and Wynne 1985). *Harveyella mirabilis* appears to the unaided eye as small hemispherical cushions on stipes of *Rhodomela subfusca* (Ceramiales) and some other red seaweeds. The plant consists of an interior portion made up of much-branched filaments which absorb food from the host cells, and the hemispherical exterior portion made up of a small-celled peripheral layer, four or five cells deep. In the exterior portion, there is a large central mass of larger cells containing the carpogonia and trichogynes. Sperm and eggs are produced in the central mass of the exterior portion in late fall. The entire exposed part of the thallus is enclosed in a fairly thick gelatinous membrane. *Harveyella pachyderma* is similar to *H. mirabilis,* differing mainly in gonimoblast details. It parasitizes *Gracilaria confervoides* (Gigartinales). Because of the reduced structure of parasites, their taxonomy is difficult; *Harveyella* has been classified by different botanists in the Gelidiales, the Cryptonemiales, and at least three families of the Gigartinales, but is now usually placed in the Gigartinaceae.

Another widespread parasite is *Choreocolax polysiphoneae,* parasitic on species of *Polysiphonia* (Ceramiales). *Choreocolax* is usually classified in the Cryptonemiales although there is difference of opinion among botanists. Melchior and Werdermann (1954) placed *Harveyella* and *Choreocolax* in the same family.

Economic Significance

Seaweeds have been used by man since ancient times and today are of much greater economic significance than most people realize. The Rhodophycopsida contribute human food, livestock feed, food supplements, dyes, stabilizers, fertilizers, and commercial gels. Japan, a leading country in the utilization of seaweeds, harvested 285,000 tons (dry weight) of marine algae for extraction of agars, carrageenins, gelans, and other colloidal polysaccharides in 1960. Approximately 50 species of red algae are used for food in Japan; the porphyras are especially important. *Chondrus crispus* (Irish moss) and *Rhodymenia palmata* (søl or dulce) have been used directly as food since ancient times and are now used commercially in the production of gels. Agar is obtained from species of *Gelidium* and other Floridae and is used in bacterial media and for desserts and other foodstuffs. Species of *Porphyra* are the source of dyes as well as food supplements. Many red seaweeds have been used over the years for fertilizer, but their commercial value is too great nowadays to allow this.

Porphyra species, commonly called "laver", although the name is often applied to other edible Rhodophycopsida, are especially important commercially. Purple laver, *Porphyra tenera,* is the species of greatest economic significance. The sulfated carbohydrates extracted from the lavers are used in the textile industry as well as in food processing. Lavers are also an important source of dyes.

In Japan, the growth of purple laver is encouraged by placing bundles of bamboo or brush in the shallow water near shore about the first of October. Spores of *Porphyra* lodge against these and by the middle of November, sporelings may be observed covering the stakes. The crop is harvested in January or February. A second crop may be harvested in early spring, but by the first of May, the algae have completely disappeared. Since the discovery that the so-called species of *Conchocelis* are actually a stage in the life cycle of *Porphyra* species, conchocelis spores have been cultured in botanical laboratories and released in the water adjacent to the nori nets during October. Laver farmers no longer need rely on the whimsical patterns of natural reproduction for the spores they need.

The bundles of brush are known as "nori", but have been largely replaced by nets stretched between bamboo poles just above the low water line and even by nets attached to buoys in deeper water. When used as food, the lavers are also known as nori. A *sushi* is a popular type of sandwich prepared by wrapping fish, meat, or cheese rolled with rice in a *nori.*

Seaweeds are at least as important a source of food in China as in Japan. Most families in northeastern China eat a seaweed dish at least once a day, according to Xia and Abbott (1987), who estimate that over 100 million pounds are consumed annually in China. Of the 73 species of seaweeds listed by Xia and Abbott as food items in China, 47 are red algae, 15 are green algae, and 11 are brown algae. Of the 28 genera of red seaweeds that contribute to the Chinese diet, four—*Porphyra* (8 spp.), *Gracillaria* (6 spp.), *Gelidium* (5 spp.), and *Hypnea* (4 spp.)—account for half of the species. Although many of the same species are used in Japan, some are different. The total number of seaweed species used in the two countries is slightly over 100 (Xia and Abbott 1987). The Chinese prefer seaweeds as a hot dish, broth or soup, for example, whereas the Japanese prefer to cook them ahead of time and then eat them cold as sandwich wrappings or a gelatinous dish.

Seaweeds are also widely used in oriental folk medicine. *Porphyra* is especially important. It has been used for centuries to control blood pressure, to prevent heart attacks, and to aid in recovery from heart ailments. In recent years, it has been the subject of nutritional research in Chinese and Japanese universities. Like *Spirulina* and other blue-green algae, *Porphyra* is a good source of vitamin B_{12}, an essential vitamin absent in most plants. Red seaweeds, especially species of *Porphyra,* are the only plants known to contain vitamin D_2 (Xia and Abbot 1987), another important vitamin. It is also high in vitamin A, vitamins B_1, B_2, B_3, B_6, and Vitamin C. *Porphyra,* and possibly other red seaweeds, contain an anticholesterase, which may account for the low rate of heart attacks in China and Japan compared to the U. S. (including Japanese Americans).

In addition to the lavers and Irish moss, dulce or søl has been an important food for centuries and can still be purchased in grocery stores in Iceland. Formerly, favorite Icelandic meals included buttered söl, dried fish, and milk. Boiled for five minutes and eaten with cream and sugar, it makes a tasty, nutritious dish that even today's American students find pleasant.

Rhodymenia palmata, or søl, is a cold water species and one of the most widespread and best known of all the red seaweeds. In the northern part of its range, e.g., at the northern end of Hudson Bay, *R. palmata* is found both in shallow and in deep water, but farther south it occurs only in deep water (Taylor 1957).

Søl (or söl) is referred to in the Icelandic sagas and in Iceland's oldest lawbook. From at least 961 A.D. until the beginning of the 20th century, søl was the Icelander's main vegetable and was prized as a palatable and nourishing food as well as a cure for seasickness, indigestion, and other maladies. Its value depended in part on its rich supply of minerals and vitamins. Rights to collect søl, as well as the right to eat it fresh when collected on another man's property, are well circumscribed by Icelandic laws dating from 1118 or earlier. Søl was collected when mature, usually late August, at low water, dried much like hay, and often packed in barrels. Because it was easily transported, it was an important food item inland as well as along the coast and was eaten by rich and poor alike. Today, it is used in Europe for making delicious candies.

Gelatinous substances, especially the sulfated polysaccharides extracted from red seaweeds, are used in many different ways. Agar-agar, or vegetable isinglass, the oldest of these, is the gelatin most easily extracted, and is probably the most difficult to harvest. It is obtained from *Gelidium amansii* and its relatives, including members of the Gigartinales and Ceramiales. Its great value as a bacterial medium depends in large measure on its peculiar melting properties. Agar must be heated to almost 70°C before it melts, but when cooled, it does not gel until a temperature slightly over 40°C is reached. Liquids or suspensions suspected of containing bacteria can be poured into the liquid agar at temperatures low enough that they are not injured. After thorough mixing, the agar can be allowed to gel and growth of the bacteria can be studied. Some of the cultures can be kept in incubators at temperatures above 40° or even 50°C, if desired, to study the effects of heat, and the agar will remain firm. Agar is also used in making desserts, puddings, and other foodstuffs and in sizing of textiles. The agar itself has no nutritional value for humans or most bacteria or fungi, which is also a considerable advantage when studying the nutritional needs of bacteria and other organisms.

Figure 4–10 Japanese woman diving for *Gelidium amansii.* The water is deep and sometimes cold; goggles protect the eyes from the effects of the high pressure in the deep water.

Obviously, therefore, *Gelidium amansii* is economically very important. A deep-sea species, it grows on the outer face of reefs, such as the barrier reef around Yoron Island at the southern end of Japan. Water temperatures vary from 19° to 30°C along this reef, which is completely immersed at high tide but forms, at low tide, a broken island about 50 m wide and 6 km long (Summerhays 1981). Divers harvest the plant from depths of several meters (Figure 4–10). Most divers are women. They tolerate cold water better and apparently can hold their breath longer than men. In other parts of the world, depending on the nature of the sea bottom, some species of agar-producing *Gelidium* are obtained by dredging rather than by diving.

Divers' sickness, or "bends", and pneumonia are always risks that divers take when collecting in cold, deep water. To prevent problems, the divers need frequent rest periods during which the chest must be kept warm. Clothing that covers the torso would not only interfere with the diver's movements under water but would keep the chest wet and cold during the rest periods, increasing the danger of pneumonia (Figure 4–10). Warm blankets that can be thrown over the shoulders and chest are always kept available.

In the Gigartinales, *Furcellaria* is a genus of bushy plants, common in colder areas. Polish workers have experimented with the possibility of commercial extraction of agar from *F. fastiqiata,* a relatively common alga of the Baltic Sea, with encouraging results. Species of *Ceramium* are also used in the commercial production of agar.

Carrageenins extracted from *Chondrus crispus,* commonly called Irish moss, are important in modern industry. For centuries Irish moss was an important food plant over much of northern Europe, although never as popular in Iceland as søl. It was used, for example, in Great Britain in making "Scottish broth." Its present use is almost entirely commercial, with much of its production coming from eastern Canada. It is the chief commercial source of carrageenins, although other Gigartinaceae are also important.

The carrageenins are sulfated polysaccharides extracted from *Chondrus* and other red seaweeds by hot water and used extensively in the dairy industry as stabilizers. The chemical make-up of various carrageenins was a topic of discussion at the 1955 and 1967 Seaweed Symposia (Margalef 1968). According to one speaker, cocoa will not remain suspended in milk unless a stabilizer is added. The wrong kind of carrageenin, or too much of it, will cause the curd and whey to separate, as will heating the mixture too rapidly. The dairy industry is therefore very selective as to the kind of carrageenin purchased; carrageenins extracted from Irish moss are of especially high quality.

As demand for carrageenins, agar, and other seaweed products has increased, some stands of *Chondrus, Gelidium,* and other red algae have become endangered. Eventually, it will be necessary to cultivate seaweeds extensively, or at least to learn how to manage the natural beds for maximum production. Laboratory studies of the nutritional requirements of *Porphyra* have been conducted in Japan and elsewhere. Nitrates and/or ammonia are essential for good growth of the early filamentous stage of *Porphyra.* Continuing studies should result in good knowledge of how and when to fertilize *Porphyra* and other seaweeds for maximum yields.

Red algae are also at times economically significant in a negative way. A common freshwater alga of warm areas is *Compsopogon coeruleus* (Bangiales), which has been reported to grow in such profusion in Arizona as to constitute a nuisance in irrigation ditches (Smith 1950). Reports of red algae causing disease in vascular plants are apparently false. The common name of one of the diseases, the red rust of tea, which is caused by a green alga, has apparently become confused with the common name of the Rhodophycopsida, or red algae.

Research Potential

The red seaweeds have not been used extensively in scientific research, beyond that essential to their commercial development, even though they seem very well suited to certain types of basic study. It has long been suggested that Rhodophycopsida are able to survive and grow well at great depths in the ocean because of the presence of red and blue photosynthetic pigments. Few, if any, experiments have been conducted, however, to test this hypothesis. In fact, very little is known about the adaptive value of color in marine habitats; the red seaweeds should be of real value in exploring this important field of research.

Rhodophycopsida are also of great potential value in solving problems of differentiation and morphogenesis. Within some taxonomic groups of red algae, three kinds of plant bodies occur in closely related species: unicellular, filamentous, and complex multicellular types. In the genus *Lithothamnium,* for example, both crustose and small shrub-like species occur; the chemical or other forces which cause such differences in habit to develop can be studied in this group of plants.

Challenging taxonomic problems in the Rhodophycopsida also await the development of research interest. *Falkenbergia* is a genus of filamentous red algae usually classified in the Ceramiales. Like many of the Rhodophycopsida, its complete life cycle has never been analyzed. The Feldmans have interpreted *F. rufolanosa* as the tetrasporophyte phase of *Asparagopsis armata,* a member of the Nemalionales. It is possible that, as further studies are made concerning *Falkenbergia,* the genus will have to be abandoned, just as *Conchocelis* was abandoned when we learned it was a stage in the life cycle of *Porphyra.*

SUGGESTED READING

"Seaweed Symposium" on the industrial uses of seaweeds edited by A. D. Virville and J. Feldman, covers species of seaweeds, methods of harvesting them, the products they produce, and how these products can be extracted, processed, and used. Brown and green seaweeds, in addition to the reds, are covered.

An article in the June 1987 issue of *Economic Botany,* entitled "Edible Seaweeds and Their Place in the Chinese Diet," by Xia and Abbott, tells of foods from the sea used in China and Japan and discusses briefly their medicinal as well as nutritional value. "Ama, Sea Nymphs of Japan" in the July 1971 issue of *National Geographic,* tells about the women who dive for abalone, pearls, and red seaweeds in the deep waters of the Pacific off the east coast of Japan.

For seaweed identification, two books by William Randolph Taylor, *Marine Algae of the Northeastern Coast of North America* and *Marine Algae of the Eastern Tropical and Subtropical Coasts of the Americas,* and one by Gilbert Smith, *Marine Algae of the Monterey Penninsula,* have detailed keys and descriptions and excellent drawings. Less complete in its coverage of species native to either of the North American coasts, but more cosmopolitan in coverage, is Yale Dawson's *How to Know the Seaweeds.* Its keys are easy to use and the drawings are very helpful in identification of species.

To find more information about red seaweeds, look up some of the genus names (Table 4-1) in *Biological and Agricultural Index.*

STUDENT EXERCISES

1. What are the stages in the life cycle of *Polysiphonia variegata*? Diagram all the stages in the sequence in which they occur.

2. How does the life cycle of *Porphyra tenera* differ from that of *Polysiphonia variegata*? Diagram the life cycle of *Porphyra tenera.*

3. If the supporting cell, or basal cell, of the carpogonial filament of *Rhodymenia palmata* has 19 chromosomes, how many chromosomes would you expect to observe in each of the following types of cells?
   (a) zygote;
   (b) any of the other cells in the carpogonial filament;
   (c) a carpospore;
   (d) a tetraspore;
   (e) a cell in the trichoblast of a tetrasporophyte embryo;
   (f) the original nucleus in an auxiliary cell;
   (g) each nucleus developing from mitosis of the auxiliary cell;
   (h) each cell in the carposporophyte;
   (i) each cell in the gametophyte.

4. If the monospore from which a gametophyte of *Porphyra umbilicalis* has 12 chromosomes in its nucleus, how many chromosomes would you expect to observe in each of the following types of cells?
   (a) a spermatid;
   (b) a zygote;
   (c) a cell in the lamina (the "leaf" or nori) you wrapped your sushi (sandwich) in;
   (d) the nucleus in a cell of a *Conchocelis rosea* plant;
   (e) a monospore produced by a *Conchocelis rosea* plant;
   (f) a monospore produced by a *Porphyra umbilicalis* plant.

5. The conchocelis stage of *Porphyra umbilicalis* was formerly known as *C. rosea.* What do you suppose was the former name of the conchocelis stage of *P. tenera*?
   (a) *P. tenera;*
   (b) *C. rosea;*
   (c) *C. tenera;*
   (d) it was known as *Conchocelis* something, but there is not enough information in the book to tell what the species epithet was;
   (e) it is impossible even to make an intelligent guess from the amount of information available.

6. If you were using a book on seaweeds published in 1940, or earlier, in what order of Rhodophycopsida would you find *Porphyra umbilicalis*? In what order would you find *Conchocelis rosea*?

7. If it is learned that most "species" of *Falkenbergia* are the tetrasporophyte stage of *Asparagopsis,* why do botanists say that one of these two genera must be abandoned? If one is abandoned, which one should be? Why? Should the remaining genus be classified in the Nemalionales or the Ceramiales? Why?

SPECIAL INTEREST ESSAY 4-1

Marine Ecology

Over 70% of the earth's surface is ocean, but we know very little about the ecology of this vast area. Over 90% of the ocean is a great desert, but the continental shelves are relatively productive, and the ocean reefs are among the most productive of all ecosystems.

Our best estimates of plant productivity indicate that despite the vast size of the oceans, only about half of the global productivity is contributed by marine plants (Table 2-5). The 7 or 8% of the ocean surface lying over the continental shelves or occupied by reefs contributes at least 33% of total ocean productivity. These are the areas dominated by red and brown seaweeds with diatoms, dinoflagellates, cryptophytes, cyanophytes, and silicoflagellates very abundant.

The most productive parts of the ocean are the rocky shores and reefs where brown, red, and siphonaceous green seaweeds abound, with the browns especially important in cold waters and reds and greens in the tropics. The reefs have been built in large measure by calcareous algae but also by animals (corals) feeding on soft algae and plankton or nourished by zooxanthellae-dinoflagellates and other unicellular algae in symbiosis with the coral.

Distinct vegetational zonation is typical of rocky shores, the kinds of vegetation and width of the zones depending on whether the shore is exposed or sheltered, how high the tides are, and the temperature of the water. On exposed coasts, algae are abundant in the superlittoral zone where only the highest tides reach, their presence made possible by the spray there. At the same elevation on sheltered shores, where there is no spray, lichens grow. Below the lichen zone, in the sheltered part of the midlittoral, a "fucoid fringe" is nearly always found.

The extremes in environmental variation occur in the intertidal zones. Some seaweeds, such as *Fucus* among the browns and *Endocladia* among the reds, are especially tolerant of the combination of heat, drying, and high light intensity here. *Rhodochorton purpureum* (Rhodophycopsida) forms velvety carpets in such habitats, and many of the coralline reds also grow in these places as well as in very deep water where light intensity is low. Many shade-loving plants such a *Delesseria* and *Membranoptera* (both reds) grow in shallow water in the winter time when light intensity is low, but when summer comes they are either killed by the high light intensity or overgrown by kelps and other sun-loving plants.

In tropical waters, light penetrates better because the angle of the sun is more direct. At higher latitudes, most of the light that strikes the ocean surface is reflected; this is especially true in the winter months, and even in the clearest water, no algae are found below about 60 meters depth. In clear, transparent water in the Mediterranean, on the other hand, some red algae grow as deep as 200 meters.

Feldman's ecological classification of marine algae is based largely on the form in which attached algae survive the dormant season, in addition to general growth habit (Fritsch 1965). **Eclipsiophytes** are annuals which overwinter as minute filamentous growth (e.g., Ectocarpales), and **Hypnophytes** are annuals which overwinter as zygotes or spores (*Vaucheria* and others). **Phanerophytes** (e.g., *Codium* and *Pterogophora*) are perennials in which the entire plant overwinters; **hemiphanerophytes** (e.g., *Cystoseira* and *Sargassum*) are perennials in which the base and part of the erect thallus overwinters; and **hemicryptophytes** are perennials in which only the basal crust overwinters (e.g., *Acetabularia* and *Cladostephus*). *Hildenbrandia, Melobesia,* and other crustose seaweeds are called **chamaeophytes. Plankters** are passively floating or weakly swimming algae which are readily carried about by water currents; in many parts of the ocean, even where attached seaweeds are abundant, the plankters are ecologically the most productive algae.

Temperature is a major factor in determining latitudinal distribution of marine algae; the Setchell system employs surface isotherms of 5°C to divide the oceans into nine climatic zones. This may be an oversimplification, even though the zones are useful, since light intensity and duration are probably as important as temperature in determining the limits of growth of algal species. Light intensity and duration follow latitudinal lines exactly, but isotherms do not.

As in other plants, the climatic factor, important as it is, is only one of the factors involved in determining geographic distribution of species. Introduction is another important factor. The tropics and the land masses form barriers that few algae can cross. In early experiments in which algae from one hemisphere were tied to ships crossing the equator, few specimens survived the prolonged exposure to higher temperature and lower oxygen and carbon dioxide concentration of the tropics. However, with the advent of rapid modern transportation, some algae have been conveyed on the bottoms of ships and introduced into new areas, creating potential ecological problems.

5 TERRESTRIAL FUNGI: MOLDS AND MUSHROOMS

For many people, the word "mold" calls to mind dark, damp dungeons, rundown houses with blinds over the windows, and old food covered with greenish or black scum; while "mushroom" brings back a childhood memory of evil umbrella-shaped "toadstools", sometimes brightly colored, springing up on a lawn or in a wooded area over night, coming apparently from nowhere for they were not there yesterday, and bringing with them the stern warning, "Don't eat them, they're poisonous."

But for others, a mold is a useful organism from which various "wonder drugs", such as penicillin, and hundreds of industrial chemicals, like citric acid, are extracted. To the millions of **mycophagists** around the world, the word **"mushroom"** conjures up visions of fancy restaurants with gourmet foods, or fun family outings in the woods hunting for morels, boletes, or chanterelles. In this chapter, we will examine the characteristics of terrestrial fungi, including the fleshy fungi called mushrooms and discuss some of the industrial uses and other attributes of common molds.

The division Eumycophyta (*eu* = good or true, *mycos* = mold or fungus, and *phyta* = plant), commonly called the higher fungi, is large, with about 80,000 species distributed among three classes: the Ascomycopsida, Zygomycopsida, and Basidiomycopsida, all three consisting entirely of heterotrophic plants. The plant body is made up of two parts, the **mycelium,** or vegetative portion, and the **sporocarp,** or reproductive portion.

There are two basic types of sporocarps, or spore fruits, in the Eumycophyta: **ascocarps** in the Ascomycopsida, and **basidiocarps** in the Basidiomycopsida (Figure 5-1). Large, fleshy ascocarps and basidiocarps are commonly called mushrooms. Their equivalents in the Zygomycopsida are minute **coenozygotes** containing numerous paired nuclei (Figure 5-1f).

The mycelium is a tangled mass of filaments, often netlike as a result of numerous secondary pit connections (anastomoses), through which the plant obtains its nourishment from the substratum. Each filament is called a **hypha.** Growth is from the hyphal tip and generally radiates from a common center in all directions, depending on food supply and moisture. In **saprophytic** species, the mycelium is typically white, filamentous, and very extensive, developing underground wherever there is abundant organic matter, such as leaves or manure, and where water and oxygen are also abundant. In Ascomycopsida, the mycelium is mostly haploid; in Basidiomycopsida it is dikaryotic (Figure 5-2).

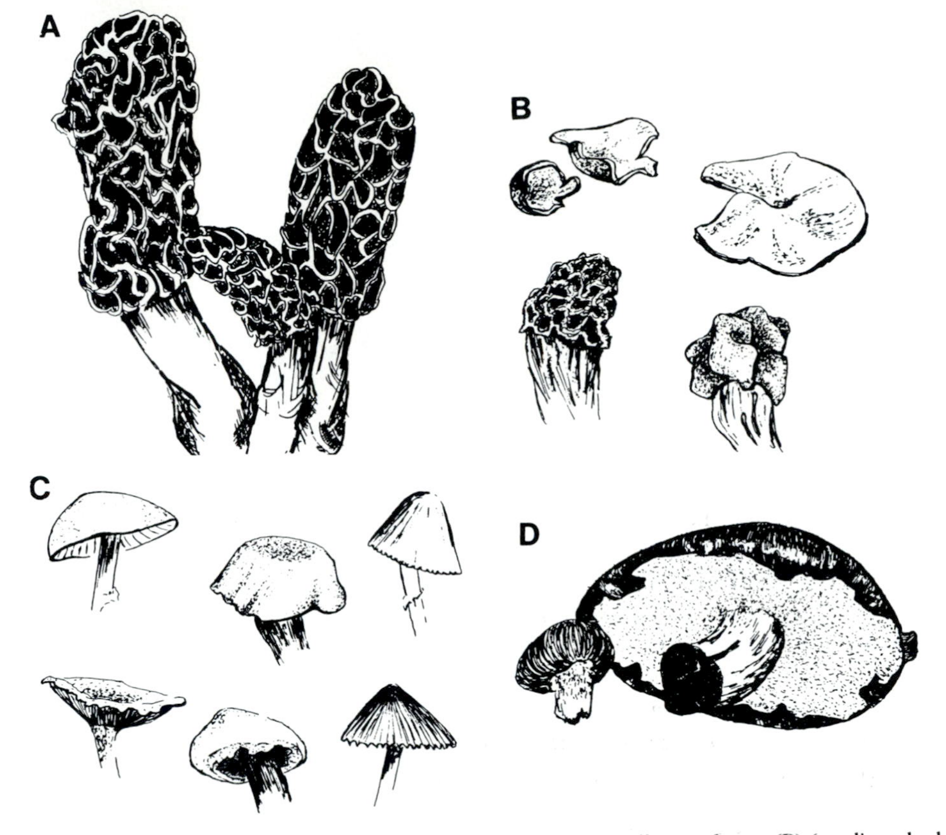

Figure 5–1 Fruiting bodies of Eumycophyta. (A) Apothecium of *Morchella esculenta;* (B) (reading clockwise) apothecia of *Peziza repanda, Discina ancilis, Helvella lacunosa,* and *Morchella conica;* (C) basidiocarps of gill mushrooms, illustrating diversity in cap margins; (D) basidiocarp of tubular type mushroom, a bolete. (From Pearson, L. C. *The Mushroom Manual,* Naturegraph, Happy Camp, CA, 1987. With permission.)

In **parasitic** species, the mycelium is less extensive and consists of intercellular hyphae and intracellular **haustoria.** In symbiotic fungi, the vegetative body is more complex. Often the basic filamentous structure is not apparent and pseudoparenchymatous tissues of different types occur.

Foliose lichens, for example, may have **rhizines** anchoring a complex thallus, made up of three or four layers of tissues, to the substrate. In **mycorrhizae,** the fungal tissues become intimately intertwined with the root tissues of the tree or other host plant, resulting in unique, stubby, dichotomously branched roots.

Most botanists agree that the Ascomycopsida and Basidiomycopsida are closely related to each other. The position of the Zygomycopsida is less certain. Traditionally, the Zygomycetes were allied with the Mastigomycopsida in a class called Phycomycetes, primarily because the plant body in both is coenocytic (Table 5-1), but recent research indicates that this alliance is artificial. Denison and Carroll (1966) suggested that the true relationships of the Zygomycopsida lie with the Ascomycopsida; physiology and biochemistry support this claim.

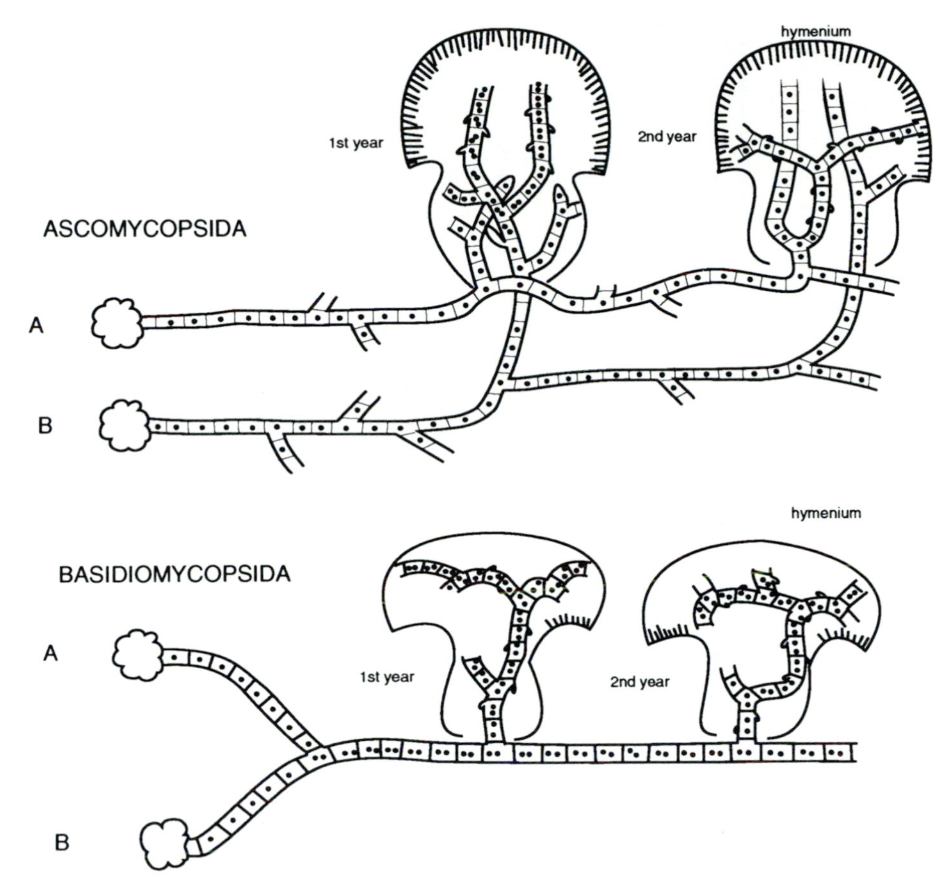

Figure 5–2 Development of haploid, dikaryotic, and diploid mycelia in Ascomycopsida and Basidiomycopsida. Meiospores (ascospores and basidiospores, respectively) of different mating types are labeled A and B. In both classes, karyogamy takes place in the hymenial tissue, followed soon by meiosis to produce meiospores. (Redrawn from Lange, M., *Systematisk Botanik,* 2, 1, 1964. Courtesy *Svampe,* Munksgaard, Copenhagen.)

Numerous similarities between Rhodophyta and Eumycophyta led Sachs (1874), Bessey (1950), Denison and Carroll (1966) and others to the hypothesis that the Eumycophyta evolved from red algae that had colonized driftwood during late Devonian. Dead trees and shrubs, washed out to sea where red seaweeds were abundant, provided a suitable medium for establishment of saprophytic organisms. Later, mutations enabled the descendants of some of these earliest fungi to invade the saprophytic niche on land. Parasitism and lichenization evolved much later, each arising independently many times.

MICROKARYOPHYTES

Because the Rhodophyta and Eumycophyta share so many characteristics, the two divisions may be referred to together as **microkaryophytes,** the name referring to the very small nuclei common to both divisions. The nucleus of microcaryophytes is not only different from and

TABLE 5–1

SIMILARITIES AND DIFFERENCES AMONG THE CLASSES OF MICROCARYOPHYTES AND FLAGELLATED FUNGI

	Rhodophycopsida	Ascomycopsida	Zygomycopsida	Basidiomycopsida	Mastigomycopsida
Sporocarp	Cystocarp or gonimoblast	Perithecium or apothecium	None (coenozygote)	Basidiocarp	None
Meiospores	4 Tetraspores	8 Ascospores	1 Zygospore	4 Basidiospores	(4 Gametes)
Ploidy of pericarp	Haploid and diploid	Haploid and dikaryotic	—	Dikaryotic	—
Filaments or hyphae	Septate	Septate	Coenocytic	Septate	Coenocytic
Vegetative body	Filamentous or tissue oriented	Filamentous or tissue oriented	Filamentous	Filamentous or tissue oriented	Filamentous
Asexual reproduction	Monospores (rare)	Mostly conidia	Mostly sporangiospores; conidia	Conidia (relatively rare)	Zoospores
Habitat	Mostly marine	Moist to dry terrestrial	Moist to dry terrestrial	Moist to dry terrestrial	Moist to aquatic
Nourishment	Autotrophic	Saprophytes, plant parasites, lichens	Saprophytes, animal parasites	Saprophytes, plant parasites, mycorrhizae	Animal parasites, aquatic decomposers

more advanced than the nucleoids of prokaryotic organisms, it is also distinctly different from the nucleus of all other eukaryotic organisms.

Rhodophyta and Eumycophyta resemble each other and differ from other plants in numerous details including

1. Presence of minute nuclei with conspicuous nucleoli
2. Nuclear membranes intact and spindle intranuclear during mitosis
3. Naked or thin-walled sperm produced one per antheridium
4. Production of a receptive filament, the trichogyne, projecting from the egg-producing body in many of the Rhodophycopsida and Ascomycopsida and in some Basidiomycopsida
5. Close chemical similarity between floridean starch and fungal glycogen
6. Close chemical similarity between floridean cellulose in the cell walls of red algae and fungal chitin in the Eumycophyta
7. Chemical similarity of water retaining gelatins which prevent desiccation of tissues
8. Absence of flagellated cells at all stages of the life cycle
9. Filamentous organization of the plant body with apical growth
10. Primary pit connections with cytoplasmic streaming through the central pore
11. Secondary pits and anastomoses
12. Retention of the zygote in the archegonium following fertilization and subsequent development of a sporocarp usually composed of both haploid and diploid tissue
13. Complex, multiphase life cycles
14. Absence of a centrosome

For a detailed discussion of these points, see Sachs (1874), Bessey (1950), Denison and Carroll (1966), and Demoulin (1974).

The plant body of microkaryophytes is often complex with distinct differentiation of specialized tissues. Tissues are basically filamentous, but the filamentous nature is frequently obscured as individual cells become rounded and closely packed to form **pseudoparenchyma** tissue. If the filamentous nature of the tissue is recognizable, it is called **prosenchyma.**

Mitosis takes place in the apical cells of each filament. In both Rhodophyta and Eumycophyta, interphase and prophase nuclei are small, typically 1 to 3 μm in diameter. They elongate as prophase draws to a close; at that time the chromosomes seem to be linked together in a chromosomal chain, similar to the single chromosome of bacteria except that each chromosome has its own centromere. In metaphase and anaphase, the nuclear membrane remains intact. Following mitosis, the new cell wall forms from the periphery inward. Septation does not become complete, however, and a small central pore, the primary pit, remains (Figure 4-5). The cytoplasm is thus continuous from cell to cell and large organelles such as mitochondria and nuclei may migrate from one cell to another through the primary pits.

Secondary pit connections also allow migration of nuclei and other organelles from one cell to another. Auxiliary cells in Rhodophycopsida, croziers in Ascomycopsida, and clamp connections in Basidiomycopsida are examples of secondary pits. These are illustrated in Figure 5-3.

Cell walls of fungi are composed of fungal chitin, which is chemically similar to floridean (or rhodophycean) cellulose. Animal chitin, which occurs in insects and some other animals, consists of much longer chains of acetyl-glucosamine only, whereas fungal chitin consists of one molecule of glucosamine attached to three molecules of acetyl-glucosamine from which four molecules of water have been removed (Waksman 1952). In floridean cellulose, glucose

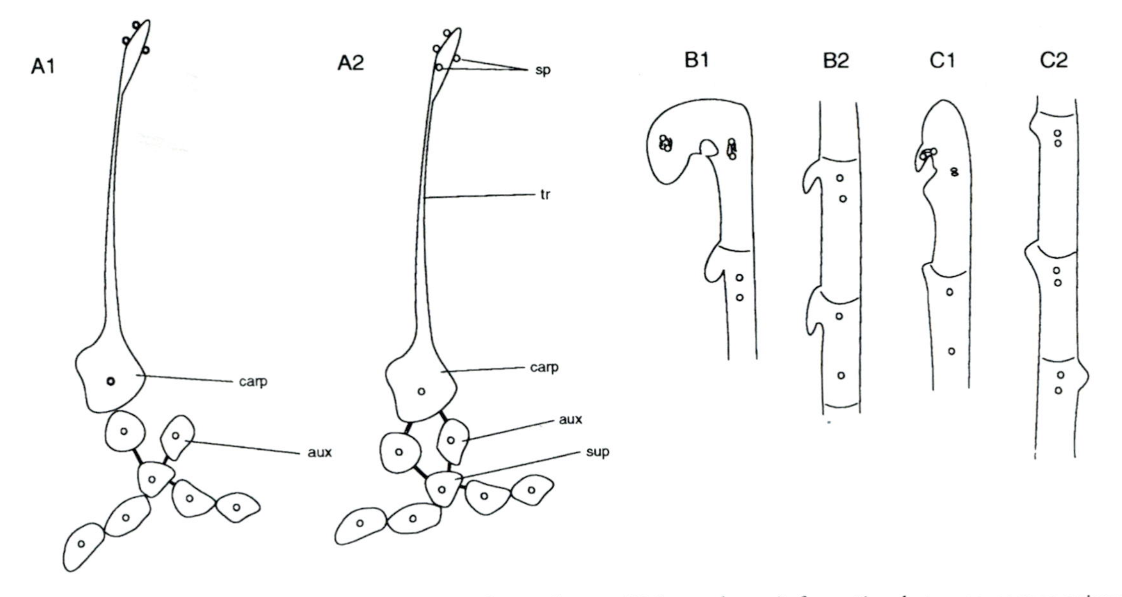

Figure 5–3 Secondary pit connections in microkaryophytes: (A) Secondary pit formation between carpogonium and auxiliary cell in Rhodophycopsida; (B) crozier formation in Ascomycopsida; (C) clamp connection formation in Basidiomycopsida; (A_1, B_1, C_1) pit connection beginning to form; (A_2, B_2, C_2) mature pit connection. (Open circle = haploid nucleus; solid circle = diploid nucleus; dotted circle = disintegrated nucleus; sp = spermatium; tr = trichogyne; carp = carpogonium; aux = auxiliary cell; sup = supporting cell.)

molecules are linked in the same way. *N*-acetyl-glucosamine differs from glucose only in having an acetylamine group attached to the number 2 carbon atom instead of OH (Figure 5–4). This difference is apparently due to the difference in mode of nutrition between red algae and higher fungi (Denison and Carroll 1966). In heterotrophic organisms, the nitrogen given off in protein metabolism must be gotten rid of, and hence there is an adaptive advantage in being able to use this excess nitrogen in structural compounds; this is especially true in the case of nonmotile, terrestrial organisms in which excretion in the form of ammonia or nitrates would present problems. In autotrophic organisms, on the other hand, nitrogen is an important and often limiting raw material and is seldom incorporated into structural compounds.

The chief food reserve in the Eumycophyta is fungal glycogen. Essentially identical to floridean starch, it is also similar to cyanophycean and bacterial starches, which are all highly branched carbohydrates built up of 14 or 16 glucose units. Animal glycogen is similar, but the chains are shorter: 11 to 13 glucose units. Fungi obtain their basic glucose units from complex cellulose and lignin molecules in the plant materials they decompose.

Life cycles of microkaryophytes are unique (Figure 5–5; see also Figures 4–7, 4–8). They tend to be variations of a three-phase cycle consisting of

- an independent gametophyte generation,
- an often dependent diploid or dicaryotic generation, and
- an independent or dependent meiosporophyte generation.

One or more of the generations may be reduced or lacking.

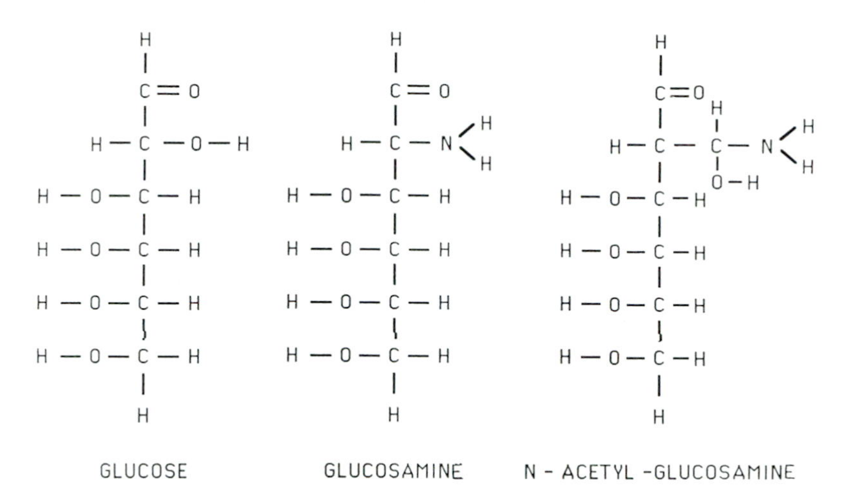

Figure 5–4 Structure of glucose, glucosamine, and *N*-acetyl-glucosamine; they differ at the no. 2 carbon position. Fungal chitin is a relatively short-chain polymer with each unit consisting of one molecule of glucose amine and three molecules of *N*-acetyl-glucosamine. Floridian cellulose is made up of short chains of glucose and acetyl-glucose with linkages like those in fungal chitin. Animal chitin contains no glucosamine and is made up of much longer chains of *N*-acetyl-glucosamine.

Fertilization, or **syngamy,** occurs at the end of the gametophyte generation when the nucleus of a spermatium, which has come in contact with a **trichogyne** (if one is present) enters the **oogonium.** In most red seaweeds, the two nuclei immediately fuse and the zygote migrates into an auxiliary cell. In most fungi and some red seaweeds, only a fusion of cytoplasm occurs at this time; the branched filaments growing out of the fertilized egg are thus **dikaryotic,** with two haploid nuclei in each cell. **Karyogamy,** or fusion of the nuclei, occurs later, just prior to meiosis (Figure 5-5).

Following syngamy, fruiting bodies, or **sporocarps,** are formed. In the Rhodophycopsida, the sporocarps are small, urn-shaped, fleshy conceptacles about 1/4 to 1 mm in diameter on the surface of the lamina of the gametophytes. These conceptacles contain diploid cystocarps which on maturity form masses of diploid carpospores (Figures 4-3, 4-7). In some species, meiosis occurs at this stage and the carpospores are haploid. Similar conceptacles occur on the diploid tetrasporophytes and produce tetrads of haploid meiospores.

In the Ascomycopsida, two basic types of sporocarps, **perithecia,** flask-shaped structures which somewhat resemble the sporocarps of red seaweeds, and cup- or disk-shaped **apothecia** are produced (Figure 5-1). **Apothecia** may be large and fleshy, as in morels and truffles, or small and leathery and sometimes hard and brittle, as in many **plant pathogens.** In the Basidiomycopsida, five or six different kinds of sporocarps may occur (Figure 5-1 g-k), the most common of which is large, fleshy, and umbrella-shaped with basidiospores produced on gills or within tubes on its undersurface.

Some of the sporocarp tissues are typically pseudoparenchymatous in microkaryophytes. The distinctive characteristics of the sporocarp makes it useful in identification of species in both the Rhodophyta and Eumycophyta. Table 4-3 summarizes some of the spore and sporocarp characteristics of both divisions.

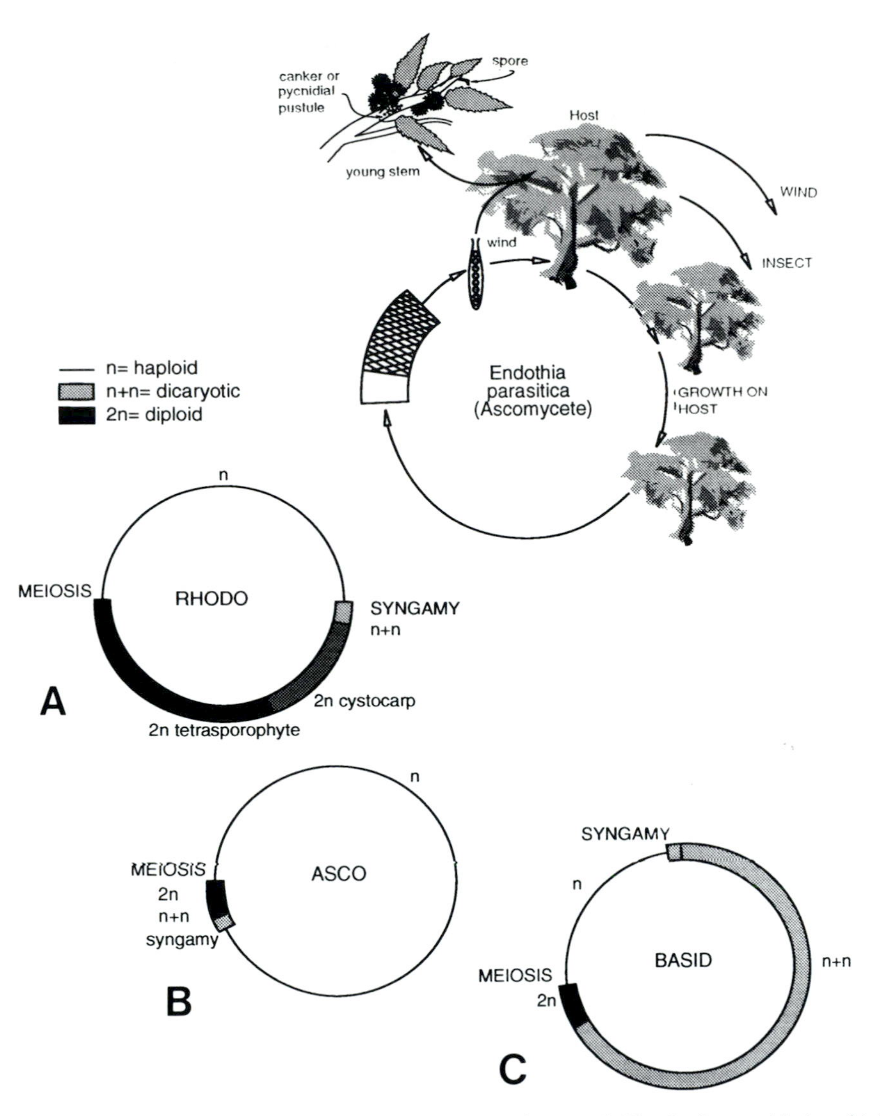

Figure 5–5 Three-phase life cycles typical of many microkaryophytes. (A) Rhodophycopsida in which dikaryotic phase is short or missing; the independent haploid gametophyte is followed by two diploid states, a dependent cystocarp and an independent tetrasporophyte. (B) Ascomycopsida with a short dikaryotic phase following the haploid gametophyte and followed by a short diploid stage. (C) Basidiomycopsida with a long dikaryotic state following a short haploid gametophyte and followed by a short diploid meiosporophyte. (D) *Endothia parasitica*, or chestnut blight, in which the disease is spread either by long-lived wind-borne ascospores or by short-lived insect-spread conidia.

CLASS 1. ASCOMYCOPSIDA

The Ascomycopsida (*ascos* = sac and *mycos* = mold or fungus), or sac fungi, are a widespread group, occurring in all terrestrial ecosystems and many aquatic systems. Kohlmeyer, for example, reported finding a member of the Sphaeriales on a piece of wood paneling at 2000 meters depth off the California coast. The class is large, with over 56,000 species, including lichens and Fungi Imperfecti, distributed among 12 orders and 84 families (Table 5–2). There are more known species of Ascomycopsida than of any other class except the Anthopsida, and species of sac fungi tend to be more widely distributed than species of any other class. Therefore, within any compact geographic area, there will generally be more kinds of sac fungi than any other plants: Canadian ecologists found 5000 species of ascomycetes in Manitoba, compared to 1200 species of flowering plants.

Saprophytic ascomycetes are found wherever there is organic matter—leaf litter, manure, plant residues of various types—that can be decomposed, and parasitic species are found on almost every species of Anthopsida and many other plants. If a small amount of horse manure or well-rotted compost is placed on the surface of a petri dish containing potato dextrose agar or corn meal agar, several species of ascomycetes can usually be recovered in a few days. Saprophytic ascomycetes may also be observed by lifting the partially decomposed litter or duff that accumulates on the soil surface of most ecosystems, but especially forests, and examining the undersurface of this material with a hand lens. Placed on a suitable medium, several species of fungi, some of them ascomycetes, usually develop from pieces of such mycelia.

Parasitic species are found on almost every species of Anthopsida and many other plants. They can be observed by examining carefully the leaves and stems of several trees, shrubs, or herbaceous angiosperms for leafspots, stem blotches, or small outgrowths. Further examination with a hand lens frequently reveals small, typical apothecia or perithecia.

Fungi Imperfecti are ascomycetes, and occasionally basidiomycetes, in which a **perfect** or **sexual** life cycle has never been observed. In some Fungi Imperfecti the ability to reproduce sexually may have been lost; in many species it either has never been discovered or else has never been linked to its sexual phase of the life cycle. Some botanists classify these fungi in a separate class called the **Deuteromyceteae** or Deuteromycopsida; this practice has not been followed in this book primarily because it fails to show the true relationships between the imperfects and the rest of the Ascomycopsida.

Lichens are symbiotic ascomycetes which can be found in most terrestrial habitats. Each lichen is a composite organism consisting of an ascomycete and either a chlorophyte or a cyanophyte. The thallus is usually differentiated into distinctive organs of food production, anchorage, and reproduction, each of which is built up of specialized prosenchyma and pseudoparenchyma tissues. They favor well-illuminated habitats subject to considerable variation in moisture and temperature. They often color vast areas of cliffs and rocks bright yellows, oranges, chartreuse, or other colors; they also grow on soil, on the bark of trees, and even on leaves. Lichens will be further discussed in Chapter 6.

Some of the choicest mushrooms are members of the Ascomycopsida. Morels and truffles are the fruiting bodies, or apothecia, of ascomycetes which live symbiotically with vascular plants or saprophytically on leaf litter or decaying wood.

Students sometimes confuse sac fungi with zygomycetes, or occasionally with club fungi or even flagellated fungi. If sporocarps are present, distinctions are easily made: the **hymenium,** or spore-bearing tissue of Ascomycopsida is nearly always on the upper surface of the sporocarp, but in Basidiomycopsida it is more frequently on the under surface. When dissected, the sporocarp of sac fungi contain asci and paraphyses with the spores inside the asci (Figure 5–6), usually eight spores per ascus. The basidiospores in a basidiocarp, on the other hand, are not enclosed in sacs but are attached to club-shaped structures called basidia, usually four per basidium. If sporocarps cannot be found, distinctions among the fungal groups are more

TABLE 5-2

CHARACTERISTICS OF THE ORDERS OF ASCOMYCOPSIDA

Orders and Included Taxa (Fam–Gen–Spp)	Sporocarp Characteristics	Other Characteristics	Representative Genera or Species
Sphaeriales (18-600-9,200)	Round, carbonaceous perithecia with hymenium, paraphyces, and ostiole; trichogyne present	Conidia are common, sometimes borne in pycnidia, more often externally on hyphae; mostly saprophytes; imperfect stage often parasitic; many lichens	*Sordaria, Neurospora, Spathulospora, Xylaria, Pyrenula, Verrucaria, Dermatocarpon, Strigula, Mycocarpon* (extinct genus)
Hypocreales (2-120-1,600)	Brightly colored, fleshy perithecia; otherwise very similar to Sphaeriales	Mostly parasites or facultative saprophytes on angiosperms and other fungi	*Claviceps, Gibberella, Nectria, Hypomyces*
Pseudosphaeriales (7-300-2,500)	Many-chambered stromata with "pseudothecia"; no paraphyses or distinctive perithecial wall	Especially abundant in the tropics; many "imperfects"; several lichens; well-developed trichogynes	*Pleospora, Venturia, Roccella, Mycosphaerella, Arthonia, Ophiobolus, Hysteria, Dothidia, Stemphyllium*
Diaporthales (2-20-2,800)	Round, carbonaceous perithecia with hymenium, paraphyses, and ostiole; no trichogyne	Mostly saprophytes, thallus around perithecium often pseudoparenchymatous	*Valsa, Diaporthe, Valsella, Gnomonia, Endothia, Melanospora, Microthecum*
Erysiphales (3-70-1,600)	Hard, black cleistothecia with projecting hyphae	Superficial white, powdery mycelium; parasexual reproduction common; mostly obligate parasites	*Erysiphe graminis, Oidium, Sphaerotheca, Sphaceloma*
Aspergillales (5-40-1,500)	Inoperculate perithecia in which the asci walls autodigest, leaving spores free in the perithecial cavity	Ubiquitous blue and green molds on cheese, jam, fruit, leather, etc.; somatic crossing-over common	*Penicillium, Aspergillus, Ophiostoma, Gymnoascus, Ceratostomella, Lilliputia*

Saccharomycetales (5–50–650)	Asci produced singly on the much-reduced mycelium; no sporocarps produced	Unicellular or filamentous	*Saccharomyces cerevisae, Rhodoturula, Ashbya, Spermophthora, Nadsonia*
Helotiales (25–565–34,400)	Hard or leathery, small apothecia with operculate asci	Most of the species are symbionts with green or blue-green algae (lichens); many are plant parasites	*Pseudopeziza, Dermatium, Ostropa, Helotium, Usnea, Lecanora, Lecidia, Peltigera, Collema*
Pezizales (5–56–950)	Large, fleshy apothecia with operculate asci	Mostly saprophytes and mycorrhizal symbionts; many excellent mushrooms	*Morchella, Peziza, Verpa, Underwoodia, Discina, Gyromitra, Helvella*
Tuberales (3–25–300)	Cup-like apothecia in some species; mostly closed apothecia with convoluted hymenium and many chambers	Mostly subterranean saprophytes and mycorrhizal fungi of oak and beech	*Genea, Tuber magnatum, Pseudobalsamia, Genabea, Terfezia, Vibrissea, Ostropa, Elaphomyces*
Taphrinales (2–2–120)	Small apothecia lacking paraphyses and with poorly developed hymenium	Yeast-like plant parasites causing hypertrophic growth	*Taphrina deformans, Ascocorticium*
Laboulbeniales (3–50–1,600)	Thick-walled unique perithecia filled with gummy mass of spores; asci autodigest	Obligate parasites on aquatic and other insects; trichogynes very prominent	*Peyritschiella, Zodiomyceta, Laboulbenia, Ceratomyceta*

Note: The number of families (Fam), genera (Gen), and species (Spp.) in each order is indicated in the first column.

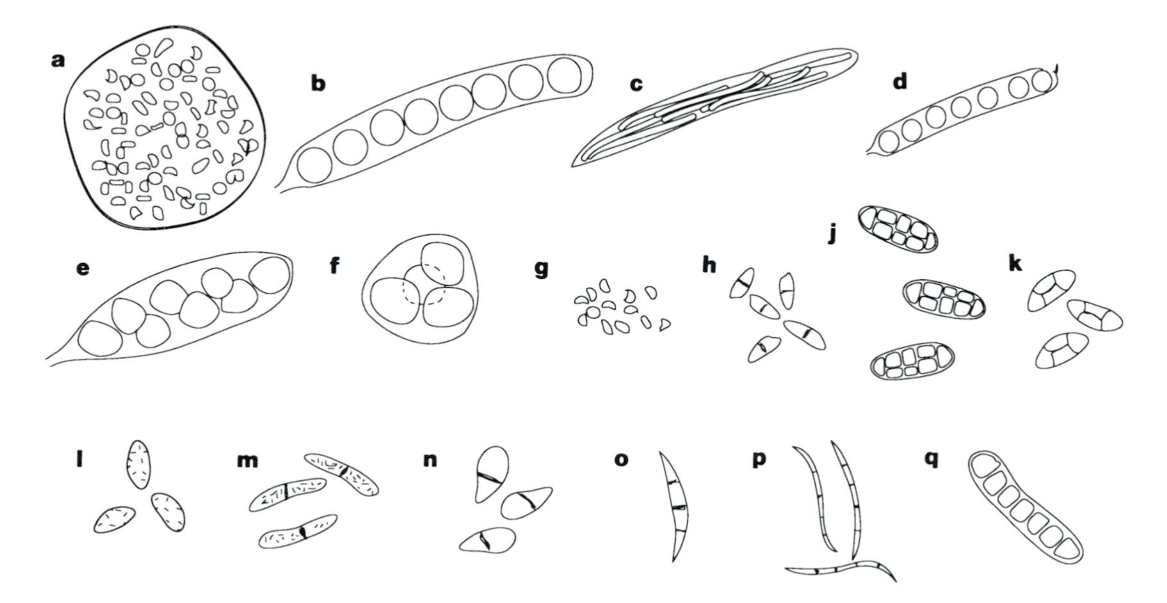

Figure 5–6 Diversity in spore sacs and spores in Ascomycopsida: Ascus diversity: (a) *Acarospora fuscata;* (b) *Lecanora rupicola;* (c) *Claviceps purpurea;* (d) *Sarcosphaeria crassa;* (e) *Diplocarpon earlianum;* (f) *Sacharomyces cerivisae.* Ascospore diversity: Nonseptate spores in (g) *Acarospora fuscata;* septate spores in (h) *Hypomyces lactifluorum;* (j) muriform spores in *Rhizocarpon geographicum;* (k) polarilocular spores in *Caloplaca trachyphylla.* Conidia diversity: (l) *Sphaeropsis malorum;* (m) *Ascochyta pisi;* (n) *Diplodia zeae;* (o) *Fusarium roseum;* (p) *Septoria apii;* (q) *Helminthosporium sativum.*

difficult to make, even with the aid of a microscope. The filaments of Zygomycopsida and Mastigomycopsida are **coenocytic** or nonseptate, but the filaments of Basidiomycopsida are septate like those of Ascomycopsida. The mycelium of sac fungi is mostly haploid; however, if dikaryotic cells are present, there will probably be **croziers** connecting them (Figure 5–3). The mycelium of Basidiomycopsida is mostly dikaryotic and may have **clamp connections,** which the mycelium of Ascomycopsida will not have. Absence of clamp connections is no guarantee that the fungus in question is an ascomycete, however. Asexual spores, or **conidia,** will frequently be present in Ascomycopsida, often in special structures called **pycnidia** and **acervuli,** which superficially resemble sporocarps but lack paraphyses and never contain asci; conidia are relatively rare in most groups of Basidiomycopsida as are pycnidia and acervuli. **Sporangiospores,** which are common in Zygomycopsida, resemble conidia but are never enclosed in either pycnidia or acervuli. Mastigomycete mycelium is usually more cotton-like than that of either ascomycetes or zygomycetes and neither sporocarps nor conidia are ever present. The asexual spores of Mastigomycopsida are called **zoospores** because they are flagellated and motile. Flagellated fungi favor more moist habitats than either Zygomycopsida or Ascomycopsida and are often aquatic.

Morphological and Physiological Characteristics

Both morphologically and physiologically, the extremes in this class seem to have little in common. The yeasts are unicellular and nutritionally resemble animals almost as much as plants;

the morels, on the other hand, are multicellular, have a very distinctive macroscopic form, and are dependent on the same mineral nutrients as other plants. Most of the class share some distinctive characteristics, however, including somewhat similar sporocarps, presence of asci and ascospores, possession of small nuclei, the chemical nature of the cell wall, and rather similar life cycles. A spectrum of variation, with one group of species resembling the next in numerous ways, ties the class extremes together.

Organelles of sac fungi are similar to those of red algae. Nuclei are small; cytoplasmic ribosomes are numerous and, as in all eucaryotes, larger than mitochondrial ribosomes; mitochondria in both classes have numerous, flat, plate-like cristae oriented parallel to the long axis of the organelle. **Cisternal rings** or vesicles often occur in lichenized ascomycetes and are believed to function like the dictyosomes of other plants; neither dictyosomes nor Golgi bodies occur in lichens or other sac fungi. **Peroxisomes** are also common and rather prominent in lichenized ascomycopsida; they are characterized by densely packed, lattice-like crystals (Figure 6-6). Cell walls are normally composed of three layers containing chitin, protein, and glucan. Both primary and secondary pits, similar to those of the Rhodophycopsida and Basidiomycopsida, occur. Electron-dense **Woronin bodies** may be found near the primary pit connections. They occur only in Ascomycopsida and are readily recognized in electron micrographs by their spherical form, crystalline structure, and large size (typically 150 to 200 nm). They will be mentioned again in the section on Fungi Imperfecti.

The vegetative mycelium begins growth with the germination of either an ascospore or a conidium. Each cell in the resulting hypha is generally haploid, although sometimes multinucleate cells develop. Under suitable environmental conditions, which vary among species, some of the cells in the mycelium develop into reproductive structures. The apical cell of a hypha divides and the two daughter cells differentiate, the basal cell becoming an ascogonium and the apical cell a trichogyne. Apical cells in other hyphae may become antheridia in which spermatia are cut off one at a time. A spermatium comes in contact with a trichogyne and its nucleus migrates through the trichogyne into the ascogonium where it pairs with the ascogonial nucleus. In many species, nuclei of undifferentiated cells function as ascogonia and spermatia.

Pairing of nonmotile gametes (spermatium and ascogonial nucleus) from two compatible mycelia is followed by the development of a more or less fleshy **stroma** which rapidly takes on the form of the fruiting body for the species. The fruiting body or sporocarp is made up of both dicaryotic and haploid hyphae. As the stroma is forming, fusion of the paired nuclei in the apical cell of each dicaryotic filament gives rise to a **spore mother cell** or zygote. The resulting diploid nucleus enlarges and undergoes meiosis followed by a single mitosis. The cell wall of the spore mother cell continues to grow, becoming a sac or ascus.

The upper surface of a sporocarp is the **hymenium,** made up of upright cells resembling a palisade. Some of these hymenial cells, formed from haploid filaments, are sterile **paraphyses.** The others are **asci.** In the Pezizaceae, differentiation occurs in the ascus wall, and the apex of the ascus develops an **operculum,** or lid, which often remains attached at one side like a trapdoor (Figure 5-6d). In other families, the end becomes soft and weak as the spores increase in size, building up pressure within the ascus which may become so great that they are shot from it with considerable force when the wall eventually ruptures. In some species, the entire ascus wall undergoes digestion at maturity and the spores are released in a mucilaginous fluid.

The ascus is usually cylindrical, though in species with no well-developed hymenium it may be pear-shaped or subglobose (Figure 5-6). Each ascus typically contains 8 meiospores, but there are only 4 spores per ascus in some species, 16, 32, 64, or more in others. In many sac fungi, spores are encased so closely within the ascus that their position relative to each other is fixed, a characteristic useful in genetic studies (Figure 5-6b,d).

Individual ascospores vary in size from about 1 μm in diameter to more than 400 μm, averaging about 10 to 30 μm. Frequently, the spores are ornamented with various kinds of markings. Usually they are unicellular, but spores with septations of various types are also common. There may be one to several transverse septations, or the septations may be both longitudinal and transverse, resulting in **muriform** spores. **Polarilocular** spores have very thick median walls, making each spore appear distinctly two-celled. Type of ascospore aids in identification of species and is also useful in studies of relationships. Some of the common types of ascospores are illustrated in Figure 5-6.

Single meiospores removed from an ascus can be germinated on an agar medium to produce pure monosporic cultures (Figure 5-7). These in turn will produce masses of conidia or asexual spores which can also be grown on agar. When spores from different cultures are placed at opposite poles of a petri dish, each will develop a mycelium. Whether sexual reproduction will occur between the two cultures is dependent upon a series of multiple alleles, called **self-incompatability alleles.** If the two mycelia contain the same self-incompatability allele, sexual reproduction will not take place (Figure 5-7C,D); if the alleles are compatible, sporocarps (apothecia or perithecia) will be produced abundantly (Figure 5-7A,B,E,F).

An apothecium is typically a slightly concave disk- or saucer-shaped structure, but it may be a deep cup-like body. Fungi bearing apothecia are commonly called **discomycetes.** The apothecia may be fleshy or leathery and are often red, yellow, brown, or black. They vary in size from less than a centimeter to several centimeters in diameter (Figure 5-1). The **hypothecium,** immediately beneath the hymenium, is pseudoparenchymatous in some species, prosenchymatous in others. In *Morchella* and its relatives, the entire apothecium is rolled back with the hymenium on the outside surface of a convex cap (Figure 5-1c).

A **perithecium** is flask-shaped and usually has a small opening or **ostiole** at its apex. If no opening is present, it is often called a **cleistothecium.** Perithecia are nearly always dark in color and are only one or a few millimeters in diameter, generally much smaller than apothecia. A true perithecium has a cellular perithecial wall or **pericarp** and the asci are borne on a hymenium which may cover only the bottom of the perithecium or may cover the walls as well. In some Ascomycopsida, stromatic cavities, superficially resembling perithecia but lacking a cellular wall, occur. In these cavities, the asci are present in tufts without sterile paraphyses among them. These pseudothecia (false perithecia) are typical of the Pseudosphaeriales (Dothidiales) and resemble the tetrasporic conceptacles of some red seaweeds (Figures 4-3, 5-5). Because perithecia are usually small and hard, the fungi bearing them are often called **pyrenomycetes,** from the Greek word for the pit or stone in cherries and peaches.

Croziers (Figure 5-3b) are unique to the Ascomycopsida. They occur only in the dikaryotic hyphae at the base of the sporocarp and are similar to the secondary pits of red algae and clamp connections of club fungi. The dikaryotic cell at the growing tip of the filament curves back 180° to form a hook; one nucleus remains in the basal portion of the cell and the other migrates into the hook. Both nuclei divide simultaneously. Cell walls separate the daughter nuclei as the tip of the hook makes contact with the middle portion of the original cell. The

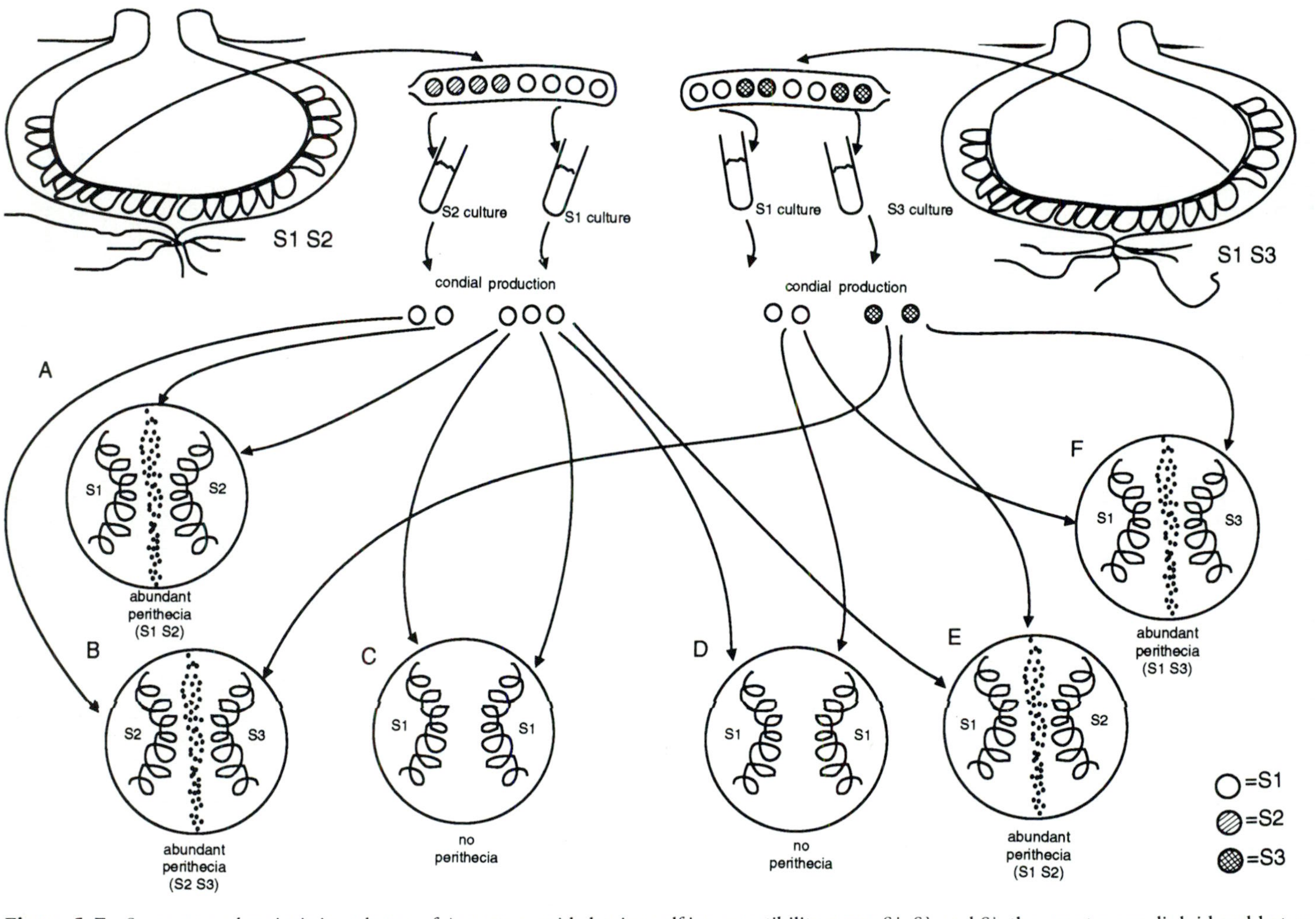

Figure 5–7 Syngamy and meiosis in cultures of Ascomycopsida having self-incompatibility genes S^1, S^2, and S^3; the zygotes are diploid and heterozygous. Four ascospores in each ascus have one self-incompatibility gene, following meiosis, and the other four have its allele. Cultures developing from each ascospore are haploid; abundant perithecia develop when cultures of different mating types are grown in the same petri dish, but not when cultures of the same mating type are grown together.

daughter nucleus from the apex of the hook migrates through the secondary pit that results from this contact. Each daughter cell is thus dicaryotic and has the same genetic makeup as the mother cell.

Asexual reproduction is by means of conidia in most terrestrial ascomycetes. **Conidia** are formed by abscission of specialized hyphae and are not, therefore, enclosed in sporangia. In some species, conidia function as sperm as well as asexual spores. Nannfeldt (1932) suggested that the trichogyne in the Ascomycopsida evolved from such conidia rather than from Floridean trichogynes, but no genetic basis for this hypothesis has been advanced. Conidia vary in size and form; they may be long and slender, often slightly curved, or roughly spheroidal, and usually range from 1 to 10 μm in diameter (Figure 5-6). They may be produced directly on the vegetative hyphae or they may be produced within special flask-shaped structures called **pycnidia** or cushion-like structures called **acervuli.**

In identifying unknown species of fungus, a student must take special care to avoid confusing pycnidia with perithecia or apothecia. Basically, the pycnidium is spherical like a perithecium, though the bottom half of the sphere may be lacking. In some species, the top half is missing, and the pycnidium resembles an apothecium. In either case, pycnidia lack a hymenium and, therefore, asci and paraphyses as well. An acervulus is a cushion-like mass of hyphae made up of **conidiophores** (the stalks from which the conidia are produced) packed closely together in a subcortical layer. Often the spores are surrounded by a gummy mass. In some species, the acervuli superficially resemble apothecia, but without asci or paraphyses.

Sexual reproduction has never been observed in many sac fungi and other Eumycophyta. In some species, which have been cultured in the laboratory under a wide range of environmental conditions without success in getting sexual reproduction to take place, it seems likely that the ability has been lost. On the other hand, many disease-producing fungi have been studied only from the point of view of identification and control with no attempts made to discover the conditions needed for meiospore production. In absence of knowledge of the complete sexual, or "perfect," life cycle, classification of these species is difficult. For convenience, therefore, these species are assigned to **form genera, form families,** and **form orders** in the **form class** *Fungi Imperfecti.* The Fungi Imperfecti will be further discussed in a later section of this chapter.

Sexual life cycles in the Ascomycopsida are three-phase like those of Rhodophycopsida. The life cycle of *Endothia parasitica,* the incitant of chestnut blight, is typical (Figure 5-5D). Ascospores of *Endothia* are transmitted primarily by wind and infect healthy plants at an optimum temperature of 20°C. The resulting mycelium is monopodially branched and made up of multinucleate cells which penetrate the cortex, phloem, and cambium tissues of young stems, branches, and even the main trunk of most species of chestnut. At first white, the mycelium becomes yellowish as it spreads through the cambium and out into the bark layers. Eventually it forms **stromata** of haploid mycelium which break through the bark as small yellow or orange papillae called **pycnidial pustules.** Very small, curved conidia, about 1.3 × 3.6 μm in size, are produced in these pustules. As they ooze out of the gummy spore mass, they form orange threads or coils called "spore horns", which are attractive to many kinds of insects and birds. As many as 100 million conidia may be present in a single average sized spore horn. After a period of activity in producing conidia (also called **pycnospores**), the pycnidial pustules become sexually active and begin to produce perithecia.

In *Endothia parasitica,* there is no trichogyne, so syngamy is between two vegetative cells, sometimes derived from the same ascospore. The dikaryotic phase is short-lived. Following karyogamy, numerous perithecia are formed, the pseudoparenchymatous pericarp being made up chiefly of haploid cells. Each perithecium opens by an ostiole which appears as a small, black pimple on the convex surface of the perithecial stroma. A very long neck connects the perithecial cavity to the ostiole. Karyogamy is followed by meiosis and a single mitosis; the asci are oblong or club shaped with eight oval ascospores in each ascus. The ascospores are somewhat larger than the conidia, averaging 4.5×8.6 μm.

Endothia subsists as cankers on live trees and as a saprophyte on dead branches. Both conidia and ascospores are produced throughout spring and summer, the former in pycnidia, the latter in perithecia. Conidia are carried by insects, birds, or rain and infect healthy plants when temperatures are around 30°C. Ascospores are carried by wind and have an optimum infective temperature of 20°C. Penetration is through open wounds. Conidia overwinter in yellow gummy masses called **cirri** on the diseased wood. In the summertime they remain viable for about 4 months, whereas ascospores remain viable for a year or more; otherwise infection by the two kinds of spores is essentially identical.

Phylogeny and Classification

Denison and Carroll (1966) proposed that the first Eumycophyta were **pyrenomycetes,** similar to modern Sphaeriales, which mutated from ancestors similar to Nemalionales and/or Gelidiales, about 350 million years ago. As evidence, they call attention to the similarities between the Sphaeriales and the Rhodophyta, the fact that most marine Eumycophyta are Sphaeriales, and the deduction that after autotrophic life had developed on land, there would be a new ecological niche in the oceans where debris from land would accumulate, and the red algae would be in the best position to supply the mutant types to fill this niche. The fossil record is in accordance with their hypothesis: forests of vascular plants covered much of the Earth's land surface by mid-Devonian, red seaweeds were abundant in the ocean at that time, and the oldest fossils of either sac or club fungi date from late Devonian at the earliest. From early Sphaeriales, evolution proceeded along three lines, one leading to the blue molds, yeasts, and Zygomycopsida by way of the pyrenomycetes, one leading to the club fungi by way of the discomycetes, and one to a group of highly specialized insect parasites, the Laboulbeniales (Figure 5-8).

The first record of a Paleozoic septate fungus was made by W. C. Williamson in 1880 when he discovered and described *Sporocarpon pachyderma* in Carboniferous coal formations from Halifax, England. This is still among the oldest good fossils of Eumycophyta. Detailed studies of this and similar fossils from England and Belgium failed to reveal any evidence of asci. They report The wall structure appears to be precisely similar to that of a fungal cleistocarp both in transverse section and in surface view. Reclassified in the genus *Mycocarpon,* they most nearly resemble modern pyrenomycetes of the order Sphaeriales. Spore-like cells in the perithecia seem to be arranged like tetraspores in a red alga conceptacle (Figure 4-3) rather than in asci, suggesting extreme primitiveness. *Protosalvinia furcata,* is a somewhat similar, chitin-containing fossil found in upper Devonian formations in Ohio. First described as *Sporocarpon furcatum,* it resembles both the Sphaeriales in the Ascomycopsida and Polysiphonia in the Rhodophycopsida; it was even assigned to a new class called Algomycetes by one paleobotanist because of its intermediate nature. However, the poorly preserved *Protosalvinia* material is difficult to interpret and may or may not be related to Williamson's *Sporocarpon.* Although

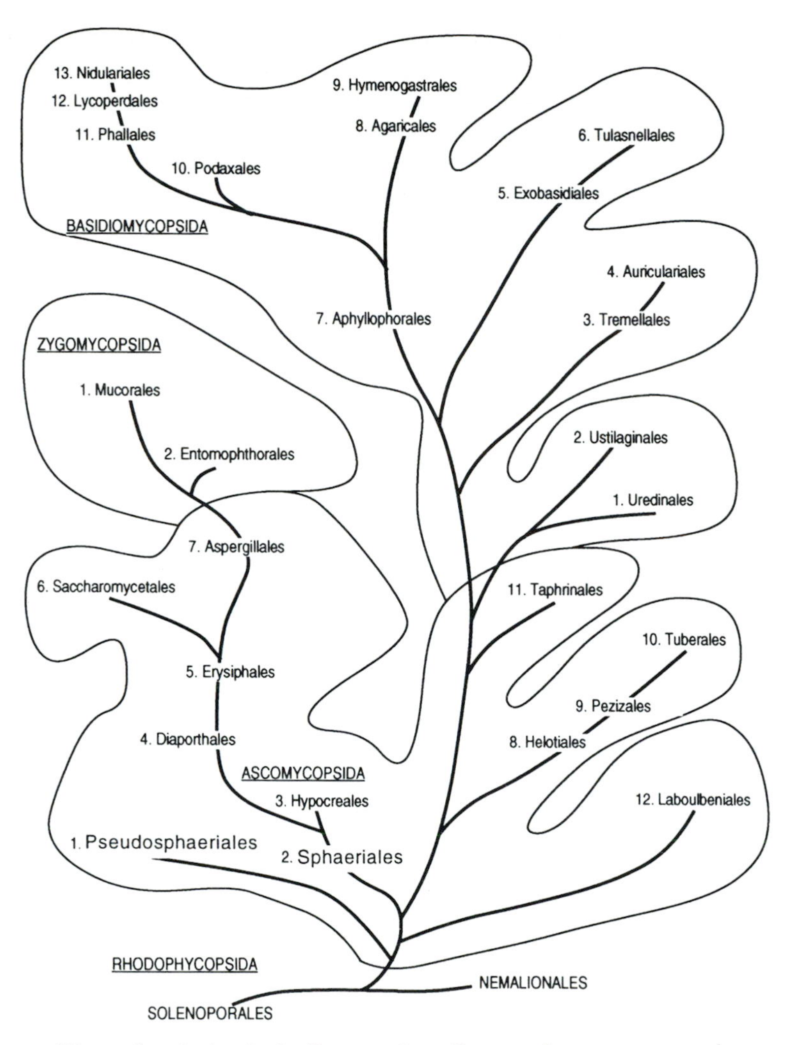

Figure 5–8 Suggested lines of evolution in the Eumycophyta. Presumed ancestors are red seaweeds in the extinct order Solenoporales.

good fossils of higher fungi are quite rare prior to mid-Pennsylvanian, they are relatively abundant in upper Pennsylvanian and Permian formations, suggesting rapid evolution in this division at that time.

Sachs (1874) very early suggested a close relationship between the Ascomycopsida and Rhodophycopsida and many modern mycologists agree. Other mycologists, however, regard the similarities between the red algae and the sac fungi to be the product of parallel evolution in two unrelated lines, and look to the phycomycetes (Mastigomycopsida), or even the Animal Kingdom, for the closest relatives of the sac fungi. Reasons for postulating a close relationship between the Ascomycopsida and Mastigomycopsida were argued by Atkinson (1915) and havebeen summarized by Demoulin (1974). They include the following observations: (1) the

Zygomycopsida seem to be transitional between the two classes, (2) fungal chitin occurs in the chytrids and *Phytophthora* of the Mastigomycopsida as well as in the Ascomycopsida but not in the Rhodophycopsida, (3) *Dipodascus* of the Ascomycopsida is coenocytic like the Mastigomycopsida, (4) lack of an auxiliary cell in both the Eumycophyta and the phycomycetes, and (5) lack of chlorophyll in any of the fungi of either group. This last point is the only one stressed by many biologists, including most advocates of the currently popular five-kingdom systems of classification.

The fact that the trichogyne in the sac fungi is two-celled, the lower cell being the ascogonium, whereas the trichogyne in the red algae is an extension of the carpogonium, has been pointed out as a flaw in the algal origin hypothesis by proponents of the phycomycete origin hypothesis. They also point out that if we accept the hypothesis that the Ascomycopsida have descended from rhodophycean ancestors, we must relate the fungi which are structurally most complex with relatively complex red algae and then assume that evolution has proceeded within the Ascomycopsida from complex to simple. On the other hand, if we accept the hypothesis that the Ascomycopsida evolved from the Mastigomycopsida, we relate simple ascomycetes to the phycomycetes and assume evolution within the Ascomycopsida from simple to complex. Many biologists feel more comfortable with the latter alternative, but are disturbed by the suggestion of retrogressive evolution on a large scale as suggested by the former.

However, if we examine other heterotrophic taxa in which parasitism is common—e.g., the Convolvulaceae, Loranthaceae, and Scrophulariaceae of the Tracheophyta in the Plant Kingdom and the Platyhelminthes and Aschelminthes of the Animal Kingdom—we find that evolution from complex to simple is the rule in parasitic groups, rather than the exception. Proponents of the rhodophycean origin hypothesis therefore contend that the simple Ascomycopsida could very likely have evolved from structurally more complex Ascomycopsida which in turn evolved from the Rhodophycopsida. They also point out that upon close examination, it appears that the ascogonium of the sac fungi is homologous to the auxiliary cell of the red algae (Demoulin 1974) being the product of fusion of sperm and egg. In summary, the weight of evidence, in the opinion of many modern mycologists, favors the algal origin hypothesis.

A variation of the Sachs' hypothesis relates primitive club fungi rather than sac fungi to the red seaweeds from which the Eumycophyta have evolved (Demoulin 1974). According to this version of the algal origin hypothesis, the first Eumycophyta were parasites on algae and were morphologically similar to the rusts and smuts in the Basidiomycopsida and the Laboulbeniales and lichenized Helotiales of the Ascomycopsida. Since a number of modern red algae are **adelphoparasites** (parasitic on closely related species), it seems plausible that the first ascomycetes were parasites on red algae if we accept an algal origin hypothesis of the higher fungi. The ancestral red alga, according to Demoulin, would be similar to modern members of the Helminthocladiaceae of the Nemalionales and/or the Chaetangiaceae of the Cryptonemiales. To date, no ancient fossils of either rusts or lichens have been discovered to add support to the Demoulin hypothesis. Examination of Figure 5-8 indicates that the Denison-Carroll and Demoulin hypotheses are not mutually exclusive: some of the earliest Eumycophyta could have been saprophytic Sphaeriales, and others could have been parasitic sac fungi and rusts.

The taxonomy of the Ascomycopsida is difficult. Factors contributing to this include (1) lack of pigments, such as chlorophyll, which are often very useful in classification of autotrophic plants; (2) lack of sexual reproduction in many ascomycetes; (3) classification of the

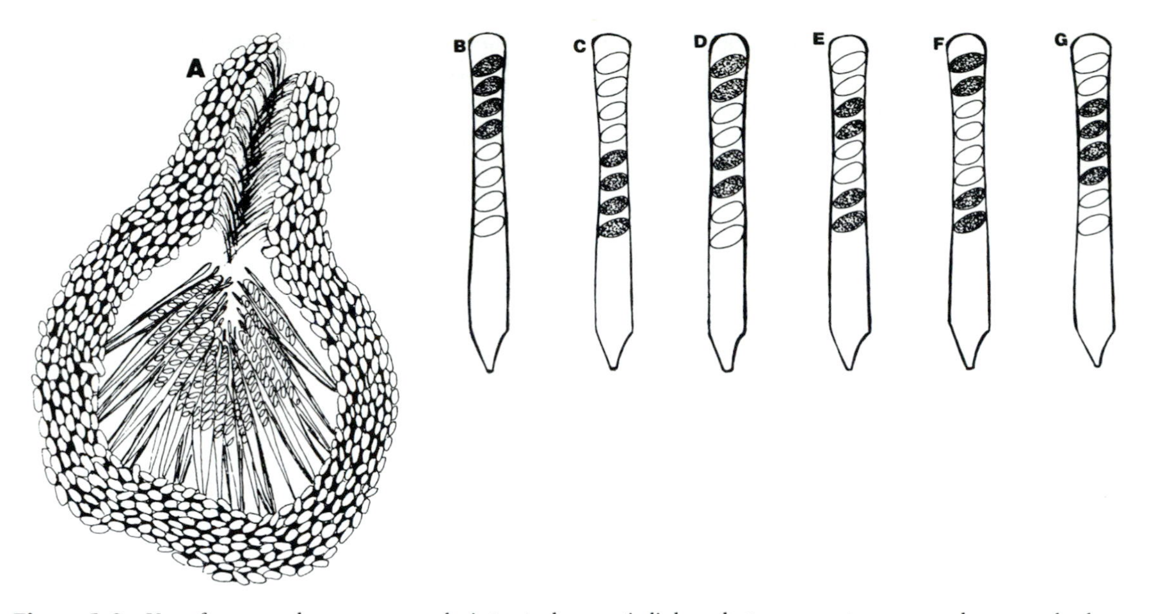

Figure 5–9 Use of ascus and ascospore analysis to study genetic linkage between centromere and spore color locus in *Neurospora crassa* (see also Figure 5-7). (A) perithecium; (B-C) segregation of the gene for dark spore color from its allele for hyaline spores occurs most frequently at first anaphase, resulting in spores of the same color being together in groups of four; (D-G) a chiasma between the centromere and the spore color gene results in segregation at second anaphase; spores of the same color are still together in pairs because a single mitosis follows meiosis as the ascus matures.

lichens, until relatively recently, in the Bryophyta and hence difficulty in incorporating them into the taxonomic scheme of the Ascomycopsida; (4) occurrence of very complex life cycles in which two closely related species do not always resemble each other in all of the stages of the life cycle; and (5) the fact that many disease-producing and other sac fungi are only partly described.

Because sexual reproductive structures are less affected by environment than vegetative structures, they have been especially useful in studying plant relationships. Mycologists, however, have overemphasized reproductive structures, employing them almost exclusively in their taxonomic systems, and forgetting that everything is important in plant classification. More cytological and biochemical data are needed if we are ever to deduce the true relationships among the higher fungi. This is especially true for the Fungi Imperfecti and the lichens.

Representative Genera and Species

The pink bread mold, *Neurospora crassa* in the Sphaeriales, is characterized by small, hard perithecia and straight, cylindrical asci in which the spores are held so tightly that they cannot change position. This feature made it invaluable to Lindegren (1948) in ascertaining when reduction division takes place and in calculating the genetic map distance between any given gene and the centromere (Figure 5-9). The number of asci in which the spores are never arranged in sets of four (Figure 5-9D–G) divided by the total number of asci examined equals the percent crossing over, and hence the map distance, between the gene being studied and the centromere.

Because *N. crassa* grows and reproduces rapidly, Beadle and Tatum (1945) used it to study the physiological effects of genes. They discovered that the production of any one vitamin or other nutrient often requires the joint action of several genes, each responsible for a specific enzyme involved in the final synthesis of the vitamin. From these studies they developed the "one gene one enzyme" concept which has led to our present understanding of how DNA functions.

Closely related to the Sphaeriales, and often combined with it, is the order Hypocreales (Figure 5-8). *Claviceps purpurea,* the best-known species, is a parasitic fungus of great economic significance, causing a serious plant disease on rye, barley, wheat, and other grasses. ***Ergot,*** the disease caused by *Claviceps purpurea,* has been known since ancient times and not only causes severe loss to the grain crops it attacks but also causes poisoning in humans and animals that eat the diseased grain. St. Anthony's fire, one form of ergot poisoning, causes convulsions and a burning sensation in the extremities which is often followed by gangrene. Another form causes severe hallucinations. As recently as 1951, an outbreak of ergot poisoning caused sickness and death in France. Some of the affected people, believing they could fly, jumped out of windows; others thought they were on fire. The miller who had produced the flour admitted observing ergot, which is easily identified but not as easily removed, in the grain and was indicted for involuntary homicide.

Considerable evidence suggests that the witnesses responsible for the death of 19 victims of the infamous Salem witch hunt in 1692 were themselves victims of ergot poisoning. Because of the effects of the ergot, the children who later served as the witnesses in the witch trials were not able to distinguish between reality and their hallucinations. Weather records reveal that environmental factors were perfect for a heavy ergot infestation in 1691. The rye harvested that fall would be used for bread the following spring and summer, resulting in ergot poisoning. Court records reveal that the witnesses all lived in the low-lying areas where rye was grown most and where the rye would most likely be infested with ergot.

The word "ergot" applies both to the disease caused by the fungus, *Claviceps purpurea,* and to the **sclerotia** that are produced by the fungus in place of the grain kernels. *C. purpurea* attacks only the developing seeds of grasses. The first symptom of infection is the exudation of "honey dew" from the grass flowers. This sweet substance is attractive to insects and contains conidia (Figure 5-10). Infection is thereby readily spread by insects from one grass head to another. Splashing rain drops also serve as a source of infection in this stage. As the grass plant matures, the fungal mycelium in each flower develops into a hard sclerotium (Figure 5-10a,b). Sclerotia vary in size, depending on the species of grass infected, in general being slightly larger than the fruit or caryopsis of that species. Studies have indicated that the specific gravity of the sclerotium varies, always being almost exactly that of the fruits of the infected species. Sclerotia remain dormant until the seed they are mixed with is planted, at which time they germinate to produce slender, purplish stalks, with aggregate sporocarps, or stromata, full of perithecia at the apex (Figure 5-10). Ascospores are produced in these perithecia; they are carried by wind and lodge against the feathery stigmas of grass flowers, where primary infection occurs.

Chestnut blight (Figure 5-5), caused by *Endothia parasitica* (Diaporthales), was apparently introduced into the New York City area on nursery stock from the Orient. It infects and rapidly kills all parts of the chestnut tree except leaves and roots. Infected plants are readily identified from afar by the wilted leaves hanging on to the dead branches throughout the winter; the roots continue to send up clusters of new stems each spring, but within a year or

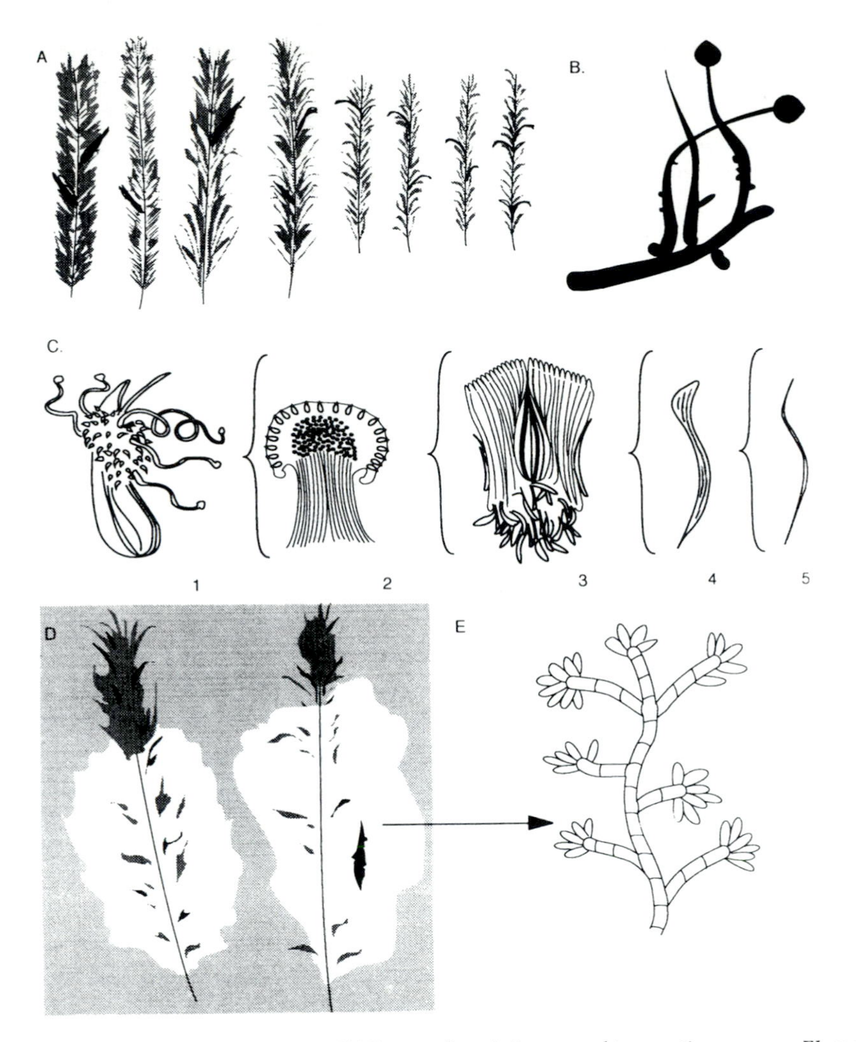

Figure 5–10 Life cycle of *Claviceps purpurea.* (A) Ergot sclerotia in rye and two native grasses, *Elymus canadensis* and *Agropyron smithii;* (B) germinating ergot sclerotium with two mature perithecial stromata; (C) successive enlargements of ergot: (1) sclerotium with six stromata; (2) enlargement of a single stroma showing 20 perithecia; (3) enlargement of one perithecium containing several asci together with sterile paraphyses; (4) a single ascus containing eight ascospores; (5) a single ascospore; (D) production of honey dew containing conidia in immature heads of rye; (E) honey dew mycelium in grass flower with several conidiophores and conidia.

two, these suckers have also succumbed. The American chestnut was the most important hardwood tree in North America at the turn of the century, but is now only a memory and the subject of an oft quoted poem by Henry W. Longfellow. Other plant diseases are listed in Table 5-3.

Helotiales is an extremely large order of discomycetes, or fungi in which the fruiting bodies are apothecia. Several plant pathogens, like *Pseudopeziza medicaginis,* which causes common leafspot in alfalfa, and *Sclerotinia minor,* the incitant of various leaf and crown rots of vegetable crops, belong here, as do many of the lichens. *Lecidea, Lecanora,* and other lichens will be discussed in Chapter 6. *Lecidea* with almost 3000 species and *Lecanora* with over 2000 are the largest genera in the Plant Kingdom. They are both members of the Helotiales, and proposals to split both genera into smaller, more manageable **segregate genera** have been made. Closely related to the Helotiales is the order Pezizales which contains several species of fleshy fungi or mushrooms such as the eminently edible morels.

Mushrooms are the fleshy sporocarps of Eumycophyta. In the opinion of many people, the choicest of all mushrooms, and the goal of almost every spring mushroom hunt, is *Morchella esculenta,* the common morel (Figure 5–1c). The stems are stout, rather short, and hollow, and the caps are rounded or oval, flesh-colored to gray with numerous pits (not wrinkles) on the surface, making this a very distinctive species that can hardly be mistaken, even by amateurs, once they have seen a specimen or even a picture.

Even more famous than the morels are the **truffles,** members of the Tuberales, and perhaps the most expensive of all gourmet foods. Truffles grow underground and are hunted with trained pigs or dogs which are able to sniff them out. *Tuber melanosporum,* the black truffle of France, and *T. magnatum,* the white European truffle, are especially famous. American truffles, on the other hand, some of which may also be very tasty, are not as well known.

Other Ascomycopsida of special interest include the Laboulbeniales, a group of fungi parasitic on insects; *Penicillium rocquefortii* which imparts a delicious flavor to blue cheeses; *Gyromitra esculenta* or false morel, which has been praised as the most delicious of all mushrooms and condemned as one of the most poisonous; *Evernia prunastri* or oak lichen ("oak moss" in the perfumery trade) and *Penicillium chrysogena* or common blue mold, both of which produce valuable antibiotics; and *Aspergillus nidulans* or black jam mold, a species that has become valuable in genetic studies and in industry. People who enjoy Coke, Sprite, 7-Up, rootbeer, and other soft drinks should appreciate *A. nidulans,* the organism responsible for the tangy flavor of their favorite beverage. The range of diversity in the Ascomycopsida is suggested in Table 5–2, in which some of the salient characteristics of each order are presented.

Ecological Significance

The Ascomycopsida are ecologically important in all terrestrial biomes and most aquatic ecosystems. As saprophytes they are important decomposers in almost all ecosystems, breaking down even cellulose and lignin into carbon dioxide, water, and various minerals, as parasites they affect, directly or indirectly, the survival and abundance of both plants and animals. As symbionts they are important in xerarch succession and in improving moisture and mineral relationships of many plants. Both saprophytic and symbiotic species provide food for herbivorous animals, and some symbiotic combinations are important in nitrogen fixation.

Almost any crop species, weed, or native plant has some type of leaf spot, canker, or other evidence of disease on it. While the diseases on native vegetation are of little economic significance, they are undoubtedly important in the ecology of the affected plants and their ecosystems. While little is known about most of these disease organisms, a majority of them are obviously ascomycetes.

The ecology of parasites is often complex. *Erysiphe graminis,* the incitant of powdery mildew in wheat and other cereal crops, is more damaging during periods of drought than during wet periods. *E. graminis* conidia can germinate at 0% relative humidity, but this is not

TABLE 5-3

EXAMPLES OF PLANT DISEASES CAUSED BY EUMYCOPHYTA

Disease	Causal Agent	Class	Order	Symptoms or Signs	Control
Apple bitter rot	*Glomerella singulatum*	ASCO	Sphaeriales	Flesh turns dark beginning under skin	Remove dead wood, use of copper sprays
Ergot (cereal crops)	*Claviceps purpurea*	ASCO	Hypocreales	Kernels replaced by sclerotia	Clean seed, rotation
Pseudoplea leaf spot (alfalfa)	*Pseudoplea trifolii*	ASCO	Pseudosphaeriales	Pin-point leaf spots with halo	Plant breeding is offering hope
Chestnut blight	*Endothia parasitica*	ASCO	Diaporthales	Wilting, cankers	Hopefully plant breeding
Powdery mildew (cereal crops)	*Erysiphe graminis*	ASCO	Erysiphales	Powdery "residue" on grain leaves	Plant breeding; dusting with sulfur
Blue mold on fruit (storage disease)	*Penicillium notatum*	ASCO	Aspergillales	Blue mold; softening of tissues	Store with good ventilation, cool
Pink ear rot (corn, barley)	*Gibberella zeae (Fusarium roseum)*	ASCO	Hypocreales	Rotting of kernels beginning at tip; pink discoloration	Resistant hybrids; crop rotation
Red laboratory mold	*Rhodotorula* spp.	ASCO	Saccharomycetales	Pink discoloration of lab medium	Complete sanitation in transferring cultures
Common leafspot (alfalfa)	*Pseudopeziza medicaginis*	ASCO	Helotiales	Yellowish leaf spot; cup-like apothecia	Resistant varieties
Take-all (cereal crops)	*Ophiobolus graminis*	ASCO	Pseudosphaeriales	Leaves, stems, and heads bleach at heading time; black mold at crown	Crop rotation, balanced fertility
Culm rot (rice)	*Leptosphaeria salvinii*	ASCO	Pseudosphaeriales	Large irregular lesions; small black sclerotia	Drain the fields but keep the soil saturated

Blackleg of cabbage (all crucifers)	*Phoma lingam*	*ASCO	*Sphaeropsidales	Inconspicuous pallid areas on seedlings; later, linear lesions at soil line	Rotation and removal of all debris
Onion smudge	*Colletotrichum circinans*	*ASCO	*Melanconiales	Black or yellowish blotches on bulbs	Red and some yellow varieties are resistant
Fusarium wilt (flax)	*Fusarium oxysporum*	*ASCO	*Moniliales	Yellowish leaves; thick terminal leaves	Rotation; resistant varieties
Helminthosporium leaf blight (corn)	*Helminthosporium*	*ASCO	*Moniliales	Narrow brown spots on leaves; black mold on kernels	Resistant hybrids
Storage rot (sweet potatoes)	*Rhizopus stolonifer*	ZYGO	Mucorales	Roots turn soft	Let roots ripen before storing; good ventilation
Corn smut	*Ustilago maydis*	BASID	Ustilaginales	Shiny hypertrophic galls; black spores	Resistant varieties; crop rotation
Loose smut (wheat)	*Ustilago nuda*	BASID	Ustilaginales	Black, bare rachis	Resistant varieties
Bunt or covered smut (wheat, rye)	*Tilletia caries*	BASID	Ustilaginales	Dwarfing of grain; foul odor	Crop rotation; seed treatment
Stem rust (cereal crops)	*Puccinia graminis*	BASID	Uredinales	Chlorotic leaf spots; masses of red spores	Resistant varieties; control of barberry
Snow mold (winter wheat)	*Typhula itoana*	BASID	Aphyllophorales	Yellow spots becoming brown blotches	Fall application of fungicides
Black scurf (potatoes)	*Rhizoctonia solani*	*BASID	*Mycelia Sterilia	Black "dirt" on tubers that won't wash off	Rotation; use of certified seed

* Refers to Fungi Imperfecti.

the total explanation for powdery mildew epidemics following periods of dry weather. *E. graminis* is parasitized by a *Cincinnobolus* species of Fungus Imperfectus which cannot grow and reproduce during dry weather but grows rapidly and seems to be very effective in limiting the growth of *Erysiphe* during wet weather. In the absence of *Cincinnobolus,* powdery mildew epidemics are sometimes severe during wet, rainy summers.

Crustose lichens are important pedogenic agents and hence pioneers in xerarch successions. Lichens with blue-green algae phycobionts, such as *Peltigera* and *Collema* species, contribute nitrogen to the ecosystem. Caribou, snails, squirrels, chipmunks, and other small animals feed on lichens. Lichens also contribute to starting and spreading of forest fires. Fleshy ascomycetes, such as *Peziza, Morchella,* and *Tuber,* also provide food for many animals including deer, elk, rodents, and insects.

Economic Importance

The annual cost of crop diseases in the U.S. at midcentury was estimated at 3 billion dollars (Pearson 1967); this had probably increased to over 10 billion by 1990. Fungi, bacteria, nematodes, protozoa, and even algae cause plant disease, but the Ascomycopsida, including Fungi Imperfecti, are the most significant (Table 5-3). To help prevent the recurrence of disasters similar to the destruction of the American chestnut, the U.S., and most other countries now enforce very strict quarantine laws.

Signs and symptoms of plant disease include (1) discoloration, or a change in color from that which is normal for the species; (2) wilting; (3) necrosis, or the death of some plant tissue; (4) increase in size, or hypertrophy of plant parts; (5) dropping of leaves, blossoms, fruits, or twigs; (6) production of excrescences and malformations, such as galls, cankers or witches brooms; and (7) rotting of tissue (See Special Interest Essay 10-1).

Some fungi are specific to one host; others are highly promiscuous. *Phymatotrichum omnivorum,* the incitant of Texas root rot or cotton root rot disease, has more than 1700 host species, more than any other known pathogen. Its hosts include cotton, alfalfa, red clover, lespedeza, soybeans, potatoes, sweet potatoes, peaches, apples, raspberry, cocklebur, wild morning glory, jimsonweed, lamb's-quarter, sunflowers, and many others; most monocots, however, and members of the cabbage, spinach, squash, and celery families are generally not affected. Crop rotation with these resistant plants is the chief means of control of this very serious plant disease. Other diseases and methods of control are suggested in Table 5-3.

Ascomycopsida are of economic significance for other reasons: the drugs, penicillin, ergotine, and usno, are all produced from sac fungi; yeasts are important in industry; *Aspergillus* and its relatives are sources of a variety of chemicals including citric, glucomic, and butyric acids (Special Interest Essay 5-1); the choicest mushrooms, the truffles and morels, are the sporocarps of sac fungi; other Ascomycopsida impart to cheeses their characteristic flavors.

Ergotine, extracted from the sclerotia of *Claviceps purpurea,* is important in stopping hemorrhaging following childbirth and is also used to control migraine headaches. The pharmacology industry purchases only rye for ergotine extraction, even though ergot from other grains is chemically identical to that from rye. Because grain produced in humid areas is often infected with *Fusarium,* a fungus that causes problems in purification of the drug, the industry avoids rye produced east of the Dakotas or Minnesota and also barley, which is more frequently infected with *Fusarium* than is rye. The toxins which occur in *Fusarium* affect the respiratory and other automatic muscular action. Ergot is also the source of a hallucinogenic drug commonly known as LSD.

Commercial yeast, *Saccharomyces cerevisiae,* is economically one of the most valuable plants cultivated by man. Its products include bread, wine, industrial alcohol, vitamins, enzymes, glycerol, and fats. It is used directly for human food and as livestock fodder. Some strains of yeast contain up to 50% protein, having the same amino acid composition as beef and pork, and it has been asserted by some that yeasts could replace animal protein in the human diet. *Torulopsis utilis, T. lactosa,* and *S. lactis* are other species of yeasts that have been considered potential human food sources since all can be grown on cheap sources of carbohydrates and none of them have the "yeasty" flavor most people find objectionable in *S. cerevisiae.*

Industrial alcohol is the most important organic chemical used in modern industry. It is used as a solvent in manufacturing plastics, dyes, safety glass, photographic materials, soaps and shampoos, explosives, varnishes, synthetic flavorings, medicines, shoe polish, and numerous other products. It is used as a raw material in manufacturing vinegar, acetic acid, ether, and chloroform. It is widely used as fuel for cooking and to provide light, heat, and power, not only in small camp stoves by vacationers but on domestic and industrial scale in many parts of the world. As a fuel, it produces a clean flame free of particulate matter and toxic fumes, and it produces more heat and about three times as much light as an equal volume of petroleum fuel.

Use in Scientific Research

Because they are easily cultured and grow rapidly, *Aspergillus nidulans* and *Penicillium notatum* have been widely used in studies of metabolism and, more recently, in genetic studies. *Neurospora crassa* has also been widely used in genetic studies (Figure 5-9).

Gibberella zeae (*Fusarium roseum*), a parasite of both corn and barley, has been useful in studying the ecological concept of limiting factors. When temperatures during the early seedling stage are relatively high, severe infections occur in barley; when germination temperatures are unseasonably low, severe infections occur in corn. In both cases, there is a three-way interaction involving host, parasite, and environment. Barley is a cool season crop and outgrows the fungus when temperatures are slightly below optimum for barley and therefore considerably below optimum for *G. zeae.* Corn, on the other hand, is a warm season crop and will outgrow the fungus when temperatures are slightly above optimum for corn and well above optimum for the fungus. At optimum temperature for barley, there will be some infection and loss in yield if the fungus is present, and the same is true in corn when temperature is optimum for it.

G. fujikuroi is the fungus that causes "crazy plant" disease in rice in which infected plants grow much taller than normal. A chemical isolated from this fungus causes "crazy" symptoms in uninfected plants. From this chemical, named gibberellin by its discoverers, we have learned a lot about differentiation and morphogenesis in plants. In many species of plants, red clover, for example, both dwarf and tall strains are known. The dwarf plants are short because the internodes do not elongate. Administration of gibberellin to such plants will cause them to grow as tall as the tall strains. If tall plants, on the other hand, are treated with gibberellin, there is no increase in size. A practical application of this observation is the treatment of first-year sugar beet plants with gibberellin to cause them to go to seed.

The sugar beet is normally a biennial. During its first year of growth, its stem is very short and it produces a large root high in sugar. After a winter dormancy period, the second year of growth begins and the stems develop long internodes, using up the sugar stored in the roots, and flowers and seeds are produced. The need for the dormancy period is eliminated in plants that are treated with gibberellin. This way a seed crop, or even two seed crops if greenhouse culture is employed, can be produced each year to speed up plant breeding programs. When the treatment is discontinued, the new generation of plants is biennial and produces a good sugar crop instead of going to seed.

The contribution of Ascomycopsida to basic research in plant physiology, biochemistry, and genetics vies with that of the Schizomycopsida and Anthopsida. The nature and function of mitochondria, enzyme action, basic morphogenesis, monitoring of air pollution, effects of microclimate, and micronutrient assay are some of the areas in which the study of sac fungi is making important contributions to our understanding of basic concepts.

FUNGI IMPERFECTI

Sexual reproduction has never been observed in the 24,000 species of sac and club fungi called **Fungi Imperfecti,** and it has therefore been difficult to assign these species to correct family and order. In some cases, it has even been difficult to be certain of the class to which a particular fungus belongs. Most Fungi Imperfecti are ascomycetes and reproduce asexually by means of conidia just as other sac fungi do.

The Fungi Imperfecti are so named because the **perfect stage** is unknown; only **imperfect stages** have been observed. By imperfect stage is meant a stage in the life cycle in which asexual reproduction takes place. If either meiosis or fertilization is observed, the life cycle is said to be perfect and the species can be removed from the Fungi Imperfecti and assigned to its correct genus, family, and order in either the Ascomycopsida or Basidiomycopsida.

Although the spores and other gross morphological features of a fungus in an imperfect stage may be quite different from those in the perfect stage, we would expect to find many biochemical and cytological characteristics constant throughout all stages. Good evidence indicates that the staining properties of the nuclei and biochemical properties of the cell walls are highly constant (Pearson and von Hofsten, unpublished data) and could have value in assigning Fungi Imperfecti to their proper families and genera. Serology also shows promise in connecting the imperfect and perfect stages. Electron microscopes reveal structures not visible with the light microscope and as transmission electron microscopy (TEM) and scanning electron microscopy (SEM) studies of fungi gradually accumulate, our knowledge of relationships among the Fungi Imperfecti and between them and other Ascomycopsida, will grow. As more and more data are gathered, employing all of the criteria used in classification of plants (Table 1-2), the form class Fungi Imperfecti can theoretically be eliminated. There are two major obstacles to achieving this: (1) the extremely large number of species that must be studied biochemically and cytologically before they can be assigned to correct genera, families, and orders, and (2) the reluctance on the part of many mycologists to consider as taxonomic characteristics anything other than sporocarp ontogeny and structure and ascospore morphology. However, if and when all Fungi Imperfecti are classified in their natural taxonomic units, their Imperfecti names may still have value and will probably continue to be used.

Saccardo (1899) classified the Fungi Imperfecti into four **form orders** on the basis of conidia production. In the **Sphaeropsidales** (with 12,500 species), the conidia are produced in **pycnidia** (Figure 5-11A). In the **Melanconiales** (2100 species), they are produced in **acervuli** (Figure 5-11B). In the **Moniliales,** or hyphomycetes (9000 species), the conidiophores are produced externally on the mycelium (Figure 5-11C,D). In the **Mycelia Sterilia** (350 species), conidia have never been observed.

Based on careful studies of many representative species, it is believed that most of the Sphaeropsidales belong to the Helotiales. Others are Diaporthales, some are Uredinales (Basidiomycopsida), and a few belong to other orders. The Moniliales belong mostly to the Sphaeriales and Aspergillales, but some Zygomycopsida also produce conidia externally and are often included in this form order. The Melanconiales include members of the various pyrenomycetous orders, the Helotiales, and some Uredinales. Most, perhaps all, Mycelia Sterilia are Basidiomycopsida. A summary of the characteristics of the form orders of the Fungi Imperfecti is presented in Table 5-4.

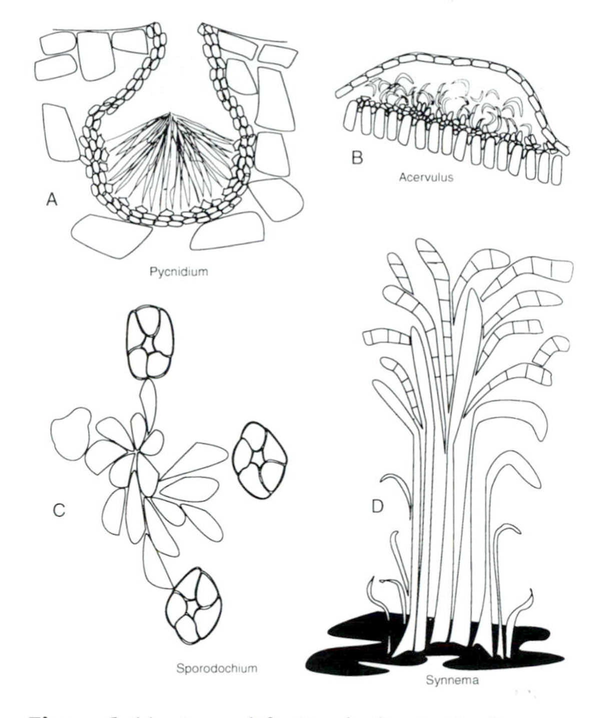

Figure 5–11 Asexual fruiting bodies in the Eumycophyta. (A) Pycnidium in *Septoria* (Sphaeropsidales); (B) acervulus in *Marssonina* (Melanconiales); (C) sporodochium in *Epicoccum* (Moniliales); (D) Synnema in *Arthrobotrium* (Moniliales). Pycnidia Superficially resemble perithecia but lack both asci and paraphyses; acervali sometimes resemble epothecia, but also lack asci and paraphyses.

General Morphological Characteristics

Following mitosis in Fungi Imperfecti, as in other Eumycophyta, new cell wall material is laid down from the periphery of the cell toward the center, leaving a primary pit through which rather large molecules and even organelles can migrate from one cell to another. The plasma membrane does not cover the pit connection; however, **Woronin bodies** are frequently observed adjacent to the pit connections; their presence is strong evidence that the fungus is an ascomycete. They are crystalline in nature and are believed to originate from microsomes; however, neither the exact chemical structure nor the function of these bodies is known. In older hyphae, they become exceptionally dense and appear to plug the pits connecting adjacent cells.

Closely related species of Ascomycopsida invariably have similar sporocarps and similar ascospores, but their asexual spores are often quite different. Within each taxonomic group, there is some correlation between the type of sexually produced spores and sporocarps and the type of conidia, but there are many exceptions. In the form genus *Gleosporium* in the Melanconiales, for example, some species are imperfect stages of *Gnomonia,* others of *Glomerella,* both in the Diaporthales, while another is the imperfect stage of *Pseudopeziza ribes* in

TABLE 5-4

CHARACTERISTICS OF THE FORM ORDERS OF FUNGI IMPERFECTI

Form Order and Included Taxa (Fam–Gen—Spp)	Spore-Producing Structures	Most Common Perfect Stage	Representative Genera or Species
Sphaeropsidales (4-568-12,500)	Pycnidia	Helotiales; also Diaporthales and Uredinales	*Phoma, Septoria, Pitidium, Aschochyta, Diplodia*
Melanconiales (1-92-2,100)	Acervuli	Sphaeriales and other pyrenomycetes; also Helotiales and Uredinales	*Colletotrichum, Gloeosporium, Myxosporium, Marsonia, Melanconium*
Moniliales (4-651-9,100[a])	Synnemata, sporodochia, or none	Sphaeriales and Aspergillales; also Zygomycopsida and some Basidiomycopsida	*Hainesia, Fusarium, Trichophyton, Verticillium, Aspergillus, Penicillium*
Mycelia Sterilia (1-20-425)	No asexual spores	Aphyllophorales	*Rhizoctonia, Typhula*

Note: The number of families (Fam), genera (Gen), and species (Spp.) in each order is indicated in the first column.
[a] Not including species of *Zygomycopsida* sometimes included here.

the Helotiales. Another member of the Helotiales, *Allophylaria oenotherae,* rarely produces ascospores but produces two kinds of conidia abundantly; at times this species is parasitic on leaves, stems, and roots of strawberries, reproducing asexually by means of externally produced conidia, and is known as *Hainesia lythri* in the form order Moniliales. At other times, it parasitizes the evening primrose and reproduces by conidia borne in pycnidia; in this stage it is known as *Pilidium concava* of the Sphaeropsidales.

It is not surprising that two closely related species often have quite different types of asexual spores. Asexual reproduction involves only one individual so there is no pressing need for uniformity in reproductive structure. In sexual reproduction, on the other hand, there must be a coupling of two individuals out of the same population; consequently, their sexual organs must be compatible. If a mutation giving rise to abnormal or incompatible sexual organs should occur it could not persist in that population because the plant or animal possessing it could not mate with other members of the species. The result is essentially perfect uniformity in the form and function of reproductive organs within each interbreeding population and similar reproductive organs in closely related species.

Sometimes there may be an ecological advantage to a species in having two or more kinds of conidia. Such is the case in *Allophylaria oenotherae.* One must keep in mind, however, that every adaptation has its genetic cost; therefore, most fungi have only one kind of conidia.

Parasexual Processes of Reproduction

Geneticists at the University of Glasgow discovered parasexual reproduction in the black jam mold, *Aspergillus nidulans,* in the early 1950s. A parasexual process is one in which assortment of genes into new combinations occurs without the aid of meiosis and syngamy. In Chapter 3, a parasexual process called meromixis, which occurs in the Schizophyta, was described (Figure 3-9). Since 1952, fungal parasexuality has been observed in many species of Ascomycopsida including Fungi Imperfecti. This phenomenon explains the presence of genetic homeostasis in the Fungi Imperfecti and their ability to evolve rapidly even in the absence of true sexual reproduction.

Fungal parasexuality occurs when the following events take place: fusion of haploid nuclei in dikaryotic cells followed by somatic crossing-over during mitosis and haploidization of the resultant diploid cells. The phenomenon is possible because microkaryophytes have a filamentous type of growth, and primary and secondary pit connections commonly occur (Figures 5-3, 5-12).

In Eumycophyta, filaments, or hyphae, developing from many spores, grow side by side in a tangled mycelium (Figure 5-12). Secondary pits (anastamoses) frequently form between cells on adjacent hyphae and a haploid nucleus from one cell occasionally migrates through the pit resulting in a dikaryotic cell. Nuclei may also migrate through primary pits between two cells of the same hypha. The two nuclei in a dikaryotic cell will sometimes fuse to become one diploid nucleus. If the two filaments originated from the same spore (Figure 5-12 E1) or if the diploid cell resulted from migration of a nucleus through a primary pit (Figure 5-12 F1), the resulting diploid nucleus will be homozygous (Figure 5-12 E2,F2); but if the two filaments originated from different meiospores, it will probably be heterozygous for many gene pairs (Figure 5-12 D2). Synapsis does not occur in mitosis, but where two homologous chromosomes come in contact with each other during prophase or metaphase of mitosis, and possibly even during interphase, one chromatid of each homolog may break, and crossing-over may occur at that point, resulting in new linkage groups (Figure 5-12 D3-7). Such **somatic crossing-over** has been observed in several species of Eumycophyta. Haploidization in some of the diploid nuclei results in dikaryotic and haploid daughter cells possessing new genetic combinations.

The chief advantage of sexual reproduction is to provide ecological or genetic **homeostasis,** which is the ability of a population to survive fluctuating environmental conditions. Sexual reproduction also speeds up the evolutionary process. However, asexual reproduction is more efficient than sexual in increasing numbers and extending the range of a species. Through their parasexual method of reproduction, many of the Eumycophyta are able to have the advantages of asexual reproduction while still maintaining good homeostasis.

Representative Species and Genera

Plant diseases caused by the Eumycophyta cost farmers (and consumers) billions of dollars each year. Most plant diseases are caused by Fungi Imperfecti, including the imperfect stages of ascomycetes and basidiomycetes. In order to combat these diseases, identification is necessary, and for efficient identification, names are needed. The creation of the admittedly artificial Form Class Fungi Imperfecti was guided by the need to identify and catalog fungi even though it was not yet possible to classify them in a natural way because their sexual life cycle was not known.

Ascomycetes and basidiomycetes that were described and named before the perfect stages were discovered have two or more names, at least one of which is its Fungus Imperfectus name. In addition, there are many Fungi Imperfecti for which no perfect stage has yet been discovered. Some of these have undoubtedly lost their ability to reproduce sexually. There are also many Ascomycopsida which produce sporocarps and ascospores only under very exacting environmental conditions. Such conditions may occur only once a year, or once in 5 years. For

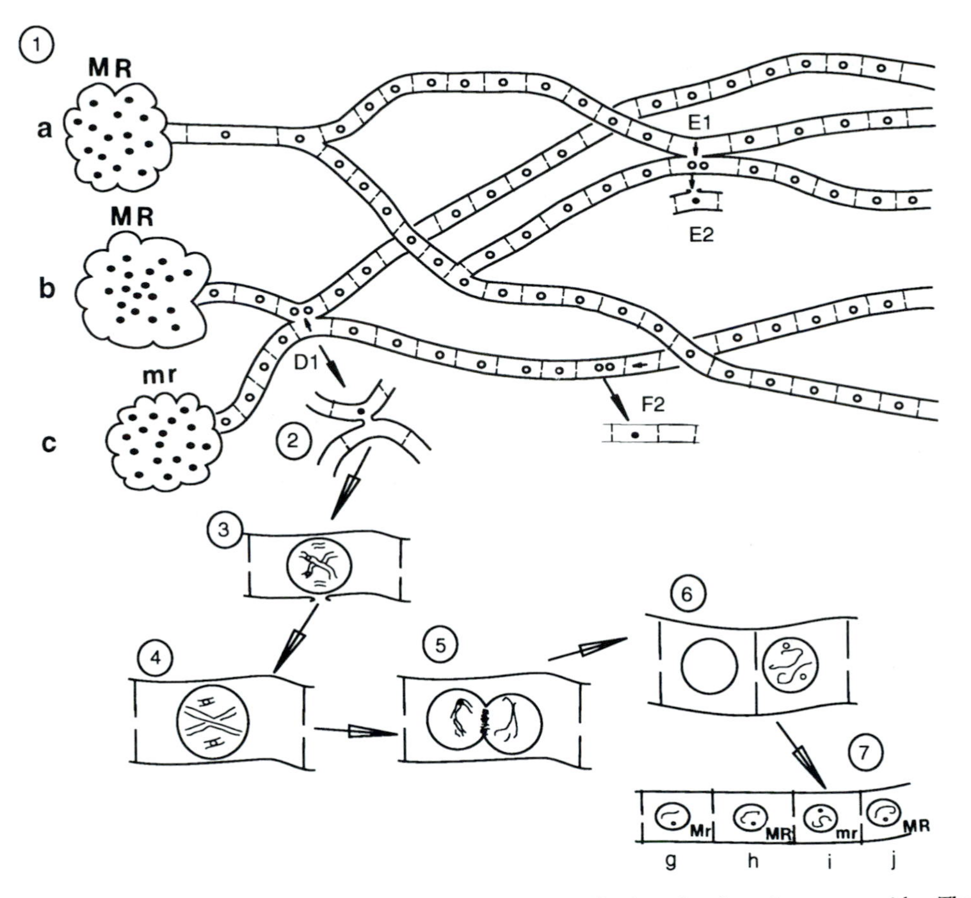

Figure 5–12 Parasexual genetic recombination in Fungi Imperfecti and other Ascomycopsida. The genes *M* and *R* are linked in spores *a* and *b* and their alleles, *m* and *r*, are on the homologous chromosome in spore *c*. (Migration through secondary pits at D and E or primary pit at F is indicated by arrows; open circles = haploid nuclei, solid circles = diploid nuclei.) 1. Migration of a haploid nucleus through a secondary pit at *D1* results in a dikaryotic cell with nuclei from different sources. At E1 and F1, the two nuclei in each dikaryotic cell are identical. 2. Diploidization results in a heterozygous nucleus at D2 and homozygous nuclei at E2 and F2. 3–4. Crossing-over occurs in D3. 5–6. Mitosis results in new linkage combinations. 7. The final step is haploidization, which may occur immediately without meiosis or can be delayed until a normal meiosis occurs. Daughter cells *g* and *j* have new gene combinations not present in parental spores *b* and *c*.

practical purposes, a name is still needed, and the form species name is used. The plant pathologist cannot afford to wait for the perfect stage to appear before beginning his control of the disease.

The largest of the form orders, in number of species, is the Sphaeropsidales with 12,500 species. Conidia are borne in flask-shaped structures called pycnidia (Figure 5-11A). The Sphaeropsidales are traditionally divided into four form families—the Sphaeropsidaceae, the Leptostromataceae, the Zythiaceae, and the Excipulaceae—which differ from each other in color and firmness of the pycnidia. The pycnidia of Sphaeropsidaceae resemble typical perithecia, with an ostiole or opening at the top, whereas in the Excipulaceae, the pycnidia open up to form a cup-shaped structure, somewhat resembling an apothecium. Some representatives of the Sphaeropsidales are listed in Table 5-3.

Second among the form orders in number of species is the Moniliales, also known as the Hyphomyceteae, with 9000 species. The conidia are borne on the surface of vegetative hyphae, at the apex of a conidiophore, or in some manner other than in organized structures like pycnidia and acervuli. There are four form families: Moniliaceae, Dematiaceae, Stillbellaceae, and Tuberculariaceae. Representative species include several incitants of plant diseases. Also in this form order are *Penicillium notatum,* a common laboratory weed and source of the antibiotic, penicillin; and *Aspergillus nidulans,* the common black jam mold which is not only a nuisance as a laboratory weed but is a valuable research organism for geneticists and the chief commercial source of citric acid used in the production of soft drinks and numerous industrial products. *Cercospora apii* causes a destructive leafspot in celery and is believed to be the cause of ugly, disfiguring facial lesions in humans (Figure 5-13); other Moniliales known to cause human diseases are species of *Trichophyton, Microsporon, Epidermophyton,* and *Blastomyces.*

The 2500 species of fungi in which the conidia are borne in acervuli are included in the form order Melanconiales. Three genera, *Colletotrichum, Gloeosporium,* and *Myxosporium* contain some of the most destructive plant parasites that exist. There is only one form family. Representative species include *Gloeosporium ribis,* the incitant of anthracnose of currants, a very widespread disease which is often locally very damaging; *Colletotrichum gloeosporiodes,* a ubiquitous species which causes anthracnose in hundreds of field and horticultural crops; *Melanconium fuligineum,* causal agent of bitter rot of grapes; *Marssonina populi,* incitant of leaf and twig blight of poplar; and *M. rosae,* which causes the very destructive and widespread black spot disease in roses.

The taxonomy of Fungi Imperfecti tends to be rather arbitrary; therefore, most mycologists nowadays combine some of the form genera where distinctions seem especially artificial, such as *Gloeosporium, Myxosporium,* and *Colletotrichum.* When not combined, *Colletotrichum* can be recognized by the stiff bristles around the acervulus. If there are no bristles, it is one of the other two: *Gloeosporium* if it grows on herbaceous plants and *Myxosporium* if it parasitizes woody plants. When conidia from the same source are inoculated on the mango, the artificiality of these distinctions are readily apparent: *Colletotrichum* grows on the twigs and leaves and *Myxosporium* grows on the fruits. There are over 1000 species of *Colletotrichum* described in the mycological literature, but based on cultural studies, von Arx (1957) recognized only 20; he concluded all of the others were synonyms. There are about 600 synonyms for *C. gloeosporiodes* alone, typically at least one for each host species it parasitizes.

The form order Mycelia Sterilia, also known as the Agonomycetales, consists of only 350 species. Included are *Rhizoctonia solani,* the imperfect stage of a basidiomycete, *Thanatephorus cucumeris,* which causes black scurf in potatoes and damping off disease in several crops. *Typhula idahoensis,* also a basidiomycete, causes snow mold in winter wheat and other cereal crops and grasses, and *Sclerotium cepivorum* causes white rot of onions and garlic. The perfect stage of *Sclerotium* is not known, but it is probably a basidiomycete.

Ecological and Economic Importance

Most of the plant diseases that cost American farmers billions of dollars every year are Fungi Imperfecti or asexually propagated Ascomycopsida (Table 5-3). A few Fungi Imperfecti cause

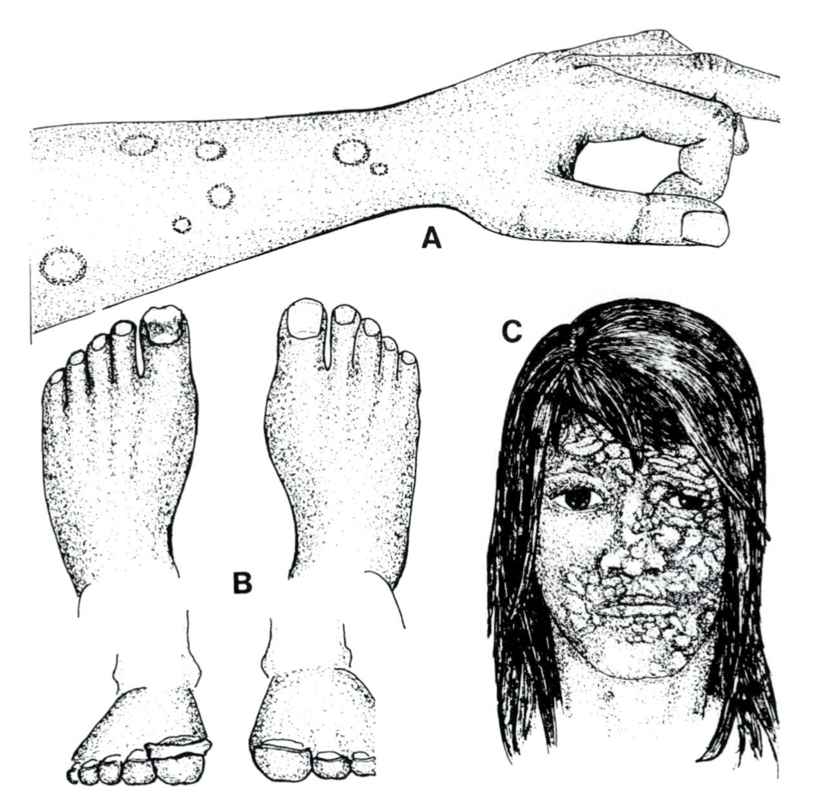

Figure 5–13 Several species of *Trichophyton, Microsporon, Epidermophyton, Blastomyces,* and other Fungi Imperfecti cause skin diseases in humans. (A) Ring worm; (B) a tropical fungus infection which causes malformation and thickening of the toe nails; (C) Lesions on the face of a child believed to be caused by *Blastomyces brasiliensis;* however, *Cercospora apii,* the incitant of a leafspot disease in celery, is suspected of causing similar lesions.

diseases in humans and others are "weeds" in biological laboratories. Geneticists, working with embryos or tissue cultures in the laboratory, must exercise special care as the plant or animal tissue is transferred to the sterile medium, or a spore from one of these weeds may enter the test tube or petri dish and soon destroy the organisms he is growing there.

Fungi Imperfecti in the form genus *Verticillium* are especially costly to farmers. Verticillium wilt often causes serious losses in several crops, especially in potatoes, where it is also known as "early die", and in cotton. Figure 5-14 is a transmission electron micrograph of a Norgold potato plant damaged by *V. albo-atrum.* The plasma membrane has been pulled away from the potato cell wall and is completely ruptured in places as a result of wilting stress and possibly toxins given off by the parasite.

CLASS 2. ZYGOMYCOPSIDA

The Zygomycopsida (*zygo* = a yoke, and *myceto* = mold or fungus) resemble the Aspergillales in many respects, both morphologically and physiologically, but have coenocytic hyphae. Traditionally allied with the Mastigomycopsida in a group called phycomycetes; they differ from

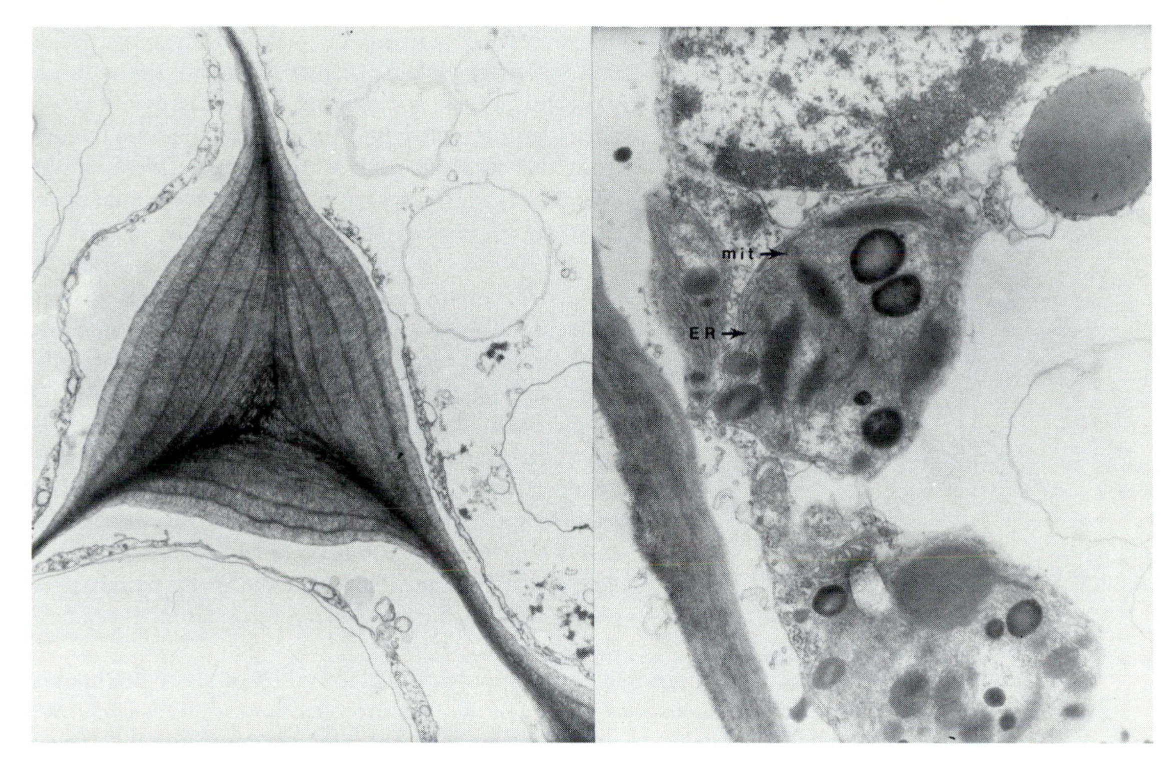

Figure 5–14 Wilted Norgold potato infected with *Verticillium* wilt fungus. Note how plasmalemma has drawn away from cell wall. ER = endoplasmic reticulum; mit = mitochondrion. (Courtesy W. M. Hess, Brigham Young University Microscopy Laboratory, Provo, Utah.)

them in having a more complex structure, never producing flagellated cells of any type, and favoring drier habitats, as well as in many biochemical characteristics.

The Zygomycopsida are among the most abundant organisms encountered when isolating microorganisms from soil, air, dung, or decaying plant material. In culture, they resemble many of the Ascomycopsida, whereas microscopically they resemble Mastigomycopsida. They may be cultured by placing a small amount of soil, partially rotted manure, or compost in a petri dish on corn meal agar and incubating at 25° to 35°C for 3 to 4 days. They are readily recognized by their cottony mycelium made up of nonseptate hyphae and their abundant asexual spores which resemble conidia but are borne within **sporangia.** One of the most common of all the Zygomycopsida is *Rhizopus stolonifer* (also known as *R. nigricans*), the black mold common on stale bread (Figure 5-15). Table 5-5 summarizes the chief characteristics of the two orders of Zygomycopsida, and Table 5-1 compares them with the other three classes of fungi.

The two classes students most frequently confuse with Zygomycopsida are Ascomycopsida and Mastigomycopsida. With the aid of a stereoscopic microscope or a good hand lens, sporangia, sporangiophores, conidiophores, and occasionally even spores can be seen. Stolons and rhizoids, typical of some Zygomycopsida, are often apparent. Although conidia and sporangiospores are essentially identical in appearance, how they are borne helps to distinguish

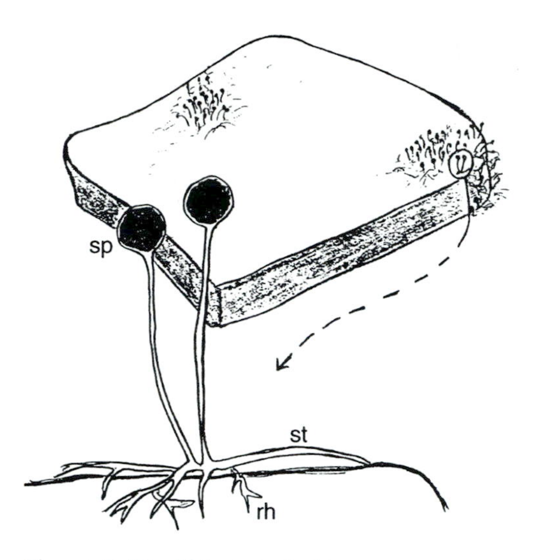

Figure 5–15 Rhizopus nigricans: sp = sporangium; st = stolon; rh = rhizoid.

between them and hence between Zygomycopsida and Ascomycopsida. Sporangiospores are borne inside a usually globose sporangium; conidia start out as ordinary hyphal cells which are then cut off, one by one, from the terminal end of the hypha. Most of the common zygomycetes have sporangia containing sporangiospores; most sac fungi have conidia.

With a compound microscope, the nonseptate hyphae of Zygomycopsida are easily distinguished from the septate hyphae of Ascomycopsida, providing one has good light and focuses the microscope carefully. Care must be taken in preparing material for microscopic examination because the hyphae of most Zygomycopsida and many other fungi are delicate and fragile. A good method is to take a piece of "double-sticky" tape, attach it to a glass cover slip, and place it gently over the mold. A drop of lactophenol can then be placed on the glass slide, the cover slip with hyphae and sporangia attached laid over this, and the mold examined microscopically.

Mastigomycopsida are coenocytic, like the Zygomycopsida, but they produce neither conidia nor the conidia-like sporangiospores typical of the zygomycetes; instead, they often produce **zoospores** (motile, flagellated asexual spores). They also tend to favor aquatic habitats, whereas Zygomycopsida are usually terrestrial.

TABLE 5–5

CHARACTERISTICS OF THE ORDERS OF ZYGOMYCOPSIDA

Orders and Included Taxa (Fam–Gen–Spp)	Reproductive Structures	Other Characteristics	Representative Genera or Species
Mucorales (8-25-400)	Thick-walled zygotes or coenozygotes produced when cross-compatible strains contact each other; life cycles haplontic	Mostly saprophytes, often on herbivore dung, also on stored vegetables or grains; sporangiospores produced abundantly	*Rhizopus, Mucor blakesleanna, Phycomyces, Blakeslea, Mortierella, Pilobolus, Kickxella, Zygorhynchus*
Enteromophthorales (2-70-80)	Thick-walled zygospores or parthenogenetic azygospores	Mostly insect parasites; also nematode-trapping soil fungi and fern parasites	*Enteromophthora muscae, Zoopage, Harpella, Basidiobolus, Conidiobolus, Amoebidium*

Note: The number of families (Fam), genera (Gen), and species (Spp.) in each order is indicated in the first column.

General Morphological and Physiological Characteristics

The plant body of the Zygomycopsida is a **coenocytic mycelium** or mass of filaments without organization into either prosenchyma or pseudoparenchyma tissue. By coenocytic is meant a cell-like structure with many nuclei not set apart by septations or crosswalls. However, crosswalls do commonly occur in older filaments and at the base of **sporangiophores;** they also often originate in places where there has been mechanical injury to the mycelium.

Hyphae are often rather long, delicate, and much branched; however, many species produce stout hyphae. **Rhizoids** may anchor the horizontal hyphae, or **stolons,** to the substrate. Conidia are abundantly produced, usually at the apex of a vertical **sporangiophore** which typically arises from a stolon opposite the rhizoids (Figure 5-15). Septations often occur in the sporangiophores. In species parasitic on insects, the hyphae are delicate, shorter, and less branched, and the conidiophores emerge where the host integument is thin.

Electron microscopy has revealed both similarities to and differences from the Ascomycopsida. Where septations occur, primary pits and septal pore plugs similar to those of Ascomycopsida are present; however, Woronin bodies have not been observed in any of the Zygomycopsida. It has been observed that mitochondria in the two classes are very similar when the cells are fixed with $KMnO_4$, but when fixed with OsO_4, the Zygomycopsida mitochondria appear tubular as in the Mastigomycopsida. Cisternal rings have been reported in a few species and are said to be similar to those of the Ascomycopsida; dictyosomes like those in Mastigomycopsida do not occur in the Zygomycopsida or other Eumycophyta.

The Zygomycopsida, like the Aspergillales, are able to produce a variety of organic compounds in large quantity. These include fats, enzymes, solvents, and organic acids. Both the Aspergillales and Zygomycopsida, therefore, are useful in the biochemical industry. Of special significance is the ability of some Zygomycopsida to convert cellulose and other complex carbohydrates into simple sugars which can then be used to produce other industrial compounds such as commercial alcohol.

Some Zygomycopsida resemble the Moniliales in their mode of asexual reproduction, having asexual spores produced externally on hyphae. Neither pycnidia nor acervuli are found in the Zygomycopsida. Sporangiospores, or in some species conidia, are produced in abundance. Sporangiophores are often quite long with globose sporangia at the apex (Figure 5-15). In some species, including many of the insect parasites, the mycelium is contained entirely within the substrate with only conidiophores extending above it (Figure 5-15); in other species, the mycelium is an extensive, white, cottony mass with black sporangia arising here and there.

Sexual reproduction occurs when two compatible hyphae send out protuberances, called **zygophores,** which join or conjugate to form multinucleate **coenozygotes.** The tips of the zygophores are called **progametangia** and their nuclei function as isogametes, or egg and sperm in species in which one nucleus (the male) migrates to the other (female) progametangium. Within the coenozygotes the nuclei fuse in pairs. Both homothallic and heterothallic species are known in the Zygomycopsida.

In heterothallic species, the progametangia must be of different mating types. It has been customary to refer to the mycelia of heterothallic species as "+" and "−" strains, or worse, as different "sexes", or male and female strains. However, genetic studies suggest that there may be several mating types, probably controlled by multiple alleles as in Anthopsida and Ascomycopsida (Figure 5-7). The hormone necessary for sexual union, or syngamy, to take place is **trisporic acid,** an organic acid which apparently exists in several slightly different forms. Each mating type gene is responsible for a precursor of trisporic acid. If conidia of the same mating type are germinated in a petri dish or other container, the resulting mycelia will all produce the same precursor and no coenozygotes will form, but if conidia of different mating types are germinated in a petri dish, the precursors will combine chemically to form trisporic acid and abundant coenozygotes will be produced (Figure 5-16). When the coenozygotes germinate, half of the daughter cultures will be of one mating type and half of the other type.

It is believed that in most Zygomycopsida, meiosis takes place within a few days after syngamy, each diploid nucleus giving rise to four haploid nuclei, all but one of which disintegrate. A thick, dark wall, often highly ornamented with little bumps or short spines, forms around the coenozygote transforming it into a **zygosporangium** which is capable of lying dormant for a long time under cold, dry conditions. When the single zygospore in each zygosporangium germinates, it usually sends out a short **sporangiophore** with a **sporangium** at its apex. **Sporangiospores** escape from the sporangium and initiate new colonies of the mold.

Phylogeny and Classification

The Zygomycopsida consist of two orders, the Mucorales with seven families and the Entomophthorales with two. The latter is believed to have evolved from the former; however, the fossil record is very scanty and without direct fossil evidence to aid us in solving the mystery of their origin, we can only compare morphological, physiological, and biochemical characteristics of these fungi with other organisms and try to infer by this means what the pattern of evolution has been. Taking all available data into account, it now appears that the first Zygomycopsida were probably similar to modern Mucoraceae; from the Mucoraceae have probably descended the other families of the Mucorales and the insect parasites of the Entomophthoraceae. The nematode-trapping and other specialized fungi of the Zoopogaceae are probably the most advanced members of the class.

This hypothesis is neither conclusive nor widely accepted. Some botanists believe that the Zygomycopsida are ancestral to the Aspergillales and other Ascomycopsida; others believe the two classes are entirely unrelated. Some look to the Mastigomycopsida for the origin of the Zygomycopsida, others to some of the green algae, especially ***Cladophora*** with its multinucleate cells or ***Spirogyra*** because of its mode of zygote formation. Referring in part to the lack of a fossil record, Hesseltine and Ellis (1973) commented, "We believe there is little or no evidence now available to support even a tentative conclusion on the mysterious origin of the Zygomycopsida." Nevertheless, a beginning point from which to conduct research is needed before any conclusions can be reached. Denison and Carroll (1966) provided the working hypothesis followed in this chapter. Characteristics of the two orders are outlined in Table 5-5.

Representative Species and Genera

Typical of the Mucorales, family Mucoraceae, is the common black bread mold, *Rhizopus stolonifer* (Figure 5-15). It is a ubiquitous saprophyte and facultative parasite that can be obtained in culture almost any place in the world simply by leaving a petri dish containing nutrient agar open for half an hour, or by allowing bread to become old. It is readily identified by its stout stolons and branching rhizoids with the conidiophore arising opposite the rhizoids. Because of its abundance, it is one of our worst laboratory "weeds," along with *Penicillium notatum* and *Aspergillus nidulans;* but it is also a useful organism in industry because of the enzymes it produces and the organic compounds it can make. At times, it becomes a serious nuisance

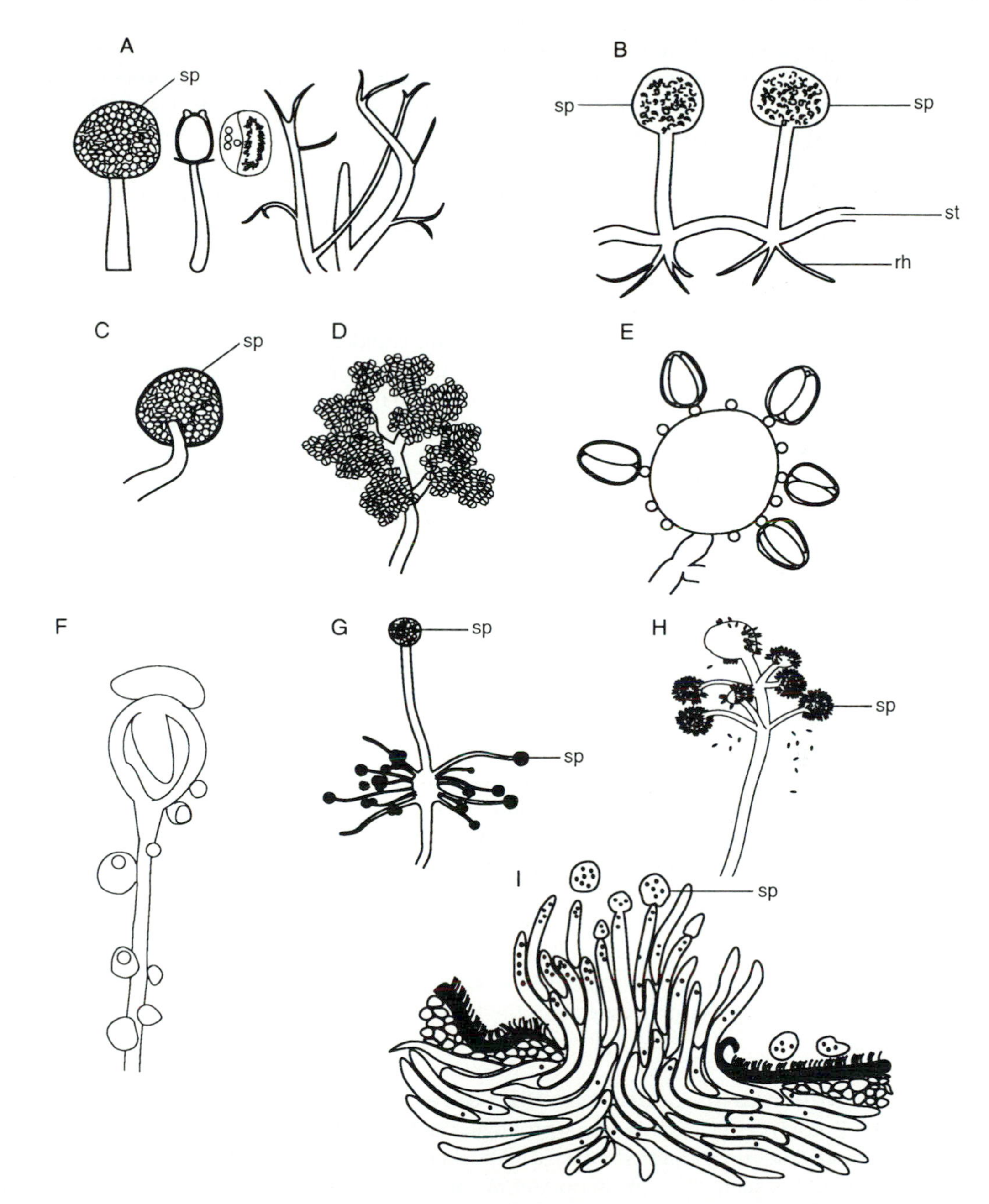

Figure 5–16 Representative Zygomycopsida species. "Double sticky tape" method was used for microscopic observation of these fungi; see text for description of method: (A) *Mucor spinosa:* mature sporangium on the left, columella after the sporangiospores have been shed in the middle, and immature mycelium on the right; (B) *Rhizopus stolonifera:* (C–H) diversity of sporangia in the Zygomycopsida; (I) *Entomophthora muscae* mycelium and conidiophores emerging from trachea of a housefly. (sp = sporangium, st = stolon, rh = rhizoid.)

in storage cellars where vegetables, especially sweet potatoes, are kept. It causes a rot that renders the roots completely useless in a very short period of time. It reproduces prolifically by means of conidia; it is also capable of sexual reproduction when cultures of different mating type are grown together in a common petri dish (Figure 5-15). The life cycle is believed to be haplontic.

Phycomyces blakesleanus, Mucor mucedo, and *M. hiemalis* are other common Mucoraceae found in soil that has been heavily fertilized with horse or cow manure. Several *Mucor* species have an antagonistic effect on other fungi. Also in the Mucorales, family Pilobolaceae, are *Pilobolus crystallinus, P. kleinii,* and several species of *Pilaira,* all commonly present in horse dung. They and other species of Pilobolaceae have phototropic conidiophores. *Pilobolus* conidiophores are relatively short; there are many conidia in each sporangium which is attached to a **subsporangial vesicle** full of a sticky, mucilaginous material. The sporangia are forcibly discharged toward the light from this vesicle. The mucilage enables the sporangia to stick to any object they strike. Some species of *Pilobolus* are reported to shoot their sporangia vertically to a height of about 2 meters (Alexopoulos and Mims 1979). *Pilaira,* on the other hand, has very long slender conidiophores; the mucilage-coated sporangia stick to any object with which they come in contact, such as a blade of grass, a leaf, or a passing herbivore. The sporangia are ingested by herbivores and pass unharmed through the digestive tract.

Typical of the Entomophthorales is a parasite of the housefly, *Entomophthora muscae* (Figure 5-16). *E. muscae* is a major cause of death of house flies in the autumn. Affected flies become very sluggish and cling to mirrors and window panes or other objects, and die there. The glass around the dying flies is visibly whitened by a halo of conidia. Postmortem examination shows the internal organs of the flies, especially the intestines, completely dissolved, apparently by action of fungal enzymes. A related species, *E. gryllii,* is common on grasshoppers on the American Great Plains. In wet weather, rings of conidiophores can be seen emerging from between the segments of the insects' abdomens. Shortly after this, the animals crawl up onto blades of grass or other plants and die. In dry weather, the signs of the disease are less obvious even though it is present. In both of these species, as in other Entomophthorales, conidia gain entrance into the insects' bodies through the trachea along their abdomens as they breathe.

Zoopage and *Gonimochaete,* both in the Zoopogaceae, trap and destroy amoebae, nematodes and other soil-inhabiting microanimals. When the conidia of *Gonimochaete horridula* germinate, they produce a small sticky knob that attaches to any nematode that happens to be passing by. A hypha penetrates into the body of the nematode where it develops into thin-walled thalli.

Ecological and Economic Significance

The Zygomycopsida are mostly saprophytes, obtaining their energy from dead plant material, including the partly digested material in the dung of herbivores and the processed plant material in bread. As such, they are important decomposer organisms. Some are weakly parasitic on stored fruits and vegetables. Two genera produce endotrophic mycorrhizae with vascular plants. About 40 species are parasites in the bodies of insects and seem to be a limiting factor in their spread.

Compared to the Ascomycopsida, the Zygomycopsida are not important plant pathogens. Nevertheless, some economic loss is occasioned by these fungi, especially by *Rhizopus stolonifer* in sweet potatoes during storage. To prevent the disease, the potatoes should be allowed to mature fully before harvesting and should be stored with adequate ventilation.

R. stolonifer is also important in the production of industrial alcohol. Before yeasts can convert complex carbohydrates, like cellulose and starch, to alcohol, they must be broken down into molecules of simple sugars. *R. stolonifer* is widely used in this saccharification process.

Potential in Scientific Research

Because of their commonness, the ease with which they are cultured, and their rapid growth rates, many of the Zygomycopsida are potentially excellent research organisms. Although the plant body is coenocytic, each nucleus behaves as though it were part of an individual cell, giving these fungi an advantage for future genetic studies. Ecological studies involving the Mucorales have the potential to improve our knowledge of decomposition processes. Probably no area of ecological research has been neglected as much as ecological niche and habitat requirements of decomposer organisms.

CLASS 3. BASIDIOMYCOPSIDA

The Basidiomycopsida (*basidi* = a small pedestal, *myceto* = mold or fungus), commonly called the club fungi, are strictly terrestrial fungi living almost exclusively on plant material, and are mostly saprophytes and facultative parasites (saprophytes which become parasitic under certain conditions), although a number of species are very destructive obligate parasites on various farm crops and other plants. The class is divided into two subclasses, the Heterobasidae with two orders and the Homobasidae with eleven. Included in the former are the rusts and smuts which are parasites on many angiosperms and other vascular plants; a variety of "mushrooms," "conks," and other fungi make up the Homobasidae. Approximately 25,000 species of basidiomycetes are known; they are distributed among 13 orders, 62 families, and 465 genera.

The most distinctive feature of the Homobasidae is the spore fruit or basidiocarp (Figure 5-17). Often it is fleshy and more or less umbrella-shaped, with spore-bearing tissue on its undersurface. In a majority of species, the spores are produced on plates or *gills* (Figure 5-1D–F), but in others they are produced in pores or on tooth-like projections on the lower surface of the sporocarp or in the interior of a globular fruiting body (Figure 5-17). In the Heterobasidae, basidiocarps do not occur and the spores are produced in small groups called *sori* just under the epidermal tissue of the host plant. Like the sac fungi and zygomycetes, many Heterobasidae produce conidia, but asexual reproduction by conidia is rare in the Homobasidae.

Basidiomycopsida are most commonly found in forested areas from early summer to late fall. They are also common in lawns and pastures. The mycelium can be found year round with the aid of a good hand lens, but the fruiting bodies—commonly called mushrooms—are seasonal, depending on the species, and appear most abundantly 2 or 3 days after a rain. Species of smut may be found by examining the ripening heads of most wild grasses. They appear as black masses of spores which replace the kernels. Some smuts, like corn smut, appear as white blisters on the stems or fruits of grasses which enlarge to massive proportions before bursting open and releasing the black spores. Rusts appear as smaller blisters which produce reddish masses of spores as they open up.

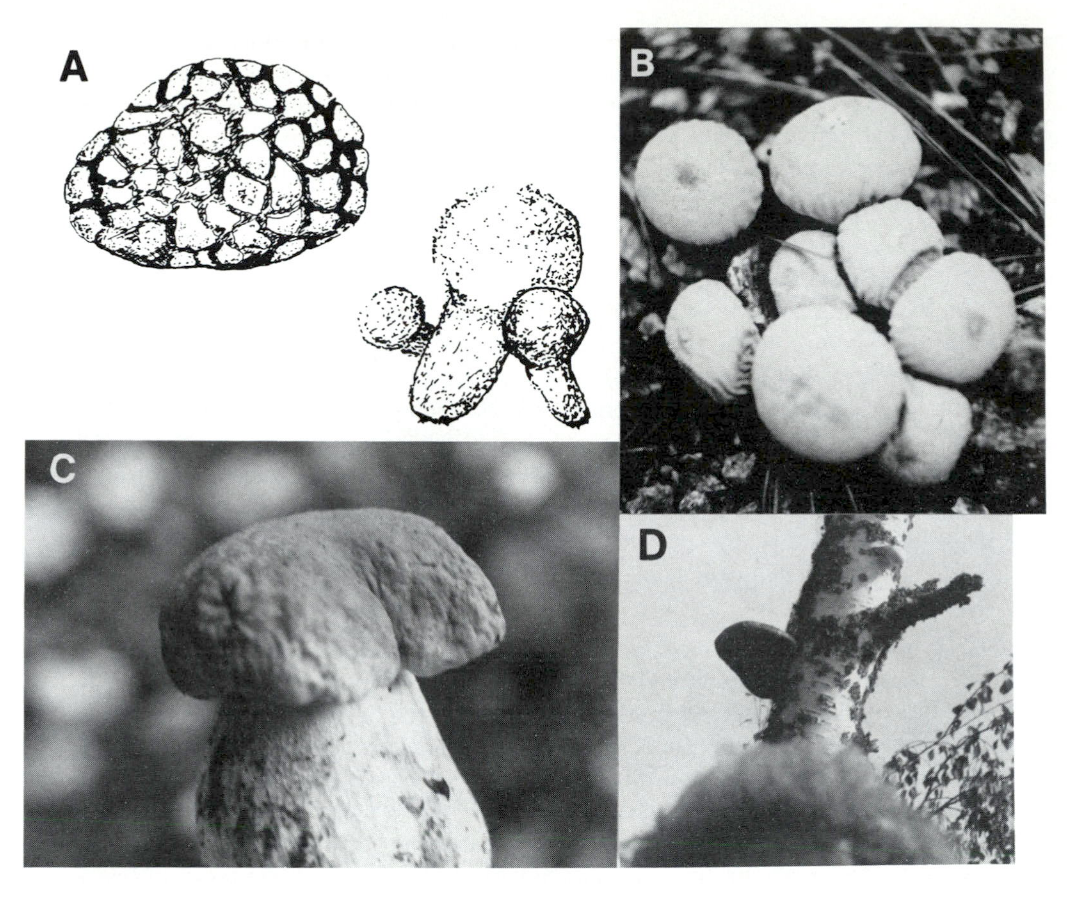

Figure 5–17 Representative Basidiomycopsida: (A and B) puffballs; (C) bolete; (D) "conk" or polypore shelf fungus in a Swedish birch. (From Pearson, L. C., *The Mushroom Manual,* Naturegraph, Happy Camp, CA, 1987. With permission.)

Students most frequently confuse Basidiomycopsida with Ascomycopsida; this is true both for the mycelium and for the sporocarps. The basidiomycete mycelium is mostly dikaryotic and clamp connections are frequently present; ascomycete mycelia are mostly haploid and never have clamp connections; where dikaryotic, there will often be croziers. The presence of either type of connection between cells is an excellent guide to which class one is examining, but the lack of both clamp connections and croziers is not a sure sign that the specimen is a sac fungus. The presence of septations in the hypha indicates that the specimen is either a sac fungus or a club fungus; zygomycetes and mastigomycetes are coenocytic. If sporocarps are present, identification is easier. In the Basidiomycopsida, there are usually four unicellular spores produced on a club-like terminal cell and attached to it by means of a short stalk or pedestal. In the Ascomycopsida, there are usually eight spores, often multicellular, produced within an ascus or sac. Most of the time, ascomycete spores are on the upper surface, basidiomycete spores on the lower surface of the sporocarps.

General Morphological and Physiological Characteristics

The mycelium in Basidiomycopsida is very similar in general appearance to that of Ascomycopsida: it is usually white, it can be found by lifting the duff or partly decomposed litter from the forest floor and examining it with a hand lens, and it is septate. It differs in being primarily dikaryotic, whereas ascomycete mycelium is mostly haploid (Figure 5-2). There are also numerous biochemical differences.

At times, parallel hyphae join together to form thick strands of mycelium in which cortical and medullary tissues may be differentiated. These are called **rhizomorphs,** suggesting that they resemble roots; they are also called "shoe strings" in some places. In mycorrhizal species, it may be difficult to tell where the root or mycorrhiza ends and the rhizomorph begins.

Like the Rhodophycopsida and Ascomycopsida, the Basidiomycopsida are characterized by primary pit connections or **septal pores** connecting adjacent cells in each hypha. Electron microscope studies have revealed a complex structure in the septal pores of many club fungi. The edge of the pore is thickened and is covered on either side by a membranous cap or **parenthesome.** The endoplasmic reticulum is continuous through the septal pore. This is in contrast to the simpler structure in Ascomycopsida and most Fungi Imperfecti, both characterized by Woronin bodies. The Woronin body is crystalline and apparently originates from microbodies, not ER.

Secondary pits in the form of **clamp connections** occur in the dikaryotic stage of many, but not all, Basidiomycopsida. A clamp connection is formed when the two haploid nuclei undergo simultaneous mitoses and one of the daughter cells of the more apical nucleus migrates into a protuberance from the side of the cell. The formation of the protuberance resembles the initiation of the secondary pit between the carpogonium and auxiliary cell in Rhodophycopsida, but in the Basidiomycopsida, it turns back on itself and forms a secondary pit connection with the basal part of the same cell (Figure 5-3).

In order for the mycelium to grow, there must be a good supply of organic matter, available nitrate and sulfate salts, other minerals, and water. Enzymes in the mycelium break the organic matter down to simple sugars; from these and the inorganic minerals, the plant builds new carbohydrates, proteins, and lipids. As the mycelium continues to grow outward in all directions from the point of origin, fruiting bodies are formed a short distance behind the growing point, providing temperature, moisture, day length, and other environmental conditions, are exactly right.

Fungus physiology is not greatly different from angiosperm physiology. Kitamoto and Gruen (1976), for example, found that stipe elongation in *Flammulina velutipes* (Agaricales: Tricholomataceae) depends on (1) a supply of water for good vegetative (mycelium) growth, (2) glucose and/or other low molecular weight carbohydrates, also supplied by the vegetative mycelium, and (3) an unidentified diffusate, analagous to the plant hormones needed by vascular plants, produced by the **lamellae** or gills. The diffusate ("D") is obtained for laboratory research by placing maturing caps, gills down, on plates of agar. Similar physiological studies of fungi are being conducted at many colleges and universities around the world; among the foremost are Tokyo University in Japan and the Institution of Physiological Botany at Uppsala University in Sweden.

In many species, the basidiocarps erupt through the soil and duff in a ring. Year after year, slightly larger concentric rings erupt at about the same time of year. Commonly called "fairy rings," these are often almost perfectly round and can frequently be observed over a period of

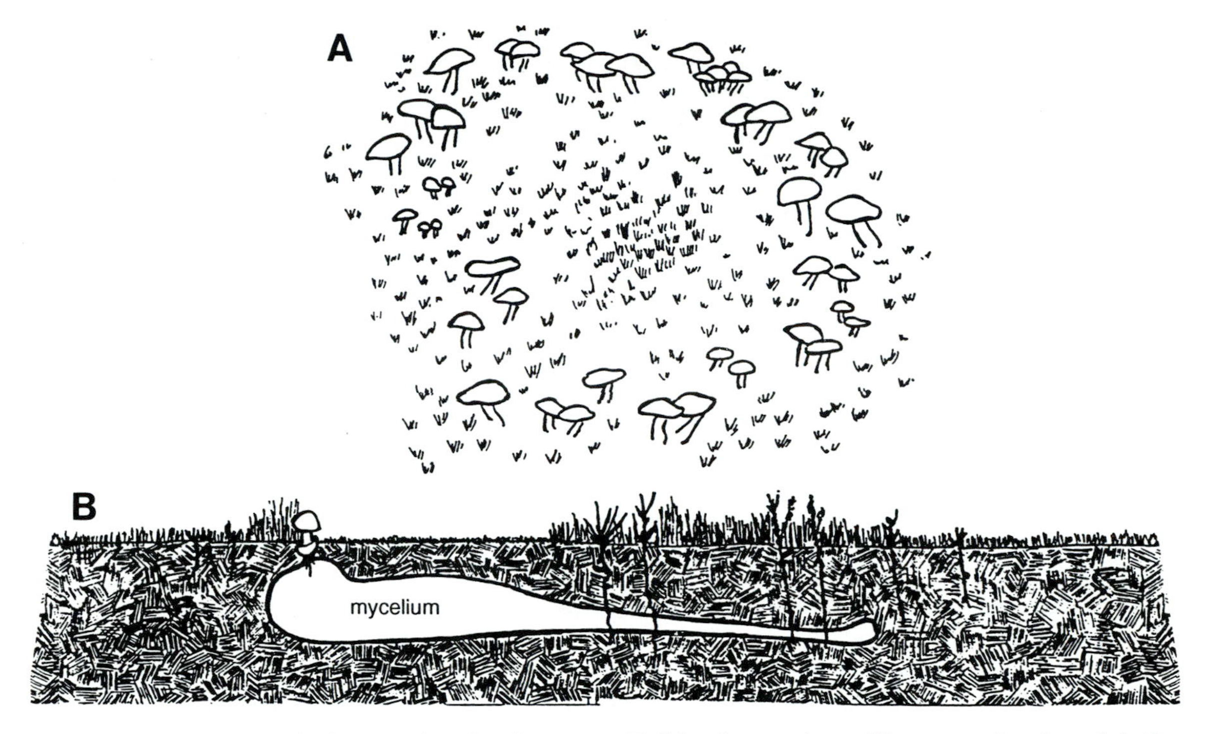

Figure 5–18 Formation of a fairy ring by a basidiomycete. (A) Fairy ring on a lawn; (B) cross section through half a fairy ring. Note the two zones of stimulated growth just ahead of and well behind the ring of mushrooms. Where the mycelium is most concentrated, the grass is stunted and yellow due to competition with the mycelium for nitrates and ammonia. (From Pearson, L. C., *The Mushroom Manual,* Naturegraph, Happy Camp, CA, 1987. With permission.)

several decades in lawns or pastures (Figure 5-18). A ring of paler colored grass, caused by competition for nitrates between fungus and lawn, signals the location of the ring before the sporocarps erupt. It was formerly believed to have been caused by fairies dancing on the turf during the night.

In a survey of the species of club fungi which grow in turf conducted in New Jersey over a 5-year period, Halisky and Peterson (1970) found ten species of fairy ring producers and six species which did not produce rings (Table 5-6). Three kinds of rings were found: those which had a stimulating effect on the grass and other plants, those which had an inhibitory effect, and those which had no apparent effect.

Asexual reproduction in the fleshy fungi is commonly by means of hyphal fragments, often called *spawn.* In commercial production of mushrooms, the spawn is strewn on soil containing straw, horse or chicken manure, and commercial fertilizer. Temperature and moisture are carefully controlled. Before seeding with spawn, the soil is heat sterilized to eliminate any molds, nematodes, or bacteria that could otherwise cause a disease in the mushrooms, as well as the spawn of all undesirable species of mushrooms.

In some species of Basidiomycopsida, especially in the Heterobasidae, asexual spores, similar to the conidia of the Ascomycopsida, are produced. These are known by a variety of names—**uredospores, chlamydospores, conidia, oidia,** etc.—depending on origin and

TABLE 5-6

BASIDIOMYCOPSIDA IN LAWNS, GOLF COURSES, PASTURES, AND ROADSIDE

Dispersal Type[a]	Effect on Turf	Number of Species	Number of Collections
S	Scattered sporocarps; no noticeable effect on turf	6	94
R-1	Rings marked by zone of dead or dying grass	5	68
R-2	Rings marked by zone of stimulated grass	2	54
R-3	Rings producing no noticeable effect on turf	3	44

Note: Summary of a 5-year study of all turf basidiomycetes found in New Jersey. S = scattered sporocarps; R = sporocarps occurring in fairy rings.

[a] Classification of Shantz and Piemeisel 1917.

Data from Halisky and Peterson 1970.

structure, but they all have the same function: asexual reproduction. The oidia also function as male gametes.

Sexual reproduction involves three phases: (1) meiosis and production of basidiospores which germinate and give rise to a haploid, gametophyte mycelium, (2) fertilization or syngamy to form a dikaryotic mycelium, and (3) karyogamy or diploidization of the dikaryotic cells. Most of the vegetative mycelium in the Basidiomycopsida is dikaryotic (Figure 5-2).

Sporocarp production begins when environmental conditions stimulate dikaryotic, vegetative mycelium to form masses of differentiated tissues. The fertile tissue is called the **hymenium.** In it, development of dikaryotic basidia and karyogamy can take place. The first stage in basidiocarp formation is the button stage (Figure 5-19). At this time, the developing spore fruit is enclosed in two membranes, an outer **universal veil** and an inner or **partial veil.** As growth of the sporocarp continues, the universal veil ruptures; remnants of the veil may remain as white patches on the top of the cap, as a **cortina** hanging from the edges of the cap, or as a **volva** or "stocking" at the base of the stem or stipe of the sporocarp. The sporocarp continues to increase in size, and as it does, the partial veil ruptures; remnants of this veil may also remain as an **annulus,** or ring, or as a cortina. The presence or absence of remnants of the two veils are useful identification characteristics; some species have both an annulus and a volva, some have an annulus but no volva, a few have a volva but no annulus, and many have neither. Only a few species have a cortina, which is a spider web-like "curtain" reminiscent of the veils once worn by Spanish noblewomen in their church worship. Patches of white or yellow tissue on the upper surface of the cap are found in some of the *amanitas* like the poisonous fly agaric or "common toadstool" (Figure 5-20) and a few other mushrooms.

As the sporocarp is developing, karyogamy, or fusion of the two haploid nuclei in the dikaryotic basidial cells, takes place. Karyogamy is almost immediately followed by meiosis. Four protuberances form on the basidium as meiosis begins, and after the four nuclei have migrated into these, they become basidiospores. Basidiospores vary in shape, color, and ornamentation; some are borne on short stalks, others are sessile; some turn blue when treated with an iodine solution, others turn red, and others do not react at all with iodine; most are unicellular, but a few are septate. The basidia on which they are borne also vary greatly in size,

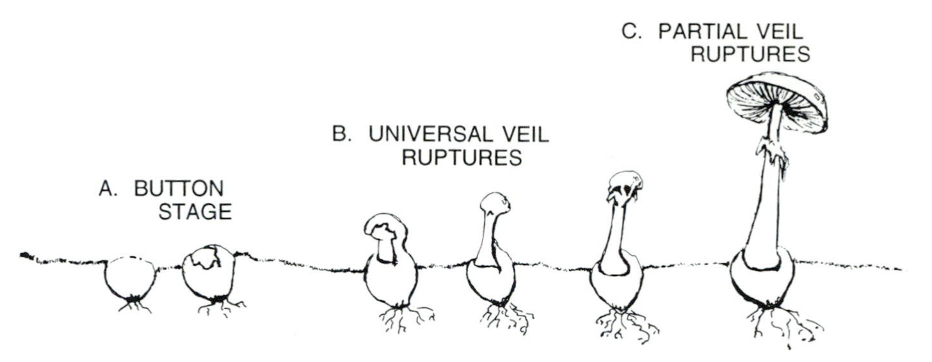

Figure 5–19 Development of a basidiocarp from button stage to mature mushroom in *Amanita phalloides.* Rupturing of the universal veil may leave a cup, the volva, at the base of the stipe or patches of tissue on the pileus or cap. Rupturing of the partial veil may leave a ring of tissue, the annulus, high on the stem or sometimes a little curtain or *cortina* hanging from the margin of the pileus. Most mycophagists (mushroom eaters) avoid species like this one that have both annulus and volva, white spore print, and free gills. (From Pearson, L. C., *The Mushroom Manual,* Naturegraph, Happy Camp, CA, 1987. With permission.)

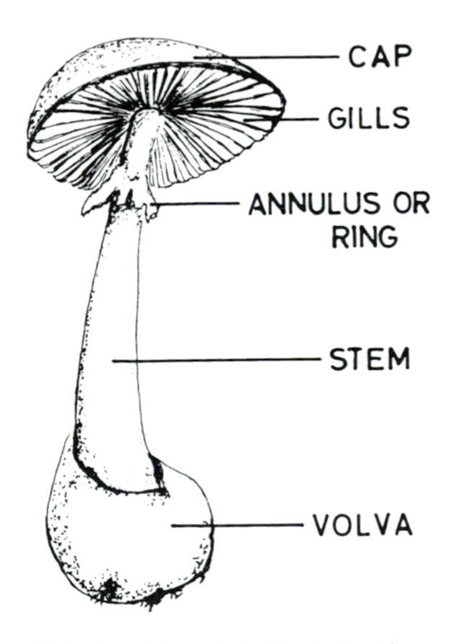

Figure 5–20 *Amanita verna,* the "destroying angel".

septation, and shape. There are generally four spores on each basidium, but a few species, such as *Amanita bisporigera,* are characterized by two spores per basidium, and *Coprinus trisporus* has mostly three-spored basidia. In many of the chanterelles, the basidia have six to eight spores each. In addition to the basidia, sterile cells called **cystidia** are present. In some ways they resemble paraphyses in the Ascomycopsida.

Basidia develop on a hymenium, just as asci do. The hymenium overlays a sterile **hymenophore** tissue; it is smooth in the simplest Homobasidae, but more frequently the development of shallow pits or slight projections provides greater surface area for spore production. In the most advanced members of the class, the hymenium and hymenophore form gills, pores, tooth-like projections, branched or unbranched coral-like structures, or a porous, fleshy gleba. The cells in this **hymenial trama** sometimes form pseudoparenchymatous tissues; at other times, the tissues are prosenchymatous.

Three types of prosenchyma are common in club fungi. **Monomictic** tissues consist of thin-walled, narrow, branched **generative hyphae** only; **dimictic** tissues consist of generative hyphae together with thick-walled, stout, unbranched, nonseptate **skeletal hyphae;** and **trimictic** tissues (Figure 5-21) consist of generative hyphae, skeletal hyphae, and much-branched, thick-walled **binding hyphae** connecting other hyphae. Clamp connections are found only in the generative hyphae.

The "typical" basidiocarp consists of a **pileus** (cap), with gills on the undersurface, and a **stipe** or stem (Figure 5-20). The pileus is usually centrally located on the stipe, but eccentric pilei also occur. The gills may be **free** so that the pileus separates readily from the stipe, or the gills may be **decurrent,** holding the two firmly together. Gills may also be **adnexed** (notched)

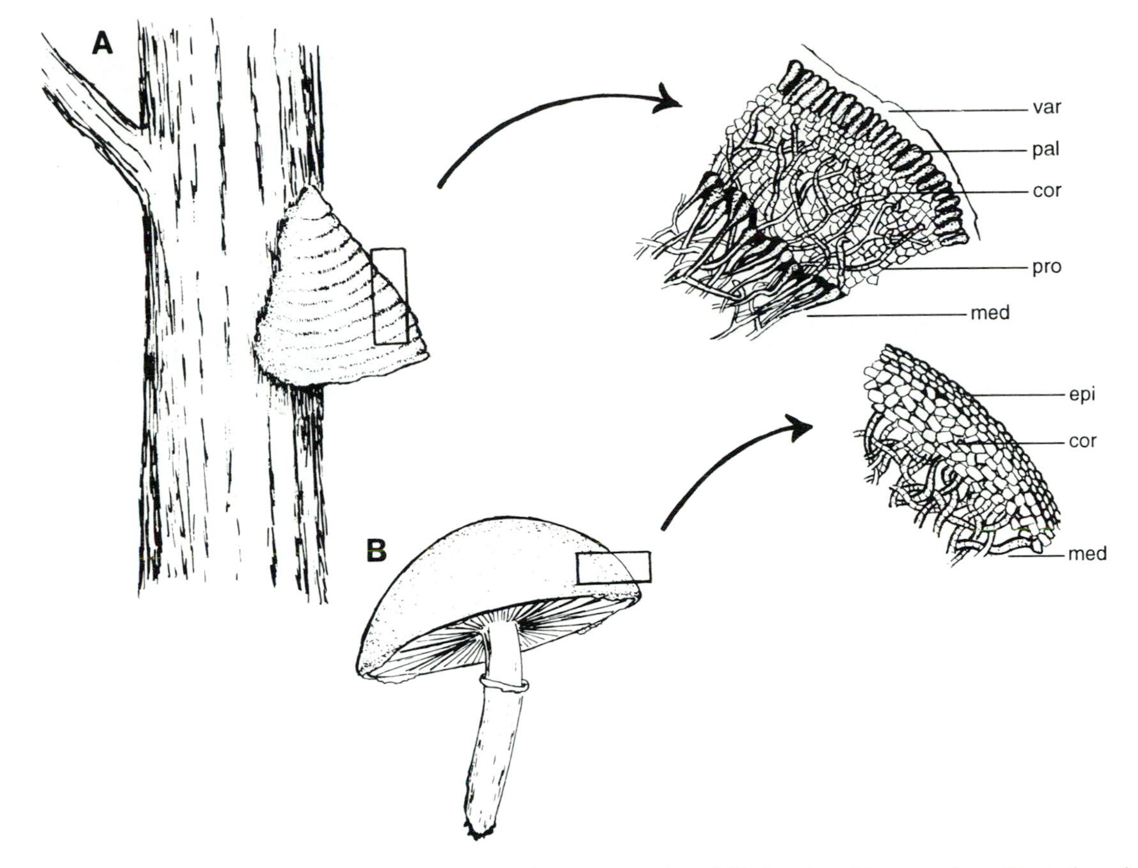

Figure 5–21 Tissues of (A) *Ganoderma lucidum,* the artists' conk and (B) *Agaricus bisporus,* the cultivated mushroom. Note the complexity of the tissues in *Ganoderma,* especially in the medulla where binding hyphae, skeleton hyphae, and generative hyphae are mingled together. (epi = epidermis, pal = outer cortex or palisade layer, cor = middle or parenchymatous cortex, pro = prosenchymatous inner cortex; med = medulla.)

or **adnate.** The base of the stipe is often either tapered or swollen. Basidiocarps also vary in color, in size, in the shape of the pileus, closeness of the gills to each other, ring and volva characteristics, spore color, presence or absence of scales, and several other ways. Color and size are seldom completely reliable characteristics; however, the color of the basidiospores tends to be highly consistent within species and usually entire genera. Color of the gills or pores often, but not always, reflects spore color. Careful observation of these characteristics is essential to correct mushroom identification.

Basidiospore color may be observed by preparing a **spore print** (Figure 5-22). The sporocarp cap is placed gill or pore side down on a piece of paper and a glass tumbler is put over it and left in place for several hours. Spore prints vary in color from white and yellow through pink, brown, purplish brown, and smoky gray to black, and are very useful in keying out species correctly (Figure 5-22B). The species of a single genus usually have the same color spore print and often, but not always, the spore color is indicated by the gill color, especially in older

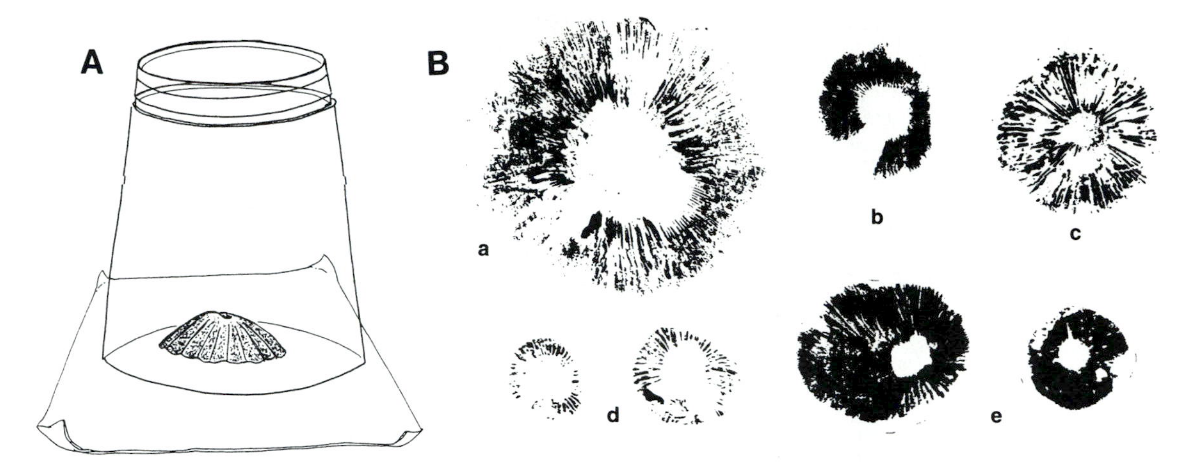

Figure 5–22 Spore prints are important in identifying mushrooms. (A) Preparing the spore print by placing the cap gill side down on a piece of white paper and covering with a tumbler. After several hours, the cap is removed and a print of the spores will remain on the paper, where their color can be observed or slides prepared to examine them microscopically. (B) Spore prints: (a) *Stropharia coronilla,* yields a purplish brown print; (b) *Pholiota terrestris,* yellowish brown; (c) *Bolbitius vitellina,* brown print; (d) *Inocybe geophilum,* ochraceous brown print; (e) *Coprinus comatus,* black print.

sporocarps. As in the Ascomycopsida, there is frequently present a mechanism to aid in forceful discharge of the basidiospores. In many cases, a droplet of liquid begins to appear on one side of the **sterigma** where the spore attaches to it as the spore matures. The droplet attains full size in 5 to 20 seconds and the spore is then forcefully discharged. It has been noted (Buller 1923) that in every case in which spores are forcefully discharged, they are attached obliquely to the sterigmata.

In the Rhodophycopsida and in many, or perhaps most, Ascomycopsida, there are specialized female and male reproductive structures—**oogonia** (more specifically called carpogonia and ascogonia), in which female gametes are produced, and **antheridia,** in which male gametes, or spermatia, are produced. In some Ascomycopsida and in most Basidiomycopsida, on the other hand, syngamy occurs by a union of vegetative cells that are not morphologically different from other vegetative cells. However, in the Uredinales, and in several species in other orders, typical trichogynes and spermatia occur as in the red algae and sac fungi.

In *Pholiota, Coprinus,* and some other genera of Homobasidae, cells which may function either as gametes or as asexual spores, called **oidia,** are produced on short, usually branched, haploid hyphae, the **oidiophores.** In *Coprinus,* droplets of sticky liquid are exuded by the oidophores and attract insects which carry the oidia to other undifferentiated haploid hyphae; if these are of a compatible mating type, syngamy takes place. The resulting hyphae are dikaryotic. If syngamy does not occur, the oidia may germinate and produce a haploid mycelium.

Phylogeny and Classification

Fossil fungal mycelium showing distinctive clamp connections have been found in Pennsylvanian formations and mycorrhizal associations of vascular plants, and fungi, probably basidiomycetes, have been observed from the same period (Andrews and Lenz 1943). This suggests that the Basidiomycopsida originated about the same time as the Ascomycopsida or slightly later. Fossil hyphae and mycorrhizae become relatively common in more recent formations, but the sporocarps of basidiomycetes are typically soft, fleshy, and susceptible to rapid decay, and therefore fossil basidiocarps are quite rare; consequently, our ideas of the patterns of evolution within the class are rather speculative and will undoubtedly remain so until we have made considerably more progress in our studies of comparative vegetative morphology and anatomy.

It is believed that the first basidiomycetes were saprophytic, or possibly parasitic, Heterobasidae similar in appearance to modern species of Uredinales (rusts) that are parasitic on ferns and gymnosperms (Demoulin 1974). Obviously the rusts that are obligate parasites of angiosperms are not primitive species: the flowering plants made their first appearance in the fossil record only a hundred million years ago, whereas the oldest club fungus fossils are about 300 million years old. Gymnosperm fossils, on the other hand, date from 375 million years, and fern fossils are slightly older than that. Nearly all of the modern fern rusts are heteroecious—meaning that they require two different species as hosts—whereas angiosperm parasites are more frequently autoecious. The fern rusts also have more complicated, macrocyclic life cycles, similar to rhodophycean life cycles.

These primitive rusts all produce sessile teliospores, while the more advanced species that parasitize flowering plants have pedicillate teliospores. The fern parasites also have trichogynes similar to those in the Rhodophycopsida and Helotiales, produce clamp connections that often resemble croziers, and have septate basidia. The high degree of correlation among these primitive traits increases our confidence in the hypothesis that the first Basidiomycopsida were heteroecious Uredinales similar to the rusts on ferns today.

The Ustilaginales (smuts) appear to have descended relatively recently from ancestors similar to the primitive Uredinales. In both of these groups of parasites, evolution seems to have led to increasingly simple morphological forms, the "basidiomycetous yeasts" of the Ustilaginales representing the ultimate in simplification: unicells which reproduce by budding. Suggested lines of relationships among the club fungi are shown in Figure 5-8.

The most primitive of the Homobasidae orders appear to be the Aphyllophorales. From early Aphyllophorales, evolution progressed along four lines, one leading to the Tremellales and Auriculariales, one leading to the Exobasidiales and Tulasnellales, one leading to the Agaricales and Hymenogastrales, and the other leading to the Nidulariales by way of the Phallales and Lycoperdales. Characteristics of each order are presented in Table 5-7.

Basidiocarp morphology and spore characteristics were used almost exclusively in early systems of basidiomycete classification. In recent years, mycelium structure and physiology, basidiocarp anatomy, biochemistry, and karyology have been added to the mycotaxonomist's collection of tools. For example, older taxonomy systems placed all mushrooms with pores on the undersurface of the cap, *viz.*, the polypores and the boletes, in a single order. Careful observation of basidiocarp anatomy and ontogeny, by revealing that pores or tubes may arise in several different ways, has caused us to abandon these older systems. As the great body of data now being accumulated is analyzed, better systems of classification will result. The system illustrated in Figure 5-8 is taken from Lange (1964), modified by suggestions presented in Ainsworth et al. (1973) and Peterson (1971).

TABLE 5-7

CHARACTERISTICS OF THE ORDERS OF BASIDIOMYCOPSIDA

Orders and Included Taxa (Fam–Gen–Spp)	Sporocarp Characteristics	Other Characteristics	Representative Genera or Species
Uredinales (2-40-2,000)	Sporocarp reduced to 4-celled septate basidium; complex life cycles with both diploid and dikaryotic generations	Asexual spores typically red; obligate parasites on all groups of vascular plants; some species parasitize two unrelated hosts	*Puccinia, Uromyces, Melampsora, Gymnosporangium, Cronartium*
Ustilaginales (3-40-1,500)	Sporocarps reduced to 4-celled septate basidium; simple 3-phase life cycles	Asexual spores typically black; faculative saprophytes on grasses	*Tilletia, Ustilago maydae, Urocystis, Entyloma, Sphacelotheca*
Tremellales (3-20-200)	Smooth, spheroidal or disk-like, gelatinous basidiocarps; basidia forked or longitudinally septate	Saprophytes on rotting wood; some very mild mushrooms (witches' butter, and some jelly fungi)	*Exidia, Tremella, Phlogiotes, Pseudohydnum, Dacrymyces, Cerinomyces, Dacryopinax*
Auriculariales (1-5-45)	Jelly-like to leathery basidiocarps; basidia transversely septate	Saprophytes on rotting wood; very mild mushrooms (jelly fungi—like the Tremellales)	*Auricularia auricula, Jola, Ecronantium, Herpobasidium*
Aphyllophorales (10-400-3,000)	Well-developed hymenium which may consist of teeth, pores, veins, or coralloid branches	Saprophytes on wood; some of the choicest mushrooms are in this order	*Clavaria, Ramaria, Fomes, Sparassis, Dentinum, Ganoderma, Hydnum, Cantharellus, Polyporus*
Exobasidiales (1-3-75)	Basidiocarps much reduced with nonseptate basidia	Parasites on Ericaceae and other angiosperms	*Exobasidium*

Tulasnellales (2-5-90)	Waxy, web-like sporocarps; similar to Exobasidiales	Parasites and saprophytes; many Fungi Imperfecti of the Mycelia Sterilia group	*Ceratium, Tulasnella, Brachybasidium*
Agaricales (14-200-8,000)	Fleshy basidiocarps with well-developed hymenium consisting of gills or pores on undersurface	Mostly saprophytes; many mycorrhizal species, also several parasites, some tropical lichens	*Agaricus, Amanita, Russula, Entoloma, Coprinus, Lentinus, Panaeolus, Boletus, Galerina, Pleurotus, Suillus, Inocybe*
Hymenogastrales (4-150-3,000)	Well-defined hymenium only in immature sporocarps; some species hypogeous	Mostly saprophytes; many mycorrhizal species; some "false truffles"	*Melanogaster, Leucogaster, Rhizopogon*
Podaxales (2-6-75)	Hymenium is well developed but not exposed; margin of cap permanently attached to stipe	Mostly saprophytes; anatomy similar to Agaricales, but spore discharge as in the Lycoperdales	*Endopticum, Podaxis, Nivatogastrum*
Phallales (1-10-150)	Globose sporocarps which open into a shaft and bulb arrangement; no hymenium	Mostly saprophytes; some are very foul smelling (stink-horns)	*Phallus, Mutinus, Dictyophora*
Lycoperdales (5-40-2,000)	Hymenium lacking, replaced by a homogeneous gleba from the dark-colored spores	Mostly saprophytes; some very choice mushrooms (puffballs)	*Calvatia, Bovista, Lycoperdon pyriforme*
Nidulariales (6-20-600)	Miniature "puffballs" (peridioles or "eggs") surrounded by a peridium ("nest")	Mostly saprophytes or parasites on grasses and forbs	*Crucibulum, Nidularia, Cyathus*

Note: The number of families (Fam), genera (Gen), and species (Spp.) in each order is indicated in the first column.

Representative Species and Genera

The Uredinales and Ustilaginales, with two families each, are widespread plant parasites. Worldwide, probably every important crop has one or more species of rust that infects it. Often the different stages of the life cycle of a single species of rust will parasitize different species; these are the **heteroecious** rusts. **Autoecious** rusts have a single host. The economic damage is often great. Epidemics of black stem rust destroyed much of the North American wheat crops, for example, in 1923, 1925, 1935, 1937, 1938, 1941, and 1950 (Pearson 1967). It continues to take its toll.

Puccinia graminis, the black stem rust, infects young wheat or grass plants by either **aeciospores,** formed on barberry plants, or by **uredospores,** formed on wheat plants. Both are dikaryotic, as is the mycelium developing from them. The mycelium grows through the stem and leaf tissues of a host plant and soon forms dozens of pustules that release masses of red conidia called **uredospores.** These red spores have given the disease its common name. Uredospores are very light and can be carried long distances by wind, but they are also delicate and short-lived and cannot overwinter except in very mild climates. Later as the wheat matures, patches of black mycelium called **telia** form and give rise to two-celled, dikaryotic **teliospores.** The teliospores overwinter on straw and stubble, where karyogamy takes place, and germinate in the springtime. Meiosis occurs in the germinating teliospores to produce a four-celled **basidium** which bears four **basidiospores.** Basidiospores cannot infect grasses, but do infect barberry where they develop an extensive haploid intercellular, primary mycelium having numerous haustoria penetrating the cells in the mesophyll tissues of the leaves. There, fertilization, or cytogamy, occurs as a haploid spermatium (or **pycnidiospore**) comes in contact with the trichogyne of a receptive cell. The pycnidiospores function not only as spermatia but also as conidia, carrying the infection to other barberry plants. The secondary mycelium developing from the fertilized "egg" is dikaryotic and soon forms pustules, called **aecia,** on the undersurface of the host leaf. The dikaryotic aeciospores can infect wheat or other grasses but cannot infect barberry. Similar life cycles are typical of many other rusts and also of the smuts. Many rusts, however, and all smuts are autoecious, i.e., they can infect only one species of host. The different stages in the life cycle of black stem rust are illustrated in Figure 5-23.

Except in isolated areas having severe winters, the barberry is not essential to the survival of the rust fungus. In the absence of the alternate host, black stem rust spreads by means of uredospores and can rapidly, in fact, reach epidemic proportions. The development of rust-resistant strains of wheat very effectively prevents this unless the alternate host is present. In most parts of North America and in Europe, genetic recombinations are created at the time fertilization takes place on the barberry host, and plant breeding is much less effective in rust control. In fact, nature often breeds new strains of rust as rapidly as plant breeders are able to develop new, rust-resistant strains of wheat, and epidemics are a constant threat. In Australia, where quarantine has kept out the barberry, rust-resistant strains of wheat are much more stable.

Tilletia caries is the incitant of bunt or stinking smut, which causes millions of dollars' worth of damage to wheat and other cereal crops each year in the U.S. Dark brown spores are produced in pustules which have replaced the tissues inside the developing kernels. Unlike the rusts, which are obligate parasites, the smuts can be grown and studied on laboratory media. Figure 5-24 shows two SEM views of teliospores of *Tilletia caries.*

Corn smut, caused by *Ustilago zeae,* is another costly disease in corn-growing areas. Closely related to these two species are the interesting "basidiomycetous yeasts", such as

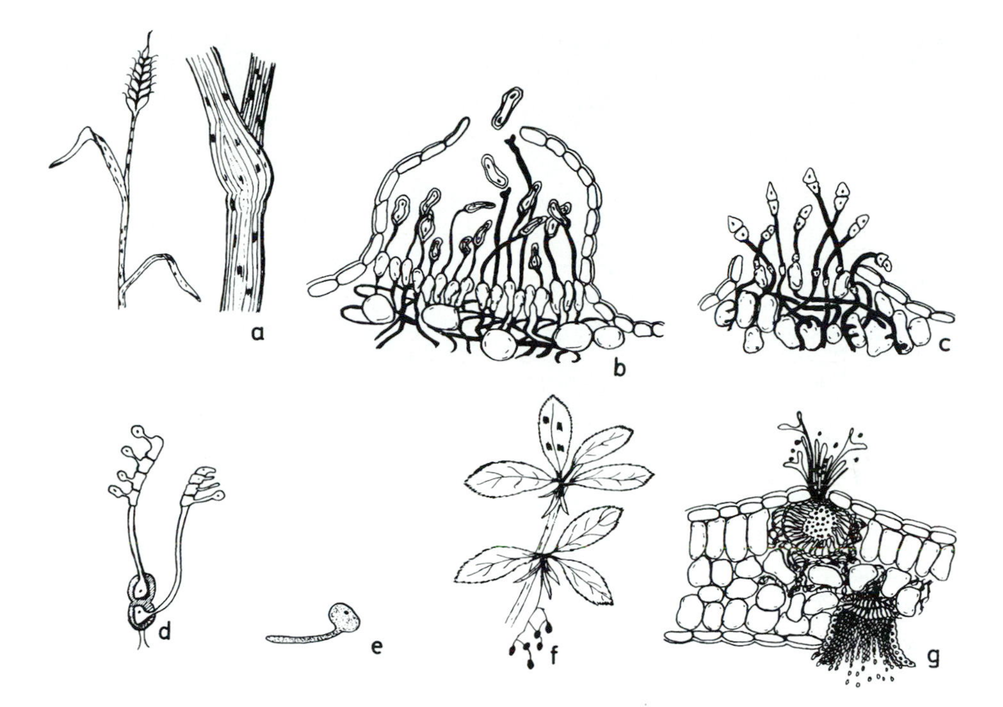

Figure 5–23 Stages in the life cycle of *Puccinia graminis,* the causal agent of black stem rust in wheat and other cereal crops. (a) Infected wheat plant with stem covered with pustules containing masses of red, oval conidia called uredospores; (b) enlargement of a uredospore pustule; (c) teliospores on stubble and straw where they survive the winter; (d) germinating teliospores with septate basidia and basidiospores; (e) germinating basidiospore on leaf (f) of barberry; (g) extensive haploid mycelium in the mesophyll of the barberry leaf with pycnidia and pycniospores on the upper surface and acervuli and aeciospores on the undersurface.

Rhodosporidium, Leucosporidium, and *Aessosporon,* which are unicellular fungi that reproduce by budding and are difficult to distinguish from the "true yeasts" of the Ascomycopsida (Kreuger-van Rij 1973).

Typical of the Aphyllophorales are the species which produce "conks" on trees. *Fomes ignatius* is a common shelf fungus which parasitizes hardwood trees. Like many Aphyllophorales, it has pores on the undersurface of the shelves rather than the gills typical of most Agaricales. Several related species form similar growths on pines and other softwoods. Two of these are *Polyporus sulphureus,* the sulfur polypore, a fleshy shelf fungus which is rated as a choice mushroom, and *Ganoderma applanatum,* the artist's conk. Also in the same order, the Aphyllophorales, are several other common mushrooms such as the coral fungi (Clavariaceae), nearly all species of which are edible, the much-sought after chanterelles (Cantharellaceae), and the tooth mushrooms (Hydnaceae). Because *Clavaria gelatinosa* and some species of *Cantharellus* are poisonous, it pays to know each species well if one would collect and eat wild mushrooms.

In many parts of the world, especially Scandinavia, eastern Europe (the republics of the former U.S.S.R.), and Japan, hunting mushrooms for the table is a favorite pastime. Mushroom eaters, or **mycophagists,** must not only know how to identify the species they are stalking,

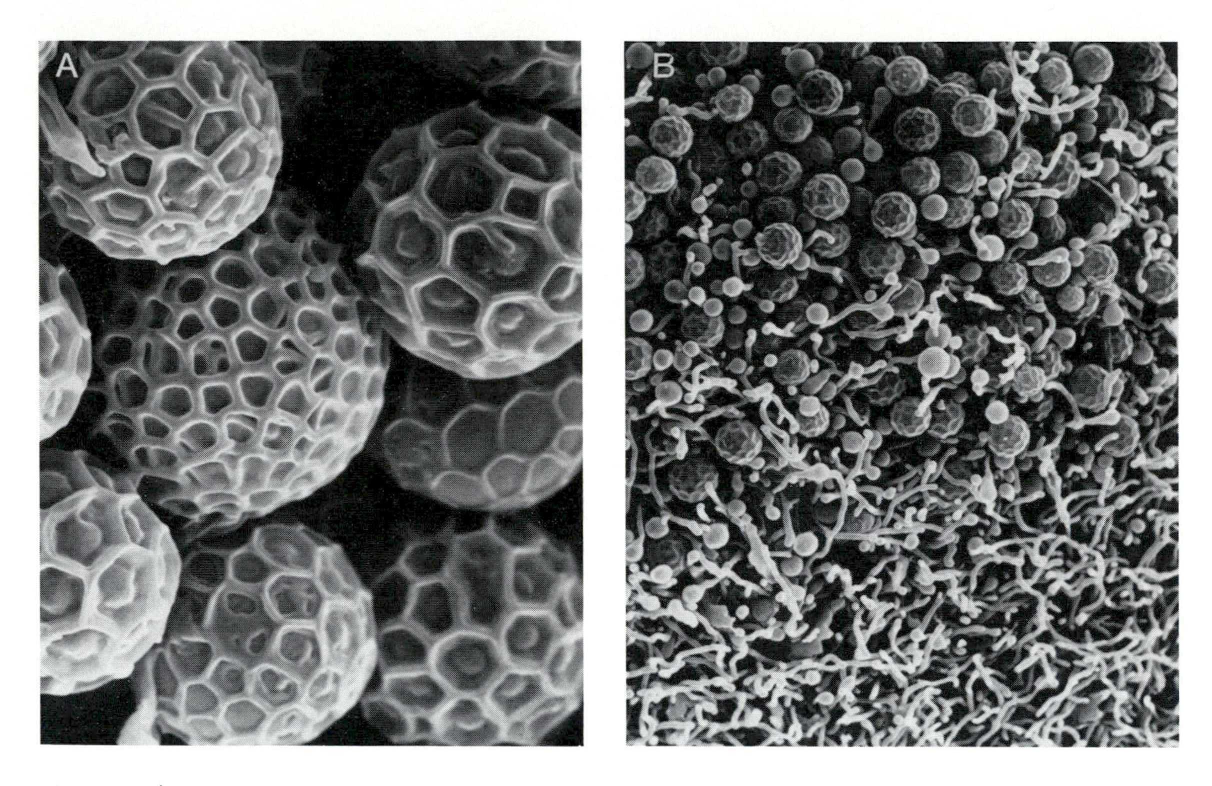

Figure 5–24 Ustilaginales. (A) Scanning electron microscope (SEM) picture of teliospores from *Tilletia caries,* the incitant of bunt or stinking smut; (B) germinating teliospores. Courtesy W. M. Hess, Brigham Young University Microscopy Laboratory, Provo, Utah.

but need to understand ecological principles, for mushrooms, like most other organisms, associate with certain trees or other vegetation and generally not with anything else. Most of all, the earnest mycophagist must have a love of nature and the out-of-doors. Mushrooms, of course, always taste best when prepared and eaten where collected (Figure 5-25).

Among the choicest of all mushrooms are several species of boletes, especially *Boletus edulis,* the cep or king bolete, and *B. aurantiacus,* the orange cap bolete. Members of the Agaricales, to which the common gill mushrooms belong, the boletes are characterized by having a tubular hymenium with pores on the undersurface, thus superficially resembling polypores. *B. satanus,* as the name suggests, is very poisonous; it is easily recognized, however, by its scarlet tube openings and flesh that turns blue when broken and exposed to the air. All of the poisonous species of **boletes** have either red pores or flesh that turns blue or both (Figure 5-26). These characteristics apply *only* to the boletes; to eat wild mushrooms safely, it is absolutely necessary to know the characteristics of each individual species!

Sporocarp tissues of *Agaricus bisporus,* the commercial mushroom commonly found in stores, and *Ganoderma applanatum* are compared in Figure 5-21. *G. applanatum* is a perennial; in addition to the complex, trimictic tissue structure illustrated, it has annual growth rings. It is called the artist's conk because detailed drawings can be etched on its smooth, white pore surface which, upon drying, are permanent. A close relative, *G. lucidum* or "ling chih",

Figure 5–25 Mycophagy. Some of the boletes are among the choicest of all mushrooms and they taste best when collected high in the forest when it's hot down in the valley and prepared out in the open with onions and butter.

the "mushroom of immortality", is supposed to have special healing properties. A candy made from ling chih is sold in some Chinese markets in the U.S.

Mushroom hunting and mycophagy have always been an important aspect of Japanese culture. A story is told of a man several centuries ago on his way to visit a temple in the mountains. Suddenly he heard boisterous laughter and shortly came upon a group of nuns singing and dancing as they came down the path; every little comment brought peals of laughter from them. They had accidentally mistaken the famous Japanese laughing mushroom for one of their favorite gourmet items. It is harmless, but causes anyone who eats it to laugh involuntarily as though being tickled or listening to humorous anecdotes.

Closely related to *Agaricus* is the genus *Amanita,* containing some of the most poisonous plants known to man, along with some edible species and some hallucinogenic species. *Amanita phalloides* is especially poisonous, and there are reports of people being fatally poisoned from eating a piece of cap no larger than a pea. The common "toadstool," *Amanita muscaria,* while not as poisonous as *A. phalloides* and some other species, has also caused many deaths. Would-be **mycophagists** with a cowardly streak, including the author of this book, simply avoid all amanitas; it has been said that there are *bold* and *old* mycophagists, but none who are both bold and old.

Of course, not all wild mushrooms are poisonous; hundreds of species of Aphyllophorales, Agaricales, and Lycoperdales are not only edible but choice, as pointed out in Special Interest Essay 5-2. To be able to recognize with certainty a few of the best unlocks the door to some delightful gastronomical experiences.

Figure 5–26 "Oh, darn it, Dennis! I had it just backwards. In the boletes it's the ones with red caps and yellow pores that are edible. The ones with the yellow caps and red pores are poisonous."

Thanatephorus cucumeris, also known as *Ceratium solani,* is a member of the Tulasnellales, and the causal agent of black scurf disease of potatoes. It is more commonly known as *Rhizoctonia solani* and classified in the Mycelia Sterilia of the Fungi Imperfecti. Closely related is the genus *Exobasidium,* in the Exobasidiales, with many species of parasites of blueberries, cranberries, and other members of the Ericales.

The puffballs and bird's nest fungi are characterized by basidiocarps in which the spores are produced in an amorphous, porous, fleshy gleba. As the puffball spores begin to develop, the gleba darkens in color until at maturity the interior of the globular basidiocarp is one mass of black or dark greenish brown spores. In many species, an opening at the apex of the spore fruit allows the spores to escape. Any pressure exerted on the tough, elastic fruit covering produces a bellows nnaction which causes puffs of black spores to be forcibly discharged.

In the Nidulariales, the peridioles or individual sporocarps resemble eggs surrounded by the nest-like peridium (Figure 5–27). Raindrops striking the "eggs" cause them to splash out of the "nest," and as they do, the funicular cord is released so that it trails after. When the peridiole strikes a twig or leaf, momentum causes the cord to continue moving until it reaches its limit; it then wraps around the twig or leaf much as an Argentine bolo wraps around an animal's legs.

Ecological Significance

The Basidiomycopsida are the most effective of all decomposer organisms involved in breaking down cellulose and lignin. Many Basidiomycopsida, along with a few Ascomycopsida, are the only organisms capable of decomposing lignin, and are therefore an indispensable part of every forest ecosystem. Most mycorrhizal fungi are also basidiomycetes; abundant evidence indicates that they are essential to the survival of many species of vascular plants, although the exact mechanism of how they are beneficial is not well understood.

Many species of club fungi, especially those in the orders Uredinales, Ustilaginales, Exobasidiales, and Tulasnellales, are widespread plant pathogens. Collectively, they attack almost every family of vascular plants and thus have far-reaching ecological effects in almost every terrestrial ecosystem.

Many animals, including man, feed on the sporocarps of fleshy basidiomycopsida. Elk are frequently observed feeding on the stout russulas, while numerous insects feed on boletes, and various snails and slugs feed on other mushrooms. While some species of invertebrates depend largely or entirely on basidiocarps for food, to vertebrates, mushrooms are generally only one of many food items. On the other hand, some species of club fungi appear to depend entirely on mammals or other animals for dissemination of their spores. Other club fungi are dependent on wind for spore dispersal.

Microclimate is more important than macroclimate to basidiomycopsida distribution. Nidulariales are found in deserts, on prairies, and in forests; however, in the forest they occur on attached branches high in the trees or on exposed twigs of fallen branches where it is relatively sunny and dry, while in prairie areas they are found only on rotting twigs or in the leaf mold of relatively deep, damp depressions, and in the desert, they grow on driftwood in shady areas along the edges of lakes or ponds.

In general, Basidiomycopsida, like Ascomycopsida and Zygomycopsida, favor much drier climate than do either the Mastigomycopsida or the Myxomycopsida. Like the Ascomycopsida, individual species tend to be very widely distributed, and the number of species found in any given ecosystem is often greater than the number of species of Anthopsida or other vascular plants.

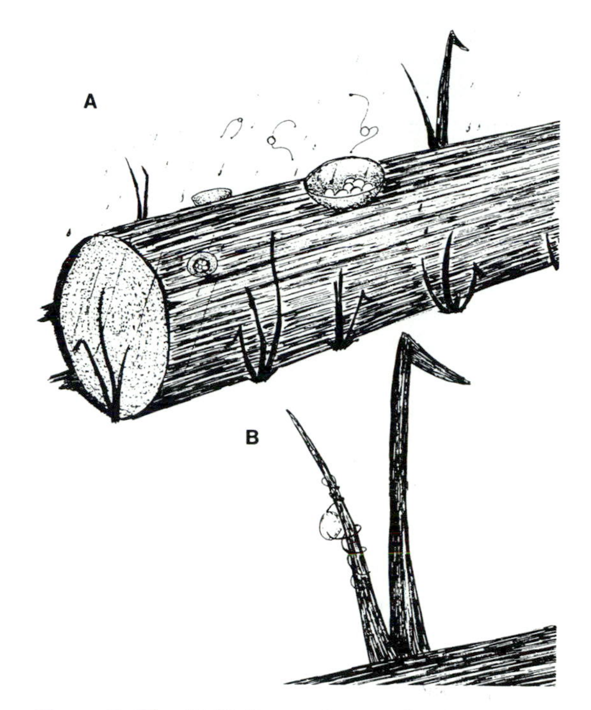

Figure 5–27 (A) Bird's nest fungus, *Cyathus striatus*, Nidulariales. When raindrops land in the "splash cup" or bird's nest, the peridiole (egg) is ejected from the "nest" or peridium. A fine thread, the funiculus, attaches the "egg" very loosely to sticky material at the bottom of the "nest"; it breaks loose and if the peridiole strikes a blade of grass or other object (B), it wraps around the object. The heavy, sticky end where it was attached to the peridium now helps it stick to the grass; when the grass is eaten, the spores pass through the digestive tract of the herbivore unharmed; those which land on dead sticks or logs develop into new "birds' nests".

Economic Significance

Puccinia graminis, the incitant of black stem rust in wheat and related cereal grasses, is the most costly of all plant parasites, costing farmers the world over hundreds of millions of dollars every year in reduced yields. Other rusts and smuts are also of great economic significance. In North America, black stem rust is damaging from Mexico to northern Canada. In northeastern Mexico, the disease is constantly present on grass and grain. Spring winds carry the uredospores northward into Texas just as the wheat crop there begins its spring growth. From Texas, the almost continual south winds carry successive crops of spores to Oklahoma, then to Kansas, to Nebraska, to the Dakotas, and finally to Saskatchewan and Manitoba. As the Canadian crop of spring wheat matures, the winter annual grasses are sprouting in the Dakotas, and the north winds of late summer and fall carry successive crops of uredospores southward. Along the way, pockets of Oregon grape or other barberries grow in woodlots and along river bottoms, and basidiospores carry the fungus to them where genetic recombination can occur, creating new combinations of virulence genes capable of infecting varieties of wheat that were previously resistant to rust.

TABLE 5-8

DOMESTIC PRODUCTION OF MUSHROOMS AND MUSHROOM IMPORTS TO THE U.S.

Type of Mushroom	Value in Dollars × 10^3		
	1980	**1985**	**1990**
Domestic production			
Commercial *Agaricus*	368,597	470,969	622,485
Shiitaki	—	19,258	16,616
Oyster mushrooms	—	2,167	3,964
Other specialty types	—	1,254	1,689
Total value	—	**493,645**	**644,754**
Imports in metric tons (1990)			
China			26,267
Taiwan			9,326
Hong Kong			7,834
Indonesia			4,235
Spain			3,300
India			1,325
Netherlands			957
Republic of Korea			813
Other major importers			1,571
Total, all imports			**56,338**
Domestic production in metric tons, 1990			325,005

Data from *1992 Agricultural Statistics,* U.S. Department of Agriculture, Washington, D.C.

Another club fungus of considerable economic significance is *Agaricus bisporus* (often referred to as *A. campestris*), the champignon or common commercial mushroom. Sales of *A. bisporus* in North America exceed $500,000,000 annually, making it a major crop (Table 5-8). Formerly, most mushrooms in the U.S. were grown in old potato cellars in Pennsylvania. As a major potato-producing as well as horse-producing state, Pennsylvania was ideally suited for this crop, which is grown in the dark under conditions of high humidity on a mixture of horse manure and straw. The **spawn,** mushroom mycelium which has been grown under sterile conditions in flasks in the laboratory, is sprinkled over this medium. Temperature is carefully controlled, ideally between 65° and 70°F, until an abundant, healthy mycelium is formed. In 2 to 3 weeks, the first basidiocarps break the surface. For several weeks, until the carbohydrates in the straw are well used up, successive crops of mushrooms are harvested every 5 or 6 days. At the present time, abandoned mines in West Virginia have replaced many of the old potato cellars of Pennsylvania and hand labor has been replaced by machines. In poultry producing areas, mushroom growers have found that chicken and turkey manure grow just as good mushrooms as horse manure, and the mushroom industry is much broader-based now than it once was.

In Japan, the shiitake mushroom, *Lentinus edodes,* is a favorite. It is cultivated on hardwood, usually oak. The spawn is produced on sterilized wood chips. When the mycelium has developed, chips containing the spawn are placed in holes that have been bored in logs of oak or other hardwoods and abundantly watered. The mushrooms may be eaten fresh or they may be dried and stored. When reconstituted, the flavor is as good as in the fresh product: rather mild, pleasant, and very different from the American champignon. It is now also grown extensively in the U.S. and Canada (Table 5-8).

The American shiitake, *Lentinus lepideus,* is generally considered inferior to the true shiitake; however, some of the Rocky Mountain ecotypes have excellent flavor not unlike the Japanese species. It is easily identified by its eccentric cap attachment and sawtooth to sharply crenulate gill edges; it grows on stumps of lodgepole pine, douglas fir, and occasionally other trees. Being drought resistant, it is one of the most reliable species in dry years.

The shiitake has been used for centuries in Japanese folk medicine. It is claimed to cure or to prevent colds, flu, rickets, and cancer, to reduce blood pressure, reduce cholesterol, accelerate recovery from fatigue, and enhance the beauty of women. The shiitake, like many other mushrooms, is high in vitamins B_{12}, and D_2, both of which are rare in plants other than mushrooms and cyanophytes. Research in the 1950s and 1960s demonstrated the presence of anticarcinogens in the shiitake. More recent research indicates that the anticarcinogenic agent is a virus which is usually, but not always, present in the mushroom spores. This virus destroys a number of other viruses, including those responsible for some types of cancer.

Research Potential

Like other Eumycophyta, basidiomycopsida produce a variety of enzymes and are of great potential value in biochemical research. Especially promising are the possibilities of analyzing the chemical nature of lignin. Mycorrhizal Basidiomycopsida, along with lichens, show promise in helping us to understand the true nature of symbiosis, the mutually beneficial process of two organisms growing in an intimate relationship together. This, in turn, could have bearing on our knowledge of how the eukaryotic cell originated.

The genetics of basidiomycopsida has been studied extensively in recent years. Rusts and smuts have been useful in showing how virulence, the ability of a pathogen to infect a host organism, is inherited. Some of the basic knowledge of DNA function was gained from studies of guanine/adenine ratios in club fungi. Further studies of this nature may help us solve some of the riddles of phylogeny in the fungi.

Research involving mushrooms show promise in helping us understand better the way in which various minerals and vitamins, especially vitamins of the B complex, function. Research with mushrooms has already helped us understand how hallucinogens and some other toxins affect the human body and further research should reveal much more. At one time, it appeared as though fungal hallucinogens might have a great potential in psychological and psychiatric treatment, but this area of research seems less promising now, although it has not been completely abandoned. However, in the process of studying psychological effects, we learned that many hallucinogens are also powerful mutagens, causing chromosomes to break. In some cases, the mutagenic effect was seen to be chromosome specific (MacKey, 1956). Thus, the first of the restriction enzymes were discovered in club fungi. The use of mushroom mutagens in plant breeding and other areas of biotechnology is a valuable tool which has led to the development of the techniques involved in such revolutionary procedures as gene splicing.

SUGGESTED READING

Several excellent textbooks on fungi are available; *Introductory Mycology* by Alexopoulos and Mims is one of these. E. A. Bessey's *Morphology and Taxonomy of Fungi* is an older classic containing excellent descriptions and keys.

Orson Miller's *Mushrooms of North America,* Lorentz Pearson's *Mushroom Manual,* and David Arora's *Mushrooms Demystified* are easy-to-use guides to wild mushrooms. The *Mushroom Manual* presents safety rules for enjoying wild mushrooms along with detailed descriptions of the "foolproof four" and the "fatal five" that enable rank amateurs to eat wild mushrooms safely. It is probably the best guide available for amateur mycophagists. For mushroom hunters who are collecting for herbarium purposes, Miller's and Arora's books have the advantage of excellent photographs. For anyone collecting seriously in order to build up a good herbarium, several guides should be available; Edmund Tylutki's five volume *Mushrooms of Idaho,* Alexander Smith's *Mushroom Hunter's Field Guide* and *Field Guide to Western Mushrooms,* and the Audubon Society's *Field Guide to North American Mushrooms* (Lincoff 1981) are among those highly recommended.

Chester's *Plant Pathology* is a good introduction to the fungi that cause plant diseases, how to recognize them, and how they can be controlled. J. C. Walker's *Plant Pathology* and J. G. Dickson's *Diseases of Field Crops* are among the other books on plant diseases that students can find interesting and useful. Linda Carporeal's articles on the Salem witch hunt, published in 1976 in several news magazines as well as *Science,* (Vol. 192) and in 1982 in *American Scientist,* are both interesting and factual accounts of the ecological effects of ergot.

STUDENT EXERCISES

1. In lifting up the damp "duff" from the forest floor and examining the underside with a good hand lens, you observe numerous white threads. Mounting these on a glass slide and examining under high power with the compound microscope, you observe the presence of cross walls or septations in these threads. Which of the classes of algae or fungi could these threads belong to? Which could they not belong to?
2. You have observed pine seedlings in the laboratory and in the greenhouse many times and noted that the main roots are relatively straight with many side roots branching off almost at right angles. In digging up a pine seedling in the forest, however, you observe that the roots are dichotomously branched or forked and there are no root hairs present. What is the cause of this? Is this pine seedling infected by some disease organism?
3. What are the two main hypotheses as to the origin of the Eumycophyta? What evidences favor each of these hypotheses? After reading the comments by Alexopolous and Bessey, which of these hypotheses do you tend to favor?
4. How old are the oldest fossils of Ascomycopsida yet found? How does this compare with the oldest fossils of Basidiomycopsida and Mastigomycopsida that have been found?
5. Using no notes, only your understanding of the cytology of microkaryophytes, show how croziers and clamp connections are formed. Compare with the formation of secondary pit connections between the oogonium and auxiliary cell in Rhodophycopsida.

6. If you were to examine some mycelium from just under the duff layer in a forest, how could your knowledge of croziers and clamp connections enable you to distinguish between ascomycete and basidiomycete hyphae? Sketch croziers and clamp connections as you would expect them to appear under natural conditions.
7. Devise a simple experiment to ascertain which of the following mushrooms your colleagues in botany class like the most: the grocery store champignon, shiitake, *Lentinus lepideus, Boletus aurantantiacus, Coprinus comatus, Coprinus atramentarius,* and *Lycoperdon perlatum.* Do all students prefer the same species of mushroom?
8. Prepare a chart showing exactly how you would replicate, randomize, and analyze your data from the above experiment in order to evaluate whether the students really can tell if there is a difference in flavor among the species of mushrooms.

SPECIAL INTEREST ESSAY 5-1

Industrial Uses of Molds

Microbial fermentation has been employed for centuries to produce bread, beer, wine, and fermented dairy products and was apparently discovered independently by ancient Egyptians, Chinese, Incas, and others long before the dawn of recorded history. Today, fermentation is of greater significance than ever; molds are used not only to produce cheese, alcoholic beverages, bakery goods, and sauerkraut, but also industrial alcohol, citric acid, vitamins, antibiotics, gluconic acid, acetone, cortisone, diastase, and many other substances. These, in turn, are used in the manufacturing of plastics, synthetic fibers, dyes, ink, cleaning agents, camera film, shoe polish, soft drinks, candy, and skin remedies, to mention only a few products.

While growing molds for commercial use involves the same principles as growing them in petri dishes or Ehrlenmeyer flasks in the biological laboratory, they must be grown on a very large scale in order to cut the costs of production to a competitive level, and special problems always arise with large-scale production. To sterilize a few hundred test tubes or petri dishes in a modern laboratory autoclave or a home-canning pressure cooker is a simple task. Even the beginning student has no trouble inoculating a hundred test tubes with cultures of fungi in such a way that uniform and rapid growth results. But to sterilize a 50,000-gallon vat full of wort, maintain it at a uniformly constant temperature, and inoculate it with a fungus in such a way that uniform growth is obtained throughout is a challenge that has been met only through innovative thinking and much experience. Optimum conditions are essential not only to achieve maximum yields but because many microorganisms produce entirely different products when environmental conditions are changed.

In the factory, as in the laboratory, four conditions must be met in order to obtain the desired products: (1) a constant source of energy in the form of carbohydrates must be available; (2) mineral and other nutrients must be present at optimum levels; (3) temperature, moisture, pH, osmotic concentration, and other environmental factors must be kept within very narrow limits, and (4) sterile conditions must be maintained.

The most versatile and widely used of all fungi are the yeasts, especially *Saccharomyces cerevisae* with its many strains and ecotypes. Its products include bread, wine, industrial alcohol, vitamins, enzymes, glycerol, and fats. It is used directly for human food and as livestock fodder, and (as has been mentioned) some yeasts contain up to 50% protein having the same amino acid composition as beef and pork. Industrial alcohol is the most important organic chemical in modern industry and is used as a solvent in producing plastics, dyes, safety glass, photographic materials, soaps and shampoos, explosives, varnishes, medicines, and many other products. It is used as a raw material in producing vinegar, acetic acid, ether, chloroform, etc. It is widely used as fuel for cooking and to provide light, heat, and power, and it produces a clean flame free of particulate matter and toxic fumes. Chemically, it is known as ethanol or ethyl alcohol and is identical to the alcohol in beer, wine, and other intoxicating beverages.

Two stages are involved: saccharification and fermentation. When waste products are used, they are converted to sugars by cooking, a mechanical process which destroys cell structure and frees starch grains from their membranes, followed by a biochemical process in which enzymes, obtained from fungi or from wheat or barley, convert the starch to sugar. In the Amylo process, pure cultures of *Rhizopus stolonifera, Aspergillus oryzae,* or *Mucor* spp. are used to convert starch to sugar. The saccharified material is called wort.

In the fermentation stage, yeast converts simple sugars to alcohol. A starter is prepared from a pure laboratory culture, allowed to grow for a few hours, and then used to inoculate a flask of wort with about 20 times the volume of the starter culture. Successively larger containers of wort are inoculated in this manner until the final culture of several hundred gallons is ready to **pitch** the 50,000-gal fermentation vat. Aeration is necessary in preparing the starters, since the object is to obtain rapid growth and reproduction of yeast cells; in the fermentation vat, on the other hand, anaerobic conditions are maintained so that maximum yield of alcohol can be achieved. When fermentation is complete, the liquor in the fermentation vat is about 20% alcohol. This is distilled to obtain 95% alcohol. To produce 100% alcohol, special drying agents are used to remove the water. For most biological purposes, 95% is superior to 100% alcohol, since water has no adverse biological effects, whereas a trace of a few parts per billion of drying agent often does.

Altering environmental conditions results in different products. For alcohol, anaerobic conditions and low pH at a temperature of 20 to 25°C are maintained. The same pH and temperature are used in yeast cake and dried yeast production, except that good aeration is required. Nitrates or ammonia are added to the wort to produce B-complex vitamins. Glycerol is produced by fermenting at high temperature (30 to 35°C) and high pH (7.5 to 8.0).

Citric acid, formerly produced from cull citrus fruit, is now produced primarily by *Aspergillus nidulans,* the black jam mold, well known to the housewife who regards it as a nuisance. Her children are generally not aware of the fact that the soft drinks they buy are produced from the same mold. While regarded by the bacteriologist as one of the worst weeds in his laboratory, it is the second most important of the industrial molds and its economic value far exceeds that of many agricultural crops.

Citric acid is produced in shallow aluminum trays which allow for a large surface area of fungal mat and hence good aeration, rapid diffusion of sugar into the cells, and rapid diffusion of acid out of the cells. The aluminum, which possibly has some catalytic effect on the process, must be of a very pure grade because slight traces of copper, manganese, nickel, or lead adversely affect fungal growth. After 7 to 9 days of growth, the fungal mat is removed and lime added to the remaining medium in order to precipitate the citric acid as calcium citrate. The liquid is poured off and sulfuric acid added to the precipitate to dissolve it; the resulting citric acid is purified by crystallization.

SPECIAL INTEREST ESSAY 5-2

Mushrooms and Toadstools

People always ask botanists how to tell poisonous mushrooms from edible ones. There seems to be a general belief that "toadstools" are botanically different as a group from edible mushrooms. Actually, the name "toadstool" has no botanical meaning whatsoever, except when used as the common name of the fly agaric or *Amanita muscaria,* the fungus frequently pictured in story books. Poisonous mushrooms occur in almost every family of fleshy basidiomycopsida, just as other poisonous plants—berries, roots, or greens—can be found in almost every family of flowering plants. In either case, there is only one reliable way of telling whether a plant is poisonous or edible: identify it, usually with the aid of a botanical key, and then look it up in a catalog of poisonous and edible plants. Excellent field guides, containing keys, descriptions, and pictures of the common mushrooms, along with information as to their edibility, can be obtained for almost any part of the U.S., Canada, or Europe, and for many other areas.

The word "toadstool" has come to us from ancient Scandinavian and has nothing to do with toads. The original word was "*dödstol*" which means *death chair,* and was applied to the genus *Amanita.* The name fits, for several species of *Amanita* are deadly poisonous, and even the common toadstool, *A. muscaria,* which is often used as a hallucinogen or as a substitute for alcohol, causes numerous deaths every year, either to people who take too large a dose while planning a "trip", or by people who mistake it for the wild *Agaricus campestris,* which it resembles in many ways, and prepare it for a meal. Even experienced mushroom hunters occasionally make a mistake; the first rule of mushroom hunting cannot be overemphasized: never eat a wild mushroom unless you are 100% positive of its identity. The same rule, of course, applies to eating wild berries or herbs.

There are two groups of mushrooms that every **mycophage** (mushroom hunter and eater) should know. These can be called the "foolproof four" and the "fatal five." The "fatal five" are those mushrooms which have been responsible for almost all of the deaths from mushrooms. There are five groups or complexes of mushrooms that are extremely poisonous: (1) the *Amanita phalloides* complex, made up of at least four species of fungi having slow-acting but deadly poisonous protoplasmic toxins, (2) the *Entoloma lividum* complex of two or three species containing very powerful gastrointestinal toxins, (3) the *Galerina venenata* complex with two species of small mushrooms containing the same toxins as *A. phalloides,* (4) the *Amanita muscaria* complex with three or four species containing fast-acting nerve toxins some of which are hallucinogenic, and (5) the *Gyromitra esculenta,* complex in which deadly amounts of a volatile, water soluble, protoplasmic toxin is sometimes present.

The mushrooms in the first two groups are very attractive and tempting to eat, but deadly even if only one or two—or sometimes only part of one—sporocarps are eaten. The *Galerinas* are so small that they are seldom picked for table use, but they are deadly poisonous and pose a constant threat to small children. The last two complexes contain water-soluble toxins that can be removed, in most of the species, by parboiling. However, the parboiling process is very exacting and does not apply to all of the species within both of these groups. Deaths have resulted when the parboiling water has been accidentally drunk.

The "foolproof four" are species that are so distinctive that there is little danger of confusing them with any harmful mushroom. They are (1) the shaggy mane, *Coprinus comatus;*

(2) the morels, *Morchella esculenta* and its close relatives in the Pezizales; (3) the sulfur polypore, *Polyporus sulphureus;* and (4) the puffballs, *Lycoperdon* spp. and their relatives in the Lycoperdales. All of the "foolproof four" species are very common and are among the most delicious of the edible mushrooms. A budding mycophagist should learn them well and confine himself to these four as he gradually gains more knowledge.

The food value of mushrooms is not great. They are low in calories and in most vitamins and minerals, but they are high in some of the B vitamins and in iron and copper. Some mushrooms, especially the puffballs, are high in proteins and the balance of amino acids is excellent. The great value of mushrooms, however, lies in their flavor. They are gourmet foods. And in this respect, the wild mushrooms are considerably superior to the cultivated species, largely because of the great variety of flavors to be encountered. Once a person has tasted a morel, or a king bolete, or the white chanterelle, he is never fully satisfied with store-bought mushrooms.

6 LICHENS AND OTHER SYMBIOTIC PLANTS

In November of 1952, a fog enveloped London and for 3 weeks hung heavy over the city. During those 3 weeks, the number of deaths per day averaged 40% higher than normal, most of the increase being due to respiratory diseases. It is officially estimated that 4000 people died as a direct result of the "killer smog." A careful analysis, however, of the day-by-day records, continuing through the following 3 weeks, during which time deaths from respiratory diseases were still abnormally high, suggests that the number killed by that pollution-laden smog may actually have been greater than 15,000.

Unfortunately, the London killer smog is not an isolated incident. Similar examples can be found in New York City, Chicago, St. Louis, Paris, and other cities. Frequency in the occurrence of such incidents seems to be increasing, even though none have been as sensational as the London smog. Pollution of the air we breathe is rapidly becoming the most serious health problem faced by the human race. It is known to cause lung cancer, emphysema and other respiratory ailments, damage to the liver and other internal organs, and even brain injury. The quest for clean air became an obsession to numerous urban dwellers during the closing decades of the 20th century.

Urban areas are not the only sites affected by air pollution. Atmospheric pollutants damage crops and kill livestock. Acid rain destroys fish in mountain lakes and ponds. Analysis of growth rings in trees near sources of pollution reveals a much reduced rate of timber growth correlated with the human activity which produced the pollution. Some pollutants act as powerful catalysts which accelerate conversion of ozone to oxygen, thus removing the protective ozone layer that surrounds the earth. Even "harmless" carbon dioxide, which is essential to photosynthesis, is coming under condemnation as concern about the "greenhouse effect" increases.

Direct measurement of the level of pollution in the air involves the use of sophisticated and expensive chemical equipment. Biological monitoring is often used in conjunction with chemical monitoring and has several advantages: it is inexpensive and it provides a history of pollution effects, whereas chemical monitoring can only tell what the levels are at the moment of measuring, and it reveals information on the biological effects of the pollution. Lichens are especially valuable biomonitors.

Most species of lichens serve this purpose well. They are often extremely sensitive to air pollution, they are long-lived, they vary greatly from species to species in their level of sensitivity, and they are easily transplanted from nonpolluted areas to areas of heavy pollution. In addition to their sensitivity to pollutants that directly harm or kill, many lichens accumulate heavy metals and other polluting elements in and around their cells. Analysis of lichen tissue can thus reveal which pollutants are especially abundant in a given area.

Contributing to the sensitivity of lichens to air pollution is the fact that they are **symbiotic** organisms. In a lichen, two unrelated species, one an alga and the other a fungus, live together in such close symbiosis as to appear to be a single individual.

The dictionary definition of symbiosis is

> "The living together in more or less intimate association or even close union of two dissimilar organisms . . . . Ordinarily it is used in cases where the association is advantageous, or often necessary, to one or both, and not harmful to either. When there is bodily union (in extreme cases so close that the two form practically a single body, as in the union of algae and fungi to form lichens . . .), it is called **conjunctive symbiosis,** if there is no actual union of the organisms . . . , **disjunctive symbiosis**."
>
> *Webster's New International Dictionary* (Unabridged)

To maintain a conjunctive symbiosis requires very exacting environmental conditions. Unless the symbiotic partners maintain a healthy balance, the system breaks down. It is not surprising, therefore, that lichens are not found in big cities or other areas of significant industrial development where any type of air pollution could be harmful to one or the other of the symbionts.

Two groups of symbiotic organisms, lichens and mycorrhizae, are of special botanical significance. Both are examples of conjunctive symbiosis. In both, one of the symbionts is a fungus and a member of the Eumycophyta discussed in the previous chapter. In lichens, the **mycobiont,** or fungal partner, is nearly always an ascomycete; in most mycorrhizae, the mycobiont is a basidiomycete. The other partner in a lichen is an alga and is called the **phycobiont**; in mycorrhizae, it is a vascular plant. Both of these relationships are so intimate that the dual nature of the associations was not discovered until relatively recently. In each of them, the composite organ or organism is vastly different from its individual parts. Therefore, a special chapter is devoted to lichens and mycorrhizae.

LICHENS

Lichens are readily recognized by their unique growth form, often distinctive colors, and the presence, in most cases, of ascocarps. The habit, or growth form, may be crust-like or crustose (Figure 6-1), leaf-like or foliose, or fruticose (vine- or shrub-like). The distinctions are not always clear and there are subdivisions of each habit type. To identify crustose lichens, it is often necessary to dissect the sporocarps, examine their structure, measure the asci and spores, and possibly run some simple chemical tests (color tests). Foliose and fruticose lichens (Figure 6-2), on the other hand, can usually be identified to species with the aid of a hand lens and simple color tests.

Prior to 1852, when Tulasne recognized the fungal nature of their hyphae, the lichens were classified in the Bryophyta as a class coordinate with but separate from the Musci (Bryopsida) and Hepaticopsida. About 15 years later, their dual nature was discovered independently by two great German botanists, de Bary in 1866 and Schwendener in 1867. Schwendener's studies were especially thorough, and his reports were elegantly detailed, yet many biologists were reluctant to accept his revolutionary hypothesis. Several studies soon demonstrated that the two symbionts, alga and fungus, could be readily grown separately in test tubes or on petri dishes. Over the years, hundreds of species of lichen fungi have been isolated and studied by botanists. Hundreds of attempts have also been made to synthesize complete, reproductive lichens from the isolated symbionts, but few have been even moderately successful.

One of the first to report success in synthesizing lichens was Bonnier in 1886; however, the validity of his work is questioned by some botanists. Thomas in 1939 and Ahmadjian and coworkers, (1980; 1983) in recent years have achieved at least partial synthesis. Considering the ease with which the isolated bionts can be grown under laboratory conditions, it comes as a surprise to most students that only half a dozen botanists have been successful in recreating complete lichen thalli.

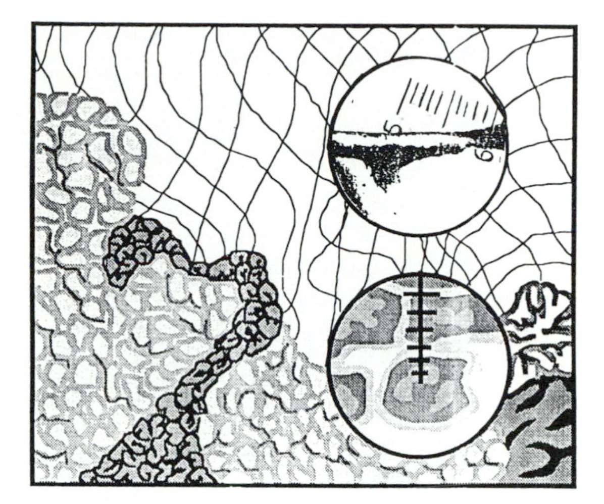

Figure 6–1 Crustose lichens growing on rock; bare basalt above, areolate lichen below, overgrown in part by a squamulose species; foliose lichen (*Lecanora melanophthalma*) at lower right. Upper inset, microscope photo of *L. melanophthalma* cross section showing loose medulla, dense upper and lower cortex and umbilicus; lower inset, areoles of crustose lichen. (Scale: upper inset, each unit = 10 μm, 100 μm from 5 to 6; lower inset, each unit = 0.5 mm.)

Lichens are among the most widespread of all plants. They can be found in almost every terrestrial ecosystem, but they are especially abundant in the forest biomes, the far north, and the high mountains. They are cold tolerant and drought resistant and are important components of and well adapted to tundra, both arctic and alpine, and deserts. On the other hand, few lichens will be found in cities, or around smelters and industries where atmospheric pollution is a problem.

Species identification demands paying careful attention to habit (or growth form) and sporocarp characteristics in addition to the obvious things, like color, size, and substrate preferences, that students immediately see. Habit of some lichens resembles that of liverworts; therefore, students sometimes confuse thalloid liverworts and foliose lichens. Lichens, however, have more highly organized tissues and often have apothecia or perithecia, similar to the sporocarps of other fungi, and liverworts never do. Most people call lichens "moss"; however, mosses have stems and leaves and lichens do not. Color distinguishes them from nonlichenized fungi; they may be orange, yellow, chartreuse, or other bright color or various shades of brown or mineral gray, and some are grassy green like mosses, liverworts, and vascular plants, but all lichens contain chlorophyll and nonlichenized fungi do not. There are over 20,000 species of lichens classified in approximately 40 families and in five of the 27 orders of Eumycopsida.

General Morphological and Physiological Characteristics

A lichen is created when a green or blue-green alga combines with a sac fungus or occasionally a club fungus to produce a plant body entirely different in appearance from that of either

Figure 6–2 Some common fruticose and foliose lichens. (A–D) Foliose: (A) *Lecanora melanophthalma* on bas taken in November; (B) same lichen from same location and distance taken in February in study of rate of growth lichens; (C) *Xanthoria polycarpa;* (D) *Parmelia exasperatula.* (E) Fruticose: *Usnea subfloridana;* (F) unidenti West Coast specimen.

component growing alone. Each kind of lichen is so distinctive that we are inclined to think of it as one specific plant, equivalent to a tree, an herb, or a moss plant. This is why lichens were once classified among the Bryophyta as a distinct class. In lichens, the total is always much greater than the sum of its parts.

The phycobiont, or algal part of the lichen, carries on photosynthesis. In the absence of available moisture, it has the ability to become dormant and thus survive dry periods. The fungal part of the organism probably contributes to the moisture balance and, during periods of dormancy, helps protect the chlorophyll in the algal cells from the bleaching effects of light.

The walls of the fungal cells are heavily gelatinized; when the thallus is wetted by rain, it can imbibe 3 to 35 times its weight in water (Hale 1979). This is apparently one of the reasons lichens are able to survive and grow under arid conditions where moisture from either rain or dew comes infrequently.

Attempts to grow lichens in the botanical laboratory have mostly resulted in failure: either the fungus takes over and destroys the alga, or the alga takes over and destroys the fungus. By varying the environmental conditions, especially moisture, Pearson (1970) and others, e.g., Bryant and Benson (1973) and Dibbin (1971) have obtained measurable growth in lichen thalli under laboratory conditions. Their studies demonstrate the importance of "adverse" conditions in order to keep first the one biont and then the other in check. Unlike single organisms, symbiotic or dual organisms cannot be grown under "optimum" condition. These studies suggest that the synthesis of lichen thalli from separate alga and fungus cultures must be dependent on fluctuating environmental conditions.

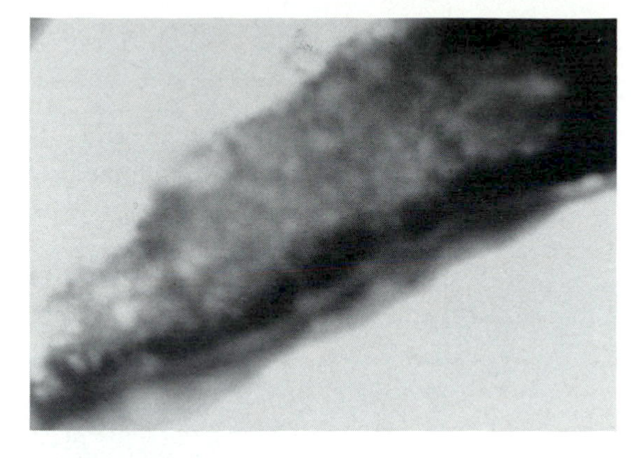

Figure 6–3 Anatomy of *Psora globifera*. Upper cortex is pseudoparenchyma, medulla is prosenchyma; there is no lower cortex.

Ahmadjian (1982), Ahmadjian and Jacobs (1983), and Ahmadjian et al. (1980) have synthesized lichens by placing fungus spores on cultures of algae. They found that the same species of mycobiont could lichenize different species of alga, resulting in essentially identical lichens. Their success confirms the need for alternating moist and dry periods in order to synthesize whole lichens.

Lichens are complex, highly evolved plants. The plant body in most lichens consists of either three or four distinct layers of differentiated cells or tissues (Figure 6–3). Four strata of cells are present in a typical foliose lichen: (1) The **upper cortex** is a protective tissue composed entirely of fungal cells and having a pseudoparenchymatous structure. The upper cortex is heavily gelatinized with substances that are often dark in color when dry but translucent when moist. Like the epidermis of leaves, it protects the inner tissues from drying out, and, in addition, its dark color protects the chlorophyll from the bleaching effect of light during dormant periods. Respiration rates are generally high in moist cortex (Pearson and Brammer 1978). (2) The **algal layer,** or **inner cortex,** as seen by scanning electron microscopy (SEM) is made up of clumps of algal cells and loosely interwoven hyphae of fungal cells (Figure 6–4). Formerly, this layer was often called the **gonidial** layer, stemming from the belief among early botanists that the algal cells were reproductive structures. When abundant moisture is available, photosynthesis takes place at surprisingly high rates (Pearson and Skye 1965, Pearson and Brammer 1978), considering how slowly lichens grow in nature. (3) The **medulla** is a thick tissue composed only of fungus cells; it may be pseudoparenchymatous in a few lichens, but is usually prosenchymatous. The cell walls are weakly gelatinized and much of the water imbibed during rainstorms is held here. Metabolically, the medulla appears to be relatively inactive (Pearson and Brammer 1978). Most of the unique lichen substances, many of which are powerful chelating agents, are stored in the medulla. (4) The lower cortex is anatomically similar to the upper cortex but is usually thinner, and may be darker in color. Lichens may also have reproductive structures on the upper surface (either disk-like **apothecia** or flask-shaped **perithecia**) and organs of anchorage (usually **rhizines**) on the lower surface.

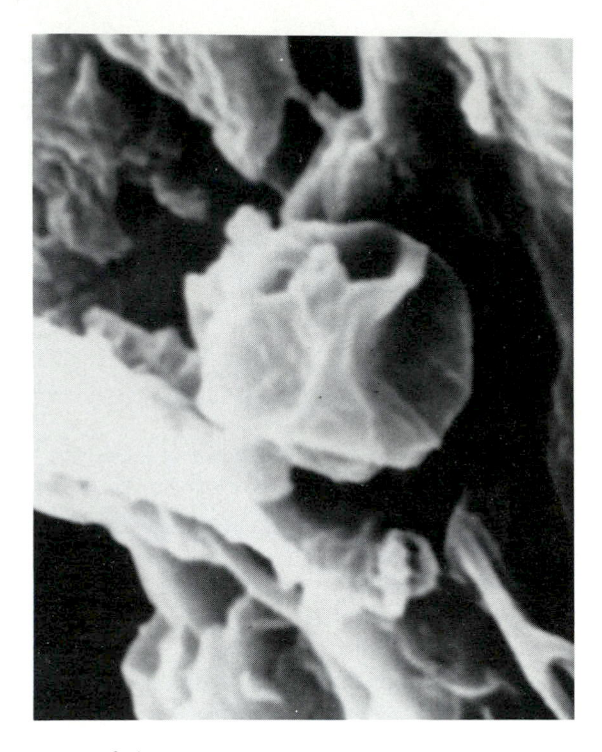

Figure 6–4 SEM photo of inner cortex in *Lecanora melanophthalma.*

Studies of lichen ultrastructure performed with the transmission electron microscope (TEM) show that the fungal cells resemble those of free-living fungi (Chapter 5) and the algal cells resemble those of free-living algae (Chapter 13). Like other Ascomycopsida, lichen mycobionts lack dictyosomes, or Golgi bodies, but possess other organelles typical of the Ascomycopsida; the phycobionts possess the same organelles as other green or blue-green algae (Figure 6–5).

Although lichenized fungi lack dictyosomes, they possess a unique organelle not found in other organisms, the *concentric bodies* or *concentric rings* (Figure 6–6A,B). These are proteinaceous bodies characterized by a large central vessicle surrounded by two or more dense concentric bands. There is often a halo surrounding the entire body. The band closest to the central vessicle often appears to be made up of concentric subbands of slightly varying density. The outer band appears to be made up of units which radiate from the center. It has been suggested that concentric bodies perform functions of secretion and protein purification accomplished by the Golgi apparatus in other organisms. They are found in nearly all species of lichens and have been found in a few species of nonlichenized ascomycetes but nowhere else. They are not found in isolated lichen mycobionts, and their presence in the mycobiont cells being cultured with green algae has been regarded as proof that the attempted synthesis was successful and lichenization complete (Ahmadjian et al. 1980).

In photosynthesis studies conducted by Pearson and Skye (1965), there was evidence of photorespiration occurring in lichens bleached by air pollution; TEM observations reveal the presence of particles in inner cortex cells characteristic of peroxisomes (Figure 6–5).

Although lichens show such a remarkable degree of differentiation, neither the fungal partner (mycobiont) nor the algal partner (phycobiont) is capable of differentiation when grown alone. The complete or intact lichen owes its distinctive morphological and anatomical characteristics to the interaction between alga and fungus, but primarily to the latter. The same species of alga can apparently be the phycobiont of greatly different lichens (Ahmadjian 1982), while closely related fungi tend to produce similar lichens, even when associated with entirely different phycobionts. For these reasons, lichens are classified among the fungi, not the algae, and the lichen name may refer either to the total lichen or to the mycobiont alone. Phycobionts are named independently of the lichens and classified among the Chlorophycopsida and Cyanophycopsida. The phycobionts of most of the more common lichens are species of *Trebouxia* of the Chlorococcaceae (Chlorophycopsida); *Nostoc* species and other Cyanophycopsida are the phycobionts of gelatinous lichens and some other moisture loving lichens.

Crustose lichens consist of a thin layer of cells closely attached to soil, rock, bark, or other substrate; often they resemble splashes of paint on the desert rocks. Upper cortex, algal layer, and medulla are usually well developed in crustose lichens, but they lack a lower cortex, and the medulla typically grows down into the stone or wood to which they are attached. **Foliose lichens** resemble leaves in general appearance and also anatomy, all four tissues being well developed in most foliose species. **Fruticose lichens** resemble the leafless stems and branches of small vascular plants. Some are erect, like miniature shrubs, growing on soil or on other plants; others are pendulous, like delicate vines hanging from the branches of trees, and occasionally reaching a length of a meter or more. The cortex completely surrounds the thallus; inside it is a ring of photosynthetic tissue; the medulla occupies the center of the thallus.

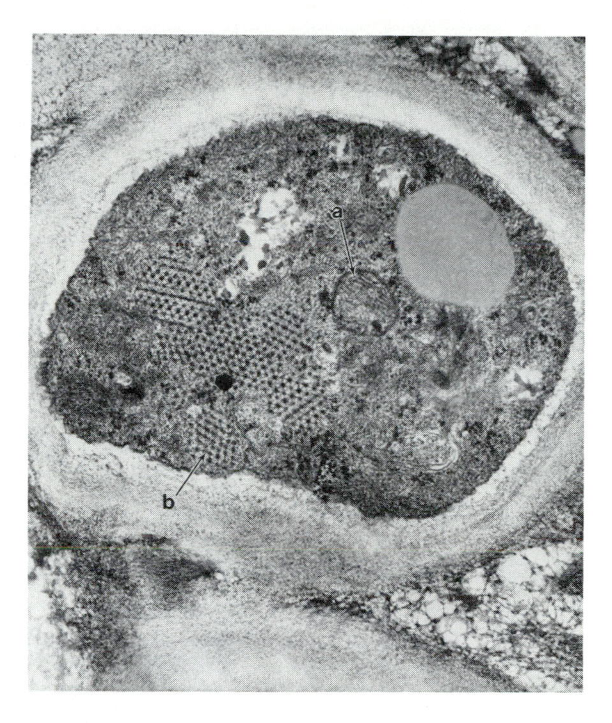

Figure 6–5 TEM of a healthy algal cell from the inner cortex of *Lecanora melanophthalma*, X62,000. The plasma membrane has pulled away slightly from the cell wall as a result of dehydration during the fixing and staining process. Mitochondria are visible (a), one near the center of the cell, below and to the left of a large starch grain; at lower left several peroxisomes are clumped together (b). Dark hexagonal bodies, carboxysomes, believed to be involved in the Calvin cycle of photosynthesis, are scattered throughout the cell. (From Pearson, 1988, unpublished INEL report; micrograph courtesy W. M. Hess and Brigham Young University Electron Microscopy Laboratory, Provo, Utah.)

Crustose lichens may be either areolate or squamulose (Figure 6–1). **Areolate lichens** usually have a "chinky" thallus which is typically broken up into small squares or polygons (called areoles) that are firmly attached to the substrate at all points by **hyphae,** or fungal filaments, growing out from the medulla. In some species, the thallus is **granular**: the areoles are reduced to small grains which may be widely spaced. Sometimes the margin of the thallus is lobed, or **effigurate,** and thus approaches foliose; crustose lichens are **endolithic**. These "*fenster flechten*" consist only of an apothecium with little or no thallus surrounding it, most of the thallus being buried in the rock. Light reflecting on the rock crystals penetrates into the substrate, where photosynthesis takes place. **Squamulose lichens** consist of multicellular scales, or **squamules,** each of which has an upper cortex, algal layer, and medulla. Along its raised margin, there is also a lower cortex; otherwise the medulla grows into the substrate as in areolate species. Therefore, squamulose lichens appear to be an intermediate type between foliose and crustose.

Foliose lichens may be either gelatinous or stratified. In **gelatinous lichens,** algal and fungal cells are not differentiated into distinct tissues but are mingled together throughout the thallus. The **phycobiont,** or algal component, is always a cyanophyte; consequently, the thallus is a characteristic bluish black color. When wet, the gelatinous materials which surround cyanophyte cells swell and give the entire thallus a gelatinous appearance. **Umbilicate lichens**

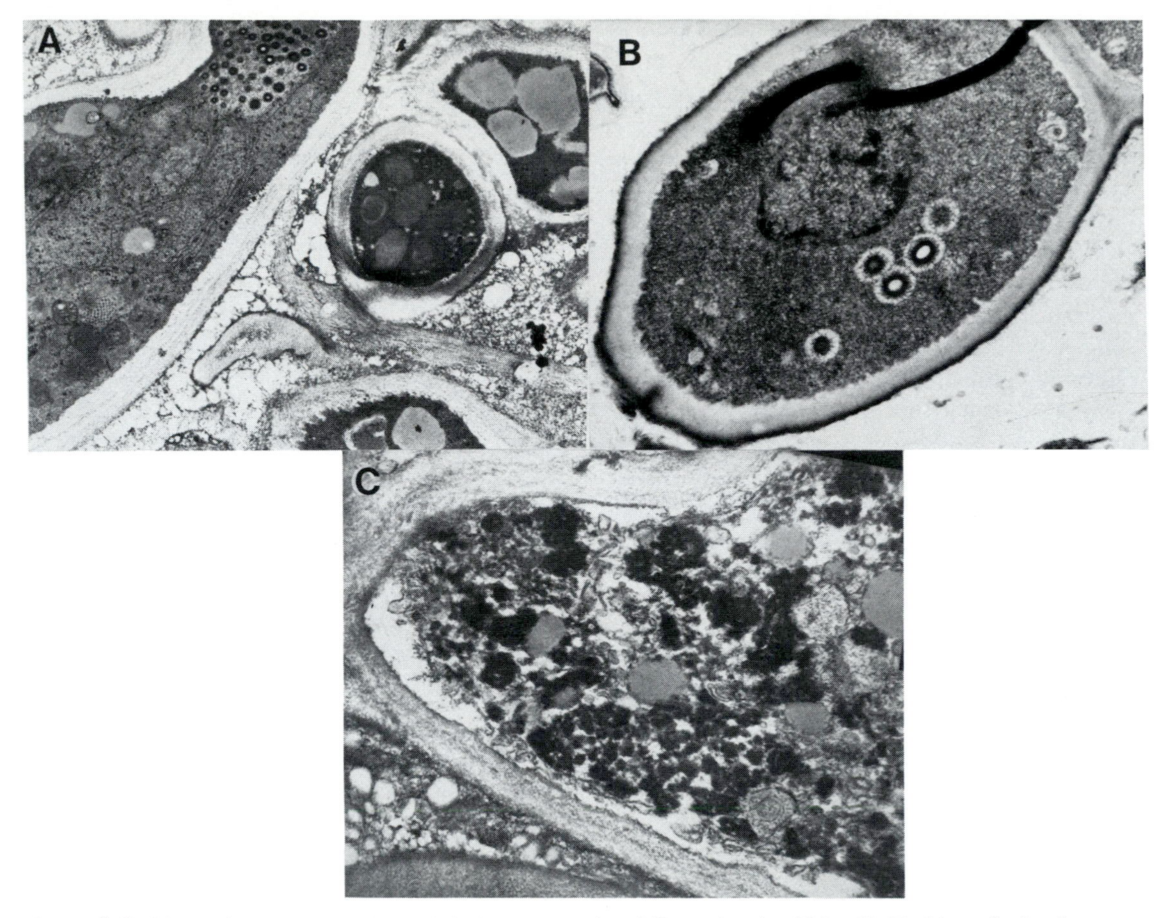

Figure 6–6 TEMs of *Lecanora melanophthalma* exposed to different levels of SO_2. (A) Healthy cells that have not been exposed to SO_2 (c = concentric rings which presumably function as dictyosomes or Golgi bodies; p = presumed peroxiosome); (B) cells after exposure to 0.0003% SO_2 for 10 days; (C) cells after 2 hours exposure to 1200 ppm in laboratory. (Arrows point to some of the ruptures in the plasma membrane.)

(Figure 6–2A, B) are stratified into the four tissue layers. The entire thallus is attached to the substrate, which is nearly always rock, by a multicellular strand called an **umbilicus**. When moist, the thallus can usually be "popped" off the rock in one piece by prying gently with a knife blade at the single point of attachment. Most foliose lichens (Figure 6–2C, D), however, are attached to the substrate by numerous simple or branched organs of anchorage called **rhizines**. Like the umbilicus, a rhizine is a complex organ, usually made up of prosenchymatous tissue, which grows out of the lower cortex and medulla. Many lichenologists reserve the term "stratified lichen" for rhizinate species plus the lichens which have neither rhizines nor umbilicus but are attached by hyphae extending from the lower cortex.

Fruticose lichens (Figure 6–2E, F) may be either erect (**tufted**) or **pendulous,** may be hollow in the center or possess a solid medulla, and may or may not have a dense, distinct central cord. They may be round in crosssection, or angular and more or less flabby. Most are limited in their distribution to habitats having relatively high humidity. Like foliose lichens, they are generally easy to identify. Some fruticose species have two kinds of thallus, a primary squamulose thallus and a secondary thallus, called a **podetium**. Podetia are erect or shrub-like.

Vagrant lichens are usually umbilicate or fruticose species which are loosely attached to the substrate, often soil, so that they readily break loose and are blown along the ground like tumbleweeds. *Parmelia chlorochroa* and a vagrant form of *Lecanora melanophthalma* are common on rangeland (steppes) of the American West. *Lecanora esculenta,* the Biblical manna, is a vagrant on the Arabian desert. Normally foliose, during periods of prolonged drought the thallus curls up and is easily blown by wind.

Foliose and fruticose lichens are easily identified, in most cases, by gross morphological features alone, but it is often necessary to examine the spores in order to identify crustose lichens. Ascospores may be hyaline (colorless) or dark in color. Some have a dark center with a hyaline halo surrounding it. They may also be simple, septate one or more times, **polarilocular** (two-celled with a thick wall between the cells and an isthmus connecting them), or **muriform** (divided transversely and longitudinally into many cells). These are illustrated in Figure 5–6.

As in other Ascomycopsida, lichen meiospores are produced in sporocarps, either disk-like apothecia or flask-shaped perithecia. Hard, dry apothecia, similar to those produced by *Pseudopeziza* and other Helotiales, are the most common. Each sporocarp results from the union of two haploid nuclei. One comes from the ascogonium, which is frequently provided with a trichogyne, and the other is the spermatium. Since karyogamy is delayed, most of the cells in the developing sporocarp are either haploid or dikaryotic. The uppermost tier of dicaryotic cells forms the **hymenium**. Karyogamy takes place in the hymenial cells, followed by meiosis and (usually) one mitosis. This results in 8 ascospores in each ascus of most lichens, although some species have 4, some 16, and one very common genus, *Acarospora,* typically has at least 64 spores per ascus.

The sporocarps, both apothecia and perithecia, are persistent and usually produce spores throughout the year. In many species, however, only spores produced between January and May are viable. When a viable fungal spore lands on a colony of algae growing on a moist piece of bark, rock, or soil, it begins growing, invades the algal cells, and develops into a lichen thallus in the same way that pathogenic fungi invade the cells of vascular plants. We hypothesize that this is a common occurrence, although no one seems to have ever observed the complete process in nature.

The nature of the the apothecia or perithecia, if present, is a very important characteristic to note when studying lichens. Sporocarps tend to be the most uniform and hence reliable of the morphological and anatomical features of a species. A lichen may produce either apothecia or perithecia but never both.

In some species, the algal layer extends up into the rim or **exciple** of each apothecium. In others, the exciple is "proper," meaning that the apothecium contains no algal cells. Apothecia with proper exciples are called **lecideine** or **biatorine,** and those having thalloid exciples, with algal cells in the rim, are called **lecanorine**. In most lichens, the disk of the apothecium is different in color from the vegetative body or lichen thallus (Figure 6–2). Lecanorine apothecia, therefore, tend to have a contrasting ring, often green in color, surrounding each apothecium whereas lecideine apothecia generally do not. However, even in lecideine species, all or part of the exciple may be raised and/or colored differently from the disk even though there are no algal cells in it, while in lecanorine species, the green rim is not always apparent. In field studies, therefore, it is often necessary to cut a section through a moistened apothecium with a razor blade or a pocket knife and examine the section with a hand lens to see if algae are present in the rim. When identifying lichens in the laboratory, thin sections of the apothecium can be examined with either a dissecting or a compound microscope.

The ostiole, or opening at the top of the perithecium through which the spores escape, is usually slightly to noticeably raised above the surface of the thallus; at times, however, it may be level or sunken. An **involucrellum** may envelop the perithecium, or sometimes only the neck or upper portion of the neck and the ostiole. Involucrella are black, carbonized structures typical of many nonlichenized fungi as well as lichens. In some genera (*e.g., Microthelium*), the involucrellum is so massive that it is easily mistaken for an apothecium.

Some species reproduce primarily, or possibly exclusively, by asexual processes. Asexual reproduction may be by **fragmentation** and **reattachment** of a portion of a lichen thallus, by isidia, by soredia, or by conidia. **Conidia** were discussed in the previous chapter. **Soredia** look like spores but contain both algal and fungal cells. Soredia are produced in pustules, called **soralia,** which somewhat resemble the **acervuli** of nonlichenized ascomycetes. Being small and light, soredia may be carried many miles by the wind. **Isidia** are finger-like, plate-like, or cylindrical growths on the upper surface of the thallus. Each isidium consists of differentiated tissues and contains both algal and fungal cells. If apothecia are abundant on a lichen, there are generally no soredia or isidia present; if soredia are abundant, ascocarps tend to be rare or lacking.

When either conidia or ascospores germinate, they develop into fungal mycelia which must thean form an association with algae in order to produce a lichen. Both sexual and asexual reproduction of the mycobiont is generally independent of the phycobiont. However, ascospores of some lichens having lecanorine apothecia will sometimes have algal cells adhering to them.

Some evidence suggests that several species of lichen may all have the same algal species as phycobiont and that the same species of lichen may, at different times or places, have different, though closely related, species of algae as the phycobiont. Nature seems to have no end of tricks with which to confuse taxonomists!

Conidia in lichens are always produced in pycnidia. Since pycnidia superficially resemble perithecia, care must be taken when identifying lichens not to confuse the two. Perithecia produce ascospores which are always enclosed in asci interspersed within the perithecium among slender filaments called **paraphyses**. Pycnidia lack paraphyses, and their spores, the conidia, are never produced in asci or other sacs.

Four simple chemical tests are widely used in studying lichens: the K test, C test, KC test, and Pd test. Unique chemical substances, commonly but inaccurately referred to as "lichen acids" in the older literature, are found in the cortex and/or the medulla of most lichens. Many have a specific reaction with one or more of the special chemical reagents used by lichenologists. For example, the C test may be used to distinguish between *Lecanora melanophthalma* and *Lecanora novomexicana.* The upper cortex and algal layer are first scraped away. A small amount of C reagent, a pungent solution of bleaching liquid, is then applied to the exposed white medulla. *L. melanophthalma* is "C+ yellow," meaning that the medulla slowly changes in color from white to yellow; whereas *L. novomexicana* is "C−," meaning no color change takes place. The K and Pd tests are conducted in a similar manner. If a specimen tests K− and C−, the KC test may be made by applying first "K" and then "C" and watching for a change in color. The K reagent is a saturated solution of potassium hydroxide or other strong base. To prepare it, dissolve a few flakes of KOH in 20 ml of distilled water. It will keep indefinitely.

The Pd reagent is prepared by dissolving a few grains of *para*-phenylenediamine in alcohol to produce an approximately 5% solution. Neither C nor Pd solutions keep well, but must be made up fresh every few days. All of these reagents, especially Pd, are very corrosive and will damage clothes if they come on them.

Other simple tests often used by lichenologists to aid in the identification of species are the UV test, the starch-iodine test, and the limestone test. The UV test is made by exposing the lichen to ultraviolet light in a dark room and watching for fluorescence. The starch-iodine test is made by placing a drop of Gram's iodine solution on the medulla or the inner tissues of the apothecium and watching for development of a blue, or occasionally a maroon, color. The limestone test is made by putting a few drops of dilute hydrochloric acid on the rock the lichen is growing on to see whether it is calcareous (HCL+) or noncalcareous (HCL−). If it effervesces—that is, if bubbles of carbon dioxide gather on the surface of the rock−it is calcareous.

Much of the scientific interest in lichens since about 1965 was stimulated by a growing awareness among ecologists and plant physiologists of the sensitiveness of most species of lichens to air pollution. Lichens were first used to monitor air pollution about 1900 by F. G. C. Arnold, who transplanted healthy lichens into the industry-polluted atmosphere of Munich. Prior to that time, Nylander (1866) and others had observed that lichens were scarce in cities and attributed their scarcity to atmospheric polluton (Ferry et al. 1973). Since the 1940s, lichens have been widely used as air pollution monitors. Lichens are excellent monitoring organisms because they are very sensitive, as a group, to pollution, because they are long-lived, and because there is considerable variation among species in sensitivity.

Most of the first studies using lichens to monitor pollution evaluated only presence or absence of species. Statistical analysis often revealed significant differences associated with distance and direction from a point source of air pollution. A recent study illustrates this. *Xanthoria polycarpa* is a bright orange foliose lichen found on nearly 100% of sagebrush and rabbitbrush plants in cold desert and steppe ecosystems over much of western North America; if it is present, it can be seen from far away. Table 6–1 shows how the abundance of this lichen was affected by distance and direction from a source of pollution. Plants along the transect toward which the wind blows in early morning hours when the lichens are wet with dew and hence most susceptible to injury had the fewest lichens on them. Instead of only presence or absence, lichens are now also used to study rates and patterns of photosynthesis and respiration, nitrogen-fixation rates, injury to cell membranes (Figure 6–6), and accumulation of toxic elements in lichen tissues. Methods and equipment have included transplanting lichens from pollution-free areas to polluted areas, electron microscope observation of cell organelles, oxygen electrode measurement of gas exchange, microrespirometers, gas chromatography, energy dispersive spectroscopy (EDS), and conductivity meters to measure the efflux of electrolytes from lichen thalli. The Cartesian diver (Figure 6–7) is a microrespirometer so sensitive that it can measure photosynthesis in slices of lichen tissue weighing less than 50 μg or in individual dinoflagellate cells.

Figure 6–8 illustrates the effect of air pollutants on photosynthesis as measured with the Cartesian diver (Pearson et al. 1981) in laboratory experiments. When photosynthesis and respiration were measured in cortex and medulla separately, the cortex was far more active metabolically than the medulla, suggesting that the latter is primarily a storage tissue (Pearson and Brammer 1978).

Most lichens are sensitive to air pollution, but there are several species which tolerate relatively high levels of pollution. A good biomonitor species needs to be widespread and

TABLE 6-1

FREQUENCY OF *XANTHORIA POLYCARPA* ON 2,400 SAGEBRUSH AND RABBITBRUSH PLANTS RELATIVE TO DISTANCE AND DIRECTION FROM THE IDAHO CHEMICAL PROCESSING PLANT

		Nighttime Downwind		Daytime Downwind		Ave.
Distance		**215°**	**135°**	**55°**	**295°**	
(km)	**AQI:**	**1**	**2**	**3**	**4**	
1		39	77	63	55	58 ab
2		26	22	82	74	51 ab
5		31	38	61	47	44 a
10		67	89	47	68	68 b
Average		41 a	57 b	63 b	61 b	55

ANOVA

	df	ms	F
Among transects	3	302.5	3.36*
Regression component	(1)	(673.4)	7.49**
Remainder	(2)	(117.1)	ns
Among distances	3	300.9	3.35*
Error	41	89.9	—

Note: Averages not followed by the same letter of the alphabet are significantly different from each other at $P \leq 0.05$ (AQI = Air Quality Index, based on previous evaluations of electrolyte leakage data, 1 indicating poorest air quality; frequency = percent of plants with the lichen on them; df = degrees of freedom, ms = mean square or variance, F = Fisher ratio for testing statistical significance).

*Significant at $P \leq 0.05$; **Significant at $P \leq 0.01$.

abundant in the area where it will be used, easily identified in the field, and sensitive to pollution. The best quality control program makes use of several indicator species, some highly sensitive and others relatively tolerant. Table 6-2 presents some advantages and disadvantages of nine species or groups of species as biomonitors of air quality in arid regions.

An article was recently published in which the reader was informed that fruticose lichens are more sensitive to air pollution than are foliose species, and crustose are the least sensitive. A perusal of the existing data on lichen tolerance indicates that all three categories are approximately equal in detecting and developing pollution control programs (Table 6-3).

Phylogeny and Classification

At least two hypotheses have been advanced as to the origin and evolution of lichens. One holds that lichens are essentially monophyletic, at least within each of two or three major groups. The other proposes that lichens are polyphyletic. Until more detailed studies have been made, employing pure mycobiont cultures, of the biochemistry, anatomy, and other characteristics listed in Table 1-2, it will not be possible to decide to every lichenologist's satisfaction which hypothesis is correct.

Considerable evidence supports polyphyletic origin of lichens. Comparison of spores and sporocarp characteristics of lichens with those of nonlichenized fungi reveals pronounced similarities; fatty acid content and protein composition of pure cultures of mycobionts compared with that of similar cultures of nonlichenized fungi grown under essentially identical conditions suggest some relationships not apparent from observation of the lichen thallus and fungal mycelium alone. Cell wall chemistry and structure, nature of the nuclei and chromosomes, and ultrastructure of cellular organelles (as revealed by electron microscopy) show promise in elucidating lichen taxonomy (Pearson and von Hofsten unpublished data). Cell shape and cell wall characteristics of mycobiont cultures of *Lecidea coarctica,* for example, resemble those in cultures of the non-lichenized fungus, *Patellaria,* more than those of *Xanthoria parietina, Rhizocarpon geographicum,* or other mycobionts of the lichens supposedly closely related to *Lecidea.* These differences and similarities correlate with those of the spores and sporocarps and clearly suggest a polyphyletic origin of lichens.

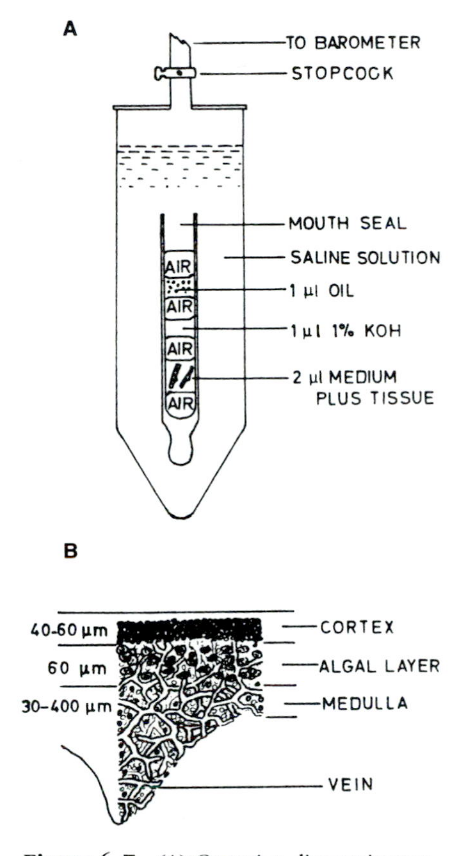

Figure 6–7 (A) Cartesian diver microrespirometer used to meassure respiration and photosynthesis in small amounts of plant or animal tissue. (B) Thin section of lichen tissue in (A) *Peltigera Canina.*

Unfortunately, we do not have a good fossil record to help us understand when and where lichens originated and how lichens are related to each other or to other ascomycetes, nor are we able to hybridize lichens experimentally in studying relationships. Lichens have not fossilized well; when placed in a humid environment, they quickly become soft and pliable. Soon the mycobiont and phycobiont begin to dissociate from each other and decompose. Another contributing factor explaining the poor fossil record is the lack of knowledge and interest, among paleobotanists, in lichen morphology. In the absence of a good fossil record, it is necessary to infer relationships from the similarities and differences observed in lichens compared to those of other sac fungi, as well as each other, as suggested in the preceding paragraph.

One of the ongoing arguments among lichenologists is whether or not "chemical strains" are separate species. Lichen species often occur in pairs (occasionally in sets of three or more). Morphologically, the two species making up the pair are very similar, often indistinguishable. One specieswill consistently have a particular chemical in either the cortex or the medulla, and its twin will not. In the morphological species *Cladonia chlorophaea,* for example, five distinct chemical strains are known. One is characterized by having fumaroprotocetraric acid in the medulla, the second by the presence of grayanic acid, the third by cryptochlorphaeic acid, the fourth by merochlorphaeic acid, and the fifth by perlatolic acid. Each has its own species name; but, are they really separate species or are they merely chemical strains of the same species?

A related problem has to do with the presence or absence of soredia and/or isidia. Often one member of a pair will consistently have soredia and seldom form apothecia whereas the other member will never produce soredia but produces apothecia profusely. Although the tendency today is to combine pairs of species that differ from each other only in their chemistry or in the presence or absence of soredia, not everyone agrees.

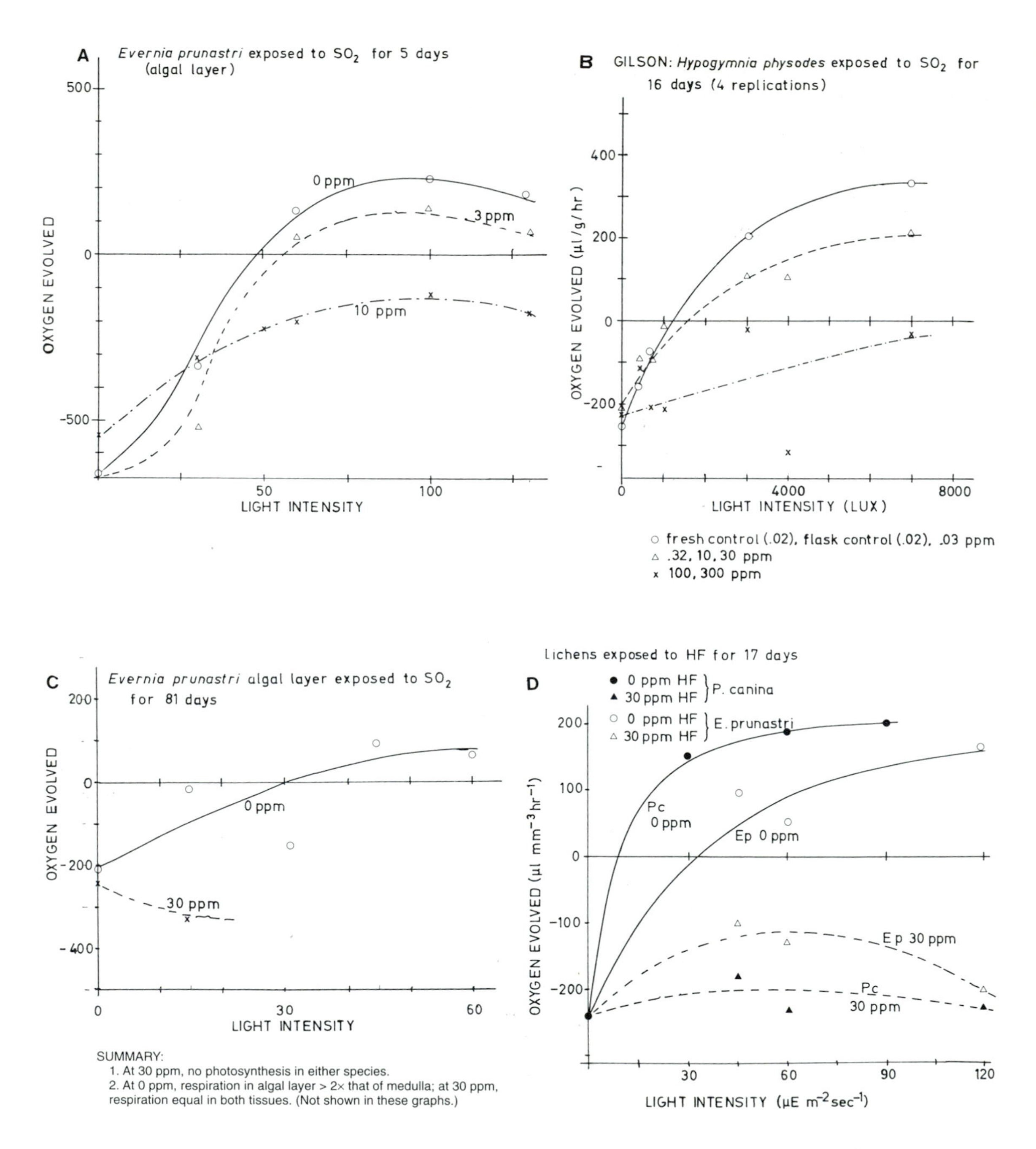

Figure 6–8 Effects of pollution on photosynthesis. Experiments conducted in laboratory flasks (stagnant air) with pure gases at exact concentrations; photosynthesis (rate of O_2 evolved) measured by Gilson respirometer (B) or by Cartesian diver (A, C, D). Data from Pearson, Henriksson, and Brammer (1981).

TABLE 6-2

CHARACTERISTICS OF SELECTED LICHENS CONSIDERED AS POTENTIAL BIOMONITORS IN ARID AND SEMIARID HABITATS

Species	Advantages	Disadvantages
Lecanora melanophthalma	Easily collected; umbilicate habit assures freedom from substrate; thallus thick, easily sectioned; pollution sensitive	Limited in distribution primarily to basalt outcroppings
Parmelia exasperatula	Corticolous, easily harvested, common; relatively sensitive to some pollutants	Rhizines adhere to bark tightly making it difficult to get clean specimens free of bark
Xanthoria polycarpa	Corticolous, easily obtained, common; bright color makes it easy to spot	Rhizines adhere tightly to bark; seems to be relatively tolerant to most pollutants
Fulgensia desertorum	Easy to spot and relatively easy to identify	Granular habit difficult to work with
Xanthoria fallax, *X. candelaria*	Relatively common, easily spotted and easily obtained; bright color	Difficult to distinguish from each other without good hand lens
Psora globifera	Widespread and abundant	Minerals in medulla damage microtome blades; seems to be pollution tolerant
Collema spp.	Abundant, foliose, easy to separate from substrate	Small, thin thalli may be difficult to work with
Caloplaca spp.	Abundant; bright color makes them easy to spot	Crustose; extremely difficult to get clean specimens; difficult to distinguish among species

Culberson et al. (1969) studied the chemical nature of partially lichenized mycobionts of the *Cladonia chlorophaea* complex and observed hybridization in taxa which morphologically and chemically appear to be separate species. In the *Ramalina siliquosa* complex of seashore-cliff fruticose lichens, similar results were observed. By studying the chemistry of single spore mycobiont cultures from the apothecia of each of the five chemical races of *R. siliquosa*, gene flow among the five **chemotypes** which dominate the supralittoral zones (the zones immediately above the intertidal or littoral zone) of northern Wales was analyzed (Culberson et al. 1993). Hybrids were identified by the presence of **depsidones** (a group of lipid chemicals common in lichens but rare in other organisms) or depsidone precursors in the cultures grown from the isolated sporelings that were not present in the apothecia from which the spores were isolated. With one exception, the chemotypes, or chemical races, observed in this study have evolved to the point that the lichenologists studying them regarded them as sibling species. This study also demonstrated that **spermatia** (nonmotile sperm) functioned only over short distances.

There are approximately 20,000 species of lichens classified in 30 families and 5 orders of Eumycophyta. As pointed out in the previous chapter, the Sphaeriales and Helotiales were probably the first of the Eumycophyta to evolve. They appear to have originated as saprophytes on driftwood in the ocean and migrated to land where they were able to colonize dead trees. Most lichens seem to be closely related to the nonlichenized fungi in these two orders and the Pseudosphaeriales. Another half dozen genera of tropical lichens are classified in two families

TABLE 6–3

RELATIONSHIP BETWEEN GROWTH FORM AND SENSITIVITY TO AIR POLLUTION IN 84 NORTH AMERICAN SPECIES AND 43 EUROPEAN SPECIES OF LICHENS

Species Sensitivity[a]	Crustose	Foliose	Fruticose	Total
Toxitolerant Species				
North America	7	9	6	
Europe	5	4	1	32
Moderately Sensitive				
North America	10	21	13	
Europe	6	9	3	62
Sensitive Species				
North America	11	13	4	
Europe	2	9	4	43
Total	41	65	31	

Chi-squared test for homogeneity: $\chi^2 = 2.05$, $P > 0.10$. (No evidence that any habit type is more or less sensitive than the others.)

[a] Sensitive North American species include *Lecanora dispersa* (Cr), *L. impudens* (Cr), and *Physcia aipolia* (Fo); moderately sensitive species include *Candelariella vitellina* (Cr), *Xanthoria parietina* (Fo), *Alectoria nidulifera* (Fr), and *Usnea* spp. (Fr); toxitolerant species include *Parmelia sulcata* (Fo), *Physcia dubia* (Fo), and *Cladonia coniocraea* (Fr). (Cr = crustose; Fo = foliose; Fr = fruticose.)

Data from several sources including LeBlanc and associates 1970–1972 and Ferry et al. 1973.

of Aphyllophorales and one of Agaricales in the Basidiomycopsida. Nevertheless, many lichenologists believe that most lichens are more closely related to each other than to the nonlichenized fungi. Henssen and Jahns (1974) classified lichens in 11 orders, several of which consist primarily or exclusively of lichens, and 46 families.

Another problem, common to all branches of botany, is the naming of a new species on the basis of a single specimen. At times, the specimen, or the entire population of which it was a part, owes its uniqueness to soil, temperature, wind, lightning, or some other environmental factor. Weber, curator of a lichen herbarium in Boulder, concluded a presentation on lichen taxonomy with a plea to botanists to try to see how few, not how many, species are comprised in the flora of their districts, and a hope that the one-specimen species will become a thing of the past. By taking into account synonyms and misidentifications, Sam Shushan and Roger Anderson (personal communications) reduced their check list of 780 species of lichens reported from Colorado to 448. Hale (unpublished report), at the 1984 Lichenology Meetings estimated that although 23,000 species of lichens have been described worldwide, the real number is probably less than 20,000.

Another problem in modern lichen taxonomy is the large number of species in some genera. Two, *Lecanora* and *Lecidea*, are each reported to have over 2000 species, a very unwieldy number for anyone to work with. Proposals to divide these two, along with *Parmelia*, *Alectoria*, and some other large genera, have recently been made. Not all botanists agree.

However, proposals to split *Lecidea*, the largest genus in the Plant Kingdom in number of species (approximately 2900), seem to be well accepted.

Ecological Significance

Lichens are important ecologically in at least four ways: (1) as pioneers in ecological succession, (2) as feed for some wild animals, especially some large game animals and some small rodents and snails, (3) in the spread of fire, and (4) in nitrogen fixation. Lichens in turn, like other organisms, are influenced by the factors of their environment. Many lichens are valuable as indicators of microclimates.

Long recognized as pioneers in xerarch succession, some crustose lichens are able to break down even the most resistant granites and quartz. This was formerly attributed to the corrosive action of "lichen acids." But the acids in lichens are weak organic acids not able to disintegrate rock. Schatz and his co-workers in the late 1950s demonstrated that many lichen substances are chelating agents which can break down rock. The process is slow, both in the laboratory and in nature. For example, it took 20 years for the first lichen to become established on boulders exposed by a gold dredge operation on the Yankee Fork of the Salmon River in Idaho. However, within another 20 years, pockets of crumbled rock and organic matter had allowed other plants to begin growth (Pearson unpublished data). The study of ecological succession is an important, but generally ignored, area of research.

Alectoria sarmentosa, A. fremontii, and *A. jubata* are the most important food sources for the woodland caribou of northern Idaho, Alaska, and western Canada. As climax forests have been cut down, the caribou have disappeared from the rest of the U.S. and from eastern Canada. Even when the climax species reappear, following the long ecological succession cycle, caribou are not able to survive. Forests must be old, as well as climax, for caribou. In an old forest, many branches fall and a certain percent of the trees themselves are downed by wind each winter; the lichens, which grow high in the crowns of conifers, primarily spruce (*Picea* spp.), become available to the caribou by this means. Ground lichens are also utilized by the woodland caribou, but they are not as important to them as they are to the barren ground or tundra caribou. *Cladonia* species, especially the "reindeer moss", *Cladonia rangiferina,* and species of *Stereocaulon* make up the bulk of the feed for tundra caribou.

Squirrels and chipmunks are often seen nibbling on umbilicate lichens and even crustose species, and snails frequently feed on foliose lichens. The extent of small animal grazing on lichens and the amount of nourishment obtained has not been carefully studied, however.

The influence of lichens in spreading fire is well documented. When dry, most pendulous lichens catch fire easily. Our pioneer forefathers knew this and frequently carried lichens in their tinder boxes before the invention of "lucifer sticks" or matches. Firefighters and observers of the Yellowstone Park fires in the summer of 1988 could be a mile or more away from the fires and see blazing material float through the air and ignite the trees near by. In some cases, both the blazing material coming from the distant fire and the tinder which first caught fire in the new area were lichens.

Gelatinous lichens, such as *Collema tenax,* and other lichens with cyanophycean phycobionts, like *Peltigera canina,* are important nitrogen fixers in many ecosystems (Figure 6-9). Based on measurements of fixation rates made with the acetylene reduction method in botanical laboratories and biomass measurements made in field studies, it has been calculated that lichens in some desert ecosystems fix about 35 kg/Ha nitrogen annually (Pearson, 1973,

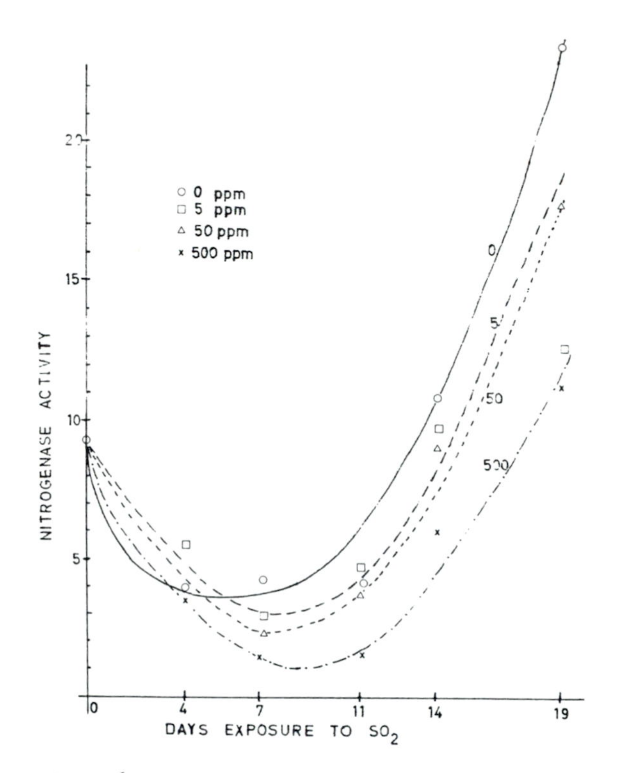

Figure 6–9 Nitrogen fixation rates in *Peltigera aphthosa,* a terricolous foliose lichen, under laboratory conditions. Rate of fixation is reduced as the lichen is brought into the laboratory, but as it becomes acclimatized to the new environment, the rate gradually increases.

IBP unpublished Desert Biome report). Studies by Elisabet Lars-Erik and Henriksson (1984) on Surtsey, an island off the south coast of Iceland, suggest that nitrogen fixation is important in the establishment of pioneer plant communities following volcanic activity.

Lichen species are relatively constant as to the substrate on which they grow. Some **saxicolous** species grow only on calcareous rocks, while others grow only on acidic or noncalcareous rocks. **Corticolous** species grow on bark, some only on what Du Rietz referred to as the "noble" trees (oak, maple, ash, etc.), and others on aspen, birch, poplars, and conifers. **Lignicolous** species grow directly on wood; **terricolous** species grow on soil, sometimes associated with terricolous mosses. However, most of these species will occasionally be found on other substrates. Saxicolous species may be found on wood or bark, for example, or on rock of different pH from the preferred type; while the corticolous lichens normally found on "noble" trees will sometimes be found on aspen, birch, or conifers, and vice versa.

In most desert areas and in the alpine tundra, crustose species are especially abundant, both on rocks and on bark. Some of the crustose lichens are probably the most drought tolerant and also the most cold tolerant of all plants. Foliose lichens are more typical of the coniferous and deciduous forests where they are especially abundant on bark but also occur on soil and on rocks. Fruticose lichens are rare in desert and alpine habitats, although they are occasionally found there, but are especially abundant in the Arctic tundra and in humid environments.

Because lichens are small and long-lived, they are ideal organisms for identifying changes in microclimate (Pearson and Lawrence 1965, Pearson 1969). Lichens, as a rule, do not do well if environmental factors are constant. Humidity and temperature, especially, must vary between day and night or over a period of a few days if lichens are to survive; otherwise, one of the partners will outgrow the other and destroy the lichen association (Pearson 1969, 1970; Pearson and Benson 1977). Nevertheless, some species tolerate relatively constant high humidity while others require weekly fluctuations in humidity; some require considerable variation in temperature between day and night, others do well when the temperature does not change appreciably.

Economic Significance

Lichens are used to dye the famous Harris tweeds in Scotland and are important in home dyeing in other countries, but the total dollar value of lichen dyes is not great. Yet at one time lichens were very important in the textile industry. At the beginning of the scientific revolution of the late 19th century, they were also an important source of the reagents used in biology to stain plant and animal tissues for microscopic examination and as indicator dyes in chemistry. Litmus, used to differentiate between acids and bases in the chemical laboratory, was extracted from *Roccella;* orcein, an excellent reagent for staining chromosomes, is found in many of the *Parmelias, Lecanoras,* and other common species. Coal tar synthetics have largely replaced them for these uses.

Our great-grandparents, if they were involved in home spinning, weaving, and dyeing, kept a wooden barrel in the back yard into which each night's accumulation of urine was poured. Lichens were collected in the woods near the home. Ammonia from the stale urine was used to extract pigments from the lichens. These pigments dyed the wool and other fabrics used for making clothes. The dyeing process, developed in Italy during the late Middle Ages, took skill and care in every step along the way. Lichens collected from conifers often contained sticky gums and were generally inferior to those collected from hardwoods. Urine from beer drinkers produced low quality, dilute ammonia having too low a pH.

Perfume manufacturers in western Europe and the U.S. import hundreds of tons of lichens annually, primarily *Evernia prunastri* or "oak moss" from the Balkan countries collectively called Yugoslavia until recently. Most soaps and bath powders contain a small amount of lichen substances as part of the fragrant essence in them.

In expensive perfumes, lichens are carriers for fragrant, aromatic compounds. At one time, lichens were even more important in the perfume industry, but coal tar synthetics, alcohol, and acetic acid have replaced them as carriers.

By the end of the 15th century, many of the species of lichens that were preferred for dyes and perfumes had become extinct in Italy. Northern Europe was emerging as an important producer of textiles. Italians were importing large quantities of lichens from the Orient for the production of dyes and perfumes, but the trade routes went through areas where the people were hostile to Europeans. Some historians believe that it was the lichen trade, more than the spice trade or the trade for medicinal herbs, that Columbus sought in 1492.

In Europe, "usno", an antibiotic obtained from *Evernia* and several other lichens, is used to treat mastitis and various skin ailments, including some forms of tuberculosis. For control of mastitis, it is applied as a salve to the nipples and is reported by Italian physicians to be completely safe for nursing infants. Usnic acid, the active ingredient, is not water soluble and has not, therefore, become popular in the U.S. for human use, but it is being used as a veterinary medicine to control mastitis and other udder infections.

Usnic, lecanoric, haematomic, and vulpinic acids show promise in controlling plant diseases. Usnic acid, for example, is effective in controlling tomato canker, a disease caused by a Gram-positive bacterium. The lichen antibiotics resemble penicillin and streptomycin in that they are ineffective against Gram-negative organisms. Usnic acid also exhibits low level control against lung carcinoma, while some other lichen substances are known to be mutagenic.

At the present time, nine substances obtained from species of *Pulmonaria* and seven from *Cetraria,* in addition to usnic acid obtained from several species of lichen, are imported for use by the American pharmaceutical industry.

Research Potential

In addition to their role in monitoring gaseous pollution, lichens also accumulate heavy metals and other polluting elements in their tissues. Because each element when bombarded by electrons gives off characteristic X-rays, the energy dispersive spectrometer (Figure 6–10) can identify elements that have been absorbed by lichens. A beam from an SEM is focused on cells in the tissue being investigated, and a detector sends the data to a computer which prints out relative quantities of the elements present. When EDS analysis was compared with traditional macroassay methods (Pearson 1993), correlation coefficients were very high ($r^2 = 0.90$ to >0.99) for some elements most of the time, though not high for others. Metabolically essential elements, like Mg, P, K, and S, often showed negative correlations as would be expected if the pollution was harmful to the cells. Figure 6–11 shows the result of one study conducted at the Idaho National Engineering Laboratory, a U.S. Department of Energy facility on the Arco Desert in eastern Idaho, in 1985.

Pollution of all types tends to be much worse in urban than in rural areas. Table 6–4 is the summary of the results obtained when lichens were collected near the two bridges connecting the east half with the west half of Idaho Falls, a city of approximately 35,000 at the time the study was made, and compared with lichens collected along each of three transects radiating from a point source of pollution at the INEL on the Arco desert approximately 65 km to the west of the city. Lead from automobile exhaust had accumulated over a period of years in the lichens near the bridges; likewise, mercurials from the seed treatment plant near one of the bridges had built up in the lichens near the bridges where farm trucks, carrying treated seed, passed by daily. Since requirements that unleaded gas be used and the banishment of mercury fungicides for seed treatment be enforced, the differences do not appear to be as great.

Table 6–5 compares the amount of each of eight pollutants relative to the position from the same Arco Desert point source. Air-borne sulfur and nitrogen oxides consistently damage cell membranes in lichens more along the nighttime downwind transects than the daytime downwind transects apparently because humidity is higher during the night and moist tissues are more susceptible to injury. Table 6–5 suggests that some polluting elements are taken in by the lichen thalli more when they are moist whereas others accumulate independently of humidity and are therefore absorbed more along the transects where the wind blows most frequently.

MYCORRHIZAE

During the 1930s, trees were planted on the Great Plains of North America to aid in the control of wind and water erosion. Extensive failures resulted because of the lack of knowledge of the importance of symbiotic fungi. Success was achieved when the seedlings were inoculated with soil from the area where the trees normally grow (Chester 1947). Mycelium of the mycorrhizal fungi, mostly basidiomycetes, needed to be present to infect the seedlings. Without the fungus, the seedlings grew poorly and lacked winter hardiness as well as drought tolerance. They were stunted, yellowish instead of a healthy green, and subject to various plant diseases.

Many species of fungi function in the formation of mycorrhizae. Some of them, such as species of *Boletus, Morchella, Russula,* and *Agaricus* are common wild mushrooms. The mycelium invades the young roots and develops a relationship so intimate that the root tissues and fungal hyphae form an organ of characteristic shape and structure in which neither the

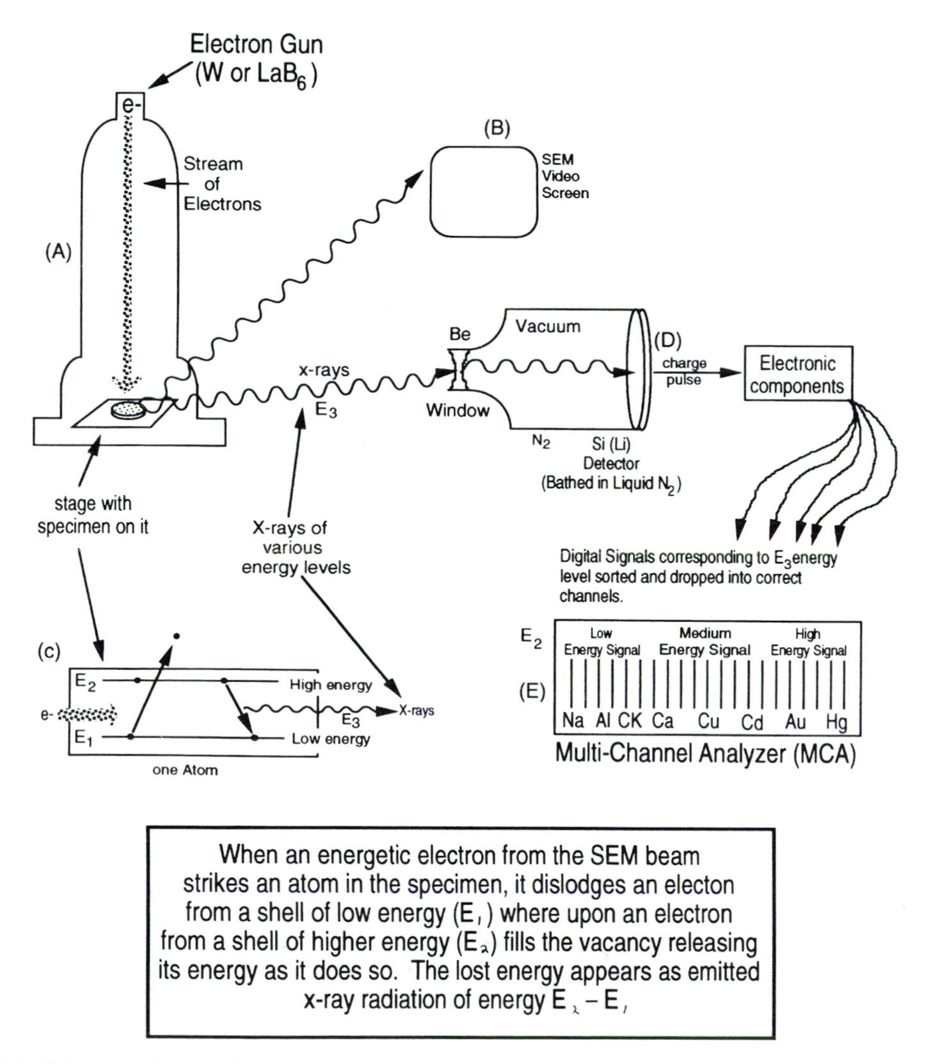

Figure 6–10 Diagram of energy dispersive spectrometer as used to measure concentration of polluting elements in lichen tissue.

host cells nor the fungal cells are readily distinguishable by the ordinary person examining them. The fungal cells apparently increase the absorbing area of the roots thus enabling the host to absorb water more readily. They also aid in absorption of phosphorus, potassium, nitrogen, and other minerals. In some cases, however, as with *Armillaria mellea,* a choice mushroom rather closely related to *Agaricus,* the relationship is pathogenic rather than mutually beneficial, at least in apple trees.

Mycorrhizae are far more common than most people realize. Trappe (1987) reported that 70% of angiosperm species are mycorrhizal. Essentially the same value (71.9%) was reported for **indigenous** species in Hawaii (species native to but not limited to Hawaii), but over 90%

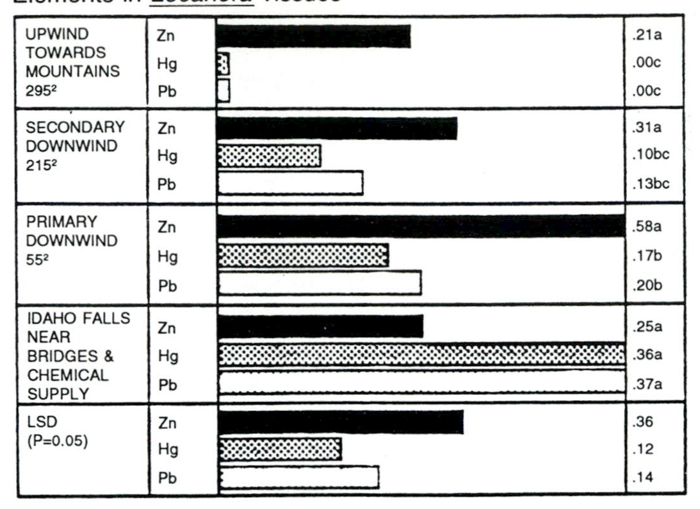

Figure 6–11 Air pollution research with lichens. Many lichens, when exposed to smokestack effluents containing heavy metals and other toxic elements, accumulate them in their tissues. Downwind from the Idaho Chemical Processing Plant, accumulation is greater than it is along crosswind transects. In this study, mercury and lead concentrations were high in areas where farm fungicides to control plant diseases were loaded and transported or where automobile traffic was considerable.

TABLE 6–4

CONTENT OF SELECTED ELEMENTS IN *LECANORA MELANOPHTHALMA* TISSUES AS MEASURED WITH ENERGY DISPERSIVE SPECTROSCOPY

Transect	Zn	Hg	Pb
295°—Upwind toward the mountains	.27 a	.00 c	.00 c
215°—Nighttime downwind	.31 a	.10 bc	.13 bc
55°—Daytime downwind	.38 a	.17 b	.20 b
Idaho Falls, near bridges	.26 a	.36 a	.37 a
Least significant difference (LSD)	.36	.12	.14

Note: Comparisons can be made among locations but not among elements. (Averages **not** followed by the same letter of the alphabet are significantly different from each other at $P \leq 0.05$.)

of **endemic** species (those native only to Hawaii) of angiosperms were mycorrhizal (Koske et al. 1992). Mycorrhizae also occur in ferns and other vascular plants in which the fungi invade rhizomes rather than roots (Andrews and Lenz 1943, Gemma et al. 1992). The morphology of mycorrhizal roots and rhizomes is unique as well as interesting (Figure 6–12). In the greenhouse and botanical laboratory we are accustomed to seeing a main root with fine secondary roots

TABLE 6-5

RELATIVE CONCENTRATION OF SELECTED POLLUTING ELEMENTS COLLECTED ALONG FOUR TRANSECTS RADIATING FROM A POINT SOURCE OF POLLUTION IN EASTERN IDAHO (1992–1993 DATA)

Group A	**Gaseous Pollutants**[a]	**Ni**	**Cr**	**Co**
215° Primary nighttime downwind	309.4 a	0.14 a	0.045 a	0.08 a
135° Secondary nighttime downwind	261.5 b	0.09 b	0.020 c	0.06 ab
55° Primary daytime downwind	247.0 b	0.07 b	0.035 b	0.04 b
295° Secondary daytime downwind	204.7 c	0.00 c	0.020 c	0.03 b
Group B	**Mg(Med)**	**Ti**	**Si**	**Se**
55° Primary daytime downwind	0.39 a	0.09 a	2.23 a	0.22 a
215° Primary nighttime downwind	0.26 b	0.07 ab	1.56 ab	0.12 b
135° Secondary nighttime downwind	0.04 c	0.05 b	0.35 b	0.12 b
295° Secondary daytime downwind	0.12 c	0.01 c	0.34 b	0.17 ab

[a] Gaseous pollutants are primarily NO_x and SO_2 and were measured by the electrolyte leakage test; Ni, Cr, Co, Mg, Ti, Si, and Se were measured by SEM/EDS.

and unicellular root hairs. The same type is seen in agricultural and garden crops. In the forest, on the other hand, mycorrhizae are common and root hairs rare.

There are two major types of mycorrhizae: ectotrophic and endotrophic. **Ectotrophic mycorrhizae** provide the more spectacular alteration of root morphology (Figure 6-12). In it, the root tissue is completely enclosed in two concentric layers of fungal tissue, the **hartig net** next to the root cortex and the dense **fungal sheath**. Cells in the outermost layer of root cortex are greatly altered and make up a palisade-like layer. Often the branching of the root is also changed, becoming thick and dichotomous rather than having fine secondary rootlets branching off at approximately 90° angles from a coarse main root. According to Harley (1968), at least 40% of the dry weight of a typical mycorrhiza is fungus.

Endotrophic mycorrhizae are formed by fungi that penetrate the roots much as pathogenic species do. Some form associations with different species of Anthopsida that are not even related to each other. In some cases they may be beneficial, in other cases not. For example, *Armillaria mellea,* which is a serious pathogen and parasite on apple trees, is a beneficial endotrophic mycorrhizal symbiont on orchids. There is little noticeable change in gross morphology of the roots as the result of endotrophic fungi. Under the microscope, mycorrhizal fungi resemble pathogenic fungi, except that the host tissues look normal and healthy. Endotrophic mycorrhizae have been found in fossil ferns from the Pennsylvanian period (Andrews and Lenz 1943) and are common in many modern herbaceous and other vascular plants.

In order for a vascular plant to develop mycorrhizae, environmental conditions must be favorable, a suitable fungus must be present, and the seedling must be in the proper stage of development. The fungi do not invade older roots. Forest biologists in the U.S. and Sweden have recently been exploring the role of root hairs in mycorrhizae. The root hair zone in laboratory seedlings begins within the first millimeter above the root apex and is typically about 20 mm long. The zone of infection begins 0.5 mm from the apex of the root and extends to 4.0 mm. When *Pisolithius tinctorius,* an edible false puffball (Basidiomycopsida), invades the roots of the black spruce, *Picea mariana,* the developing root hairs within the infection zone

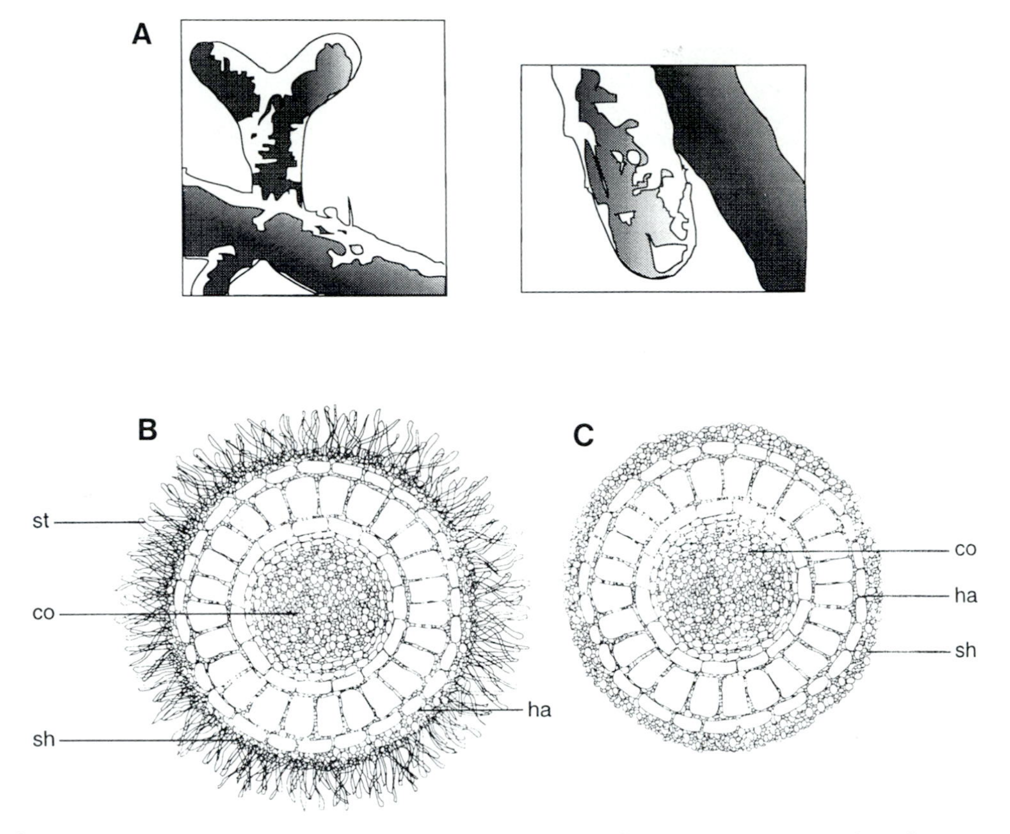

Figure 6–12 Mycorrhizae. (A) Typical mycorrhizal root as seen on a forest tree; (B) cross section of an ectotrophic mycorrhiza showing the hartig net (ha), root cortex (co), and fungal sheath (sh); (C) mycorrhiza lacking fungal strands.

become enveloped by a layer of fungal hyphae (Figure 6-12). Root hairs which have already elongated, on the other hand, have few fungal hyphae on their surface.

The evolution of mycorrhizae depended upon coevolution of fungi and vascular plants. Boucot (1990) and others have studied the evolution of this and other aspects of co-evolution, the evolution of leaves with abundant mesophyll correlated with that of moths and flies that become leaf miners, for example. But in the case of mycorrhizae, the two organisms have become mutually dependent on each other resulting in a true symbiosis.

OTHER SYMBIOSES

Many marine plants form symbiotic relationships with animals. The best known, and probably most important of these, is the association between corals and various algae known as **Zooxanthellae**. The coral obtains most of its food directly from dinoflagellates, cryptophytes, or heterokonts living within its transparent tissues where photosynthesis takes place. Presumably the alga benefits through protection from herbivores as well as through higher concentrations of carbon dioxide and nitrates than occur in sea water. The protein needed by the corals is obtained, in part, at least, by ingestion of free-living animals. As the protein is metabolized, nitrates are released to the benefit of the **phytosymbionts** or microscopic plants.

Symbioses between bacteria and legumes and between *Prochloron* and various ascidians were discussed in Chapter 3. A number of symbioses between blue-green algae and vascular plants, especially some of the cycads, are also known. Stem galls containing nitrogen-fixing actinomycetes (Schizomycopsida) have been reported in some desert shrubs, especially rabbitbrush and sagebrush (Farnsworth 1973). But the most spectacular of all symbioses, as well as the most significant, is the hypothetical endosymbiosis between colorless cells and unicellular blue-green algae that is believed by many biologists to have given rise to the plastids.

SUGGESTED READING

Three books by Mason Hale, *The Lichen Handbook, Biology of Lichens,* and *How to Know the Lichens,* are excellent references for anyone wanting to know more about these interesting dual organisms. James Lawrey's *Biology of Lichenized Fungi* (Prager 1984) is an up-to-date treatment of the morphology, physiology, and ecology of lichens. Like Hale's books, it is readable, well documented, and illustrated with good photographs, charts, and graphs.

Baddeley, Ferry, and Hawksworth's *Air Pollution and Lichens,* published in 1973, covers many aspects of monitoring air quality with lichens. In 1993, the U.S. Forest Service, in cooperation with the U.S. National Park Service, published a handbook of air-quality monitoring methods, *Lichens as Bioindicators of Air Quality,* edited by Laurie Huckaby, which describes all aspects of biomonitoring including site characterization, identification of sensitive species, transplant methods, collection, chemical analysis, floristics, and others. Harley and Smith's *Mycorrhizal Symbiosis,* published by Academic Press in 1983 explores the nature and role of these interesting and important structures in detail.

STUDENT EXERCISES

1. If you were to embark on a project to find fossil lichens, what kind of paleobotanical environment would you look for: swampy lowland, forest, grassland, sandy desert, basalt cliffs in an ancient desert, reefs, or other?

2. You have observed pine seedlings in the laboratory and in the greenhouse many times and noted that the main roots are relatively straight with many side roots branching off almost at right angles. In digging up a pine seedling in the forest, you observe that its roots are dichotomously branched and there are no root hairs present. Is this pine seedling infected with one of the pathogenic fungi discussed in Chapter 5? Or is there some other cause for its peculiar appearance?

3. To obtain maximum growth in cultures of bacteria, fungi, or algae, the cultures are kept under **optimum** conditions of light, temperature, and moisture. Why does this not work in obtaining maximum growth in lichens? Outline a plan for obtaining maximum growth in lichens under laboratory conditions.

4. Outline a program for studying the genetic diversity in a hybrid population of lichens. Could you obtain an F_1 population by crossing two homozygous strains that differ from each other in a number of ways? How would you go about obtaining such an F_1 population and also an F_2 population?

5. Outline an experiment, using lichens, to evaluate the possible environmental damage that could be caused by a large, coal-fired power plant proposed for construction in central Nevada. Spell out replication, randomization, and evaluation methods you would propose in detail.

6. *Dermatocarpon, Verrucaria,* and *Staurothele* are generally classified in the Sphaeriales. Their perithecia are similar to those of *Neurospora* and *Claviceps* and have typical parphyses. On the other hand, the sporocarps in some of the species begin development in what Alexopoulos and Mims interpret as chambers (locules) in a stroma-like tissue. If Alexopoulos and Mims are correct in their interpretation, should these genera be kept in the Sphaeriales or transferred to another order? If transferred, to what order?

7. Lichens may be crustose, foliose, or fruticose. However, subtypes exist for each of these habit types. Indicate with a *CR, FO,* or *FR* which of the three habit types each of the following subtypes belong to (some may fit more than one type):

tufted	areolate	perithecia
granular	pendulous	lecanorine
gelatinous	shrub-like	vagrant
rhizinate	lecideine	endolithic
vine-like	umbilicate	squamulose

8. Which of the names in no. 7 indicate lichens belonging to the Sphaeriales? Which indicate lichens belonging to the Helotiales? Which could fit either?

PART III

THE BROWN LINE

Pyrrophyta
Myxomycophyta
Chrysophyta
Phaeophyta

One of the marvels of plant geography is the abundance of plant life in the cold oceans where temperatures are constantly only a few degrees above freezing. No fossil evidence of life in these habitats prior to the Paleozoic era has been found, but once adaptations evolved making life at low temperatures possible, the cold oceans were soon colonized. The earliest plants possessing enzymes that make possible metabolism at low temperatures were similar to the Bracteobionta of today.

The Bracteobionta are so named because they have tinsel-like projections, called **mastigonemes,** on their flagella. They are also known as the "brown line" because most autotrophic species have brownish carotenoid pigments, such as fucoxanthin, diatoxanthin, and dinoxanthin, that aid in photosynthesis.

True flagella occur in both plants and animals and are characterized by a core of two strands surrounded by a ring of nine. Lashing of the flagella enables unicellular plants to stay slightly immersed in water without sinking to the bottom.

Five types of flagellum occur in plants: (1) whiplash flagella, which lack mastigonemes, are tapered and typically terminate with a tuft of strands at the apex; (2) pectinate flagella have a blunt apex and a single row of mastigonemes; (3) pinnate flagella have a blunt apex and a double row of mastigonemes; (4) truncated whiplash flagella have a blunt apex and no mastigonemes, and (5) tapered pectinate flagella have a single row of mastigonemes and taper to a tuft of strands at the apex. The first four occur in the Bracteobionta. Each cell typically possesses two flagella, one whiplash, the other tinsel or modified tinsel (truncated whiplash).

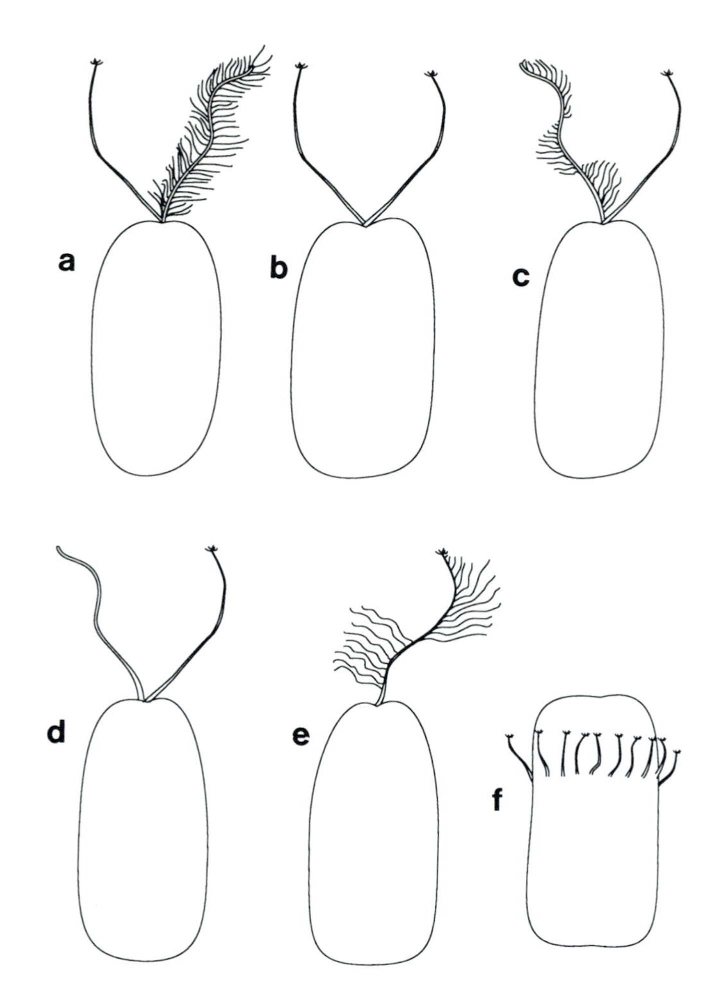

PART III. THE BROWN LINE

The Architecture of Flagella is one of the most consistent of the many features differentiating plants into three major groups or sub-kingdoms. The Brachteobionta are characterized by non-tapered flagella adorned with tinsel-like projections called *mastigonemes* as in sketches a and c. Whiplash flagella, as in b, are typical of most Lorobionta but are also found on most flagellated cells of Bracteobionta. The sketches represent *Synura* (a) in the Chrysophycopsida with one whiplash and one pinnate flagella, *Chlamydomonas* (b) in the Chlorophycopsida with two whiplash flagella, *Exuviella* (c) in the Dinophycopsida with one whiplash and one pectinate flagella, *Physarum* (d) in the Myxomycopsida with one blunt, untapered, modified tinsel flagellum, lacking mastigonemes, *Euglena* (e) in the Euglenophycopsida with a tapered, tinseled flagellum that basically resembles a whiplash, and *Derbesia* (f) in the Siphonophycopsida having a ring of short whiplash flagella or cilia.

7 FIRE PLANTS AND CRYPTOPHYTES

At the tip of the Huon Peninsula of New Guinea, black sand beaches extend for miles with occasional rocky outcroppings or reefs separating one stretch of beach from the next. From the beach, one can walk out into the ocean on a cloudy, moonless night and see every detail of one's feet while observing the pebbles and sea shells in clearest detail. The strange fire which makes this possible, lighting up the ocean floor and everything on it, is caused by unicellular plants of the Division Pyrrophyta. Similar to the eerie light of fireflies, so delightful to children wherever they see it, it never ceases to enthrall those who travel by ship through the tropics.

The Huon Peninsula is one of many places in the tropics where dinoflagellates thrive; there are also islands in the Caribbean which are famous tourist attractions because of their "burning" bays—dense with blooms of dinoflagellates—that seem to be on fire. The Huon Peninsula is also the home of fascinating forest-dwelling fireflies which flash their lanterns in a most unusual and spectacular synchronized display of bioluminescence. Their luminescing mechanism is essentially identical to that possessed by dinoflagellates, dependent upon reaction between a substrate, called luciferin, and adenosine triphosphate (ATP), catalyzed by an enzyme called luciferase. The enzyme is specific to each category of luminescing organism.

Aside from this interesting phenomenon of bioluminescence for which they are famous and from which they probably received their scientific name, the Pyrrophyta (*pyro* = fire, or *pyrrho* = reddish or orange; *phyto* = plant) are ecologically among the most important of all plants (Table 2-5). Morphologically, physiologically, taxonomically, and genetically, however, they are perhaps the least perfectly known. As circumscribed here, the division contains two classes, the Cryptophycopsida and Dinophycopsida, with more than 1100 species classified in 90 genera, 38 families, and 12 orders. It is highly possible that only a small percentage of the species that belong to this division have been discovered, described, and named.

The Cryptophycopsida probably originated in middle to late Proterozoic time from *Euglena*-like ancestors which were not yet sharply differentiated from the Cyanophycopsida. From the Cryptophycopsida have evolved the Dinophycopsida and other divisions of the Bracteobionta. Lines of relationships among the 12 orders of Pyrrophyta are suggested in Figure 7-1.

Most Pyrrophyta are autotrophic and have brown or brownish green plastids that owe their color to the presence of yellow and brown carotenoids in addition to chlorophylls *a* and *c*.

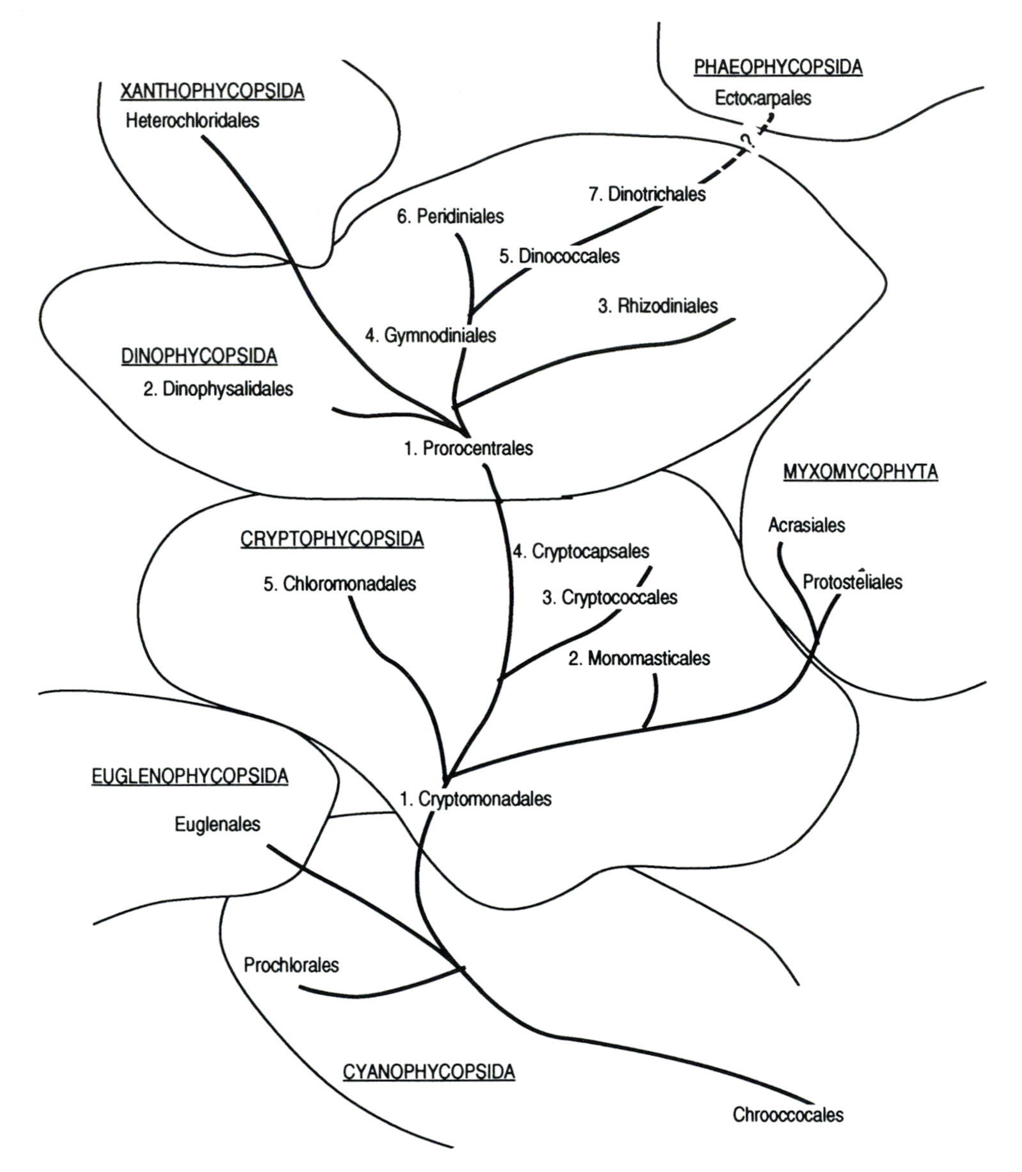

Figure 7–1 Suggested lines of relationship among the Pyrrophyta. From ancestral *Euglena*-like Chroococcales, evolution has proceeded along three lines, one leading to the Chloromonadales, one to the slime molds by way of the Monomasticales, and one leading to the dinoflagellates and heterokonts. Some botanists believe that the Phaeophycopsida evolved from filamentous dinoflagellates.

The most abundant of the carotenoids are diadinoxanthin, dinoxanthin, and peridinin. Other Pyrrophyta are heterotrophic, obtaining their nourishment either **saprophytically** or **phagotrophically.** Some autotrophic species can also ingest particulate food phagotrophically. Similarities in pigmentation, flagellation, cell wall structure, reserve foods, ultrastructure of plastids and other organelles, and other characteristics suggest a relationship to the Chrysophyta and Phaeophyta. Because of phagotrophic nutrition in some species, and other animal-like characteristics, they are often claimed by zoologists who classify them in the Phylum Protozoa, Class Mastigophora.

Like other Bracteobionta, members of the Pyrrophyta have tinsel flagella, store foods primarily as oils, but also as slightly branched, short-chain carbohydrates, and have thylakoids stacked in sets of two or three. Unlike other Bracteobionta, the tinsel flagella are **pectinate** rather than **pinnate** and are inserted more or less laterally. They arise in pairs from the base of the gullet and seem to have a single row of long, flexuous mastigonemes (Brown and Cox 1954, Pitelka 1963); in the dinoflagellate *Oxyrrhis marina,* however, no evidence of mastigonemes has been found. The cell walls are composed of cellulose, food is frequently stored as starch, their distribution is more tropical than cold ocean, and a blue pigment, similar to c-phycocyanin, may be present. Thus, in flagellation and food storage, they somewhat resemble the Euglenophycopsida, and in habitat and pigmentation the Cyanophycopsida. The Pyrrophyta, therefore, seem to form a link between the higher Bracteobionta and the other two subkingdoms.

Because of their many similarities, the Cryptophycopsida and Dinophycopsida are included in the same Division. They also differ from each other in a number of ways, including some details of pigmentation and of thylakoid structure, presence of trichocysts in most Cryptophycopsida but in relatively few Dinophycopsida, and the more complicated vacuolar arrangement of the Cryptophycopsida.

The Dinophycopsida also have a type of nucleus—moniliform or mesokaryotic—that does not occur in other plants or animals. In both classes, the chromosomes are large, easy to stain, difficult to count, and highly variable (Table 7-1). They remain condensed during interphase and can be counted then; however, they are faint in cryptophyte interphases and dinoflagellate cysts. Partly because Cryptophycopsida are usually considered eukaryotic and Dinophycopsida mesokaryotic (Table 7-2), some botanists classify them in different divisions. However, the similarities between cryptophyte and dinoflagellate nuclei are more pronounced than the differences, and separation into two divisions can hardly be justified by karyotype or other reasons.

CLASS 1. CRYPTOPHYCOPSIDA

The class Cryptophycopsida (*crypto* = hidden, and *phycus* = seaweed or alga) is a small, imperfectly known assemblage of organisms commonly found in fresh water containing considerable organic matter (especially nitrogenous material) and in the ocean. They are apparently among the most common of the tropical and subtropical marine plankters, although because of their small size they have long escaped much notice. Their morphological similarities to the chrysomonads are readily apparent (see Table 7-3) and formed the basis of early classification systems, but early in the present century, the great German phycologist Adolph Pascher was able to demonstrate that these similarities are superficial and proposed that their true relationships are with the dinoflagellates. There have been 19 genera and 74 species described, and they are distributed among 9 families and 5 orders, including one order, the Chloromondadales, often classified with the Chlorophycopsida, the Xanthophycopsida, or by itself as a separate class.

TABLE 7-1

CHROMOSOME NUMBERS IN THE PYRROPHYTA

		Chromosome Number	
Species	**Order**	**Range**	**Average**
Cryptomonas	Cryptomonadales		
ovata			86 ± 6
appendiculata			61 ± 6
sp. nov.			42 ± 2
laboratory strains		108–209	145 ± 5
Chroomonas mesostigmatica			24 ± 2
Chilomonas paramecium			35 ± 2
Cyanophora paradoxa			10
Exuviella	Prorocentrales		
baltica		18–23	20
pusilla		20–25	24
Prorocentrum			
micans		65–69	68
triestinum		20–27	24
Amphidinium klebsii	Gymnodiniales	24–26	25
Gymnodinium			
vitiligo			44
zachariasi			64
Oxyrrhis marina		c. 40–48	c. 45
Ceratium	Peridiniales		
tripos			c. 200
hirundinella		264–284	274
Peridinium trochoideum			44
Gonyaulax polygramma			c. 100

Livestock sewage lagoons are good places to examine when in search of cryptophytes. Because of obvious health hazards, students are warned never to collect from municipal sewage lagoons. Late summer and winter blooms of green and blue-green algae in lakes or ponds will usually have cryptomonads present. Being small, they often go unnoticed. As bottles of algae are kept in the laboratory, the number of Cryptophycopsida often increases.

General Morphological Characteristics

The typical cryptophyte is small, motile, unicellular, and autotrophic with a single small, non-granular plastid, or occasionally two plastids (Figure 7-2). However, "typical" cryptophytes are not often encountered; as suggested in Table 7-4, considerable diversity exists within the class in color, size, and shape. Among the smallest are *Cryptomonas vectensis*, 6 × 4 × 3 μm, and *Rhodomonas minuta*, 7 × 3 × 3 μm; among the largest is *Cryptomonas stigmatica*, 21 × 11 × 5 μm. *C. vectensis*, therefore, has a volume of approximately six picaliters. *Chroomonas vectensis* has a single blue chromatophore and *Rhodomonas minuta* a single reddish brown one, but most species have green to chocolate brown plastids (Hulbert 1965).

TABLE 7–2

DIFFERENCES BETWEEN PROKARYOTES, MESOKARYOTES, AND EUKARYOTES

Prokaryotes	Mesokaryotes	Eukaryotes
No basic protein in chromosomes	No RNA or basic protein in chromosome	Chromosomes contain RNA and protein
Chromosomes consist of fibers	Chromosomes consist of fibers[a]	Chromosomes nonfibrillar
Continuous DNA synthesis	Continuous DNA synthesis	DNA synthesis only during synthesis stage of interphase
No nuclear membrane	Typical nuclear membrane	Typical nuclear membrane
One chromosome	Many chromosomes, often 100 or more	Many chromosomes, usually 10 to 50
Chromosome number constant for any given species	Chromosome number variable within individual species	Chromosome number constant for any given species
No centromere; chromosome attached to plasma membrane	One central body to which all the chromosomes are attached	Each chromosome has its own individual centromere
No distinct metaphase or anaphase stages	Mitosis with typical metaphase and anaphase stages	Mitosis with typical metaphase and anaphase stages
Chromosome visible during "interphase"	Chromosomes countable during interphase	Chromosomes decondensed and not visible during interphase

[a] There are two types of nucleus in the Dinophycopsida: in the *Amphidinium* type, the chromosomes consist of fibers; in the *Noctiluca* type, the chromosomes are nonfibrillar.

As cryptophytes swim through the water, they rotate, revealing their flattened and rigid shape. There is typically a sub-apical **oral groove** ending in a **gullet** and lined with spindle-shaped cavities containing **trichocysts,** spine-like structures, the exact function of which is not known. However, they are discharged at times of physiological stress and may have a defensive function. Trichocysts have a greater diameter than flagella and are readily visible with the light microscope. In some species, they occur only in the oral groove; in other species they are on the cell surface (Figure 7–2). Because they vary in location and size, they are useful to the taxonomist in identifying species.

The position of the oral groove varies from apical to lateral. When the groove is lateral, the cell tends to be reniform; when the groove is apical, the base of the cell is rounded and the apex is truncated. Adjacent to the gullet is one or more contractile vacuoles. When the cells are slowed down, the pulsations of the vacuoles are especially noticeable; apparently they pump water from the cell.

Attached at the base of the gullet are two flagella, one whiplash, the other pectinate, or sometimes both pectinate, neither of them visible with the light microscope unless the cells are specially prepared. A colored stigma, or "eyespot," may be present. It apparently controls the direction of movement although the exact mechanism is not known.

If a cell wall is present, it is composed of cellulose and pectin; in many species the cells are naked. Rigidity in naked cells is supplied by a firm, sometimes striated, periplast formed on the inside of the plasma membrane. Periplasts are chemically different from cell walls and commonly occur in animal cells as well as in Pyrrophyta, Chrysophyta, and Euglenophyta.

TABLE 7–3

COMPARISON OF PYRRHOPHYTA WITH OTHER UNICELLULAR ALGAE

Class	Cell Shape	Plastics and Pigmentation	Motility	Nucleus and Chromosomes	Other
Cryptophycopsida	Flattened, ovate to subligulate or slipper-like, rigid	Thylakoids in pairs; phycocyanin-*c* may be present; green-brown	Rapid; two subapically inserted pectinate flagella	Spherical to elongate; many chromosomes, countable in interphase	Trichocysts nearly always present, usually lining the oral groove; contractile vacuoles present
Dinophycopsida	Globose to flattened ovate; rigid	Thylakoids in sets of three; brown carotenoids present	Rapid; one pectinate, one whiplash flagellum	Elongate to crescent; many chromosomes, countable in interphase	Trichocysts sometimes present; distinctive girdle groove and sulcus; moniliform nucleus
Euglenophycopsida	Long ovate to subligulate or slipperlike, nonrigid	Thylakoids forming simple grana; chlorophyll *b* abundant	Rapid; one modified pectinate, one whiplash flagellum apically inserted	Nucleus open in mitosis; mitosis as in vascular plants	Red stigma or "eye-spot" conspicuous; contractile vacuoles as in Cryptophytes
Chrysophycopsida	Globose to long ovate, nonrigid; often symmetrical	Thylakoids in sets of 3; brown to golden carotenoids	Rapid; 1 pinnate 1 whiplash flagellum apically inserted	As in vascular plants; centrioles prominent	Rigid, siliceous lorica, open at one end often present
Xanthophycopsida	Variable; globose to elongate, often curved or even coiled, rigid; often symmetrical	Thylakoids in 3s; plastids disklike with smooth margins, nongranular	Frequently epiphytic or free floating; when motile, rapidly so; flagella unequal, pinnate longer	As in vascular plants; centrioles present	Lorica occasionally present; color often green but chlorophyll *b* not present; learn to recognize representatives in each order

Bacillariophycopsida	Very rigid, symmetrical with distinct striations	Thylakoids in 3s; golden brown to yellow green	Slow, jerky; no flagella in vegetative cells	As in vascular plants; centrioles present	The distinctive pinnate or centric striations make identification easy
Chlorophycopsida	Globose to ovate, usually symmetrical; no oral groove or gullet	Thylakoids stacked in small grana	Slow, rapid, or nonmotile; 2 whiplash flagella	As in vascular plants; centriole in motile species only	The grassy green color and granular, rough plastids aid identification
Cyanophycopsida	Globose, cubical, or ovate; very small	Thylakoids single and unstacked, no plastids	No flagella; herpokinetic motility in some species	No nucleus, one chromosome attached to plasmalemma	Lack of plastids and other vesicular organelles; small size, bluish color aid identification
Myxomycophyta	Spindle-shaped swarm cells in some species	No thylakoids, no plastids, no chlorophyll	One or two modified whiplash flagella in swarm cells	Globose nucleus often closed in mitosis	Lack of chlorophyll, slimy nature make identification easy

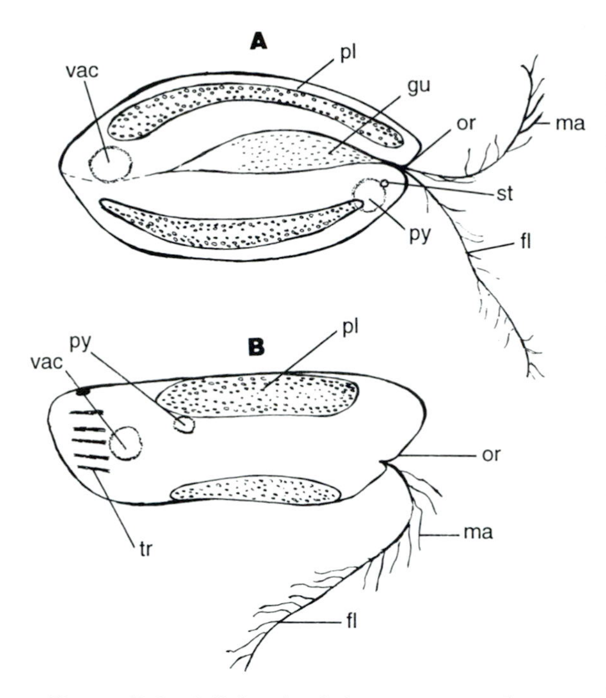

Figure 7–2 Cellular detail in two cryptophytes: (A) *Cryptomonas erosa;* (B) *Monomastix opisthostigma.* (or = oral groove, gu = gullet, st = stigma, tr = trichocyst, vac = contractile vacuole, fl = flagellum, ma = mastigoneme, pl = chloroplast, py = pyrenoid.)

Some cryptophytes have well-developed oral grooves and gullets, others have an oral groove but no gullet. The position of the trichocysts also varies among species. Plastid color ranges from bright green to various shades of brown, yellow, and even blue; some species lack plastids and are colorless. Some have more than the typical one or two plastids.

Electron microscope studies have revealed that the plastids lie along the surface of the cell and are laminated with the **thylakoids** in pairs rather than in sets of three as in Dinophycopsida and most other Bracteobionta. As in the Dinophycopsida, girdle lamellae are lacking. **Pyrenoids** are present in many cryptophytes. These are not always embedded in the chloroplasts, as in other algae, but may lie free in the cytoplasm. Starch grains are formed outside the plastids and are held within an extension of the nuclear membrane. Instead of β-carotene, as in other Bracteobionta and most plants, α-carotene and ε-carotene occur in combination with chlorophylls *a* and *c* and two brown carotenoids, diadinoxanthin and dinoxanthin. A water-soluble photosynthetic pigment, phycocyanin-*c,* occurs in some cryptophytes.

Vegetative reproduction occurs either while the cell is in a motile condition or in a resting stage. In the latter case, the mitotic products become embedded in a mucilaginous matrix, thus forming an amorphous **palmelloid** mass. The nucleus is elongated and the chromosomes are faint, but countable, during interphase (Godward 1966). They are so closely aggregated on the metaphase plate that they appear to be one continuous chromosome, however, making counting difficult. Isolated reports of sexual reproduction have not been confirmed; possibly it is common but has escaped detection; possibly it does not occur in most species.

Phylogeny and Classification

Fossil cryptophytes are extremely rare or nonexistent. The paleobotanist faces two handicaps in studying this class: being strictly unicellular, cryptophytes lack an attention-getting macroscopic structure, and possessing no hard substances in their cell coverings, they do not readily form sharp imprints, good petrifactions, or other morphologically recognizable fossils. Because fossils of Dinophycopsida are found in late Proterozoic formations, and because the cryptophytes seem to form a link between them and the Cyanophycopsida, which originated about the beginning of the Proterozoic era or earlier, we assume that the first cryptophytes evolved in middle to late Proterozoic time. Cronquist (1960) proposed that the Cryptophycopsida

TABLE 7–4

CHARACTERISTICS OF THE ORDERS OF CRYPTOPHYCOPSIDA

Orders and Included Fam-Gen-Spp	Thallus Type	Oral Groove and Gullet	Trichocysts	Plastids and Pigmentation	Representative Genera and/or Species
Cryptomonadales (5-12-70)	Unicellular	Subapical to apical or lateral	Abundant, lining oral groove	One or two yellowish green to brown; some species holozoic 1 or 2 lateral, shield-shaped	*Cryptomonas erosa, Chilomonas, Sennia, Kateblepharis, Cryptaulax, Rhodomonas*
Monomasticales (1-1-1)	Unicellular	Subapical oral groove poorly developed	1 to 7 basal, parallel oriented	1 or 2 lateral, shield-shaped	*Monomastix opisthostigma*
Cryptococcales (1-1-1)	Nonmotile vegetative unicells; zoospores	Two apical flagella on zoospores	In oral groove	One brown plastid with a large pyrenoid	*Tetragonidium verrucatum*
Cryptocapsales (1-2-2)	Nonmotile gelatinized colonies	Two unequal apical flagella	In oral groove	Brownish plastids	*Phaeococcus clementi, Phaeoplax marinus*
Chloromonadales (1-6-12)	Unicellular	Apical flagella and apical oral groove	Numerous, over entire cell	Several bright green plastids	*Vacuolaria, Reckertia, Merotrichia, Trentonia, Conyostomum*

Note: Fam = families, Gen = genera, Spp = species.

probably evolved from *Euglena*-like ancestors, inasmuch as both groups have similar flagella and similar gullets.

Pascher (1931) divided the class into three subclasses, each with one order. To these three orders may be added another, Chloromonadales, which was ignored by Pascher but had earlier been classified with the Cryptophycopsida by Oltmanns (1904). One of Pascher's orders is so variable that Melchior and Werdermann (1954) and others have split it into two orders. Pringsheim (1944) has discussed additional aspects of taxonomy in the Cryptophycopsida. Relationships among the five orders are suggested in Figure 7-1 and chief characteristics are indicated in Table 7-4.

From the primitive Cryptomonadales, with their euglenid-like gullet and cyanophycean-like pigmentation, three lines of evolution seem to have led to the chloromonads, the dinoflagellates, and the slime molds, respectively. Along the dinophycean line, starch has been replaced by oils as the chief storage product and the oral groove has become more pronounced by taking on first a spiral and eventually a transverse orientation with insertion of the flagella becoming increasingly lateral. In the line leading to the chloromonads, cell size and plastid number have increased and the quantity of brown carotenoids per cell has decreased. In the line leading to the Myxomycophyta, the oral groove has become less pronounced, insertion of flagella has become more apical, and mastigonemes have disappeared (even though the flagella have remained unequal in length).

The Chloromonadales have been a taxonomically difficult group, most frequently allied with the Chlorophycopsida because of their bright, grassy green color. However, flagellation, food storage, general cell form, smooth rather than granular plastids, and the nature of the gullet convinced Oltmanns and others that they should be allied with the Cryptophycopsida. Like the cryptomonads, they are flattened, have a subapical gullet, and possess trichocysts; however, they are much larger than most cryptomonads and have many plastids per cell. On the other hand, Luther (1899) included them in the Xanthophycopsida, as did Scagel et al. (1965). Smith (1950) relegated them to what he called "groups of uncertain systematic position." The ten species, included in four genera and two families, are all freshwater; it is possible that other species of green, unicellular flagellates now included in the Chlorophycopsida will be transferred to this order after they have been studied more thoroughly.

Representative Genera and Species

The best-known genera are *Cryptomonas* and *Chilomonas;* the former is autotrophic, the latter is colorless and saprophytic. Both belong to the family Cryptomonadaceae, order Cryptomonadales, and both are very common; Conrad and Kufferath (1954), for example, listed 30 species of *Cryptomonas* for one small area in Belgium. *Cyathomonas, Katablepharis,* and *Cryptaulax* are holozoïc. Other genera frequently encountered are *Rhodomonas, Cryptochrysis,* and *Monomastix.*

Cryptochrysis and *Monomastix* differ from other common cryptophytes in several ways. The former has an oral groove but no gullet; like most cryptophytes it has the trichocysts lining the oral groove. *Monomastix* is relatively common and has traditionally been included in the Cryptomonadales; however, it has its one to seven trichocysts situated at the posterior end of the cell, rather than in the oral groove, and the oral groove is poorly developed; unlike other cryptomonads, it has only one flagellum. Melchior (1954) placed *Monomastix opisthostigma,* the only species in the genus, in a family and order by itself.

Three species are nonmotile in the vegetative stage: *Phaeococcus clementi* lives in soil, *Phaeoplax marinus* lives in the ocean, and *Tetragonidium verrucatum* lives in fresh water. The first two build palmelloid colonies (colonies of cells held together in an amorphous gelatinous matrix), the last is a unicellular coccus-type alga.

Ecological and Economic Significance

The Cryptophycopsida appear to be of greatest ecological significance in tropical and subtropical oceans, where they probably vie in importance with Cyanophycopsida and Dinophycopsida as primary producers (Tables 2–4 and 2–5). However, more studies of primary productivity in the ocean are sorely needed and, until they have been made, it is impossible to evaluate the full significance of these and other plankters. Many of the Cryptophycopsida, like most Cyanophycopsida, are so small that they are not retained by the filters traditionally used in collecting marine algae and their numbers have obviously been underestimated in the past.

Freshwater plankton fluctuates widely and rapidly in species composition. Marine plankton probably does the same. If so, the relative significance of any plankter to total productivity cannot be ascertained until weekly studies have been made throughout the year for several consecutive years employing methods that give a truly representative sample of what is present. As an example of seasonal fluctuation, a study conducted in a British pond revealed that *Cryptomonas* was the most abundant alga from January to March but was not abundant during the remainder of the year. Unless care is taken to ensure adequate sampling throughout the year, the importance of many species might be grossly overestimated while the importance of others is ignored depending on when the study was made.

Cryptophycopsida are sometimes important constituents of "algal blooms" that often occur in ponds and streams, especially in late summer, either as the result of organic pollution or of normal seasonal fluctuations in climatic, water quality, suitable temperature, or other factors. Distinctive odors are often associated with such blooms (Kudo 1966). The odor of water containing high densities of *Cryptomonas* is reported to resemble candied violets, while most Chrysophycopsida and Dinophycopsida impart to water a fishy odor. *Cryptomonas erosa* was reported to be a good pollution indicator by 27 of 165 authors reporting on pollution-tolerant algae in the *Journal of Phycology* and other scientific publications. Some Cryptophycopsida, especially species of *Chrysidella* (Cryptomonadales), form symbiotic relationships with various coelenterates and other marine animals. The resulting plant–animal complex appears green or yellowish green in color and carries on photosynthesis. The organic matter produced by the cryptomonad cells is undoubtedly of considerable significance to the host animal, but it is not known exactly how much net productivity results from these **zooxanthellae.** Except for color, the animals are little changed in appearance, but the plant cells are often considerably altered.

Research Potential

Cryptophycopsida have not been widely employed in scientific research. Difficulty in recognizing and identifying them has probably been a greater deterrent to scientific investigational use than difficulty in culturing them, even though it has only been in recent years that success in culturing these and other algae has been achieved. With the culture barriers being broken down, they can feasibly be more widely used in the future.

It has been suggested that plastids and mitochondria have arisen in plants as symbiotic organisms which have lost their ability to live independently of the host organism (Margulis

1971). Electron microscope and biochemical studies of plastids and other organelles in cryptophytes should be valuable in testing this hypothesis, inasmuch as the Cryptophycopsida represent a primitive group of organisms. Some basic questions that need answering are: (1) do the plastids and mitochondria of cryptophytes contain DNA and RNA? (2) If so, do plastid and/or mitochondrial DNA differ from chromosomal DNA? (3) What characteristics of cryptophytes make them likely candidates as the invading organism in establishing the symbiosis, or are they more likely the result of a symbiosis? (4) If the latter, what existing species of cyanophytes most nearly resemble the invading symbiont?

Basic research in patterns and pathways of photosynthesis could profitably be conducted with cryptophytes since both pigmented and nonpigmented forms occur and since there is a wider range in photosynthetic pigments reported in this class than any other: chlorophylls, phycobilins, and photosynthetic carotenoids all occur in this one class.

Of special ecological interest is the structure and function of trichocysts. Two types of trichocysts occur in cryptophytes, a slender type and a stout type. Either, neither, or both may occur in a given species. The slender type is found in many Pyrrophyta including *Oxyrrhis marina* in the Dinophycopsida and *Chilomonas paramecium* in the Cryptophycopsida. The less common stout type is also found in *C. paramecium* but not in *O. marina* (Pitelka 1963). It is believed that trichocysts function as defense mechanisms; if this is so, survival value of different kinds of defense mechanisms or complete lack of them could be ascertained with studies involving cryptophytes. While the trichocysts of cryptophytes seem to be structured differently from those of *Paramecium* and other predatory animals, they are apparently discharged in a similar manner. Discharge of trichocysts in all organisms that have been studied is almost instantaneous upon stimulation, only a few milliseconds time being required.

Cryptomonads grow best in water containing nitrogenous organic matter and should therefore be of value in studies of processes involved in the biodegradation of water pollutants. *Cryptomonas erosa*, a common alga in sewage, must be supplied with various B vitamins when grown in pure laboratory culture, indicating that some B vitamins must be released in the biodegradation of sewage. When nitrogen is abundant as either nitrate or ammonia, *Rhodomomas lens* excretes phosphorous, the rate of excretion being proportional to light intensity and nitrogen concentration. Since many cyanophytes are able to fix nitrogen, but are limited by lack of phosphorus, it has been suggested that some cryptomonads and some cyanophytes form a synergistic relationship mutually beneficial to both; this could account for the abundance of these two classes in polluted water.

CLASS 2. DINOPHYCOPSIDA

The Dinophycopsida (*dino* = whirling or terrible, hence armored; *phycus* = seaweed or alga), commonly called the dinoflagellates, is ecologically one of the most important groups of plants. While more studies of marine ecology are needed, it now appears safe to say that with exception of the Anthopsida and Bacillariophycopsida, the Dinophycopsida are the most important producers of food and oxygen in the biosphere. About 1000 species of dinoflagellates are known, but it is highly probable that there are many species not yet discovered, many of which may be ecologically very important. The 1000 species are classified in 126 genera, 29 families, and 7 orders.

The common name, dinoflagellate, is usually applied to the entire class, but in a stricter sense refers only to motile, armored species having a typical "girdle groove." The species in two of the seven orders have naked cells and species in two other orders are nonmotile, but most of the remaining species are characterized by overlapping, interlocking plates, which are often ornamented with ridges or spines, and by two flagella, one of them pectinate.

Dinoflagellates are present in most lakes, rivers, estuaries, and shallow bays in late summer and autumn. *Peridinium* and *Ceratium* are the most commonly encountered freshwater genera and can frequently be obtained by filtering large quantities of river or lake water through filter paper or through a plankton net and examining the residue. The residue typically contains diatoms, desmids and other chlorophycean plankters, and small animals; it may or may not contain dinoflagellates.

When an area of ocean displaying bioluminescence can be found, dinoflagelates are easily obtained in large numbers by using a plankton net. Occasionally, they will be so abundant that the water is colored red, yellow, or brown. If a bottle of such seawater is placed in a dark or dimly lighted room and illuminated from one side at a relatively low light intensity, the dinoflagellates will accumulate on the illuminated side of the bottle.

General Morphological and Physiological Characteristics

A dinoflagellate cell may be **naked,** enclosed only by a plasma membrane and an outer membrane with flattened vesicles between them, or it may be **armored,** having cellulose or other structural polysaccharides deposited within the vesicles and forming a cell wall. The typical dinoflagellate cell wall is made up of **thecal plates,** varying in number from two to approximately 100. The number and arrangement of the thecal plates are distinctive for each species and are the most useful characteristics to aid in their identification.

Most dinoflagellates are characterized by cells having a girdle groove, also called a **singulum,** dividing it into two halves, an **epithecum** and a **hypothecum,** and a **sulcus** or longitudinal groove (Figure 7-3). A ribbon-like pectinate flagellum encircles the cell in the girdle groove, and a trailing, whiplash flagellum lies in the sulcus. Trichocysts have been reported in a few species.

While most dinoflagellates are both armored and motile, about 30 motile species have cell walls but are unarmored (lack thecal plates). One naked, nonflagellate species, *Dinamoebidium varians,* has only an inner cell membrane and moves in an amoeboid fashion. Two species, *Dinothrix paradoxa* and *Dinoclonium conradi,* are filamentous, having immobile, somewhat cylindrical cells joined end to end in branching filaments. Some 13 species of Dinophycopsida consist only of free-floating or attached, sometimes epiphytic, nonmotile unicells. About 330 species with naked cells and over 600 species of armored dinoflagellates, all possessing the typical girdle groove and sulcus characteristics, make up the remainder of the class.

Most dinoflagellates are brownish in color, but some are bright green. Analysis of the chloroplast pigments reveals the presence of chlorophylls *a* and *c* together with typical Bracteobionta carotenoids: *dinoxanthin,* which is especially abundant in most Pyrrophyta and has been reported in some Chrysophyta; *diadinoxanthin,* which is very abundant in Chrysophyta and occurs in some Cryptophycopsida, as well as Dinophycopsida; and *peridinin,* unique to the Dinophycopsida and responsible for the golden to chocolate brown color in many

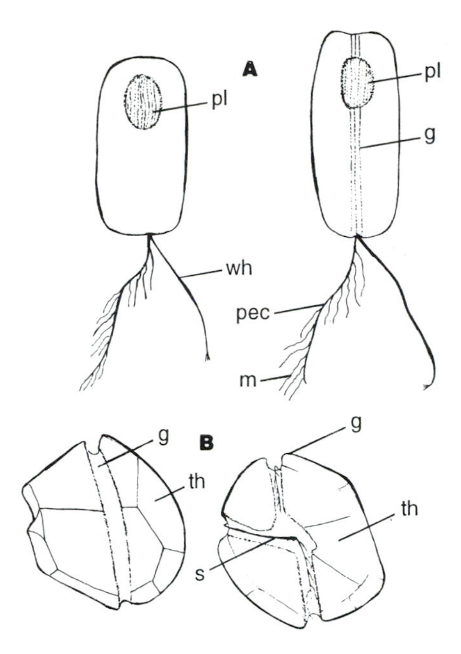

Figure 7–3 Cellular detail in two common dinoflagellates: (A) *Exuviella marina,* left, valve view; right, girdle view, (B) *Peridinium cinctum,* left, dorsal view; right, ventral view. (g = girdle or girdle groove, s = sulcus, th = thecal plate, wh = whiplash flagellum, pec = pectinate flagellum, m = mastigoneme, pl = chloroplast.)

dinoflagellates. The brown pigments are masked by chlorophyll in some species and mask the chlorophyll in others. A large number of species are colorless and holozoic; some photoautotrophic species obtain part of their food by ingestion of particles through the gullet.

The ultrastructure of chloroplasts and mitochondria resembles that of other Bracteobionta: each plastid band consists of three thylakoids and the mitochondria are microtubular. In large dinoflagellates, there are usually no pyrenoids with the many plastids, but in smaller species, pyrenoids commonly occur. In many Dinophycopsida, especially in the freshwater species, starch instead of oils is the chief storage product. In some species, the sulcus ends in a ventral chamber, which is apparently homologous to the gullet of the Cryptophycopsida, and remains of bacterial and algal cells and portions of diatom frustules have been found in it. Trichocysts have been reported along the edges of the sulcus and on the horns of *Ceratium hirundinella.*

Bodies hypothesized to be the centers of bioluminescent activity have been observed with the electron microscope in many dinoflagellates. These have been called **scintillons.** Using genera in which both luminescent and nonluminescent species occur, phycologists have recently to observed scintillons in both groups of species thus casting some doubt on the scintillon hypothesis of luminescence.

In *Prorocentrum,* spherical bodies which fluoresce green in blue light and red with ultraviolet light form a horseshoe-shaped ring around the cell center. These are believed to be dinoflagellate lysosomes (Zhou and Fritz 1994).

Both asexual and sexual reproduction have been studied in complete detail in about 30 species of Dinophycopsida out of the nearly 2000 listed by most algal taxonomists. Increase in numbers is primarily by means of the former. Meiosis was reported to be immediately postzygotic in *Glenodinium* and *Ceratium,* indicating haplontic life cycles in these two autotrophic genera, but in the holozoic genus, *Noctiluca,* Zingmark (1970) reported that meiosis was part of gametogenesis, indicating a diplontic life cycle.

In most of the life cycles that have been studied, the meiospores, asexual spores, and gametes are naked and flagellated, resembling mature *Gymnodinium* cells. For example, in *Dinamoebidium varians,* a colorless, heterotrophic, marine species, the amoeboid vegetative cell becomes spindle-shaped and forms a gelatinous sheath around it just prior to mitosis. It then divides two or three times to form four or eight zoospores, each very similar in appearance to *Gymnodinium* except that they constantly change shape as they swim away. Within 15 minutes, the flagellum is retracted and the zoospore becomes a vegetative amoeboid cell. Although the vegetative stage of *D. varians* is very similar to that of the slime mold, *Labyrinthula,* it is included in the Dinophycopsida and not the Myxomycophyta because of its *Gymnodinium*-like zoospores. Reproduction in *Tetradinium, Pyrocystis,* and other coccoid types is also by *Gymnodinium*-like zoospores, and the same is true of the filamentous and armored species.

A typical diplontic life cycle was described by Highfill and Pfeister (1992). In *Glenodiniopsis steinii,* the gametes, zygotes, and asexual spores are identical in appearance to the vegetative cells except for size and the presence of a thin mucous covering on old zygotes to which bacteria attach, giving them a warty appearance. The pattern of the thecal plates, position of sulcus and cingulum, and other characteristics are the same in vegetative and reproductive cells.

The Pyrrhophyta differ from most organisms in that the chromosomes are visible and countable during interphase. Chromosome numbers are often high in the Dinophycopsida, 260 in one strain of *Ceratium hirundinella,* for example, and approximately 274 in another (Table 7-1). Synthesis of DNA is apparently continuous. In some species, e.g., *Prorocentrum* sp., there is either no spindle, or else a unique type of spindle, since mitotic inhibitors, such as colchicine, do not interfere with mitosis. Nucleoli are usually more or less typical; however, some species lack nucleoli and in *Gonyaulax* the nucleolus lies outside the nucleus, whereas in *Noctiluca* the nucleoli are so numerous and large as to have been interpreted as chromosomes (Afzelius, according to Zingmark 1970). Meiosis is postzygotic and is typical of plants in general in *Ceratium cornatum* and other species that have been carefully studied. However, interkinesis is extra long.

Physiological studies in the dinoflagellates have centered chiefly on bioluminescence and mineral nutrition. Bioluminescence is of the unique anaerobic, flashing type unknown elsewhere among plants or animals. Many, but not all, marine dinoflagellates are bioluminescent, but no freshwater species are. Even if most or all of the marine species in a genus luminesce, none of the freshwater species do. Tropical bays and lagoons have sometimes become world famous for their luminescence. The slightest agitation triggers the mechanism that releases the light. Although luminescence in the ocean near Finschhafen, New Guinea, is intense at night, there is no evidence in the daytime of any organisms present in the perfectly clear water.

Mineral nutrition has attracted attention because it has seemed desirable to be able to grow some of the nuisance species in the laboratory in order to study their mineral requirements and develop control for them. Although they are easy to grow on laboratory media prepared from lake water, it has been virtually impossible to grow many of the species on a completely synthetic medium. One such nuisance species is *Peridinium cinctum,* which at times may grow in such rich abundance in the Sea of Galilee that the water supply for the city of Tel Aviv develops an objectionable flavor. To control this, more needed to be known about the dinoflagellate creating the nuisance.

Historical records reveal that the Sea of Galilee has always been an important source of food for much of Israel. The New Testament, for example, relates how Peter, Thomas, and their associates harvested fish in such quantity that they were astonished that their nets did not break. A "nuisance species," like *P. cinctum,* may well have provided most of the food that enabled the fish to increase in such quantity. The difference between "good" species and "bad" is seldom a taxonomic question.

Kåre Lindström (1980) found that the essential mineral missing from synthetic media but needed for growth of *P. cinctum* is selenium. Selenium is now known to be essential for animals, including humans, but how widespread the need for selenium is for plants must be ascertained by further studies with other species of dinoflagellates as well as other plants.

Phylogeny and Classification

Excellent fossils of dinoflagellates are found in Ordovician formations. What appear to be typical dinoflagellates have also been found in Hadrynian formations. These earliest fossils resemble Prorocentrales; other early fossils suggest that the Peridiniales were also among the first of the dinoflagellates to evolve. The morphologically simple Gymnodiniales do not fossilize well but may have been present prior to the origin of the Peridiniales without having left a record.

Acritarchs are organic-walled vesicles or fossils, up to 100 µm in diameter, which cannot be identified with absolute certainty as to their affinities with modern taxonomic groups; they make up a garbage can "taxon," so to speak, into which ancient rubbish is dumped. Nevertheless, there are recognizable patterns, and many acritarchs definitely resemble modern and ancient dinoflagellates. The Hadrynian Period is considered by the Russian paleobotanists, who are probably the best authorities on the fossils from that time, to consist of two periods, the **Riphean** and the **Vendian;** the number of dinoflagellate-like species of acritarchs increased rapidly during the early Riphean, leveled off during middle and late Riphean, increased again in early Vendian, and then "crashed" during late Vendian and did not build up again until Silurian times (Figure 7-4). Silurian formations are rich in abundance of excellent fossils very similar to modern dinoflagellates.

Because of their similarity to the Cryptomonadales, as well as the evidence provided by early fossils, it is assumed that the Prorocentrales of today most nearly resemble the original Dinophycopsida. From primitive ancestors, having apical cryptomonadlike flagella and resembling both *Prorocentrum* and *Gymnodinium,* evolution seems to have proceeded along three lines: one, leading to the Xanthophycopsida, in which the girdle groove and sulcus are not pronounced and in which the cell is composed of two valves; the second leading to the typical armored dinoflagellates (Peridiniales) that make up the majority of species in the class; and the

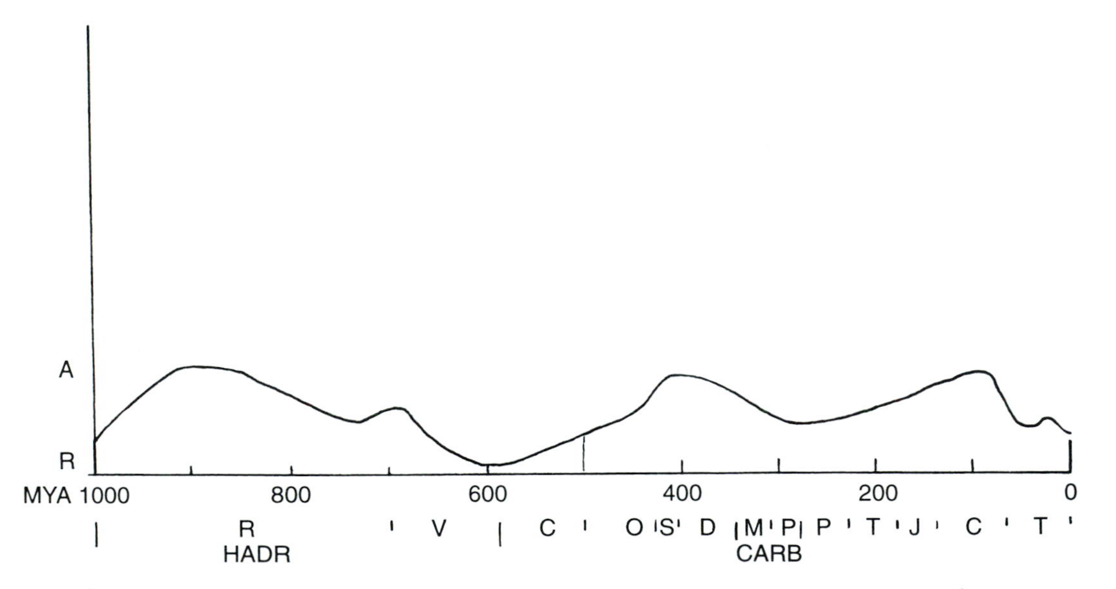

Figure 7–4 Fluctuations in abundance of Dinophycopsida, from Hadrynian (Riphean) time to the present, as revealed by the fossil record.

third, in which motility has been lost in the vegetative stage and in which filamentous types occur. Some phycologists have suggested that the third line leads to the Phaeophycopsida by way of the Dinococcales and Dinotrichales. Suggested relationships are illustrated in Figure 7-1; a list of orders with their chief characteristics is presented in Table 7-5. The total number of species in Dinophycopsida may be two or three times more than indicated in the table; according to Taylor (1993), there are approximately 2000 species in the class.

Because the nuclei of dinoflagellates seem to be intermediate between those of prokaryotes and eukaryotes, Dodge (1963), Zingmark (1970), and others have proposed placing them in a kingdom by themselves, the Mesokaryota. Table 7-2 compares Mesokaryota with Prokaryota and Eukaryota.

Representative Species and Genera

Of the freshwater dinoflagellates, *Ceratium hirundinella* (Figure 7-5) and *Peridinium cinctum* are among the most commonly encountered. Species of *Gonyaulax, Gymnodinium, Peridinium,* and *Pyrodinium* are common marine dinoflagellates. *Noctiluca miliaris,* a large, holozoic, luminescent, marine species, is easy to find in ocean water and readily available from biological supply houses. The somewhat smaller but very common *Gonyaulax polyedra* gives off light at a very high intensity compared to most bioluminescent organisms. Many marine species of *Gymnodinium, Peridinium, Ceratium,* and *Prorocentrum* are luminescent, but none of the freshwater species in these or other genera are luminescent.

Ecological and Economic Significance

Most recent surveys of marine life have indicated that the dinoflagellates are ecologically the most important plants in the warm temperate zone; at high latitudes, the diatoms become more important, though dinoflagellates are still abundant, and in the tropics, the Cryptophycopsida and Cyanophycopsida are at least as important as the dinoflagellates. Where dinoflagellates are abundant, environmental factors are often so favorable for their rapid growth and reproduction that "blooms" or "tides" of them form. It is likely that their ecological significance has been underestimated in the past.

While their year round significance is greatest in subtropical areas, bursts of rapid dinoflagellate production appear during the summer months at high latitudes. Norwegian ecologists have reported a density of dinoflagellate cells in Oslo Fjord exceeding 100,000 per liter. During red tide blooms, cell densities in excess of 40 million per liter commonly occur (Bold and Wynne 1985). Thus, the ecological significance of Dinophycopsida to fish, other marine animals, and man is considerable.

Autotrophic organisms living symbiotically in the digestive cavities of some marine animals, where they carry on photosynthesis and thus contribute directly to the nutrition of the animals, are called **Zooxanthellae.** According to the Goreaus and other writers in *National Geographic Magazine,* symbiotic Dinophycopsida may be the chief producers of coral reef ecosystems. Zooxanthellae having dinoflagellate symbionts are abundant to a depth of 100 meters; below that, the corals lack chlorophyll and exist primarily as carnivores supported ultimately by detritus from the surface. Growth of the symbiotic corals is very rapid, while the nonsymbiotic ones grow very slowly. According to Taylor, *Gymnodinium* and *Amphidinium* are the genera to which most Zooxanthellae belong. Other investigators have found *Chrysidella* (Cryptophycopsida) in most of the Zooxanthellae they have examined. According to Smith (1950), most

TABLE 7–5

CHARACTERISTICS OF THE ORDERS OF DINOPHYCOPSIDA

Orders and Included Fam–Gen–Spp	Thallus Type	Other Characteristics	Representative Genera and/or Species
Gymnodiniales (7-21-365)	Naked unicells with striated pellicles	Girdle groove and sulcus usually well developed; bioluminescence common	*Gymnodinium, Oxyrrhis, Oodina, Noctiluca, Polykrikos, Gymnaster, Warnowia*
Rhizodiniales (1-1-1)	Naked amoeboid unicells	Holozoic nutrition; Gymnodinium-like reproductive cells	*Dinamoebidium varians*
Dinoccoccales (4-8-14)	Nonmotile or free-floating unicells or palmeloid colonies	Overlapping cellulose thecal plates present; some epiphytic, others free-floating or in irregular colonies; sulcus and girdle well developed	*Tetradinium, Cystodinium, Pyrocystis, Hypnodinium, Phytodinium*
Dinotrichales (2-2-2)	Cells attached end to end in simple or branched filaments	Cell walls thick, often covered with a layer of gelatinous material to form a firm mass; filaments may be tapered	*Dinothrix paracoxa, Dinoclonium conradi*
Peridiniales (12-60-530)	Unicells with thecal plates, girdle, and sulcus	Pattern of thecal plates distinctive, often with spines, wings, or other ornamentation; bioluminescence common	*Ceratium, Peridinium cinctum, Pyrodinium, Gonyaulax, Glenodinium, Oxytoxum*
Dinophysidales (3-15-160)	Unicells with soales or plates; two apical flagella	Motile; 17 overlapping cellulose scales; cells divided vertically into two valves with a vertical groove	*Dinophysis, Dinofurcula, Amphisolenia, Histioneis, Ornithocercus, Desmocapsa*
Prorocentrales (1-6-30)	Motile unicells with two apical flagella	Cells naked or with cellulose cell wall; mostly marine, some freshwater	*Prorocentrum, Exuviella, Porella, Haplodinium*

Note: Fam = families, Gen = genera, Spp = species.

Zooxanthellae involving corals contain cryptophytes; however, other algal symbionts occur in coral and other marine animals. Luminescent coral has been observed at 400 m depth near Hawaii, but the source of the luminescence has not been ascertained. It was likely due to symbiotic dinoflagellates; however, radiolarians and some other marine animals are also luminescent.

Periodic bursts of dinoflagellate reproduction are commonly designated "red tide," "chocolate tide," "chocolate water," or similar terms. Cell density increases during these bursts to extremely high values so that the water is actually colored by the organisms. Mass mortality of marine life is often associated with these bursts. In the 1946–1947 red tide in the Gulf of Mexico, the density of *Gymnodinium brevis* was reported at 60 million cells per liter. A toxic substance produced by the algae resulted in the death of billions of fish, many of which were washed up on the beaches of Florida and other gulf states.

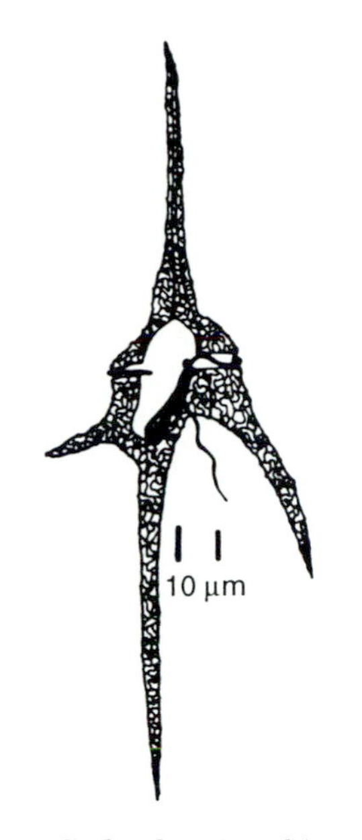

Figure 7–5 *Ceratium hirundinella,* a common freshwater dinoflagellate, order Peridiniales.

Red tide is a coastal phenomenon caused by any of several species of dinoflagellates only a few of which cause mass mortality of fish. During the period from 1901 to 1945, there were seven red tides recorded off the coast of southern California. In one of the seven, heavy mass mortality occurred and in two additional tides, there was considerable destruction of marine animal life. Apparently no ill effects occurred as a result of any of the other four. The autotrophic genera, *Gymnodinium* and *Gonyaulax,* and occasionally the holozoic genus, *Noctiluca,* are the dinoflagellates involved most often in poisonous red tide outbreaks. The poisoned fish are not toxic to warm-blooded vertebrates and in some areas—e.g., Mobile Bay in Alabama, Malabar in India, and Walvis Bay in southwest Africa—people gather the dying fish by the truckload for food.

Mollusc poisoning caused by marine dinoflagellates has occurred in many places in the world. When *Gonyaulax* spp. or other toxin-producing dinoflagellates are consumed by oysters, clams, or other molluscs, their flesh is rendered poisonous to humans, although the molluscs are apparently unharmed. The most severely affected region is the Pacific Coast of North America from southern California to Alaska, where *G. catenella* is the usual causative organism. The poison it produces is one of the most toxic substances known. In eastern Canada, *G. tamarensis* causes similar outbreaks, and in Europe, *Pyrodinium phoneus* does the same. All molluscs and many crustaceans become poisoned in the vicinity of the area where the dinoflagellate thrives. In California, the mussel *Mytilus californicus,* which is an open coast form, becomes poisonous while *M. edulis,* an inhabitant of bays, never does. In Europe, it is exactly the other way around. The explanation of this difference is that *G. catenella* multiplies in the open ocean whereas *P. phoneus,* the European dinoflagellate, occurs only in brackish water. The dinoflagellates causing mollusc poisoning are normally yellow-green to golden in color, but as they become poisonous they change to a reddish brown or orange color.

The symptoms of **paralytic shellfish poisoning (PSP)** vary, depending on the species of dinoflagellate involved and the resistance of the person affected. They may begin immediately or they may be delayed as long as 12 hours after consuming the affected shellfish. They often include a feeling of prickling numbness, constriction about the lips, a sensation of lightness, and an impression that objects have no weight. Patients often believe they can fly. The pulse rate may go as high as 160 per minute. There may be headache, backache, nausea, and difficulty of urination, and various muscles may become paralyzed. Because any warmblooded vertebrate may be affected by PSP, a mouse bioassay has been effective in detecting the presence of PSP toxins.

Research Potential

Their ability to luminesce has made dinoflagellates the subject of special research interest (Special Interest Essay 7-1). Where species of *Peridinium, Pyrodinium, Ceratium,* and *Gonyaulax* are abundant, the sea at times seems to be on fire. Often the water is colored by the high density of cells of the luminescing species. Luminous red water is usually associated with the "red tide" organism, *Gonyaulax polyedra;* other species may produce other colors or no colors at all. At the present time, no clearcut evidence of an ecological advantage to bioluminescence has been demonstrated in dinoflagellates; consequently, it is now generally assumed that it is the result of an alternate pathway for dispersal of energy which many species of dinoflagellates have independently evolved.

The ornamentations on dinoflagellates have also aroused the research interest of ecologists who have observed that distinct ecotypes occur in species of *Ceratium* and other Dinophycopsida. The forms occurring in warmer areas have considerably longer horns than those in colder areas. Apparently the longer horns slow down the rate of sinking in the less dense warmer waters. For the dinoflagellate, to maintain a favorable position near but not at the surface of the ocean while alive, slowly sinking out of the trophic zone when dead, is of considerable importance for the successful functioning of the ecosystem of which it is a part and depends upon a delicate balance involving protoplasmic density, water density, appendages (such as horns), phototactic responses, and automotility.

SUGGESTED READING

Smith's *Cryptogamic Botany,* Vol. I and Bold and Wynne's *Introduction to the Algae* have chapters on the morphology of the Pyrrophyta. *Animal Toxins,* edited by Russell and Saunders, has a chapter by D. V. Aldrich, K. A. Steidinger, and co-workers on the toxic substances associated with red tides and other dinoflagellate blooms.

E. N. Harvey, in his book *Bioluminescence,* (Academic Press, 1952), summarizes the history of scientific studies of bioluminescence and makes comparisons among the different groups of luminescent organisms. *The Journal of Phycology* is a rich source for information on these and other planktonic plants. Every botany student should frequently peruse the latest issues.

STUDENT EXERCISES

1. What possible adaptive advantage is there to the trichocysts in Cryptophycopsida? To bioluminescence in Dinophycopsida? To the presence of flagella in either class?
2. Is the absence of a fossil record proof that the Cryptophycopsida did not evolve until relatively modern times? Explain.

3. A student collects two samples of *Ceratium hirundinella* from a mountain reservoir, one from the deepest part, near the dam, where the temperature is 5°C, the other from a shallow bay where the temperature is 24°C. Examining several cells from each sample, he finds that the average diameter of the cells measured at the girdle is 53 μm in both samples. The length of the horns differs, however. In sample A, the average distance from the girdle to the tip of the longer horn is 87 μm; in sample B, the average distance from girdle to tip is 99 μm. Which sample was collected near the dam? Which in the shallow bay?

4. As a consulting botanist, you are called on by the National Oyster Farmers Society to evaluate what the society regards as a potentially dangerous situation. Samples of water from three areas have revealed densities of dinoflagellates greater than 1000 cells per milliliter: at site A, *Gymnodinium brevis* is the species responsible for the bloom; at site B, *Gonyaulax catenella* is the abundant species; at site C, the bloom is caused by *Dinamoebidium varians.* What can you recommend to the oyster growers as far as the marketing of their product is concerned? Would you expect to observe a visible "red tide" coloring the water at any of the three sites?

5. Obtain luciferase and luciferin from dried firefly lanterns and mix the two in a dark room; add two or three drops of water noting the exact time as you do so. How long does the luminescence continue? Does addition of ATP cause the luminescence to last longer?

6. A sample of luciferase (call it GPL) is obtained from a pure culture of *Gonyaulax polyedra.* You then extract luciferin from several species of plants. A brilliant luminescence results when luciferin from sample A is mixed with GPL; nothing at all happens when luciferin from sample B is added to GPL; and a very faint luminescence comes from sample C mixed with GPL. What is the source of each of these samples of luciferin? (a) *Gonyaulax polyedra;* (b) same family but another species; (c) another dinoflagellate but from a different order; (d) a luminescent bacterium; (e) a luminescent mushroo

7. What is the chemical nature of GPL (*Gonyaulax polyedra* luciferase) referred to in question no. 6? (a. carbohydrate; b. lipid; c. organic acid; d. protein.) Was it extracted by grinding cells of *G. polyedra* in cold water? or in hot water? or in cold water first, then hot water? or in hot water first and then cold water?

8. How would you go about extracting luciferin and luciferase from *Peridinium cinctum,* a dinoflagellate native to the Sea of Galilee in Israel? *Hint:* What do you know about luminescence in freshwater dinoflagellates?

SPECIAL INTEREST ESSAY 7-1

Bioluminescence

While numerous kinds of marine organisms have the ability to emit light, when the sea displays a brilliant and far-flung luminescence the most common cause is a dinoflagellate bloom. Similar blooms can be produced in the laboratory. Bioluminescence occurs in the Schizomycopsida and Basidiomycopsida, in addition to the Dinophycopsida, and also in many animals:

radiolarians, sponges, corals, nemerteans, molluscs, annelids, fireflies, echinoderms, chordates, and others. It has also been reported from the Ascomycopsida and Myxomycopsida, but these reports have not been confirmed.

The cause of bioluminescence has puzzled and excited scientists and others for centuries. In 1672, Robert Boyle wrote, ". . . and then I plainly saw, both with wonder and delight, that the joint of meat did in divers places shine like rotten wood or stinking fish . . . ." Pouring a little wine upon a portion of this meat, he discovered that alcohol extinguished the light; water, however, "would not so easily quench our seeming fires." He also discovered by means of experimentation that turpentine, hydrochloric acid, and ammonium hydroxide inhibit luminescence by meat-inhabiting bacteria, and that oxygen is essential to bacterial bioluminescence. Using a vacuum pump, he discovered that extinction of luminescence by lack of oxygen is reversible, whereas extinction by addition of toxic chemicals is not, and that a very small oxygen pressure produces maximum brightness. He wrote, "I could not by the touch discern the least degree of heat," an observation since confirmed by others.

In 1885, Dubois prepared, from a West Indies elaterid firefly, crude extracts which he called luciferin and luciferase. Later, he prepared similar substances from the boring clam, *Pholas.* To prepare luciferase, Dubois ground the photogenic tissue in cold water and left the homogenate standing until the light disappeared. To prepare luciferin, he ground photogenic tissue in hot water and quickly cooled the homogenate. When a small amount of luciferase was added to luciferin at room temperature, light was produced. Because his luciferase was heat labile and because it was still active in the mixture of luciferin and luciferase after the mixture had ceased to luminesce, as could easily be demonstrated by adding some more luciferin, Dubois concluded that the luciferase was an enzyme, luciferin the substrate upon which it acted. Dubois demonstrated that clam luciferase could not activate firefly luciferin, nor could firefly luciferase activate clam luciferin.

In the early 1900s, E. N. Harvey attempted to extract luciferin and luciferase from every type of luminescent organism known, and he cross-tested in all possible combinations cold-water and hot-water extracts from these organisms. Only when species were closely related, as in the case of *Cypridina* and *Pyrocypris,* for example, would luciferase from one species activate luciferin from another.

Since the 1960s, luciferin and luciferase have been extracted from some groups of plants and animals (e.g., jelly fish) in which Harvey had been unsuccessful. One reason success has now been attained where Harvey met failure has been the discovery that cofactors are frequently necessary for luminescence. The specific cofactors or activators are various and include adenosine triphosphate (ATP), magnesium ion, reduced flavine mononucleotide ($FMNH_2$), hydrogen peroxide, and other chemicals.

The fact that each group of luminescing organisms has its own specific substrate, specific enzyme, and specific activator, strongly suggests that in the course of evolution, a diversity of luminescing systems have evolved independently in completely unrelated lines. A number of other observations support this suggestion. In bacteria, for example, light is emitted along the entire visible spectrum, with peak emission at about 470 nm. The light emitted from fungi is not as bright as that from bacteria and extends from the orange to the blue or from 660 nm to about 470 nm with maximum emission at 520 nm. The emission spectrum from fireflies is narrower, from about 630 nm to 520 nm with peak emission at 570 to 550 nm, depending on the species. Furthermore, bacteria and fireflies require much more oxygen than do fungi and dinoflagellates; likewise, fungi and bacteria give off a steady glow, while in fireflies and dinoflagellates the light comes in bursts or flashes of rather short duration.

Although it has frequently been observed that organisms closely related to luminescent kinds may or may not themselves luminisce, there is paucity of genetic studies on luminescence. Mycologists and geneticists have studied luminous and nonluminous ecotypes of *Panus stipticus,* a basidiomycete. The European form, which is non-luminous, hybridizes readily with the luminous American form; the hybrids are completely fertile. It was concluded that the luminous and nonluminous forms differ from each other by only one gene pair, with luminous dominant over nonluminous.

The feasibility of using luminous bacteria and fungi to detect toxic substances has been demonstrated by botanists and biochemists working in the space industry. Unsymmetrical dimethyl hydrazine (UDMH) is fuel used in the space industry. It is highly toxic; if any should leak into the manned chamber of a space ship, the astronauts would be in serious straits. Cultures of *Photobacterium fischeri, Vibrio albensis,* or *Armillaria mellea* can be kept in the capsule at all times. At concentrations of 1.6 ppm, too low to be detected by other means, *P. fischeri* luminescence begins to decrease; it ceases at 64 ppm. The inhibition of luminescence is thus a measure of UDMH concentration.

8 THE SLIME MOLDS

The summer of 1973 was warm and wet in many parts of the U.S. That was the year the "Blob" made headlines. Its "red pulsating" plasmodium was photographed covering lawns and flower beds and even climbing telephone poles. Sensationalist journalists writing for the national tabloids were awed by its massive, slimy, slug-like body slowly, persistently creeping on, seemingly deterred by nothing that happened to be in its way. Speculation circulated, that rainy summer, that "the blob" was a mutant bacterium sent from another planet to take over the earth. However, the rains stopped and the "blob" passed into oblivion, or into the secluded world of a few research botanists who study it under its more euphonious name, *Fuligo septica.*

It is unusual for slime molds to attract as much attention as *Fuligo septica* did in 1973. They seldom attain the massive size of the "blob", and most people have never consciously seen a slime mold, even though they are relatively common in moist forest habitats. Few biologists study them. Taxonomically, they are an enigma: they resemble animals in many ways and are claimed by zoologists, yet they possess plant-like characteristics, especially in mode of reproduction, and are also claimed by botanists. Zoologists classify them in the phylum Protozoa, class Sarcodina, along with the amoebae, where they make up the order Mycetozoa. As plants, they do not seem to fit well into any of the existing classes or divisions, and so most botanists place them in a division by themselves, as has been done here. Many botanists, on the other hand, include them with other heterotrophic plants in the division Mycophyta, or scatter them among other plant and/or animal taxa. Still other biologists classify them in a separate kingdom, either in the Fungi, along with both the Mastigomycopsida and the Eumycophyta, or else in the Protista with unicellular algae and Protozoa. And to complicate matters even more, many biologists admit to uncertainty as to whether the three classes included in the Myxomycophyta are really closely enough related to one another to be retained in a common taxon, whether it be class, division, or kingdom.

There are six main groups of slimy organisms that have been called "slime molds" at various times: (1) the slime bacteria (Myxobacteriales), (2) the cellular slime molds (Acrasiales), (3) the net slime molds (Labyrinthulales—with three subgroups which may or may not be closely related to each other), (4) the plasmodial or true slime molds (Myxomycopsida, sensu stricto), (5) the Plasmodiophorales, and (6) the Protosteliales. The organisms in all six groups resemble each other in producing slime material that often helps in holding the cells together, in forming a net-like plasmodium or pseudoplasmodium, in taking food phagotrophically at some time in the life cycle, in producing often brightly colored sporangia with delicately sculptured spores, and in aggregation of cells into a common thallus of some type.

Inasmuch as the first of these six groups, the Myxobacteriales, are prokaryotic and resemble other bacteria in numerous biochemical and cytological details, they are included in the Schizomycopsida. We are confident that their similarities to the slime molds are superficial and the result of parallel evolution under more-or-less similar environmental conditions.

The other five groups are collectively known as the slime molds or Myxomycophyta (*myx* = slimy or mucous, *myco* = mold or fungus, phytum = a plant) and are discussed in this chapter. They are often included in a common order within a single class; however, many of their shared similarities may be analogous rather than homologous, and it seems better to classify them as three separate classes and thus emphasize our uncertainty as to their phylogenetic relationships. As more data are gathered, we should be able to ascertain more nearly their true relationships and, if necessary, remove some of the genera, or even classes, to other divisions, or combine genera that are superficially different but are actually related.

Verifiable slime mold fossils are nonexistent at the present time. However, fossil spores should exist and may someday provide important clues as to the origin of these plants. The widely held belief among botanists that they are an ancient group is purely speculative. Since the plasmodial slime molds are mostly terrestrial organisms, living in close association with vascular plants, they likely did not originate prior to the invasion of land by Tracheophyta. However, the members of the other two classes feed on bacteria and might conceivably have evolved from now extinct organisms of ancient origin. On the basis of the information available today, we speculate that the ancestral Myxomycophyta evolved from flagellated plants, unicellular or colonial, probably during or after the Devonian period. Their early evolution was along two lines leading to the three classes (Figure 8–1), the cellular slime molds (Acrasiomycopsida), the net slime molds (Labyrinthulomycopsida), and the plasmodial or true slime molds (Myxomycopsida).

The Cryptophycopsida, Chrysophycopsida, Mastigomycopsida, Euglenophycopsida, Chlorophycopsida, and Protozoa have all been suggested as the parent stock from which the Myxomycophyta evolved. Like the Cryptophycopsida, the Myxomycophyta are characterized by naked, amoeboid swarm cells and pulsating vacuoles. The flagella, where they have been examined, are truncated like those of the Pyrrophyta and other Bracteobionta, but they lack mastigonemes. In species having two flagella, the two are of different length, as in the Pyrrophyta. The slime mold flagella could easily have arisen by loss of a single row of mastigonemes while retaining the other characteristics of Cryptophycopsida flagella. In fact, the marine pyrrophyte, *Oxyrrhis marina,* has flagella exactly like those of many slime molds.

It would seem less likely that the Chrysophycopsida, Xanthophycopsida, or Mastigomycopsida, with their pinnate flagella, or the Euglenophycopsida, with their tapered pectinate flagella, would have given rise to the type of flagellum in the Myxomycophyta. Furthermore, in the Cryptophycopsida and some Dinophycopsida, naked amoeboid stages that resemble the swarm cells of many Myxomycopsida are common. The mitochondria are tubular as in the Pyrrophyta. The colorless dinoflagellate *Dinameobidium* could be classified in the Labyrinthulomycopsida except that it goes through a *Gymnodinium* stage in its life cycle. On the other hand, the apical insertion of the flagella in most Myxomycopsida is similar to that of the

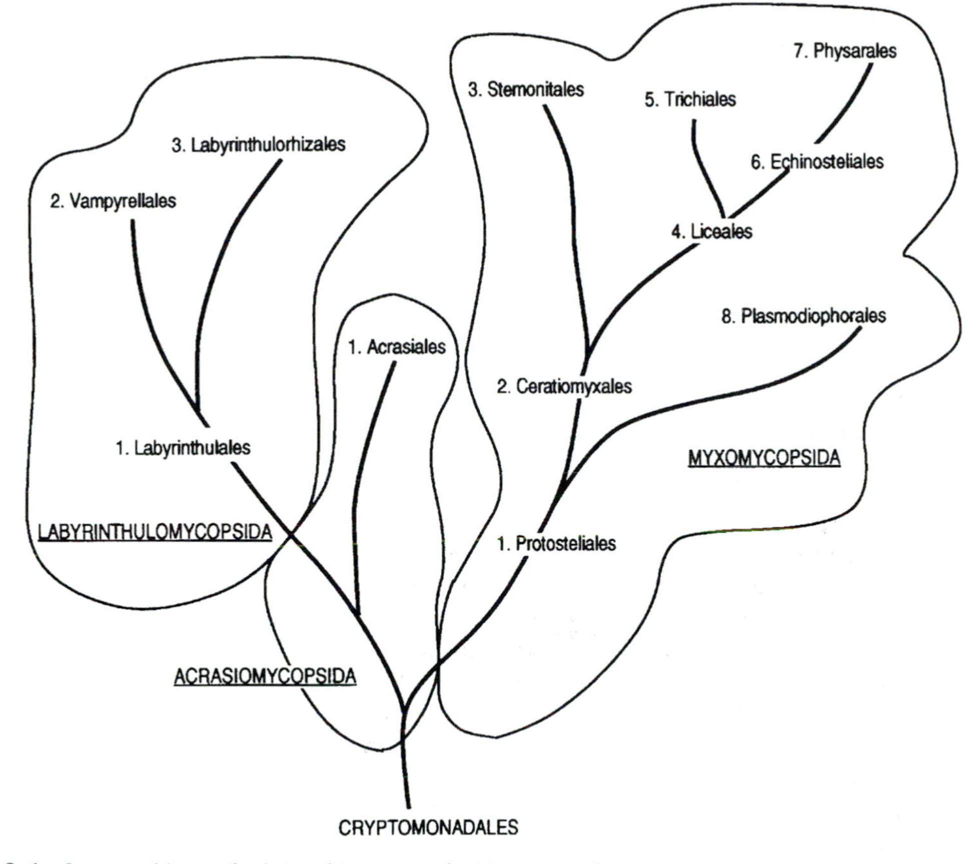

Figure 8–1 Suggested lines of relationship among the Myxomycophyta.

Chlorophycopsida and Euglenophycopsida, and the possibility of their origin from either of these groups must be considered.

Although Myxomycophyta are often described as saprophytes living on plant material in very wet and often densely shaded habitats, the species that have been studied intensively, including those that are commonly cultured, are better described as herbivores and predators. A few species are parasitic on vascular plants.

The Acrasiomycopsida are usually abundant in moist, decaying vegetable matter making up the humus of most soils. They are easy to grow in culture dishes on nutrient agar. Collecting Labyrinthulomycopsida is more of a challenge; it is generally necessary to find stands of the plants which they parasitize, such as eelgrass, *Zostera marina,* or various kelps and green algae. The plasmodial (or true) slime molds can be found in very wet forests by peeling back the bark on a stump or fallen log and looking for the net-like, usually white or yellow, plasmodium. Some species appear on the soil, often partially covering stems and leaves of grass or other plants, as white or brightly colored, typically yellow, grains which, when touched, feel very slimy and not at all granular. During dry weather, the often brightly colored fruiting bodies can be found on similar sites.

There are three morphological stages in the life cycle of most slime molds: a unicellular, amoeboid or flagellated stage, a plasmodial or plasmodial-like migratory stage, and a sporulative stage. Superficially, the three classes resemble each other in all three stages, but there are significant physiological and other differences among them that cause us to suspect they may not be as closely related to each other as was formerly believed.

The unicellular stage is primarily a growth or cell replication stage in the Myxomycopsida, but it is the assimilative or vegetative stage in the other two classes. In the Myxomycopsida, the assimilative structure is the plasmodium; in contrast, pseudoplasmodia in the other classes are primarily migratory structures. The individual cells, or the plasmodia, move slowly over the substrate ingesting bacteria, yeast, fungal spores, and other particulate matter. As long as food and moisture are both abundant, the assimilative organisms continue to grow, but in the absence of either, the assimilative stage ends and the plant goes into a reproductive stage if there is abundant light. In the dark, they tend to remain in the assimilative stage indefinitely.

Morphological differences among the three classes occur in all three stages of the life cycle. In the unicellular stage, cells may be either amoeboid or flagellated in the Labyrinthulomycopsida and Myxomycopsida, but are never flagellated in the Acrasiomycopsida with the exception of a single species, *Guttulina rosea.* In the plasmodial stage, true plasmodia are extremely rare in the Acrasiomycopsida and uncommon in the Labyrinthulomycopsida. Pseudoplasmodia, resulting from the aggregation of individual cells, occur in these two classes but never in the Myxomycopsida, which have a thallus consisting of a true plasmodium. The pseudoplasmodium of Acrasiales is usually more or less star-like with few anastomoses. In the Labyrinthulomycopsida it is abundantly anastomosed; its distinctive feature, however, is the net of nonliving material on which the individual cells glide. Superficially, the sorocarps of the three classes are very similar, but the mode of formation and their anatomy are vastly different.

Reproduction is by spores, usually produced in a sporangium, distributed by wind. The spore walls are of cellulose as in other plants. Following germination of the spores, either a unicellular myxamoeba, resembling the amoeboid cells of some Cryptophycopsida and Dinophycopsida, or a flagellated swarm cell develops. Differences in the assimilative stage as well as differences in sporangia and life cycles are taken into account in classifying the 450 species into 3 classes and 12 orders. Relationships among the orders are suggested in Figure 8-1, and differences among the three classes are summarized in Table 8-1. The division contains 27 families and 73 genera.

CLASS 1. ACRASIOMYCOPSIDA

The Acrasiomycopsida (*acra* = at the apex, *sio* = to move to and fro, *myceto* = a fungus), commonly called cellular slime molds and social amoebae, are an interesting group of phagotrophic organisms which in their assimilative stage cannot be distinguished from amoebae, but which have a fungus-like reproductive stage that in general morphology resembles that of the true slime molds. Cellular slime molds differ from slime bacteria in having larger cells which are nucleated. The 9 genera and 21 species are all classified in one order, the Acrasiales, and they are all usually included in a single family.

Acrasiomycopsida are easily isolated and grown in petri dishes. A good medium may be prepared by adding 20 grams of granulated agar, 1 g lactose, and 1 g peptone to 1000 ml double-distilled water (or spring or well water that has not been chlorinated), and autoclaving

TABLE 8-1

COMPARISON OF THE THREE CLASSES OF MYXOMYCOPHYTA

Acrasiomycopsida	Labyrinthulamycopsida	Myxomycopsida
Thallus is a pseudoplasmodium made up of individual cells	Thallus is a net plasmodium consisting of tracks of slime	Thallus is a true plasdium of anastomosed coenocytes
Few anastomoses between cells occur	Anastomoses between cells occasionally occur	Anastomoses between cells are very common
With one exception, cells are never flagellated	Cells typically have one modified whiplash flagellum	Cells typically have two modified whiplash flagella
Produce acrasin, a mammal type hormone	Do not produce any mammal type hormones	Do not produce mammal type hormones
Common in soils, especially forest soils	Mostly aquatic, often marine	Common on bark and in soil of forested areas
Feed on bacteria in soil, dung, compost	Often parasitic on aquatic plants	Feed on bacteria in soil and decaying vegetation

at 15 lb pressure for 10 minutes. Sprinkle particles of soil on this medium or on a sterile hay infusion. For denser growth, increase the amount of sugar and peptone and add some phosphate; 10 g peptone and 10 g glucose plus 1 and 1.5 g, respectively, of Na_2HPO_4 and KH_2PO_4, added to the 20 g agar and 1000 ml water, gives excellent growth. Soils vary considerably in species and abundance, deciduous forest soils generally yielding the greatest diversity and abundance of cellular slime molds; however, the cosmopolitan *Dictyostelium mucoroides* is usually present in cultivated as well as forest soil (Figure 8-2).

General Morphological and Physiological Characteristics

The assimilative stage of the cellular slime molds consists of unicellular, nonflagellated myxamoebae that are common in moist organic soil, on dung, and in other habitats where bacteria are abundant. Like the protozoan amoebae studied by zoologists, the myxamoebae are naked, changing shape continuously as they move slowly over the substrate, engulfing food particles, such as bacterial cells, by sending out pseudopodia which enclose them. Most species are easily cultured if grown on media containing bacteria. Under favorable circumstances, the assimilative stage goes on indefinitely with the myxamoebae dividing every few hours; if the food supply is depleted, if overcrowding occurs, or if moisture becomes limiting, the organism goes into the next, or aggregation, stage of its life cycle.

There is considerable diversity in the appearance of the individual myxamoebae. On the basis of unicell morphology, Bonner (1967) divided cellular slime molds into four groups; the *Hartmanella* type in which the cells are large and have long, slender (filose) pseudopodia, the *Dictyostelium* type with small cells having many filose pseudopodia (Figure 8-2), the *Sappinia* type in which the pseudopodia are broad and lobose and the cells are binucleate, and the *Guttulina* type with fat, stubby, slug-like (limax) pseudopodia. The only species of Acrasiomycopsida to produce flagellated unicells, *Guttulina rosea,* is included in this last group.

Prior to the beginning of aggregation, the myxamoebae go through an interphase period which lasts from 4 to 8 hours, during which time they stop feeding, decrease in size, and undergo other morphological changes including the appearance of granules in the cytoplasm and the disappearance of food vacuoles. As the individual cells move toward the aggregation

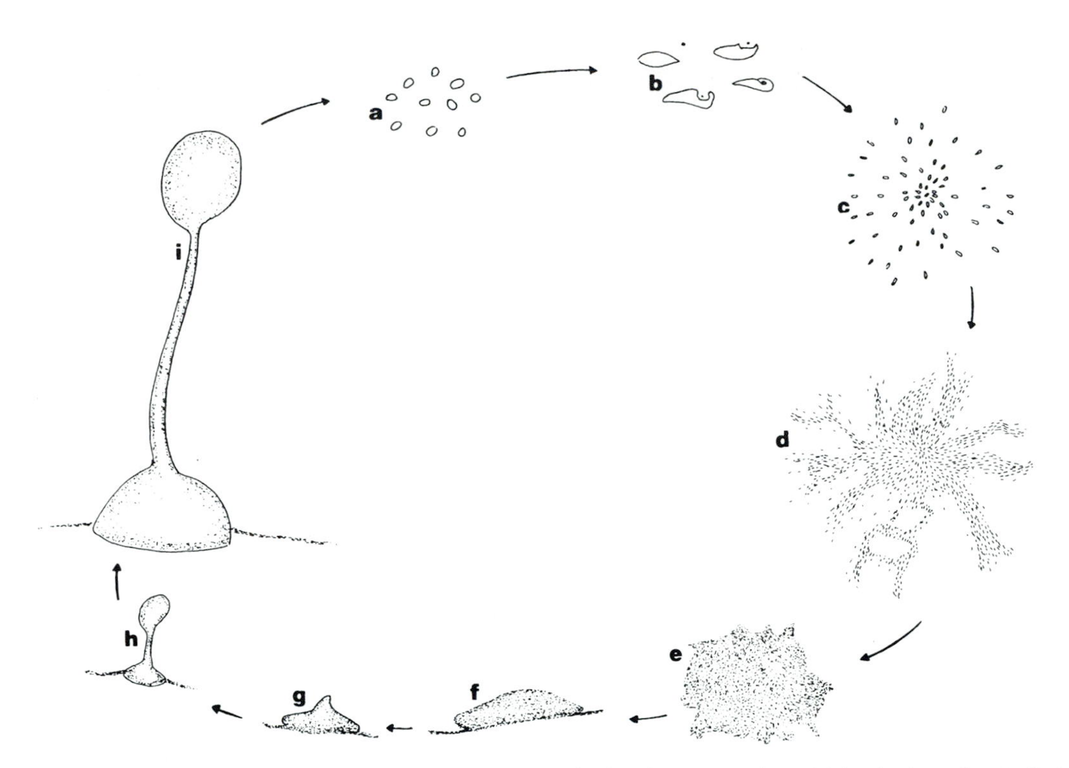

Figure 8–2 Life cycle of a typical slime mold: (a) spores; (b) free-living amoebae; (c) beginning of assimilation; (d) aggregation into a pseudoplasmodium; (e) aggregation complete and migration beginning; (f) migration; (g–h) sorocarp beginning to form; (i) mature sorophore and sorocarp. (d and e from unnamed soil collection, sagebrush-grassland, Rexburg, ID; others from published descriptions.)

center, they congregate into an elongated reticulate mass, or **pseudoplasmodium,** similar in appearance to the true plasmodium of the Myxomycopsida, but differing from it in two significant ways: (1) the pseudoplasmodium is strictly cellular except just prior to sporulation when, in some species, the cells unite into a true plasmodium; and (2) the pseudoplasmodium does not feed on the substrate.

Under conditions that lead to aggregation, the myxamoebae secrete a substance called **acrasin,** which produces a chemotactic response in the individual cells. Although their chemical composition varies slightly from species to species, acrasins are secreted by all Acrasiomycopsida. Wright and Anderson (1958) demonstrated that acrasin is steroid in nature and that urine from pregnant mammals, including humans, contains similar steroids and will evoke the same aggregation response in the assimilative myxamoeba of some Acrasiales. It was originally hypothesized that a single large gradient over the entire aggregation pattern was responsible for the phenomenon. It has since been demonstrated that many gradients of acrasin concentration are set up beginning at the center and moving outward in waves as previous acrasin-producing cells become fatigued. In each wave, myxamoebae that have not yet undergone aggregation move toward the nearest area of high concentration of steroid. Correct interpretation of test results is therefore challenging.

As the pseudoplasmodium continues its development, the cells become more densely packed and it begins to resemble the snaillike mollusk known as a slug (Figure 8–2). Considerable differentiation of cells now occurs, those that will become spores remaining isodiametric while those that will become stalk cells become elongated. In this stage, the pseudoplasmodium is commonly called a **slug,** and it creeps slowly over the substrate leaving a slimy sheath behind it much as slugs and snails do. The slug is sensitive to conditions of moisture during this **migration period,** which may last for only several hours under dry conditions (Figure 8–2), but as long as 20 days if conditions are moist. The length of period determines to some extent the morphology of the organisms in the next stage: the longer the migration period, the longer the stalk or sorophore that is produced.

Sporulation follows the migration stage. The pseudoplasmodium stops gliding and gradually becomes erect (Figure 8–2). Within the pseudoplasmodium, a cellulose cylinder forms and becomes the stalk of the fruiting body, the **sorophore,** on which the **sorocarps** with their masses of spores are borne. As the stalk lengthens, the rest of the cells making up the pseudoplasmodium migrate upward around the outside of the stalk and after 3 to 4 hours have formed one or more sorocarps. Sorocarps differ from one species to another, but in many species the sorocarp is globose and is formed at the apex of the stalk. The cells of the stalk become vacuolated, produce a rigid cellulose sheath, and then die. According to Gezelius and Rånby (1957), the cellulose in the fruiting body of *Dictyostelium* is similar to angiosperm cellulose but of a lower crystalline order. In some species, an oscillating movement may be noted at the apex of the sorocarp at this stage (hence the genus name, *Acrasis*). Each cell in the sorocarp, still retaining its individuality and without further cell division, becomes a spore. The spores are scattered by wind and upon germination release single myxamoebae, whereupon the assimilative stage is reestablished and the life cycle completed (Figure 8–2).

Sexual reproduction is reported to be by fusion of myxamoebae. Apparently, the resulting zygote does not always undergo meiosis since both haploid and diploid strains of *Dictyostelium discoideum* are known. Haploidization of diploid strains has also been reported with meiosis taking place in individual myxamoebae.

Phylogeny and Classification

If the Acrasiomycopsida have left a fossil record, it has not yet been discovered. Like the Cryptophycopsida, which they resemble in several ways and from which they likely evolved, their cells are soft and decompose rapidly. They differ from the cryptophytes in lacking flagellated cells at any stage in the life cycle, as well as in lacking photosynthetic pigments. They differ from the Labyrinthulomycopsida and Myxomycopsida in the type of assimilative stage, in how individual cells form a pseudoplasmodium, in their general lack of flagella, and in a number of biochemical characteristics; but also resemble them in many ways: all three classes possess naked assimilative cells, have cellulose in the cell walls of the spores, and are characterized by phagotrophic nutrition. Furthermore, there are many similarities between pseudoplasmodia and true plasmodia, and similar myxamoebae are produced in all three classes when the spores germinate. Table 8–1 summarizes some of the variation among the three classes.

Without a fossil record to guide us, we can only speculate as to the origin of the Acrasiomycopsida and the origin and evolution of the other two classes of Myxomycophyta as well, basing our hypotheses on observed morphological, biochemical, and physiological similarities and differences. Assuming that the first slime molds evolved from the Cryptophycopsida, they were probably both unicellular and flagellated. Mutations resulting in aggregation of cells into pseudoplasmodia imparted an adaptive advantage to species subject to desiccation when the habitat periodically dried out. Such mutations would be of great significance by making it possible for aquatic, achlorophyllous Cryptophycopsida to migrate onto moist habitats on land. Presumably, the first cellular slime molds were morphologically similar to both *Cryptomonas* in the Cryptophycopsida and *Protostelium* in the Myxomycopsida, possessed pectinate flagella similar to those of *Guttulina rosea,* and obtained their nourishment by ingesting the bacteria that were decomposing dead algal masses. As vascular plants became abundant on land, primitive slime molds found an unoccupied niche in damp forests where they were able to feed on the bacteria and fungi that were decomposing the wood and bark of fallen logs and stumps. From ancestors similar to *Protostelium* and *Guttulina,* the other genera of cellular slime molds have evolved along one line and the Labyrinthulomcopsida along another line.

Representative Genera and Species

Among the earliest cellular slime molds discovered and named was *Acrasis granulata,* found on attached dead plant parts in France. It was named and carefully described by van Tiegham in 1880 and never seen again since then. In 1962, a second species was described by L. S. Olive and C. Stoianovitch and named *Acrasis rosea.* (Bonner 1967). Both species of *Acrasis* are found on the dead parts of live plants, which may account for failure to find it for so many years after the initial discovery. Another genus, described by van Tiegham in 1884, and never seen again since then is *Coenonia.*

Acytostelium differs from other cellular slime molds in that the sorocarps are produced on noncellular stalks. In this respect, *Acytostelium* resembles *Protostelium;* both genera are sometimes classified in the Acrasiales, but in a different family from the rest of the order. Martin et al. (1983) separated the two genera, allying the Protosteliales with the Myxomycopsida and putting *Acytostelium* in the Acrasiaceae.

Several species of Acrasiomycopsida, e.g., *Dictyostelium discoideum, D. mucoroides, Guttulinopsis vulgaris,* and *Polysphondilium violaceum,* are cultured in biological laboratories. The generalized morphological characteristics described above are taken from studies of cultured Acrasiomycopsida and seem to apply to most of the known species.

Ecological and Economic Significance

What little is known of their ecological niche suggests that the Acrasiomycopsida are not an ecologically important group. In some textbooks, they are described as saprophytes significant as decomposers in some ecosystems. However, their niche actually seems to be that of herbivore, grazing upon saprophytes, and their total effect may well be a slowing down rather than speeding up of the decomposition processes. This, too, is speculative since little is known about their ecology. Hopefully, the students of today will be stimulated to investigate the ecology of decomposers in general and better ascertain the ecological contributions of the slime molds.

Economically, the Acrasiomycopsida have so far been of little value, but they have a potential for use in running pregnancy tests. The **Shaffer** test works well when environmental conditions have reached the point where the myxamoebae are capable of aggregating. Urine from a pregnant woman will then cause immediate aggregation; urine from a nonpregnant person will have no effect on aggregation. In the latter case, aggregation will still proceed once the natural supply of acrasin has built up at the aggregation center. Recognition of the point where myxamoebae are capable of aggregating is a problem not yet fully solved.

Cellular slime molds also show promise of economic value in scientific research. Several species are cultured by biological supply houses and sold for demonstration material and research purposes at colleges.

Potential for Scientific Research

Cellular slime molds are easily cultured in the laboratory and interesting experiments have been conducted with them, both in the assimilative and in the aggregation stage. The significance of aggregation in "social organisms" has been largely overlooked. Bonner (1967) has discussed some of the features which make Acrasiomycopsida valuable in studies of aggregation; Franks has reported a similar phenomenon in ants in the *American Scientist.* In both slime molds and ants, the aggregated body is able to do things that the same number of cells, as individuals, cannot possibly do.

Experiments involving steroid hormones have already been mentioned. In other experiments, portions of the pseudoplasmodium have been cut away to ascertain what effect, if any, this might have on sorocarp formation. When this is done, the fruiting body is abnormal and the abnormality is consistent for the portion of the pseudoplasmodium that was excised. These results have been confirmed with experiments involving dyes, suggesting that once aggregation is complete, the cells are actually differentiated as though the pseudoplasmodium were truly a multicellular organism. On the other hand, when the pseudoplasmodium is experimentally shaken vigorously enough to separate the cells, they can be plated out on medium and each will develop into an assimilative myxamoeba.

Since naked cells have some distinct advantages for studies of cell structure and physiology, cellular slime molds have a potential for providing information on the function of various organelles, effects of osmosis on life processes, reversal of sol-gel reactions in living cells, etc. For example, the micromanipulator, which has proven so useful in studying the physiology of isolated organelles, is more easily employed in removing organelles from naked cells than from cells enclosed in a rigid cell wall or possessing a more-or-less rigid pellicle.

The cellular slime molds are also excellent organisms for studies of morphogenesis. Even though the cells have retained their individuality in the pseudoplasmodium, there is a division of labor that is now being studied in greater detail and will hopefully give us insights into the processes involved in morphogenesis and cellular differentiation in all kinds of plants and animals, including humans. A gene in *Dictyostelium discoidium,* discovered by scientists at the Pasteur Institute in Paris, transmits the aggregation signal needed for morphogenesis. Essentially identical to a gene in humans responsible for metastasis of cancer cells, this gene may prove to be the key factor in producing a cure for cancer (Marx 1990).

Studies of acrasin, how it is formed in the cellular slime mold, the biochemical pathways involved, and the action of genes in its formation may help us in our understanding of biosynthesis of steroid hormones in more complex organisms, especially in humans.

CLASS 2. LABYRINTHULOMYCOPSIDA

The Labyrinthulomycopsida (*labyrinth* = maze of interconnected channels, and *myceto* = mold or fungus) are the least well known of the Myxomycophyta. Their inclusion in this division, or even in the Plant Kingdom, is controversial. Nevertheless, because they produce a slimy pseudoplasmodium, have naked cells, produce cellulose walled spores in a sorus which somewhat resembles that of the Acrasiales, and produce flagellated myxamoebae with a modified whiplash flagellum as in Myxomycopsida, they are often classified as Myxomycophyta. There are about 40 species of Labyrinthulomycopsida, included in 3 orders and distributed among 4 families and 17 genera (Table 8-2). To collect members of this class, one needs to find stands of eelgrass or other plants which they parasitize and examine carefully for dark spots or other signs of infection.

General Morphological Characteristics

Labyrinthula macrocystis, a parasite on an aquatic plant known as wrack or eelgrass (Figure 8-3), is the best-known species. Its assimilative stage consists of a uninucleate, spindle-shaped myxamoeba about 4 × 8 μm in size with tufts of filaments (sometimes called pseudopodia) on each end. The threads of slime secreted by filaments from adjacent cells fuse as they come in contact with each other, thus forming a nonliving, plasmodium-like network on which the myxamoebae glide along. This "plasmodium" is often called a **net plasmodium,** typical of and unique to the Labyrinthulomycopsida (Figure 8-3). Being made up of nonliving filaments, rather than living protoplasm, it is not homologous to either the true plasmodium of the Myxomycopsida nor the pseudoplasmodium of the Acrasiomycopsida.

In the assimilative stage, each cell divides mitotically to form two or four daughter cells which feed phagotrophically for a while, gliding along their slime tracks, before dividing again. Under favorable conditions, a large, elongated colony of cells sometimes develops. The colonies may fuse to form pseudoplasmodia similar to those of the Acrasiomycopsida (Figure 8-3).

At times, groups of cells will mound up and form a membrane-covered sorus from which spores are later released. In some species the spores are encased in a cellulose wall, in other species they are naked. Each spore germinates to produce one to four spindle-shaped myxamoebae with tufts of filaments. In at least one species, biflagellate zoospores are produced. The flagella are unequal in length: one is a typical whiplash flagellum, the other a truncated whiplash. In *Labyrinthomyxa,* uniflagellated zoospores are produced. In several genera, e.g., *Leptomyxa* and *Reticulomyxa,* multinucleated myxamoebae with radiating pseudopodia, which anastomose to produce a fine microscopic network, possibly represent a type of true plasmodium.

Phylogeny and Classification

The class is divided into three orders, the Labyrinthulales, Vampyrellales, and Labyrinthulorhizales (Figure 8-1, Table 8-2). The Labyrinthulales, with four genera, contains the typical net slime molds. The taxonomic positions of the Vampyrellales, with 12 genera, and of Labyrinthorhizales, with one genus, are still controversial.

The two families of parasites on freshwater algae that make up the Vampyrellales were regarded by Kudo (1966) as being closely related to each other and to the typical net slime molds. *Plakopus rubicundus,* the only species in the Plakopaceae, is a parasite on species of *Oedogonium.* There are about a dozen genera and 25 species in the other family, the Vampyrellaceae, typical of which is *Vampyrella lateritia,* a parasite on another genus of Chlorophycopsida, *Spirogyra.*

Labyrinthorhiza is often included in the Labyrinthulales, but its differences are distinctive enough to suggest that it is best classified in a separate order. It is a genus of nonflagellated

TABLE 8-2

GENERAL CHARACTERISTICS OF THE ORDERS OF ACRASIOMYCETEAE AND LABRINTHULOMYCETEAE

Order and Number of Fam, Gen, Spp	Habitat and Ecological Niche	General Characteristics	Representative Genera
Labyrinthulales (2-4-9)	Marine: parasitic on aquatic angiosperms, brown seaweeds, or green algae	Cells usually motile, two flagella; nonphagotrophic; cells make a "plasmodium" on nonliving tracks	*Labrinthula, Labyrinthulomyxa, Pseudospora, Protomonas*
Vamprellales (2-13-26)	Mostly parasites on freshwater algae like *Spirogyra, Oedogonium,* etc.; some marine saprophytes	Multinucleate plasmodium develops inside the host cells; mostly nonflagellated	*Vampyrella, Plakopus, Leptomyxa, Reticulomyxa, Arachnula*
Labyrinthulorhizales (1-1-2)	Freshwater; phagotrophic	Cells amoeboid, flagella lacking; cells make "plasmodia" on living tracks	*Labyrinthulorhiza*
Acrasiales (4-9-21)	Essentially ubiquitous in soil; phagotrophic on bacteria and yeast	Cells amoeboid, nonflagellated; produce a pseudoplasmodium following chemotropic aggregation	*Dictyostelium, Polysphondilium, Copromyxa, Guttulina*

Note: Fam = families, Gen = genera, Spp = species.

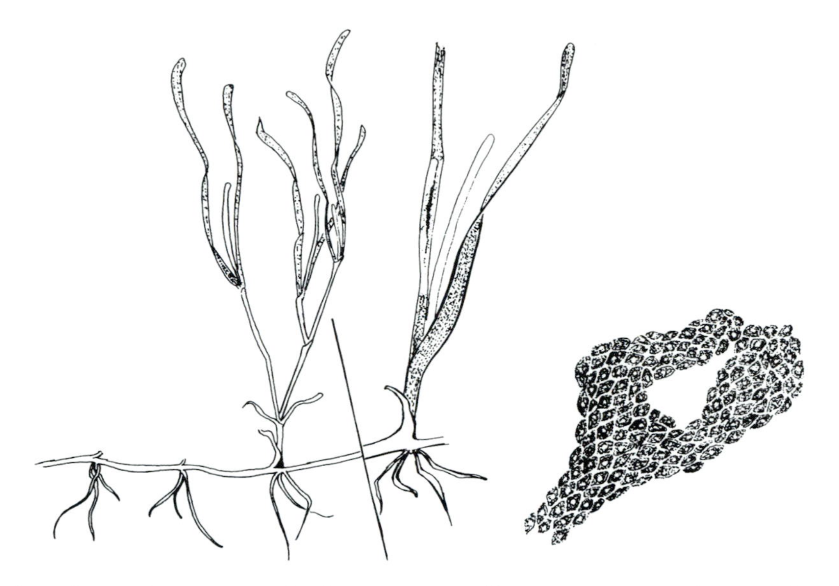

Figure 8–3 Left to right: eelgrass or wrack, *Zostera marina* (Potamogetonaceae), host species of *Labyrinthula vitellina;* closeup of rhizome leaf, showing chlorotic and discolored streaks, signs of infection; microscopic view of *Labyrinthula* net or pseudoplasmodium.

freshwater species which ingest food phagotrophically. Unlike *Labyrinthula,* its myxamoebae glide on tracks of living material. These, and some other characteristics suggest a possible link between the Acrasiales and Labyrinthulales. Detailed comparative studies of the tracks of *Labyrinthula* and *Labyrinthulorhiza* are needed in order to ascertain the true relationships of these genera to each other and to the Vampyrellales.

The fossil record has been of no help in studying relationships among the Labyrinthulomycopsida nor in discovering their origin. Flagellation suggests a relationship to the Myxomycopsida, as do the simple plasmodia in some Vampyrellales. Sorus formation and the unicellular assimilative stage suggest relationship with Acrasiomycopsida.

All species of *Labyrinthula* are parasitic on marine angiosperms and *Labyrinthomyxa* is parasitic on brown algae, suggesting origin of both of these genera from a marine phytoflagellate or possibly a protozoan. From the first *Labyrinthula*-like ancestors, which probably resembled both *Protostelium* in the Myxomycopsida and *Acrasis* in the Acrasiomycopsida, as well as *Cryptomonas* in the Cryptophycopsida, the other two orders have presumably evolved.

Representative Species and Genera

While most Labyrinthulomycopsida are microscopic in size, a few species are visible to the unaided eye. The largest of these, *Megaamoebomyxa argilloibea,* an opaque white organism found only in debris-rich marine sediments in Gullmar Fjord, Sweden, grows to 2.5 cm long. *Leptomyxa reticulata* is not as large as *M. argilloibea,* but is more common. It forms plasmodia up to 3 mm in diameter superficially resembling the plasmodia of true slime molds.

Ecological, Economic, and Research Significance

Eelgrass, *Zostera marina,* which is not a grass but a member of the Potamogetonaceae or pondweed family, is important to the production of cod, eel, ducks, and other marine animals. Dried eelgrass, or wrack, is also a high-quality packing material for glass and ceramics. An epidemic of eelgrass disease, caused by *Labyrinthula macrocystis,* destroyed 80% of the eelgrass along the north Atlantic coast of the U.S. in the late 1930s, nearly ruining the cod fishing industry in much of that area.

Several species of pond algae are parasitized by members of Labyrinthulomycopsida. Their ecological significance has not been fully assessed but may be considerable. *Pseudospora* and *Protomonas,* parasites on *Spirogyra, Volvox,* and *Eudorina* in the Chlorophycopsida, appear to be the most important.

Little is known about the physiological characteristics of the Labyrinthulomycopsida. It is therefore difficult to predict where they will be of greatest research value. Complete studies of life cycles of several of the species are needed; once these are available, it will be possible to study physiological characteristics and develop culture techniques. Studies of the ultrastructure of the filaments on which the myxamoebae glide and the biochemistry of the aggregation principal may have far-reaching results.

CLASS 3. MYXOMYCOPSIDA

The Myxomycopsida (*myxo* = slime and *myceto* = a mold), or true slime molds, are also known as the plasmodial slime molds because the plant body is a true **plasmodium** or netlike, multinucleate, naked mass of protoplasm. The plasmodium of a myxomycete differs from the pseudoplasmodium of the Acrasiomycetes in that it is not septate, and it differs from a coenocyte, as in the Zygomycopsida and Mastigomycopsida, in being much branched and **anastomosed** (with branches reuniting to form a net), and in migrating slowly like an ameba. The anterior portion of an advancing plasmodium is completely naked, but the posterior portion is enclosed in a sheath of slime.

Plasmodial slime molds are relatively common organisms found on plant litter in moist habitats. Once he has seen, or preferably felt, a few specimens of slime molds, the student is not likely to confuse them with any other group of plants. At first, he may think he has a Myxobacteriales or Mastigomycopsida, or possibly even an Ascomycopsida; however, the reticulate nature of the plasmodium is so distinctive that, with moderate care, this is unlikely. Slime bacteria produce pseudoplasmodia similar to those of Acrasiomycopsida but made up of very small cells. While slime molds and ascomycetes both produce sclerotia, the sclerotia of slime molds tend to be more brightly colored and more variable in structure and form than ascomycete sclerotia.

Plasmodia and sclerotia of slime molds may be brought into the laboratory and grown on artificial media. Placed in a petri dish on sterile agar or filter paper, many species can be kept alive and growing by providing them with oat flakes. If great care is exercised, spores may also be germinated, but this is much more difficult.

The moist chamber technique is especially useful in culturing slime molds. Petri dishes or small covered bowls are fitted with disks of filter paper or paper towels cut to size and sterilized in an oven or autoclave. The material to be cultured—which may be a piece of bark from a

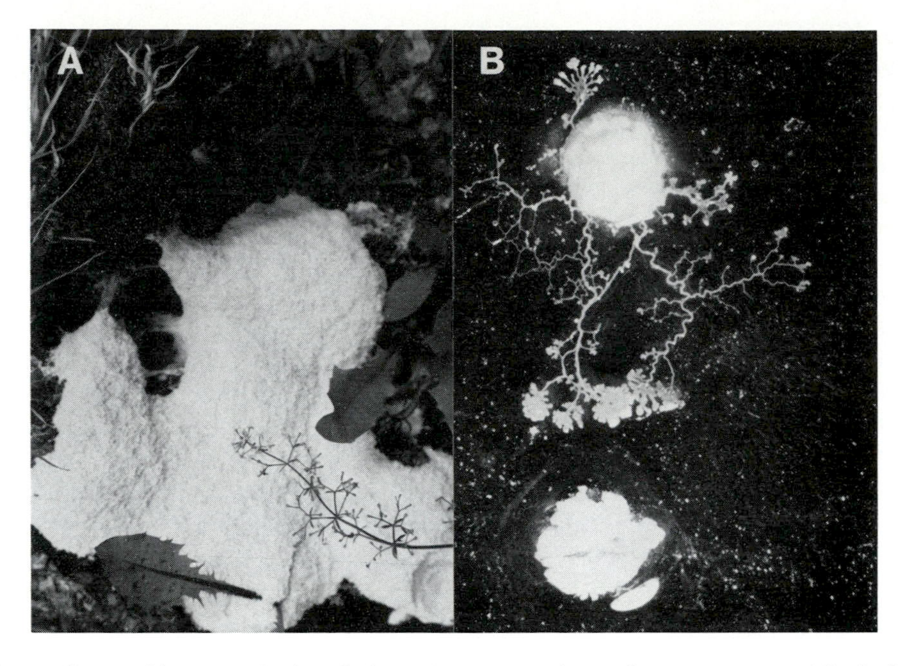

Figure 8–4 (A) Slime mold, tentatively identified as *Fuligo septica,* observed in grassy area partially shaded by *Betula verrucosa,* Sigtuna, Sweden, July 1993; (B) *Physarum* sp. on agar plate with oatmeal flakes.

living tree, some dead wood, decaying leaves, conifer needles, old seed pods, or other plant debris or animal dung—is placed on the filter paper, covered with sterile distilled water, and allowed to soak overnight. The water is then poured off and the cultures are incubated at 23 to 30°C for a week to 10 days or longer. Sterile water should be added as needed to keep the paper moist. Plasmodia and fruiting bodies develop in a large percentage of such cultures and may be detected with a good stereoscopic microscope or often with a strong hand lens. Some species, such as *Echinostelium elachiston,* produce sporangia within 24 hours; others may take several weeks; it is important therefore to examine the culture daily over a rather long period of time (Martin et al. 1983).

General Morphological and Physiological Characteristics

The assimilative structure of the true slime molds is a plasmodium, which may be white or brightly colored: yellow, red, brown, or violet. It is usually rather small and inconspicuous on the forest floor, but occasionally attains considerable size, rarely as large as a meter in diameter (Figure 8-4A). The plasmodium is typically fan-shaped, formed by usually synchronous divisions of the nuclei but without divisions of the cytoplasm. Naked, filament-like coenocytes anastomose to produce a slowly migrating, slimy amoeboid mass with interconnecting veins. The veins are large and coarse at the trailing edge, but very fine at the moving edge. In many species, the advancing edge, instead of being vein-like, consists of a jelly-like **ectoplasm** with many nuclei in it moving ahead of an **endoplasm** made up of small veins. In other species, there is no differentiation of ectoplasm and endoplasm.

As the plasmodium creeps over the decaying litter, it engulfs solid particles of food by means of **pseudopodia** ("false feet": protuberances extending from the body which later anastomose) and digests these internally. The creeping of the plasmodium is most rapid when abundant moisture is present; during dry periods or periods of prolonged high temperatures or other adverse conditions, it becomes a hard, waxy structure called a **sclerotium,** which may also be brightly colored. When adequate moisture is available, the sclerotia give rise to plasmodia again. In some species, sclerotia have been known to remain viable for at least 2 years; sclerotia purchased from biological supply houses for classroom or teaching lab purposes, however, usually should be germinated within a few months to ensure success in obtaining good plasmodial development.

Under laboratory conditions, the plasmodia continue to grow for long periods of time. In most species, mitoses are synchronized, i.e., all of the nuclei undergo mitosis simultaneously. Eventually, most plasmodia sporulate. The conditions needed to induce sporulation in nature are not fully known, but moisture, temperature, light, pH, availability of food, and plasmodial size are all known to be involved. In some species, age alone (by determining plasmodial size) may be the main initial factor inducing sporulation. In the many species which are pigmented, light seems always to be involved as an initiator.

As the food supply diminishes, or as the result of other changes in the environment, the plasmodium migrates to a more exposed, drier location, often on stems or leaves of plants, and there forms a cushion-like mass leading to the development of reproductive fructifications or **sporophores.** One type of sporophore is the **sporangium.**

In the production of sporangia, two patterns of sporulation have been observed. In the **myxogastroid type,** hemispherical masses with a nipple at the top are formed. The nipples elongate and their bases constrict to form the stalk. Food vacuoles form and extrude fibrous material outward, leaving a central vacuole in the stalk. In the **stemonitoid type,** spherical masses form and begin depositing material for a stalk in the center. The protoplasm creeps upward and at the apex continues secreting stalk material inward. In either type, the outer later of protoplasm forms an enclosing membrane called the **peridium.** Within the sporangium, meiosis takes place. Each daughter nucleus is surrounded by a cellulose wall and becomes a spore. Sometimes a network of nonliving, thread-like strands, called a **capillitium,** is interspersed with the spores (Figure 8-5).

The spores are spread by wind and, upon germination, may develop into amoeboid cells or may produce zoospores (swarm cells) which swim around by means of two flagella of unequal length and lacking mastigonemes (Pitelka 1963). The spores of slime molds are nearly always globose, though in a few species they are usually oval. Often they are highly ornamented and may be brightly colored, ranging from hyaline to almost black.

The spores are long-lived compared to sclerotia; in some species, they have been known to germinate after more than 75 years in storage. Optimum temperature for germination is 22 to 30°C; optimum pH is between 4.5 and 7.0 for most species. Upon germination of the spores, either flagellated swarm cells or naked amoebae are produced, the former if free water is present, the latter in the absence of free water. Depending on moisture conditions, swarm cells may revert to myxamoebae and vice versa.

The amoeboid cells or the zoospores may function as gametes which fuse in pairs to form zygotes. Germination of the zygotes gives rise to the vegetative reticulate plasmodia. Each zygote may develop into a plasmodium which goes on to complete the entire life cycle, or two

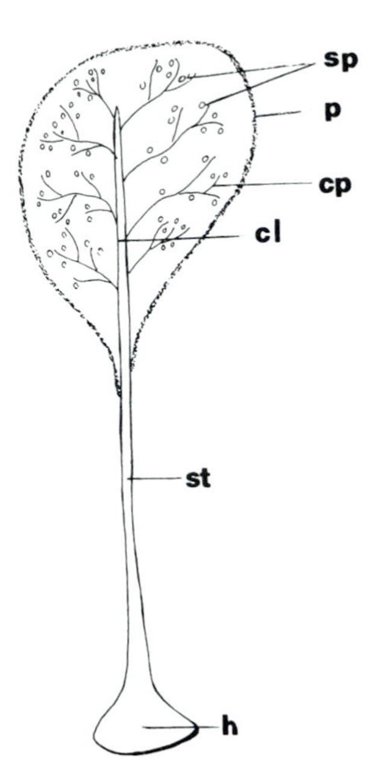

Figure 8–5 Anatomy of a typical slime mold sporangium (p = peridium; cp = capillitium; sp = spore; cl = columella; st = stalk; h = hypothallus).

or more zygotes may aggregate and then produce a plasmodium in which the nuclei are not all genetically alike. In some species, several plasmodia may fuse to form a single one. It has been suggested that aggregation of several zygotes or plasmodia into a single individual having genetically different nuclei provides homeostasis, just as sexual reproduction does, although in this case there may be several parents instead of only two (Bonner 1967). Homeostasis enables a species to survive better under changing environmental conditions.

Three types of fructification, or **sporophores,** occur in the true slime molds: **sporangia, aethalia,** and **plasmodiocarps.** In *Ceratiomyxa,* no sporangia or other fructifications are formed, and the spores are borne singly and externally on a white threadlike or columnar structure. In *Fuligo* and *Lycogala,* the plasmodium heaps up into a rounded or cushion-like mass or into several such masses, within which the spores are located. These masses are called **aethalia.** An outer sterile layer becomes the peridium, and a capillitium may or may not be formed. In *Tubifera,* the sporangia are short-stalked and so closely associated with each other that they resemble an aethalium. Such a mass of sporangia is called a **pseudoaethalium.** In some species of *Physarum* and other slime molds, the vein-like structures of the reticulate plasmodium enlarge and become hard and dry; within these structures, called **plasmodiocarps,** meiosis takes place and spores are produced. Variations of these fructifications, as well as intergradations, occur, and in some species, both sporangia and plasmodicarps occur.

Sexual reproduction has been observed in several species of Myxomycopsida; however, the complete life cycles have been studied in less than 10% of the total known species. With the exception of the Plasmodiophorales, the pattern is about the same in all of the species in which life cycles have been studied (Figure 8-6). Spores, produced by meiosis, germinate to develop into either flagellated unicells or amoeboid unicells. These reproduce asexually to form populations of genetically identical cells. In most species, flagellated unicells readily lose their flagella and become amoebae, and amoeboid unicells readily gain flagella, but the environmental conditions that trigger these changes are not fully understood. Most species seem to be **heterothallic,** meaning that self-incompatibility genes prevent conjugation of cells coming from the same spore. Some species, however, are known to be **homothallic** or self-fertile. Conjugation of two unicells results in a zygote. According to early studies, some species conjugate only in the amoeboid stage, others only in the flagellated stage; recent studies indicate, however, that in several species conjugation may be either between two flagellated cells or between two amoeboid cells. The zygote develops into a typical plasmodium and meiosis takes place in the plasmodium at the time the spores are formed. The life cycle is best described, therefore, as sporophyte prominent diplohaplontic.

Electron microscope studies have revealed that the nucleus is round and enclosed in the usual double membrane (Pitelka 1963). It contains one to several round nucleoli and numerous dark granules. Mitosis in all members of the class that have been studied carefully is **closed** or **intranuclear** in diploid nuclei and **open** in haploid nuclei. Meiosis is also intranuclear.

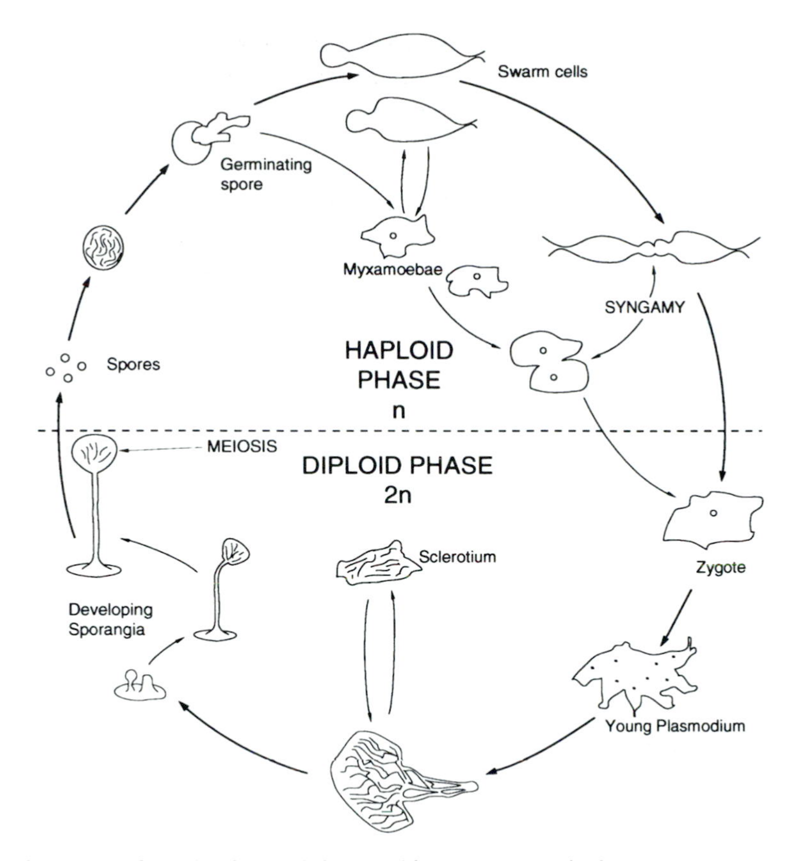

Figure 8–6 Diplohaplontic life cycle of typical slime mold. Syngamy may be between two swarm cells or between two myxamoebae. Meiosis takes place within the mature sorocarp, resulting in haploid spores. The nuclei in the unicellular stages are haploid, those in the plasmodial and sporophore stages are diploid.

Centrioles are present as haploid nuclei divide, but not in meiosis or in mitosis of diploid nuclei. In general, the nuclei seem to resemble those of the Cryptophycopsida.

In *Physarum,* there are numerous clear vesicles in the cytoplasm of assimilative plasmodia, but no evidence of reticular membranes. Cell walls of spores and sporangia contain cellulose. Mitochondria have twisted microtubules and a rather dense matrix, typical of Bracteobionta. The stages of mitosis take place within the nuclear membrane in haploid cells, but not in diploid cells where the nuclear membrane disintegrates following prophase.

The parasitic Plasmodiophorales have a more distinctive alternation of generations with plasmodia produced both in the haploid and in the diploid stage. Mitosis has been reported to be **intranuclear** in the haploid but not in the diploid stage. A unique type of cell division occurs in the Plasmodiophorales, however: the nucleolus elongates and as it does, the chromosomes line up in a ring around it. Thus, in equatorial view, the chromatin takes on a cross-like appearance. As far as is known, this **cruciforme** type of mitosis does not occur in any other organism, including any of the other Myxomycophyta (Figure 8–7).

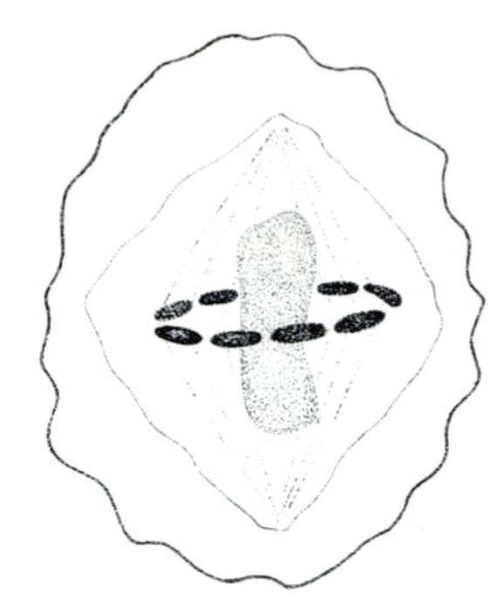

Figure 8–7 Cruciforme mitosis in *Plasmodiophora brassicae*. In prophase, the nucleolus elongates and the chromosomes form on the metaphase plate surrounding it. They are not attached to each other but almost touch each other end to end.

Phylogeny and Classification

Although there have been reports of Paleozoic slime molds, one from the Pennsylvanian and one from the Permian, the earliest confirmed myxomycophyta fossil is a *Stemonitis* found in Baltic amber, probably a few million years old at the most. The vegetative or somatic cells of slime molds, being naked, do not fossilize well, and paleobotany has not been helpful in studying relationships among the Myxomycophyta. Slime mold spores and sporangia are distinctive enough, however, that there is hope of finding and identifying myxomycete microfossils some time in the future.

Traditionally, the Acrasiales and Labyrinthulales have been included in the same class as the Myxomycopsida. The tendency at the present time is to separate them from the true slime molds as has been done in this book. Some botanists also remove the Protosteliales and Plasmodiophorales from the Myxomycopsida.

Including the Plasmodiophorales and Protosteliales, there are eight orders of Myxomycopsida. The most primitive is the Protosteliales, consisting entirely of microscopic species. Sexual reproduction has never been observed in the Protosteliales. Superficially at least, some of them resemble the Vampyrellales. In the Protosteliales, *Cavostelium* and *Protosporansium* are flagellated and suggest relationship to the Cryptophycopsida, whereas *Protostelium* is nonflagellated and seems to form a link between the Myxomycopsida and Acrasiomycopsida. Other primitive slime molds are included in the Ceratiomyxales. Unlike the more advanced slime molds, reversible protoplasmic streaming does not occur in the Protosteliales and Ceratiomyxales, and sporophores are either microscopic or nonexistent.

From flagellated ancestral types similar to *Protosporansium* and *Cavostelium* in the Protosteliales, and *Ceratiomyxa* in the Ceratiomyxales, three lines have apparently evolved within the Myxomycopsida (Figure 8–1). In one of them, leading to the Stemonitales, sporophore development is of the stemonoid type in which the stalk material is deposited inward, the plasmodia are not differentiated into ectoplasm and endoplasm, lime is not deposited in the sporangia and seldom in any part of the plant, and spores are dark brown or purplish brown in color. In the second line, leading to the Physarales, sporophore development is of the myxogastroid type in which the stalk material is deposited outward from a central vacuole, the plasmodia are differentiated into ectoplasm and endoplasm, lime is often deposited in the sporophores, and the spores are mostly light in color or brightly colored, although considerable variation exists within each order. In the third line, spores are hyaline or colorless, and all of the species are plant parasites, the best known being *Plasmodiophora brassicae* and *Spongospora subterranea*, parasitic on cabbage and potatoes, respectively. Some of the chief characteristics of the eight orders are shown in Table 8–3. Suggested relationships among them are indicated in Figure 8–1.

Representative Genera and Species

Ceratiomyxa fruticulosa is common, and usually abundant, in all parts of the world. The thallus is pink or rose colored and only a few millimeters in length. The spores are light colored and

TABLE 8–3

GENERAL CHARACTERISTICS OF THE ORDERS OF MYXOMYCOPSIDA

Order and Number of Fam, Gen, Spp	Spore and Other General Characteristics	Representative Genera and/or Species
Protosteliales (2-4-12)	Sporophores microscopic, producing few (1-4) spores; cytoplasmic streaming in plasmodia unidirectional	*Protostelium, Cavostelium, Ceratiomyxella, Protosporansium*
Ceratiomyxales (1-1-3)	White, threadlike hypothallus with numerous colorless spores borne directly on it	*Ceratiomyxa fruticulosa*
Liceales (5-11-50)	Sporophores noncalcareous, lacking true capillitium; spores hyaline to pale yellow; sporangia often grouped into pseudoaethalia, aethalia may resemble small puffballs	*Dictydium, Cribraria, Lycogala, Tubifera*
Trichiales (4-9-70)	Sporophores often brightly colored sporangia or plasmocarps with well-developed capillitia; peridium may rupture and form goblet-like structure; spores rosy or yellowish	*Hemitrichia, Trichia, Dianema, Arcyria, Margarita*
Echinosteliales (2-2-9)	Microscopic plasmodia without veins, with amoeboid motility; each plasmodium produces a single sporangium lacking peridium and capillitia; spores light colored	*Echinostelium, Clastoderma*
Stemonitales (3-8-70)	Spores black or purplish brown, borne in sporangia having both columella and capillitia; stipe, but never peridium or capillitium, sometimes calcareous	*Stemonitus, Lamproderma, Diachaea*
Physarales (2-10-162)	Spores dark colored, violet tinged, in sporangia with lime-encrusted peridium and/or capillitium; aethalia and plasmodiocarps produced	*Fuligo, Didymium, Mucilago, Physarum, Badhamia*
Plasmodiophorales (4-9-24)	Spores hyaline colorless, produced in zoosporangia; plant parasites	*Plasmodiophora, Spongospora*

Note: Fam = families, Gen = genera, Spp = species.

borne singly and externally on a thread-like structure. The primitive order, Ceratiomyxales, consists of three species of *Ceratiomyxa,* two of them limited to the tropics.

Members of the Liceales are also relatively common. In *Dictydium* and *Cribraria,* there is neither a capillitium nor a pseudocapillitium present, but in *Lycogala,* thread-like remnants of the plasmodium form a pseudocapillitium. *Lycogala epidendrum* produces aethalia which look and function like puffballs; the force of raindrops striking the peridium produces a bellows effect which forces the spores out through an irregular opening. Some species of *Lycogala* are characterized by a large (2 to 4 cm), green aethalium attached to a flat, bright red or crimson base, resulting in a fruiting body which looks like a stuffed olive. In the other orders that have evolved from early Liceales, true capillitia are present.

In *Trichia,* the capillitium is marked by regular spiral bands. Mature sporangia of *Stemonitis* have no peridium, but the persistent peridia of *Lamproderma* (Stemonitales) and *Diachaea* (Physarales) are beautifully iridescent.

Fuligo septica and *Physarum oblonga* are common species that are frequently cultured. The plasmodia and aethalia of *F. septica* are sometimes 20 cm or more in diameter. Sclerotia of *Ph. oblonga* can be obtained from biological supply houses and are easily cultured in the botanical laboratory by placing them on moist filter paper in a petri dish with a few flakes of rolled oats. The plasmodia are yellow, fan-shaped, and relatively large. Reversible protoplasmic streaming is easily observed with the aid of a dissecting microscope. Several other species which are occasionally found in forests or on the bark of trees also develop relatively large plasmodia. Plasmodia in *Echinostelium,* however, are microscopic in size.

Ecological and Economic Significance

While slime molds are relatively abundant in damp, shady places, they are typically ignored by student and investigator alike, and their ecological significance is therefore difficult to assess. Many species are reported to produce enzymes that dissolve some plant materials which they then take in osmotrophically as food. Other species have been observed ingesting bacteria and fungal spores phagotrophically. Presumably they are of some significance both as decomposers and as bacterial grazers in some forest ecosystems where there is much moisture present. The adaptive advantages, if there are any, of pigmentation in slime mold sclerotia and fructifications is an area that needs further study.

Economically, only members of the Plasmodiophorales are of real significance at the present time. The most important of the plant parasites making up this order are *Plasmodiophora brassicae,* the incitant of clubroot disease of cabbage and other crucifers, and *Spongospora subterranea,* causal agent of powdery scab in potatoes.

Records of damage by the clubroot disease go back as far as the 13th century in western Europe. By 1800 it was widespread on turnip and rutabaga in England, and by 1872 had become so destructive in Russia that the Royal Russian Gardening Society in St. Petersburg offered a prize for its control. At that time, it was believed by some that insects caused the malady, while others held that environmental disturbances, over-manuring, or "reversion to wild types" was the cause. In the early 1870s, a brilliant Russian botanist, Michael S. Voronin, began his classic studies of the disease in which he was able to demonstrate that a slime mold was the cause of the trouble. In 1872 he published a report on the cause and control of the disease, and in 1878 he published a very complete account of the life history of *Plasmodiophora brassicae.* This monographic work has become a classic in plant biology.

Clubroot occurs on both wild and cultivated species of the family Cruciferae. The symptoms may develop at any stage in the life cycle, but the disease has often progressed to a considerable extent before any symptoms are noticeable on above-ground parts of the infected plants. The infected roots rapidly enlarge and take on a variety of shapes, depending in part on species and in part on the infection site. In cabbage, a spindle-shaped club frequently results.

Secondary infections by bacteria usually do more damage than the primary infection; as the clubs decay, toxic substances are released which may be responsible for most of the wilting of the foliage associated with the disease (Figure 8-8).

Cabbage is usually started in a greenhouse or in coldframes and then transplanted to the field. The major means of widespread distribution of the disease is through the transplants. In controlling the disease, it is therefore important that the soil used in propagating the seedlings be free from the pathogen. Apparently its spores are capable of surviving in the field for years, and long rotations are necessary, once the pathogen has become established, if the disease is to be avoided. Mercuric chloride as a transplanting liquid was formerly used to prevent the disease; some other fungicides, safer to use than the mercurials, have more recently shown promise. Resistant varieties of turnips and rutabagas are now available. Physiological races of the disease organism differ in their virulence against different hosts and complicate the plant breeding approach to control of the disease (Walker, 1942).

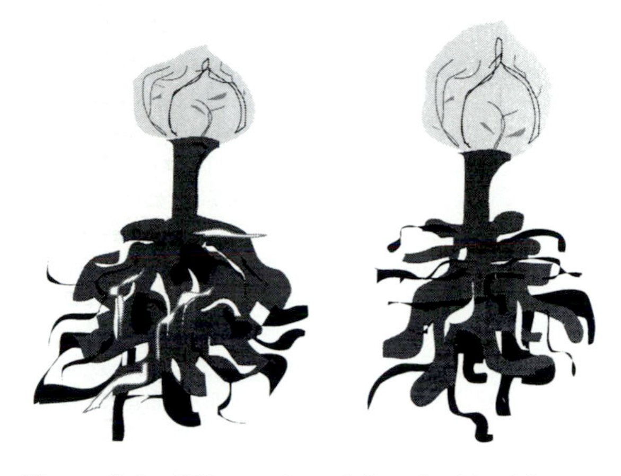

Figure 8–8 Cabbage plants infected with club root mold, *Plasmodiophora brassicae;* Multiple infections have resulted in the heavily clubbed root; single infections produce spindle-shaped clubs.

In an entirely different way, *Fuligo septica,* "the blob," has been of some economic value. *F. septica,* a common species occurring on lawns and other open areas, may become extremely abundant during prolonged periods of wet weather. In 1973, it became the subject of newspaper and magazine articles because of its abundance and unique appearance. *Fuligo septica* continues to grow in damp meadows, as it has for centuries, and in botanical laboratories where a few scientists continue to use it in their research. It is not known when first it attracted the attention of botanists; possibly, "the blob" is the same slime mold referred to in some 9th century Chinese writings as "kwei hi" (meaning "demon's droppings"), which was supposedly able to cure ulcers.

Research Potential

Because of their large, naked plant body, slime molds have long been considered good organisms for the study of the nature of protoplasm. Years before the development of the electron microscope made it possible to observe the endoplasmic reticulum (ER) in cells, its presence had been inferred from studies involving various slime molds. With the light microscope, ground cytoplasm—the part of the protoplasm that the nucleus, plastids, and mitochondria appear to "float" in—seems to be a homogeneous, jelly-like mass. Early studies, however, in which the protoplasm of a myxomycete plasmodium was drawn out with a microneedle, revealed a birefringence phenomenon which is best explained by assuming that a network of fibers or membranes was being pulled into parallel orientation. In other studies, it was discovered that undisturbed plasmodia could pass through the pores of bacterial filters without injury, whereas plasmodia that were forced through the same filters by means of a vacuum pump could not survive, again suggesting fibers that could be oriented into parallel units. The electron microscope has demonstrated the presence of an ER in all eukaryotic organisms that have been studied and also the presence of fine filaments less than 10 nm in diameter in most cells.

When the zygotes undergo mitosis to begin the development of a plasmodium, the fusion of the cells into a naked coenocyte is essential. It has been reported that several zygotes may become involved in the development of one plasmodium by means of cell fusion. The causes of cell fusion have been investigated to some extent. When chelating agents were added to cultures of young flagellated cells, cell fusion and plasmodium development were inhibited. The cells continued to divide as long as food (bacteria) was present but they did not fuse until they were transferred to a medium lacking the chelates. Obviously the chelating agents were tying up some metal ion involved in an enzyme essential to cell fusion. As this phenomenon is studied in greater detail, we should learn more about the general principles of cellular differentiation relating to all multicellular and coenocytic organisms.

The plasmodial slime molds are of special interest in the university classroom and laboratory because they can be used so effectively in observing cytoplasmic streaming. In research, they have already taught us something of the causes of this phenomenon and they promise to teach us more. Myxomyosin, a contractile protein similar to actomyosin in animal muscle, contracts when ATP is added to it. Myxomyosin has been found in all investigated slime molds, and it is believed that as it contracts in periodic patterns, the cytoplasm is forced to move through the tubules making up the plasmodial network. In most slime molds, the streaming is reversible, flowing for a while in one direction, then changing direction for a while. In the Protosteliales, streaming is vigorous, as in other slime molds, but unidirectional. Protoplasmic streaming is reported to be irregular in the Ceratiomyxales.

Under certain conditions, mitosis in some slime molds, as in some Pyrrophyta, is internal, that is, the nuclear membrane remains intact as the chromosomes divide. In most slime molds, this commonly occurs in diploid but not in haploid cells, but the reverse seems to be the case in the Plasmodiophorales. The slime molds may, therefore, be valuable organisms in teaching us more about the process of mitosis in all organisms.

SUGGESTED READING

Introductory Mycology by Alexopoulos and Mims is an excellent textbook on fungi and presents many of the latest findings on the slime molds. *The Cellular Slime Molds* by J. T. Bonner gives excellent coverage of the Acrasiomycopsida. *The Genera of Myxomycetes* by Martin, Alexopoulos, and Farr is a detailed, and interestingly written, monograph of the plasmodial slime molds.

Three articles in scientific journals, "The Mycetozoa" by L. S. Olive in the 1970 *Botanical Review;* "Isolation, cultivation, and concentration of simple slime molds" by K. B. Raper in the 1971 *Quarterly Review of Biology;* and "The etiologic agent of the wasting disease of eel-grass" by E. L. Young III in the 1943 *American Journal of Botany,* are highly recommended.

Aggregation is an important phenomenon in all social organisms, not only slime molds. "Army ants: a collective intelligence" by Nigel Franks in the March–April 1989 issue of *American Scientist* (vol. 77, pp. 138–145) discusses this topic from a different point of view.

STUDENT EXERCISES

1. Prepare and sterilize media as described in this chapter for growth of cellular slime molds. Sprinkle soil from several sources on petri dishes containing the media. Observe the development of the organisms, keeping careful notes as to length of time required for development of each culture.

2. On a field trip to a nearby woods, search for the presence of slime molds. In what kinds of habitats would you expect to find them?

3. What is the Shaffer test? What equipment and plant materials are needed for conducting the test? How is it conducted and what happens if the person tested is pregnant?

4. A married student isolated several cultures of cellular slime molds from a sagebrush ecosystem in an outdoor ecology laboratory. From one of the pseudoplasmodia, she started several subcultures. Knowing that she was pregnant, she added a sample of her urine to ten of the subcultures and a sample of her husband's urine to another ten subcultures. She detected no difference in growth or development among the twenty subsamples. Why? Suggest a course of action which might have given her better results.

5. Seed six petri dishes fitted with moist filter paper disks with *Physarum* sclerotia and add a few flakes of rolled oats to each. Incubate at several temperatures from 18°C to 35°C. Take careful notes of growth and development of the plasmodia in each dish. Plot growth (as the Y axis) against temperature and fit a smooth curve to the resulting scatter diagram. What do the minimum, optimum, and maximum temperatures appear to be?

9 THE UBIQUITOUS ALGAE: DIATOMS AND OTHER CHRYSOPHYTES

Historical records and prehistorical anthropological findings indicate that food from the sea has always been important to man. In some countries, fish is the major source of both calories and protein; in others, such as the United States, fish are eaten largely to add variety to meals, to ensure a healthy diet, or to provide an excuse for an enjoyable day in the mountains with a casting rod.

The prodigious diversity of aquatic fauna includes freshwater species; species that feed on the well-lighted bottom of estuaries, bays, and the continental shelf; species that live near the surface of the deep ocean; and species that live on the bottom beneath them, where it is perpetually dark and the temperature is a constant 4°C. Some fish in this last category carry lanterns—lens-shaped organs located just above, below, or around each eye—which enable them to see their prey: the lanterns house bioluminescent bacteria or other luminescent plants such as dinoflagellates.

Over a hundred billion pounds of fish are hauled in each year by the fisheries of the world. The base of the ecological pyramid, or food chain, which supports this vast harvest consists largely of the plants that will be discussed in this chapter. The diatoms are especially important, although anciently, before the diatoms evolved, the Chrysophycopsida must have been of greatest significance. According to Tiffany (*World Book Encyclopedia:* Diatom), most of the 30,000 species of fish that exist today would soon become extinct if all diatoms were suddenly to disappear.

Extinction of many species of fish is threatened by industrial development, causing alarm among ecologists and conservationists throughout Europe and North America. Acid rain and runoff from acid snow, both caused by air pollution, bring down the pH of many lakes and ponds to a point at which most fish, salamanders, and other aquatic animals cannot survive. Mountain lakes, as well as the lakes and ponds in other forested areas, generally have little buffering capacity; consequently, a small amount of acid rain reduces the pH of the water rapidly. When lake pH falls below 4.8, fish are almost always eliminated (Evans 1984). Although numerous excellent studies on the effects of acid precipitation on trees, shrubs, and grasses have failed to produce any clear-cut evidence of decreased productivity or extensive foliage injury to terrestrial plants, we cannot assume that aquatic plants are unharmed. We know little about the effect of pollution-derived acids on diatoms and other microscopic plants which are important sources of food for fish.

Though a small division in number of species, the Chrysophyta (*chryso* = gold, *phytum* = a plant) is a large one in number of individuals. Ecologically, only the Tracheophyta is more important (Table 2–5); in spite of this, the Chrysophyta are poorly known physiologically, genetically, and biochemically. There are four classes in the division according to the classification system used in this book: Xanthphycopsida, Chrysophycopsida, Bacillariophycopsida, and Mastigomycopsida. Most species in the first three classes are autotrophic, while the fourth contains only saprophytic and parasitic species. The three autotrophic classes are discussed in this chapter; the Mastigomycopsida will be discussed in the next chapter. The common name, chrysophyte, is applied to the autotrophic species, sometimes in a very narrow sense (Chrysophycopsida only), sometimes in a broader sense (Chrysophycopsida and Xanthophycopsida), and sometimes in reference to all three classes.

There is much difference of opinion as to relationships among the four classes. Formerly, the Bacillariophycopsida were sometimes allied with the Phaeophycopsida and the Xanthophycopsida with the Chlorophycopsida; today most botanists agree that the Bacillariophycopsida are closely related to at least the Chrysophycopsida, but there are still differences of opinion regarding the Xanthophycopsida. The relationship between the Mastigomycopsida and the other classes is also controversial. Clements (1909, 1931), in his textbook on mycology, classified the Mastigomycopsida among existing families and genera of the Xanthophycopsida, but others have placed them in the division Mycota or the kingdom Fungi along with the zygomycetes, sac fungi, and club fungi, and sometimes the slime molds. On the other hand, recent studies of fossils, plastid pigments, cell flagellation, and life cycles suggest a close relationship among the four classes and thus support the classification used here.

The Chrysophyta fossilize well and have left an excellent record. Fossils tentatively identified as Xanthophycopsida and Mastigomycopsida have been found in Hadrynian formations, about 900 to 1000 million years old. Fossils resembling modern Saprolegniales (Mastigomycopsida) have been found in middle Devonian (Chesters 1964) and younger formations, and good fossils of Chrysophycopsida and Xanthophycopsida, similar to modern coccolithophores and Heterococcaceae, are relatively abundant in many Paleozoic formations. Apparently these three classes originated anciently from similar or identical stock. The Bacillariophycopsida are much younger; the oldest confirmed diatom fossils date from early Cretaceous, some 120 million years ago. Diatoms fossilize very well and are abundant in all periods since the Cretaceous. Some of the fossil genera are now extinct, but many of the modern genera have also been recovered as fossils. Probable relationships are illustrated in Figure 9–1.

The four classes resemble each other in a number of ways: (1) flagellated cells in all four classes usually have two flagella, one whiplash and one pinnate, or else just one pinnate flagellum; (2) cellulose is a basic component in cell walls in all four classes, although it is frequently masked by other materials, most frequently silicates; (3) spiny endospores or cysts are produced by some members of all four classes; these cysts are usually called statospores in the autotrophic species, aplanospores in the heterotrophic species; (4) in contrast to the Eumycophyta and other Aflagellobionta, and also Lorobionta, mitochondrial cristae are tubular in all four classes; (5) Golgi bodies are also well developed and more complex than the dictyosomes

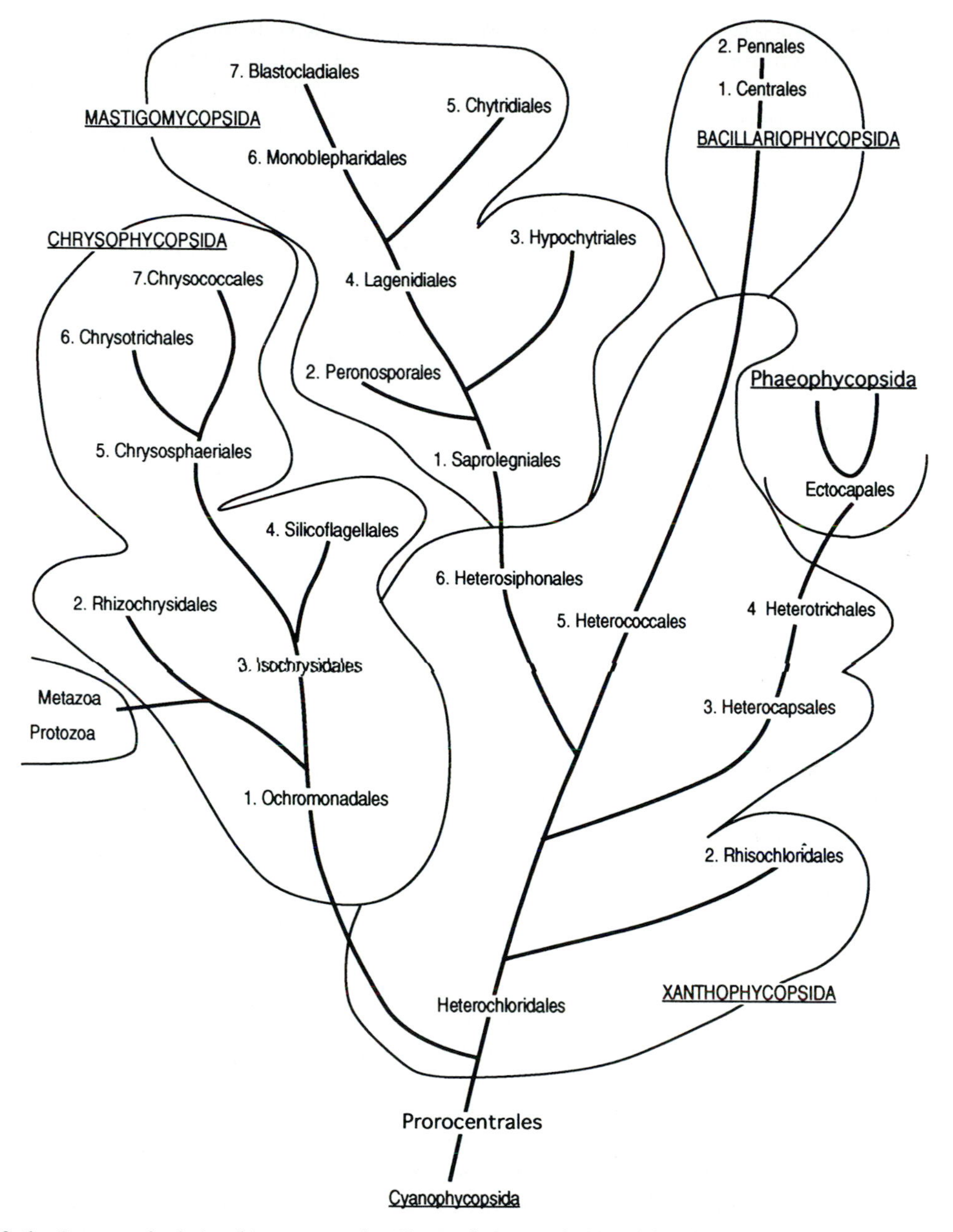

Figure 9–1 Suggested relationships among the Chrysophyta. It is believed that the first animals evolved from early Chrysophycopsida, similar to modern Ochromonadales and Rhizochrysidales, during late Proterozoic time.

of most plants; (6) in the autotrophic species, each lamella in the plastids consists of three thylakoids and the lamellae may be arranged parallel or nonparallel to each other, (7) food is stored primarily as oils but also as a slightly branched carbohydrate called leucosin or chrysolaminarin; (8) cells often consist of two **valves** that fit together like the two halves of the gelatin capsules that pharmacists use: an **epithecum,** usually the smaller of the two but having a larger diameter, and a **hypothecum,** which fits inside the other.

Chrysophyta commonly occur not only in freshwater and marine habitats, but in soil, on or in other plants, and in various other moist habitats. Many freshwater species are limited to soft water ponds and streams; others are common in both freshwater and marine ecosystems.

CLASS 1. XANTHOPHYCOPSIDA

The Xanthophycopsida (*xantho* = yellow, *phycus* = algae) are also known as the Heterokontae (*hetero* = different, *kont* = flagella). Commonly called heterokonts, because one flagellum is longer than the other, or yellow-green algae, because α-carotene is usually so abundant that its color partially masks the green of chlorophyll *a,* most of the 275 species occur in freshwater. A few are marine. Species of *Vaucheria, Tribonema,* and *Characiopsis* (Figure 9–2) are very widespread and are so common at certain times of the year as to dominate algal blooms in many pools and lakes. The typical yellow-green alga is characterized by cells having overlapping cell walls composed of cellulose, pectic materials, and silicates. In this respect it resembles both the Chrysophycopsida and Bacillariophycopsida. In filamentous species, the overlapping nature of the cell walls is especially noticeable when the filaments are broken (Figure 9–2).

Heterokonts usually make up a significant portion of the algal mats that form on ponds in late summer. Epiphytic species are abundant on green algae, such as *Rhizoclonium,* at all seasons, but the free-floating and filamentous species are less common in the winter, at least in cold climates. To study Xanthophycopsida, one needs only tear off pieces of the algal mats that occur on ponds and slow-moving streams, store these in bottles with plenty of water from the same source, and examine at leisure with a microscope in the laboratory (Figure 9–2).

Students often have difficulty in distinguishing motile heterokonts from cryptophytes and chrysomonads. Cryptophytes tend to be flat and rigid and the gullet is usually subapical; chrysomonads have less rigid cells than either cryptophytes or heterokonts and are typically golden to golden brown in color. Both cryptophytes and chrysomonads possess contractile vacuoles which heterokonts lack. Heterokonts are generally grassy green to yellowish green; the gullet is apical as in the chrysomonads.

Epiphytic, free-floating, and filamentous species are most frequently confused with Chlorophycopsida, but the latter tend to be more grassy green in color and generally have only one or two granular plastids, often of rather bizarre shapes (nets, open rings, spiral bands, etc.), whereas the Xanthophycopsida typically have many small, smooth, disk-shaped plastids (use high power or oil immersion magnification). When the cells of both are heated in concentrated hydrochloric acid, most Xanthophycopsida will turn blue-green because of the abundance of carotenoids, whereas most chlorophycopsida will remain green or turn yellowgreen. Individual cells in filamentous xanthophytes are usually swollen or barrel-shaped, and where a filament has broken, H-shaped segments indicate the overlapping nature of the two halves of the cells. The "type method" of identification—becoming well acquainted with a few "type species"—is an effective method of learning how to distinguish among these superficially similar groups of plants. Characteristics of the classes that most frequently cause confusion are summarized in Table 7-3.

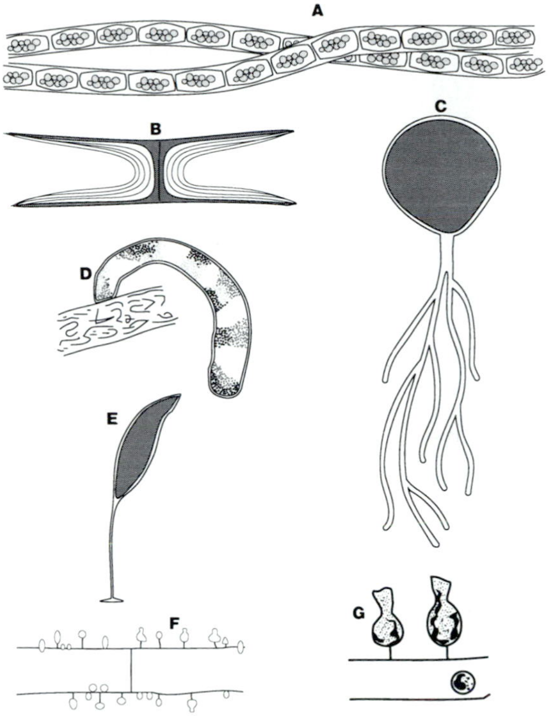

General Morphological Characteristics

As in other planktonic and filamentous algae, the Xanthophycopsida contain unicellular and colonial flagellated species, nonflagellated unicells, species in which the cells are held together in a gelatinous mass, species in which the naked cells move in an amoeboid fashion, species in which the cells are arranged end to end in simple or branched filaments, and species having coenocytic filaments (Table 9-1).

Like most other Bracteobionta, the Xanthophycopsida have small, disk-shaped plastids with smooth edges. Pyrenoids are generally absent. Each lamella contains three thylakoids. Photosynthetic pigments are chlorophyll *a* and brown carotenoids similar, but not identical, to fucoxanthin and diatoxanthin; chlorophyll *e* has been reported in some species of heterokonts, and also in some diatoms, but it has been difficult to confirm these reports. Food reserves are chrysolaminarin (leucosin) and oils, as in many other Bracteobionta.

Cell walls are composed mainly of either pectose or pectic acid often impregnated with silica, especially in the Heterotrichales and Heterococcales. Cellulose is present but not abundant. When cells are treated with concentrated potassium hydroxide, they can be seen to consist of a large hypothecum, often complex in structure, and a small, simple epithecum.

Figure 9–2 Unicellular and filamentous Xanthophycopsida: (A) *Tribonema bombycina,* two filaments with many disk-shaped plastids in each cell; (B) same filament, details of cell wall structure; (C) *Botrydium granulatum,* growing on soil with subterranean rhizoids and a green aerial vesicle; (D) *Ophiocytium* sp. epiphytic on *Rhizoclonium hieroglyphicum* (Chlorophycopsida); (E) *Characiopsis* sp.; (F) several epiphytic xanthophytes on a green alga filament; (G) unidentified epiphytic heterokont, possibly a *Maleodendron* species.

Sexual reproduction has been studied in detail in many Xanthophycopsida. Life cycles all seem to be haplontic, and induction of sexual reproduction in laboratory cultures is a simple process. Sexual reproduction in *Botrydium,* and possibly some other genera, is by fusion of isogametes produced by mitosis from ordinary vegetative cells. Germination of the zygote is by meiosis. Sexual reproduction in *Tribonema* is similar except that a nonmotile cell, which was formed in and then escaped from an ordinary vegetative cell, is fertilized by a flagellated sperm. More typical of the class, however, is *Vaucheria,* a coenocytic genus, in which small, motile, biflagellate sperm fertilize larger, nonmotile eggs within an oogonium (Figure 9-3). The sperm are produced in elongate, often slightly spiral-shaped branches called **antheridia** and eggs are produced in globose protuberances called **oogonia** arising from the same

TABLE 9-1

DIVERSITY IN THE XANTHOPHYCOPSIDA (ORDERS AFTER PASCHER 1917)

Orders (Fam–Gen–Spp)	Thallus Morphology	Other Characteristics	Habitat	Representative Genera
Heterochloridales (1-8-8)	Motile unicells with 2 or more discoid or rod-shaped plastids and a contractile vacuole; amoeboid stage sometimes sessile	One species as a pyrenoid, others do not; unicells can change from flagellated to amoeboid stage; amoebae may be phagotrophic	Freshwater and (1 species) brackish water	*Chlorochromonas, Chloramoeba, Ankylonoton, Bothrochloris, Heterochloris*
Heterococcales (9-35-170)	Nonmotile unicells or loose colonies; cells usually consist of 2 overlapping halves that may be silicified; cells often curved	Epiphytes often stipitate; plankters often have long bristles on them	Mostly freshwater but with many marine and some terrestrial species	*Characiopsis, Dioxys, Perone, Ophiocytium, Tetradriella, Gloeobotrys, Pleurogaster, Monallantus, Tetrakentron, Chlorobotrys*
Heterosiphonales (2-2-50)	Coenocytic filaments, often grassy green to olive green in color; numerous disk-shaped plastids	Sexual reproduction by means of a large oogonium with an apical pore and a helical antheridium producing flagellated sperm	Freshwater and terrestrial; some species are marine	*Botrydium, Vaucheria*
Rhizochloridales (3-7-10)	Permanently rhizopodial or amoeboid; cells solitary or joined to form net-like plasmodial masses with up to 150 cells	Some species have loricas; cells uninucleate or multinucleate; 2 of the families are plasmodial with cells joined by cytoplasmic bridges	Mostly fresh water	*Chlorarachnion, Stipitococcus, Myxochloris, Chloromeson*
Heterocapsales (2-8-12)	Produce amorphous or dendroid colonies of various shapes with nonmotile cells held together in gelatinous envelope	Colonies may be attached to rocks or other substrate by gelatinous stipes; or free-floating in plankton	*Heterocapsa* is marine, other genera are freshwater or occur in brackish water	*Heterocapsa, Gloeochloris, Malleodendron*
Heterotrichales (4-7-35)	Cells usually barrel-shaped, joined end to end in simple or branched filaments	Where filaments break, H-shaped pieces can often be observed; some epiphytic on other algae	Freshwater and terrestrial, in small pools and puddles	*Tribonema, Heterococcus, Heterothrix, Bumilleria, Monocilia*

Note: The number of families (Fam), genera (Gen), and species (Spp.) in each order is indicated in the first column.

filaments as the antheridia. Depending on species, antheridia and oogonia may be produced on the main filament or in clusters on a side branch. The antheridia begin forming first. The antheridial filament is generally slightly tapered at the apex, and at maturity a small pore develops at the tip. In some species, it may be so oriented that it points toward the oogonium. Sperm cells are produced by mitotic divisions of the original nuclei in the antheridial initial. In the oogonium, a series of mitoses produces a large number of egg nuclei. As it matures, a pore develops at the apex and through this pore the sperm swim. A chemical attractant is apparently involved in producing a chemotactic response in the sperm.

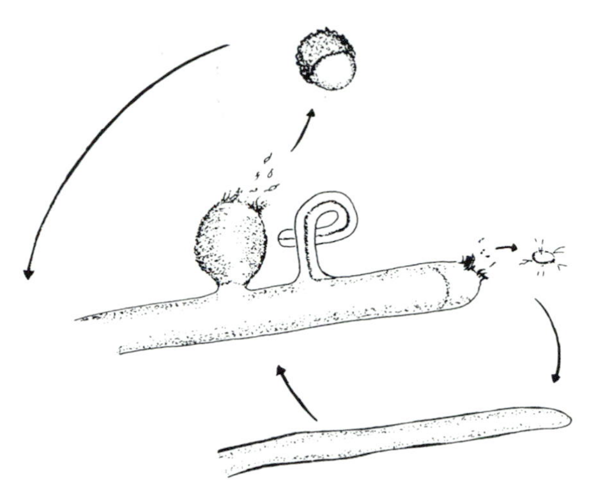

Figure 9–3 Sexual reproduction and life cycle of *Vaucheria geminata*.

Following fertilization, the zygote loses its chloroplasts and develops a thick cell wall; the resulting oospore is able to withstand adverse conditions and can survive for a long period of time before germinating by meiosis to produce four motile, biflagellated meiospores which germinate on any suitable substrate and develop into coenocytic, haploid filaments. There is difference of opinion as to whether more than one of the many egg nuclei in a given oogonium can develop into oospores.

Asexual reproduction in the coenocytic genus *Vaucheria* may be by (1) septation and subsequent constriction of a branch which then gives rise directly to a new plant; (2) swelling and septation of the tip of a branch to form a club-shaped sporangium which gradually rounds up into a spherical multinucleate zoospore with many flagella arising in pairs; or (3) fragmentation as the entire contents of the filament break up into a series of **statospores,** which are two-valved spores formed **inside** the cell, often having ornamentations such as spines on the epithecum (upper valve) or on both valves. Mature statospores are only slightly smaller than vegetative cells; each spore has a conspicuous collar and plug (Figure 9-4). The zoospores and the statospores germinate directly into new coenocytic filaments; however, the statospore has a thick cell wall and is capable of surviving desiccation or other adverse conditions for a considerable length of time. Similar methods of asexual reproduction occur in other xanthophytes. In some species, akinetes as well as statospores are produced. In akinetes, the cell wall of the vegetative plant becomes the spore wall, but in statospores the entire spore is formed inside the vegetative cell; in *Tribonema,* several statospores may form in one cell.

Phylogeny and Classification

Fossilized Xanthophycopsida have been reported in Hadrynian formations and are relatively common in mid-Paleozoic rocks. Presumably, the very first xanthophytes were unicellular, motile species which did not fossilize well but were similar to modern Heterochloridales. The ancestral stock from which they evolved was probably similar to the Gymnodiniales and Prorocentrales, judging from morphological similarities among these three orders. The Chrysophycopsida may have evolved independently from this same stock during late Proterozoic

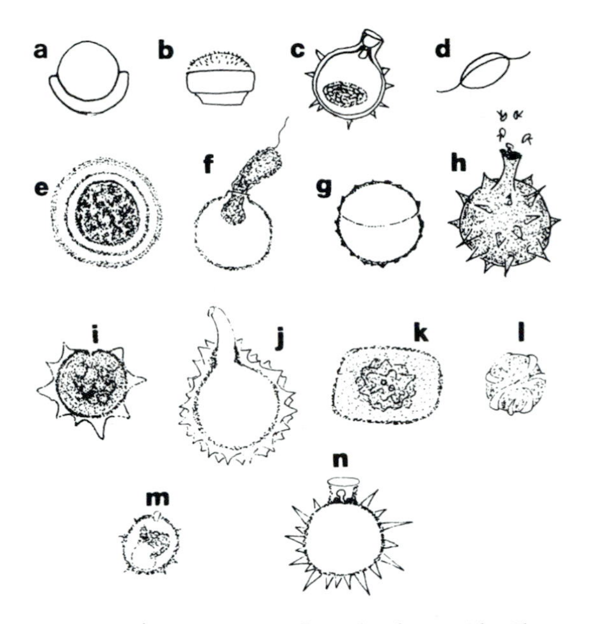

Figure 9–4 Statospores of Xanthophycopsida, Chrysophycopsida and Bacillariophycopsida, and spiny endospore of a Mastigomycete: (a–e) unidentified; (f) *Chromulina;* (g) *Chloromeson;* (j) zygote of *Monoblepharis,* Mastigomycopsida; (h and n) *Ochromonas;* (l) *Celloniella;*(k and m) unidentified.

time or directly from early Heterochloridales. A second line very early produced the coenocytic xanthophytes and from them the Mastigomycopsida probably evolved, while along a third line the overlapping, silicified cell wall characteristics became more pronounced and gave rise to the Bacillariophycopsida as well as the colonial and filamentous yellow-greens. The amoeboid xanthophytes have apparently evolved along a fourth line in which cell walls and periplasts have both completely disappeared (Figure 9–1). Although the amoeboid xanthophytes are usually pigmented and carry on photosynthesis, they can also engulf food particles by means of pseudopodia. In one marine species, *Chlorachnion reptans,* as many as 150 cells may remain attached to each other by protoplasmic strands in a large plasmodium which thus resembles a pigmented slime mold. Reproduction is by flagellated cells and the nature of the chloroplast pigments as well as the pinnate flagellum show that it is related to other xanthophytes and not the slime molds. Suggested relationships among the six orders are illustrated in Figure 9–1 and a summary of the chief characteristics is presented in Table 9–1.

Classification of the Xanthophycopsida into the six orders illustrated in Figure 9–1 is based on Pascher's system, in which only two characteristics, flagellation of the vegetative cells and cell wall characteristics, including those that result in grouping of cells into colonies or filaments, were considered. This is admittedly an artificial system and needs revising based on other characteristics such as cell ultrastructure and biochemistry. Some of the Heterococcales, for example, resemble *Vaucheria* in cellular physiology and reproductive processes more than they resemble other Heterococcales. Unfortunately there seems to have been a decrease in research interest in the Xanthophycopsida in recent years, and the information needed to make a better classification may be slow in coming.

Representative Genera and Species

In late summer and autumn, *Vaucheria* species are among the most abundant algae found in ponds. Color is typical of green algae such as *Cladophora* and *Rhizoclonium,* and *Vaucheria* is often listed in botanical keys with the Chlorophycopsida where it "keys out" under "unbranched filaments lacking septations." Species of *Vaucheria* are the only green, coenocytic algae commonly encountered in freshwater habitats. This is an example of the value of identification through the "type" method; knowing the general characteristics of the class and division does not help one in recognizing this genus unless reproductive structures can be found.

One cannot see the hidden qualities—food reserves in the form of oils, pinnate flagella, thylakoids in sets of three—which cause us to classify this genus in the Xanthophycopsida. There are over 40 species of *Vaucheria,* some living in freshwater, others in damp soil, and others in marine habitats.

Tribonema is a common freshwater genus abundant in ponds and lakes in late summer. There are several yellow-green disk-shaped plastids in each of the barrel-shaped cells which are attached end to end in long filaments. Where a break has occurred in the filament, H-shaped segments reveal the two-valve arrangement of the cells. Similar to *Tribonema,* but with more fragile filaments and almost golden plastids, is the genus *Bumillaria.* Unlike *Vaucheria* and *Tribonema, Bumillaria* is common in the winter time as well as late summer. Only 3 species of *Bumillaria* and 15 of *Tribonema* have been described.

Many heterokonts are epiphytic on other aquatic plants, especially some of the green algae like *Rhizoclonium.* Among the most commonly encountered are species of *Characiopsis,* a genus frequently confused with the chlorophycopsid epiphyte, *Characium.* The yellow-green, nongranular plastids with smooth margins (use high power or oil immersion!) make identification relatively easy, however. Less common, but readily recognized because of its curved to spiral shape, is the planktonic genus, *Ophiocytium.* Species of *Ophiocytium* and *Characiopsis* (Figure 9–2D,E) are frequently encountered growing on or near *Rhizoclonium* and other green algae of ponds and streams.

Harpochytrium is a genus of pale green and colorless algae, some saprophytic, others autotrophic, formerly classified in the Mastigomycopsida. It was transferred to Xanthophycopsida when a species having yellowish green chloroplasts was discovered; other autotrophic species of *Harpochytrium* have since been discovered, and all have pale, yellowish green plastids.

Ecological and Economic Significance

Nearly all modern Xanthophycopsida are found on soil or in freshwater streams and ponds, where blooms consisting largely of *Vaucheria, Tribonema, Characiopsis,* and other genera commonly occur in late summer. *Botrydium* also occurs mingled with moss on tree trunks. The fossil record indicates heterokonts were once much more abundant than they are now. The marine species from which modern Xanthophytes evolved are long since extinct; possibly they lacked ability to compete with diatoms and Chrysophycopsida. The ecological importance of the class is significant in some ecosystems, but on a worldwide basis, considering all biomes and all ecosystems, the ecological significance of heterokonts is relatively slight.

The class is apparently of little economic importance at the present time. *Pleurochloris pyrenoidosa,* commonly called "mustard algae," is a nuisance alga in swimming pools. The cost of controlling it makes it the most important species, economically speaking, in this class. Unlike the Dinophycopsida and Cyanophycopsida, the Heterokontae do not seem to produce toxic blooms harmful to livestock or other animals. They are reported, however, to be a major cause of the "mossy" flavor in fish that most people find objectionable.

Research Significance

The Xanthophycopsida are among the most easily cultured of all algae and therefore have considerable potential for studying metabolic pathways in photosynthesis, for studying mineral nutrition of plants, and for basic ecological studies involving limiting factors, especially in the development of algal blooms. Simplicity in inducing sexual reproduction in laboratory cultures and the occurrence of haplontic life cycles suggest that the Xanthophycopsida should be ideal for genetic studies.

Inasmuch as xanthophycean fossils are among the most ancient of all known fossils, the class should also be of special interest to paleobotanists and geologists and of great potential in furthering our understanding of evolution. If the present fossil record is accurate in its implication that the Xanthophycopsida are the most ancient of the Chrysophyta, then we need to study their ultrastructure, biochemistry, and physiology in order to understand better their origin and relationships to the diatoms, the golden algae, and the flagellated fungi, all of which are ecologically very important. To do this, we also need to look for fossil evidence of these plants wherever we can find it. In the past, fossils or suspected fossils from this and other nontracheophyte taxa have frequently been described only as "algae," a term having no real botanical significance and tending to produce only frustration in students seriously studying evolution.

CLASS 2. CHRYSOPHYCOPSIDA

The Chrysophycopsida (*chryso* = golden and *phycus* = alga), often called the golden algae, resemble both Cryptophycopsida and Xanthophycopsida in many ways and are believed to have evolved from heterokont-like ancestors in late Proterozoic time (Figure 9-1). Two groups of chrysomonads, the coccolithophores and silicoflagellates, fossilize especially well and have left an excellent fossil record, going back to the Devonian in the case of the coccolithophores.

Chrysophycopsida occur both in marine habitats, where they are especially common in the cold oceans, and in freshwater. The freshwater species also tend to favor colder habitats, such as mountain streams, waterfalls, and oligotrophic lakes, and are often abundant in such habitats. The cells of most species are very delicate and literally disintegrate (some writers say "explode") when the environment is changed drastically; consequently, they are not seen in the college laboratory as often as might be expected from their abundance. Collections of lake water generally need to be examined within a very short time, ideally at the collection site, if Chrysophycopsida are to be observed. There are approximately 400 species of Chrysophycopsida distributed among 7 orders, 26 families, and over 70 genera. Some of the morphological and ecological characteristics of the 7 orders are summarized in Table 9-2.

General Morphology and Physiology

The vast majority of the species are unicellular or colonial flagellates commonly called chrysomonads. A few of the Chrysomonads are reported to have only one pinnate flagellum, but most species are biflagellate, the anterior flagellum pinnate, the trailing one whiplash. In some species, the pinnate flagellum is much longer than the whiplash, just as in the heterokonts, but in others the two flagella are approximately equal in length. A few chrysophycean species are filamentous; others consist of gelatinous, nonmotile colonies or of nonmotile unicells. Naked, amoeba-like unicells occur in one order.

The typical chrysomonad has one or two smooth, golden plastids, an apical gullet, and no cell wall. A number of species have no plastids and are hence colorless. A pliable **periplast** formed inside the plasma membrane provides moderate rigidity to most species, but the cell nevertheless displays considerable **metaboly**—i.e., the cell changes shape as it moves through the water. In many species, the cell is inside a **lorica,** an open, rigid case commonly composed of silicates. In species having cell walls—and these are more common in the nonmotile orders than in the chrysomonad groups—the wall is composed of cellulose often impregnated with silica or carbonates and is typically composed of two valves that fit together like the two halves of a petri dish or a gelatinous pharmaceutical capsule.

Figure 9–5 *Synura uvula,* a colonial chrysophyte.

The chrysomonads can best be differentiated from cryptomonads by their color, their more pliable cells, their apical gullet, and their more or less spherical, rather than flattened, form. They differ from Prorocentrales and Heterochloridales by color and by the usually larger plastids (Table 7-3). Many chrysomonads, e.g., *Synura uvula* (Figure 9-5), form motile colonies, superficially similar to *Volvox* in the Chlorophycopsida but differing in color and in having smooth plastids.

The plastids are similar to those of Dinophycopsida and Phaeophycopsida, as well as those of other Chrysophyta, having thylakoids in sets of three, and containing chlorophyll *a,* frequently chlorophyll *c,* and brownish carotenoids. The chief carotenoid is fucoxanthin, which is a bright golden color. Also present are diadinoxanthin, diatoxanthin, dinoxanthin, B- and α-carotenes, and lutein. Ultracellular and biochemical characteristics are typical of the division.

In several species, especially marine Chrysophycopsida, a structure resembling a third flagellum is present. Called a **haptonema,** it is structurally different from flagella and is often held rigid as the organism moves; it is sometimes coiled into a tight spiral. With the advent of the electron microscope, haptonemata are more easily observed than was possible when only light microscopes were available. Because of its unique characteristics, the species of algae possessing haptonemata have been classified in a new class called the Prymnesiophycopsida or the Haptophycopsida by some phycologists; many botanists are reluctant to accept, however, a class based on a single characteristic, regardless of its uniqueness.

Chrysomonads commonly reproduce by simple mitotic fission. Under certain conditions, **statospores** are formed. In *Ochromonas tuberculata,* the vegetative cell has a single plastid, a large golgi body, and several contractile vacuoles; the beginning of statospore formation is marked by the contractile vacuoles becoming very active, followed by the appearance of a thin refractile sphere just under the surface of some of the cells. Deposition of silica soon begins in the vicinity of the Golgi body. The nucleus, plastid, and other organelles, with exception of some mitochondria, ribosomes, and contractile vacuoles, are enclosed in the developing membrane and siliceous wall.

Epiphytic species, species having gelatinous colonies, and filamentous species produce flagellated spores that resemble small chrysomonads. Frequently, these asexual spores (commonly called **zoospores** because they are motile) even contain plastids. Sexual reproduction involving isogametes has been reported in a few species of Chrysophycopsida, but most of the reports need confirmation. In the coccolithophores, life cycles are reported to be diplohaplontic.

TABLE 9–2

DIVERSITY IN THE CHRYSOPHYCOPSIDA (ORDERS AFTER PASCHER 1931)

Orders (Fam–Gen–Spp)	Thallus Characteristics	Flagella and Plastids	Other Characteristics	Habitat	Representative Genera
Chromulinales (4-12-165)	Motile unicells, occasionally rhizopodial, often united into motile colonies; lorica may enclose cells	One long pinnate flagellum; 1 or 2 large, discoid plastids that may become net-like	Cells fuse to form a plasmodium in *Myxochrysis;* a pectic sheath with small overlapping siliceous scales may enclose the cells	Mostly freshwater; a few are marine and often common; many species limited to clear lakes	*Chrysamoeba, Chrysocapsis, Chrysococcus, Chromulina, Chrysosphaerella, Mallomonas*
Isochrysidales (4-11-75)	Motile unicells or small motile colonies, cell walls often impregnated with lime (the coccolithophores)	2 equal flagella, 1 pinnate, 1 whiplash; a haptonema may be present; 1 or 2 elongated plastids	Cells obpyriform, enclosed in thin, pectic sheath containing silicious or carbonate scales. (The prymnesiophytes are included in this order)	Important component of the marine plankton in the cold oceans	*Chrysidella, Syncrypta, Synura, Coccolithophora, Pontosphaera, Ochrosphaera*
Ochromonadales (3-10-105)	Motile unicells with or without loricas, also epiphytes on diatoms	2 unequal flagella; 1 or 2 parietal plastids	Cells typically attached at base of vase-shaped loricas which may be joined to each other	Mostly freshwater, some marine and brackish water	*Uroglena, Ochromonas, Dinobryon, Chrysoxys, Eusphaerella*
Silicoflagellales (3-6-25)	Motile unicells; cell walls impregnated with silicates	1 flagellum, rarely 2; 2 to many small plastids	Excellent fossil record dating to the Cretaceous	Marine, sometimes forming major blooms	*Mesocena, Ebria, Distephanum, Cannopilus*

Rhizochrysidales (3-10-15)	Amoeboid unicells or rhizopodial colonies with or without loricas	1 long flagellum in zoospores and motile vegetative cells; 1 to 2 large golden brown plastids	Cytoplasmic strands may unite cells into plasmodial colonies; fusion plasmodia common	Freshwater; some species both autotrophic and phagotrophic	*Rhizochrysis, Chrysarachnion, Bicocoeca, Palatinella, Heliactis*
Chrysocapsales (3-10-15)	Nonmotile vegetative cells united into amorphous gelatinous colonies	Zoospores have 1 long flagellum; one large plastid	A common species in clear mountain streams, *Hydrurus foetidus,* covers rocks with slippery layer and smells like spoiled fish	Freshwater, especially waterfalls and swift rivers	*Hydrurus, Celloniella, Geochrysis, Naegeliella*
Chrysosphaerales (2-6-7)	Unicellular or non filamentous colonies never motile in vegetative stage; epiphytic cells typically hemispherical	Zoospores with 1 pinnate flagellum, 1 large golden brown plastid	Mitosis unknown except to produce gametes or zoospores; cell walls often impregnated with silica	Epiphytes on freshwater algae; some known only as fossils	*Epichrysis, Chrysosphaera, Chrysapion*
Chrysotrichales (3-4-5)	Cells joined end to end in simple or branched filaments or in pseudo-parenchymatous masses	Zoospores with 2 flagella of unequal length; basal cells lack plastids, other cells with 1 to many golden brown plastids	Superficially similar to Heterotrichales but cell structure, zoospores, and especially statospores are regular chrysophycean type	Epiphytes on filamentous green algae in bogs; *Phaeothamnion* is only genus in North America	*Phaeothamnion, Phaeodermatium, Nematochrysis, Sphaeridiothrix*

Note: The number of families (Fam), genera (Gen), and species (Spp.) in each order is indicated in the first column.

Phylogeny and Classification

All of the unicellular flagellated Chrysophycopsida were included in a single order by Pascher; his Chrysomonadales has recently been divided by different botanists into three or four orders based on additional characteristics including type of flagellum and nature of the cell wall (See Special Interest Essay 9-1). In the Ochromonadales, the group which most nearly resembles the heterokonts in plastid characteristics as well as flagellation, the pinnate flagellum is longer than the whiplash, which in some species is rudimentary or lacking. In the Isochrysidales, the two flagella are equal in length. The Silicoflagellales are a group of cold ocean marine flagellates, with silica-impregnated cell walls, which apparently evolved during the Cretaceous from *Ochromonas*-like ancestors. They have left an abundant fossil record and today make up a dominant part of the nanoplankton of the North Sea and other cold waters. Like the Ochramonadales, they have a long tinsel flagellum, but the whiplash flagellum is short or lacking.

Suggested lines of relationships among the seven orders of Chrysophycopsida are illustrated in Figure 9-1. The animal-like characteristics of *Chromulina, Ochromonas, Uroglena,* and some other chrysomonads cause many biologists to look to the Chrysophycopsida as the most logical stock from which the animal kingdom has evolved. *Chrysamoeba radians,* with its single flagellum, naked cell, and phagotrophic nutrition when conditions are unfavorable for photosynthesis, is especially animal-like, despite its two large golden plastids. It is believed by many biologists that both the Protozoa and the Metazoa evolved from organisms similar to *Chrysamoeba* in late Proterozoic time. No fossils have been found, however, which resemble either the earliest animals or the presumed chrysophycean stock from which they evolved.

The earliest really good chrysophycean fossils are found in Paleozoic formations and resemble modern coccolithophores of the order Isochrysidales. More recently, "molecular fossils," in the form of an organic residue (24-*n*-propylcholestane) found in sedimentary rock at Prudhoe Bay, Alaska, indicate the presence of Chrysotrichales during the Paleozoic Era (Moldowan et al. 1990). The same residues are produced by *Sarcinochrysis marina* and *Nematochrysopsis roscoffensis;* the formations in which they have been found are dated Ordovician to Devonian. Moldowan et al. (1990) place these species in a new order, the Sarcinochrysidales, and suggest that it is a link between the Chrysophycopsida and the Phaeophycopsida.

From primitive chrysomonads have evolved the Isochrysidales, Chrysosphaeriales, Chrysotrichales, and Chrysocapsales along one line, the Rhizochrysidales and animal kingdom along a second line, and the Silicoflagellales along a third line. Characteristics of the seven orders are summarized in Table 9-2. The recently suggested class Haptophycopsida (DeMort et al. 1972) is included in the Isochrysidales because of lack of sufficient evidence of correlated differences to warrant class distinction.

Table 9-2 is based largely on Pascher's classification, which took into account only flagellation and cell wall characteristics and is thus artificial. Hopefully, an increased interest in these ecologically important and interesting plants will soon give us the information needed to arrive at a more natural and hence better classification.

Representative Genera and Species

Synura ulva (Figure 9-5), is probably the freshwater chrysophyte most frequently encountered by college students. Its distinctive golden color and attractive form catch the eye. *Hydrurus foetidus* is a common colonial species found in cold mountain streams and often recognized merely by its bad odor. It forms a brown, slippery, gelatinous film on bottom rocks. If the film is scraped off, the colonial nature of the mass and almost spherical form of the individual nonmotile cells can be observed under the microscope. *Phaeothamnion confervicola* is a rare freshwater species which lives as an epiphyte on other algae and is easily recognized by its branching filaments. Marine chrysophytes often observed by students include numerous species of silicoflagellates, such as *Dictyocha fibula* and *Distephanum speculum,* and coccolithophores, including species of *Pontosphaera* and *Syracosphaera. Distephanum speculum* is common in all oceans of both hemispheres; both *D. speculum* and *D. fibula* occur as fossils in numerous formations.

Ecological and Economic Significance

Marine chrysomonads are among the major producers in cold ocean ecosystems, both in coastal waters where brown algae are dominants and in the open seas where only planktonic forms are found. One group of chrysomonads, the coccolithophores, are very important producers in the southern Mediterranean Sea and in the tropical Atlantic and Indian oceans. The wide distribution and relative abundance of freshwater species indicate that they may be of greater ecological significance in food chains leading to freshwater fish than has been recognized. Some studies have suggested that the so-called "mossy" odors in fish at some seasons of the year may originate from a few species of Chrysophycopsida and Xanthophycopsida.

Research Potential

If it is true, as many biologists believe, that both the Protozoan and Metazoan lines of evolution, which have led to the chordates, including man and other mammals, had their beginnings in the Hadrynian period from organisms similar to the Chrysomonads of today, then further research involving these plants should be of great interest to all zoologists and historical geologists as well as to geneticists, evolutionists, paleontologists, and even anthropologists. Chrysophycopsida appear to be especially valuable research organisms for studying the origins of phagotrophic nutrition inasmuch as both autotrophic and phagotrophic species occur in the class.

Some Chrysophycopsida are tolerant of extremely wide ranges in salinity and temperature. This is a possible explanation as to why they are more important as a class in some marine habitats than either the Xanthophycopsida or the Chlorophycopsida. *Olisthodiscus luteus,* for example, grows well both in brackish water and in water containing over 4% salt, and over a range of temperatures from less than 10°C to more than 30°C. Salinity tolerance is especially wide at higher temperatures, and temperature tolerance is especially wide at salinities around 2.5% This helps us understand the summer blooms of this species in some bays.

Because of their ability to increase in numbers rapidly, the Chrysophycopsida should be valuable in studying limiting factors affecting marine productivity. We know little about the sources of nitrogen, calcium, sulfur, and other minerals needed by plankton, or even, in many cases, which minerals are needed; furthermore, we know little about the presence of toxic minerals and how they affect productivity, or even what minerals are toxic in ocean ecosystems. Chrysophycopsida should be useful organisms for this type of research.

CLASS 3. BACILLARIOPHYCOPSIDA

The Bacillariophycopsida (*bacillum* = a little stick, *phycus* = alga) are commonly called diatoms, and are the largest in number of species of the classes of Chrysophyta as well as ecologically the most important. There are more than 5000 species of diatoms, about half of them marine and the other half freshwater. Another 4500 species of fossil diatoms have been described. All diatoms are essentially unicellular; however, in some species, loose colonies or filaments regularly form. It has been estimated that between 15% and 25% of the oxygen we breath is produced by diatoms; if so, they are ecologically among the most important of all plants. Like other members of the division, they are especially abundant in the cold oceans and in cold streams or lakes on land (See Special Interest Essay 9-2). Several species are commonly found in soil.

Diatoms can be found at any time of year in any stream, pond, lake, bay, or estuary where anything will grow. One of the easiest ways to find them is to place glass slides or other pieces of glass in a clear stream or pond and leave them there for 3 or 4 days. Mark the slides with a drop of bright paint to help locate them when it is time to recover them. Diatoms are also nearly invariably present as epiphytes on coarse filamentous green and yellow-green algae, and they can always be found on the mud or sand at the bottom of ponds and lakes.

Students have little if any trouble recognizing diatoms. Their striations and rigid form are distinctive. Each unicellular plant is bounded by angular or geometrically curved lines: diatoms obey the rules of solid geometry better than any other organisms do. Many biologists aver that diatoms are the most beautiful of all creatures; certainly they possess an elegant beauty that has attracted both scientists and amateur naturalists over the years. There is probably no group of plants in which nonprofessionals have played as large a role in publishing descriptions and names as the diatoms.

General Morphological Characteristics

Morphologically, the diatoms can be divided into two groups, the pennate diatoms and the centric diatoms. The former are characterized by bilateral orientation and often bilateral symmetry, the presence of a raphe, one or two plastids per cell, herpokinetic motility in vegetative cells, and lack of flagella at all stages of the life cycle. They are especially common in freshwater habitats. Centric diatoms are characterized by radial orientation and often radial symmetry, many plastids per cell, flagellated gametes, production of statospores, and complete lack of motility in the vegetative stage. They are especially abundant in marine habitats. Their general characteristics are summarized in Table 9-3.

The cell wall of diatoms is composed of pectic materials well impregnated with silica. On the death of the organism, the empty shell, or **frustule,** is very resistant to decay and can remain in identifiable condition for millions of years. The frustule consists of two **valves** or halves that fit together like the two parts of a petri dish (Figure 9-6). The outer valve is called the **epithecum,** the inner valve is the **hypothecum.** In some species, the two valves are essentially identical; in others they differ in various ways: for example, there may be a raphe on the epithecum but not on the hypothecum. When diatoms are examined with the microscope, they sometimes lie so that the top or bottom is in view, or they may lie on their side. Students examining a pure culture of diatoms will frequently think they have two species because in side view the cell looks so different from top view. Side view is known as "**girdle view**" and top or bottom view as "**valve view**" (Figure 9-6).

In girdle view, the overlapping nature of the cell wall is often apparent (Figures 9-6 to 9-8). In some species, the hypothecum is slightly smaller than the epithecum because in cell division the hypothecum of the mother cell becomes the epithecum of one of the daughter

TABLE 9-3

CHARACTERISTICS OF THE TWO ORDERS OF BACILLARIOPHYCOPSIDA

	Centrales	Pennales
Cell orientation and/or symmetry	Radial	Bilateral
Plastids	Many	One or two
Flagella	Present on gametes	Never present
Motility	Only in reproductive cells	Jerky motility common in vegetative cells
Asexual spores	Statospores common and highly ornamented with spines, etc.	Statospores never present; chlamydospores occur in some species
Raphe	No raphe or pseudoraphe present	Responsible for motility in some species; pseuraphe occur in others
Habitat	Mostly marine	Mostly freshwater
Origin according to the fossil record	Early Cretaceous: possible Berriasian epoch	Tertiary: Miocene or possibly Oligocene epoch

cells each time cell division occurs. If the cell walls are so highly silicified that the new epithecum is incapable of "stretching," after a few cell generations some cells are smaller than the original parent cell; the striations and other cell markings remain the same, however. When minimum cell size is achieved, production of **auxospores** (spores formed by liberation and enlargement of a protoplast followed by secretion of a silicified wall around the protoplast) makes possible a restoration to the original cell size. Unicellular cultures may, therefore, consist of cells of many different sizes. In most species, however, all of the cells remain the same size, either because the cell walls grow as cell division occurs, or because new **connecting bands,** which are present in some species and make it possible for two valves of equal diameter to be connected to each other, are formed.

The ornamentation that make diatoms so distinctive from all other plants, and also so attractive, is the result of thin areas in the cell wall bounded by siliceous ridges. Very small, thin areas arranged in rows are **punctae** and the rows of punctae are **striae,** or striations. The striae lie

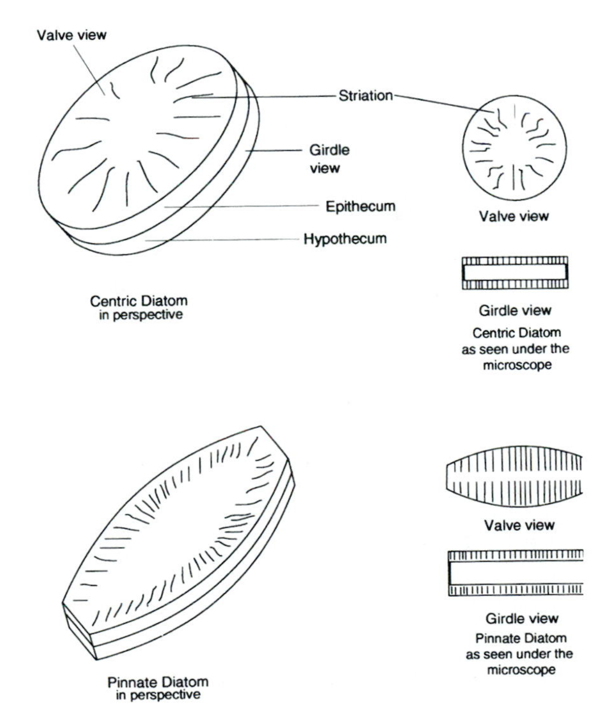

Figure 9–6 A diatom, in three dimensions at left to illustrate the distinctions between girdle view and valve view as they appear under the microscope.

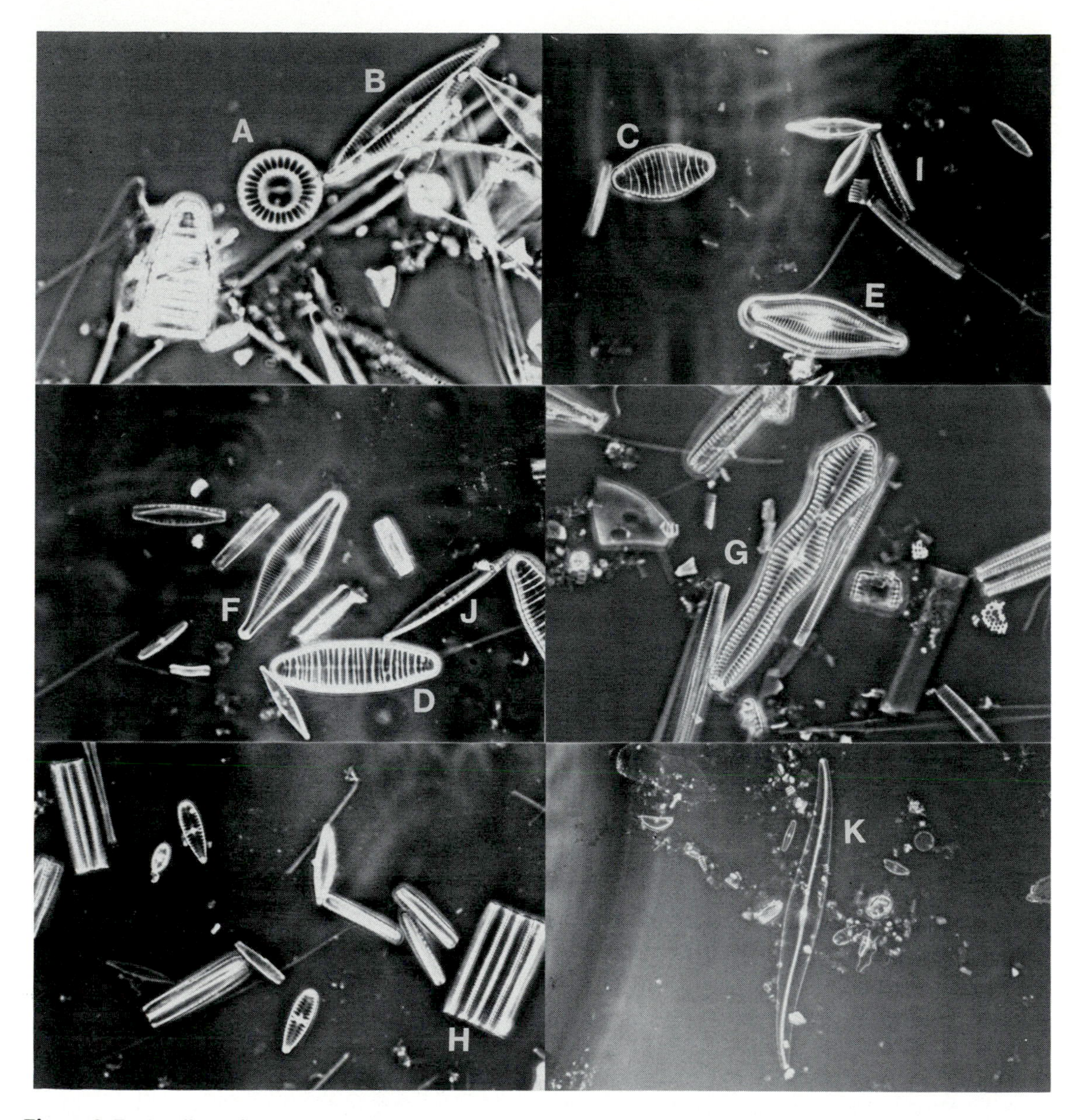

Figure 9–7 Bacillariophycopsida: Centrales, Gomphaceae, Nitzschiaceae, and others. (A) *Cyclotella meneghiana;* (B) *Navicula rhyncocephala;* (C,) *Diatoma vulgare;* (D) *Diatoma vulgare;* (E) *Gomphonema herculeana;* (F) *Gomphonema acuminatum;* (G) *Gomphonema herculeana;* (H) *Fragilaria vaucheria*, girdle view; (I) *Fragilaria vaucheria,*, valve view; (J) *Nitzschia palea;* (K) *Gyrosigma sciotense.* (Micrographs courtesy Richard Clark, Ricks College, Rexsburg, ID.)

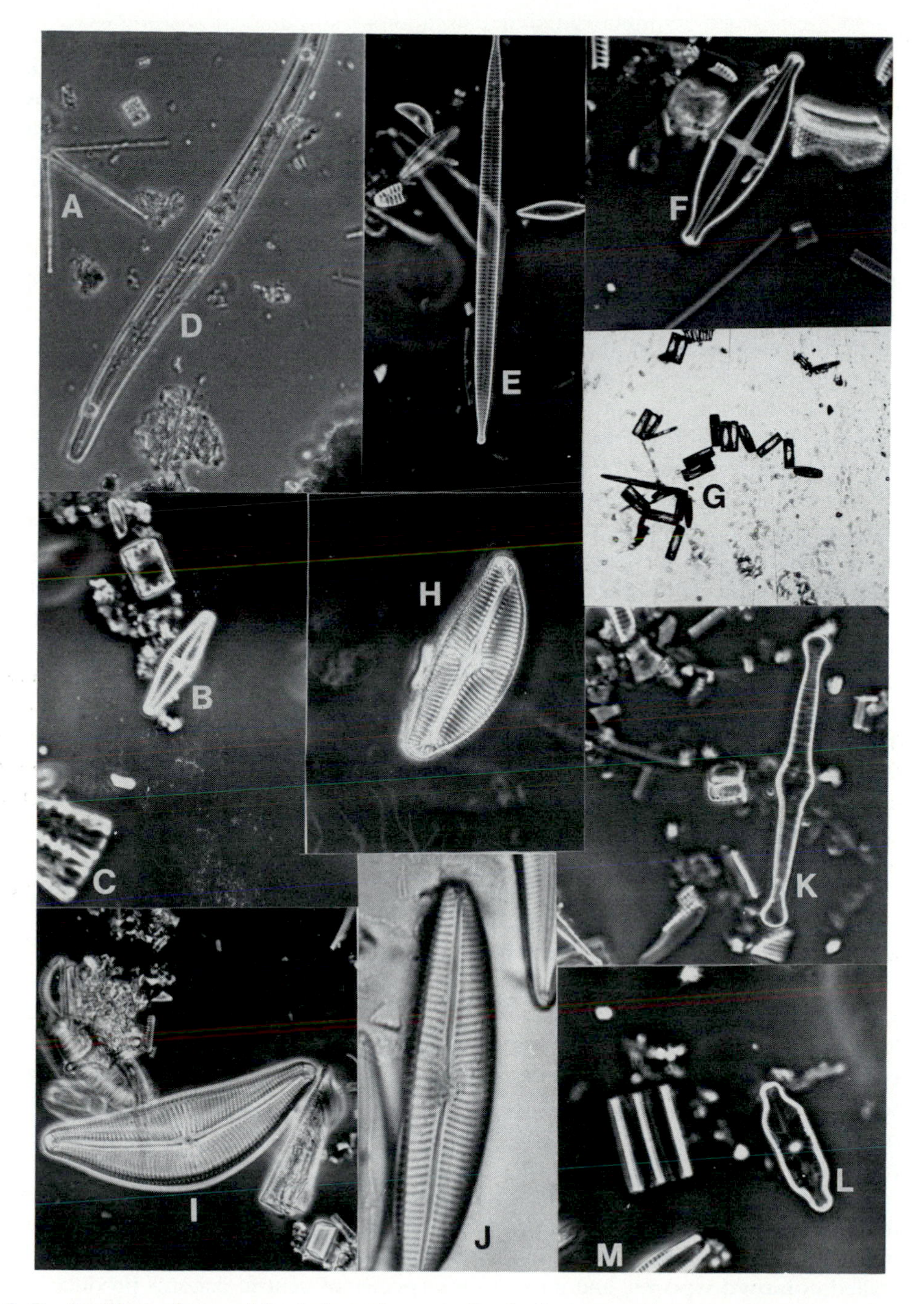

Figure 9–8 Bacillariophycopsida: Achnanthaceae, Cymbellaceae, and others. (A) *Astrionella formosa;* (B) *Achnanthes minutissima;* (C) *Meridion circulare;* (D) *Gyrosigma sciotense;* (E) *Synedra ulna;* (F) *Stauroneis anceps;* (G) *Tabellaria flocculosa;* (H) *Cymbella prostrata;* (I) *Cymbella gracilis;* (J) *Cymbella mexicana;* (K) *Tabellaria fenestrata;* (L) *Neidium iridis;* (M) *Amphora ovalis.* (Micrographs courtesy Richard Clark, Ricks College biology department Rexsburg, ID.)

between elevated ridges called **costae.** In the Centrales they radiate outward from a common center; in pennate diatoms they tend to be parallel to each other in two rows, one along each margin of the valve. Between the two rows of striations, on either the upper or the lower valve, or both, in most pennate diatoms, there is an unsilicified groove or canal called a **raphe.** Species of diatoms having a raphe are motile, while those lacking a raphe are non-motile; it is believed that protoplasm streams through pores in the valve and along the raphe and that the friction between protoplasm and water causes the motion.

In the walls of centric diatoms (Figure 9–7A), there are larger thin areas bounded by thick, siliceous ridges. These are called **areolae.** The areolae are often roughly hexagonal in shape and give a coarse appearance to many marine diatoms. Within each areola, minute vertical canals, or **pores,** occur in many species. In some species, the canals are incomplete and are called **poroids.**

The electron microscope has revealed some differences between centric and pennate diatoms in the nature of the striations. Striations in the pennate diatoms consist of a single row of punctae, while those in the centric diatoms consist of multiple rows, similar to the striations on the cell walls of many Chrysophycopsida and Xanthophycopsida.

In addition to striations, many centric diatoms have spines or other projections on their valves which aid in holding cells together in colonies or loose filaments. Mucilaginous material may also hold the cells together (Figure 9–7H, Figure 9–8G). The mucilage is secreted by cellular protoplasm where it comes in contact with the external environment, either through the pores in the areolae or through similar pores in the raphe.

Striations, raphes, and other ornamentations, and the tendency to form colonies or filaments are important in diatom identification. It is often easier to identify diatoms if they are first treated with a strong oxidizing agent, like chromic acid, to remove the chlorophyll which tends to obfuscate the characteristic ornamentation. Plastids are uniform throughout each of the morphological groups and hence of little value in species identification. However, the acid also destroys the mucilage holding the cells together, making it difficult to tell if the specimen is colonial, filamentous, or strictly unicellular (Figure 9–8G).

The plastids in the Bacillariophycopsida are typical for the division; the lamellae are usually, but not always, parallel to each other, and each lamella consists of three thylakoids. Brown or yellow carotenoids tend to mask the green of the chlorophyll.

The most abundant of the carotenoid pigments is fucoxanthin. Diadinoxanthin, α-, β-, and γ-carotenes, and chlorophylls *a* and *c* are also present. Diatoxanthin and dinoxanthin are found in several species. Chlorophyll *e* has been reported, but the reports need confirmation. Fucoxanthin is responsible for the golden color typical of many Bacillariophycopsida. Food reserves are in the form of chrysolaminarin (formerly called leucosin) and oils; fucosterol has been reported in some species. The buoyancy of diatom cells is reported to be due to the presence of oil droplets.

Asexual reproduction in diatoms is by simple mitosis accompanied in some species by a diminution of cell size. Mitosis in most species of pennate diatoms occurs around midnight and is synchronous with plastid division (Smith 1938). Pennate diatoms frequently produce thick-walled **chlamydospores,** and most marine species of centric diatoms produce statospores. Chlamydospores may be modified statospores, although they are not typical in appearance, being neither spiny nor otherwise ornamented.

Auxospores are produced by both pennate and centric diatoms. In most cases they are actually zygotes formed by the union of two gametes which resulted from meiosis. Auxospores are also called rejuvenescent cells and are considerably larger than the vegetative cells which produced them. In pennate diatoms they are produced by (1) two cells each producing two gametes which fuse to become two zygotes or auxospores; (2) two cells each producing one gamete which fuse to become a zygote or auxospore; (3) a single cell producing two gametes which fuse to become a zygote or auxospore; or (4) a single cell producing a single gamete which develops parthenogenetically into an auxospore. Some species are isogamous, others heterogamous. In centric diatoms, auxospores are apparently produced by a method similar to (3) above; possibly they are produced asexually in come centrics. Life cycles in all pennate diatoms that have been investigated are diplontic.

The so-called "microspores" in centric diatoms are apparently produced by meiosis followed by several mitoses and are actually flagellated isogametes. Some have a single flagellum, others have two flagella, one tinsel and a shorter whiplash. In some species, the microspores may be sperm and the eggs may be nonmotile, but there are conflicting reports concerning sexual reproduction in centric diatoms. Very likely both isogamous and oogamous species exist.

It is fashionable in scientific research to use only the latest tools and methods and to regard old methods as out-dated and useless. It was recently pointed out to members of the Botanical Society of America that, because of this "faddist" bias, research in diatom sexuality came to a stand still long before our knowledge in this area was satisfactory. The electron microscope, wonderful as it is, is not suited for life cycle studies. Since then, several studies of diatom life cycles, using only light microscopes, has added significantly to our knowledge of diatom sexuality. Conjugation is usually girdle to girdle. Pairing may be dorsal to dorsal, dorsal to ventral, or ventral to ventral, often in a 1:2:1 ratio. (The side with the pyrenoid is called the ventral side; the nucleus is usually located dorsally.)

Phylogeny and Classification

Diatoms fossilize well; an excellent paleobotanical record is therefore available extending to the Cretaceous. There are reports of Paleozoic diatoms, but these are not reliable, and reports of Jurassic diatoms are also suspect. In every case in which an attempt to confirm the reports of pre-Cretaceous diatoms has been made, either an error in plant identification—e.g., coccolithophores, xanthophytes, or dinoflagellates were reported to be diatom fossils—has been revealed, or an error in identification of the geological formation was found. The oldest known diatoms for which we have fossil specimens are centric marine species from the early Cretaceous. Very few have been found in early or lower Cretaceous, but they are relatively abundant in mid and upper Cretaceous formations. No freshwater species have been found prior to the Miocene epoch, with the possible exception of *Melosira granulata* in what may be an oligocene formation. Pennate diatoms date from the Paleocene, or possibly somewhat earlier. Most of the genera of living Bacillariophycopsida can be found in the fossil record.

The earliest diatoms to appear in the record were centric marine types that produced statospores and had overlapping cell walls that were strongly silicified. Present-day species belonging to the same or closely related genera are characterized by these same features, by having several plastids per cell, and by producing motile gametes with unequal flagella. They thus resemble the Heterococcales; hence it appears as though the Bacillariophycopsida evolved from Xanthophycopsida similar to modern Heterococcales during the early Cretaceous or possibly the late Jurassic period. From these first diatoms have evolved other centric diatoms and the pennate diatoms.

The 5500 species of diatoms (10,000 if fossil species are included) are classified in 2 orders, 20 families, and 170 genera. The 2 orders are easily distinguished because Centrales have radial symmetry and Pennales bilateral symmetry. Most Centrales are marine and most Pennales are freshwater.

To differentiate between two orders on the basis of only one characteristic—bilateral vs. radial symmetry, in this case—would result in an artificial classification. Actually, the Centrales are differentiated from the Pennales on the basis of many characteristics, including flagellation, plastid number and morphology, both sexual and asexual reproduction, and geological record in addition to symmetry. Characteristics of the two orders are summarized in Table 9-3 and their relationships to each other and other Chrysophyta are illustrated in Figure 9-1.

Some botanists elevate the two orders to rank of subclass and divide the Pennales into four orders based entirely on presence or absence of the two raphes, and the Centrales into three orders based entirely on cell geometry. The resulting classification is considered artificial by most botanists.

Representative Species and Genera

Navicula viridis is especially common at some times of the year. The name means "little boat," referring to the jerky motion characteristic of most motile diatoms. *Gomphonema constrictum* is another common species. Resembling an Egyptian mummy case when observed in valve view, it is simply wedge-shaped in girdle view. *Melosira* is one of the few genera of centric diatoms containing freshwater species; the individual cells are held loosely together in long chains. *Fragillaria* and *Tabellaria* also tend to form chains. The cells are long and thin in both of these genera and also in *Synedra ulna*, a very common and highly variable species. Species of *Astrionella* also have cells that attach to each other, but the individual cells radiate from a common point so that the colony resembles a star.

Marine genera are mostly centric diatoms. *Melosira granulata* is a common freshwater species; *M. varians* has been reported not only from great depths in the ocean but also from freshwater (Clark and Rushforth 1977). Closely related to *Melosira* are *Podosira*, *Coscinodiscus*, and *Thalassiosira*. The last named is often very abundant. Less closely related are *Arachnoidiscus*, with beautiful radial symmetry, *Actinocyclus*, and *Aulacodiscus*. Including fossils, there have been about 70 species of *Actinocyclus* and 100 species of *Aulacodiscus* described; most of the species in the latter genus are now extinct. Sometimes separated from the rest of the Centrales as distinct orders are the two suborders, Solenineae and Biddulphineae, represented, respectively, by *Rhizosolenia* with 35 species, and *Biddulphia* with 100. The diversity in diatom morphology is suggested by Figure 9-1.

Ecological Significance

There is no doubt of the great ecological significance of these ubiquitous organisms. It is extremely unusual to sample any aquatic habitat, freshwater or marine, without finding diatoms; frequently they are the most abundant organisms present. They are also common in soil. Not only is the world's fishing industry dependent upon diatoms as the main component of the primary producer base upon which the rest of the ecological pyramid is built, but all aerobic organisms depend upon them for continual replenishment of the atmosphere with oxygen. Marine Centrales are especially significant in this regard.

Diatom blooms are triggered by an increase in water temperature and light intensity, following a build-up of mineral nutrients during the winter months, and last for a few weeks until the minerals are used up to the extent that they become limiting. Diatom numbers then decrease rapidly from a high of tens of millions of cells per liter of water to a few hundred per liter. Dinoflagellate blooms often follow the spring diatom bloom. As minerals build up during the summer, conditions become favorable for good diatom growth again and the fall bloom occurs, lasting until the low light intensity of winter days limits the growth.

The causes of diatom blooms and subsequent decrease in numbers has been the subject of considerable physiological research. All diatoms require large amounts of soluble silica in order to grow under laboratory conditions. Silica is not essential to most organisms, and its availability is probably a major factor in the occurrence of diatom blooms in lakes. Ryther (1956) reported that diatoms are more efficient than dinoflagellates at low light intensities and that could be the explanation for the sequence of algal blooms in many lakes and ponds; others, repeating Ryther's experiments, have not been able to confirm that any difference between the two groups exists; however, Chan observed that when five species of diatoms were compared to five species of dinoflagellates, all of the former had a higher level of chlorophyll *a* per gram protein than any of the latter. Apparently the explanation of algal bloom sequences is much more complicated than a simple light efficiency relationship with the ability to rapidly produce high levels of chlorophyll *a* being one of several factors, variation in needs for minerals another, and possibly a difference in photo-efficiency still another.

All centric and many pennate diatoms are either free-floating or attached to rocks, soil, or other water plants; some pennate diatoms are motile and may be phototactic. Spiny projections from the cell walls of marine Centrales enable them to remain afloat just below the surface where conditions are best for photosynthesis, by taking advantage of wind-induced Langmuir circulation of the surface waters. Phototactic pennate diatoms in lakes and ponds are apparently able to remain in the zone of optimum photosynthetic activity in the same way Dinophycopsida and other flagellates are.

In streams, many species are epiphytic on filamentous algae. Other diatoms attach to rocks and other objects by means of gelatinous substances. This characteristic has led to the development of the "glass plate" method of studying diatoms. Glass slides or plates are immersed in a stream or pond and after a few days or a week to a month they are taken up and examined in the laboratory. Changes in species density through time or space reveal ecological trends.

Diatoms are reported to produce inhibitory substances that retard the growth of dinoflagellates and other aquatic plants and also to be inhibited in some cases, or have their growth enhanced in other cases, by similar substances. Of the 20 basic amino acids common to all biological systems, 18 have been identified in diatoms, the exceptions being glutamine and asparagine.

The influence of man's migration and commerce has become a factor in diatom ecology. *Biddulphia sinensis* was formerly found only in the Pacific, but has been carried by ships to the Atlantic, and since 1903 has become an important component in the North Sea plankton.

Ecotypic adaptation occurs in diatoms as in other plants. For example, strains of *Thalassiosira pseudofanana* which have evolved in mid-Atlantic, where nutrients are limiting, take in essential nutrients so slowly, compared with strains that have evolved in the nutrient-rich waters of estuaries, that they are at a disadvantage in survival when both are grown in a nutrient-rich medium. Presumably the mid-Atlantic strains would have survived better than the estuary strain had both been grown on nutrient-poor medium.

Economic Significance

As mentioned in the first paragraph of this chapter, the billions of pounds of fish and other seafood harvested each year are dependent upon the phytoplankton of the ocean. Chief among the plankters are the diatoms and the golden brown algae.

World fisheries each year land over 40 billion dollars' worth of produce (Tables 9-4 and 9-5). Leading countries include Japan, Russia, China, Norway, Peru, and the U. S. The economy of several nations, especially Portugal, Iceland, Norway, the Philippines, North and South Korea, and Ecuador, depend heavily on fishing. How much of this income is dependent upon Chrysophycopsida and how much upon Bacillariophycopsida and other plankton groups cannot be precisely and accurately judged at the present time, but the Chrysophycopsida are considered the most important producers in the North Sea and very important in many other waters, and the Bacillariophycopsida are considered to be of prime importance in Antarctica and the open oceans.

Japan and the republics making up the former Soviet Union are the leading nations in harvesting fish; together, they take in over 27% of the total annual catch. The ten leading countries (listed in Table 9-4) harvest 60% of the total annual take. The value of the U. S. share of this harvest is over 2 billion dollars annually (Table 9-5), roughly comparable to the total farm value of cotton, wheat, soybeans, or tobacco. Of special significance is the annual harvest of shrimps, primarily from the Gulf of Mexico, valued at almost half a billion dollars, considerably more than the farm value of several of our most important crops including rice, apples, oranges, peas, and beans, and about the same as potatoes.

In early spring each year, the rapid growth of diatoms soon depletes the dissolved silicates in the reservoirs which provide culinary water to London. If the phosphate level in the water is high, blue-green algae "bloom" and cause a "fishy" flavor about which people begin to complain. As the complaints come in, the municipal water board fertilizes the reservoirs with silica. This stimulates diatom growth and hence the fish which are dependent upon them. The fish consume the excess blue-green algae and the fishy flavor disappears. Paradoxically, it takes an increase in mass of fish, caused by the increase in diatom productivity, to eliminate the "fishy" flavor (Russell-Hunter 1970).

Not only are living diatoms of great economic significance to the whaling and fishing industries, fossil diatoms are just as important to the electronics, abrasives, and sugar manufacturing industries. One of the first uses of diatomaceous earth was as an absorbent for nitroglycerin to make an explosive that could be handled and transported safely, dynamite. Diatomaceous earth has now been replaced by wood meal for this purpose, but it still has many uses including as filtering medium in sugar factories, insulation in blast furnaces, manufacturing of cement, and as a mild abrasive in polishes and toothpaste. When 2% of diatomaceous earth is added to cement, the workability and strength of the product are markedly improved.

The siliceous cell walls of diatoms remain unaltered when the diatom dies, and as the empty frustules settle to the bottom of lakes or of the sea they build up deposits that may be

TABLE 9-4

TONNAGE OF FISH (METRIC TONS) LANDED BY LEADING COUNTRIES, 1977 AND 1986

Country	Landed Harvest	
	1977	**1986**
Japan	10,733,000	11,966,800
U.S.S.R.	9,352,000	11,260,000
Communist China	6,880,000	9,346,500
Chile	—	5,695,500
Peru	2,530,000	5,609,600
U.S.	3,102,000	4,943,200
South Korea	2,419,000	3,102,500
India	2,540,000	2,925,300
Indonesia	—	2,521,200
Thailand	1,778,000	2,119,000
Philippines	—	1,916,300
Norway	3,562,000	1,898,400
Denmark	1,806,000	1,871,300
Republic of Korea	—	1,700,000
Iceland	—	1,657,100
Canada	—	1,466,600
Mexico	—	1,303,700
Spain	—	1,303,500
Ecuador	—	1,019,300
World Total	**73,500,000**	**91,456,800**

Data from 1982 *Information Please Almanac* and *United Nations Statistical Yearbook, 1985-86.*

many meters thick, 1000 meters in the Santa Maria oil fields in California, for example. Diatomaceous earth is light in weight; a slab that appears to be too heavy to lift may weigh only a few kilograms.

Research Potential

Long ignored by plant physiologists, geneticists, ecologists, and other botanists in their scientific investigations, the Bacillariophycopsida have now become very popular research material. Their greatest contributions have been in the understanding of photosynthesis, the testing of the laws of ecology having to do with limiting factors, and the analysis of causes of ecological succession. They show promise in unraveling the tangle of contradictory data concerning efficiency in light utilization and in discovering pathways of synthesis and uses of fats and other lipids. They have a potential for genetic research inasmuch as they multiply rapidly, show considerable morphological and physiological variability within species, and are easily cultured; this is an area of research that has hardly been touched as yet, however.

Scientific research in any branch of biology is dependent upon availability of pure cultures of the organisms the researcher is interested in. The increased availability of cultures of Bacillariophycopsida in recent years is greatly enhancing their use for research. There is no longer

TABLE 9-5

PRODUCTION AND VALUE OF SOME MAJOR GROUPS OF FISH AND OTHER SEAFOOD TO THE U.S. ECONOMY, 1980, 1987, AND 1990

Seafood Group	Landed Harvest (millions of pounds)			Value (millions of dollars)		
	1980	**1987**	**1990**	**1980**	**1987**	**1990**
Pacific salmon	613.8	562.0	733.0	352.3	596.4	612.0
Pollock	64.0	598.0	3,178.0	9.0	64.0	283.0
Flounder	244.8	197.7	502.0	84.2	145.1	159.0
Tuna	500.0	100.1	62.0	289.3	95.8	105.0
Halibut	19.1	76.1	70.0	16.8	88.3	97.0
Menhaden	2,496.6	2,712.3	1,962.0	112.0	104.4	94.0
Atlantic cod	118.2	59.1	96.0	323.3	44.2	61.0
Sea herring	291.0	207.1	221.0	44.9	51.3	38.0
Whiting	35.6	34.7	44.0	6.1	11.6	11.0
Haddock	55.2	6.7	5.0	21.4	8.5	6.0
Ocean perch	31.1	24.1	1.0	6.6	6.2	1.0
Jack mackerel	44.0	26.7	9.0	4.0	1.8	1.0
Shrimp	339.7	363.1	346.0	402.7	578.1	491.0
Crabs	523.1	386.4	499.0	291.3	321.9	484.0
Lobster	43.8	45.6	68.0	90.9	133.6	178.0
Scallops (meat)	29.7	40.2	42.0	114.3	141.1	158.0
Clams (meat)	95.4	134.1	139.0	90.2	132.9	130.0
Oysters	48.0	40.0	29.0	81.0	92.0	94.0
Total (all seafood)	**6,267.0**	**6,895.7**	—	**2,534.0**	**3,114.7**	—

Data from the 1981, 1989, and 1992 *Statistical Abstracts of the United States,* published by U.S. Department of Commerce, Washington D.C.

any reason for research in diatom physiology, biochemistry, and genetics to lag as it has in the past. Starr reports 53 diatom cultures representing 32 species in 15 genera in the University of Texas culture collection.

SUGGESTED READING

Smith's *Cryptogamic Botany,* Vol. 1 has a chapter devoted to the morphology and taxonomy of the Chrysophyta. Shorter chapters on the Chrysophyta are found in various botany textbooks.

The Philadelphia Academy of Natural Sciences has published two comprehensive books on diatoms: *Synopsis of North American Diatomaceae* by Boyer, published in 1927, and *The Diatoms of the United States* by Patrick and Reimer, published in 1966.

Plankton and Productivity of the Sea, by J. E. G. Raymont, is an interesting and thorough treatise on marine productivity, including the organisms of greatest significance, how productivity is measured, and the environmental factors that affect rates of primary production in the different oceans.

Several regional monographs have been published, most of them of a purely taxonomic nature. Examples of these are J. R. Carter's *Diatoms from Andorra,* published in 1971 in volume 31 of *Nova Hedwigia;* H. E. Sovereign's *Diatoms of Crater Lake, Oregon,* published in volume 77 of the American Microscopic Society Transactions; and *Diatom Studies of the Headwaters of Henry's Fork of the Snake River, Island Park, Idaho, U.S.A* by R. L. Clark and S. R. Rushforth, published in 1977 by J. Cramer Verlag. The Clark publication is ecologically oriented.

Journal articles on diatoms and other chrysophytes can be found in *Biological and Agricultural Index* in your college library under such headings as "diatoms," "chrysophytes," "algae," and any genus names you are interested in (*Vaucheria, Synura, Navicula,* etc.). All botany students should get in the habit of perusing library copies of *Journal of Phycology* and *Protozoology* on a regular basis.

STUDENT EXERCISES

1. Prepare a table with three columns and four or more rows in which you can indicate the major differences among the three subkingdoms of plants, the Aflagellobionta, the Bracteobionta, and the Lorobionta for each of the following characteristics (and any others you may wish to add): photosynthetic pigments, thylakoid structure, food reserves, and flagella.
2. Prepare a table with four columns and five rows in which you can indicate the differences and similarities among the four classes of Chrysophyta for the following characteristics: photosynthetic pigments, earliest fossil evidence, probable origin in time, thallus organization, cell shape, and most common habitat.
3. In which of the orders of the Xanthophycopsida would you expect to find cells arranged end to end in unbranched filaments? In which order would you expect to find filament-like coenocytes?
4. A long, thin, unicellular plant is observed in a drop of water. It has two golden green plastids. Along its sides there are parallel, linear markings called striations. To what division and class does this plant belong to? To which order?
5. If you were to see two cells of the same species of diatom under the microscope simultaneously, would you immediately recognize them as belonging to the same species? Why or why not?
6. Define *plankton, plankter, phytoplankton, zooplankton.* Five of the eleven divisions of plants and eight of the 29 classes have important plankton species in them. Which are the divisions and classes with important plankters?
7. List the 10 leading countries in fish production in the most recent year for which you can find statistics.

SPECIAL INTEREST ESSAY 9-1

Parallel Evolution of Plankton Species

Unicellular, colonial, and short filamentous algae which are more or less passively carried about by water currents make up the *phytoplankton* of lakes, rivers, and oceans. Plankton species are called *plankters.* Phytoplankters (photosynthetic plankters) occur in 5 of the 11 divisions and in 8 of the 29 classes. Most marine and freshwater aquatic animals depend on plankton for food. In some countries, well over 50% of the meat consumed by humans is from animals of the sea which have fed off plankton.

When species of phytoplankton from different classes are compared, it appears as though similar growth forms have evolved independently several times. Certain prefixes and suffixes have been coined to indicate the plant body type in the algal classes containing species of plankters. The suffix **-monas** refers to flagellated, motile, unicells; **-kont** is an alternate term which means about the same; **-coccus** refers to nonmotile unicells; and **-capsa** indicates cells with gelatinous capsules that tend to attach to each other in **palmelloid** colonies. The prefix **rhizo-** refers to unicellular organisms having amoeboid motility; **siphono-**, used either as a prefix or suffix, refers to a coenocytic (or tubular) nonseptate organization of multinucleate bodies; and the suffix **-thrix,** or **-tricha,** refers to a septate, filamentous type of body organization. The prefix or suffix **gymno-** means naked.

In the Cyanophycopsida, **Chroococcales** and **Tolyptothrix** are examples of how these suffixes may be used. **Heterococcales, Heterosiphonales,** and **Rhizochloridales** are examples from the **Xanthophycopsida** (or **Heterokontae**). **Gymnodiniales, Dinotrichales, Cryptomonadales, Chrysomonadales, Chloromonadales, Chrysocapsales,** and **Ulotrichales** are further examples from other classes of algae.

Ecologically, the ideal place for phytoplankters to be is just below the surface of the water. At the surface, drying is too rapid and space too limited. At great depths, light is poor. Most plankters are slightly heavier than water and possess some type of mechanism to keep them from sinking. Because water is a noncompressible fluid, any object that is heavy enough to sink in surface water will sink all the way to the bottom of the ocean or lake unless a counteracting force can prevent this. Adaptations that prevent complete sinking are discussed by Fogg et al. (1973) and others and include: (1) Phototaxis in flagellated organisms. Many dinoflagellates swim upward toward the light during the day, sink slowly toward the depths at night, and thus maintain themselves in the ideal photosynthetic zone. (2) Spines and other projections that enable cells to take advantage of upwelling currents caused by wind or temperature differences. (3) Fat globules that hold the cell very close to the density of water; usually of little ecological advantage because their density is nonadjustable in most plankters. (4) Mucilage production which offers a relatively rapid rate of adjusting buoyancy. (5) Adjustable gas vacuoles, found only in the blue-green algae, and which quickly increase in size in the presence of light and decrease in the absence of light, enabling many Cyanophycopsida to change position as rapidly as any flagellated organism can do.

SPECIAL INTEREST ESSAY 9-2

Plankton Ecology

Ponds, streams, lakes, and oceans all contain invisible microorganisms in often rather large quantity. Occasionally they become so dense that the water takes on a greenish or other color, but more often the water appears clear. If a drop of this water is examined under the microscope, usually nothing is seen, even though organisms are there in significant numbers. Simple arithmetic explains why. Put one drop of water on a glass slide under an ordinary square cover slip. Observe the size of the cover slip relative to the field of view under low power. The cover slip will be about seven fields square. Seven squared times four divided by π equals slightly more than 60, which is the number of fields under that cover slip. An average of one diatom, or other plankter, per field $\times 60 \times 20$ (the average number of drops per ml water) $\times 1000 =$ 120,000 diatoms per liter, a very high density of organisms for most natural bodies of water.

In the ocean and in lakes, the minerals needed for plant growth—calcium, magnesium, potassium, copper, zinc, ferrous, and other cations, and nitrate, phosphate, sulfate, manganate, molybdate, and other anions—are used by the phytoplankton as rapidly as they become available. They thus become incorporated into the food chain. Many of these minerals remain in the bodies of plants or animals when they die and slowly sink to the bottom of the ocean. Since light can penetrate only 2 or 3 hundred meters into the water, there is no photosynthesis there; nevertheless, there is great activity by living organisms feeding on the detritus raining down from the ocean surface. Finally, decomposition by bacteria releases the minerals in soluble form, but diffusion is slow. Ions of calcium, nitrate, or ammonium require 5 years to migrate to the surface from a depth of 5 km, assuming diffusion at the maximum rate we observe under laboratory conditions. If it were not for the great underwater currents that can sweep the available minerals with them, life at the ocean surface would be even sparser than it is.

Filters, centrifuges, and plankton nets are used to concentrate plankton for microscope and other studies. If the density of all phytoplankters in a body of water is 10,000 per liter, which is a rather high density, the probability of finding one of them in a drop of water chosen randomly from that body and placed on a glass slide under the microscope is less than 50%, providing the entire slide is examined, since there are approximately 20,000 drops of water in 1 liter. If 20 liters of that water are poured through a filter to concentrate the algae and a drop of the concentrate, prepared by washing the filter with 20 ml water, is examined under the microscope, the probability of finding a cell has been increased one thousand fold.

A useful tool for concentrating plankton is a type of plankton net invented over 100 years ago by Charles Darwin (1854). At first employed only in qualitative surveys of the kinds of organisms present in a lake or other body of water, they are now employed in quantitative studies. A number of improvements have been made during the past century, including self-recording equipment. Lead weights of various sizes tied to the tow ropes make it possible to sample plankton at different depths. In practice, a 3-kg lead will put the net at about 100 meters and 7-kg lead will put it at 800 meters depth when the ship is moving at 15 to 20 knots and the nets are tied to a 1000-m tow line. A serious limitation of plankton nets is that the nanoplankton, the plants and animals of very small size, are not held. To use netting fine enough to filter out the nanoplankton is not practical because of the resistance the fine netting presents to water flow. Evaporation and centrifuging are used in connection with plankton nets to gain a better picture of the kinds and numbers of organisms present in open bodies of water.

Plankton nets are easily constructed using a metal hoop about 50 to 75 cm in diameter and fine mesh bolting cloth sewed to form a cone about 120 cm long with a glass jar sewed into the end of the net in which the organisms may accumulate (Wimpenny 1936; Russell-Hunter 1970).

Much of our information about plankton distribution has been obtained from passenger ships and freighters pulling plankton nets across the ocean on routine trips. Some parts of the ocean are rich in plankton, other parts have very little plankton. Where underwater currents meet physical obstacles, such as suboceanic ridges or mountains, or where two currents moving in opposite direction meet, there will frequently be upwellings rich in minerals from the ocean floor. There, zooplankton and fish as well as phytoplankton will be found in abundance. The North Sea, the coasts of Nova Scotia and Maine, the Pacific off Peru and near the Philippines, and large areas adjacent to the Antarctic Continent are famous fishing areas because of upwellings which furnish the phytoplankton with needed minerals.

To better understand the ecological relationships among organisms in the ocean and in lakes and rivers, it is desirable to grow them in the laboratory and study their physiological characteristics. Early attempts to grow planktonic species under laboratory conditions were not very successful, largely because the minerals in the mycological media that were used tended to be too concentrated for good algal growth. Later, success was achieved by adding a small amount of soil to spring water and autoclaving this. Better-defined media were developed by diluting mycological media, after eliminating the sugars and proteins, and adding a few drops of lake or pond water. At the present time, a number of good algal media are available.

The grazing of herbivorous zooplankters often affects the survival of phytoplankters. In the interior of Shark Bay in Western Australia, zooplankters are sparse because the salt content limits their growth and the phytoplankton forms blooms which develop stromatolites. Near the mouth of the bay, grazing by zooplankters prevents the formation of stromatolites. Analysis of this phenomenon has led to better interpretation of stromatolite formation and hence greater understanding of the environment in which ancient organisms evolved.

10 THE FLAGELLATED FUNGI

The potato crop in Ireland looked extra good in the summer of 1846 as harvest time approached. This was cause for rejoicing because the poor—and that was nearly everybody—relied almost entirely on potatoes for food. When wet weather set in and delayed the harvest, no one was concerned. As everyone knew, potatoes require much water. The rains ceased, and as the farmers and their families arrived in the fields and opened the first mounds, they were met by a powerful stench. Their winter food supply had been reduced to a stinking pulp.

Thus began the terrible Irish famine of the 1840s. Before it was over, a million men, women, and children—but mostly children—had died of starvation or of associated nutritional diseases. Another million and a half emigrated from their beloved emerald island, most of them to the U.S.

A flagellated fungus, *Phytophthora infestans,* is the incitant of late blight, the potato disease that brought on the famine. Late blight continues to take its toll in some parts of the world some years, but resistant varieties are now available in most potato-growing areas and the danger of late blight causing another potato famine is rather remote.

The flagellated fungi, or Mastigomycopsida (*mastigo* = a whip, *myceto* = a mold or fungus), also known as the algal fungi or water molds, is an ancient group of plants which show some pronounced affinities to the Xanthophycopsida, especially to the Heterosiphonales, and are therefore included in the Chrysophyta in this book. Other botanists have allied them with the Zygomycopsida, either in a division by themselves or in the division Eumycophyta with the Ascomycopsida and Basidiomycopsida. Still others have placed them in the animal kingdom, and others in the Kingdom Fungi together with the Eumycophyta, the Myxomycophyta, and the Schizomycophyta. Clements (1903) and Clements and Shear (1930) distributed the various genera of flagellated fungi among the families of the Xanthophycopsida.

The general characteristics which the Mastigomycopsida share with the autotrophic Chrysophyta were discussed in the previous chapter. Especially striking are the similarities in flagellation and reproduction. There are also similarities in asexual spore types, in structure of the mitochondria and Golgi apparatus, and in some biochemical pathways, especially those involved in amino acid synthesis. Fossil evidence indicates that they originated about the same time as Xanthophycopsida from similar stock. They differ from the autotrophic Chrysophyta primarily in lacking chlorophyll. Like the Eumycophyta, they are heterotrophic and have osmotrophic nutrition.

CLASS 4. MASTIGOMYCOPSIDA

Flagellated fungi are worldwide in distribution, being especially abundant in cool, moist habitats. Many species are aquatic parasites and saprophytes. Most fungi which parasitize fish, aquatic crustaceans, and filamentous algae, as well as most of those which are saprophytes on the dead vegetation in streams and ponds, including dead algae, are Mastigomycopsida.

To collect algal fungi for laboratory observation, four general methods are commonly employed: (1) collect water from pools high in organic matter, especially the scum that may be on the surface, and examine microscopically for coenocytic fungi; (2) collect plants known or suspected to be parasitized by algal fungi, especially aquatic vascular plants like water lilies, species of *Ranunculus* or *Eleocharis,* etc., and observe carefully for presence of brown dots on the leaves, black streaks on the stems, or other signs or symptoms of infection, and examine these signs under the microscope; (3) bait a pool or stream where algal fungi are expected to occur with pine pollen, sweetgum pollen, boiled paspalum grass leaves, dead insects, or other suitable bait, and after 2 or 3 days, collect and examine microscopically; and (4) prepare a suitable medium, such as potato dextrose agar or corn meal agar to which peptone has been added, and add water from a pool, soil, dead fly, or other likely source of mold. Other media that are suitable for culturing flagellated fungi are various gross water media to which hemp seed, sphagnum moss, or paspalum grass leaves have been added and boiled.

The beginning student usually has more trouble finding and identifying flagellated fungi than other fungi or algae. Lacking chlorophyll, they easily pass undetected even when present unless the light is properly adjusted for the microscope. Students most frequently confuse them with Zygomycopsida; but they are also often confused with Basidiomycopsida and Ascomycopsida from which they should be readily distinguished by their coenocytic nature. However, the septations in the filaments of sac and club fungi may not be detected if the light intensity is too high. Confusion with zygomycetes is more of a problem. If zoospores are present, identification is certain since none of the Eumycophyta have flagellated or motile cells in any stage of the life cycle. Zoospores often are not present, however. The flagellated fungi tend to be more cottony, favor moister habitats, and are simpler in structure than the Zygomycopsida and other Eumycophyta. Zygomycopsida and Ascomycopsida usually produce large numbers of sporangiospores or conidia (nonmotile asexual spores); Mastigomycopsida do not, but may produce an abundance of zoospores (motile asexual spores).

General Morphological Characteristics

The plant body of the algal fungi may consist of (1) a single cell with or without nonnucleate rhizoids, (2) a **polycentric** colony of cells joined by rhizoid-like threads of protoplasm through which nuclei may migrate, or (3) a true coenocytic or filament-like multinucleate cell similar to the coenocytes of *Vaucheria.* Plant bodies composed of prosenchymatous or pseudoparenchymatous tissues, as in most Ascomycopsida and Basidiomycopsida, never occur in the Mastigomycopsida.

Some flagellated fungi are unicellular parasites with the cell lying wholly within a single host cell and with a single nucleus. Just prior to reproduction, the nucleus undergoes mitosis to form a **globose coenocyte.** Other species are multinucleate in the vegetative state. Still others are filamentous, growing between adjacent cells of the host and sending **haustoria** into individual cells where they obtain nourishment. **Saprophytic** species typically grow in the tissues of dead plants or animals in much the same way that **parasitic** species grow in living organisms.

Mastigomycopsida vary in size from slender filaments not much larger than bacteria or blue-green algae to coarse hyphae 0.3 mm in diameter and several millimeters long. The largest fungus known is *Achlya oblongata* with hyphae twice the diameter of human hair and several centimeters long. It was found growing on hemp seed in water and when the seed was picked up the hyphae were so stiff they stood straight out for 15 mm.

Most Mastigomycopsida have flagellated cells at some stage in the life cycle (Figure 10–1). In *Saprolegnia, Peronospora, Lagenidium,* and their relatives there are two flagella present, both usually laterally inserted and often the same length, but one of them pinnate and oriented forward and the other whiplash and trailing. *Protomyces,* which appears to be related to *Peronospora,* has no flagella. In *Rhizidiomyces* and *hyphochytrium,* there is no whiplash flagellum, only a single tinsel flagellum. In *Chytridium* and other chytrids, only a trailing whiplash flagellum is present.

As in other Bracteobionta, the mitochondria are made up of tubules rather than flat plates. **Centrioles** are well developed adjacent to the dividing nucleus, and the nucleolus persists morphologically intact throughout mitosis. In contrast to the Eumycophyta, well-developed **Golgi dictyosomes** occur in all of the Mastigomycopsida that have been examined.

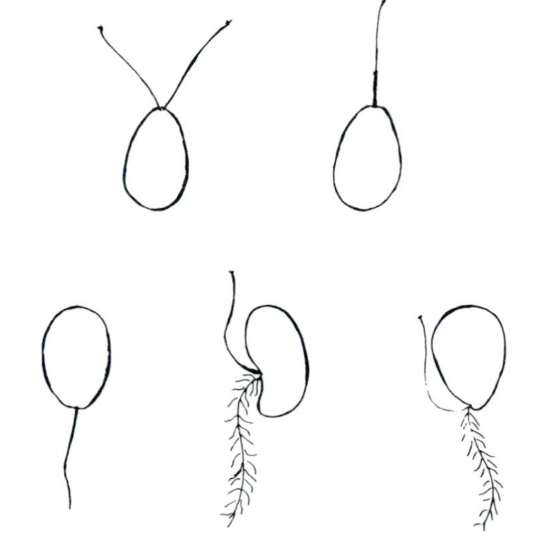

Figure 10–1 Types of flagellation in Mastigomycopsida and related algae: Upper row, left to right, two whiplash flagella; one anterior whiplash flagellum; Lower row, left to right, one posterior whiplash flagellum; two flagella attached laterally, the anterior flagellum whiplash, the posterior flagellum pennate (tinsel-type with two rows of mastigonemes), two flagella attached posteriorly, one whiplash, one trailing pennate.

Asexual reproduction is by means of either zoospores or aplanospores. Zoospores are flagellated spores, aplanospores are nonflagellated spores usually enclosed in a thick cell wall. A few species produce both kinds of spores, but most produce one or the other.

Spiny **aplanospores** or cysts, similar in appearance to the statospores of autotrophic Chrysophyta, are produced in several species (Figure 9–4). In *Saprolegnia ferax,* the cysts are double layered with the outer layer forming first by fusion of membrane-bound vesicles with the plasma membrane. Following fusion, the vesicle contents are released to the cell surface. The wall is thickened by an inner layer formed partly from Golgi vesicles.

Sexual reproduction in many of the algal fungi (Figure 10–2) is similar to that in *Vaucheria* (Figure 9–3) in which a protuberance grows out from a hypha and develops into a multinucleate oogonium. Adjacent to it, a long, slender antheridium develops in much the same way. At maturity, the oogonium develops into a wide pore through which the sperm migrate. In some species, the tip of the antheridium makes contact with the oogonium, adjoining cell walls dissolve at point of contact, and a sperm nucleus with considerable cytoplasm migrates into the oogonium where the zygote is formed (Figure 10–2). In most cases, only one zygote is formed in each oogonium, and the other egg nuclei and sperm nuclei disintegrate. The zygote secretes a thick wall, often ornamented with spines or other projections, thus forming a resistant oospore which can survive adverse conditions. Prior to germination, several nuclear divisions, the first one believed to be meiotic, results in several nuclei in the oospore. The zygote or oospore germinates directly into a hypha.

In many of the chytrids, motile, uniflagellate isogametes unite to form a biflagellate zygote which remains motile for some time until it comes to rest on a host plant and develops a thin wall. The resting zygote sends out a haustorium through which its two haploid nuclei migrate. The invading nucleus remains dikaryotic and dormant for some time; after karyogamy occurs, several nuclear divisions result in destruction of the host cell and infection by asexual zoospores of other cells.

A number of hormones are produced by Mastigomycopsida (Table 10–1). The hormone produced by the female organs has been called **sirenin,** although some botanists identify it and other hormones simply by letters of the alphabet. Sirenin has been purified and its molecular structure has been worked out, but it is not known how it attracts male gametes. RNA is probably involved in sex determination, and in *Allomyces* it is possible to increase the ratio of male to female gametangia by interfering with the production of RNA in a hypha. Carotene is found only in male gametes, and some plant physiologists think it may be involved in sex determination in mastigomycopsida.

Phylogeny and Classification

Formerly the Zygomycopsida were included in the same class (Phycomycetes) as the flagellated fungi; cellular ultrastructure and other characteristics indicate they are more likely related to the Ascomycopsida. The Plasmodiophorales are also included in the Mastigomycopsida by some biologists; however, their mode of reproduction, the nature of their flagella, and other characteristics indicate a relationship with the slime molds, and they are included in the Myxomycopsida in this book. This leaves three groups of fungi, the Oomycetidae, the Hyphochytridae, and the Chytridiomycetidae, in the Mastigomycopsida. The three are generally treated as subclasses; however, the taxonomic position of the last group, commonly called chytrids, is very uncertain; they show affinity to the Zygomycopsida in some ways (Bessey 1950) as well as to the Oomycetidae. Several mycologists have suggested a relationship between some of the algal fungi and the green algae (Bessey 1950), and it is certainly possible that either the Zygomycopsida or the Chytridiomycetidae, or both, evolved from the Chlorophycopsida. More evidence is needed, however, before realigning the chytrids with any other group of fungi or algae.

Good fossils of nonseptate fungi consisting of both hyphae and spores and closely resembling *Saprolegnia* and *Peronospora* are found in middle-Devonian chert (Chesters 1964). Identification of earlier fossils is less certain; however, those found in the Australian Bitter Springs formation and identified as mastigomycetes seem quite convincing (Barghoorn and Schopf 1965). Reports of fossil fungi in the Neohelikian Belt series of Montana are less convincing. It is known that many reports of fungi have turned out to be nothing more than worm burrows, and what at first appear to be very good nonseptate fungi have turned out to be the gelatinous sheaths of cyanophytes. (In stromatolites that are being formed at the present time, it is quite common for *Lyngbya* and other Hormogonales to leave empty sheaths which resemble fungi).

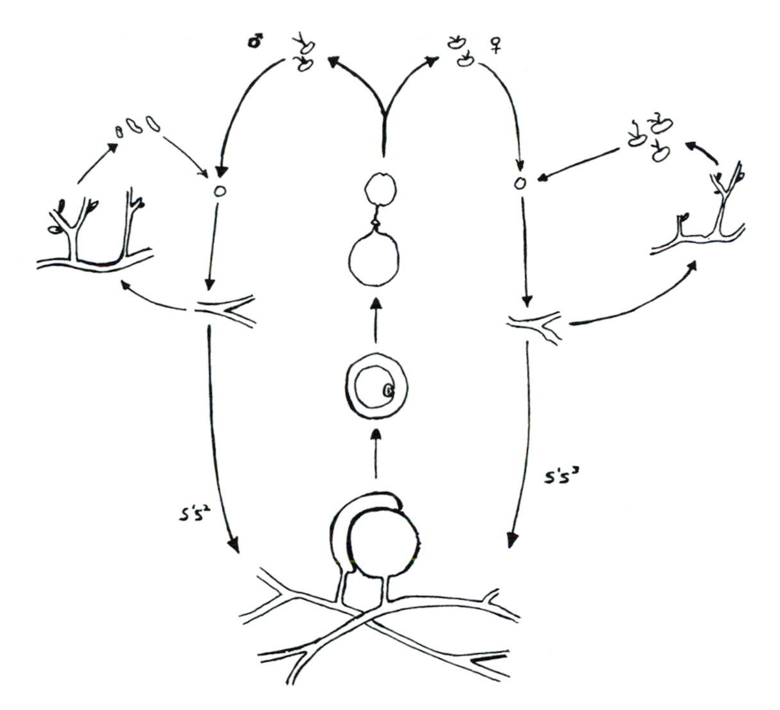

Figure 10–2 Sexual reproduction and life cycle of *Saprolegnia;* compare with Figure 9-3.

TABLE 10-1

SEX HORMONES IN FUNGI

Hormone	Molecular Formula	Molecular Weight	Effect on Fungi	Comments
RNA	$(C_{10}H_{15}O_8N_2PA)_n$	100000+	2:1 = female 1:1 = male	$A=C_{0-1}H_{0-2}N_{0-3}$
Trisporic Acid C	$C_{18}H_{28}O_4$	306	Control of mating type	
Sirenin	$C_{15}H_{24}O_2$	236	Chemotaxis of male gametes	Produced by female
Antheridiol	$C_{22}H_{42}O_4$	470	Chemotropism of antheridia	Produced by female

Function of Antheridiol in Dioecious Species of *Achlya*

A. Produced by female plants; initiates development of antheridia

A_2—initiates

A_1—augments

A_3—inhibits

B. Produced by male plants; initiates production of oogonia

C. Produced by female plants; chemotropism, antheridia to oogonia (C is now believed to be a different manifestation of A)

D. Produced by male plants; delimitation and differentiation of oogonia (septation between oogonium and hypha; oogonium becomes a cyst or endospore, often ornamented or spiny)

Note: The upper part of the table lists the hormones that have been studied in detail. The bottom half shows how the different types of antheridiol function.

In the past, many so-called paleobotanists called everything they were not sure of "a fossil alga" or "a fossil fungus," making these two terms a "refuge for the destitute."

Presumably, the first Mastigomycopsida were species similar to modern Saprolegniales which evolved from *Vaucheria*-like Xanthophycopsida near the end of the Proterozoic Era or possibly earlier. Like their xanthophyte ancestors, they had two flagella, as do the Saprolegniales, Peronosporales, and Lagenidiales of today. Two lines of primarily aquatic fungi have evolved from them: along one line, consisting of the Hyphochytriales, the pinnate flagellum has been retained but the whiplash flagellum has been lost; along the other line, leading to the Chytridiales, the whiplash flagellum has been retained and the tinsel flagellum lost.

The Saprolegniales most nearly resemble the Xanthophycopsida and like them are able to produce amino acids from simple carbohydrates, nitrates, ammonia, and sulfates. In tests of protein similarities, Mez reported cross-compatibility between *Vaucheria* (Xanthophycopsida) and *Saprolegnia* on serological tests (Bessey 1950). The cell walls of Saprolegniales and other Oomycetidae are made of glucans and cellulose; chitin is seldom if ever present. Both chitin and cellulose occur in the cell walls of the Hyphochytridae, but only chitin, with small amounts of cellulose rarely present, occurs in the Chytridiomycetidae.

Although the Saprolegniales are not able to produce sugar from carbon dioxide and water, as the Heterosiphonales do, they are able to produce all of the essential amino acids entirely from simple sugars and minerals in the medium on which they are growing. A medium containing nitrates, ammonia, sulfates, and other essential elements that will support good growth of *Vaucheria* will support good growth of *Saprolegnia* if some sugar is added. The Lagenidiales, Peronosporales, Hyphochytridae, and Chytridiomycetidae have lost this ability and must have some of the amino acids supplied.

Suggested relationships among the seven orders of Mastigomycopsida are illustrated in Figure 9-1, and a summary of their chief characteristics is presented in Table 10-2. There are 25 families, 144 genera, and 1125 species in the class.

Representative Species and Genera

The Saprolegniales are commonly called "water molds," even though they are now known to be as common in wet soil as in water. (The term "water mold" is also commonly applied to all Mastigomycopsida.) They are parasites on algae, aquatic insects, fish, and some flowering plants, or saprophytes on plant or animal matter in freshwater and occasionally brackish water. *Saprolegnia* and *Achlya* are the best known genera.

Saprolegniales grade into the Peronosporales, all of which are obligate parasites, many of them on crop species. *Phytophthora infestans* is the incitant of late blight disease in potatoes, the disease responsible for the Irish Famine of the 1840s. Closely related are the downy mildew, damping-off, and root rot organisms, *Peronospora, Pythium,* and *Aphanomyces.* Both orders are characterized by large, coarse, filamentous species having cellulose in the cell walls and reproductive structures similar to those of the Heterosiphonales.

The Lagenidiales, sometime called the Ancylistales, are mostly parasites, closely related to the Saprolegniales but much smaller. *Lagenidium giganteum* parasitizes mosquito larvae and has shown promise in controlling *Culex tarsalis,* the mosquito which causes encephalitis (Washino 1980).

TABLE 10-2

GENERAL CHARACTERISTICS OF THE ORDERS OF MASTIGOMYCOPSIDA

Orders (Fam–Gen–Spp)	Thallus Characteristics	Flagella	Other Characteristics	Representative Genera
Saprolegniales (3-35-160)	Filamentous coenocyte; septate gametangia and injured areas	Anterior flagellum pinnate, posterior flagellum whiplash	Cell walls contain cellulose; saprophytes, or parasitic on small animals	*Achlya, Saprolegnia, Aplanes, Aphanomyces, Sapromyces, Rhipidium, Pythiopsis*
Lagenidiales (5-25-100)	Filamentous coenocyte; holocarpic (entire thallus becoming reproductive structures	Anterior flagellum pinnate, posterior flagellum whiplash, both attached laterally	Parasites within cells of algae and of some aquatic animals	*Lagena, Olpidiopsis, Woronina, Sirolpidium, Aphanomycopsis*
Peronosporales (3-20-500)	Slender filaments with well-differentiated oogonia and antheridia	As in the above orders	Parasites on angiosperms; also saprophytes and parasites on algae and aquatic animals.	*Phytophthora, Pythium, Peronospora, Albugo, Sclerospora, Zoophagus, Bremia, Plasmospora*
Hyphochytriales (3-6-15)	Unicellular to short filamentous; cellulose often present	One pinnate flagellum anteriorly oriented, attached anteriorly	Parasites on seaweeds, algae, and fungi	*Anisolpidium, Reesia, Rhizidiomyces, Cystochytrium*
Monoblepharidales (1-3-10)	Multinucleate, coenocytic filaments with terminal oogonia and lateral antheridia	One whiplash flagellum, attached posteriorly	Saprophytes in freshwater ponds, lakes, or streams; nonmotile female, motile male gametes	*Monoblepharis, Gonapodya, Monoblepharella*
Blastocladiales (3-5-40)	Club-shaped to branched or unbranched, slender coenocytes; sexual reproduction by isogametes	One whiplash flagellum, or, occasionally, 2 whiplash flagella attached posteriorly	Cell walls contain chitin; large, nuclear cap adjacent to nucleus; parasites on worms, fungi, and mosquito larvae	*Allomyces, Catenaria, Coelomyces, Blastocladia, Blastocladiella*
Chytridiales (7-50-300)	No true mycelium, unicellular or having a rhizomycelium (haustoria lacking nuclei which radiate from a central cell)	One whiplash flagellum, attached posteriorly	Cell walls contain chitin; parasites on vascular plants, algae, fungi, eggs and larvae of invertebrates	*Olpidium, Synchytrium, Micromyces, Physoderma, Chytridium, Karlingia, Rhizidium, Rozella, Entophlyctis, Siphonaria*

Note: The number of families (Fam), genera (Gen), and species (Spp.) in each order is indicated in the first column.

The Hyphochytriales, formerly included in the Chytridiales because of a number of similarities in gross morphology, are characterized by a single tinsel flagellum and the same ultrastructural and cell wall features as in the Oomycetidae. *Hyphochytrium infestans*, the species for which the order was named, was found growing on the apothecia of a species of *Helotium* (Ascomycopsida) in 1884 and described and named by Zopf, a famous German mycologist. It has never been seen since.

The Blastocladiales seems to be the link between the Oomycetidae and the chytrids. Most members of the order are aquatic or soil saprophytes. The Monoblepharidales are also saprophytes in fresh water, differing from the Blastocladiales in having heterogametic sexual reproduction. The Chytridiales are structurally the simplest of all the Mastigomycopsida and believed by many mycologists to represent the stock from which the other flagellated fungi have descended. Here they are considered to represent the end product along a line of increasing parasitism accompanied by simplification. They are largely parasites on algae, other fungi, worms, leaves, stems, or roots. Many of the species are saprophytes on plant or animal material in wet soil or in aquatic habitats. *Olpidium* and *Rozella* in the Chytridiales and *Allomyces* in the Blastocladiales are among the best-known genera in the Chytridiomycetidae. *Olpidium viviae* is parasitic on the leaves and stems of some species of vetch, and *Rozella allomyces* is an obligate parasite of several species of *Allomyces*.

Ecological Significance

Ecologically, the Mastigomycopsida are important as saprophytes in aquatic habitats, especially in freshwater, and in damp, cold soil. As parasites, they influence the survival of many species of plants and animals, especially algae, but also flowering plants, insects, crustaceans, and nematodes. The genus *Olpidium*, for example, has species that are (1) parasites on freshwater algae, (2) parasites on marine algae and seaweeds, (3) parasites or saprophytes on pollen grains that fall in the water, (4) parasites on small aquatic animals, (5) parasites on insect eggs in the water, and (6) parasites on the roots of various angiosperms, such as cabbage and mustard.

The "crayfish plague" is caused by a member of the Saprolegniales, *Achlya prolifica*. During the first half of the 20th century, it spread in the streams of central and southern Europe until the crayfish became extinct in much of its former habitat. It reached Scandinavia about 1970 and has now, therefore, effectively eliminated commercial crayfish harvesting from all of Europe.

There are a number of species of Lagenidiales and Saprolegniales which are parasitic on mosquito larvae, but the extent to which they actually limit the survival and distribution of these pests is not known. However, 100% kill of larvae has been reported in some cases of manmade epidemics (Federici et al., in California Agriculture, 1980), and they show promise in controlling some species which cause human disease.

Economic Significance

Plant diseases annually take a toll of several billion dollars in the U.S. alone or over $20 for every man, woman, and child. (See Special Interest Essay 10–1.) In some countries the toll is considerably higher on a per capita basis. To an individual farmer, the cost may be ruinous, running into the tens of thousands of dollars and causing bankruptcy. To the rest of us, the cost is reflected in higher prices for farm products. The Mastigomycopsida are important contributors to this wastage. Some of the diseases caused by flagellated fungi include late blight of potato and tobacco wilt, both caused by species of *Phytophthora* (Figure 10–3); black wart of potato; crown wart of alfalfa; brown spot of maize; white rust of crucifers; damping-off of greenhouse plants; downy mildew in grapes, grasses, lettuce, onions, peas, clover, and melons; and root rot in peas (Table 10–3).

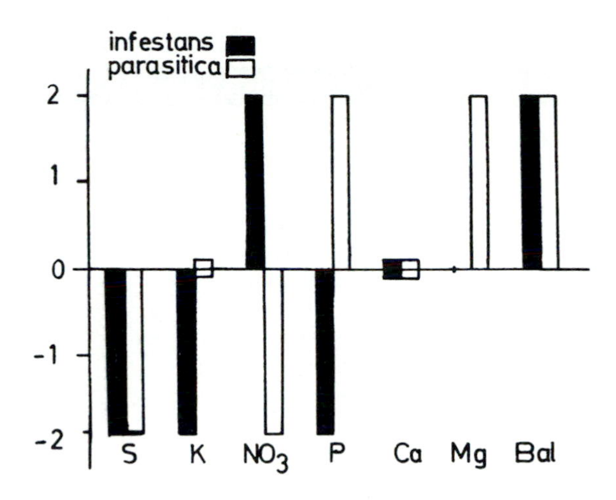

Figure 10–3 Results of laboratory tests on mineral requirements of two fungi as part of a program to learn more about their control. *Phytophthora infestans* is the incitant of late blight disease in potatoes; *Phytophthora parasitica* causes a damaging disease in tobacco.

Phytophthora infestans, the causal agent of late blight, was simultaneously reported in Europe and the U.S. in the early 1830s. This disease was the main cause of the "Irish Famine" of 1846 through 1848, which has probably had a greater effect on American history than any other event in Europe of that period, including wars and all of the acts of kings and prime ministers. For at least 15 years, it was the main contributor to the "America Fever" epidemic that swept the continent and brought millions to America's shores. Late blight spread from Ireland to Poland, northern Germany, and Denmark during the late 1840s and 1850s, causing famine and hardship wherever it went. Attempts to control it were largely unsuccessful until resistant varieties were developed, through application of the theory of evolution, midway through the 20th century. Success in growing the fungus in laboratory cultures has been instrumental in developing disease-resistant varieties and disease control methods (Figure 10–3)

Damping-off disease, usually caused by species of *Pythium* and *Aphanomyces,* especially *Pythium debaryanum,* cause widespread damage in greenhouses and, sometimes—though fortunately very rarely—in other crops, and rarely even in lawns (Figure 10–4). It is usually controlled, if caught early enough, by holding off water, opening windows, and improving ventilation in general. It is very destructive, causing a blackening of the lower stem and roots with death of roots quickly resulting. Frequently, it spreads and does its damage even though all the foregoing precautions are taken. Once the organism becomes well established in the wooden trays commonly used in greenhouses, it seems to be impossible to eliminate it even when fungicides are applied and the trays are thoroughly steam sterilized.

Of special interest to plant pathologists is a disease that nearly ruined the French wine industry in the mid 1800s, grape mildew caused by *Plasmopara viticola.* According to a popular legend, a farmer in Bordeaux whose grapes were dying was whitewashing his outbuildings one day and some whitewash was left over. He dumped the excess into a bucket containing blue vitriol which he had used to treat his wheat seed the day before. This mixture of copper sulfate and calcium hydroxide he threw on to his dying grapes. To his surprise, the grapes recovered, and Bordeaux mixture, the first important fungicide, was invented. Regardless of the truthfulness of the legend, Bordeaux mixture has been a boon to grape growers for well over 100 years.

TABLE 10-3

SUMMARY OF SOME SERIOUS PLANT DISEASES

Crop	Disease	Causal Agent	Control
Corn	Corn smut	*Ustilago maydis* BAS	Res, Rot
	Dry rot	*Diplodia zeae* ASC	Res, Rot, San
	Pink ear rot	*Gibberela zeae* ASC	Res, Rot
	Stewart's disease	*Bacterium stewartii* SCH	Res
Rice	Helminthosporium leaf blight	*Helminthosporium turcicum* ASC	Res
	Culm rot	Drain fields	
	Leptosphaeria salvinii ASC	Insect-transmitted virus VIR	Quarantine, leafhopper control
	Stunt		
	Kernel smut	*Tilletia horrida* BAS	Res (?)
Cotton	Soreshin	*Pellicularia filamentosa* BAS	Rot, Chm
Wheat	Stem rust	*Puccinia graminis* BAS	Res, Control of barberry
Rye, and Barley	Bunt or covered smut	*Tilletia caries* BAS	Chm, Rot
	Loose smut	*Ustilago nuda* BAS	Res
	Scab	*Gibberella zeae* ASC	Res, Rot
	Take-all	*Ophiobolus graminis* ASC	Rot, Fer
	Snow mold	*Typhula itoana* BAS	Chm
	Ergot	*Claviceps purpurea* ASC	Certified seed
Oats	Stem rust	*Puccinia graminis* BAS	Res, Control of barberry
	Victoria blight	*Helminthosporium victoriae* ASC	Res
Sorghum	Bacterial streak	*Xanthomonas holcicola* SCH	Chm, Rot, Res
	Milo root rot	*Periconia circinata* ASC	Rot, Res
Alfalfa	Bacterial wilt	*Corynebacterium insidiosum* SCH	Rot, Res
	Witch's broom	Insect and dodder transmitted virus VIR	San, leafhopper control
Tea, coffee, and citrus	Red rust	*Cephaleures virescens and C. parasitica* CHL	San Fer
Apple	Fireblight	*Erwinia amylovora* SCH	Res, San, Fer
Potato	Ringrot	*Corynebacterium sepedonicum* SCH	San, certified seed
	Common scab	*Streptomyces scabies* SCH	San, Fer, Rot
	Powdery scab	*Spongospora subterranea* MYX	Chm, Res
	Early die	*Verticillium alboatrum* ASC	Rot
	Late blight	*Phytophthora infestans* MST	Res, Rot
	Golden rootknot	*Heterodera rostochiensis* NEM	Quarantine, long Rot
Nursery plants	Damping-off	*Pythium debaryanum MST*	Good ventilation, San

Note: In 1985, the estimated cost of all plant diseases in the U.S. was $10,000,000,000. Costs include reduced yields, expenses of spraying and other control methods, and increased prices of farm products. (VIR = virus, SCH = bacterium, ASC = sac fungus, BAS = club fungus, ZYG = zygomycete, MYX = slime mold, MST = flagellated fungus, CHL = green alga, NEM = nematode; Res = use of resistant hybrids or varieties, Rot = crop rotation, Chm = chemical sprays, Bio = biological control, San = sanitation, Fer = good fertility.)

Research Potential

The Mastigomycopsida have been useful in studying how the factors in the disease complex interact to produce disease. In some species, a wide variation in virulence has been noted among strains of the fungus; consequently, a potential exists for studying the inheritance of virulence in pathogens to go along with our studies of inheritance of resistance in the host species. It has been proposed, for example, that some varieties of mustard are resistant to white rust because of the presence of sulfur compounds that inhibit the causal organism, *Albugo candida.* The hypothesis is supported by the observation that strains high in sulfur compounds are more resistant than those with average or low levels. Nevertheless, the hypothesis does not explain all of the observations made in connection with this disease organism. When mustard is grown on soil deficient in sulfate, for example, all strains were low in these sulfur compounds, yet the differences in resistance were the same as before.

Figure 10–4 Damping-off disease caused by *Pythium* sp. in a lawn. Upon observing the first symptoms, wilting and yellowing of the lawn, the owner watered it more heavily, which only caused the disease to spread faster. It was controlled by digging up the entire lawn, applying a strong fungicide, waiting a year for the fungicide to take effect, and then planting a new lawn. The disease is very rare and did not spread to any neighbors' lawns, stopping when it reached a concrete curb or driveway.

In recent years, growth factors, including the so-called plant hormones, have been extensively studied. Plant hormones are mostly relatively simple organic acids and are responsible for a variety of tropic and tactic responses as well as differentiation of the developing proembryos and embryos. Animal hormones, on the other hand, are of a quite different nature, mostly belonging to a family of lipids called steroids, and are primarily involved in differentiation of sexual organs. Sex hormones in fungi were first investigated in *Saprolegnia.* Since then, several hormones have been studied in some detail (Table 10–1). From these studies have developed hypotheses as to the mode of sex differentiation and development of reproductive organs in plants. In contrast to the growth factors or hormones of the Eumycophyta, which are like those of other plants, mastigomycete hormones seem to show similarities to animal hormones.

SUGGESTED READING

Introductory Mycology by Alexopoulos and Mims (1979) is an interestingly written and authoritative book about all groups of fungi and is highly recommended to all botany students. Several textbooks on plant pathology are available: Walker's *Plant Pathology* (1950) and Chester's *Nature and Prevention of Plant Diseases* (1952) are especially recommended as is the 1953 U.S.D.A. Yearbook of Agriculture, *Plant Diseases.*

STUDENT EXERCISES

1. Prepare a flask of hemp seed medium by boiling half a hemp seed in a weak solution of minerals, sugar, and peptone. When the flask has cooled, add a dead fly, wait 2 or 3 days, and examine for the presence of fungi on the body of the fly or on the hemp seed. For variation, divide the medium into several portions and try different sources of infection to each.

2. From your local county agricultural agent or other source, such as direct conversation with farmers of the area, find out what the most destructive plant diseases are in the area in which you live. From your textbook and other sources, classify these to class and order, and where possible, family.

3. The price farmers receive for a crop increases when the supply decreases and/or the demand increases; with increasing supply and decreasing demand, prices go down. The following formula is a rough approximation of the relationship:

$$P = (1 + D) / (1 + S)^n$$

   where P = change in price, D = change in demand, S = change in supply, and n = measure of the inelasticity in demand for a product (if demand is completely elastic, n = O).

   Assume that an epidemic of late blight disease has destroyed 20% of the potato crop in Poland (S = −0.2). Demand for potatoes, in Poland, is very inelastic (n = 2, D = O). Will the price of potatoes go up or down? How much? If potatoes cost 33 zloty on the Warsaw market a year ago, before the epidemic, what do you predict they will cost this year?

4. What is the best way to control damping off disease in the tomato seedlings you have growing in your home garden hotbed or coldframe?

5. In terms of the number of crops subject to different categories of plant diseases, what is the relative importance of Basidiomycopsida, Ascomycopsida, Mastigomycopsida, Myxomycopsida, Schizomycopsida, and viruses as causal agents?

6. Compare the life cycle of *Vaucheria* with the life cycle of *Phytophthora* as outlined in this chapter. In what ways are they similar? In what ways are they different? Draw your own diagram of the life cycle of *Phytophthora,* assuming that meiosis occurs as the zygote germinates as indicated by Olive (1965) in his description of the Saprolegnialean life cycle. How does the *Vaucheria* life cycle compare with this diagram of *Phytophthora*?

7. Several species of *Coelomyces* and *Lagenidia* are known to parasitize mosquito larvae. (a) Which order or orders do these genera belong to? (b) Outline an experiment to test the value of a species of *Coelomyces* in controlling the anopheles mosquito (which spreads malaria) under field conditions. Be sure to indicate how you would replicate and randomize your experiment, what precautions you would take to prevent *confounding,* and how you would measure your results.

SPECIAL INTEREST ESSAY 10-1

Plant Pathology

Plant pathology, or phytopathology, is the scientific study of the causes, nature, and control of plant diseases. Plant productivity is essential to the maintenance of human civilization, and indeed all animal life. Plant pathology is primarily concerned with maintaining and increasing plant productivity.

The causes of plant disease can be summarized under the following heads: (1) soil, weather, and other environmental factors; (2) viruses; (3) bacteria; (4) parasitic green algae; (5) sac fungi and fungi imperfecti; (6) club fungi; (7) slime molds; (8) flagellated fungi; (9) protozoa; (10) nematodes; and (11) parasitic angiosperms. Injury caused by insects, rodents, birds, or competition from autotrophic plants is generally not considered plant disease or within the scope of plant pathology. Examples of plant diseases and their causal agents are presented in Table 10-3. One of the most destructive of these is the flagellated fungus, *Phytophthora infestans,* the incitant of late blight disease of potatoes.

A plant parasite is an organism that lives in or on a plant and obtains some or all of its nourishment from the host. Not all parasites according to this definition, are pathogenic or harmful. There are several degrees of pathogenic parasitism. Some parasites are destructive, others are balanced parasites, The former, when inoculated into a host plant under the right environmental conditions, multiply rapidly and are so damaging that they soon kill the host; the latter allow almost normal development of the host plant, multiplying and growing in harmony with the growth and development of the host. Most parasites are somewhere between the extremes.

The distinction between **parasite** and **saprophyte** is not sharp. Many club fungi, for example, live only on the dead tissue of woody angiosperms and are thus saprophytes. However, even though most of the wood in a living tree stem is dead xylem tissue, it is essential to the health of the total tree since it provides strength and support. Furthermore, the outermost dead cells conduct water and minerals to the photosynthesizing cells in the leaves. A saprophyte living in dead xylem tissue in a live tree stem is therefore pathogenic. Many saprophytes, such as the common bread mold, *Rhizopus stolonifera,* can become weakly parasitic at times, and are called facultative parasites. Dormant organs, such as roots, are the structures they most readily parasitize. Other pathogens, such as corn smut, can grow on dead organic matter if conditions are just right even though they are normally parasites; they are called facultative saprophytes. Then there are other parasites that cannot be cultured on non-living materials, such as the rusts, many of which are destructive on cereal crops. These are obligate parasites. At the other extreme, some soil organisms never grow on living tissue and are called obligate saprophytes. Obligate parasites and obligate saprophytes are rare; most fungi live somewhere on the broad spectrum between.

People often are confused as the meaning of the terms **parasite** and **pathogen.** A parasite may or may not produce disease symptoms in its host organism; if it does, it is a pathogen. A pathogen is not *the* **cause** of a disease, but is part of the **causal complex,** along with temperature, humidity, mechanical injury, and several other environmental factors. It is not strictly correct to call *Phytophthora infestans* the cause of late blight disease; rather it is an **incitant** or a **causal organism.** If *P. infestans* is present, late blight will develop *if* humidity and temperature are both high and other environmental factors are right for the development of the disease. The study of the combination of factors that lead to an outbreak of a disease is called the **etiology** of the disease.

The fact that living organisms are part of the causal complex of plant disease is a relatively recently discovered concept. Until the middle 1800s, most biologists attributed disease in plants to the weather, the soil, or "acts of God." Farmers who believed that fungi caused plant disease were labeled superstitious by most botanists. One of the leaders in developing the germ theory of disease was Anton de Bary who, in the late 1800s, demonstrated that the fungus commonly seen on barberry leaves was, in fact, the cause of stem rust disease in wheat, just as "superstitious" farmers had believed for years.

Methods of disease control include (1) use of chemicals, as in the control of bunt disease by treating the seed with copper carbonate, (2) elimination of an alternate host, as in the case of barberry for wheat rust control, (3) crop rotation, as in corn smut or flax wilt control, (4) quarantine, as in the control of citrus canker in Texas, Arizona, and California, (5) plant breeding, as in the control of most strains of smut and rust in cereal crops and bacterial wilt in alfalfa, (6) control of insect vectors, as in elimination of the white fly to control curly top disease in sugar beets, (7) ecological control, as in choice of early planting for control of Gibberella rot in barley or late planting for control of the same disease in corn, and (8) protection methods, as in proper fertilization and irrigation for control of scab in potatoes.

When plant diseases are not controlled, crop losses can be very serious. Fire blight has destroyed valuable orchards over a period of a few years; millions of dollars are spent annually to control rust and other diseases of cereal crops. But the costs of plant disease often go beyond the purely economic cost. Downy mildew of grape once threatened to destroy the wine industry of France and Germany, and late blight of potatoes led to widespread famine and human suffering in several countries in the mid 1800s.

The challenges which face plant pathologists can often be discouraging and even frustrating. Inspired by Charles Darwin's presentation of the theory of evolution, a young biologist by the name of Torbitt developed a plan for creating varieties of potatoes resistant to the late blight organism. Late blight was still endemic and destructive over much of northern Europe in the 1860s when Torbitt began his potato improvement projects. However, neither he nor any other biologist or agronomist of the 19th century had any success in controlling late blight. Until the early 1930s, when the University of Maine finally released a resistant variety, all a potato farmer could do was plant in a field where no plants of the nightshade family had grown for a few years and pray that the season would not be too wet (Stevenson and Clark 1937).

Torbitt's plan was good, except for one detail. He (and other biologists of the time including Darwin) lacked an understanding of the basic principles of genetics and the importance, therefore, of sexual reproduction to the establishment of new races, or new breeds of livestock, or new varieties of crops. Potato varieties are clones. They are asexually propagated. Except for rare somatic mutations, every individual member of a clone is genetically identical to every other member of that clone. But the little berries that form from the blossoms of a few potatoes in every field contain seeds that have resulted from sexual recombination. Potato berries from all the known varieties and from wild plants growing in Mexico and in the northern Andes were collected by agronomists in Maine. From the thousands of seedlings grown, a few were found to survive in greenhouse flats infected with *Phytophthora infestans.* Some of these survivors were "escapes" and of no value to the plant breeders, but some possessed resistance genes. Hybrids between the latter and high yielding, high quality varieties led to the release of good, disease-resistant potatoes.

11 KELPS AND OTHER BROWN SEAWEEDS

Among the largest of all plants, or at least the *longest,* are the giant kelps of the northeast Pacific Ocean. On the Alaska coast, kelp plants 50 meters long are frequently found. Early explorers reported finding plants over 100 meters long; either human activities have altered the environment to the extent that the giant kelps no longer grow that big, or else the explorers were exaggerating. Many conservationists believe the former. Either way, there is no question that exploitation of marine resources by greedy humans, harvesting fish, whales, and fur-bearing animals without regard to the total environment, has had its effect on the Phaeophycopsida.

When it was discovered that the fascinating forests of kelp off the coast of California were a lucrative natural resource, harvesting of these plants rapidly increased until they were threatened with total destruction in the 1950s as the result of human activities. Even now, they are not out of danger. It is not clear as to whether overharvesting, water pollution, or elimination of the sea otters which prey on kelp grazers was the major contributing factor, but it is clear that man's exploitative activities were involved.

The Phaeophyta (*phaeo* = dusky, *phytum* = a plant), or brown algae, are the most complex and highly evolved of the Bracteobionta, and are among the most efficient of all plants in photosynthesis and conduction of food. The level of cellular differentiation and physiological specialization of their structures is comparable to that in the Tracheophyta.

As in the Pyrrophyta and Chrysophyta, brown pigments, especially fucoxanthin, mask the green of chlorophylls *a* and *c;* colors in brown seaweeds consequently range from olive green, occasionally almost grassy green, through rich golden browns, to almost black. The plastids are usually small, smooth, and discoid, but some genera have large, flat plastids which may be irregular in shape. The thylakoids are stacked in threes or fours and the resulting layers, or lamina, tend to be more consistently parallel oriented than in the other two divisions. Pyrenoids are seldom present.

Flagellated gametes and meiospores are common; the flagella are laterally inserted with a forward-oriented tinsel flagellum and a trailing whiplash. Cell walls are two layered; the inner layer is of cellulose and the outer of pectin-like phycocolloids such as algin and fucoidin. Golgi bodies, tubular mitochondria, centrosomes, and centrioles are typical of the subkingdom.

Brown seaweeds are the dominant plants in marine habitats along the continental shelves and in estuaries and bays of the colder parts of the temperate oceans. Although as a general rule they become less and less conspicuous as one travels from Arctic and Antarctic areas toward the equator, they do occur in many topical waters; for example, the gulf weed, *Sargassum,* is abundant in a great eddy in the Atlantic just north of the equator known as the Sargasso Sea. Suggested lines of relationship among the eleven orders are shown in Figure 11-1 and general characteristics of the orders in Table 11-1. The 900 species are all included in one class.

CLASS 1. PHAEOPHYCOPSIDA

The brown algae, with the exception of four rare freshwater species found in Europe, are all marine organisms. They are among the largest and structurally most complex of all plants, many species having stipes 50 m long or longer anchored by strong holdfasts to the rocks on the ocean bottom and with large, flat fronds kept afloat by unique structures called **pneumatocysts** (Figure 11-2). The largest of the brown seaweeds, and structurally the most complex, are the kelps. Kelp is harvested commercially in many areas, including the west coast of North America.

Students have little difficulty in recognizing brown seaweeds. Occasionally they may confuse them with red seaweeds or some of the Chlorophycopsida or Siphonophycopsida, but the color is distinctive enough that this is seldom a problem. Any plant possessing pneumatocysts is a phaeophyte, but not all phaeophytes have pneumatocysts. Occasionally, brown seaweeds are so green as to cause confusion with chlorophytes but none of the green algae attain the level of tissue and organ differentiation commonly found in the browns.

The best time and place to observe brown seaweeds is at low tide on a rocky shore. Many species grow in the intertidal zone, and others can be found anchored to the rocks of the subtidal zone. Sandy beaches and mud flats have very few seaweeds on them except for those that have been torn loose by wind and wave from other places and washed ashore. Many brown seaweeds are annuals; most of the perennials are deciduous, losing most of their fronds during the winter storms; consequently, spring and early summer are the ideal seasons for collecting. Simple dredges can be used to collect seaweeds at greater depths, and scuba diving makes many species available that the "hip-boot algologist" never sees. Some areas where the best seaweeds grow have treacherous tides and currents that the novice needs to be aware of; it is easy to become engrossed in collecting and fail to observe the passing of time and flow of another tide.

Morphological and Physiological Characteristics

Although a few brown algae—e.g., species of *Ectocarpus, Pylaiella,* and *Ralfsia*—are filamentous, the plant body of most Phaeophycopsida is differentiated into broad, flat **lamina** or **blades** which are efficient food producers, much-branched **holdfasts** which anchor the plant firmly to the ocean bottom, and cylindrical or sometimes flattened **stipes** which connect the two and conduct food from the organs of productivity to the organs of anchorage (Figure 11-2). Fertile blades are sometimes called **fronds.** Unique gas-filled floaters or **pneumatocysts,** found only in Phaeophycopsida, aid in holding the blades at or near the surface. Morphologically, blades, stipes, and holdfasts very nearly resemble leaves, stems, and roots,

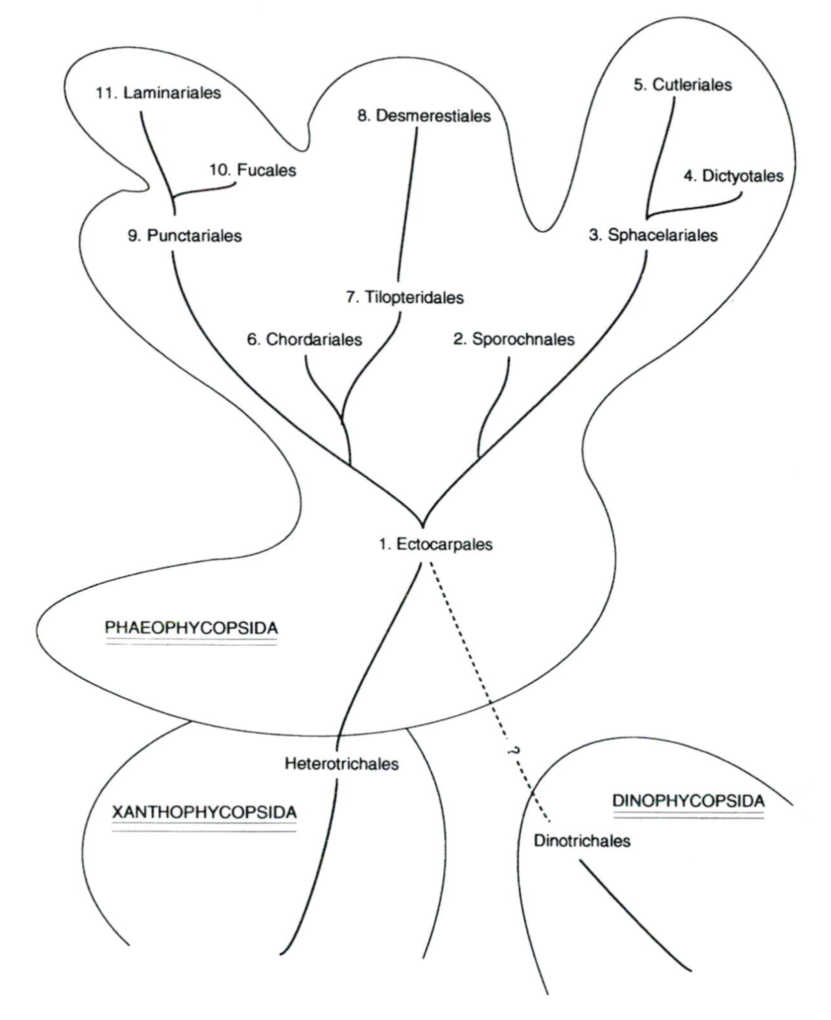

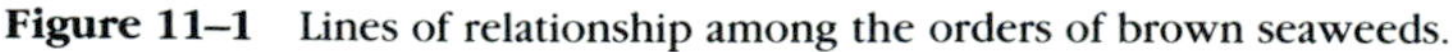

Figure 11–1 Lines of relationship among the orders of brown seaweeds.

respectively, of vascular plants; physiologically they function in an almost identical way. Anatomically, however, the tissues are arranged differently, and they have had a different pattern of development. Structures having different origin and/or development but similar functions are called **analogous** structures to distinguish them from **homologous** structures, which have the same origin and pattern of development, whether the function is the same or not.

Tissues in many of the brown algae are **pseudoparenchymatous;** in many others, they are predominantly **prosenchymatous.** Both of these types of tissue are basically filamentous. In the kelps and fucoids, true **parenchyma** tissues, resulting from cell divisions in all three dimensions, are found.

Anatomical structure is most complex in the Fucales and Laminariales. The two orders resemble each other in many ways including anatomy of the mature plants as well as

TABLE 11–1

GENERAL CHARACTERISTICS OF THE ORDERS OF PHAEOPHYCOPSIDA

Orders (Fam–Gen–Spp)	Life Cycles	General Morphological Characteristics	Habitat	Representative Genera
Ectocarpales (2-30-120)	Isomorphic diplohaplontic; syngamy of biflagellate isogametes	Hemispherical "fuzzy" cushions on rocks, shells, etc., made up of uniseriate, branching filaments sometimes compacted into pseudoparenchyma tissue	Tropical and temperate oceans; 4 species in European streams and freshwater lakes	*Ectocarpus, Pylaiella, Bodenella, Rolfsia, Lithoderma*
Sporochnales (1-6-24)	Sporophyte prominent diplohaplontic; oogamous; parthenogenetic development of sporophyte	Growth from intercalary meristematic cell; sporophyte a dichotomously branched, flattened cylinder with prominent midrib, ending in tuft of hairs; parenchymatous cortex and medulla	Tropical oceans; southeastern U.S. shores, Caribbean, Atlantic coast of Europe, Australia	*Neiria, Sporochnus, Carpomitra*
Sphacelariales (4-8-35)	Isomorphic diplohaplontic; isogamous; asexual reproduction by propagula	Growth polysiphonous with large apical cell; rows of cells in distinct tiers	Tropical oceans; warmer waters of the Atlantic coasts of North America and Europe	*Chaetopteris, Cladostephanus, Batteria, Sphacelaria*
Dictyotales (1-18-100)	Isomorphic diplohaplontic; oogamous	Flattened, erect stalks and numerous blades; parenchymatous. Male gametes uniflagellate	Mediterranean, Brazil, Indian Ocean, and other tropical seas	*Zonaria, Padina, Dictyota*
Cutleriales (1-2-4)	Gametophyte prominent diplohaplontic; oogamous. Sporophytes small cushions up to 1 mm thick	Erect plants, typically about 20 cm tall with flat, dichotomously branched blades with epidermis-like outer cortex, small-celled inner cortex, and large-celled medulla	Warm waters along Atlantic coast of Europe in other tropical waters	*Cutleria, Zonardia*
Chordariales (8-20-100)	Sporophyte prominent diplohaplontic, isogamous; gametophytes microscopic	Loose prosenchymatour sporophyte, often crustose or forming globose cushions, or sometimes erect and branched	North Atlantic, Baltic, North Pacific and other temperate oceans; some species epiphytic on green algae	*Myrionema, Chordaria, Compsonema, Acrothrix, Mesogloi, Haplogloia, Leathes, Chordariopsis*

Tilopteridales (2–5–10)	Isomorphic diplohaplontic; oogamous	Trichothallic growth with *Sphacelaria*-like base and *Ectocarpus*-like apex	North Sea, Baltic, and other cold temperate oceans	*Haplospora, Tilopteri, Acinetospora, Scaphspora, Masonophycus*
Desmerestiales (2–3–22)	Sporophyte prominent diplohaplontic; oogamous with motile sperm; syngamy within oogonium where embryo begins development	Growth trichothallic with single filament at each growing apex and pseudoparenchyma below	North Pacific, North Atlantic, and antarctic waters	*Arthrocladus, Phaeurus, Desmerestia*
Punctariales (6–19–50)	Sporophyte prominent diplohaplontic, usually isogamous	Some species foliose, others cylindrical to globose; cell division intercalary but not localized in meristems, producing true parenchyma	Mediterranean, North Atlantic, North Pacific, and other cold and warm temperate oceans; some species epiphytic	*Giraudia, Punctaria, Stictyosiphon, Dictyosiphon, Myrotrichia*
Fucales (7–23–430)	Sporophyte prominent modified diplohaplontic, usually described as diplontic; oogamous, unicellular gametophytes	Parenchyma tissue differentiated into distinct meristoderm, cortex, and medulla tissues; growth from an apical meristem; microsporangia contain 64 sperm; megasporangia 1–8 eggs	Mostly temperate and arctic seas; some tropical species	*Ascophyllum, Fucus, Pelvetia, Hesperophycus, Cystoseira, Sargassum, Durvillea, Ascoseira*
Laminariales (4–20–85)	Sporophyte prominent diplohaplontic with embryo developing in the oogonium of a minute gametophyte	Parenchymatous cortex with cambium-like meristoderm covered with mucilage, parenchymatous outer cortex, phloem-like inner cortex, and loose medulla; stipes up to 50 m long or more	Cold temperate and arctic oceans	*Nereocystis, Egregia, Macrocystis, Laminaria, Postelsia, Pelagophycus*

Note: The number of families (Fam), genera (Gen), and species (Spp.) in each order is indicated in the first column.

Figure 11–2 *Nereocystis lutkeana,* a kelp. Note the branched holdfast and the single, large, gas-filled pneumatocyst that keeps the fronds afloat at the surface where there is abundant light.

morphogenesis (Figure 11-1). Much of the growth of species in these groups results from activity in two embryonic tissues, both of which are meristematic; the **primary meristem** and the **meristoderm.** The primary meristem in the Laminariales is called the **transition zone.** It is located between the embryonic stipe and blade and gives rise to new cells in both structures. Its counterpart in the Fucales is the **apical meristem.** The meristoderm is the outermost tissue of blades and stipes, thus resembling the epidermis of vascular plants.

Differentiation of the *Fucus* embryo begins as the side of the zygote away from the light grows out into a rhizoid, which soon forms a narrow multicellular strand or basal stalk. Mitotic divisions in the illuminated side of the zygote result in a pear-shaped embryo, wide at the apex and narrow at the base. As growth continues, one of the apical cells develops an especially conspicuous nucleus. A fine hair grows out of this cell, the base of the hair lying in a funnel-shaped depression created as surrounding cells continue dividing and growing. A second, much longer, hair is produced as growth continues. The apical meristem originates in this depression. From it, both stipes and blades develop.

In *Laminaria,* the zygote begins to develop while still in the oogonium of the microscopic gametophyte. A downward-oriented rhizoid typically penetrates the oogonial wall and forms a narrow stalk, while a vertical growth takes on the appearance of a small leaf or flattened club. As growth continues, a transition zone develops in the basal portion of the club-like structure; a cylindrical stipe begins to develop below it and a flat blade above it. From the embryonic stipe, additional rhizoids develop and anchor the germling to the substrate. Rhizoids may be septate or nonseptate, branched or unbranched, depending on species.

As the embryo grows and differentiates into juvenile and mature plants, four distinctive tissues develop. In the blades of *Fucus, Laminaria, Macrocystis,* and other Laminaria-like seaweeds, these are the **meristoderm, outer cortex, inner cortex,** and **medulla.** The meristoderm, a meristematic tissue analogous in some ways to the epidermis of vascular plants, consists of palisade-like cells with thickened outer walls and contains numerous flattened, disk-shaped plastids. A thick mucous layer covers the meristoderm and aids in protecting the blade against desiccation. On the inner side of the meristoderm, a cortex develops. The outer cortex is made up of parenchyma cells and also contains plastids. The inner cortex is mostly parenchymatous, but also contains thick-walled sieve cells. It serves primarily in food transport and probably storage of food and other materials. The innermost tissue is the medulla, which is more or less prosenchymatous.

Blade and stipe anatomy in Fucales and Laminariales is illustrated in detail in Taylor (1956, 1957, 1960) and in Smith (1944). As in the blades, four distinct tissues can usually be discerned in stipes of annual and deciduous kelps and fucoids. The meristoderm typically consists of two layers, a palisade layer and an inner layer, both containing numerous chloroplasts. The outer cortex also contains chloroplasts; it is made up of larger cells originating from the meristoderm. Running through it are mucilaginous ducts surrounded by small, thicker walled secretory cells. The inner cortex is made up of thick-walled sieve cells through which food is conducted from the blades to all other parts of the plant body. The medulla appears to be prosenchymatous, but actually is made up of three kinds of cells: parenchyma cells, connecting cells, and hyphal cells. The last two develop from protuberances that grow out of the cortical cells. The first of these become the connecting cells: as they grow, they come in contact with other protuberances, and the adjoining walls dissolve and fuse, but nuclear fusions do not occur. Later protuberances continue growing into long filaments or hyphae which penetrate throughout the medulla. Some of these have swellings at the septations and are called trumpet hyphae. Trumpet hyphae occur in many Phaeophycopsida and are believed to be active in food transport in species that do not have sieve cells. In *Pterogophora* and other kelps with perennial stipes, a secondary meristem develops between the outer and inner cortex and new cortical tissues are produced from it, as growth begins each spring, resulting in distinct growth rings in the stiff, "woody" stipes.

The anatomy of holdfasts is less complex than that of stipes. Branches, called haptera, contain two or three tissues: meristoderm, cortex, and, in a few species, medulla. The cortex contains mucilage ducts and broad parenchyma cells; numerous rhizoids grow out from the meristoderm, curl around objects, and penetrate into crevices, and thus attach the plant to the substrate. In most species, the new haptera grow out from the lower portion of the stipe.

In the kelps and many other phaeophytes, portions of the stipe, in the vicinity of the transition zone, are modified into gas-filled bladders, called pneumatocysts or floaters, made up mostly of cortex tissue with some sieve tubes present. Often they are located at the apex of the stipe between it and the blades. These bladders sometimes have a capacity of several liters of gas, mostly carbon dioxide, although carbon monoxide is believed to be present in some species. Both gases are products of respiration. Copious production of mucilage prevents water from entering or gases from escaping.

Growth in most other groups of brown algae is **trichothallic** (filamentous in origin). Modification of simple and branched filaments results in stipes and blades made up of complex pseudo parenchyma tissues. In *Desmerestia,* structure is especially complex and three distinct tissues can be observed: a meristoderm-like outer layer, a cortex, and an internal system. The cells of the internal system are relatively thick-walled. They originate from tufts of hair-like filaments at the apex of the main stipe and its many branches, they continue elongating throughout the life of the plants, and they give rise to the pseudoparenchymatous cortex which surrounds them. At the center of the internal system is a single tube-like series of axial cells which are believed to conduct food. Plastids are abundant in the cells making up the tufts of apical hairs and in the outer cells of the internal system.

The anatomy and physiology of organisms are always closely related to each other and in the Phaeophycopsida represent millions of years of selection and evolution. The geometry and anatomy of blades, stipes, and holdfasts are marvelously adapted to the functions of food production, food conduction, and anchorage, respectively, in a marine habitat, just as leaves, stems, and roots, are adapted to these same functions on land. Anatomical details are different in land plants and brown algae largely because the gene pools of the parent stocks from which they evolved were different, but the inner cortex of the brown algae and the phloem of vascular plants are remarkably similar in general form and they perform very efficiently the same function. Nothing similar to xylem tissue exists in the seaweeds; it isn't needed and therefore never evolved, because the plants are almost continuously immersed in water. Sexual reproduction by means of flagellated gametes, on the other hand, is a real advantage in brown seaweeds, although it would be a drawback in land plants which grow in more xeric environments. In many land plants, some of the stems and leaves are modified into colorful flowers which attract pollinating insects; such an adaptation would be useless in seaweeds, just as pneumatocysts would be useless in land plants.

Other adaptations especially valuable to brown seaweeds in their struggle to survive, but which would be of less value to land plants, include those that make metabolism—respiration and photosynthesis—possible at very low temperatures and those that make production of amino acids and hence rapid growth possible at a time of year when there is virtually no available nitrogen in the water.

Plastids in most Phaeophycopsida are small, flat disks, lacking pyrenoids, situated on the periphery of the cells. In the Fucales and Laminariales, they are most abundant in the meristoderm. They vary little from species to species throughout the class, with the greatest diversity in groups in which the plants are filamentous or prosenchymatous, especially the Ectocarpales. Most Ectocarpales have typical discoid plastids, but star-shaped axial plastids occur in *Pylaiella fulvescens,* and twisted or spiral, ribbon-shaped plastids occur in some species of *Ectocarpus.* Some species of *Pylaiella* have pyrenoids. Branched ribbon-shaped plastids occur in *Phloeospora,* a member of the Punctariales sometimes included in the Ectocarpales. Large, single plastids, instead of small multiple ones, occur in *Ectocarpus,* in *Petalonia, Scytosiphon,* and *Colpomenia* of the Punctariales, and in *Ascocyclus* and several other species in the Chordariales. In the Sporochnales, *Carpomitra* sporophytes have the typical large number of discoid plastids, while the gametophytes have a single, much dissected plastid in each cell.

Physiologically, brown seaweeds are well adapted to their environment. Levels of nitrate and other minerals build up in surface layers of the sea during the winter but are in limited supply during the warmer months; nevertheless, the brown seaseeds grow most rapidly and produce the greatest amount of protein in late spring and summer. How they can do this has been something of a mystery, and hence the object of considerable physiological research in recent years. We have learned that brown algae differ from other plants in a number of ways related to nitrogen metabolism: (1) they can use either nitrate or ammonia equally well and their utilization of either one is not reduced by a high level in their tissues of the other; (2) they can utilize nitrites (which are toxic to most other plants) if these are present; (3) the uptake of nitrate and its conversion into ammonia is not light dependent; (4) they can take up nitrates and other minerals at temperatures near and even below 0°C; and (5) they can take up excess nitrogen in a form of luxury consumption during the winter months, when neither temperature nor light is suitable for rapid photosynthesis, and store it either as simple amino acids or as nitrates in their medullary or other tissues.

Ultrastuctural and biochemical characteristics are similar to those of other Bracteobionta. Nuclei are especially large and have prominent centrosomes associated with them. Flagella are usually paired, one pinnate and one whiplash, and occur only in reproductive cells: meiospores and gametes. Mitochondria are tubular. Food reserves are oils and related lipids, such as fucosterol, a slightly branched carbohydrate called laminarin, and sugar alcohols such as mannitol. The most abundant plastid pigments are chlorophylls *a* and *c*, β- and α-carotenes, fucoxanthin, flavoxanthin, lutein, and violaxanthin.

Reproduction is primarily sexual; however, many species of *Sargassum* in the open sea and *Macrocystis* along the Pacific coast reproduce primarily, if not exclusively, by asexual means. Flagellated asexual spores, called mitospores, are produced in these and a number of other species.

Sexual reproduction in *Ectocarpus, Pylaiella,* and other filamentous brown algae involves flagellated isogametes, flagellated meiospores, and an isomorphic alternation of generations. Mature sporophytes and gametophytes are typically about 2 cm long and appear as soft, light brown tufts or plumes on rock, wood, or other algae. Distinguishing between gametophytes and sporophytes can only be done when the sporangia or gametangia have developed and, like identification to species, requires the use of a microscope.

Most Phaeophycopsida have a diplohaplontic type of life cycle (Figure 11-3). In many species, the generations are isomorphic or nearly so, but in others the sporophytes are large and conspicuous and the gametophytes are small and inconspicuous. In the Fucales, life cycles are usually interpreted as diplontic. In many brown algae, the gametophytes can be cloned simply by dividing them into smaller pieces and providing each piece with the suitable nutrients. The life cycle of *Macrocystis pyriforme* is illustrated in Figure 11-3.

Phylogeny and Classification

Although pre-Devonian fossils have been identified as Phaeophycopsida and referred to the Ectocarpales (Fritsch 1965), the resemblance is very superficial. The earliest fossils similar in appearance to modern brown seaweeds appeared in late Devonian, but Phaeophycopsida-like fossils are extremely rare prior to the Triassic when they became abundant. Unfortunately, paleontologists have not yet shown the interest in Phaeophycopsida that they have in the Siphonophycopsida and coralline Rhodophycopsida, partly because Phaeophycopsida do not fossilize as well. Recent calcareous formations, however, often contain excellent fossils of *Padina* and related genera.

Probably the first brown algae were similar to modern *Ectocarpus* and originated in temperate, or possibly subtropical, seas during the Devonian period, either from Dinotrichalean-like ancestors or, more likely, from the Xanthophycopsida. Already present in this parent stock were enzymes which make respiration at rather low temperatures possible, and these early plants were thus able to survive brief periods of low temperature that occur in such seas. As the genes which code these enzymes became better established, some lines were able to carry out their entire life cycles in cold water. Other lines became adapted to tropical habitats.

From the filamentous, highly variable, physiologically diverse and unspecialized Ectocarpales, evolution seems to have progressed along three lines. One line, leading to the Cutleriales, became well adapted to tropical conditions; in it, life cycles are largely isomorphic diplohaplontic, although in *Cutleria,* differentiation has led to the gametophyte generation being more prominent than the sporophyte. The other lines are characterized by the development of sporophyte dominating heteromorphic life cycles and specialized adaptation to cold ocean habitats. In one of these, leading to the Desmerestiales, organs built of very complex

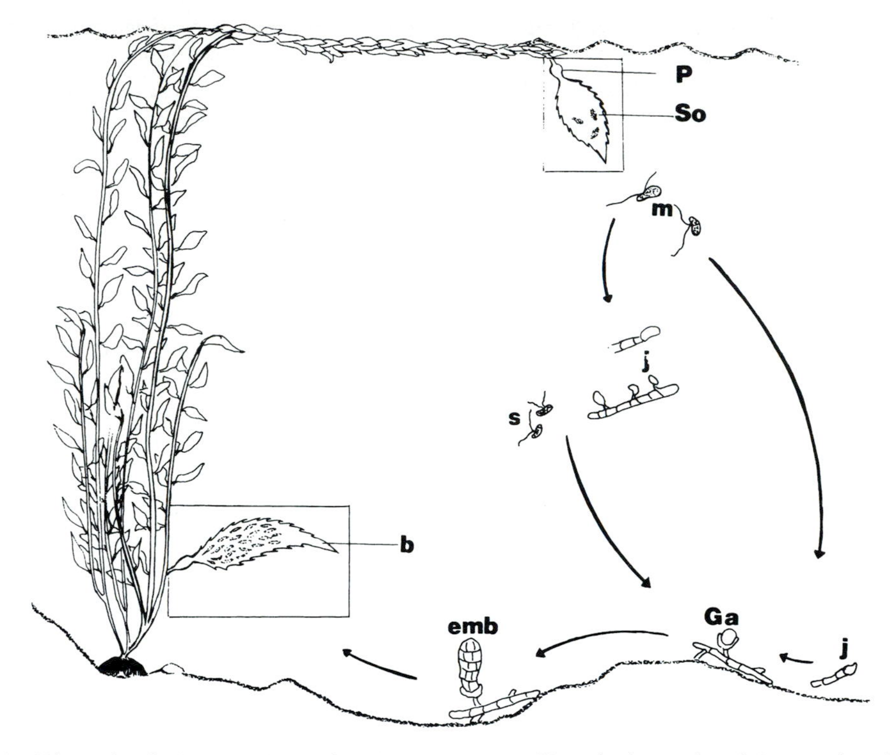

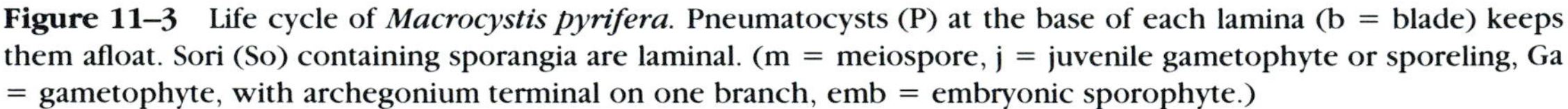
Figure 11–3 Life cycle of *Macrocystis pyrifera*. Pneumatocysts (P) at the base of each lamina (b = blade) keeps them afloat. Sori (So) containing sporangia are laminal. (m = meiospore, j = juvenile gametophyte or sporeling, Ga = gametophyte, with archegonium terminal on one branch, emb = embryonic sporophyte.)

pseudoparenchyma tissue based on trichothallic growth, of a type unique to the Phaeophycopsida, have evolved. In the third line, organs composed of true parenchyma have evolved. Complex organic compounds such as algin, which is made up of units of *d*-mannuronic and *l*-glucuronic acids and is found primarily if not exclusively in the Laminariales and Fucales, make it possible for many of the species to withstand the effects of drying that organisms adapted to the littoral zone must cope with. The chief characteristics of the orders are summarized in Table 11-1. The 900 species are classified in 31 families and 190 genera.

Plastid diversity, differences in tissue structure and arrangement, variation in life cycles and sexual reproduction, and many other characteristics have been taken into consideration in the classification of the Phaeophycopsida. In the classification used here, the Ectocarpales includes only species with filamentous structure; all of these have isomorphic diplohaplontic life cycles. This is the most variable of the orders in a number of respects, ecologically and morphologically. *Ectocarpus siliculosa* and *E. confervoides* are cosmopolitan species found both in the tropics and in cold oceans. *Bodanella lauterbornii, Heribaudiella arvernensis, Lithoderma fontanum,* and *L. fluviatile,* are found in freshwater, the first two in France, the *Lithodermas* in lakes and streams in Scandinavia. *B. lauterbornii* lives in shady or deep lakes, *H. arvernensis* in swift alpine streams.

The gametophytes of the Sporochnales show similarities to the Desmerestiales and Laminariales, but the mature sporophytes show little evidence of relationship to any specific group within the browns, although they are definitely phaeophytes. On the basis of morphogenesis and ecological preferences they have been allied here with the Cutlerian line.

In *Cutleria,* the gametophytes and sporophytes are sufficiently different from each other that for years the sporophytes were classified as *Aglaozonia* and the gametophytes as *Cutleria.* By observing the development of zygotes and the germination of meiospores, botanists learned that *Aglaozonia parvula* is the sporophyte of *Cutleria multifida.* Since then, the life cycles of the other species have been traced.

Although the tissues of the Punctariales are pseudoparenchymatous, a number of similarities can be seen between them and the kelps, and they are believed to form a link between the Ectocarpales and the Laminariales. In some classifications, the Chordariales, Tilopteridales, and Sporochnales are included in the Ectocarpales; the Punctariales, as circumscribed here, are also often included in the Ectocarpales, but they are just as often divided into two orders, Punctariales and Dictyosiphonales.

Tissue similarities, life cycle similarities, and similarities in ecological preferences all suggest a close relationship between the Laminariales and the Fucales. Special evidence of a close relationship is found in some chemical substances abundant in many species of both orders but rare or absent in other orders of the Phaeophyta. Among these are the alginates, laminarin, and fucoidin. The alginates seem to be limited to the Fucales and Laminariales and are very abundant in both; laminarin and fucoidin are more widely distributed, the former having been reported in diatoms and possibly Chlorophycopsida, although it is absent from most phaeophycopsida other than the kelps and fucoids.

Representative Species and Genera

Although the kelps and the rockweeds are the best known Phaeophycopsida, other brown algae are also interesting, attractive, and ecologically important (Figure 11–4). The Ectocarpales, for example, are abundant in most marine ecosystems but are often overlooked. They are small filamentous plants which form little brown "pin cushions" on rocks, pilings, and seashells along the coasts of Europe, North America, eastern Asia, and many Pacific islands.

Coilodesme is an epiphyte which consists of a single vesicle attached to the stipe of other seaweeds. If one is careless in his examination of marine plants, one may believe he has some new species of Phaeophycopsida with slightly flattened pneumatocysts, when in reality he has a red seaweed, or even a vascular plant, with a brown epiphyte attached to it.

In *Cutleria,* the sporophyte is a crustose plant only half a dozen cells thick, firmly anchored to the substrate by numerous rhizoids and bearing many sporangia on its upper surface. Spores are produced by meiosis. The meiospores germinate and develop into vertical, columnar structures with a tuft of hair-like filaments at the apex. As growth continues, a much-branched thallus about 20 cm tall consisting of strap-shaped segments made up of prosenchymatous tissue develops. There is no differentiation of stipe and blade in this dichotomously branched thallus, which is usually dioecious. Branches result where filaments fail to fuse. Oogonia are produced on dark brown tufts of unbranched filaments or hairs on the female plants and antheridia are produced on bright yellow tufts of much branched hairs on male plants. Both eggs and sperm are motile, but the eggs are larger and have many chloroplasts while the sperm have few or no chloroplasts. The eggs lose their flagella prior to fertilization, and penetration of the sperm occurs in a colorless area near where the flagella were attached. The zygotes germinate immediately to produce crustose, parenchymatous sporophytes.

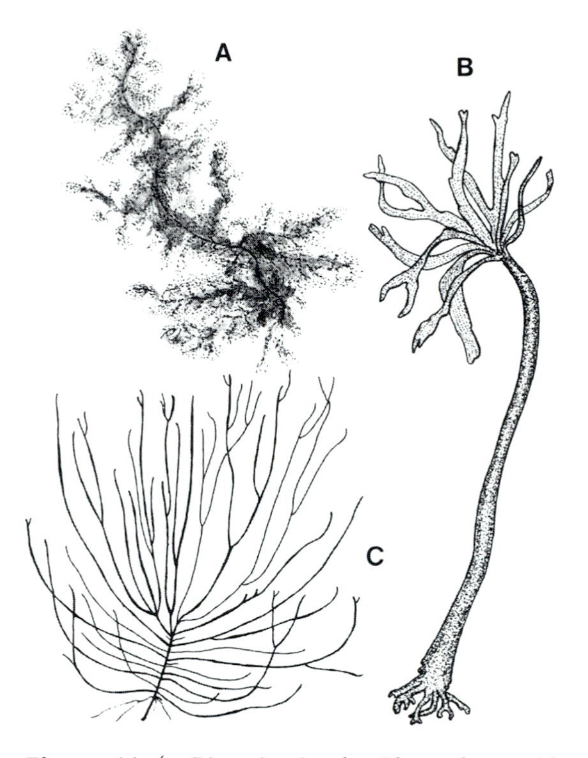

Figure 11–4 Diversity in the Phaeophycopsida: (A) *Pylaiella littoralis,* Ectocarpales; (B) *Postelsia palmaeformis,* Laminariales; (C) *Chordaria flagelliformis,* Chordariales.

Laminaria, type genus of the Laminariales, is a taxon of about 30 species that occurs throughout the colder seas of the northern hemisphere. In northeast Asia several species are important sources of human food and are also widely used in making candy (Xia and Abbott 1987). The blades are usually long and slender and attached to the rocky substrate by short stipes. Meiospores are produced in numerous sporangia which occur in large patches called **sori** located on both surfaces of the blades. The sporangia start out as club-shaped mononucleate cells intermingled among elongate sterile cells or **paraphyses.** Meiosis, followed by three or four mitoses, results in 32 or 64 nuclei per sporangium. The cytoplasm undergoes cleavage and each nucleus acquires a plasma membrane and two laterally attached flagella. After the meiospores are released, they swim around for a while and then settle on a suitable substrate where half of them become short, unbranched, filamentous female gametophytes and half of them become many-branched, filamentous male gametophytes. Several globose, or sometimes elongate, oogonia develop on each female gametophyte, and several cylindrical antheridia develop on each male gametophyte. Each oogonium produces a single nonmotile egg, and each antheridium produces many biflagellated sperm. The sperm escape from the antheridia and swim to the oogonia where fertilization takes place. The zygote, still held in the oogonium, begins development immediately. The first three or four divisions are all in the same plane resulting in a short filament; then divisions in other planes begin and a parenchymatous embryo results. Differentiation into meristoderm, cortical, and medullary tissues soon takes place.

Growth rates in the Laminariales are highly variable, depending both on species and on environmental factors. In some annuals, like *Nereocystis lutkeana,* a length of 30 to 40 meters may be attained within 4 months following fertilization, making this species one of the fastest-growing plants known. In a slow-growing perennial like *Pterogophora californica,* on the other hand, it may take 10 to 12 years for the plant to attain a length of 2 meters and a stipe diameter of 4 or 5 centimeters.

Life cycles in the Fucales are similar to those in the Laminariales except that the gametophyte generation, according to the usual interpretation, consists only of gametes. In *Fucus,* (Figure 11–5A), gametes are produced by meiosis, followed by several mitoses, in small cavities called conceptacles which are located on the swollen tips of the blades. In some species of *Fucus,* male and female gametes are produced by separate plants which can be distinguished from each other by the color of the conceptacles. In other species, for example *Fucus furcatus,*

Figure 11–5 Diversity in the Phaeophycopsida: (a) *Fucus vesiculosus,* Fucales; (b) *Sargassum filipendula,* Fucales.

sperm and eggs are produced on the same plant within the same conceptacle; in yet other species, conceptacles may produce only one kind of gamete but both male and female conceptacles occur on the same plant. There is no alternation of generations according to the usual interpretation of reproductive events in *Fucus.*

The first division of the fertile nuclei within the conceptacles is meiotic. The four daughter nuclei are meiospores according to those botanists who do not accept the usual interpretation but regard *Fucus* life cycles as representing an alternation of generations. They do not develop spore-like walls, however, but remain naked and almost immediately divide by mitosis. In the male structures, mitoses continue until there is a mass of 32 nuclei. Plasma membranes develop, resulting in 32 uninucleate protoplasts, each of which then divides mitotically one more time. This alternate interpretation of life cycles implies that the mass of 32 protoplasts is the male gametophyte which then produces 64 male gametes or sperm. Eggs are produced in a similar manner with 8 eggs formed within each oogonium.

Both eggs and sperm are motile, but the eggs are much larger than the sperm and are sluggish, whereas the sperm are rapidly motile. When eggs and sperm are mixed in a drop of water in the laboratory, fertilization can be observed under the microscope. Chemical attractants cause the sperm to migrate to the eggs where they attach and begin the process of penetrating the membrane surrounding each egg. As hundreds of attached sperm continue moving their flagella, the eggs begin spinning rapidly, but as the first sperm penetrates the egg membrane, all of the others quickly detach and swim away. The zygote develops into a complex, parenchymatous embryo before becoming the dichotomously branched mature rockweed common along rocky sea shores.

Ecological Significance

The Phaeophycopsida occupy in cold oceans much the same niche that Rhodophycopsida occupy is warm oceans. However, the brown algae are not as tolerant of shade as are the reds and therefore, with the exception of species with extremely long stipes, are limited to somewhat shallower water.

Three or four characteristics are responsible for the great ecological success of brown algae in all of the oceans of the world, but especially in the cold oceans. (1) Having long stipes and a very efficient food transport system, many species are able to grow anchored in deep water and still have their blades displayed to abundant light in the photic zone. (2) Several species produce large amounts of mucilage which enables them to live in the intertidal zone and tolerate periods of drying. (3) In common with many other bracteobionta, they are able to live and carry on metabolism at very low temperatures. (4) Some species, at least, are able to accumulate nitrogen during the winter months and store it for use during the summer growing season.

Distinct zonation of seaweed vegetation is typical of rocky coasts, especially where tides are pronounced (Figure 11-6). Algae of the "fucaceous girdle," which is the highest zone, must be able to withstand long periods of drying, exposure to both very low and very high temperatures, and considerable variation in salt concentration caused either by dilution of rainwater or by evaporation of water from the tidal pools. *Ralfsia* and *Fucus* are commonly found in this zone, the former on more exposed sites and often in the spray zone. The sea palm, *Postelsia palmiformis* (Figure 11-4B), is found on exposed sites at a lower tide level than *Ralfsia.* It grows best on bare rocks where wave action is strong and spray abundant, but it also grows on barnacles. It may, however, kill the barnacles, and then a storm will sweep away both the barnacles and the palms and the rock will be left bare, an ideal site for further *Postelsia* colonization. Below the *Postelsia* zone on exposed sites, an *Alaria* zone is common. Both on exposed and protected sites, *Laminaria, Lessoniopsis,* and *Sargassum* grow in the lowest littoral zone, which is only exposed during the very lowest tides, or in the subtidal zone. *Macrocystis* and *Nereocystis* are anchored in relatively deep water far beyond the *Laminaria* zone. According to early accounts, Arctic specimens of *Nereocystis* were sometimes anchored in water over 100 m deep; with their massive bulb-like pneumatocyst and strap-shaped fronds floating on the surface, and the thick, tough stipe tying the fronds to the holdfast, they were very impressive. Further south, they were never that long.

If care is taken to avoid cutting the basal part of the plants where new shoots will develop, some species of Phaeophycopsida can be successfully transplanted. Transplant studies are useful in elucidating the effects of toxins discharged into the bays and estuaries where brown seaweeds grow, in studying the ecological effects of sewage on seaweeds, and in studying growth, development, and phenology of species.

All producer organisms are normally grazed upon by herbivorous animals; the brown seaweeds are no exception. In the giant kelp forests off the southern California coast, sea urchins are the chief grazers. During the 1950s, these forests were faced with total destruction. Four causes were suggested (1) over-harvesting by commercial users, (2) water pollution creating either poor light conditions or adding poisonous chemicals to the water, (3) a warming trend deleterious to the seaweeds, or (4) increase in sea urchins due to the destruction of their natural predators, the sea otter. Because the commercial harvesters were taking the mature fronds on a schedule that had never been harmful in the past and leaving the perennial holdfasts to produce a new crop the next year, the first possibility seemed to be eliminated. After several years' careful study it was concluded that excessive trapping of the sea otter had allowed numbers of urchins to increase to the point where they were completely destroying the live beds. The studies revealed that sea urchins prefer old, and even dying, fronds of *Nereocystis* to all other food, but will consume other algae when necessary, and even eat *Agarum,* an unpalatable kelp,

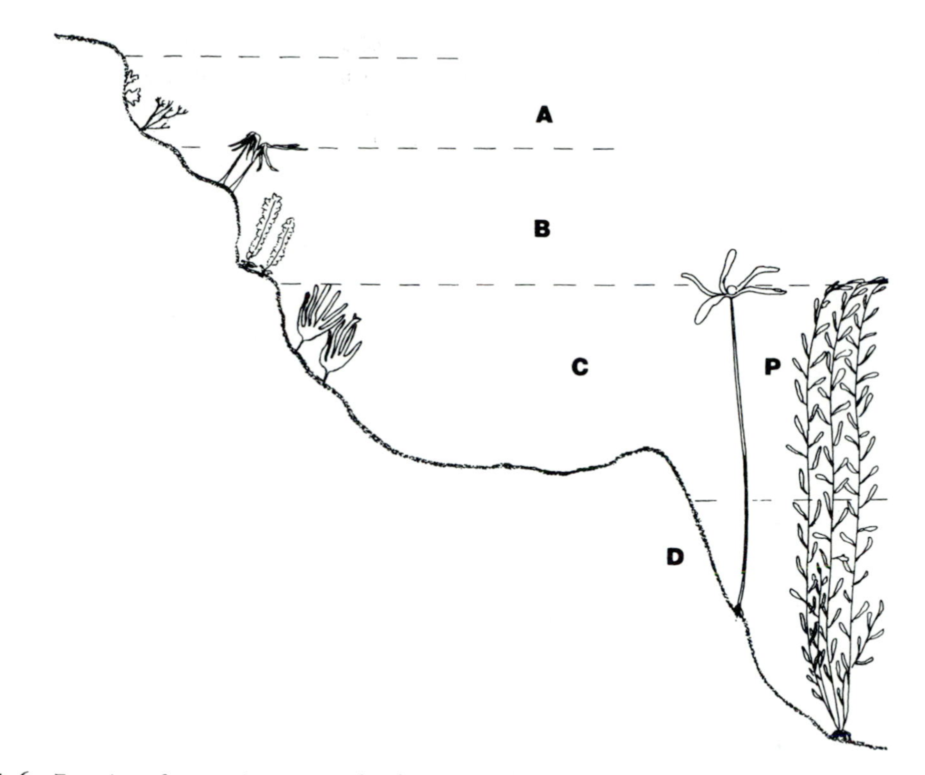

Figure 11–6 Zonation of vegetation on a rocky shore. (A) Salt spray zone and fucaceous girdle; (B) intertidal zone; (C) subtidal zone; (P) pelagic zone; (D) benthic zone.

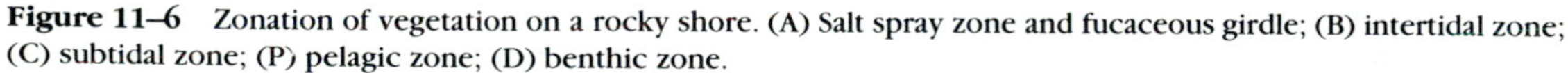

as a last resort. Lime was used to control the urchins, and the forests have now come back almost to their pristine condition. At the present time, the urchin population is kept small enough that the animals are able to feed on the forest detritus in their normal manner rather than on the young kelp plants. Unless or until sea otter populations build up, however, periodic lime application will be required to maintain the balance between producer and herbivore normally maintained by the secondary consumers.

In additions to herbivores which depend on phaeophytes for food, many species of fish and other sea animals find shelter and protection from their predators in the dense beds of kelp and other brown algae. Species of red algae grow in the shade of brown seaweeds in ecosystems in which light intensity is at a much higher level than they could otherwise tolerate. Xanthophytes, diatoms, red algae, and other plants live as epiphytes on many of the brown seaweeds. Bacteria and various fungi, mostly Mastigomycopsida, live on the detritus which forms as the fronds and old stipes die in the late autumn and are battered loose by winter storms.

Economic Significance

The brown algae are of great economic value, and their importance in this regard has rapidly increased in recent years as more and more uses have been developed. The commercial use

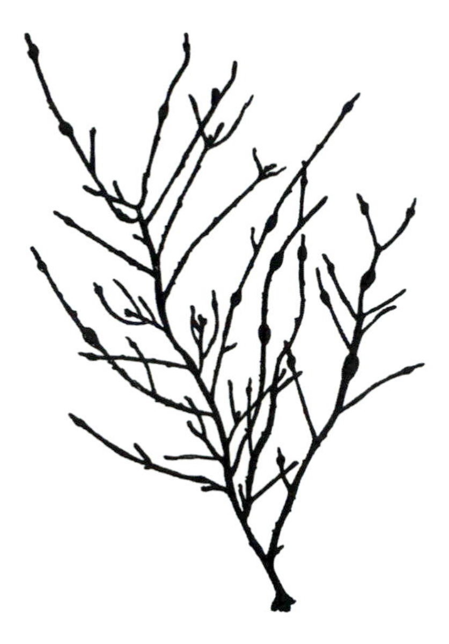

Figure 11–7 *Ascophyllum nodosum,* a commercial source of vitamins.

of alginates is now valued in the hundreds of millions of dollars annually. The interest in brown seaweeds for physiological research has been sparked to a large degree by the increase in their economic value.

Some brown seaweeds have been used for food for hundreds of years, although the red seaweeds have historically been more important in this regard. *Ascophyllum nodosum* (Figure 11-7 is extremely high in vitamin C (ascorbic acid) and is an important source for the commercial processing of vitamins and minerals. Recently, kombu, dried *Laminaria japonica,* which for centuries has been a favorite in the Orient, has become a popular food item in the U.S., Canada, and Europe where it is used in soups and with various meats, as in kombu maki. Xia and Abbott (1987) list 11 species of brown seaweeds as important sources of food in China. They estimate over 100 million pounds of seaweed are consumed annually. Probably 10 to 20% of this is from brown seaweeds.

Algin, obtained from most kelps and fucoids, is used as a stabilizer in ice cream and other desserts, and is the most important seaweed food consumed in the U.S. today (Table 11-2). Barges with reciprocating power scythes, similar to the mowing machines used by farmers to cut hay, harvest millions of pounds of kelp along the coast of southern California each year. This is processed to produce alginates and other organic compounds (Figure 11-8).

TABLE 11-2

SOME INDUSTRIAL USES OF SEAWEEDS

Use	Type of Seaweed	How Processed
Human food	Rhodophycopsida	Breakfast cereals; soup, sandwiches, etc.; candy; gelatins and stabilizers
	Phaeophycopsida	Soup, warm dishes, kombu; stabilizers in ice cream
	Chlorophycopsida	Soup, warm dishes
Livestock feed	Phaeophycopsida	Fresh fodder or pasture; silage
Fertilizers	Phaeophycopsida (kelp)	Freshly collected and applied as green manure; burned and ashes applied to provide potassium, iodine, trace minerals
Vitamins and minerals	Rhodophycopsida	Trace and rare vitamins: B_{12}, D_4, etc.
	Phaeophycopsida	Vitamin C (ascorbic acid), vitamin A precurser, (β-carotene), iodine, trace minerals
	Cyanoplycopsida	Vitamin B_{12}, and anticarcinogens
Medical industry	Rhodophycopsida	Agar and other gelatins; stabilizers
	Phaeophycopsida	Alginate for plastics and artificial limbs; emulsifiers and disintegrants for medicines; degradable gauze; relieve gas

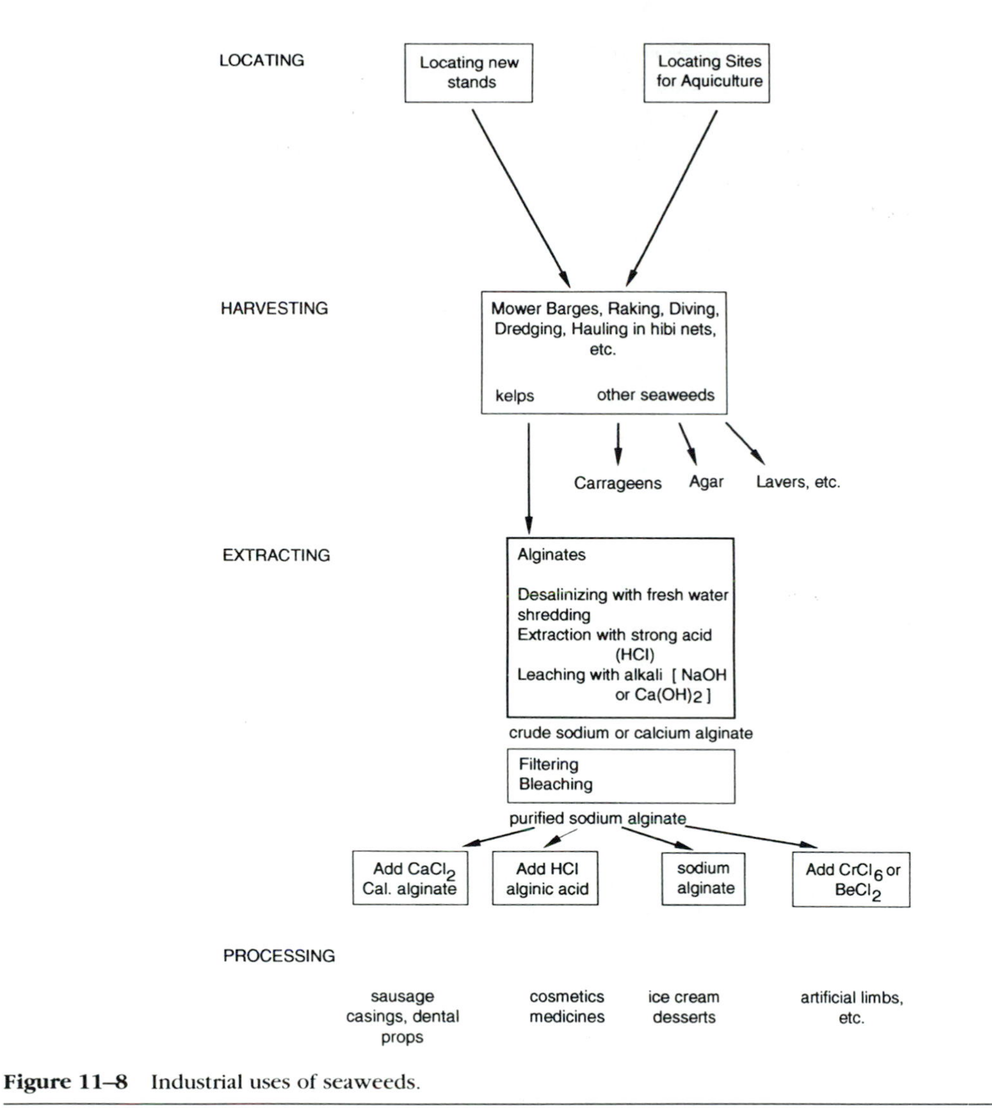

Figure 11–8 Industrial uses of seaweeds.

In Scandinavia, Great Britain, and along the coasts of North America, farmers have collected kelp for fertilizer for at least a hundred years. Now, however, kelp is too valuable a resource to be used this way to any great extent. Kelp was either applied to the fields directly in a partially decomposed condition and plowed under, or it was dried and burned and the ashes, which are high in potassium, used for fertilizer. Kelp can still be purchased as a soil additive; in addition to adding nitrogen and potassium to the soil, it improves soil tilth, just as barnyard manure and compost do. It also supplies several of the trace elements needed by plants and the animals that feed on them.

Kelp and other seaweeds have also been used extensively for chicken feed and for supplements to the rations for sheep, cattle, and horses. In the Orkney Islands, the sheep live on kelp about 10 months of the year, and in several other areas, kelp is ensiled for feeding to sheep. In New Zealand, dairy cattle fed on kelp silage are reported to produce more butterfat. Where iodine, copper, or cobalt are deficient, kelp can help provide these essential elements; however, cattle and horses fed kelp sometimes get toxic doses. Food faddists, aware of the value seaweed has had in correcting nutritional disease in livestock, make use of kelp and other seaweeds extensively. There is some evidence that for some individual cows or sheep, copper or iodine in organic form, as in kelp, is more effective in correcting nutritional diseases than the same element in purely inorganic form. The same may be true for people. However, for most people and most animals, it makes no difference whether these minerals are in an organic or an inorganic form.

Kelp products are widely used for medicinal purposes. One of the major purchasers of commercial alginates is the vast pharmaceutical industry. Algin is used to thicken liquid medicines, as a disintegrant in the coating of pills, and as a dispersing agent for antibiotics. Alginic acid seems to have direct medical value where an emulsifier is needed to reduce formation of gaseous pressures. Gauze made of alginates does not have to be removed from wounds because it can be absorbed directly into the body. Plastics made from alginates of the heavy metals are ideal for making artificial limbs.

In contrast to alginates, other organic chemicals that are easily extracted from various brown seaweeds, such as laminarin, fucoidin, and mannose, have not found extensive commercial use as yet. Small amounts are used in the biochemical supply industry.

The Native Alaskans used the stipes of *Nereocystis* for fishing lines and the pneumatocysts as syringes to bail out their boats. Children in many societies down to the present day have used kelp stipes as jump ropes.

Research Potential

Phaeophycopsida appear to be ideal organisms for solving some basic problems of plant physiology and ecology. Ecology as a science has been heavy on theory with many untested hypotheses often treated in textbooks and taught in classrooms as facts. At the present time, research is being conducted on the effects of grazing on various phaeophytes, on rates of translocation of food from photosynthetic sites both up and down the plant, on utilization of minerals by brown algae, and on mineral nutrition in general.

Because of the variety of tissues in the Phaeophycopsida, they have an excellent potential for studying morphogenesis and differentiation. The nuclei show promise for cytogenetic studies. Differences between green plants and brown algae in structure of plastids and mitochondria suggest that they could complement each other in studies of photosynthesis and respiration that would help us understand the processes better. For example, in green plants (Lorobionta), hexoses and trioses are early products of photosynthesis, but it has been reported that in the Phaeophycopsida, pentoses are the first sugars produced.

The Phaeophycopsida have been largely ignored by paleobotanists. There is a real need for increasing our knowledge and understanding of evolutionary patterns in this group of plants, not only for their own sake but to further our total knowledge of evolutionary principles. Paleobotanists working with any group of algae, however, walk on treacherous terrain because the literature abounds in both incorrect and incomplete identifications and descriptions. The "it's only an alga" attitude has created serious problems. Much of the paleobotanical work of the near future must of necessity be based on new collections because of the unreliability of much of the old reports.

SUGGESTED READING

Gilbert Smith's *Marine Algae of the Monterey Peninsula* and Taylor's *Algae of the Northeastern Shore of the United States* are excellent sources for students desiring to identify the seaweeds of these two areas. Yale Dawson's *How to Know the Seaweeds* is an easy-to-use pictured key. *Between Pacific Tides* by Ricketts and Calvin and Colman's *The Sea and It's Mysteries* are interesting books with ecological emphasis.

STUDENT EXERCISES

1. With the aid of a hypodermic needle and large (50 to 100 ml) syringe, with the plunger at the halfway position, find out whether the gases in a pneumatocyst of a freshly collected brown seaweed (e.g., *Nereocystis*) are under greater or lesser pressure than the atmosphere.

2. Make a free-hand section of the stipe of a brown seaweed and describe the cells and tissues that you see.

3. Make a free-hand section of the blade of a brown seaweed and describe the cells and tissues in it.

4. Compare the anatomy of kelp (Figure 11-2) with that of a vascular plant (Figure 18-8). Label the tissues in each. What is the function of each tissue? Are they analogous or homologous structures? (If prepared slides are available, make your own sketches from the prepared slides rather than from illustrations in the book.)

5. Test a kelp stipe for the presence of sugars, starches, proteins, and lipids, using simple assay techniques such as the iodine test for starch, Benedict's solution for reducing sugars, concentrated nitric acid for proteins, etc. How does the kelp compare with a common vegetable such as carrot, potato, or onion for each kind of substance?

6. Store a sample of fresh kelp or other brown seaweed at room temperature for 48 hours, store a second sample in a refrigerator at 2°C, and a third at 8°C for the same length of time. Compare the samples for morphological and chemical changes that may have occurred during the 48 hours.

7. Compare four or five species of seaweed for the presence of sugars, starch, protein, or other organic substances.

8. For each of the following genera of brown algae, indicate the order to which it belongs and the type of life cycle (haplontic, isomorphic diplohaplontic, gametophyte dominant diplohaplontic, sporophyte dominant diplohaplontic, or diplontic) you would expect to observe: *Ectocarpus, Laminaria, Cutleria, Pylaiella, Sargassum, Leathesia, Coilodesme, Lithoderma.*

SPECIAL INTEREST ESSAY 11-1

Industrial Uses of Seaweeds

Four groups of seaweed products are of especially great economic value today: the alginates, agar-agar, carageenins, and lavers. The use of algin and its derivatives is the basis of the fastest-growing of the industries utilizing seaweeds and probably the most valuable in terms of dollars received.

Commercial use of seaweeds involves (1) locating, (2) harvesting, (3) extracting the desired product, and (4) processing. The first three are performed by a small number of individual processing companies, such as Marine Colloids, Inc., and General Mills Chemicals, Inc. The fourth is performed by a large number of pharmaceutical, plastics, and foods companies of all sizes.

Probably most major beds of kelp, Irish moss, dulce, and other economically important seaweeds have already been located, so the location process at the present time consists of finding minor beds, less extensive than the California kelp forests, for example. This step will soon consist of establishment of beds under human control, and the improvement of existing beds by fertilizing or other management practices. It is projected that **aquiculture** will be an important area of employment and research within a few years. At the present time, aquiculture is most advanced in Japan where nori, or laver, *Porphyra* spp., is cultured on tidelands by suspending **hibi nets** from bamboo poles about the first of October. In about 6 weeks, the growth on the hibi nets is easily visible and by early January is mature and ready for harvesting. By May first, the growth has disappeared. A discovery made about 1950 that a species of *Conchocelis* was a stage in the life cycle of a *Porphyra* species, rather than a separate species as previously believed, has made it possible to increase the yield of *Porphyra* by growing the conchocelis stage asexually under laboratory conditions and seeding the hibi nets with conchocelis spores (Figure 4-8).

Harvesting procedures depend on the species and the locality. Rowboats or small motor boats are commonly used to harvest nori in Japan from hibi nets, although some aquiculturists use larger boats on to which the nets are hauled. A type of mowing machine, mounted on large, self-propelled barges, is used to harvest kelp off the American west coast. *Chondrus crispus, Gigartinia stellata,* and *Iridaea laminarioides,* the chief sources of carrageenin, are harvested by hand or with small rakes at low tide in eastern Canada. In Japan, divers harvest much of the *Gelidium* used for agar-agar production. *Gelidium* grows deep, often at 20 m or more. Women are much better divers, on the average, than men, and most of the *Gelidium* is harvested by women. *Gelidium* harvesting in North America is largely by dredging, which tends to be more damaging to the environment, however.

As the harvested kelp or other seaweed is brought into the processing plant, it is first desalinized by leaching with fresh water and then shredded. The shredded material is leached with hydrochloric acid and treated with either sodium carbonate or calcium chloride to yield a crude sodium or calcium alginate containing considerable laminarin, mannitol, and other impurities. The crude product is filtered and bleached. Purified sodium alginate can be converted into calcium alginate by adding calcium chloride, into alginic acid by treating with hydrochloric acid, or into heavy metal alginates by treating with salts of chromium or beryllium.

Algin derivatives are used in the textile industry, in plastics, in the production of cosmetics, and in other industries. Because of their nontoxicity and desirable colloidal properties, they are used widely in the food industry. Sodium alginate is considered the best stabilizer and creaming agent for ice cream. Sausage casings, which were formerly made of intestinal membranes from sheep and cows, are now made of alginates. Alginates are widely used as pill coatings and they make especially accurate dental impressions. Artificial limbs made from heavy metal alginates possess a resilience not found in other plastics or materials and hence are less annoying to the user than wooden, metal, or plastic limbs.

Other seaweed products include the lavers, used in the paint and textile industries and as food, agar-agar used in the food industry and in biological laboratories, and carrageenin, used to stabilize chocolate milk and other dairy products and elsewhere in the food industry.

Agar has no food value, but most of the agar production is used in the food industry because it is nontoxic and has properties that make it desirable in thickening some foods and in canning. Much of the agar production goes to biological laboratories as a nutrient carrier for growth of microorganisms. High quality agar is insoluble in cold water, but soluble in hot water and at about 90°C it goes into solution. As it cools, it remains liquid to about 40°C, at which point it hardens. To liquefy it again, it must be heated to at least 65°C. This property is very useful in microbiological research, for example, in studying the heat tolerances of microorganisms.

Carrageenins are especially valuable stabilizers. In chocolate milk, the cocoa solid will not settle out if the right kind and amount of carrageenin are added under the right conditions. However, if the wrong kind of carrageenin is used, or not enough, or too much, or at too high a temperature as the drink was prepared, the cocoa will settle out. Carrageenins are also used in the textile, leather, pharmaceutical, brewing, and food industries.

PART IV

THE GREEN LINE

Euglenophyta
Chlorophyta
Bryophyta
Tracheophyta

Judging from the fossil record, the rapid increase of large, unicellular eukaryotes, similar to modern Chlorophycopsida, near the close of the Proterozoic era was accompanied by a decline in both numbers and diversity of small-celled prokaryotes, suggesting an ecological advantage for eukaryotic species. Grazing by herbivorous animals that originated in late Proterozoic time was less likely to completely destroy entire populations of large unicellular plants than populations consisting only of small unicells. By the beginning of the Cambrian period, primitive Lorobionta were well established in all of the oceans. During the Cambrian, multicellular Siphonophycopsida, many of them with heavily calcified cell walls, evolved and became abundant.

Later in the Paleozoic era, mutations occurred which resulted in adaptation of some Lorobionta to terrestrial life. The ability to become dormant during dry periods was of special significance. Equally important were the mutations that resulted in differentiation of organs such that one part of a plant could live in an extremely dry habitat (air), in which excellent conditions of light and oxygen prevailed, while another part of the same plant was living in a moist, but dark, habitat (the soil).

The Lorobionta are so called because they have whiplash flagella, typically two long ones or many short ones per cell. They are also referred to as the "green line" because of the grassy green color imparted by chlorophylls *a* and *b*. They differ from Aflagellobionta and Bracteobionta in flagellation, plastid pigmentation, and numerous other details of cellular biochemistry, ultrastructural characteristics, and ecological preferences.

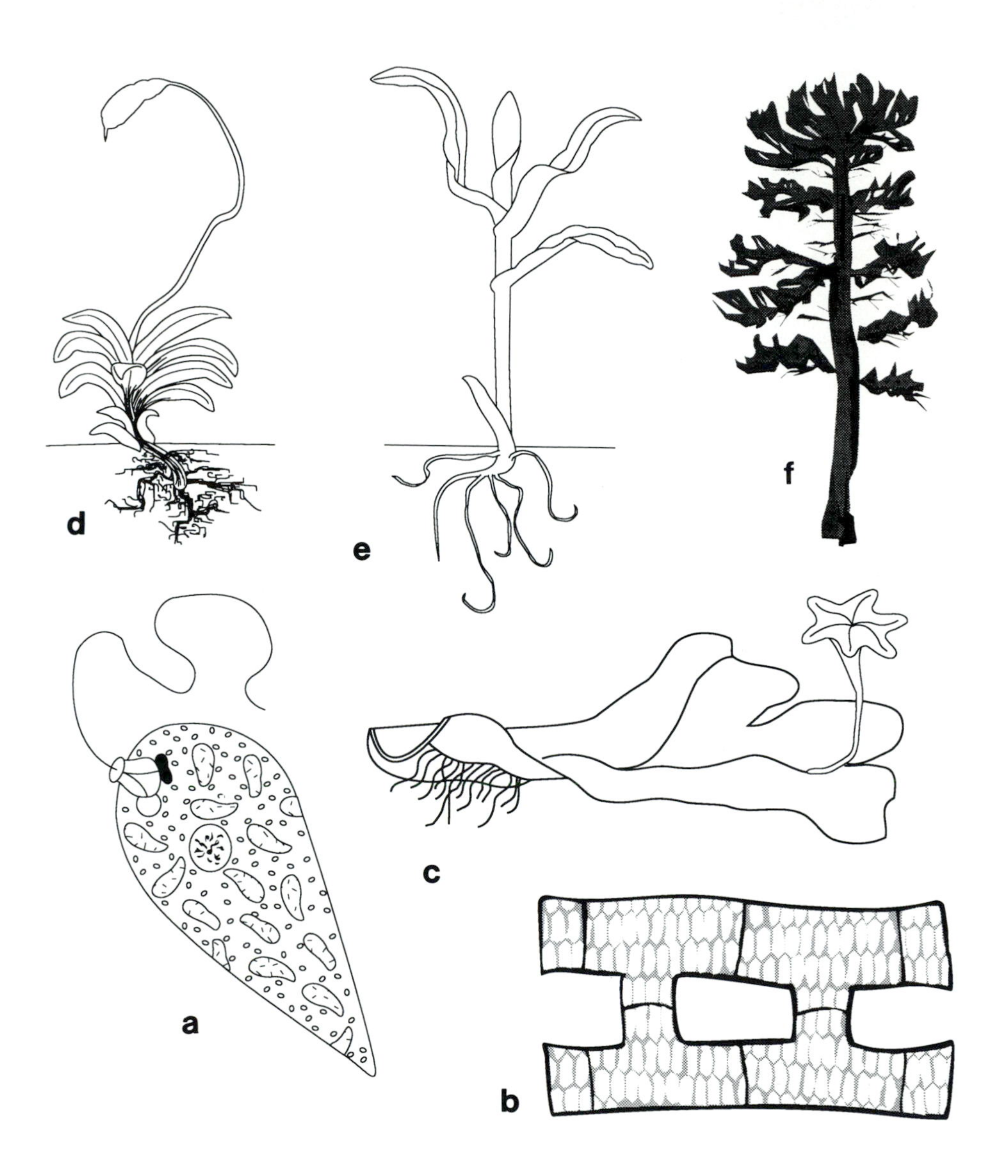

PART IV. THE GREEN LINE

Diversity in the Lorobionta from simplest and most primitive to the most highly evolved: (a) *Euglena* showing nucleus and several plastids (Eugelenophyta); (b) conjugation between two filaments of a green alga (Chlorophyta); (c) *Marchantia,* male plant with antheridiophore and rhizoids (Hepaticopsida); (d) *Tortula,* leafy gametophyte supporting an elongated sporophyte with capsule (Bryopsida); (e) a grass (Anthopsida); (f) *Ginkgo* (Pinopsida).

12 EUGLENIDS

Among the minute organisms described by Antony van Leeuwenhoek in 1675 were green creatures which moved swiftly through the water, "upwards, downwards, and round about." They had "divers colours, some being whitish and transparent; others with green and very glittering scales. . . . And the motion of these animalcules in the water was so swift, and so various, . . . that 'twas wonderful to see." So wrote the world's first microscopist, fascinated by the new world his microscope had opened up to him—and ultimately to us. Judging from his description, we are certain that some of these swift, green "animalcules" were euglenids.

The Euglenophycopsida (*eu* = good or true, *gleno* = the socket of a joint, *phyta* = plant), commonly called euglenids or euglenoids, are essentially ubiquitous, grassy green, mostly unicellular algae found chiefly in fresh water containing considerable organic matter. Although claimed by zoologists as well as by botanists, they are characterized by the same pigments that occur in pine trees, ferns, corn plants, and maple leaves. The ultrastructure of their cells resembles that of other Lorobionta more than any animal group or the animal-like plants of the Bracteobionta: the cryptophytes, dinoflagellates, and chrysomonads.

Like other Lorobionta, they are green in color due to an abundance of chlorophylls *a* and *b* in the same ratio as occurs in mosses and vascular plants, together with the typical Lorobionta carotenoids, β-carotene, lutein, and neoxanthin; they have mitochondria with plate-like cristae, the thylakoids of the plastids tend to be stacked into grana or at least show the beginnings of such stacking; the carbohydrate food reserve, paramylon, is unbranched, like amylose; and the flagella are tapered like whiplash flagella. On the other hand, the presence of contractile vacuoles, the condensed chromosomes during mitosis, the β-(1-3) linkages in paramylon, and the presence of a single row of mastigonemes on the flagella suggest relationship with the Pyrrophyta. Thus euglenids resemble both Bracteobionta and Lorobionta. In addition, the euglenids differ from other Lorobionta and most other plants in not being able to utilize nitrates in amino acid synthesis, in preferring acetate and alcohol over glucose in metabolism, in complete absence of cell walls, and in some basic differences in light and dark reactions in photosynthesis that caused Lynch and Calvin in the early 1950s to speculate that *Euglena* is the type of organism from which both plants and animals evolved (Wolken 1967).

At times, the euglenids have been included in the Chlorophyta as a class distinct from both Chlorophycopsida and Charophycopsida. In recent years, the tendency has been to classify them as a separate division. On the basis of their many differences from other green algae, it seems most logical to list them as a separate division with one class, the Euglenophycopsida. They are included in the subkingdom Lorobionta, where they appear to represent a primitive group which shows evidence of relationship to the other subkingdoms, especially the Bracteobionta. Their immediate ancestors were probably prokaryotes similar to extant species of *Prochloron.*

CLASS 1. EUGLENOPHYCOPSIDA

The name, *Euglena,* which means "having good joints," refers to the ability of these organisms to bend easily, and consequently change shape, as they move. This characteristic is often referred to as "euglenid motility" or "metaboly." It occurs in a few other plants besides euglenids, notably the golden algae.

Euglenids can be found in almost any pond or pool having an abundance of green, blue-green, or yellow-green algae in it, but the best place to look for them is in small pools near barnyards where there is an abundance of organic nitrogen in the water. They can often be found in bottles of algae in the biology laboratory that have been allowed to decompose due to too much plant material and not enough water and oxygen. If a little ammonium sulfate is added to such bottles, the probability of finding euglenids is increased. They can also be found in sewage lagoons in rich numbers, but there are obvious health hazards in collecting from such habitats.

General Morphological Characteristics

Euglenids are rather large, rapidly moving, naked, slipper-shaped green plants (Figure 12-1). Many of them continually change shape as they move through the water in a rotary manner. Usually several disk-shaped plastids are present, but some species are colorless. Cell shape varies from species to species and is also dependent on environmental factors. Some euglenids are almost spherical, whereas others are more or less cigar-shaped, but they all tend to be smaller and more rounded when grown in the dark. A red "eyespot" or stigma is often prominent at the anterior end; near it is a gullet and near the base of the gullet there may be a contractile vacuole. In one genus, *Colacium,* the cells are nonmotile and grow as epiphytes on various aquatic animals.

Because of their color and size, euglenids are not often confused with other plants or animals. The apical gullet is rather distinctive. The cells tend to be more symmetrical than those of Cryptophycopsida, Chrysophycopsida, and Xanthophycopsida, but less symmetrical than Chlorophycopsida. Colorless species might be confused with some animal flagellates like *Trypanosoma,* but can be distinguished by the apical gullet and general form of the cells.

Most euglenids have a single, modified pectinate flagellum originating at the base of the gullet. A swelling near the flagellar base, called the **paraflagellar** body, is light sensitive. Good evidence indicates it has something to do with phototactic responses. A rudimentary flagellum frequently has its base near that of the true flagellum but extends only to the paraflagellar body. A true second flagellum, sometimes the same length as the first and sometime much shorter,

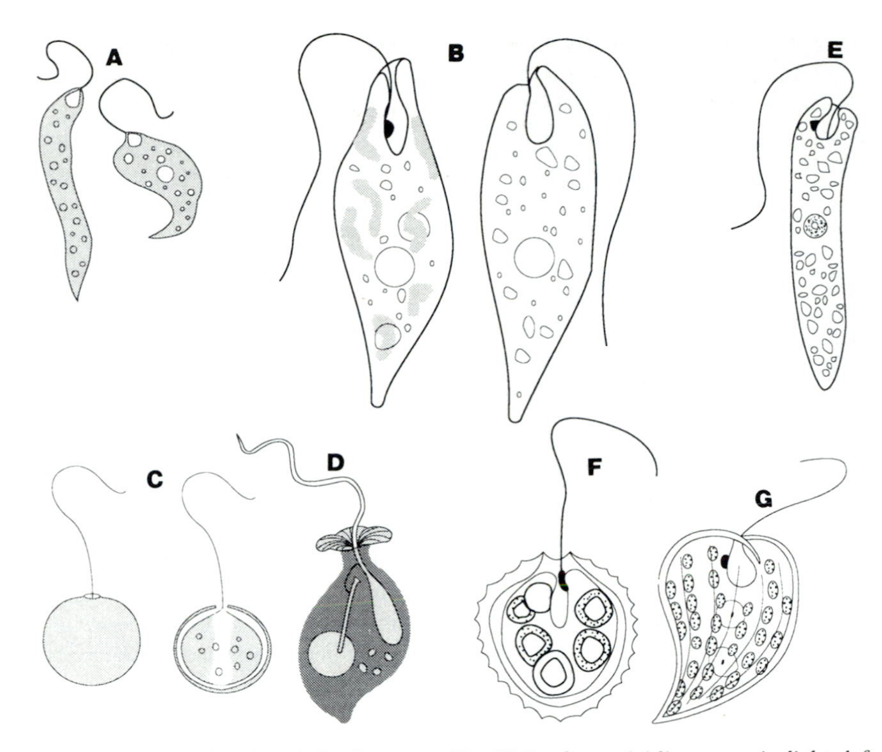

Figure 12–1 Some typical euglenids. (A) *Euglena gracilis;* (B) *Euglena viridis,* grown in light, left, and in dark, right; (C) *Euglena* sp.; (D) *Urceolus cyclostomus;* (E) *Astasia* sp.; (F) *Astasia dangeardii;* (G) *Phacus longicauda.*

has been reported in a few species. As in other flagellated plants, the flagella are not easily observable, but can be seen if the organisms are slowed down by applying heat or certain chemicals or if flagellar stains are applied (Figure 12-2).

The function of the gullet is unknown. In contrast to Cryptophycopsida and Chrysophycopsida, where ingested food particles are often seen in the gullet and ingestion itself has been observed, there are no accurate reports of ingestion in euglenids (Doyle 1970). Likewise, the exact mode of phototaxis is not known. It has been suggested that a shadow of the stigma cast periodically on the paraflagellar body causes it to contract or expand and thus alter its orientation, slightly changing the course of the movement. When shadows no longer strike the paraflagellar body, the organism has to be moving straight toward the light. Attractive as the hypothesis sounds, there are reports of strains of *Euglena* which have no stigma but have as strong phototactic responses as those with stigmata.

Chlorophylls *a* and *b* in a ratio between 1:1 and 3:1 are responsible for the green color. Lutein and β-carotene are also abundant; neoxanthin is relatively abundant. Traces of cryptoxanthin, echininone, and α-carotene have also been reported. Reports of astaxanthin, both in euglenids and Chlorophycopsida, are controversial (Wolken 1967).

Electron micrographs show that the plastids of euglenids are very similar to those of Chlorophycopsida. The thylakoids show some tendency to be stacked in sets of four, or sometimes

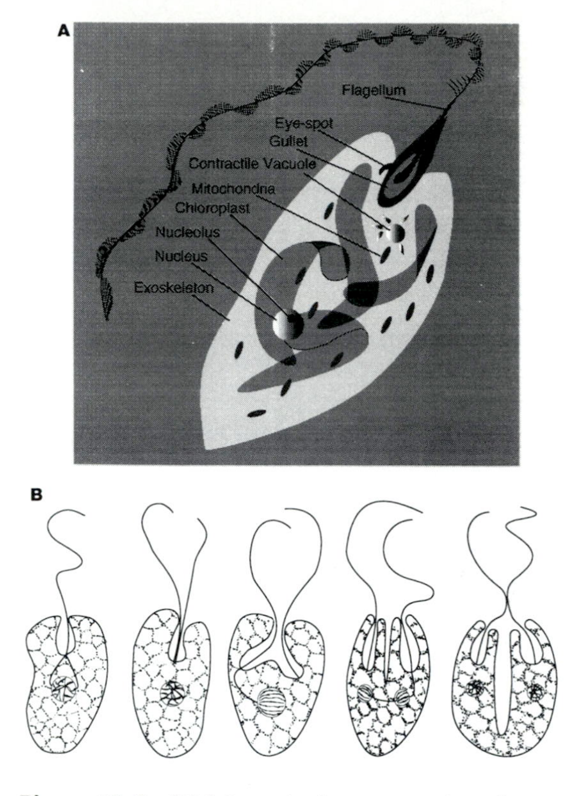

Figure 12–2 (A) Schematized representation of a euglenid; (B) cell division in a euglenid.

two, with stacking periodically in grana-like sets. Based on microspectrophotometry measurements (Figure 12-3), it has been calculated that there is at least one carotenoid molecule for every three chlorophyll molecules in the plastid (Wolken 1967). Also associated with chlorophyll in the plastids are DNA and ribosomes. Plastid DNA is different from nuclear DNA, however, and plastid ribosomes are different from ER ribosomes. According to Edelman, Schiff, and Epstein (1965), plastid DNA in *Euglena* has the same density as cyanophycean DNA, and according to Doyle (1970), plastid ribosomes have the same sedimentation rate as cyanophycean ribosomes; this suggests that euglenids could have evolved, as theorized by Margulis (1968), from an endosymbiotic relationship between early eukaryotes and blue-green or green prokaryotes, with the prokaryotic cells evolving into chloroplasts.

Different species of euglenids vary in rigidity; some have completely unskeletonized cells and move by means of pseudopods in amoeboid fashion, others have a firm periplast—also called a pellicle or an exoskeleton—that allows very little change in cell shape, and others are enclosed in a lorica. Most species, however, have a flexible periplast made of proteins lying just *inside* the plasma membrane. Such cells are still called "naked" because they have no cell wall or other structure *outside* the plasma membrane. Electron microscope studies have shown that the periplast is made up of a series of rings which partly fit inside each other (as in a collapsible cup) and thus allow for flexibility. Many species have striated pellicles which are usually spiral. Species identification is largely dependent on the pattern of the striations.

Three different kinds of structures occur at or near the cell surface of plants and provide rigidity in varying degrees to their cells: **periplasts** or pellicles, which are produced by the organism *inside* the plasma membrane and are usually proteinaceous in nature; **cell walls,** which are produced by the organism *outside* the plasma membrane as part of the cell and usually consist of cellulose but may contain chitin, lignin, silicates, carbonates, pectinates, or a host of other substances; and **loricas,** which are produced as a housing or shell completely free from the cell and are usually made of carbonates or silicates. Loricas are always open at the anterior end. Of the three, cell walls are typical plant structures and are often two or more layers in thickness. Only periplasts and loricas are found in the Euglenophyta. Both occur in animals as well as in plants.

The nucleus in euglenids is similar in appearance to nuclei of other eukaryotic organisms. As in most Lorobionta, it is relatively large and has a prominent nucleolus. There are also one or more **endosomes** associated with the nucleus; these resemble nucleoli in size and general

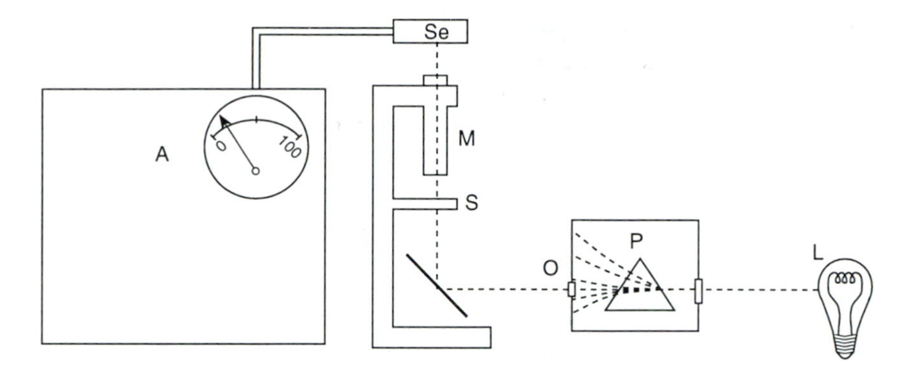

Figure 12–3 Microspectrophotometer used to measure the concentration of pigments in cells of *Euglena* and other algae and to measure phototactic and other light responses. L = light source; P = monochromator with prism which separates the different wave lengths making up white light; O = narrow opening or slit which permits light of only one color (very narrow range of wave lengths) to continue to the microscope, M; S = movable mechanical stage with specimen in clear liquid; Se = light-sensitive photoelectric cell which converts light into electricity which is then carried to the amplifier and oscilloscope, A, where the percent transmittance of light passing through the specimen can be measured. By moving the specimen, S, the amount of light absorbed at any wave length by the specimen can be compared with the light absorbed at the same wave length by the clear liquid it is in. By comparing the two values, percent transmittance can be obtained.

appearance, but show different staining characteristics. Mitosis is different from that in many eukaryotic organisms in that the chromosomes remain condensed during interphase, as in the Dinophycopsida. The nuclear membrane is intact during metaphase and anaphase, no spindle fibers form, and it takes 2 to 5 hours to complete the two stages in contrast to 2 hours or less in most organisms and as little as 30 minutes in some. Colchicine does not have any influence on mitosis in euglenids, presumably because there are no spindle fibers. Anaphase in *Euglena* averages about an hour, in other organisms it averages about 15 minutes, with a maximum of 30.

Asexual reproduction is by simple mitotic fission of vegetative cells. Thick-walled cysts are formed in some genera. Sexual reproduction has never been observed in any species of *Euglena* or most other Euglenophycopsida. Conjugation of gametes has been reported in one genus, *Scytomonas,* but meiosis has never been observed; apparently, neither sexual nor parasexual modes of reproduction are common in the Euglenophycopsida. Considering the great amount of laboratory research work that has been undertaken with *Euglena* and the peculiarities in euglenid mitosis, it seems odd that if meiosis does occur, it has never been observed. It is possible that sexual reproduction has never evolved in this genus, and possibly in the entire division.

Phylogeny and Classification

Euglenids are soft-celled organisms that do not fossilize well; that does not mean that we will never find any fossil evidence of their origin. Since the discovery about 1955 that protein molecules can be preserved for millions of years, molecular or biochemical paleobotany has made some important contributions to our understanding of evolution. Some lipids seem especially able to persist in geological formations for extremely long periods of time. We may yet find evidence of euglenids in the fossil record even if we never find morphologically identifiable cells.

There are some 520 species of euglenids; these are classified in two orders, the Euglenales with 2 families and 13 genera and the Colaciales with 1 genus and 4 species. *Euglena* with 155 species, *Trachelomonas* with 150 species, and *Phacus* with 130 species are the largest and best-known genera. *Astasia* with 10 species seems to be identical to *Euglena* except that the cells are colorless; however some significant physiological differences were recently noted when laboratory-induced colorless strains of *Euglena* were compared with their counterpart *Astasia* species, and so the genus has been retained instead of merging it with *Euglena,* as proposed by some.

The Euglenaceae probably evolved from Prochloralean stock during the Paleohelikian about the same time the Chlorophycopsida originated. Wolken (1967) pictures the euglenids as descended from unicellular Chlorophycopsida; more likely, both descended from the same ancestral stock along slightly different lines. The Cryptophycopsida may have also evolved from the same or similar stock soon after. From the Euglenaceae have evolved the other two families of Euglenophycopsida. Suggested lines of relationship are illustrated in Figure 12-4 and characteristics of the two orders are summarized in Table 12-1.

Ecological and Economic Significance

The euglenids are probably not significant producer organisms, except in sewage lagoons where, with the Chlorophycopsida and Cyanophycopsida, they provide much of the oxygen needed to break down the organic matter (Figure 3-24). As the sewage enters a lagoon, bacteria act on it, releasing nitrates, ammonium, potassium, sulfates, phosphates, and other minerals, but depleting the oxygen supply. *Chlorella, Scenedesmus, Euglena, Oscillatoria,* and other green and blue-green algae are important in restoring the oxygen so that the breakdown can continue.

Because euglenids require organic nitrogen to grow, they are among our best indicators of water pollution, a characteristic having considerable practical significance. C. M. Palmer listed 60 genera and 80 species of freshwater algae according to tolerance to pollutants. *Euglena* was one of the five most tolerant genera. The five most tolerant species were *Euglena viridis, Nitzshia palea, Oscillatoria limosa, Scenedesmus quadricauda,* and *Oscillatoria tenuis.* But not all euglenids were tolerant of water pollutions; the five most sensitive species in Palmer's list were *Tetraedron muticum, Pyrobotrys gracilis, Euglena proxima, Gonium pectorale,* and *Cryptomonas ovata.* Samples of water containing any or all of the most tolerant species and none of the most sensitive species are more seriously polluted than those containing many of the sensitive species and few of the tolerant species. A sample containing only one or two tolerant species, and no other species, is more seriously polluted than a sample containing all of the most tolerant species; species diversity is less in polluted water than in clean water.

Research on the structure and function of the flagella and of motility in euglenids has revealed that, at low light intensity, a very slight increase in light will invariably cause a quick response and the illuminated cells will suddenly increase their velocity many-fold. This is most pronounced in the blue-violet part of the spectrum (Figure 12-5).

Research Significance

Although euglenids were probably among the first microorganisms to be viewed with the microscope, they were never studied to any great extent, beyond very basic taxonomic observations, from the time van Leeuwenhoek discovered them in 1675 until the second half of the

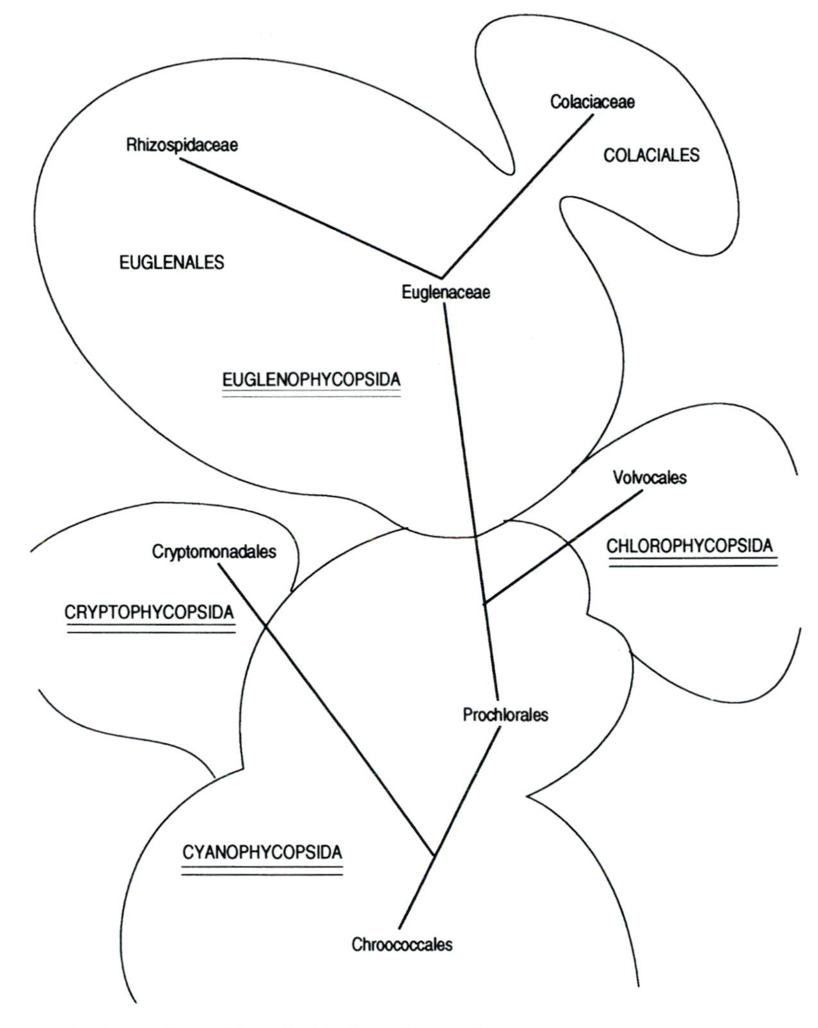

Figure 12–4 Lines of relationship within the Euglenophycopsida.

20th century. Few research papers on Euglenophycopsida were published prior to about 1950, but there have been hundreds since that time.

Euglena gracilis and some other euglenids have emerged as especially useful in studies of general cell morphology, cellular biochemistry, growth curves and patterns, pigment synthesis, photosynthesis, phototaxis, photoexcitation, respiration, and related phenomena (Figure 12-5). Euglenids promise to be of continuing value in scientific research since they are easy to culture, there is considerable variability in both morphological and physiological characteristics, the cells can be bleached of chlorophyll in a number of ways, some of which are reversible, growth rate is rapid, and they contain chemical substances also found in organisms not as easy to work with. In one important area, however, *Euglena* has not yet made significant contributions to scientific research: lacking sexual reproduction, they have not been used in genetic

TABLE 12–1

GENERAL CHARACTERISTICS OF THE ORDERS OF EUGLENOPHYCOPSIDA

Orders (Fam–Gen–Spp)	General Morphological Characteristics	Habit	Habitat	Nutrition	Representative Genera
Euglenales (2-13-520)	Discoid, band-shaped or stellate plastids with pyrenoids; pyrenoids not associated with plastids also occur; cells with 1-3 flagella	Strictly unicellular	Mostly freshwater in water high in organic matter; a few marine	Autotrophic, saprophytic, or holozoic	*Trachelomonas, Euglena, Astasia, Phacus, Ottonia, Peranema, Urceolus, Rhizaspis, Rhynchopus,*
Colaciales (1-1-4)	Discoid chloroplasts with or without pyrenoids; vegetative cells lack flagella; spores uniflagellate	Palmelloid colonies of up to 20 or more cells	Epizoic on copepods, rotifers, and other fresh water plankters	Autotrophic	*Colacium*

Note: The number of families (Fam), genera (Gen), and species (Spp.) in each order is indicated in the first column.

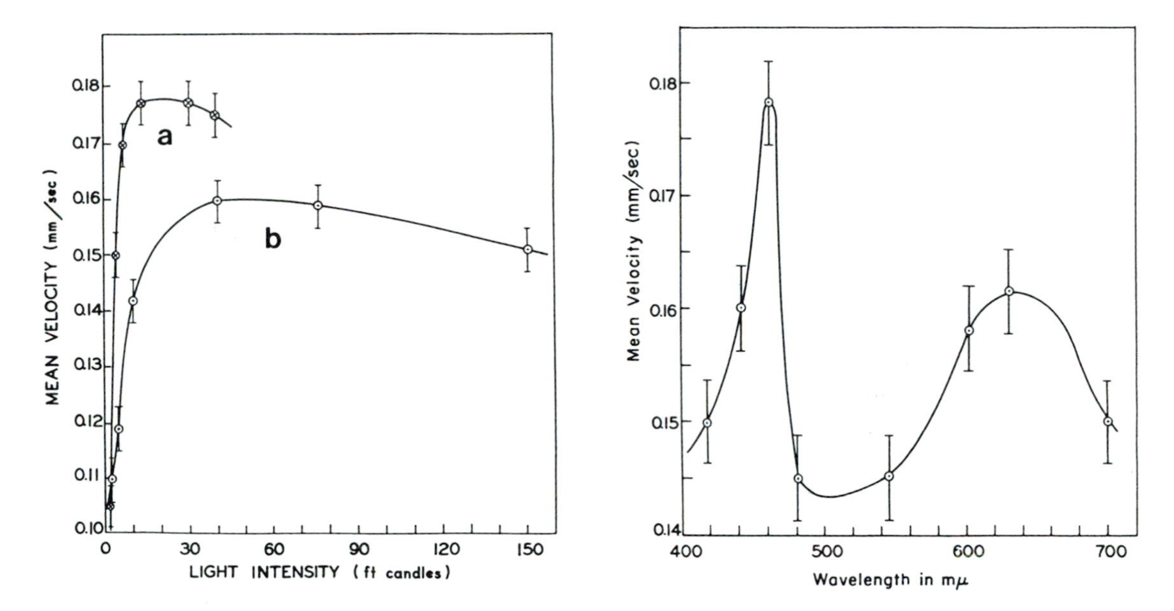

Figure 12–5 Phototactic responses in euglenids. Left: maximum cell velocity is reached at very low light intensity, especially with polarized light; a = response to nonpolarized light, b = response to polarized light. Right: phototactic response is greatest in blue light within a very narrow range, slightly over 460 nm, and second greatest in red light within a much wider range centering around 625 nm. (From Wolken, J. J., *Euglena, An Experimental Organism for Biochemical and Biophysical Studies,* Appleton-Century-Crofts, New York, 1967. With permission.)

studies. However, mutations frequently occur, and there is therefore considerable genetic variability in strains of *Euglena gracilis,* a species widely employed in physiological research at the present time.

Euglenids have shown promise in recent years as bioassay organisms in nutrition research. Vitamin B_{12}, for example, does not occur in very many foods but is important in human nutrition. It is a cofactor in numerous reactions involving shifts of carbon and hydrogen atoms within organic compounds. A nucleotide reductase reaction utilizes a B_{12} coenzyme in *Euglena gracilis* to reduce ribose to deoxyribose in the synthesis of DNA. *E. gracilis* cells grown in a medium deficient in vitamin B_{12} become abnormally large because nuclear division, and hence cell division, cease while RNA and protein synthesis continue (Shubert 1984). A small amount of the food to be tested for presence of this vitamin can be added to a medium containing the minerals necessary for growth of *E. gracilis* but lacking B_{12}. If the euglenid grows normally, we know that the food being tested contains B_{12}, but if the cells grow abnormally large, we know that B_{12} is lacking in it.

Some of the earliest scientific research utilizing euglenids also involved nutrition. It was soon learned that green euglenids as well as colorless ones require organic sources of nitrogen, and that green ones become colorless when grown in the dark for several generations; however, color can be restored after a few generations in the light. It was then learned that when *Euglena* is grown at higher temperatures, cell division outstrips plastid division and after a few generations, colorless strains of *Euglena,* completely devoid of chloroplasts, can be obtained.

These cannot generate new chloroplasts and remain, therefore, forever colorless, completely dependent on an organic food supply in the medium. The two kinds of colorless or bleached Euglenas differ in significant ways. Cells bleached as a result of growing in the dark contain a type of satellite DNA not found in cells bleached as a result of not having any plastids, indicating that the satellite DNA is located in the plastids.

Streptomycin, ultraviolet light, and other agents also bleach *Euglena gracilis* cells. Streptomycin-bleached cells lack plastid ribosomes although they contain cytoplasmic (ER) ribosomes; streptomycin also prevents the formations of ribosomes in bacteria, a fact of considerable interest to proponents of the endosymbiosis-origin hypothesis of eukaryotic cells.

Because of the special characteristics of euglenid cells, they will continue to be used extensively in scientific research. They should be of special value in studying the cytochrome system in the mitochondria, eyespot pigments, periplast nature, and water pollution. Even though periplasts occur in a number of plants and animals, we know very little about their structure. Water pollution problems are now very real in almost all parts of the world and promise to become more serious.

SUGGESTED READING

Euglena, an Experimental Organism for Biochemical and Biophysical Studies, by Jerome J. Wolken, is an excellent source of information on these interesting unicellular plants. *Algae as Ecological Indicators,* edited by L. Elliot Shubert, includes information on all groups of algae, including the euglenids.

Every student of botany who has any interest in the euglenids—whether as research material, ecological indicators, or other purposes—should become well acquainted with two scientific journals, *Journal of Phycology* and *Journal of Protozoology,* which carry the latest research on algae in general and unicellular algae, respectively.

STUDENT EXERCISES

1. Obtain several samples of pond or mud puddle water, preferably from a barnyard area, and examine for euglenids. How many species can you find?

2. Take the sample from exercise 1 with the most euglenids in it and divide it into 10 or 12 parts. Put a small amount of ammonium nitrate in two of these aliquots, a slightly larger amount in another two, and so on, leaving two units unchanged as controls. Randomize the units within each of the two replications, place in two well-lighted locations, and leave them for a week. At the end of that time, sample each unit by placing a drop of water on a microscope slide, cover with a cover slip, and calculate the number of cells in each unit. Which of the five or six treatments resulted in the greatest number of euglenid cells per drop of water?

3. Why do you suppose there was so little scientific research conducted with euglenids prior to the 1950s?

4. There is probably a copy of C. Dobell's book, *Antony van Leeuwenhoek and his "Little Animals,"* in your college library. Find it and read the entire section in which

van Leeuwenhoek describes the green flagellates. Does his descriptions of any of these "animalcules" match your observations of common euglenids?

5. Lists of the five most pollution-tolerant and the five most pollution-sensitive species of algae, as described by Parker, is presented in the ecological section of this chapter. To which class and division does each of these ten species belong?

13 THE POND "MOSSES," SIPHONOPHYTES, AND STONEWORTS

One of the most interesting plants on Planet Earth is the "mermaid's wineglass," *Acetabularia mediterranea,* sometimes common in shallow water in the Mediterranean and other tropical and semitropical seas. As the common name, "mermaid's wineglass," implies, it resembles a miniature green goblet, typically 2 to 5 cm tall. Like wineglasses made of fine crystal, herbarium specimens of *Acetabularia* are fragile and must be handled with tender care. The species in its natural environment is also ecologically fragile and in danger of becoming extinct as human activities alter marine ecosystems; hence it attracts the attention of ecologists and conservationists.

It has also attracted the special interest of plant anatomists, morphologists, and morphogeneticists. In common with some other members of the Siphonophycopsida, it seems to defy the cell theory, for the entire plant, with its "roots," "stems," and photosynthetic crown, is a single cell with the nucleus in one of the rhizoids, or "roots," which anchor the plant to the ocean bottom. How can an organism have such distinctive differentiation of its body parts yet not be built up of differentiated cells and tissues?

The attention of plant physiologists, too, has been attracted to the mermaid's wineglass. Much of our understanding of circadian rhythms has come from research with this interesting plant. A **circadian rhythm** is the repetition of a physiological process approximately every 24 hours independent of outside environmental forces; "jet lag" is the manifestation of such a rhythm in humans. When these rhythms were first observed in a variety of organisms, it was hypothesized that they were controlled by something in the nucleus. Because of the ease with which its nucleus and other cellular structures are removed, *Acetabularia* has been ideal for studying circadian rhythms and the organelles involved in producing them.

"Landlubbers" seldom have the opportunity to observe either the mermaid's wineglass or any of its close relatives in their natural habitats, for the Siphonophycopsida are almost entirely marine. Members of the other two classes of Chlorophyta, on the other hand, occur in freshwater habitats and can be studied directly by anyone. In ponds and streams, for example, "pond mosses" (Chlorophycopsida) are always abundant, as every fisherman knows. Among the most common are several species of "mermaid's tresses," members of the genus *Spirogyra,* frequently found tangled in fishing tackle as it is reeled in.

The Division Chlorophyta (*chloro* = green, *phyta* = plants), with its three classes, is one of the most widespread of all the plant divisions. Possibly members of this division can be found on every square kilometer of the earth's surface with exception of the most extreme polar and desert environments. They are not limited to streams, ponds, and tropical seas; the genus *Trebouxia,* for example, as the phycobiont of many lichens, has been found on rocky cliffs within a few miles of the South Pole and commonly grows on desert rocks, while *Chlamydomonas gloeocystiformis* thrives in the Great Salt Lake and on the deserts to the west. Green algae make up much of the marine plankton, and many species are instrumental in the building of tropical reefs.

The division is an ancient one, dating from early neohelikian time or earlier. The oldest eukaryotic fossils which have been found resemble modern unicellular Chlorophyta such as *Chlamydomonas.* The green algae slowly increased in diversity and abundance, judging from the fossil record, throughout the remainder of the Proterozoic era, but they were never as abundant as the Cyanophycopsida until late Hadrynian. At that time, unicellular animals evolved and grazed many of the prokaryotic species to extinction; however, the larger green algae were apparently able to resist the destructive onslaught of the herbivores. The larger size of the eukaryotic cell is possible because of compartmentalizing of enzyme and other chemical systems into separate and distinct organelles, each with its specific function to perform.

A perusal of Figure 13-1 suggests that the Chlorophyta have evolved along three lines, one leading to the Siphonopycopsida or calcareous green algae, one to the Charophycopsida or stoneworts, and one to the Bryophyta and possibly the Tracheophyta.

The division has much in common with the mosses, liverworts, and vascular plants. Green algae contain the same chloroplast pigments, store food in the same form, build cell walls of the same materials, have the same kind of flagella, have similar chromosomes, nuclei, mitochondria, ribosomes, dictyosomes, and plastids, and possess similar adaptations which enable them to survive periodic dry conditions. Because of these similarities, Doyle (1970) included the Chlorophyta, Bryophyta, and Tracheophyta in one division which he called Chlorophyta. On the other hand, Bold (1957) and others have divided the Chlorophyta, as circumscribed in this book, into two divisions, the Chlorophyta and Charophyta, each with a single class. The Siphonales are usually treated as an order of the class Chlorophycopsida; however, their pigmentation, cell wall chemistry, fossil history, and structure are so different that it seems best to treat them as a separate class. As discussed in this book, therefore, the division consists of three classes, the Chlorophycopsida, the Siphonophycopsida, and the Charophycopsida. Suggested relationships among the three classes and nine orders are illustrated in Figure 13-1 and some representative genera in Figure 13-2.

CLASS 1. CHLOROPHYCOPSIDA

The Chlorophycopsida (*chloro*-green, *phycus*-alga) is a relatively large class with approximately 6000 species. Common names include green algae, pond moss, ditch moss, frog soup, and just plain "moss" which they are not. To most people they are something of a mystery, to fishermen they are often a nuisance, and to limnologists they are a very important group of plants. There are 320 genera classified in 36 families and 6 orders. The diversity found in the class is suggested in Table 13-1.

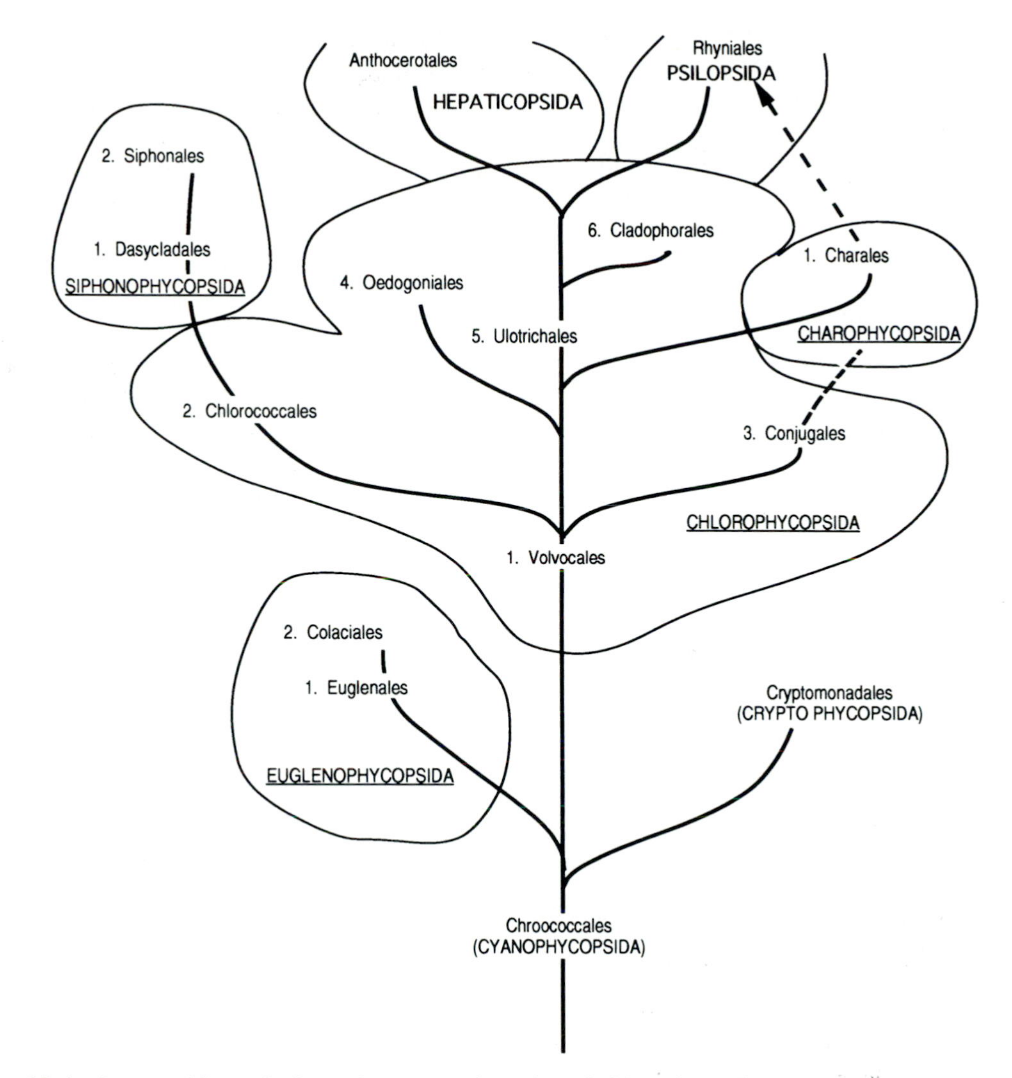

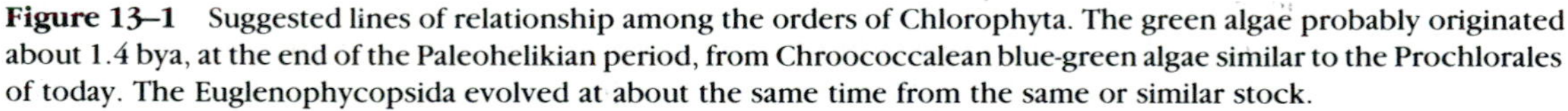

Figure 13–1 Suggested lines of relationship among the orders of Chlorophyta. The green algae probably originated about 1.4 bya, at the end of the Paleohelikian period, from Chroococcalean blue-green algae similar to the Prochlorales of today. The Euglenophycopsida evolved at about the same time from the same or similar stock.

Green algae can always be found in abundance in ponds, lakes, and even mud puddles, wherever a green mat or scum covers the water surface. In late summer, the water of ponds and slow-moving streams will often take on a greenish tinge because of the abundance of unicellular chlorophytes, and they also occur in the cooler parts of hot springs, in saline ponds and lakes, and in the ocean. They are not limited to aquatic environments, however, but often form green patches on the shady side of trees or rocks and grow profusely on damp cliffs. Their green mats are common on damp soil; the phycobiont of most lichens is a green alga, and the

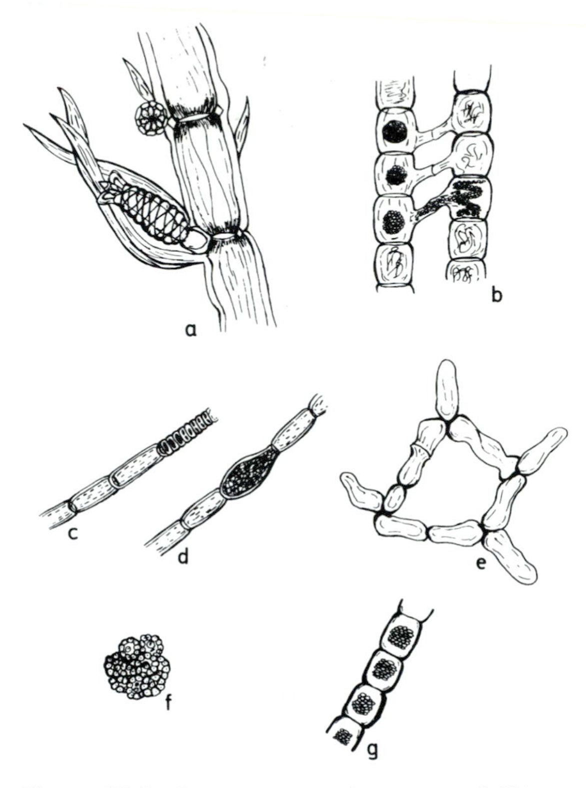

Figure 13–2 Some representative genera of Chlorophyta. (a) *Chara* (Charales); (b) *Spirogyra* (Conjugales); (c) *Oedogonium* with antheridia (Oedogoniales); (d) *Mougeotia,* reproductive cell (Conjugales); (e) *Hydrodictyon* (Chlorococcales); (f) *Palmella* (Volvocales); (g) *Microspora* (Ulotrichales).

"watermelon snow" seen in late summer on snowbanks and glaciers in the mountains is most commonly a member of the Chlorophycopsida.

A single drop of water from a pond will often contain a dozen species of Chlorophycopsida plus a few species of diatoms, blue-green algae, and other small plants and animals. Most chlorophytes are readily identified by their bright green color, but students sometimes have difficulty distinguishing them from other microscopic plants, expecially Xanthophycopsida. Chloromonads and epiphytic heterokonts also cause confusion. If the plastids are large and unusual in shape, ribbon-like for example, the plant in question is almost certainly a chlorophyte, but if overlapping cell halves can be discerned, or H-shaped segments be seen where filaments have broken, it probably is not. Chlorophycean plastids are usually granular, have rough or irregular margins and prominent pyrenoids, and are usually a brighter and clearer green than those of other algal groups.

General Morphological Characteristics

Chlorophycopsida may be unicellular, colonial, filamentous, or rarely parenchymatous. Some unicellular species are motile, others are nonmotile and free-living, epiphytic, or endophytic. Some free-floating species have horns or other ornamental projections on them (Figure 13-3): these apparently aid in maintaining buoyancy. The algae in Figure 13-3 belong to the Volvocales and Chlorococcales.

Motile colonies are usually spherical with nonflagellated cells on the inside and flagellated cells on the outside (Figure 13-3D), but they may be flat sheets or cubical masses.

Among the most beautiful of all algae are the desmids, which are members of the Conjugales, as is *Spirogyra,* a very common genus known as mermaid's tresses (Figure 13-4D). Mermaid's tresses is a filamentous alga; the desmids (Figure 13-4) are nonmotile unicellular organisms.

Flagella, which vary from one per cell to several, usually two, are of the whiplash type. As in the Euglenophycopsida, these arise from a **centriole** lying near the nucleus. Division of the centriole is correlated with mitosis. A **stigma** or photosensitive "eyespot" is also associated with the flagella in many species, especially in the zoospores. In addition, there are one or two **contractile vacuoles** in each cell of some species. They apparently aid in controlling osmotic concentration by pumping out fluid. They contract quite rapidly, but distend very slowly. When two are present, the contractions generally alternate with each other.

The plastids, usually one or a small number in each cell, vary in size and shape more than in other plants. Some are in the form of large sheets or nets that lie just under the plasma membrane; others are in the form of plates, disks, or ribbons. Associated with each plastid, often in the center, is a membrane-bound structure called a **pyrenoid** (Figure 13-4). In some large plastids, there may be several pyrenoids. Each pyrenoid consists of a proteinaceous central core surrounded by starch plates. In some species, pyrenoids arise only from previously existing pyrenoids, in other species, they arise *de novo,* their origin controlled by genes in the nucleus.

Many Chlorophycopsida have a single, net-like plastid in each cell. The strands of the net are typically very fine except where the pyrenoids occur. As the cell gets older, the strands of the nets become increasingly coarse until the plastid has become a large perforated plate. At this stage, many students have a difficult time recognizing that there is a plastid in the cell, but describe the cell as having chlorophyll merely dissolved in the cell sap as in the Cyanophycopsida. Careful examination of such cells with the high power or the oil immersion objective of a microscope is often necessary to reveal the true nature of the plastid. As the cell continues to age, the perforated plate often breaks up into small disks with ragged edges and usually with one pyrenoid in each disk.

In other Chlorophycopsida, the plastid may be a single open ring, or there may be one or more ribbon-like plastids that form spirals around the protoplast, or two ragged, star-like plastids may be present (Figure 13-4). *Spirogyra* has ribbon-like spiral plastids. This genus is widespread and often very common in freshwater ponds. There are many species and it may be difficult to distinguish among them; however, identification is frequently possible through a combination of chloroplast characteristics: number of plastids per cell, whether they are wide or narrow, and whether the margin is entire or wavy (Table 13-2).

The ultrastructure of the organelles is similar to that of the euglenids (Figure 13-5). Figure 13-5 is a TEM microphotograph of *Trebouxia parmeliae,* the phycobiont of a lichen, *Lecanora melanopthalma;* note how the thylakoids tend to be stacked in pairs or in sets of four, but here and there the beginnings of grana-like stacking can often be observed. The mitochondria contain plate-like cristae and are thus typical for the subkingdom. Structure of the endoplasmic reticulum and nature of the ribosomes are like that of other Lorobionta.

Chlorophylls *a* and *b* (in a ratio of about 1:1), β-carotene, neoxanthin, violaxanthin, and zeaxanthan are the principal plastid pigments: α-carotene, astaxanthin, and lycopene have also been reported. Food is stored as starch, built up of branched amylose molecules that consist of about 20 glucose units each, and unbranched amylopectin molecules that consist of about 1000 glucose units each. In chlorophycopsida, as in other Chlorophyta, these are in the ratio of about 5:1. Cell walls are made up of very orderly arranged cellulose units.

Both sexual and asexual reproduction are common. Asexual reproduction is often by means of **zoospores.** In *Cladophora* and *Ulothrix,* for example, the contents of some of the cells will become flagellated asexual spores which swim around for a while, then settle on a suitable substrate, withdraw their flagella, and germinate into new filaments. In the Conjugales, asexual spores are never formed, but vegetative reproduction by means of fragmentation of mature filaments provides an effective means of asexual reproduction.

Sexual reproduction has been observed in many species in all of the orders. **Haplontic life cycles** (Figure 2-9), in which the only diploid stage is the zygote, are the rule in four of the six orders. The zygote forms a thick wall and serves as a resting cell capable of tiding the

TABLE 13-1

GENERAL CHARACTERISTICS OF THE CHLOROPHYCOPSIDA

Orders (Fam–Gen–Spp)	General Morphology	Plastids	Life Cycles	Habitat	Representative Genera
Volvocales (14-51-570)	Unicellular, motile or nonmotile, or colonial, rarely branched filaments; 1 family with naked cells; contractile vacuoles and red stigma often present	Mostly cup-shaped, one per cell, with single pyrenoid; rarely stellate, disk-like or parietal; pyrenoids may be absent, may be 2 to many	Mostly haplontic; biflagellate meiospores and gametes	Many freshwater, also marine and in saline lakes; a few species in alpine summer snow banks	*Tetraspora, Volvox, Palmella, Eudorina, Chlamydomonas, Pedinomonas, Prasinocladus, Caryosphaeroides, Dunaliella*
Chlorococcales (11-90-700)	Unicellular, coenocytic, or in nonfilamentous colonies; lack cell division except in reproduction	Large, cup-shaped with single pyrenoid or ragged disks with 1 to many pyrenoids	Both haplontic and diplontic species occur; meiospores are biflagellate	Mostly freshwater, many marine species; one genus parasitic, others symbiotic with fungi in lichens; many epiphytic on bark	*Chlorella, Trebouxia, Hydrodictyon, Pediastrum, Characium, Scenedesmus, Selenestrum, Chlorangium*
Conjugales (3-38-2750)	Filamentous in one of the families, unicellular in the others; cell walls stratified with gelatinous outer layer	Ribbons, bands, or ragged star-like disks with 1 to many pyrenoids; margins ragged, plastids granular	Mostly haplontic; meiospores and gametes nonmotile	Mostly freshwater	*Spirogyra, Zygnema, Mougeotia;* also *Spirotaena, Cosmarium, Closterium, Xanthidium,* and other desmids

Oedogoniales (1-3-350)	Branched or unbranched filaments with formation of unique cell caps; cell wall of 3 layers, outermost of chitin	Net-like sheets with pyrenoids at net intersections	Haplontic with stephanokont meiospores and sperm; male filaments epiphytic on oogonia	Freshwater and terrestrial; a few species epiphytic on other algae, mosses, and liverworts	*Oedogonium, Oedocladium, Bulbochaete*
Ulotrichales (10-89-565)	Filamentous, or flat sheets 1 or 2 cell layers thick, or tubular sheets, or parenchymatous; Cell walls 2-layered, pectinate outer layer	Ring-like plastids encircling ⅔ of the cell; pyrenoids 1 to many; also parietal plates; 1 stellate plastid in 1 family	Haplontic and diplohaplontic with biflagellate meiospores and gametes	Marine and freshwater	*Ulothrix, Ulva, Enteromorpha, Microspora, Hormidium, Chaetophora, Prasiola, Fritschiella, Schizogonium*
Cladophorales (2-12-340)	Coenocytic filaments, often branched profusely, some genera unbranched; cell walls as in the Ulotrichales	Net-like, becoming plate-like in age; often breaking up later into ragged disks	Both diplontic and diplohaplontic, with isogamous chlamydomonad gametes and biflagellate spores	Marine and freshwater	*Cladophora, Rhizoclonium, Pithophora*

Note: The number of families (Fam), genera (Gen), and species (Spp.) in each order is indicated in the first column.

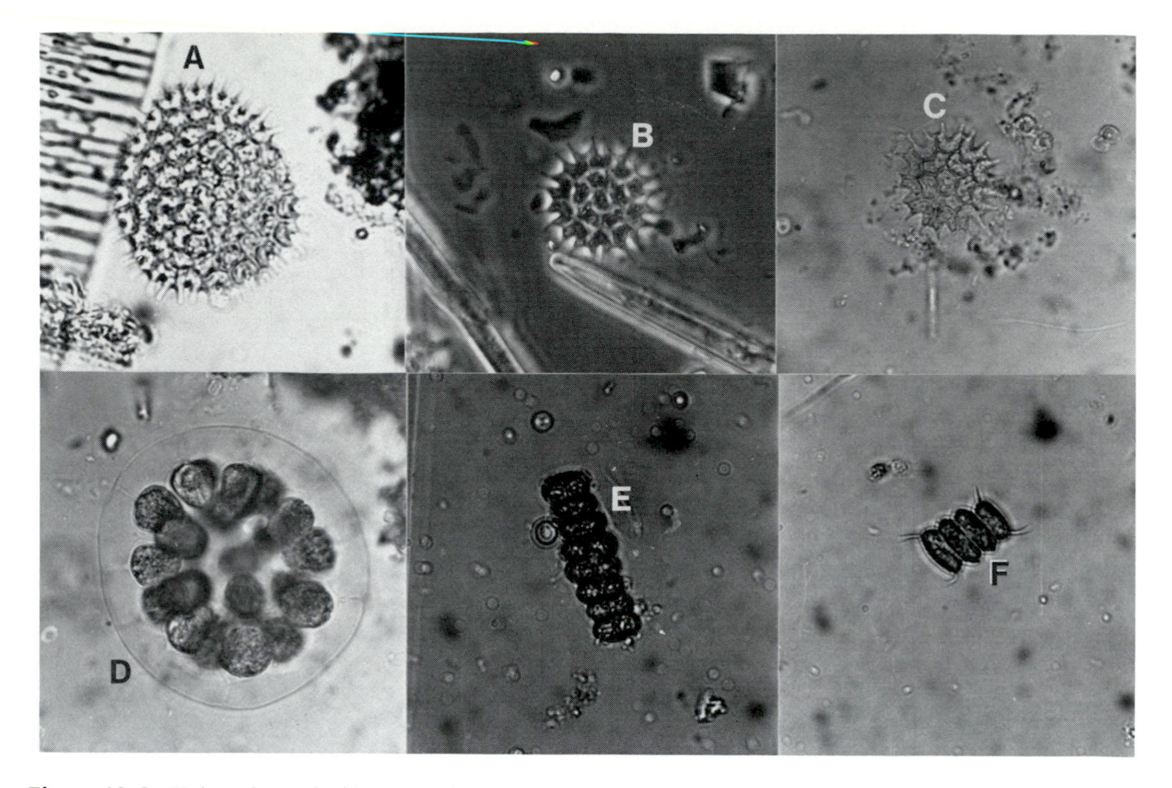

Figure 13–3 Volvocales and Chlorococcales. (A) *Pediastrum boryanum;* (B) *Pediastrum duplex;* (C) *Pediastrum tetras;* (D) *Sphaerocystis schroeteri;* (E) *Scenedesmus bijuga;* (F) *Scenedesmus quadrisina.* (Micrographs courtesy of Richard Clark, Ricks College, Rexburg, ID.)

population over periods of drought, as when a mud puddle dries up. In some species of *Chlamydomonas,* flagellated vegetative cells divide to form motile, biflagellate isogametes. Conjugation occurs at a fixed time each day, at 2 a.m., for example, in one species. The zygotes remain dormant for several weeks. Upon germination, they produce motile meiospores which gradually grow into vegetative cells almost identical in appearance to the meiospores. The zygotes of *Ulothrix* typically lie dormant for 5 to 9 months after formation. They then divide meiotically, producing biflagellate meiospores that resemble the meiospores of *Chlamydomonas.* These swim around for a while, then lose their flagella and settle to the bottom of the stream or pond. They immediately divide mitotically to produce two daughter cells: the lower one elongates, becoming narrow at the basal end, and serves as a holdfast, anchoring the filament to the substrate. The upper one divides transversely to become a photosynthetic, multicelled filament.

In one group of chlorophytes, the Zygnematales or Conjugales, flagellated spores or gametes never occur. When environmental conditions are conducive to sexual reproduction, filaments lying in proximity to each other orient themselves in a parallel manner. Conjugation tubes form between adjacent filaments, if they are of different mating types (Figure 13-2b). In some species, syngamy takes place in the conjugation tube. In most species, the contents of each cell in the "female" filaments round up and become eggs, each cell an oogonium; the cellular contents of the "male" filaments migrate, by amoeboid motility, into the oogonia,

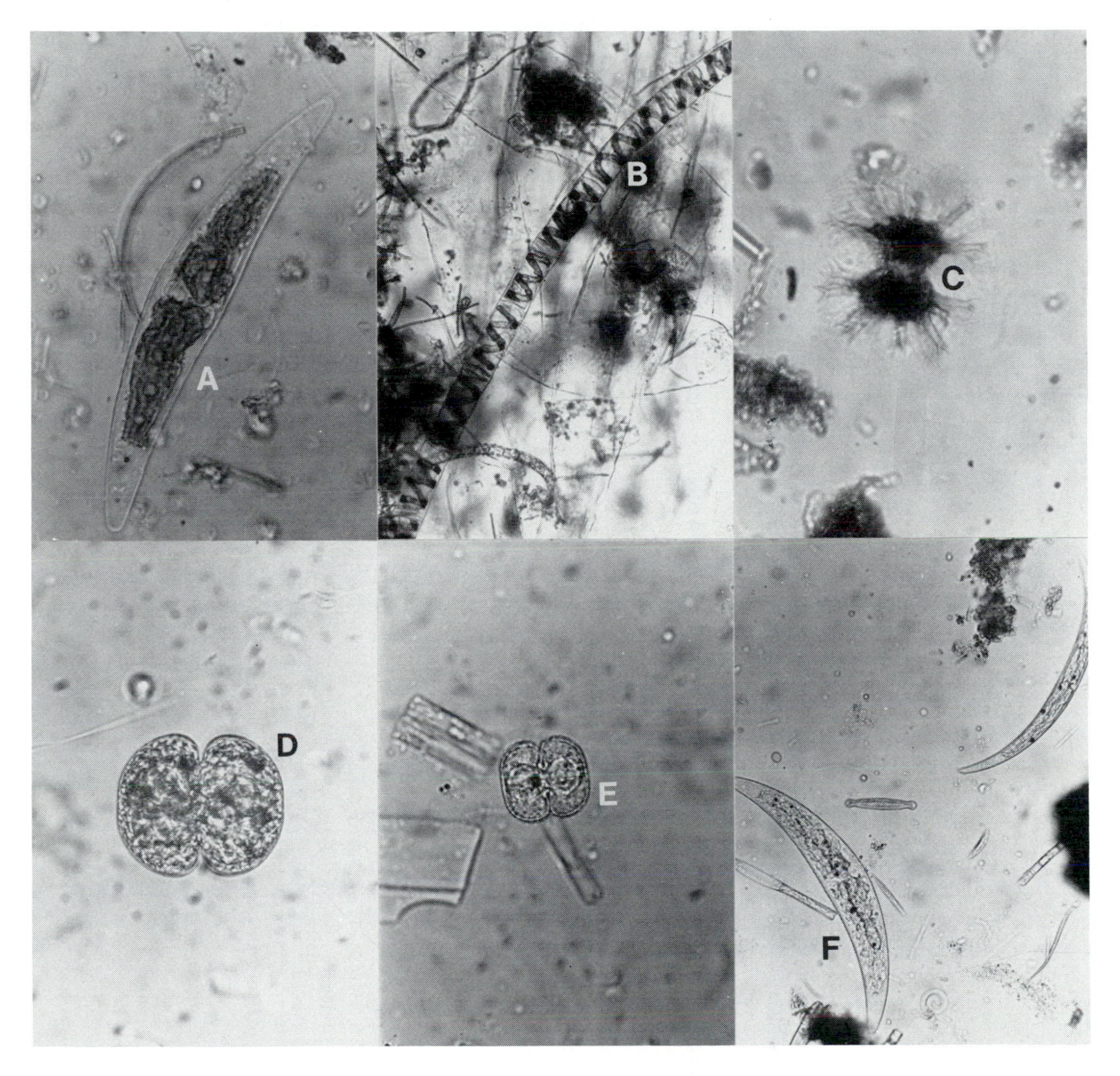

Figure 13–4 Conjugales. (A) *Closterium acerosum;* (B) *Spirogyra communis;* (C) *Staurastrum desmidium;* (D) *Cosmarium intermedium;* (E) *Cosmarium circulare;* (F) *Closterium moniliforme.* (Micrographs courtesy Richard Clark, Ricks College, Rexburg, ID.)

where syngamy takes place (Figure 13-6A). The zygotes remain dormant, often for a year or more, and finally divide by meiosis. Three of the potential meiospores disintegrate; the other germinates and grows into a vegetative filament.

In many Chlorococcales and Cladophorales, **diplohaplontic life cycles** are the rule. In most species of *Cladophora,* for example, biflagellate meiospores, very similar in appearance to those of *Chlamydomonas,* are produced by the sporophytes and germinate to give rise to a gametophyte generation (Figure 13-6B). The meiospores are produced in **meiosporangia,** and similar appearing biflagellate isogametes are produced by the gametophytes in **gametangia.**

TABLE 13-2

CHARACTERISTICS OF SOME COMMON SPECIES OF *SPIROGYRA*

Species	Cell Ratio: Diameter (μm)	Diameter to Length	Chloroplasts: Number and Morphology	Are Fertile Cells Swollen?	Zygotes (Spores)	
					Shape	**Diameter (μm)**
S. communis	20-25	1:3-5	1, slender, wavy	No	Elliptical	19-23
S. porticalis	30-48	1:2-6	1, broad, dentate	No	Ovoid-globose	42
S. varians	33-40	1:2-3	1, broad, dentate	1 Side	Oval to elliptic	38
S. catenaeformis	24-27	1:2-5	1, broad, dentate	Yes	Elliptical	30
S. decimina	34-40	1:2-4	2-3, broad, entire	No	Globose to ovoid	38
S. crassa	60-150	1:1-2	6-10, slender	No	Ovoid, flat	45-140
S. lutetiana	30-36	1:3-7	1, broad, dentate	Sometimes	Variable	30-43
S. neglecta	50-65	1:2-5	3, slender, wavy	Yes	Ovoid	50-60
S. fluviatilis	34-38	1:5-6	4, slender, wavy	Yes	Ovoid	50-80
S. dubia	43-50	1:1.5-3	2, slender, wavy	Slightly	Ovoid to ellipsoid	40
S. inflata	15-18	1:3-8	1, broad, tight	Yes	Ellipsoid	30-36
S. weberi	22-28	1:6-16	1, slender, loose	Slightly	Ovoid	26-30
S. grevilleana	28-33	1:3-10	1 (2), broad, crenate	Yes	Ovoid to rounded	30-36
S. insignis	38-45	1:4-12	3 (2), slender	Yes	Ellipsoid	48

From Flowers, S., *Algae of Utah,* University of Utah, Salt Lake City, 1948. With permission of Mrs. Seville Flowers.

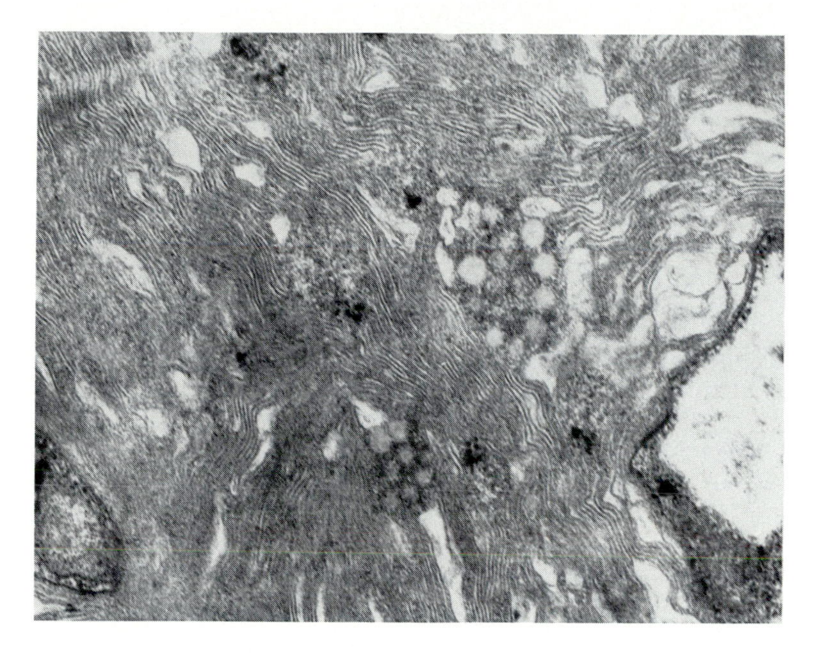

Figure 13–5 Electron micrograph of *Trebouxia parmeliae.* (Micrograph courtesy W. M. Hess, Brigham Young University Microscopy Lab, Provo, Utah.)

The flagella are at the anterior pole of the meiospores, and they escape from the sporangium through a pore, usually posterior end first. After swimming around for a while, they develop into the much-branched filaments of the young *Cladophora* gametophytes. The gametophytes of some species are **homothallic,** others are **heterothallic.** In heterothallic species, the gametophytes of different mating types are identical in appearance, but conjugation can occur only between gametes of different mating type. The zygotes germinate by means of mitosis and develop into branched, filamentous sporophytes identical in appearance to the gametophytes.

Although most species of *Cladophora* have isomorphic diplohaplontic life cycles, *C. glomerata* is **diplontic** and gametes are produced in gametangia on diploid plants directly by meiosis (Figure 13-6C). Unlike species having an alternation of generations (or diplohaplontic life cycle), the only haploid cells in species having diplontic life cycles are the gametes.

Phylogeny and Classification

Stromatolites from the Paleohelikian Bitter Springs formation in Australia have yielded fossils of both Cyanophycopsida and Chlorophycopsida. Thin sections from the black chert reveal cells with nuclei, some of which appear to be dividing. Representative fossils include *Glenobotrydion* and *Caryosphaeroides,* both of them similar in appearance to modern *Pedinomonas, Chlamydomonas* and *Dunaliella. Sphaerocystis* (Figure 13-3D) and *Trebouxia* (Figure 13-5) are common modern genera which also resemble these ancient fossils.

Our knowledge of the early evolution of plants has come almost entirely from morphological and biochemical studies of such fossils. While other classes and divisions are represented in ancient stromatolites, the most abundant and most readily identified Proterozoic eukaryotic

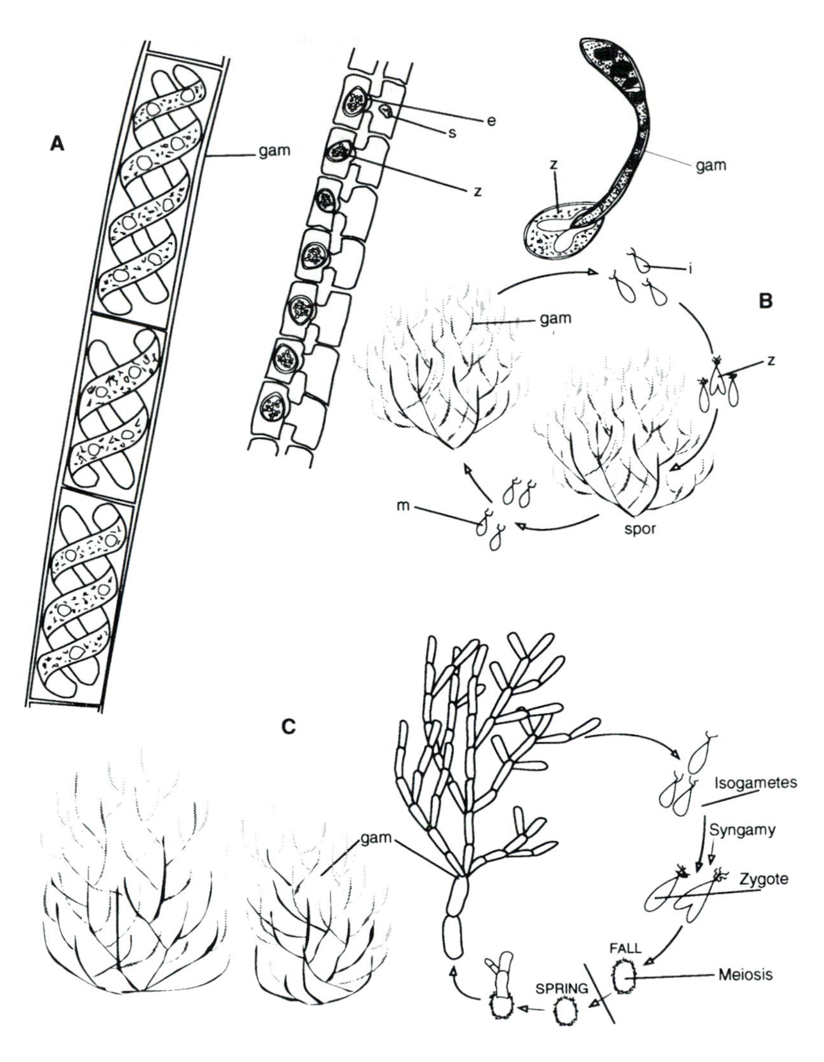

Figure 13–6 Life cycles of green algae. (A) *Spirogyra inflata,* haplontic, no motile gametes or spores; (B) *Cladophora keutzingiana,* diplohaplontic, spores and gametes motile; (C) *Cladophora gracilis,* haplontic. (s = sperm, e = egg, z = zygote, m = meiospore, gam = gametophyte, spor = sporophyte.)

fossils are green algae. We need to keep looking for other fossils, both of Chlorophycopsida and of all other classes that might possibly be present, yet we must be careful of our identifications, not only of the fossils but also of the geological formations. A few years ago, a challenge was issued to paleobiologists and geologists to find more Proterozoic fossils. The challenger stated that our knowledge of the early evolution of plants was not hampered by the lack of fossilization in Precambrian periods as much as by our failure to find the fossils that are there. Soon reports of Proterozoic fossils began pouring in; unfortunately, most of the reports were based on faulty identifications, resulting in even greater confusion than existed before.

Better understanding of the processes involved in fossilization has greatly enhanced our knowledge of ancient microscopic forms of life in recent years. Carbon and hydrogen atoms are taken up in the process of photosynthesis and are present in all cells in the form of sugars, proteins, lipids, and other organic compounds. These carbon atoms exist in three inorganic forms in seas and lakes: carbon dioxide, calcium bicarbonate, and calcium carbonate. The first two are soluble in water, the last is insoluble. They readily convert from one form to the other as indicated in Formula 13-1:

$$CaCO_3 + CO_2 + H_2O \rightleftharpoons Ca(HCO_3)_2 \qquad (13\text{-}1)$$

Plant cells use carbon dioxide as a raw material in photosynthesis, drawing it from the water and causing the reaction to proceed to the left. The calcium carbonate thus formed precipitates, forming solid calcium carbonate in the cell cavities and around the cells. **Phototactic responses,** the swimming of cells toward the light, sometimes cause unicellular algae to congregate into dense masses or mats where the carbonates they precipitate enhance their preservation. In order to find and correctly identify fossils, especially microfossils, paleobotanists must obviously be specialists in plant morphology; they must also be good students of plant physiology and biochemistry.

Since about 1975, a number of studies have been conducted on the green prokaryote, *Prochloron,* in order to ascertain its relationships to the eukaryotic green algae as well as to other prokaryotes. Pigmentation is similar to that of the Chlorophycopsida; chlorophylls *a* and *b* are both present, though in a wider ratio than in the Chlorophycopsida and other Lorobionta, and the cells lack the phycobilins that are typical of the Cyanophycopsida. Otherwise, it is biochemically more like Cyanophycopsida than Chlorophycopsida. Ultrastructurally, *Prochloron* seems to be intermediate between the Cyanophycopsida and Chlorophycopsida. At least four or five species of *Prochloron* are known, all of them obligate symbionts living with various species of colonial ascidians (marine chordates of the family Didemnidae). *Prochloron didemni,* the type species, lives in grooves on the outside of the testis of a *Didemnum* species in Baja California, Mexico. Other species live in the cloacal cavity of other ascidians.

Because *Prochloron* is prokaryotic and resembles the Cyanophycopsida in many ways, it is classified in this book in the Cyanophycopsida and discussed in Chapter 3. The genus probably descended from now extinct free-living prokaryotes that once linked the Cyanophycopsida to the Lorobionta (see Stackbrandt 1989). Presumably, the first Chlorophycopsida were Volvocales which evolved from *Prochloron*-like Cyanophycopsida almost a billion and a half years ago. Among their earliest descendents were the unicellular types, like *Glenobotrydion,* which have been preserved in stromatolites.

The Volvocales not only resemble the earliest eukaryotic fossils that have been found, they also resemble the Euglenales, Prochlorales, Chroococcales, and Cryptomonadales in numerous ways. Many are unicellular, have a single red stigma or eyespot and, like *Euglena* and *Crytomonas,* have contractile vacuoles. Unlike *Euglena,* the stigma in Volvocales is inside the plastid membrane. Other Volvocales are colonial: *Volvox* is a spherical colonial type in which the outermost cells are flagellated and the inner cells are not. In *Palmella,* the cells are nonflagellated and held together in a gelatinous mass. *Palmella* reproduces by isogametes and meiospores which are flagellated and resemble *Chlamydomonas.* Many green algae go through *Chlamydomonas*-like and palmelloid stages in their life cycles, even though the vegetative stage may be filamentous or even parenchymatous.

From the Volvocales, three lines of evolution can be traced (Figure 13–1). The first of these is characterized by increasing coenocytism and in it there is a pronounced accumulation of lime in the cell walls. Vegetative growth is by cell enlargement only, never by cell division, along this line leading to the Siphonophycopsida or calcarous green algae. Flagellation has been lost along the second line, leading to the Zygnemataceae and desmids (Conjugales); plastids are large and often lobed or ragged, and the cell wall consists of an inner cellulose layer and a soft outer pectic layer. Stewart and Mattox (1975) suggested that the Charophycopsida and Tracheophyta both evolved from this line; however, the more traditional view is that the third line, which leads to the Ulotrichales and Cladophorales, has given rise to the Charophycopsida, Bryophyta, and Tracheophyta. Along this line, flagellation has been retained and most species go through chlamydomonad and palmelloid stages in their life cycles.

The first line evolved mainly in marine habitats, the other two primarily in freshwater. Characteristics of the orders of Chlorophycopsida are presented in Table 13–1.

Representative Genera and Species

Several species of *Chlamydomonas* are frequently encountered by students. *C. nivalis* is the species most frequently observed in the red snow ("watermelon snow") of late summer on mountain snow banks. *C. gloeocystiformis* thrives in extremely saline conditions. *C. eugametos* and *C. reinhardtii* are cultured in the laboratory and used in genetic and other research.

Hydrodyction reticulatum (Figure 13–2e) is a common species in shaded pools in the summertime where it forms nets strong enough to hold small fish. The meshes of the net are usually six-sided, sometimes five- or seven-sided in places, and increase in size and coarseness as the individual cells enlarge. Closely related to *Hydrodyction* are *Characium, Trebouxia (Cystococcus), Pediastrum,* and *Scenedesmus,* all with haplontic life cycles, and *Chlorchytrium* with a diplontic life cycle. *Pediastrum* (Figure 13–3A–C) and *Scenedesmus* (Figure 13–3E–F), in the Chlorococcales, confuse many students, who think they are desmids; they are common in the summer plankton of ponds and lakes. According to Smith (1950), practically every body of standing water contains at least one species of *Scenedesmus.*

Characium is a common unicellular epiphyte in ponds and lakes growing on many different kinds of algae but especially on *Rhizoclonium,* often in association with *Characiopsis,* of the Xanthophycopsida, which it resembles in gross morphology. The genus *Trebouxia* or *Cystococcus* contains most of the common lichen phycobionts and is therefore extremely common all over the earth (Figure 13–5). It is not known how many species of *Trebouxia* there are; some lichenologists have assumed that almost every species of lichen has a different phycobiont; others believe that many different fungi parasitize the same species of phycobiont. When grown in pure culture in the laboratory, *Trebouxia* strains taken from different species of lichens are generally indistinguishable by either morphological or physiological tests. Engler's *syllabus* indicates only 20 species of described and named *Trebouxia.*

Two especially common genera of freshwater algae are readily recognized by their slimy nature. *Spirogyra,* or "mermaid's tresses" (Figures 13–2a, 13–4B, 13–6A), and *Zygnema* both form greenish yellow mats on pond surfaces. Haploid filaments conjugate in a **scalariform** (ladder-like) manner, the cell contents becoming gametes (Figure 13–2b). Zygotes are resistant to drying and may lie dormant for a year or more.

The desmids are unicellular freshwater plants which occupy a niche very similar to that of the pinnate diatoms, often forming dense blooms in late summer in lakes and ponds. Because of their beautiful cells, they are common objects of study of artists and photographers, as well as amateur botanists. Each cell typically consists of two semicells joined by an isthmus within which the nucleus is situated (Figure 13-4). In mitosis, the nucleus divides and then the two semicells separate and generate new halves.

Some botanists divide the Ulotrichales into two or three orders; others include the Cladophorales and/or Oedogoniales in the Ulotrichales. In *Ulothrix,* the vegetative plant is a haploid filament attached to the substratum, usually rocks in fast flowing streams in late summer or fall, by a funnel-shaped basal holdfast cell. Each of the other cells of the filament contains a single plastid with one or more pyrenoids, depending on the species. The plastid is a curved band that resembles an old-fashioned napkin ring or a Navajo bracelet. The haploid nucleus is in the center of the cell. The plant can reproduce sexually by means of biflagellate isogametes or asexually by means of quadriflagellate **mitospores.** The zygotes are also quadriflagellate and swim about after fertilization before settling down, losing their flagella, secreting a thick wall, and undergoing a rest period. Germination of the zygote results in four Chlamydomonas-like meiospores. Closely related are *Ulva lactuca,* or sea lettuce, an edible foliose marine alga, and *Enteromorpha intestinalis,* an ecological pioneer in streams and ponds. Both consist of flat sheets of cells which are rolled into hollow tubes in the latter species. In the same family is a common soil alga of India, *Fritschiella tuberosa,* in which differentiation is even more pronounced. A perennial paranchymatous basal thallus gives rise annually to tufts of slightly branched filaments in which most of the photosynthesis takes place. It is easy to imagine that plants like *Anthoceros* in the Bryophyta could have evolved from an alga similar to *Fritschiella tuberosa* (Figure 13-7).

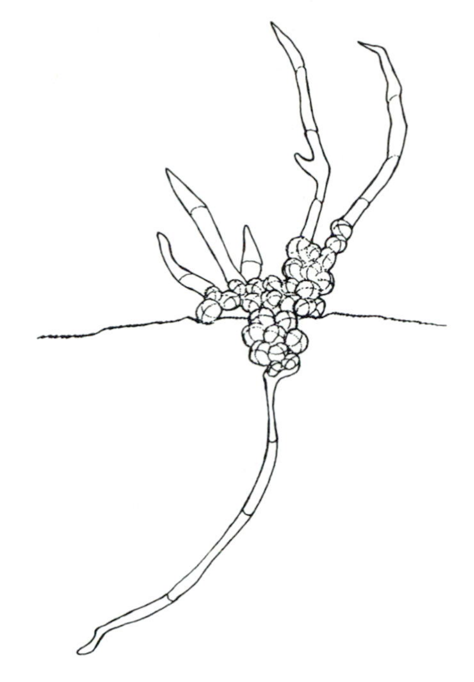

Figure 13–7 *Fritschiella tuberosa,* parenchymatous base with filaments growing out from it. A green alga similar to this one is believed to be the ancestral stock from which came the Bryophyta and possibly the Tracheophyta.

Oedogonium is characterized by a unique type of cell cap found nowhere else in the plant or animal kingdoms. In some species of *Oedogonium,* the male filaments are much smaller than the females, but in other species, there is no difference in size. In either case, motile sperm swim through an opening in the oogonium to fertilize the single egg.

In macrandrous species of *Oedogonium,* an antheridial mother cell divides by mitosis to form a series of antheridia, each of which produces one or two stephanokont antherozoids, or sperm, which are at first enveloped in a vesicle. The sperm invade the oogonium through a fertilization pore; following syngamy, the zygote develops a thick cell wall and may lie dormant for over a year before undergoing meiosis and germinating to produce four stephanokont meiospores, similar in appearance to sperm cells and also to the zoospores involved in asexual reproduction. In species which have been carefully investigated, two of the meiospores produce male filaments, two female.

In nannandrous species, some of the cells in a filament become androsporangia and produce haploid androspores which are chemically attracted to an oogonial mother cell on the same (in homothallic species) or a different (heterothallic species) filament where they attach and germinate, producing a delicate, epiphytic male filament called a nannandrium. The terminal cell of the nannandrium then divides to form a series of antheridia. The oogonial mother cell divides and produces an oogonium containing a single egg. Each antheridium produces a single antherozoid, or sperm, and fertilization occurs in a manner similar to that of macrandrous species.

Because of their coenocytic nature, the Cladophorales have been included in the Siphonales by some botanists. Their close resemblance to *Ulothrix* in general morphology, life cycles, and ecological preferences, along with pronounced differences from the Siphonales in pigmentation, morphology, and physiology, argues against this. Two genera, *Rhizoclonium* and *Cladophora,* are among the most frequently encountered green algae. Most species of *Cladophora* have diplohaplontic life cycles; *C. glomerata,* however, is diplontic. The occurrence of two kinds of life cycles in the same genus is unusual and has prompted some botanists to suggest dividing the genus into two genera, or even separate families, on the basis of life cycle. Life cycle differentiation is genetically controlled, just as other characteristics are, and in less highly evolved taxa, like the Chlorophycopsida, the genetic basis is simple. Phenotypes determined on a simple genetic basis are easily altered, either by single mutations, by genetic recombination, or by environmental factors, and do not justify the splitting of otherwise uniform taxa.

Ecological and Economic Significance

Ecologically, the Chlorophycopsida is probably the most important class of freshwater organisms. In arctic and desert ecosystems, they are also of great relative importance as phycobionts of lichens, and thus grow under conditions of extreme cold and/or dryness. They also make up an appreciable portion of marine plankton. On a world basis, they are not of as great ecological significance as some groups of plants, but in some ecosystems they are of the greatest significance and have often been underestimated in the past).

The zooplankton of streams and lakes is dependent on planktonic forms of Chlorophycopsida, especially the desmids and some of the Chlorococcales, for their survival. To a lesser extent they depend also on the filamentous green algae both for food and protective cover. All freshwater animal life—snails, crustaceans, fish, and even birds are, therefore, to a large measure dependent on the Chlorophycopsida. In the management of farm fishponds, for example, it is very important to fertilize for good growth of desmids and diatoms if a good harvest of fish is to be taken. Numerous studies have shown that it is probably impossible to decrease the annual yield of a farm fishpond by overfishing with line and hook, but that annual yield is very readily affected by the amount of fertilizer applied to the ponds and hence density of the phytoplankton.

For rapid and relatively odor-free breakdown of raw sewage, good oxygenation is essential. Species of *Scenedesmus* and *Chlorella* are reported to be especially important oxygen producers in sewage lagoons (Figure 3–17); other Chlorophycopsida, especially many of the desmids, also supply large amounts of oxygen; in brackish water, *Ulva* species are important.

Early ecological studies revealed that *Chlorella pyrenoidosa* is one of the most rapidly growing of plants. This species was consequently widely employed in research by the National

Aeronautic and Space Administration (NASA) to develop self-contained ecosystems for distant space travel. It was hypothesized that *C. pyrenoidosa,* a common sewage lagoon "purifier," could be grown on the waste products of the astronauts and thus provide both oxygen and food. Since *Chlorella* species are easy to culture, they are especially desirable for this type of project. Recipes for cookies and bread containing up to 10% *Chlorella* flour content were tested by thousands of people all across the country, and were found to give tasty and apparently nutritious products. However, 10% is far from the 50 to 95% needed in a practical space trip, and when the *Chlorella* flour content was increased much beyond 10%, digestive disturbances were soon noted. Schwimmer and Schwimmer (1968) have reported the isolation of toxins from this and related species, which is now not as popular in space research as it once was. Nevertheless, the idea is sound, even if the choice of species is not yet right. Perhaps some other easily cultured alga, possibly one that does not grow quite as fast as *Chlorella* or have as simple nutritional requirements, will turn out to be the species needed for sustained journeys through space. Space scientists might well consider the desmids.

Many Chlorophycopsida affect man's economy in a negative way. Dense mats of *Rhizoclonium, Microspora, Zygnema,* and other filamentous green algae consume large quantities of oxygen from the water in which they are growing while discharging the oxygen they are producing into the air, with the result that anaerobic conditions develop. Many animals and plants die for lack of oxygen in such ponds or lakes, and foul smelling products of anaerobic decomposition become a nuisance. In some cases, "ditch moss" may clog up irrigation systems. To control this problem, copper sulfate and other algicides are used. This can be expensive. Dense blooms of filamentous green algae also impede boat traffic in streams and lakes, especially in recreational areas. These algal mats often harbor animals which cause painful stings or rashes on some people. The Chlorophycopsida are not the only creators of problems of this nature, of course, but in freshwater ponds, lakes, and streams, they are the most abundant of the filamentous forms and hence the most common trouble makers.

A few species of green algae are incitants of plant disease, the most serious of which is red rust of tea, caused by *Cephaleuros virescens* and *C. parasitica,* members of the Trentepohliaceae in the Ulotrichales. Red rust, also called orange rust, is a very serious disease causing millions of dollars of damage to tea crops of Sri Lanka (Ceylon) and other tea-producing countries some years. *C. virescens* also damages coffee, citrus fruits, mango, pecan, jasmine, guava, rhododendron, jujube, and occasionally some greenhouse plants.

Research Potential

Because they are easily obtained any place on earth where people live and work and are easily grown in the laboratory, the Chlorophycopsida are especially promising for many kinds of scientific research. Having the same pigments as vascular plants, information gained from experiments with the Chlorophycopsida has been used in studies to learn more about photosynthetic pathways in crop plants. From such studies, we hope to learn how to manage crops better in order to feed more people. Since the thylakoids are stacked in a manner intermediate between the unstacked condition of the Cyanophycopsida and the dense grana of the Angiosperms, green algae may prove to be of great significance in studying physiological and ecological advantages of grana.

Much of our basic knowledge of cellular respiration has come from research with green algae. It is estimated that approximately 100 individual chemical reactions, each controlled by an enzyme, are involved in converting the chemical energy in a molecule of carbohydrate into kinetic energy to drive the life processes of a plant or animal. This energy is released in very small "bundles" which convert adenosine diphosphate (ADP) into adenosine triphosphate (ATP) in a manner analagous to how an electrical current converts water and lead sulfate in an automobile storage battery into lead, lead oxide, and sulfuric acid. Atoms are rearranged in the molecules involved in both cases and also, in both cases, the new chemical substances are stable until the right signal is given—the closing of a switch in the case of the storage battery—at which time they promptly release energy through reverse reactions, dephosphorylation of ATP to ADP in the case of mitochondrial respiration.

Most of the reactions in respiration take place in the mitochondria. Recent studies by plant cytologists employing radioactive nuclides and electron microscopy have shown that the enzymes involved in respiration are arranged on the cristae of the mitochondria in exactly the same order as had been hypothesized independently by chemists studying the chemical reactions that would need to take place in order to break down carbohydrates to carbon dioxide and water with the release of kinetic energy. Figure 13-8 is a diagram of a mitochondrion illustrating the chemical reactions of the oxidative phosphorylation phase of respiration.

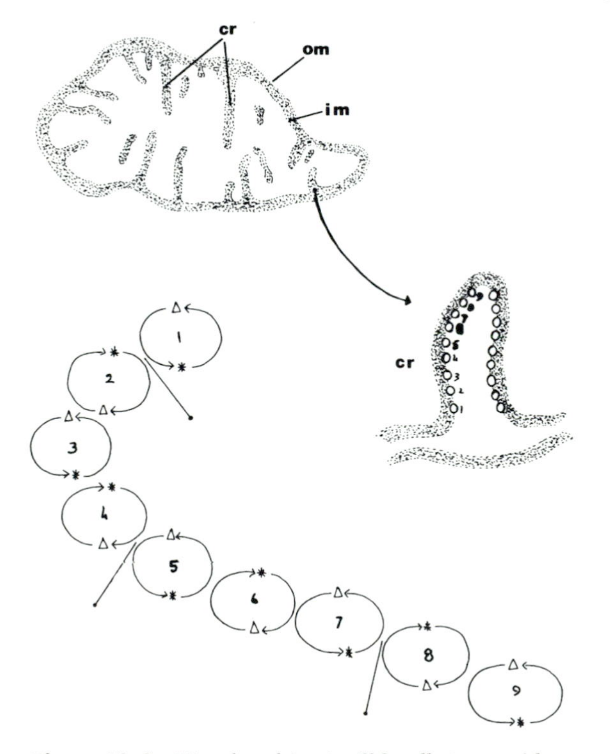

Figure 13-8 Mitochondrion in *Chlorella pyrenoidosa,* one of the green algae used in working out the processes involved in respiration. The enzymes are located on the cristae in the order indicated: 1 = NADH, 2 = FAD, 3 = CoQ, 4 = Cyt b, 5 = Cyt c-1,6 = Cyt c, 7 = Cyt a, 8 = Cyt a-3, 9 = O_2; * = oxidized form; om = outer membrane, im = inner membrane, cr = crista, cyt = cytochrome.

In recent years, green algae have been "discovered" by the geneticists. Since about 1960, *Chlamydomonas reinhardtii* has emerged as a major species for genetic studies. Considering that *Chlamydomonas* possesses some of the same characteristics that made *Neurospora* and *Aspergillus* so useful in genetic studies, it is surprising that it was not discovered much sooner. Actually, it was discovered during the early days of genetic research, but it did not become popular research species at that time. Between 1916 and 1920, Adolf Pascher published the results of a number of genetic experiments with *Chlamydomonas,* carefully pointing out the advantages this plant possesses for genetic research, including tetrad analysis (as in *Neurospora crassa,* Figure 5-9). Presumably his taxonomic research, which made him the world's outstanding algologist of the first half of the century, prevented him from continuing the genetic studies begun prior to 1916. Nevertheless, his understanding of genetics is unquestionably the solid base on which his taxonomic research still stands. One of Pascher's students, Franz Moewus, conducted genetic experiments with *Chlamydomonas, Polytena, Protosiphon* (Siphonales), and *Botrydium* (Xanthophycopsida) between 1930 and 1955, but despite their excellent quality and real significance, mention of them never got into the

TABLE 13-3

CHARACTERISTICS OF THE ORDERS OF SIPHONOPHYCOPSIDA

Order (Fam–Gen–Spp)	Chief Carotenoids	Thallus Structure	Representative Genera
Dasycladales (2-15-90)	β-Carotene, neoxanthin, zeaxanthin, violaxanthin	Septate filaments with 1 to several nuclei per cell	*Acicularia, Acetabularia, Valonia*
Siphonales (4-45-350)	α-Carotene, siphonein, siphonxanthin	Coenocytic filaments with numerous nuclei in each coenocyte	*Bryopsis, Codium, Derbesia, Caulerpa*

Note: The number of families (Fam), genera (Gen), and species (Spp.) in each order is indicated in the first column.

popular textbooks on genetics. Other Chlorophycopsida used effectively in genetics research include *Eudorina, Gonium, Dunaliella, Microcystis, Spirogyra,* and *Ulva.*

CLASS 2. SIPHONOPHYCOPSIDA

The Siphonophycopsida (*siphono* = tube or pipe, *phyco* = seaweed or alga), also called the Bryopsidophycopsida (Round 1969), are marine plants fairly common in shallow tropical waters. The cell walls often contain large amounts of calcium carbonate and they have therefore left an excellent fossil record, dating from Cambrian to recent formations. The class contains about 500 species included in some 75 genera and 6 families, with at least 58 additional genera known only as fossils. Ancient siphonophytes have built many calcareous geological formations since early Paleozoic time, and modern siphonophytes provide much of the organic matter and lime in tropical reefs that are still forming. The two orders, Dasycladales with two families and Siphonales with four, are traditionally included in the Chlorophycopsida, often as a single order. Characteristics of the two orders are summarized in Table 13-3.

Siphonophycopsida are almost exclusively marine organisms of shallow, tropical or subtropical seas. One species, *Dichotomosiphon tuberosum,* lives in freshwater in both Europe and North America, but is rare. They are readily distinguished from other seaweeds by the combination of color and lime encrustations that is so common in the class. They have more advanced structure than the Chlorophycopsida, they lack the pneumatocysts common in the Phaeophycopsida, and they are easily distinguished from lime-encrusted red algae by color and usually by lack of distinctive articulation. Several species are very abundant and can be obtained from the intertidal belt on almost any rocky shore. "Landlubbers" can be almost certain of obtaining *Codium fragile* and other siphonaceous green algae by sending to a biological supply house for a random sample of littoral seaweeds.

General Morphological and Physiological Characteristics

The Siphonophycopsida are morphologically rather complex, often-lime encrusted seaweeds which are almost entirely restricted to tropical and subtropical waters. The basic structural unit is usually a coenocyte rather than a cell, but in some species, it is a structurally complex cell containing a single nucleus and resembling a complex, multicellular plant.

In the Siphonales, the thallus is a branched, tubular coenocyte with the branches arranged on an axis or intertwined in such a way as to create a structure of definite macroscopic form. In most Dasycladales, the thallus starts growth as a siphonaceous tube, but cell walls divide it,

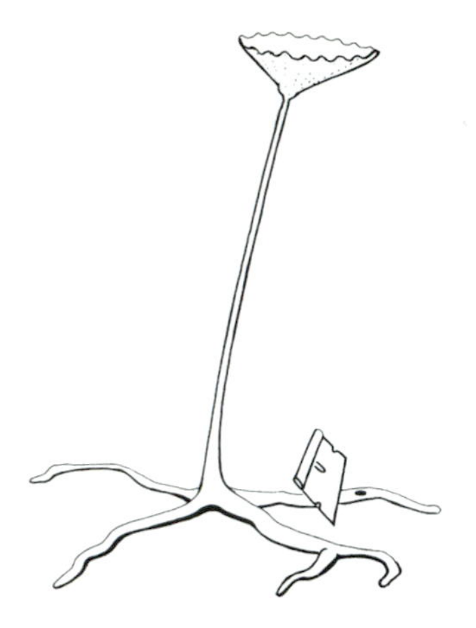

Figure 13–9 *Acetabularia mediterranea*, as used in experiments which indicated that circadian rhythms are not located in the nucleus.

as it grows, into several coenocytes. However, the entire plant in some Dasycladales, e.g., *Acetabularia*, is a morphologically complex uninucleate cell in apparent defiance of the cell theory (Figure 13-9).

On the basis of their own studies, as well as those of other botanists and zoologists, M. J. Schleiden and Theodor Schwann, in 1839, concluded that the bodies of all organisms are built up of cells and products of cell activity, thus formulating the Cell Theory. According to this theory, the anatomy, the overall morphology, the physiology of every plant or animal body is determined by how individual cells are put together to form tissues, organs, and organ systems. Cells of all living organisms are remarkably uniform; diversity is dependent on how those cells are arranged. This is a concept on which all biologists seem to agree: the story of life in all its manifestations is basically the story of the cell.

Who is bold enough to challenge such a universally accepted and widely acclaimed concept? Who or what would suggest that there might somewhere be an exception to the rule that every live body large enough to be seen without the aid of a microscope is composed of millions or billions of cells? The exception is, of course, the Siphonophycopsida.

One need only observe how the cells of oak xylem are arranged relative to each other compared to those of maple wood, to appreciate the significance of the cell theory. Yet, *Acetabularia* is also complex in structure and function even though the plant body is not built up of differentiated cells organized into functional tissues and organs.

There is considerable morphological variation among species of Siphonophycopsida. In some, the plant body is nothing more than a bladder-like vesicle attached by unicellular rhizoids or multicellular holdfasts to the substratum (e.g., *Derbesia* gametophytes and *Valonia*); in others, the plant body is macroscopically quite complex and highly differentiated. In *Acetabularia,* branched rhizoids anchor the plant to a rocky substrate, and a vertical axis supports a branching or umbrella-like crown (Figure 13-9). In *Derbesia* sporophytes, in *Codium,* and *Dasycladium,* filamentous coenocytes intertwine to form structures of distinctive form and often considerable size. *Caulerpa* and *Bryopsis* consist of single coenocytes with stolons, rhizoids, "stems," and "leaves" arranged in such a way that they superficially resemble vascular plants. Like *Acetabularia,* they seem to defy the cell theory.

Electron microscope studies reveal an ultrastructure similar to that of other Chlorophyta. Flagella, where present, are whiplash type. Thylakoids in the plastids show simple arrangement into grana-like structures. Mitochondria cristae are plate-like as in other Lorobionta.

Pigments in the Siphonales are different from other Lorobionta: α-carotene is more abundant than β-carotene, which may not be present at all, and the main xanthophyll pigments are siphonein and siphonoxanthin, found nowhere else in the plant kingdom, but present in all Siphonales that have been examined. Food reserves and cell wall structure are typical for the division and similar to the Chlorophycopsida (Table 1-3). In many of the siphonophytes there is an unusually large amount of plastid DNA present, causing Green (1967; 1973) to comment, "Perhaps *Acetabularia* is a 'living fossil' which has not had its chloroplast DNA pruned down by selection or by transfer to the nucleus."

Asexual reproduction seems to be rare in the Siphonophycopsida; when it occurs it is primarily by vegetative fragmentation. Sexual reproduction involves flagellated gametes; the zygotes usually remain dormant for long periods of time. Life cycles are heteromorphic diplohaplontic in *Derbesia marina* and modified diplontic in *Acetabularia* and several other genera.

Gametophyte and sporophyte generations of *Derbesia marina* are so unlike in general appearance that they were long believed to be different species and were even classified in different families (Smith 1938). The gametophyte, formerly known as *Halicystis ovalis*, is a perennial epiphyte growing on various coralline red algae, especially species of *Lithophyllum* and *Lithothamnion*. The multinucleate thallus is differentiated into a peg-like, starch-filled, colorless holdfast which penetrates into the host between adjacent articulated segments, and a large, dark green, globose vesicle a centimeter or more in diameter. The vesicles are shed each fall, but the holdfasts become more deeply embedded in the host each year as the host tissues grow up around them. Male and female plants are similar in appearance except for the color of the gametangia. In the summer, during spring tide of each lunar month, biflagellate gametes are discharged. Shortly after sunrise, a cloud of green "smoke" emerges from the dark green oogonia of the female plants; intermittent puffs of eggs continue for several minutes. About the same time, puffs of sperm are discharged from the yellow gametangia of the male plants. Eggs and sperm resemble each other, but the eggs are larger and contain several plastids, while each sperm contains a single plastid. Both swim around for a while until contact is made at anterior ends of both gametes, and syngamy takes place. Zygotes germinate within a week and develop first into a branched coenocyte (superficially similar in appearance to *Vaucheria* of the Heterosiphonales with which the species was once thought by many botanists to be closely related) and then gradually into the typical *Derbesia* sporophyte consisting of a dense basal portion of intertwined coenocytes with tuft-like erect filaments. The upright filaments are much branched, some of the branches being short, ovoid sporangia in which meiosis takes place. The egg-shaped meiospores resemble the spores of *Oedogonium* with a ring of short flagella, or cilia, near the tapered end. These **stephanokont** meiospores contain numerous disk-shaped plastids; they swim around for a while and finally germinate on a branch of red alga where they first send out a rhizoid which develops into the holdfast and then develop the photosynthetic vesicle.

There are about 15 species of *Acetabularia*, the best known of which are *A. mediterranea* and *A. crenulata. A. mediterranea* is commonly called the "mermaid's wineglass" (Figure 13-9). It is a perennial species which takes about 3 years to mature in nature, but will mature in about 10 months in laboratory culture. When a zygote of *A. mediterranea* germinates, which is in early summer, it forms a branched filamentous coenocyte-like cell containing only one nucleus. One of the branches penetrates the rocky substrate. This rhizoid is colorless. In the autumn a crosswall forms between the starch-loaded branched rhizoid and the green upright filament, and the upright portion of the plant dies back. In the spring of the second year, a new upright axis is formed with one or more whorls of branches at its apex; this axis also dies back at the end of the season. In the spring of the 3rd year, the axis forms an umbrella-like crown at the apex. The nucleus divides during the summer, and the daughter nuclei, all of them diploid, migrate through the axis and into the crown where cross walls form between them. The cell wall thickens and each cell becomes an ovoid, uninucleate **aplanospore.** Meiosis takes place in each aplanospore, followed by several mitoses, and each aplanospore thus becomes multinucleate. In many species, the multinucleate aplanospores remain dormant until spring, when a lid-like opening releases biflagellate isogametes, each with a single nucleus and many plastids. Gametes from the same aplanospore will not fuse with each other, although gametes from aplanospores produced by the same plant will, suggesting the presence of self-sterility alleles that segregate during meiosis. Following syngamy, the zygotes germinate to begin the next generation. In some species of *Acetabularia,* the aplanospores develop within a few days after formation and the zygotes overwinter.

Acetabularia has proven to be valuable material for scientific research in many areas. During the 1950s, for example, it was suggested that DNA was present and functional in the cytoplasm of plants and animals; this idea was hotly contested for several years. Species of *Acetabularia,* which had been popular and important research objects since the late 1930s, were ideal for settling this type of argument: all that was needed in addition to the plants themselves and a method for analyzing DNA was a pair of scissors or a razor blade (Figure 13-9). A single snip, and the nucleus was gone, while the rest of the plant lived on for several months, though it had no reproductive future. Biochemical tests demonstrated the presence of DNA in such enucleated plants.

Experiments with *Acetabularia* also demonstrated that the nucleic acid in cytoplasm is functional. When *A. crenulata* stalks and crowns were grafted onto *A. mediterranea* bases, as illustrated in Figure 13-10, growth was normal. If the crown was removed, a new crown, somewhat intermediate between the two, was generated. If the second crown was removed, the third one was typical of the *mediterranea* base rather than the *crenulata* stalk. If nucleus and crown were both removed, a crown similar to that of *A. crenulata* was formed on the *crenulata* stalk. If this second crown was removed, no third crown was generated. Apparently DNA in the nucleus was able to transfer its message to nucleic acids in the stalk, but these were used up in regenerating one cap and had to be resupplied from nuclear DNA.

Phylogeny and Classification

The Siphonophycopsida have left an excellent fossil record, which has been described by Herak et al. (1977) as the best fossil record of any algal group. At least 12 genera of Siphonales and 31 genera of Dasycladales have been studied in considerable detail; some of these extend back to the Cambrian Period (Figure 13-11). Although fossils of the two orders show many similarities to each other, they have been clearly distinct since Paleozoic time, supporting the claim that the Siphonales and Dasycladales should not be combined into a single order. Fossils similar to *Acetabularia* can be found throughout the Mesozoic and into the Paleozoic era; those with whorled branches extend back only to the Cretaceous period, while older fossils have alternate branching more like the Chlorophycopsida. Proposed relationships between the orders of Siphonophycopsida and Chlorophycopsida are illustrated in Figure 13-1. Having the same pigmentation as the Chlorococcales, but ecologically, morphologically, and physiologically resembling the Siphonales, the Dasycladales seem to form a link between the two classes.

Ecological and Economic Significance

Very limited in their geographical distribution, the Siphonophycopsida are of little ecological significance on a worldwide basis, as far as productivity is concerned, even though they are important in some tropical waters, and a few species are important in more temperate habitats. In another way, however, they are of great importance. Among the most productive of all ecosystems are tropical reefs, and the Siphonophycopsida are among the most important of all reef-building organisms (Figure 13-11). They have been important reef builders ever since the

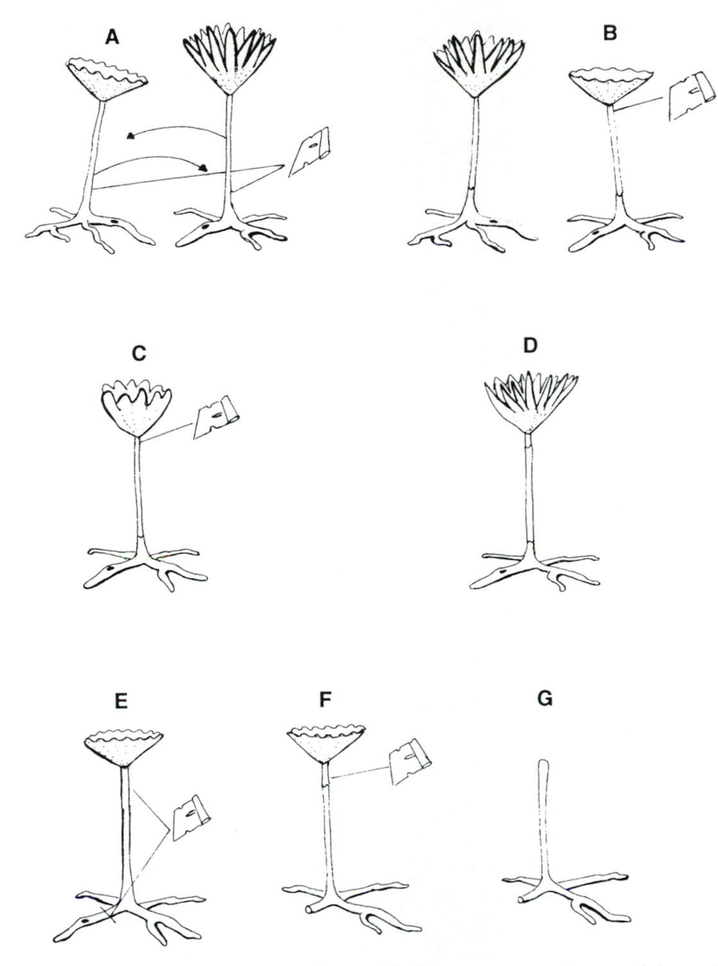

Figure 13–10 Grafting experiments with *Acetabularia.* (A) When stems and caps of *A. mediterranea* and *A. crenulata* are exchanged, the grafts continue growing normally (B). When the *mediterranea* cap is removed (B), a new cap, intermediediate between *mediterranea* and *crenulata* is produced (C). When this second cap is removed (C), a crenulata cap is formed (D), as would be expected, inasmuch as the nucleus is *A. crenulata.* If the nucleus and cap are both removed from a mature plant (E), a new cap, identical to the severed one, is produced (F). If it is removed, the plant apparently lacks sufficient nucleic acid to produce a new cap of any type (G), even though the stem and rhizoids continue living, with the stem carrying on limited photosynthesis.

Cambrian period of the Paleozoic era. At the present time, they are especially important as reef builders in the Caribbean.

In the building of a reef, two types of material are essential, the skeletal structure, or superstructure, which provides the support for the entire reef, and the fill materials that hold the structure together. Three groups of organisms are important superstructure builders: coralline red seaweeds, coral animals, and calcareous green algae. All kinds of marine algae are important in providing the fill materials that hold the reef together.

Limestone formations in the Rocky Mountains, as in other mountainous areas all over the world, often contain hundreds of species of plant and animal fossils; in the Grand Canyon of

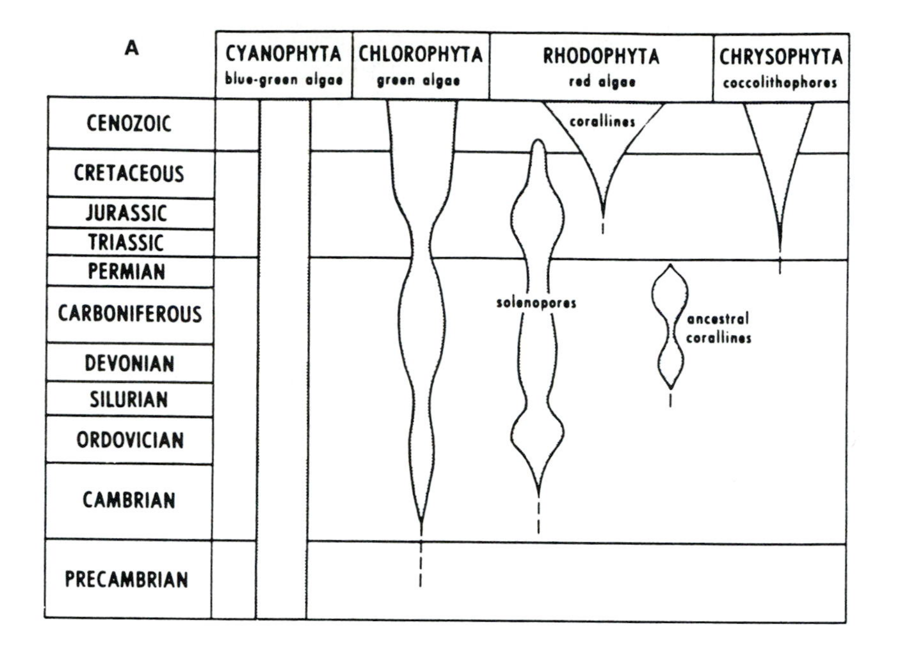

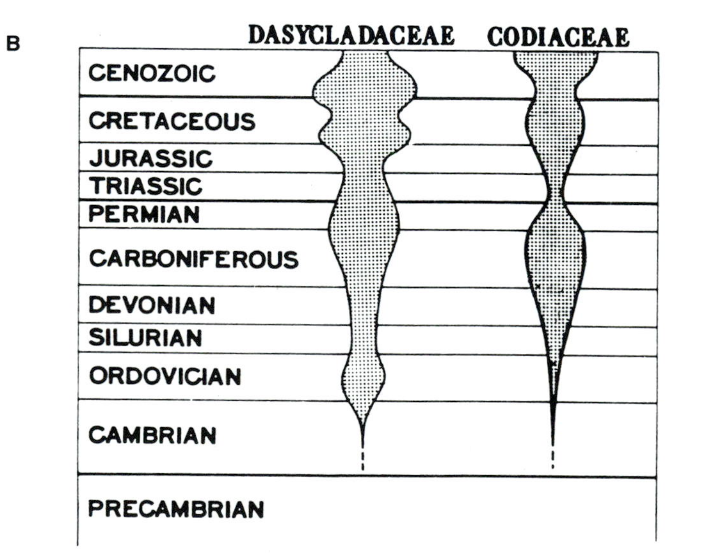

Figure 13–11 Fossil algae important in reef building. (A) Geologic distribution of all calcareous algae including all three classes of Chlrorophyta; (B) geologic distribution of Siphonophycopsida. Width of columns based on abundance of fossils from each geologic period. (From Ginsburg, R., Rezak, R., and Wray, J. L., *Geology of Calcareous Algae*, University of Miami, Miami, FL, 1972. With Permission.)

the Colorado River, for example (Figure 2-1), much of the biological history of that area, which has been submerged below the ocean at least seven times in the last 2 billion years, has been laid bare. Many of these limestone formations originated as reefs, the oldest of which were formed about 600 million years ago at the beginning of the Paleozoic era. The superstructure of reefs built during the Cambrian period consist entirely of calcareous green algae. Rhodophycopsida did not evolve until early in the Ordovician period and corals during the Devonian. At the present time, the Siphonophycopsida are less important than the other two groups for superstructure construction, but are very important in providing fill materials needed to bind the structure together. The importance of different groups of plants in reef building during geological time is illustrated in Figure 13-11.

Reef building is still going on. The distribution of Siphonophycopsida in comparison with other groups of plants around small islands and atolls is illustrated in Figure 13-12. Most species of Siphonophycopsida grow in the littoral or shallow sublittoral zones. Beyond this, relatively little is known about their autecological requirements, unfortunately.

Economically, Siphonophycopsida are of little direct value. In contrast to the Rhodophycopsida, Phaeophycopsida, and Chlorophycopsida, none of the calcareous green algae are important sources of either food or industrial products. Their only economic significance at the present time, in addition to the selling of specimens for educational purposes or for research, is as marine aquarium plants: *Bryopsis* and *Caulerpa* are especially attractive in aquaria. *Codium fragile* is easy to culture and grows very well under aquarium conditions. A siphonophyte with more bizarre appearance is *Penicillus,* Neptune's shaving brush (Figure 13-13), also a reef builder.

Research Potential

Several features contribute to the great value of Siphonophycopsida as research organisms. One of these is the advantage of giant cells for many kinds of research (Table 13-4). Because of its large cells, *Valonia* was chosen by Lark-Horovitz (1929) for experiments to see if ions could cross plasma membranes. Thus was ushered in the atomic age of botanical research. He placed cells in solutions of lead nitrate containing radioactive lead and observed that nearly all of the lead was concentrated in the cell walls; little or no lead had entered the cell vacuole at the end of 4 hours. When the plants were placed in an atmosphere containing radon gas, radioactivity was picked up inside the vacuoles within a few minutes. In dead cells, lead did not move into the vacuole. Apparently gas molecules but not lead ions could pass through the membrane. Although his research gave no final answer to the question he was investigating, his research was of great significance because this was the first time radioactive isotopes were used in biological research. It also demonstrated the value of plants with very large cells for some types of biological research. It took another 10 years before radioactivity was again experimented with; today, radioactive isotopes are widely employed in physiological, genetic, and ecological research. (See Poisoux-Deo 1970; Jacobs 1984.)

Research into the causes of **circadian rhythms** (*circa* = approximately, *dial* = day or 24 hours) has been accomplished with the aid of calcareous green algae. Such rhythms are common in most plants, and also in animals, but are especially pronounced in the Siphonophycopsida. *Acetabularia* grown in continuous light carries on photosynthesis most rapidly

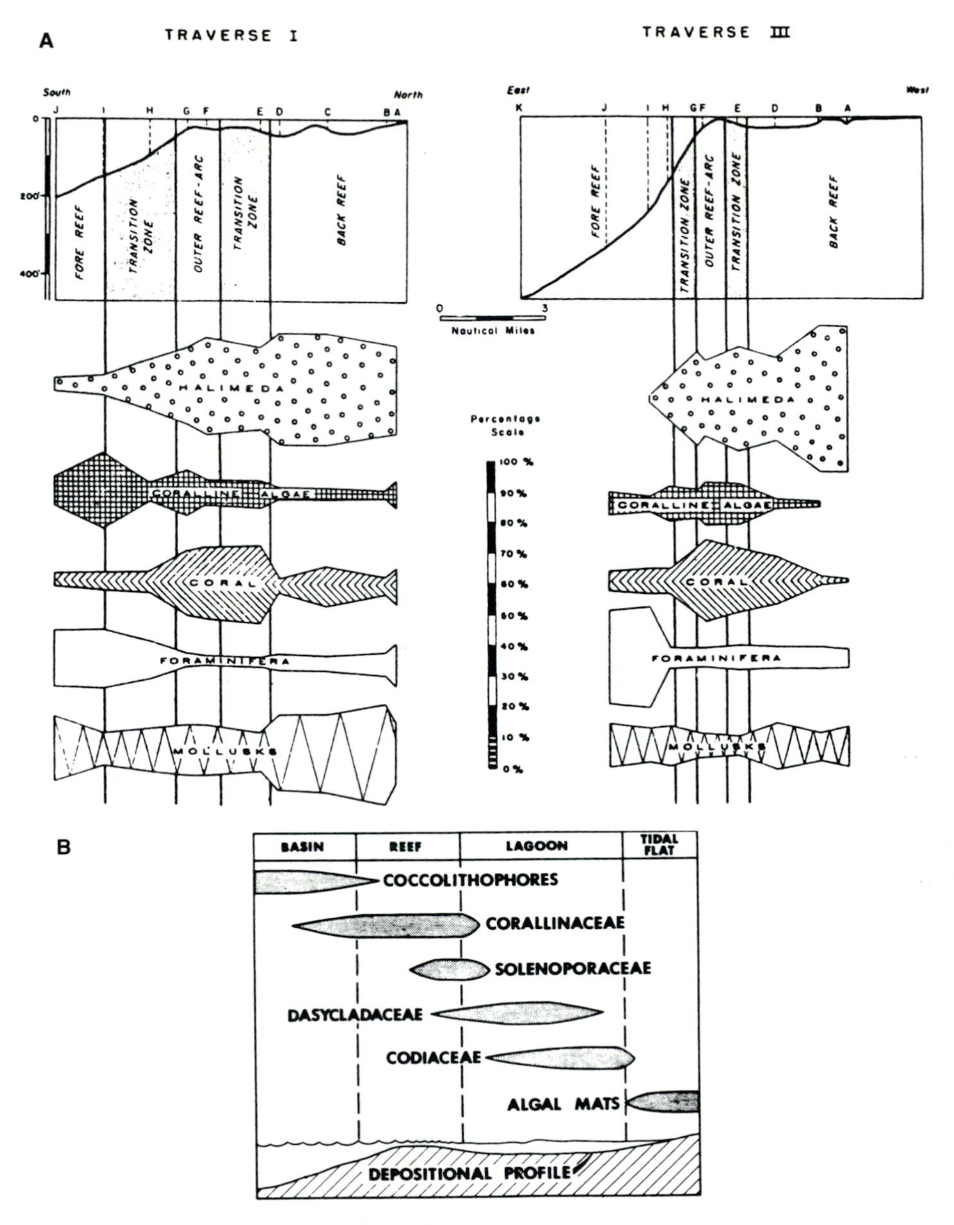

Figure 13–12 (A) Distribution of important reef building plants and animals in waters surrounding atolls and tropical islands. The two transects (or traverses) are at right angles to each other. In the shallow waters of the lagoons, outer reefs, and back reefs, *Halimeda,* in the Siphonales, is by far the most important provider of both superstructure and binding materials. (B) Fossilized plants found along the depositional profiles including stromatolites in the algal mats. (From Ginsburg, R., Rezak, R., and Wray, J. L., *Geology of Calcareous Algae,* University of Miami, Miami, FL, 1972, . With permission.)

during the hours of the day when light is normally intense and slows down during the hours that would normally be night. In continuous light, the interval between peaks of maximum photosynthesis is approximately 25 hours, rather than 24, which helps explain how organisms are able to adjust to new light-dark regimes. Assuming that circadian rhythms are controlled in a similar manner in other organisms, including mammals, this could also explain why most people recover from "jet lag" more rapidly when they travel from west to east than when they travel from east to west. We do not know where the "clock" that controls these rhythms is located, but it is somewhere inside the cell. We know it is not in the nucleus, as at one time hypothesized, because enucleated cells behave the same as cells with nuclei and can be reset by exposure to a new light-dark regime exactly the same way as a cell with a nucleus (Figure 13-9). The fact that nuclear DNA is not involved in the clockwork of *Acetabularia* does not prove that this is the case in all plants and animals; nevertheless, it suggests that cellular substances other than nucleic acids may be of great importance in establishment of circadian rhythms.

Figure 13–13 *Penicillus capitatus,* Neptune's shaving brush.

CLASS 3. CHAROPHYCOPSIDA

The Charophycopsida (*char* = graceful, *phyco* = alga), commonly called the stoneworts, are lime-encrusted green algae that show relationships with *Ulothrix, Ulva, Cladophora,* and possibly *Spirogyra.* Because of their distinctive macroscopic structure, they are sometimes included in the Bryophyta; however, differentiation of cells is much simpler in the stoneworts than in the mosses or liverworts, and so are the reproductive structures. There are over 200 species of Charophycopsida, many of them relatively common and widespread; the single family contains six genera. Table 13-5 summarizes the main differences between the Charales and the other Chlorophyta.

Most stoneworts are freshwater algae growing submersed in lakes and streams that are high in calcium and low in phosphorus. In clear, slow-flowing streams they may form dense "forests" on the muddy bottom, several meters below the water surface; in fast-flowing streams they will usually be found in the shallow water near the edges. They are easily recognized by their vertical axes with whorls of "leaves" at the nodes, by their harsh texture due to the lime encrusted in the cell walls, and by the musky odor as they begin to dry. Because of these distinctive characteristics, students seldom confuse stoneworts with any other group of plants. To some amateurs they seem to resemble *Equisetum,* but the similarities are purely superficial. Stoneworts are submersed aquatics, very delicate and graceful in appearance, and lack the vascular tissues characteristic of *Equisetum* and other Tracheophyta. Reproductive structures are also very different.

TABLE 13-4

PLANTS HAVING EITHER LARGE CELLS OR LARGE COENOCYTES USEFUL FOR SEVERAL KINDS OF PHYSIOLOGICAL RESEARCH

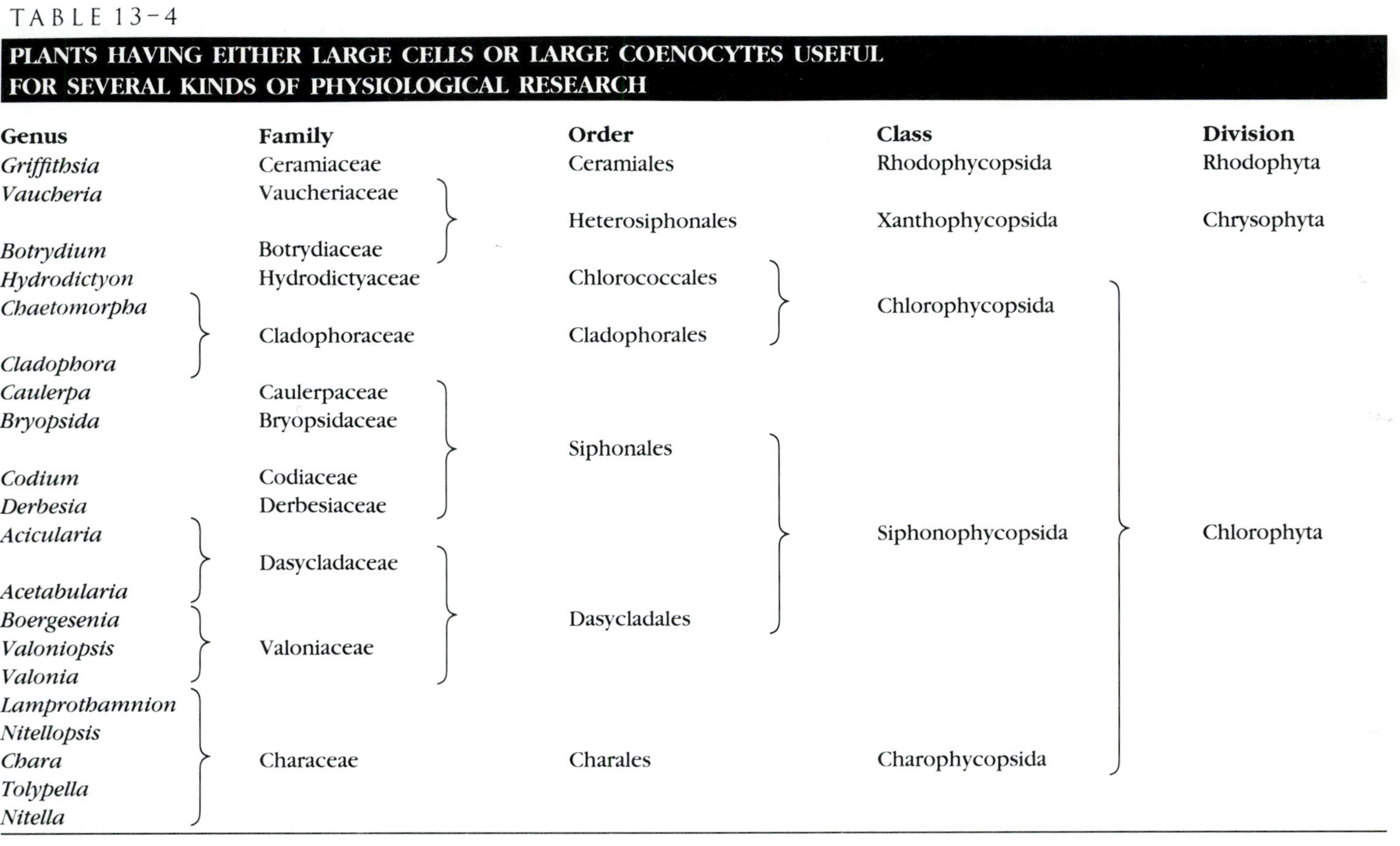

Genus	Family	Order	Class	Division
Griffithsia	Ceramiaceae	Ceramiales	Rhodophycopsida	Rhodophyta
Vaucheria	Vaucheriaceae			
		Heterosiphonales	Xanthophycopsida	Chrysophyta
Botrydium	Botrydiaceae			
Hydrodictyon	Hydrodictyaceae	Chlorococcales		
Chaetomorpha			Chlorophycopsida	
	Cladophoraceae	Cladophorales		
Cladophora				
Caulerpa	Caulerpaceae			
Bryopsida	Bryopsidaceae			
		Siphonales		
Codium	Codiaceae			
Derbesia	Derbesiaceae			
Acicularia			Siphonophycopsida	Chlorophyta
	Dasycladaceae			
Acetabularia				
Boergesenia		Dasycladales		
Valoniopsis	Valoniaceae			
Valonia				
Lamprothamnion				
Nitellopsis				
Chara	Characeae	Charales	Charophycopsida	
Tolypella				
Nitella				

From Hope, A. B. and Walker, *The Physiology of Giant Algae Cells*, Cambridge University Press, London, 1975. With Permission.

TABLE 13-5

COMPARISON OF CHARALES WITH OTHER GREEN ALGAE

	Charales	**Chlorophycopsida**	**Siphonophycopsida**
Number of species	215	5,300	440
Growth habit	Upright stem with whorls of leaves at each node; internode a single cell	Variable, seldom upright; never with stems or leaves	Variable, often upright; no whorls of leaves
Cells	Stems multinucleate, leaves may be uninucleate or multinucleate	Uninucleate or multinucleate	Mostly multinucleate
Tissues	Simple cortex in stems or some genera; always lime encrusted	Simple parenchyma in one genus; plants may be unicellular, filamentous, or made up of thin sheets; seldom lime-encrusted	Prosenchymatous in some genera; usually lime-encrusted
Oogonia	Enclosed by a jacket of sterile cells	Always unicellular	Always unicellular
Gametes	Biflagellate, motile	Biflagellate, stephanokont, or nonmotile	Biflagellate or stephanokont
Probable origin	Devonian	Paleohelikian	Cambrian

General Morphological Characteristics

The stonewort thallus consists of a delicate, upright axis with whorls of "leaves" at the nodes, and rhizoids anchoring it to a muddy or sandy substrate. The internodes consist of giant, elongated cells or coenocytes, sometimes surrounded by a cortex of smaller cells; the leaves consist of series of smaller, uninucleate cells. The axis continues growing almost indefinitely and may attain a length of 50 cm or more, although a plant height of 15 to 20 cm is more common. The leaves, on the other hand, are limited in growth, generally consisting of three to eight cells, depending on the species. Chloroplasts are abundant in the cells of the leaves and axis. In some species, branches of indeterminate length form in the axils of the leaves. The rhizoids are filaments of colorless, uninucleate cells.

In *Nitella* and *Tolypella,* the stem is one cell thick. Each internode consists of a single cell (a coenocyte) up to 5 cm long and 2 or 3 mm in diameter and containing several nuclei. In *Chara,* modified branches form just above and below each node. These grow parallel to the internode cell to form a corticating tissue usually one cell layer thick, but sometimes with two or more cell layers on the older portions of the stem. As some of the branches grow up, the others grow down, and they meet halfway along a zigzag line.

Cells of charophytes are large, with a relatively thick cytoplasm between the thin cell walls and the large vacuoles. Except in the giant cells or coenocytes of the vertical axis or "stem," there is usually one haploid nucleus in each cell. Plastids are small and disk-shaped and there are several in each cell. Pyrenoids are prominent. Thylakoids show only a slight tendency to form grana. In addition to β-carotene and chlorophylls *a* and *b,* γ-carotene and lycopene are generally abundant. Flagella occur only on male gametes. They are long, slender, and tend to coil. Male gametes are biflagellate, both flagella being the whiplash type.

Gametangia are multicellular and are produced at the nodes. The egg is enclosed in an archegonium—often called an oogonium even though it is multicellular—which is always situated just above the antheridium, or sperm-producing structure. In older literature, the archegonium was always called a **nucule,** the antheridium a **globule.** These terms are still frequently used in referring to the gametangia of Charophycopsida. The antheridium or globule is bright orange which helps make identification of charophytes sure even at a distance.

The archeogonium or nucule contains a single egg which is surrounded by five two-celled filaments attached to the base of the antheridium by a unicellular pedicel. The five cells in the upper tier do not become very large and ultimately form the corona of the nucule. The cells of the lower tier elongate to many times their original length, and as they elongate, they twist into a tight spiral (Figure 13-14).

Numerous biflagellate sperm or antherozoids are produced in each antheridium; these escape through a pore in the antheridium wall and enter the archegonium through slits between adjacent cells just below the corona. The fertilized egg develops within the archegonium into a thick-walled hardy zygote. The archegonium soon falls free of the vegetative plant and lies for several weeks on the bottom of the pool or stream. Germination of the zygote is by meiosis; two cells, one with three nuclei, the other uninucleate, result. The three-nucleate cell disintegrates. The uninucleate cell forms a protuberance at the coronal end of the nucule, and this develops into two filaments, a rhizoidal initial and a protonemal initial. The protonemal initial develops into a green **protonema,** or branched filament, from which a vertical axis is produced and grows into the typical stonewort plant.

Asexual reproduction in the Charophycopsida is by (1) **amylum stars,** or star-like aggregates of starch-filled cells developed about a lower node, (2) **bulbils** which develop on the rhizoids, or (3) **protonema-like outgrowths** from a node. Asexual spores are not produced by any of the stoneworts.

Cell walls consist of an inner cellulose layer and an outer pectic layer which, in many species, becomes heavily encrusted with calcium carbonate. Beyond this, very little is known about the chemistry or the structure of stonewort cell walls. As the protoplasts and the cellulose layer of the cell walls decompose, carbonate casts remain (Figure 13-14). These accumulate on the lake or river bottom, sometimes in thick layers, and may be preserved for millions of years. Identification of species from casts of nucules is not difficult.

Figure 13–14 Fossilized nucules, extinct speces of Charophycopsida.

Physiological studies have shown that stoneworts do not tolerate high levels of phosphorus; even levels that would be normal or below normal for other plants are not tolerated (Forsberg 1963). Stoneworts are also less demanding of oxygen than most plants. These characteristics make them useful indicator organisms in studies of water pollution (See Special Interest Essay 13-1).

Phylogeny and Classification

Marine species of Charophycopsida apparently evolved during early Paleozoic time, presumably from chlorophycean ancestors similar to either *Ulothrix* or *Oedogonium* or possibly *Spirogyra.* Because of the ability of stoneworts to survive under almost anaerobic conditions, together with their ability to accumulate limestone in the cell walls of both vegetative and reproductive structures, they have left an excellent fossil record. Well-preserved fossils very similar in appearance to modern *Oedogonium* occur in early Paleozoic formations and could represent the stock from which the Charophycopsida evolved. Judging from the present fossil record, stoneworts were the first plants to colonize freshwater habitats, near the end of the Silurian or early Devonian. From these first Charophycopsida have evolved the six genera we know today, plus several genera which have become extinct. Extant stoneworts are all included in one order, the Charales, and one family, the Characeae. Fossil Characeae go back to the early Devonian. Three families, the Clavatoraceae from the Jurassic, the Palaeocharaceae from the Pennsylvanian, and the Trichiliscaceae from the Devonian, are known only as fossils. Suggested relationships to other Chlorophyta are illustrated in Figure 13-1.

Biochemical and cytological similarities shared with the Tracheophyta have led some botanists to suggest that the charophycopsida represents the stock from which the vascular plants evolved. However, the oldest fossils of vascular plants are older than the oldest known charophyte fossils. Furthermore, there are no striking morphological similarities between the ancient fossils of the two groups.

Ecological and Economic Significance

Although Charophycopsida probably originated in marine habitats, most species today are freshwater habitants and very few are marine. In some ecosystems they are important producer organisms and provide food for a number of invertebrate animals and also for ducks and other birds. The relationship with ducks is apparently a symbiotic one, the ducks helping to provide the high lime conditions needed by the stoneworts, the stoneworts providing part of the food consumed by the ducks.

Shallow lakes where both ducks and stoneworts once were abundant now have no stoneworts and fewer ducks as the result of pollution by sewage containing high levels of phosphate from phosphorus detergents. Although still very abundant in many mountain lakes and streams, *Chara vulgare,* the most common of all the stoneworts, appears to be an endangered species in many parts of the world, especially in North America and western Europe where most lakes and rivers contain phosphate pollutants.

Research Potential

The giant cells of stoneworts are especially suitable for some kinds of research (Special Interest Essay 13-2). Having a thicker layer of protoplasm between the cell walls and the vacuole than many giant cells, they have been a favorite with physicists and plant physiologists studying rates of diffusion across cell membranes, electric potential differences, cytoplasmic streaming, sugar-starch conversion, differences among individual cells in rates of photosynthesis and respiration, and other botanical phenomena that are best studied on a single-cell basis.

The Charophycopsida have been used and are still being used to study patterns of evolution. Three basic questions pondered by biologists ever since Darwin suggested the evolution of species are still unanswered: (1) does evolution have definite direction? (2) what is the motor of organic change—is it primarily internal or external? (3) what is the tempo of evolution and how constant is it? (Gould 1977). The excellent fossil record provided by the Chlorophyta may help in answering these questions more satisfactorily.

The physiological adaptations of the Charophycopsida—especially their sensitivity to phosphate, their tendency to accumulate calcium carbonate, and their ability to function well at low oxygen levels—give them additional advantages in some kinds of research, such as that having to do with water pollution and with fossilization.

SUGGESTED READING

Gould's article, "Eternal metaphors of paleontology," in Hallen's *Patterns of evolution* is an interesting summary of the conflicts that evolutionists face as they try to unravel the patterns of evolution and reconstruct the fossil history. Wray's *Calcareous Algae* summarizes the findings of paleobotanists who work with the Siphonophycopsida and Charophycopsida. *The*

Physiology of Giant Algal Cells, by Hope and Walker, summarizes some of the work done by physicists exploiting the advantages of giant cells in their research.

STUDENT EXERCISES

1. What is the difference between periphyton and plankton?
2. Classify as periphyton, plankton, seaweed or symbiont each of the following species of green alga: (a) *Spirogyra dubia,* (b) *Zygnema insignis,* (c) *Cosmarium ehrenbergii,* (d) *Trebouxia parmeliae,* (e) *Chlorella pyrenoidosa,* (f) *Rhizoclonium hieroglyphicum,* (g) *Ulva lactuca.*
3. Indicate where you would go to collect each of the above.
4. Filamentous algae are ideal organisms to use in studying osmosis and plasmolysis in cells. Suggest a replicated and randomized experiment for ascertaining the isotonic solution of a species of Chlorophyta. Would you expect the isotonic solution of a freshwater alga to be more concentrated or less concentrated than that of a marine alga?
5. Prepare a series of five or more sugar solutions varying from 0% (the control) to 4% or 5% and find isotonic concentration of solutes in each of four or five species of algae that you can collect near your home.
6. What are the laws of diffusion that are involved in osmosis? List five basic laws of diffusion.
7. Citric acid and isocitric acid have the same empirical formula, $C_6H_8O_7$. What is the chemical difference between them? What do we call the type of chemical reaction in the Krebs cycle that converts citric acid to isocitric acid?
8. Why is citric acid more sour than isocitric if they both have the same number of atoms of each type in their molecules?
9. When a bottle of water from a saline lake is placed in subdued light, as near a north-facing window, *Chlamydomonas gloeocystiformis* responds by swimming toward the light. When brightly illuminated, on the other hand, the cells swim away from the light. What are these responses called?
10. Some kinds of algae are popular research organisms. Look up *Chlamydomonas, Chlorella, Acetabularia, Cladophora,* and *Chara* in a recent issue of *Biological and Agricultural Index.* Are any of these genera being used in research today? If so, how?
11. From the written description of reproduction in *Ulothrix* in this chapter, sketch a life cycle of *Ulothrix zonata.*
12. How many kilometers (or miles, if you prefer) would you have to travel in order to find a specimen of Chlorophycopsida? Of Siphonophycopsida? Of Charophycopsida? Indicate specifically where you would go to find the specimen closest to your home.

SPECIAL INTEREST ESSAY 13-1

Water Pollution

Botanists are interested in water pollution for a number of reasons: (1) part of the problem is caused by some plants, (2) water pollution affects how plants grow, (3) plants are useful in controlling water pollution, and (4) some plants are reliable indicators of water pollution.

To be a good pollution indicator, a plant species needs to be widely distributed, easily recognized, and rather exacting in its nutritional requirements. In using algal indicators in pollution studies, it is best to use several species simultaneously; a few of these should be very tolerant of pollution, a few very intolerant. Among the most reliable indicators of high levels of pollution are species of *Oscillatoria, Euglena,* and *Cryptomonas.* Some of the Charophycopsida, Bacillariophycopsida, and Chrysophycopsida are good indicators of pollution-free water.

Pollution indicators fall into two categories: (1) though harmless themselves, they indicate the presence of problem substances which they can tolerate or use in their metabolic processes; (2) they produce substances that are undesirable, either toxic or nuisance factors. In the latter category are such species as *Peridinium cinctum, Aphanizominon flos-aquae,* and *Gonyaulax catenella. P. cinctum* produces undesirable, although harmless, odors in culinary water supplies, *A. flos-aquae* produces large quantities of nitrogen which in turn stimulate other algae—this may be good or bad depending on other factors. *G. catenella* produces a toxin that renders shellfish that have fed on it deadly poisonous to humans who eat them.

Categories of water pollution include (1) occurrence of pathogenic bacteria resulting from untreated sewage being discharged into a stream, often at such low levels that the water looks, smells, and tastes perfectly good: (2) occurrence of large quantities of organic matter, with associated bacteria, from barnyards, pulp mills, canning plants, and the like; (3) occurrence of heavy metals or other poisonous chemicals, either inorganic or organic; (4) occurrence of phosphates, sulfates, and other nontoxic fertilizer substances; and (5) an increase in temperature of the water as a result of heating in slow-moving water of an impoundment or of the water being used to cool a nuclear power plant or other factory. As far as direct effect on human health is concerned, the first and third problems are by far the most hazardous; the others, however, affect the overall environment and may have far-reaching effects, not necessarily bad in every case.

Some of the problems of water pollution are linked to **eutrophication,** which is the process of aging that occurs naturally in all lakes and ponds. Young (or oligotrophic) lakes are generally deep, cold, and of low productivity compared to old (or eutrophic) lakes. Adding fertilizers to a lake, either directly or by organic pollution, increases its productivity and hastens the aging process. It also changes the faunal composition of the lake: trout, salmon, and stone flies, for example, are found in oligotrophic lakes; carp, leaches, frogs, and salamanders are found in eutrophic lakes. Through eutrophication, the total amount of fish that a lake will support may be greatly increased, but if you prefer trout to carp, you may not be pleased by this increase. Knowing that eutrophic lakes tend to be warmer than oligotrophic lakes, a water skier may search out eutrophic lakes for his favorite sport; as dense water blooms of *Spirogyra, Rhizoclonium, Vaucheria,* and *Hydrodyction* develop after midsummer, he may lose his enthusiasm for eutrophication, however.

One of the most serious problems in North America today, and one of the most talked about, is eutrophication. There are at least two popular misconceptions that need clarification, however: (1) eutrophication is *not* a synonym for pollution, and (2) phosphate detergents do not *always* cause eutrophication. Pollution *does* generally hasten eutrophication, and one of the major problems has been the discharge of phosphate detergents into lakes and streams by way of household sewage. If phosphorus is a limiting factor in the body of water into which the detergents are discharged, various kinds of algae will grow much better. If calcium, nitrogen, potassium, or other mineral nutrients are limiting, discharge of phosphorus has little or no effect on the growth of most plants, unless it is in such extremely large quantities as to be toxic. Often, both nitrogen and phosphorus are limiting, but since the blue-green algae can utilize atmospheric nitrogen, they are stimulated by the addition of phosphates alone. The resulting cyanophycean blooms are often characterized by strong odors and flavors and sometimes by production of toxins. They are generally followed by blooms of other algae that thrive on the nitrogen fixed by the blue-greens. If an alga bloom consists largely of filamentous species, oxygen may become a limiting factor and both plants and animals will die in large quantity, resulting in foul odors. This may seem paradoxical, since filamentous algae produce far more oxygen than they consume; the problem, of course, is that the filamentous algae accumulate in thick layers at the surface of the lake or pond and the oxygen they give off, being relatively insoluble in water, is released to the atmosphere, while the shaded cells at the bottom of the layer are continually using up large quantities of oxygen from the water in respiration. The fish that survive in such habitats are sluggish and hence low oxygen-consuming species.

Not all fish and other aquatic animals that live in eutrophic lakes are undesirable and not all animals that live in oligotrophic lakes are desirable, from the human utilitarian point of view. In Indonesia and some other areas, controlled eutrophication is being practiced in order to provide more food for man and eliminate starvation and malnutrition.

What are some danger signs that indicate water pollution may become a problem? Although any pond or lake may develop a degree of algal blooms at some time of the year, large numbers of *Euglena, Cryptomonas,* and *Oscillatoria* suggest that appreciable organic pollution is entering the lake. *Cryptomonas* or *Oscillatoria* alone do not mean much, but the presence of the three together is indicative of a problem. Conversely, the presence of *Chara* or *Nitella* indicate that the level of soluble phosphates is low since these plants cannot tolerate high levels of phosphate. The presence of *Asterionella formosa, Synedra acus, Tubellaria ferestrata,* and *Melosira granulata* suggest water relatively free of organic pollution.

One method of controlling water pollution is by construction of sewage lagoons (Figure 3-18). As raw sewage enters the lagoon, bacteria act on it, releasing various minerals and simultaneously depleting the oxygen supply. Both aerobic and anaerobic bacteria are active. The latter convert proteins to hydrogen sulfide and other "smelly" substances in some or their metabolic activities. For continued aerobic decomposition of the sewage, oxygen must be added. Some systems use pumps for aeration; others rely on algae. Species of *Chlorella, Chlamydomonas, Scenedesmus, Euglena, Oscillatoria, Cryptomonas, Rhodomonas, Chrysomonas, Aphanothece,* and *Aphanizominon* are all important in this aeration process. Frequently the water will take on a greenish tinge because the plankton is so dense. As the water from the treated sewage is discharged, a large biomass of living algae is discharged with it. They will soon die as they float on down the stream, and their decomposition may seriously deplete the stream of oxygen sometimes resulting in fish kill. To remedy this, some botanists have proposed harvesting the algae from the lagoon outlet and using it for animal feed, for "green manure," or as fertilizer.

SPECIAL INTEREST ESSAY 13-2

Giant Cells in Biological Research

A few species of plants and animals have extremely large cells; these present opportunities for research in areas that would be difficult using typical cells. Imagine, for example, trying to inject a small quantity of sulfur dioxide or other chemical into individual cells of *Oscillatoria spirulina* or *Escherichia coli,* with a microhypodermic needle. Like other bacteria and blue-green algae, their cells are only 0.5 to 1.5 μm in diameter. Even with the aid of a good microscope and micromanipulators, the task probably would be impossible. On the other hand, chemicals can be injected into specific cells of *Valonia, Acetabularia, Chara, Nitella, Cladophora,* and other algae having giant cells. Some genera possessing such cells are listed in Table 13-4.

An example of research in which giant cells were utilized is Englemann's classical study of the action spectrum of light and photosynthesis. Using *Cladophora glomerata,* Englemann set up a prism so that different colors of light fell on different parts of a cell mounted on the stage of his microscope. From this experiment, he learned that red and blue light were most effective in photosynthesis, as measured by oxygen evolution, while green and yellow light caused little oxygen to evolve. The use of *Valonia* in experiments involving radioactive tracers and of *Acetabularia* in studies of possible relationships between DNA and biological clocks are other examples of how large cells have been very useful.

14 MOSSES AND LIVERWORTS

During World War I when cotton gauze became scarce in England, Canadian and British troops used sphagnum moss, obtained mostly from Ireland, for dressing wounds. Although used as an emergency substitute, it was soon learned that the sphagnum dressings had some advantages over cotton dressings. Apparently moss contains antibiotics which were helpful in reducing infection. In a similar way, 19th century pioneers applied poultices made of moss, with or without the addition of juices of berries or roots, to scratches, bruises, and cuts to hasten the healing process; how to prepare them was part of the education they gained from their American Indian neighbors (See Coon 1963, McCleary et al. 1960).

Liverworts, too, have figured prominently in medical history. In the 1500s, when the "doctrine of signatures" was held in high esteem, plants were believed to reveal their medicinal powers through gross morphology. The lichen *Parmelia sulcata,* for example, is convoluted like a human brain and was used to treat the mentally ill; the wild lettuce, *Lactuca sativa,* has a milky sap and was given to women suffering from mastitis. But especially respected among all herbs were several species of thalloid liverworts which resemble the human liver in form and were believed to be capable of correcting all types of liver ailments. Although there has never been a shred of evidence presented to substantiate this belief, nor is there any important drug on the market derived from liverworts, one occasionally, even in the 20th century, comes in contact with people who believe that liverworts are valuable for medical purposes. Some superstitions die slowly. In retrospect, we must sadly admit that the "doctrine of signatures" greatly retarded, rather than advanced, medical science. Nevertheless, this peculiar doctrine stimulated medieval herbalists to study several groups of plants, including the two classes discussed in this chapter, the mosses and liverworts, and thus helped advance the science of botany.

Mosses and liverworts are recognized by nearly all botanists as being closely related to each other. Most taxonomists classify them as separate classes in a common division. For example, the "Eichler system," which classified plants into four divisions and ten classes, was the system of plant classification from the late 1870s to 1947. For almost three quarters of a century, it appeared in most biology textbooks published in the U.S. and many other countries. It is still being published in encyclopedias and popular books on botany. Today, the Bryophyta (*bryo* = moss, *phytum* = plant) is the only division of the four proposed prior to 1880 by Schimper and popularized by Eichler and others still accepted as a natural taxon. Schimper's Thallophyta is now recognized to be an assemblage of several unrelated groups and his Pteridophyta and Spermatophyta have been combined into the Tracheophyta and rearranged into new class groupings. But mosses and liverworts still stay together.

Although a large taxon with much variation in it, the bryophytes all tend to share in a number of characteristics: (1) all have a gametophyte-dominant diplohaplontic life cycle; (2) the gametophyte consists of simple leaf and stem-like or thalloid food production organs anchored to the soil by rhizoids; (3) the sporophyte is attached to the gametophyte and is usually completely dependent on it for nourishment; (4) fertilization takes place in a multicellular archegonium and the zygote develops into an embryo within the archegonium immediately following fertilization; (5) the sperm are little more than elongated nuclei with two anterior whiplash flagella; and (6) pigmentation, ultrastructure, food reserves, mitochondria, cell walls, etc., are typical for the subkingdom Lorobionta.

Of the two classes, the Hepaticopsida is the more ancient, is made up of simpler plants, is more tropical in its distribution, and is more hydric in its ecological preferences. Bryopsida are always leafy, Hepaticopsida may be leafy or thalloid. Bryopsida are more erect, Hepaticopsida tend to be prostrate. A list of characteristics of the two classes is presented in Table 14–1. Comparison of the growth form of typical liverworts and mosses and of the differences in structure of the sporangia is shown in Figure 14–1.

Although the fossil record suggests that the first bryophytes evolved at least 50 million years later than the first vascular plants, we will discuss them first, primarily because they are not as morphologically complex or fully as well adapted to terrestrial environments as are many Tracheophyta. At one time, it was believed by many botanists that the Bryophyta evolved from green algae and, in turn, gave rise to the vascular plants. This hypothesis has been abandoned by most biologists. Modern hypotheses generally acknowledge that the Tracheophyta is the older of the two terrestrial divisions of Lorobionta. One modern hypothesis presumes independent origin of the two divisions of land plants with the bryophytes evolving from green algae after the Tracheophyta had become well established. Another suggests that the Bryophyta evolved from early Tracheophyta.

Modern views of bryophyte origin are based largely on three kinds of evidence: morphological and physiological similarities between bryophytes and other Lorobionta, especially some Chlorophycopsida, ecological niche studies, and the fossil record.

Table 14–2 compares the characteristics of a primitive liverwort (Anthocerotales) and a primitive moss (Funariales) with the Chlorophycopsida presumed to be ancestral to both (Ulotrichales) and an advanced, more highly evolved liverwort (Marchantiales).

Proponents of the modern hypotheses point out that mosses and liverworts today, as in the past, are usually ecological subordinates in environments dominated by vascular plants. Removal of the dominants almost invariably results in elimination of the bryophytes from the ecosystems in which they were abundant. Thus ecological considerations support the hypothesis that the Bryophyta evolved later than the Tracheophyta, utilizing an ecological niche that was first created when vascular plants became well established on land.

In the Silurian formations in which fossils of the oldest terrestrial plants have been found, there are no recognizable bryophytes. According to Chaloner (1967), the number of species of fossil spores increased, over a 30 million-year span of time, from three in terrestrial Silurian formations to eight in the Gedinnian epoch (lowest Devonian) to 41 in the Emsian and 56 in

TABLE 14–1

COMPARISON OF THE GENERAL CHARACTERISTICS OF HEPATICOPSIDA AND BRYOPSIDA

	Hepaticopsida	Bryopsida
Growth form	Dorsiventral orientation	Radial orientation
Leaves	If present, unequally spaced around the stem	Always present, equally spaced around the stem
	Usually notched; if not, then rounded	Never notched, usually tapering to a slender tip
	Typically 1 cell layer thick, sometimes 2–3 layers thick	Typically many cells thick
	Cells isodimetric	Cells elongate, spindle-shaped or isodiametric
Rhizoids	Unicellular	Multicellular with oblique septations
Protonema	Spherical mass of cells or short filament	Alga-like branched filament differentiated into erect photosynthetic axis and colorless rhizoids
Archegonium	Usually surrounded by a perianth	Arises from a whorl of regular leaves
Capsule	Divides into 4 linear segments	Cup-like with peristome usually present
Variation within species	Varieties few, species very distinct	Many varieties within each species; species grade into each other
Chromosomes	Multiples of 9; polyploidy rare	Multiples of 5, 6, or 7; polyploidy common
Habitat	Abundant in tropics and warm temperate zone; mostly very hydric	Abundant in polar and cool temperate zone; both hydric and xeric species occur
Oldest known fossils	Upper Devonian, becoming common in Pennsylvanian	Permian, becoming common in Triassic

the Givetian; significantly, none of them resembled Bryophyta spores. Yet moss and liverwort spores have thick walls, are resistant to decay, and are quite distinctive in appearance (Figure 14-2). In the 1980s, however, were reported very small spores from Ordovician formations, which were believed by some botanists to be small liverwort spores, thus reviving the hypothesis that the first land plants were Bryophyta.

The Bryophyta have achieved a much higher level of cellular organization and differentiation than the Chlorophyta from which they appear to have evolved during the mid-Paleozoic Era. They possess separate and distinct organs for food production, for anchorage, for reproduction, and sometimes for foliage display. Some species possess differentiated tissues which increase their efficiency in photosynthesis and other processes.

In number of species, this is a large division with over 20,000 species distributed among 47 families. There are two classes, the Hepaticopsida and the Bryopsida, each with eight orders (Figure 14-3). All of the species are terrestrial or freshwater, none are marine. Although most favor wet sites, there are a number of desert species of Bryopsida. Many mosses are well adapted to survive the very harsh conditions of Arctic and alpine tundras where they make up a conspicuous part of the flora. Liverworts, on the other hand, are most abundant in moist tropical ecosystems.

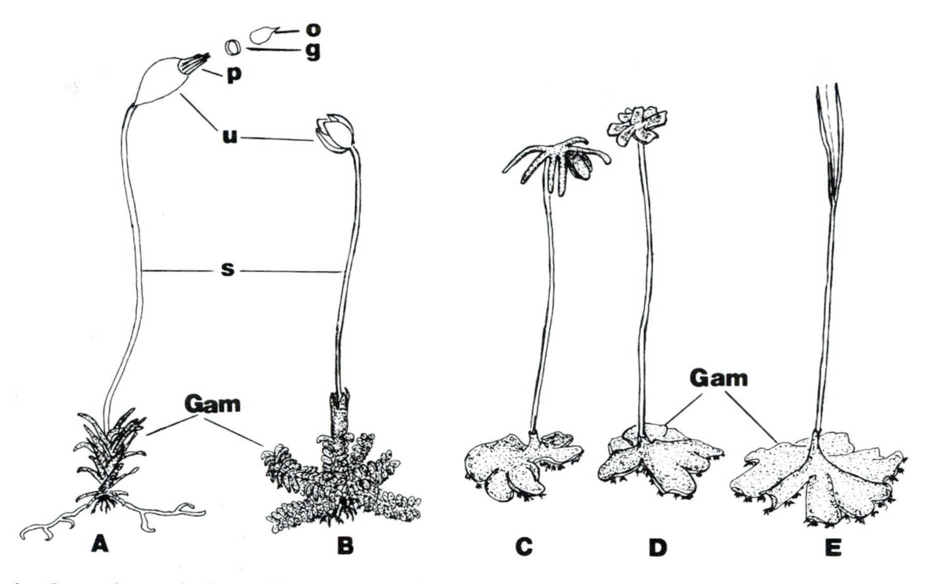

Figure 14–1 General morphology of liverworts and mosses. (A) A typical moss, *Mnium marginatum;* (B) a leafy liverwort, *Jungermannia pumila;* (C) female and male *Marchantia polymorpha,* a thalloid liverwort; (D) a hornwort, *Anthoceros crispulus.* (o = operendum, g = gasket, p = peristome, u = urn or capsule, s = seta, Gam = gametophyte.)

CLASS 1. HEPATICOPSIDA

The Hepaticopsida (*hepato* = liver), commonly called the liverworts or livermosses, may be either leafy or thalloid. By thalloid is meant that there is no distinct differentiation into leaf and stem as there is in the mosses and leafy liverworts. Thalloid liverworts superficially resemble lichens, which at one time were included in the same division by many botanists. There are about 9000 species and 250 genera of liverworts classified here in 23 families and 10 orders, but by some others in as few as 3 or 4 orders and 50 or more families (e.g., Reimers in Melchior and Werdermann 1954). Chief characteristics of the orders are summarized in Table 14-3. One order, Anthocerotales, is sufficiently different from the other nine that it is sometimes placed in a class by itself, the Anthoceropsida.

Liverworts can be found in wet, densely shaded forests, especially along the edges of streams and near waterfalls, and on the exposed rocks of waterfalls. They are more abundant in warmer climates than in cold, but a few species live in arctic and subarctic areas; they are seldom, however, very abundant outside the tropics. The most commonly encountered species in North America and Europe is a thalloid species, *Marchantia polymorpha* (Figure 14-4E). Since this is the only liverwort many biology students ever see during their college experience, it is not surprising that most people seem to think all liverworts are thalloid, when, in fact, about 80% of all species are leafy.

Thalloid liverworts are often confused by students with some of the foliose soil lichens, especially peltigeras. They are easily distinguished from lichens by the absence of apothecia, frequent occurrence of gemmae cups and sometimes archegonia or antheridia, the large cells on the surface of the liverworts as opposed to small, gelatinous, cortical cells on lichens, by lack of stratified tissues including a distinct chlorophyllous layer characteristic of most lichens, by color, and by habitat.

TABLE 14-2

COMPARISON OF SOME OF THE CHARACTERISTICS OF FOUR ORDERS OF GREEN ALGAE, LIVERWORTS, AND MOSSES

Ulotrichales	Anthocerotales	Funariales	Marchantiales
Gametophytes and sporophytes have single large plastid with prominent pyrenoids	Gametophyte with single large plastid and a prominent pyrenoid; sporophyte with large plastid and pyrenoid	Gametophyte with large plastid; sporophyte also with green plastids having prominent pyrenoids	Gametophyte with many small plastids; sporophyte without plastids or pyrenoids, not green
Cells short, cylindrical; no tissues or organs except in one genus having parenchymatous thallus; plants filamentous	Cells short, cylindrical, organized into parenchymatous thallus	Cells short, cylindrical, gametophyte differentiated into erect stem with many green leaves; sporophyte green, lacking leaves	Cells mostly isodiametric and often elongated; sporophyte inconspicuous, colorless
Gametophyte and sporophyte same size and form; both independent and autotrophic	Gametophyte more prominent and slightly larger than sporophyte; both autotrophic	Gametophyte more prominent and slightly larger than sporophyte; both autotrophic	Gametophyte much more prominent and much larger than sporophyte; sporophyte heterotrophic
Gametes with 2 anterior whiplash flagella	Gamete with 2 anterior whiplash flagella	Gametes with 2 anterior whiplash flagella	Gametes with 2 anterior whiplash flagella
Isomorphic alternation of generations	Alternation of generations with gametophyte slightly more prominent than the sporophyte	Alternation of generations with gametophyte slightly more prominent than the sporophyte	Alternation of generations with gametophyte decidedly more prominent than the sporophyte

Leafy liverworts are frequently confused with mosses; if reproductive structures are present, telling the two apart is simple. If there are no reproductive structures or sporophytes present, identification will have to be based on growth form and leaf characteristics. If the leaves are thin, delicate, and notched, and unevenly spaced around the stem in three ranks, the ventral rank often smaller than the others, the plant is a liverwort. If the plant is upright in habit with relatively thick leaves spaced evenly on all sides of the stem and with a midrib but no notch, the plant is a moss (compare Figures 14-5 and 14-6). It always amazes the freshman student how many bryophytes there are with characteristics somewhat intermediate between the typical moss and typical liverwort. As in all plant identification, one must exercise his best judgment in deciding the taxon some unknown specimen belongs in; with experience, judgment improves.

General Morphological Characteristics

The plant body is thalloid in four of the ten orders of hepatics, meaning that there is no differentiation of stem and leaf, but a large, leaf-like or plate-like thallus is attached directly to the substratum by rhizoids. In some of the small Sphaerocarpales, the cells making up the thallus are undifferentiated and all of them contain plastids and carry on photosynthesis and other metabolic activities. In the Metzgeriales (Figure 14-5D), the thallus tissue is also undifferentiated

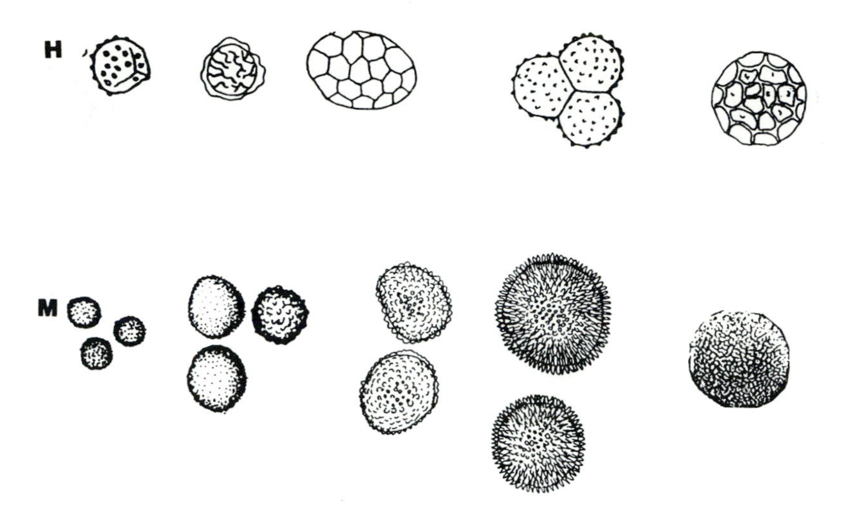

Figure 14–2 Moss and liverwort spores (spore diameter in μm indicated in parentheses). (H) Hepaticopsida: *Anthoceros laevis* (40), *Riccia frostii* (ca. 40), *Riccia sullivantii* (ca. 70), *Riccia curtisii* (ca. 45), *Riccia fluitans* (ca. 65). (M) Musci or Bryopsida: *Distichium capillaceum* (17–23), *Distichium inclinatum* (30–45), *Physcomitrium californicum* (30–34), *Physcomitrium pyriforme* (27–52), *Tortula ruralis* (10–14).

except that the upper cells often contain more chloroplasts. In the Marchantiales, on the other hand, the thallus is differentiated into (1) an upper epidermis layer, (2) a loose green region having one or more layers of air chambers, and (3) a compact, colorless, parenchymatous ventral region; if moisture becomes limiting, collapse of the cells surrounding the air chambers partially seals off the chambers and thus reduces evaporation. There are also internal cavities in many Anthocerotales, but these are filled with mucilage, not air.

In the other orders, the plant body consists of a prostrate, or occasionally ascending, stem usually attached to the substrate by rhizoids, and bearing two or three rows of leaves. In the four species of Calobryales, the basal part of the plant is a branched, creeping "rhizome," lacking rhizoids, from which arise vertical, leafy stems; the leaves are simple and entire. The stems of the only species of Takakiales are erect like those of the Calobryales, but the leaves are cylindrical with a central series of cells surrounded by approximately eight peripheral series. In the circa 7000 species traditionally included in the Jungermanniales, but here separated into three orders, considerable variation in leaf size and shape as well as in stem and rhizoid characteristics exists; however, nearly all species have the following characteristics in common: (1) the leaves are almost always only one cell layer thick, although a few species have leaves two or three cells thick; (2) a true midrib or nerve is never present; however, a series of elongate or rounded cells called a **vitta** occurs in a few species; (3) the uppermost whorl of leaves forms a **perianth** which surrounds the archegonium or antheridium; (4) as each leaf grows, there are nearly always two distinct apical growing points resulting in a bilobed or cleft leaf; and (5) the stems are usually prostrate with the ventral row of leaves considerably modified and smaller than the two dorsal or lateral rows. In *Lepidozia* and its relatives, the leaves tend to be divided into three or four lobes. In *Herberta* and a few other genera, the stems are erect.

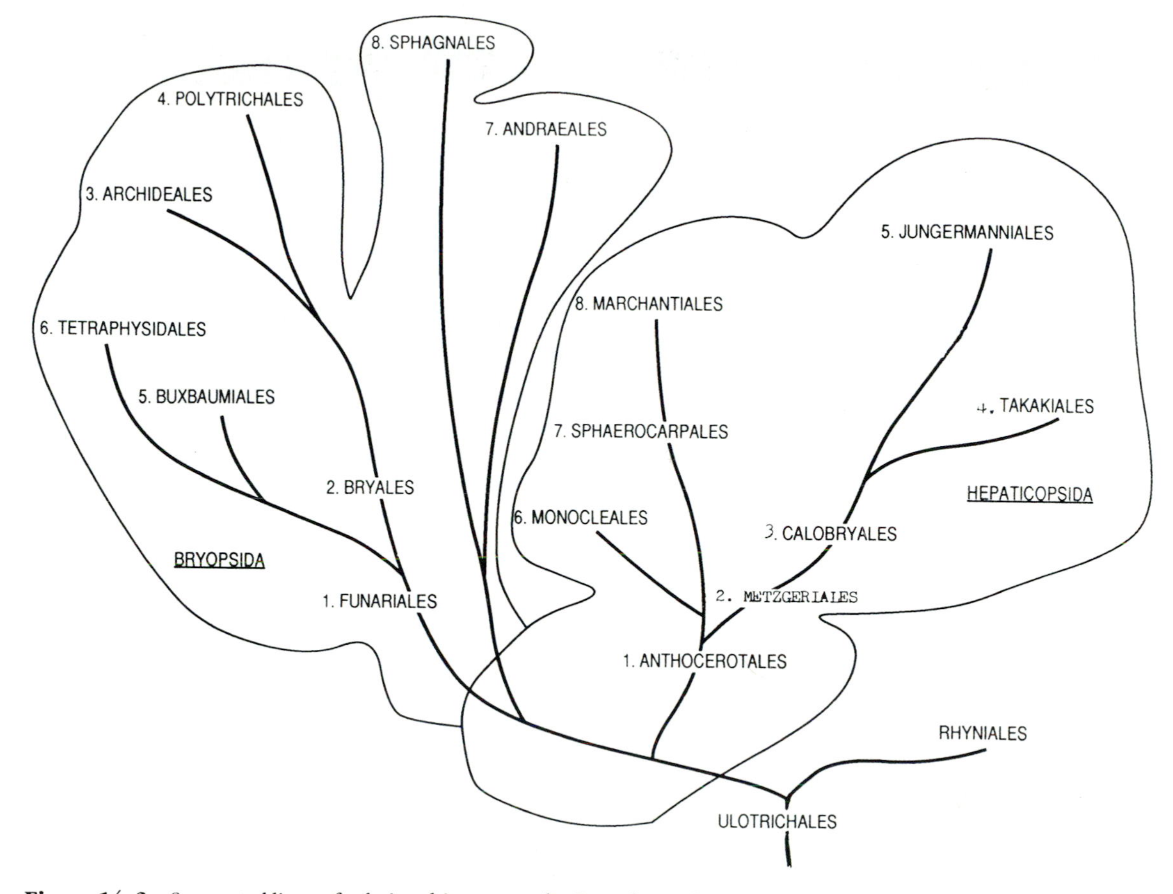

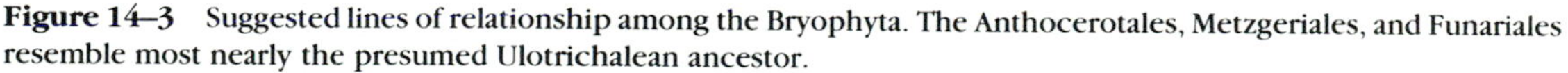
Figure 14–3 Suggested lines of relationship among the Bryophyta. The Anthocerotales, Metzgeriales, and Funariales resemble most nearly the presumed Ulotrichalean ancestor.

Stems of leafy liverworts range from about 0.1 to 0.6 mm in diameter and anatomically are parenchymatous with little or no differentiation of cells. They develop from a single apical cell resembling an inverted isosceles tetrahedron. In primitive genera, differentiation is so slight that the leaves are radially three ranked, with the ventral row of leaves only slightly smaller than the laterals. In most species, however, the ventral row of leaves, called **amphigastria,** is considerably smaller than the laterals. Although none of the liverworts have lignified cells, *Marchantia* possesses the enzyme system necessary to produce lignin from eugenol, a lignin precursor found in some mosses and all vascular plants.

In both leafy and thalloid liverworts, the rhizoids are nearly always white, slender, unicellular threads that may become brown in age. Multicellular rhizoids occur in two leafy genera, *Plagiochila* and *Schistochila,* and purplish rhizoids are common in the Metzgeriales.

Following germination, leafy liverworts enter a primary leaf stage, then a juvenile leaf stage, before becoming mature gametophytes with production of archegonia, antheridia, and sporophytes (Figure 14-1). Four patterns of development were described by Fulford, in the most

TABLE 14-3

GENERAL CHARACTERISTICS OF THE HEPATICOPSIDA

Orders (Fam–Gen–Spp)	**Growth form**	**General morphology**	**Representative genera**
Anthocerotales (1-4-320)	Thalloid	Thallus orbicular with dichotomously lobed margins; cells undifferentiated; single plastid per cell with pyrenoid; sporophyte green; capsules long, horn-like	*Anthoceros, Notothylas, Megaceros, Dendroceros*
Metzgeriales (1-20-1,000)	Mostly thalloid, lobed	Growth indeterminate; internal structure smooth, simple	*Pellia, Fossombronia, Metzgeria*
Calobryales (1-2-4)	Leafy, unlobed stems from "rhizome" w/o rhizoids	Small, erect with round to oval 3-ranked leaves, ventral rank slightly smaller than laterals; archegonial neck long, twisted	*Calobryum, Haplomitrium*
Takakiales (1-1-2)	Modified leafy	Small, erect, delicate, with leaf-like cylindrical branches called "phyllids"	*Takakia*
Jungermanniales (8-170-7,000)	Mostly leafy	Growth determinate, apical cell becoming archegonial initial; leaves 3-ranked with very small ventral leaves	*Jungermannia, Frulania, Porella, Scapania, Gymnocolea, Lejeunea, Lophozia*
Sphaerocarpales (2-3-20)	Thalloid	Simple gametophyte lacking internal differentiation of cells	*Sphaerocarpos, Riella, Geothallus*
Monocleales (1-1-2)	Thalloid	Large ($\approx 5 \times 20$ cm), made up of homogenous parenchyma cells lacking air spaces; fungi often present; some contain *Nostoc* and fix nitrogen	*Monoclea*
Marchantiales (3-15-150)	Thalloid	Differentiation into epidermal and cortical tissues; sporophytes embedded in the gametophyte; air spaces present, close when humidity is low	*Riccia, Ricciocarpos, Corcinia, Raboulia, Asterella, Lunularia, Marchantia, Conocephalum*

Note: The number of families (Fam), genera (Gen), and species (Spp.) in each order is indicated in the first column.

common of which germination results in an alga-like filament which produces the primary leaf. In the thalloid liverworts, the development stages are more condensed, the primary leaf stage being replaced by a disk stage which develops into the broad, flat thallus.

Sexual reproduction is by fertilization of nonmotile eggs by biflagellate sperm produced in gametangia. The first evidence that an archegonium or antheridium is about to appear is the development of an **involucre** in some thalloid hepaticopsida such as *Sphaerocarpos,* or of a **perianth** in leafy liverworts. In *Marchantia,* special structures, the **archegoniophore** and **antheridiophore,** extend above the thallus and bear the gametangia (Figure 14-1). In the Jungermanniales, the perianth consists of large, modified leaves at the apex of the branches.

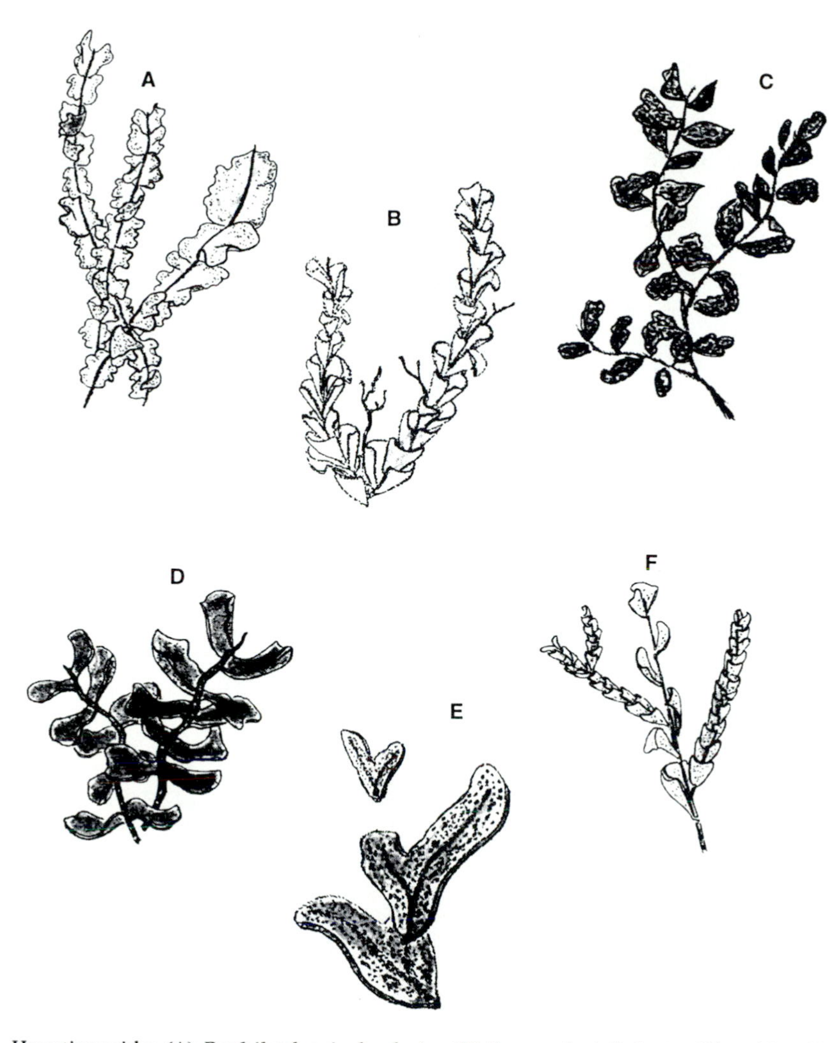

Figure 14–4 Hepaticopsida: (A) *Barbilophazia barbata,* (B) *Bazzania trilobata,* (C) unidentified species from Fern Falls, Rocky Mountain National Park, CO, (D) unidentified leafy liverwort from Cow Creek, Larimer County, CO, (E) *Marchantia polymorpha,* (F) *Porella platyphylloides.*

Bryophyte gametangia are much larger and more complex than gametangia of either Chlorophyta or Tracheophyta. In most thalloid liverworts, the archegonium is elevated above the thallus surface, but in Anthocerotales, it is embedded in the thallus. In leafy liverworts, it is at the apex of the stem enclosed loosely by the perianth. Above the chamber containing the egg, specialized canal cells are surrounded by thicker walled neck cells, the whole structure having the appearance of a long-necked flask. The archegonium in all thalloid liverworts except *Anthoceros* develops from a single cell on the surface of the thallus and consists of two parts, the venter or egg chamber and the neck. When the egg is ready for fertilization, the canal cells of the neck break down into a mucilaginous liquid through which the sperm are able to swim in

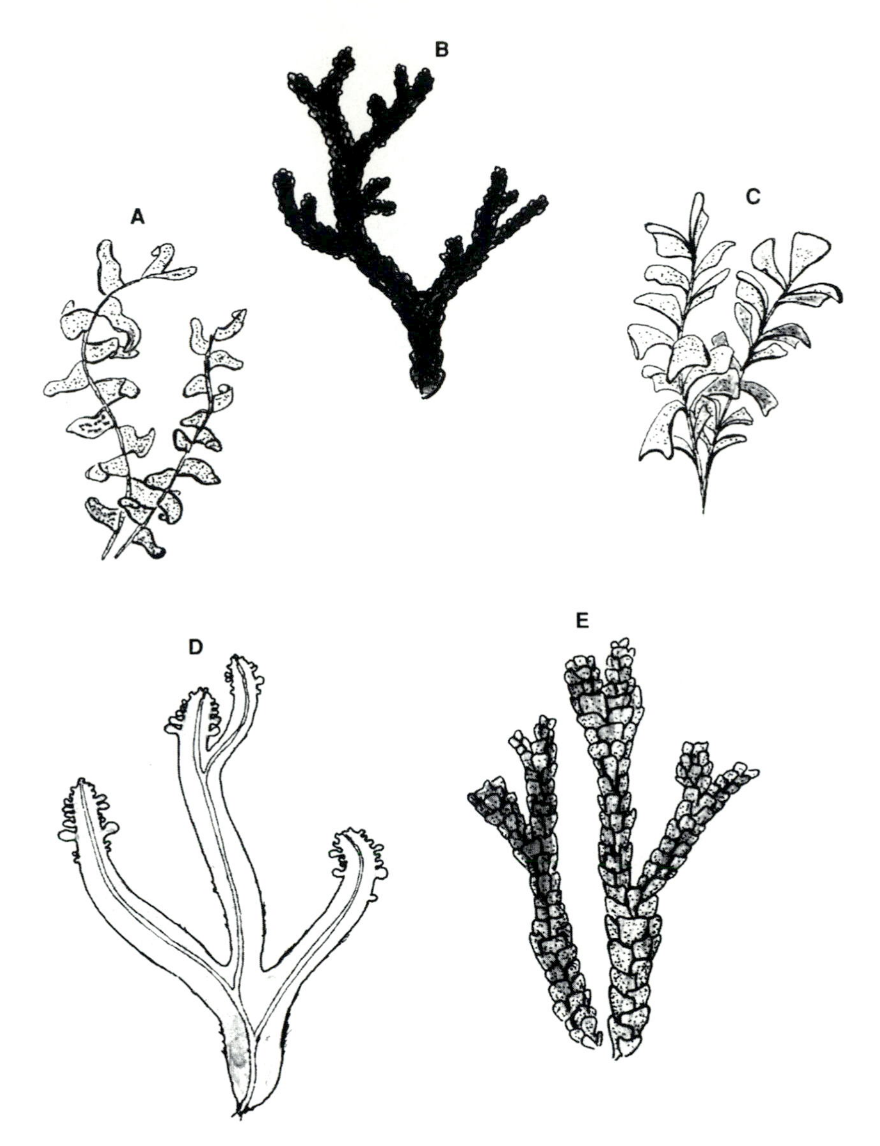

Figure 14–5 Hepaticopsida. (A) *Jungermannia exsertifolia,* (B) *Frullania asagrayana,* (C) *Scapania nemorosa,* (D) *Metzgeria fruticulosa,* (E) *Bazzania Trilobata.*

effecting fertilization. In *Anthoceros,* there is no specialized neck, but only four or five neck canal cells directly above the venter. The egg is nonmotile.

Like the archegonia, the antheridia develop from a single cell and are elevated on receptacles except in the Anthocerotales, in *Asterella,* and in a few other species in which they are sunk in the upper surface of the thallus. In *Marchantia,* the receptacle is elongated into an

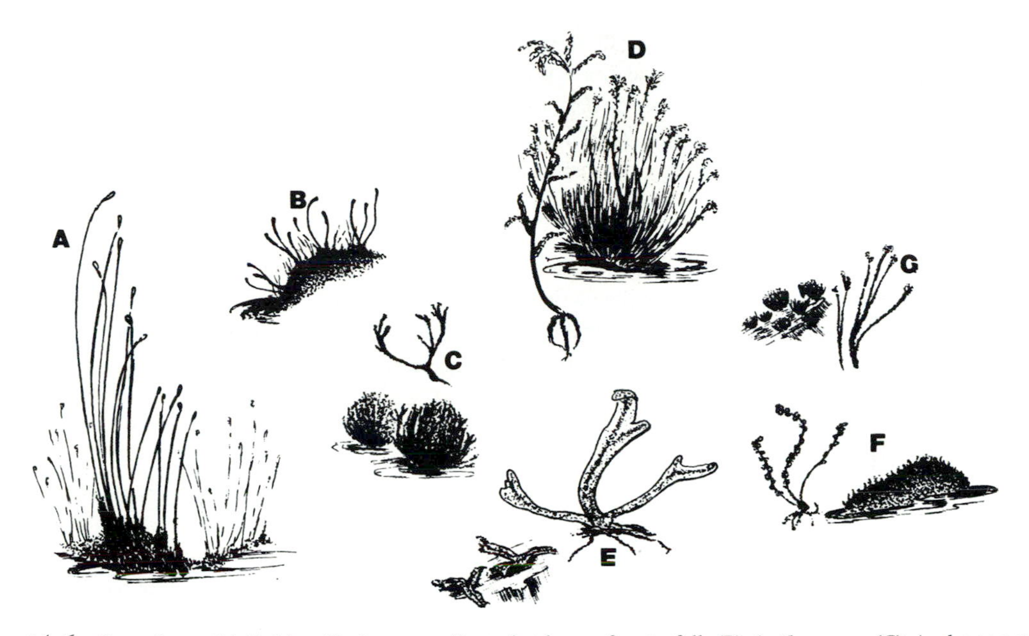

Figure 14–6 Bryophyta. (A) Unidentified moss collected at base of waterfall; (B) *Anthoceros;* (C) *Andreaea rupestris;* (D) *Spagnum squarrosum;* (E) *Conocephallum conocum;* (F) *Pellia endivifolium;* (G) *Grimmia apocarpum.*

antheridiophore (Figure 14-1D) with the antheridia on its ventral surface. The sperm are enclosed in delicate, multichambered sacs attached to the receptacle by short stalks. Above each antheridial sac, small openings allow for the escape of the sperm from the antheridial chambers. Each sperm is essentially an elongated, often curved, nucleus with two flagella. When the antheridium is mature, the sperm are released and air currents help to spread them. In the Marchantiales, the antheridia are embedded in a special tissue (14-3d) and on maturity, the sperm are forcibly discharged, an adaptation which makes it possible for air currents to disperse them more effectively. Water is essential to fertilization, but a drop of dew or a light rain provides sufficient moisture for this purpose.

Following fertilization, the zygote develops immediately into an embryo within the archegonium. Two patterns of development have been observed; in several families, the zygote develops into a four-celled filament and then begins dividing longitudinally to produce the parenchymatous sporophyte tissue, and in other families the zygote develops into an eight-celled ball first. In nearly all liverworts, the mature sporophyte consists of three parts, the foot which anchors the sporophyte and absorbs the nutrients needed for growth and development, the seta or stalk, and the capsule which contains spores and elaters. In *Anthoceros,* there is no seta, but an intercalary meristem between the large foot and the elongate capsule results in capsules that continue growing until they sometimes achieve a height of 20 cm or more. Other liverworts lack sporophyte meristems and the capsules and setae are of limited size. In *Riccia,* the sporophyte is reduced to a simple capsule only slightly embedded in the surface layers of gametophyte tissue.

Asexual reproduction in liverworts is by (1) growth and branching followed by death and decay of older parts, (2) separation of whole organs and regeneration of new plants from them, and (3) specialized propagules called gemmae. In *Riccia,* repeated branching of the thallus followed by death of the original body is common. In *Frullania, Plagiochila,* and other leafy liverworts, new plants commonly develop from detached leaves, and leaf cuttings can be used in the laboratory to propagate *Frullania fragilifolia.* Gemmae are cup-like or disk-like structures which grow on the upper surface of bryophyte thalli or leaves. The gemmae of *Marchantia* and *Lunularia* are large and prominent, but the gemmae of most leafy liverworts are microscopic.

The ultrastructure and pigmentation of liverworts are typical of the Lorobionta. The basic chromosome number of most liverworts is 9. In the Anthocerotales, however, it is 5; in *Takakia,* it is 4; and in some genera it is 8 or 10. Polyploidy is rare; however, polyploid races occur in *Marchantia* (9, 18, and 27) and some other genera.

Phylogeny and Classification

The oldest bryophyte fossils that have been found to date are liverwort fossils in the upper Devonian; the oldest moss fossils appear 100 million years later in the Permian. According to the independent origin hypothesis, the first bryophytes evolved from Ulotrichalean stock during the Devonian and were probably similar to *Anthoceros.* The same or similar stock had already given rise to vascular plants.

The presumed ancestors of the Bryophyta, the Ulotrichales, are characterized by short cylindrical cells containing a single, large chloroplast with a prominent pyrenoid, gametes having two anteriorly inserted whiplash flagella, cells with prominent starch grains, and cell walls primarily of cellulose with relatively little pectic compounds. Some of the Ulotrichales produce parenchymatous tissue and survive well on damp soil, although they require free water for reproduction (Figure 13-7). The Anthocerotales, or hornworts, resemble the Ulotrichales in all of these characteristics. Life cycles are also similar. Ulotrichales often have isomorphic diplohaplontic life cycles. Life cycles in the Anthocerotales are heteromorphic diplohaplontic with gametophytes somewhat larger than sporophytes, but sporophytes that have been separated from the mother gametophytes can carry on photosynthesis and continue to live independently. In most Hepaticopsida and Bryopsida, on the other hand, the sporophyte is reduced to a simple structure completely dependent on the gametophyte for food. In both the Anthocerotales and Funariales, the sporophtye is relatively large, structurally complex, and photosynthetic. Thus, the life cycle in these two orders is only slightly different from the isomorphic life cycles of the Ulotrichales. Genetically, life cycles in which the two generations are similar to each other are simpler than those in which they are vastly different from each other, regardless of whether the gametophyte or the sporophyte dominates. The Ulotrichales, Anthocerotales, and Funariales are compared in Table 14-2.

Morphologically, *Anthoceros* seems to form a link between the presumed ancestors among the Chlorophyta and the rest of the Hepaticopsida. Our present, very incomplete fossil record fails to confirm this assumption, however, inasmuch as no indisputable fossils of Anthocerotales, or **hornworts,** have been found in Paleozoic formations. A few well-preserved liverworts have been found in upper Devonian and lower Carboniferous formations, most of them similar to modern Sphaerocarpales and Metzgeriales. In slightly younger formations, liverwort fossils are much more abundant, and all of the modern orders are represented in late Mesozoic and early Cenozoic records.

The earliest bryophytes, like their modern descendants, were probably well adapted to growing in the shade of vascular plants in relatively wet environments. From these first bryophytes have evolved the modern Anthocerotales and other liverworts along one line and the Bryales and other mosses along a second line.

Spores of *Anthoceros* are similar to the spores of some of the early vascular plants, especially *Psilophyton.* This has led some botanists to propose that the Anthocerotales evolved from the Psilopsida. *Pallavaciniites devonicus,* the earliest known fossil liverwort, was found in a formation dominated by *Psilophyton* and associated also with lycopod, fern, and gymnosperm remains.

Because so little paleobotanical work has been done with bryophytes, it is difficult to develop a classification system based on the fossil record as has been done in some groups of plants such as the Rhodophycopsida, Siphonophycopsida, and Tracheophyta. According to Schuster (1966), the oldest bryophyte fossils now known, including *Pallaviciniites devonicus, Treubiites kidstoni, Blasiites lobatus,* and *Hepaticites metzgerioides,* are members of the Metzgeriales. Morphologically, the Metzgeriales show a superficial similarity to the Anthocerotales, but bryologists who have studied the oldest fossils carefully seem confident of their classification in the Metzgeriales.

At least four arguments can be advanced to explain why the hypothesis that the Anthocerotales is the most primitive of the Hepaticopsida and the order from which other liverworts have evolved has not been supported by fossil evidence: (1) the hypothesis itself may be wrong; (2) fossil hornworts may exist in relative abundance but have not yet been found; (3) environmental conditions may not have been suitable for fossilization in the habitats where the earliest *Anthocerotales* lived; many modern hornworts, for example, are epiphytes, and epiphytes do not fossilize well; and (4) fossil hornworts may have been found but misidentified, thrown into the portmanteau genera *Metzgeriites* and *Hepaticites,* for example.

Suggested lines of relationship among the 16 orders are illustrated in Figure 14-3.

Modern Metzgeriales are primarily thalloid liverworts, but Watson (1971) also listed several genera of leafy liverworts among the earliest bryophyte fossils, suggesting a rather wide range in taxonomic groups suddenly appearing in the Devonian and Mississippian periods. Relationships suggested from studies of comparative morphology of both extant and fossil liverworts are indicated in Figure 14-3.

In the Hepaticopsida, and also in the Bryopsida, evolutionary trends seem to have resulted in greater and greater differentiation between the two generations, the sporophyte and gametophyte. In *Anthoceros,* the two generations are not as different from each other as is the case in the Marchantiales and Jungermanniales where differences are very pronounced. Evolutionary trends in the Bryophyta have been toward increasingly simple sporophytes and increasingly complex gametophytes.

Most botanists have regarded thalloid liverworts as primitive and leafy species as advanced; however, two problems prevent our accepting this hypothesis without reservation: (1) some of the leafy species, e.g., *Trichocolea tomentella* found in northern Europe in habitats where the humidity is constantly high, resemble some algae quite closely, and (2) among the most ancient fossils yet found, some are thalloid and others are leafy. Watson suggested that the leafy habit, if it is advanced, may have evolved independently several times. In this case, the orders in current systems of classification may need to be split and the species recombined in more natural orders. Characteristics of the eight orders are summarized in Table 14-3.

In the early classification systems there were very few genera of liverworts and also few families. About 1900, several taxonomists, e.g., V. F. Brotherus (Melchior and Werdermann 1954) began splitting these genera (which in many cases had grown into giant taxa with many hundreds of species), into more manageable units. This splitting trend continued until very few large genera and many monotypic ones existed. New families with fewer genera were also created en masse. Thus Reimer (Melchior and Werdermann 1954), following Brotherus' outline, divided the bryophytes into 4 orders and 48 families of liverworts and 3 orders and 80 families of mosses. At the present time, many botanists feel this has gone too far. For example, about half of the 48 families in Engler's Syllabus (Reimers 1954) have only one genus and almost a quarter of the families have five or fewer species. The classification used here follows Watson with the families roughly equivalent to Reimers' superfamilies.

Representative Genera and Species

Anthoceros is different from most other Hepaticopsida in many ways. It is the only liverwort to have pyrenoids; both generations, haploid and diploid, are autotrophic; each cell of the gametophyte has a single large plastid and the sporophyte cells have two large plastids each; the sporophytes can exist by themselves under laboratory conditions and come close to independent living quite often in nature; the sporophyte lacks a seta; the elongated capsule continues growing in an indeterminate manner for some time, and there is elaborate internal organization of the capsule. These differences prompted Hayes in 1898 (Fulford 1964) to propose a separate class for the Anthocerotales, and his example has been followed by several other botanists. On the other hand, other Anthocerotales—*Notothylas, Dendroceros,* and *Megaceros*—are much like other liverworts and form a link between *Anthoceros* and the rest of the class. The Anthocerotales, in turn, seem to be a primitive order which forms the link between the Chlorophycopsida and the rest of the Hepaticopsida.

Typical of the thalloid liverworts is *Marchantia polymorpha* (Figure 14-4), probably the most widespread of all liverworts. It seems to thrive any place where humidity is constantly relatively high, but is especially abundant in humid forests where burns have recently occurred. It grows in arid regions in shaded spots near waterfalls, on the shady banks of fast-flowing creeks, and near sprinkler outlets on the shady side of buildings or trees. Individual plants grow rapidly, often attaining a length of 10 to 15 cm. They are dichotomously lobed and branched and exhibit the apical growth and posterior decay typical of many liverworts. Other common genera of thalloid liverworts are *Riccia* and *Sphaerocarpos.*

Porella, Frullania, and *Scapania* (Figures 14-4F, 14-5B, C, respectively) are examples of leafy liverworts. Unlike *Herberta,* which is erect with three ranks of equally sized leaves, these genera are prostrate and have two ranks of larger leaves plus a ventral rank of amphigastra or smaller leaves. They superficially resemble prostrate mosses, but the leaves are thinner and they lack a midrib. The flattened, much branched stems of *Porella* have scattered rhizoids emerging from the under surface and anchoring them to the soil. The leaves overlap each other (Figure 14-4F) and the plants form densely interwoven mats closely appressed to the ground with only the youngest branches ascending in a semi-erect position. The leaves lack a cuticle and rapidly dry out when picked. *Porella platyphylloidea* is one of the most common species of leafy liverworts in North America. It is heterothallic, with the male plants somewhat narrower than the female. The perianths are swollen and each encloses eight to ten archegonia.

Splitting in recent years has reduced the size of most of the giant, heterogeneous genera that formerly existed, replacing them with a large number of smaller, more homogeneous genera, many of them monotypic. Nevertheless, a few giant genera, in terms of number of species, still exist, including *Plagiochile* with 1200, *Frullania* with 700, and *Bazzania* with 450 species. Among temperate zone genera, *Lophoclea* (380 species), *Riccia,* and *Anthoceros* each has over 200 species.

The largest families in number of species are the Lejeuneaceae (1800), Lepidoziaceae (1750), and Plagiochilaceae (1470), all in the Jungermanniales.

Ecological and Economic Significance

Liverworts are plants of warm, humid areas, being most common in tropical rain forests. In general, liverworts are not important elements, in terms of total productivity, of any ecosystem, but they occasionally become abundant in special habitats. They tend to be subordinate to other plants in all ecosystems, even to mosses when associated with them. Nevertheless, *Diplophyllum albicans* is an important element in Great Britain in some ecological successions, growing abundantly in the chinks between rocks, and in North Carolina, *Sphaerocarpus texanus* is a dominant pioneer species in ecological successions of abandoned fields. *Frullania dilatata* occurs in moss-dominated bark ecosystems in Britain. *Scapania* species (Figure 14-5C) are associated with mosses in bog communities in the Black Forest of Germany.

In addition to temperature and moisture, light is important in the survival of bryophytes. The onset of reproduction in most temperate zone liverworts is triggered by changes in photoperiod. Few mosses, on the other hand, are sensitive to daylength (Chopra and Bhatla 1983). Increase in light intensity and temperature are more important in triggering reproduction in the Bryopsida.

Liverworts are of little economic value, either positively or negatively. At one time they were widely used in medicine, and a few naturopaths, hanging on to the *doctrine of signatures,* still recommend them for liver ailments. *Pulmonaria* spp. in the Anthopsida, so named because of their morphological similarity to the human lung, was recommended by medieval herbalists for tuberculosis and pneumonia, and a common lichen, *Peltigera canina,* which smells like dog urine, for rabies. Other species were mentioned earlier in the chapter.

Research Potential

Liverworts have great potential for scientific research and have been used in the past in significant ways. The first studies of sex determination in plants were made with *Sphaerocarpos.* Studies of nitrogen fixation have been made with *Anthoceros* and other liverworts containing symbiotic cyanophytes in their tissues. In *Scapania* growing under extreme environmental conditions, morphological changes occur and the plant takes on the appearance of other species that grow there; this phenomenon needs further investigation and could lead to a better understanding of morphogenesis in general.

Some Hepaticopsida are easily cultured. Since the dominant phase of their diplohaplontic life cycle is haploid, they offer some special advantages in genetic studies. The genetics of photosynthesis, the genetics of respiration, and the genetics of cellular differentiation could possibly be studied more effectively with liverworts than with the traditional garden varieties of Anthopsida.

Gametophytes of *Cryptothallus mirabilis*, in the Metzgeriales, are very similar in appearance to *Riccardia* except that they lack chlorophyll and form a mycorrhizoidal relationship with a fungus. When first discovered in Austria about 1920, *Cryptothallus* was thought to be a *Lycopodium* gametophyte. This interesting species should be studied more in order to improve our understanding of mycorrhizal relationships.

CLASS 2. BRYOPSIDA

The Bryopsida, or Musci, differ most noticeably from the Hepaticopsida in having an upright, leafy stem with radially arranged leaves which usually have prominent midribs. They also differ in several other ways (Table 14–1): mosses have multicellular, branched rhizoids, develop from a protonema, tend to be better adapted to xeric habitats than liverworts, and are much more abundant in temperate and colder climates. The 12,000 species are included in 27 families; general characteristics of the 8 orders are outlined in Table 14–1, and suggested relationships among them are presented in Figure 14–3. Three representative moss species are illustrated in Figure 14-6 and compared with three common liverwort species.

Mosses are easily found in all terrestrial ecosystems, from bogs to deserts and from arctic and alpine tundra, where their only living associates are lichens, to tropical forests. There are no marine mosses, however.

Not everything that is called moss is moss. Fishermen call green algae moss, many people call the fruticose lichens hanging from trees moss, and even foliose and crustose lichens are sometimes called moss. In the southern states, a flowering plant of the pineapple family is called Spanish moss, and on the Atlantic coasts of Europe and North America, a red alga used in making broth is called Irish moss.

Students seldom confuse mosses with other plants. In contrast to algae and lichens, mosses have leaves and stems. Leafy liverworts and lycopods cause some trouble, but the former can be recognized by their dorsoventral orientation and thin leaves with notched apices, in contrast to the radial orientation and thick acuminate leaves, of mosses, and the latter have a complex water conducting tissue, xylem, in their stems, whereas moss stems are seldom more than 2 or 3 mm in diameter with no xylem and little differentiation of tissue in them.

Morphological and Physiological Characteristics

The commonly observed moss plants, like those of liverworts, are gametophytes and therefore haploid. They consist of a thin stem anchored to the substrate with multicellular rhizoids and with leaves arranged radially about the stem.

Some mosses are prostrate, others erect, and they vary in size from a few millimeters tall to half a meter or more in some species of *Sphagnum, Polytrichum, Dawsonia,* and *Fontinalis.* Most mosses are normally bright, grassy green, but when they dry, they sometimes become black or dark brown. Often the mosses on the desert pass almost completely unnoticed until moistened by a rain, when they change from dull brown to bright green and suddenly seem to be everywhere.

At the apex of mature male plants, antheridia are formed. On female plants, archegonia develop at the apex and, following fertilization by sperm produced in the antheridia, erect sporophytes grow from the archegonia. As the sporophytes mature, a unique type of sporangium forms at its apex (Figure 14–1). Meiosis takes place and four spores, two of which will develop into male gametophytes and two into female, result. These are disseminated by splashing rain or by wind and eventually germinate.

Growth of the moss gametophyte begins when a meiospore germinates and develops into a **protonema** which is an algal-like filament, with transverse crosswalls and numerous chloroplasts. Mitosis takes place in the terminal cell, but the other cells continue to elongate in an uneven manner and the crosswalls become oblique. At first, only the terminal cell divides, but after several cells have been formed, some of the older ones begin bulging at the anterior end, and from these bulges, branches are formed. The branches rebranch and as the protonemata continue to grow, some of the cells produce **bud primordia.** These are pear shaped, with a tetrahedral cell resembling an inverted pyramid at the apex. As the apical cell divides and forms additional cells on its three faces, a bud results, which soon develops into the juvenile gametophyte.

Moss stems are frequently more complex than liverwort stems. Although some stems have no differentiation of tissues, in many species there is a central strand, a cortex, and an epidermis (Figure 14-7). Cells in the central strand may be adapted to conduction of food from the leaves to the rhizoids and other plant parts. Branches often arise, but from beneath the leaves rather than in the axils of the leaves as in the Tracheophyta.

Leaves are mostly simple, consisting of a single layer of cells making up the leaf blade and a costa or midrib two or three cell layers thick resembling a vein. Depending on species, moss leaves vary from less than a millimeter in length to over a centimeter. They vary in shape from ovate through lanceolate to acicular (Figure 14-8B). In many species, the leaves terminate in a hair-like projection or awn; in others they are either obtuse or terminate in a sharp angular point. The base of the typical moss leaf is broad and sessile, but some species have leaves with tapered bases. Leaves are usually in sets of three,

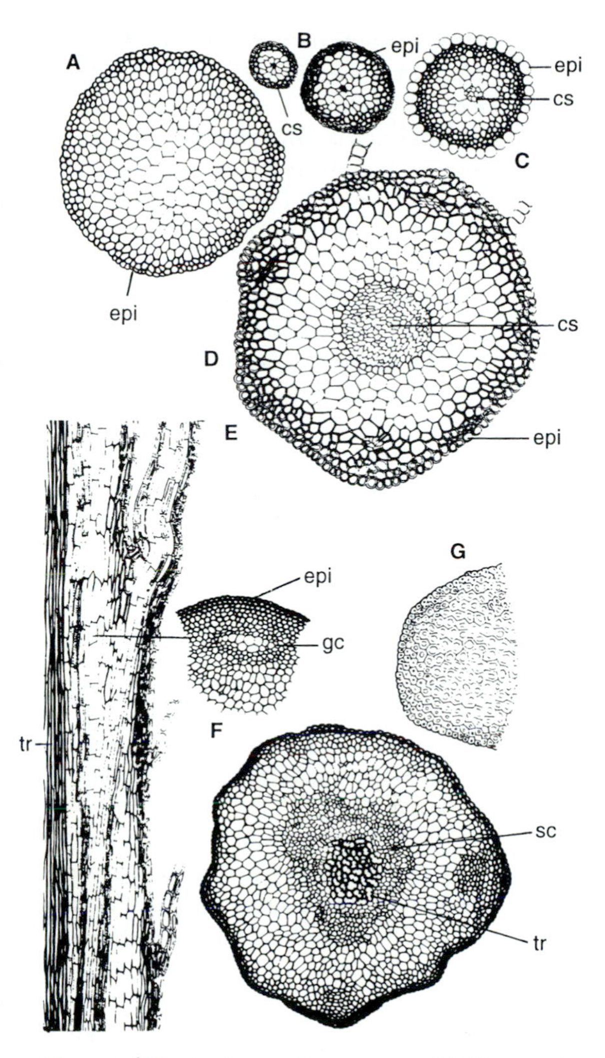

Figure 14–7 Anatomy of moss stems. (A) *Merceya latifolia,* no central strand; (B) *Amblystegium serpens,* small central strand and thick epidermis; (C) *Mnium medium,* larger central strand and epidermis of enlarged, thin-walled, hyaline cells; (D) *Hygrohypnum ochraceum* with large central strand; (E) longitudinal section of *Polytrichum commune* stem with thick-walled tracheid-like cells in addition to sieve cells and companion cells in the large central strand; (F) transverse sections of *P. commune* stem; (G) *Orthotrichum affine,* older stem with uniformly thick-walled cells throughout. (cs = central strand; epi = epidermis; sc = sieve cells; tr = tracheid-like cells; gc = girdle cells) (From Flowers, S., *Mosses: Utah and the West,* Brigham Young University Press, 1973. Courtesy of Mrs. Seville Flowers, with permission of Brigham Young University Press.)

spirally arranged on the stems; phyllotaxies of 2/5 and 3/8 are common, but in *Polytrichum*, it is 5/13, in *Leskia* 8/21, and in *Fontinalis* 1/3. Two-ranked alternate leaves occur in *Fissidens* and *Distichium.*

Rhizoids in Musci are multicellular and often branched. In many species, the protonemata become transformed into rhizoids. The oblique crosswalls characteristic of protonemata also occur in rhizoids. The crosswalls of newly forming cells are perpendicular to the cell axis, but as the cells continue to grow, the upper and lower surfaces grow at different rates resulting in oblique septation.

At the apex of the stem of mature female plants is an archegonium or egg-bearing structure. This may be surrounded by a specialized whorl of leaves called a **perichaetium** (Figure 14–8A). Antheridia occupy a like position at the apex of mature male gametophytes. Ontogeny of archegonia is essentially uniform in all mosses. From the archegonium, a yellowish or brown leafless stalk, the sporophyte, develops. The sporophyte consists of a foot, which is embedded in gametophyte tissue, the seta or stalk, and the vase-like or globose capsule in which the spores are produced (Figure 14–1). Sporophyte characteristics are especially useful in identifying the mosses; the nature of the peristome and the teeth of which it is composed are of major significance (Figures 14–9 to 14–12).

In the Sphagnales and Andreaeales, the sporophyte is reduced to a foot and capsule, and a gametophyte structure, the pseudopodium, performs the function of the seta in raising the capsule above the rest of the plant for effective dispersal of spores.

Cellular differentiation is more diverse in mosses than in liverworts. Leaf cells vary in shape from square to much elongated. At the base of the leaves are cells which in some species are much different from the other leaf and stem cells; these are the **alar** cells. Cells on the margins of leaves are also sometimes different and the leaf margins in many species are two or more cell layers thick.

Pigmentation in the Musci is typical of the Lorobionta (Strain 1958). Electron microscope studies show that plastids, mitochondria, and other organelles are also typical of the subkingdom. Chromosome numbers in the Musci are mostly multiples of 11, 12, or 13 (Figure 14–13). In *Bryum* the basic number is 10 and in *Mnium* and many other genera it is either 6 or 7; in *Sphagnum* the basic number is 21. Polyploidy is common in the Musci, in contrast to the Hepaticopsida where it is rare. Frequently a species of moss which is dioecious will have polyploid strains which are monoecious.

Asexual reproduction in the mosses, as in the liverworts, is by (1) branching accompanied by death of older parts so that each branch becomes a new plant, (2) generation of new plants from organs or parts of organs, and (3) gemmae. In many mosses, new plants commonly form from old protenemata. Vegetative reproduction from fragments of old plants is very common and is apparently the main mode of reproduction in many species of moss. Gemmae in Tetraphidales develop into a thalloid structure in a manner similar to that of the thalloid liverworts.

Sexual reproduction involves fertilization of nonmotile eggs by motile sperm. Gametangia are similar to those of liverworts, but the archegonia tend to have longer necks. The young sporophytes have a dagger-like foot, which penetrates the gametophyte tissue, instead of the broad foot characteristic of liverworts. The setae elongate by means of intercalary meristems; thus the moss sporophyte is indeterminate in length. Setae longer than 5 cm are, nevertheless, rare, although they reach a length of 7 to 10 cm in a few species. Anatomical structure of the setae is generally identical to that of the gametophyte stems (Flowers 1973).

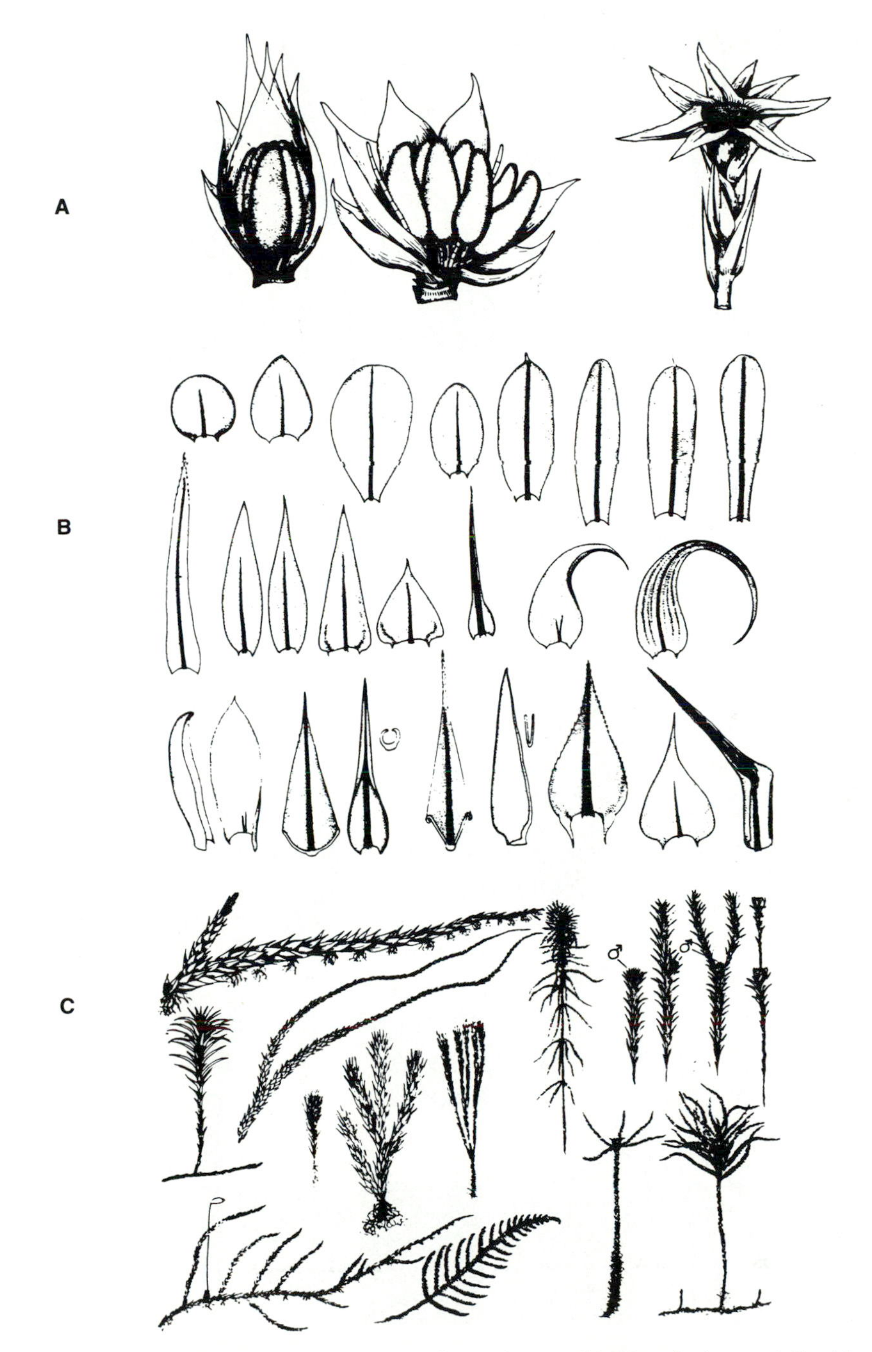

Figure 14–8 General morphology and anatomy of moss leaves. (A) Diversity in specialized leaves surrounding archegonium; (B) diversity in stem leaves; (C) diversity in habit. (From Flowers, S., *Mosses: Utah and the West,* Brigham Young University Press, Provo, Utah, 1973. Courtesy of Mrs. Seville Flowers, with permission of Brigham Young University Press.)

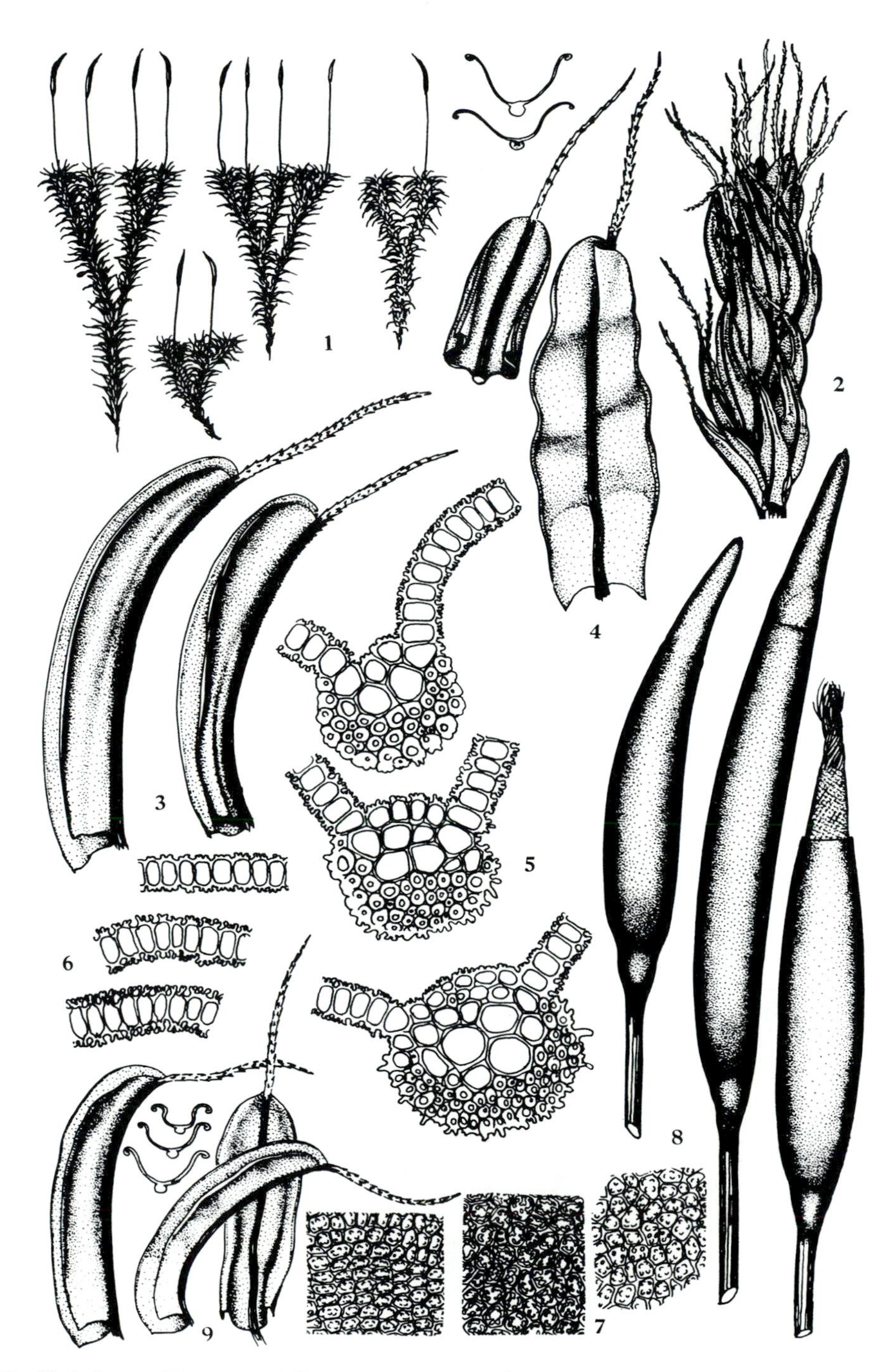

Figure 14–9 *Tortula ruralis,* an especially common moss in desert areas. (1-2) Habit, sketches × 4; (3-4) typical leaves × 3; (5-6) leaf cross-sections × 120; (7) surface of leaves × 120; (8) capsules × 50; (9) leaves of a desert ecotype. (From Flowers, S., ***Mosses: Utah and the West,*** Brigham Young University Press, Provo, Utah, 1973. Courtesy of Mrs. Seville Flowers, with permission of Brigham Young University Press.)

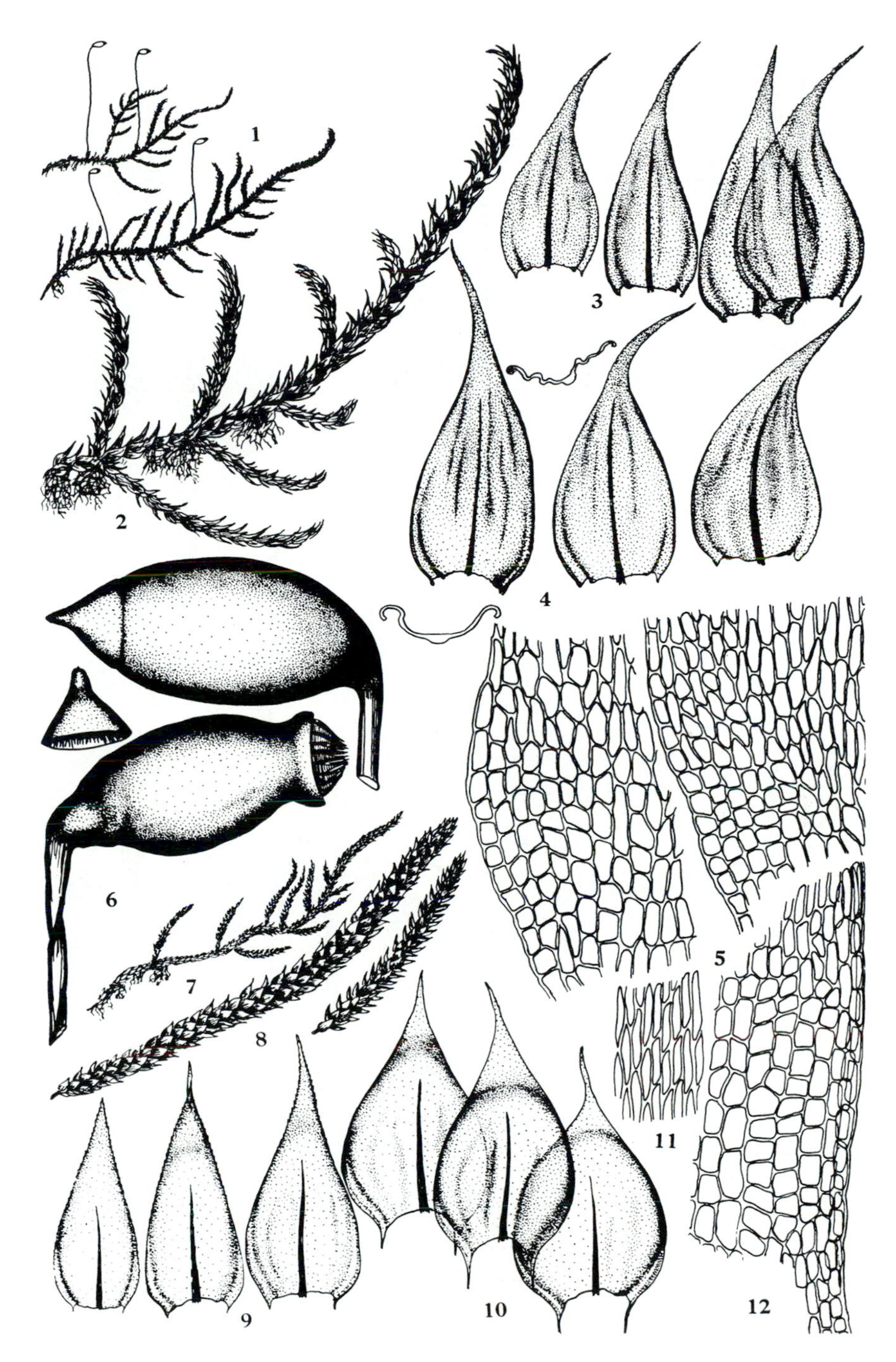

Figure 14–10 (1-6) *Brachythecium erythrorrhizon;* (7-12) *Brachythecium digastrum.* (From Flowers, S., *Mosses: Utah and the West,* Brigham Young University Press, Provo, Utah, 1973. Courtesy of Mrs. Seville Flowers, with permission of Brigham Young University Press.)

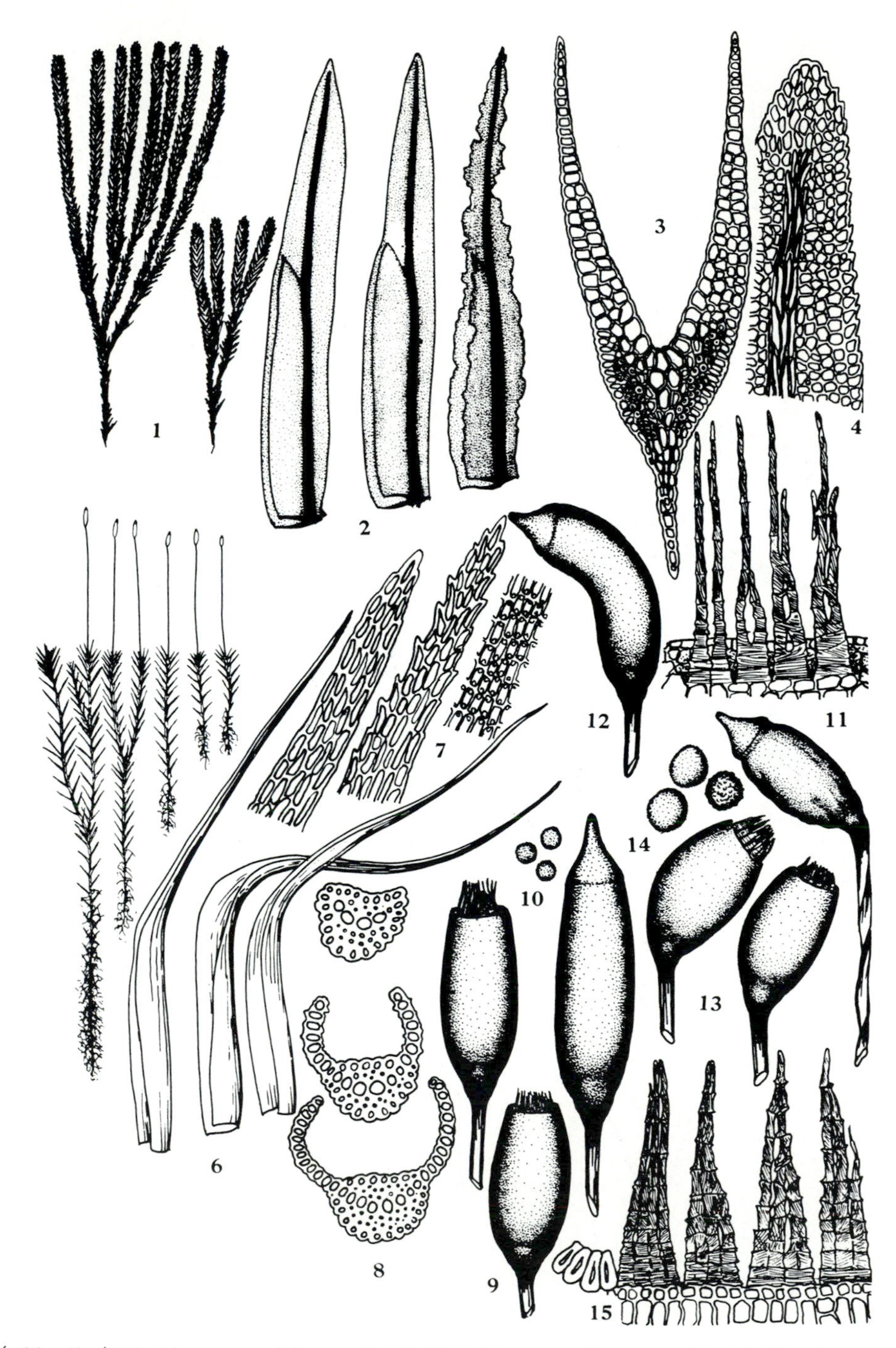

Figure 14–11 (1-4) *Fissidens grandifrons;* (5-12) *Distichium capillaceum;* (13-15) *Distichium inclinatum.* (From Flowers, S., *Mosses: Utah and the West,* Brigham Young University Press, Salt Lake City, 1973. Courtesy of Mrs. Seville Flowers, with permission of Brigham Young University Press.)

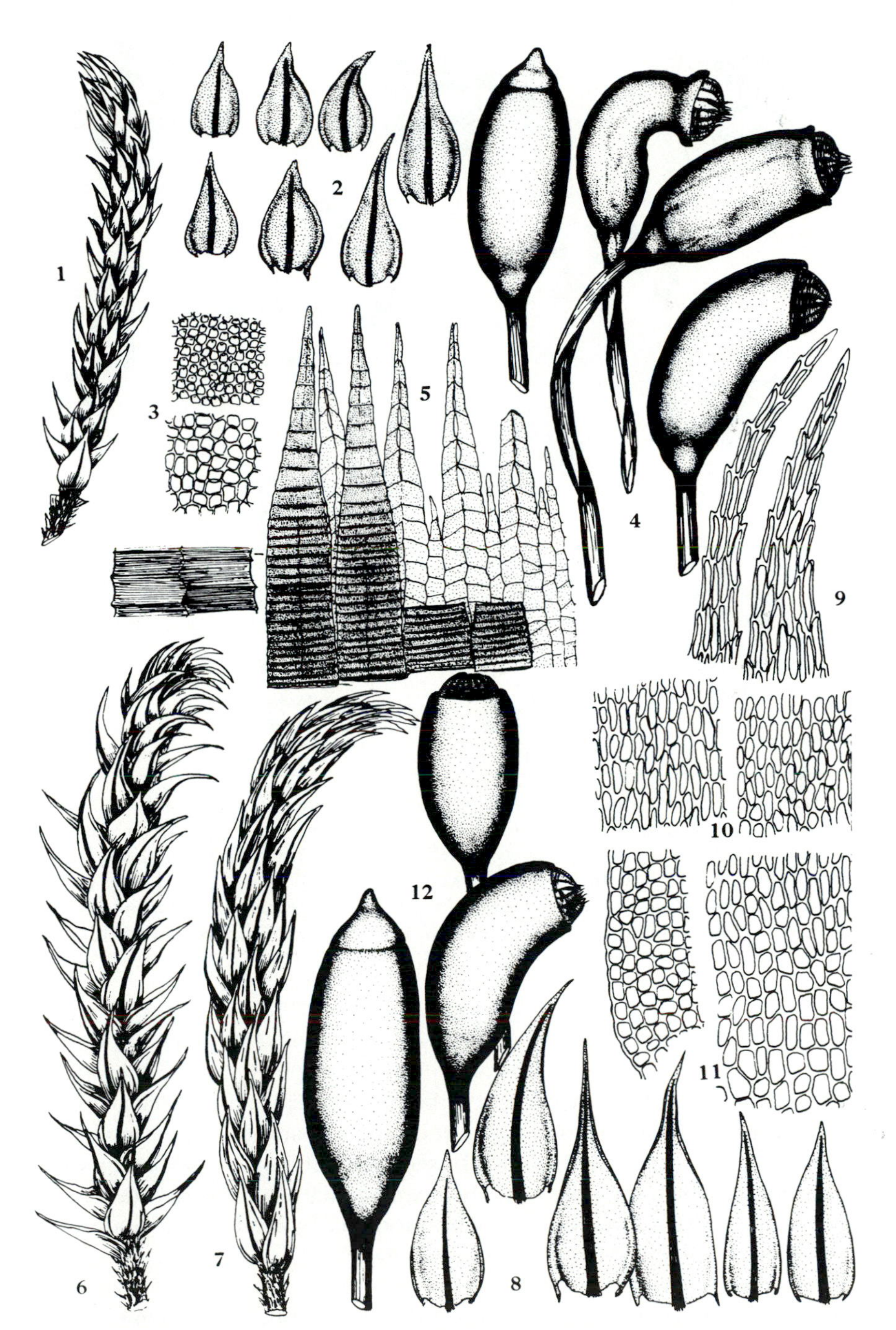

Figure 14–12 (1-5) *Lescuraea incurvata;* (6-12) *Lescuraea radicosa.* (From Flowers, S., *Mosses: Utah and the West,* Brigham Young University Press, Provo, Utah, 1973. Courtesy of Mrs. Seville Flowers, with permission of Brigham Young University Press.)

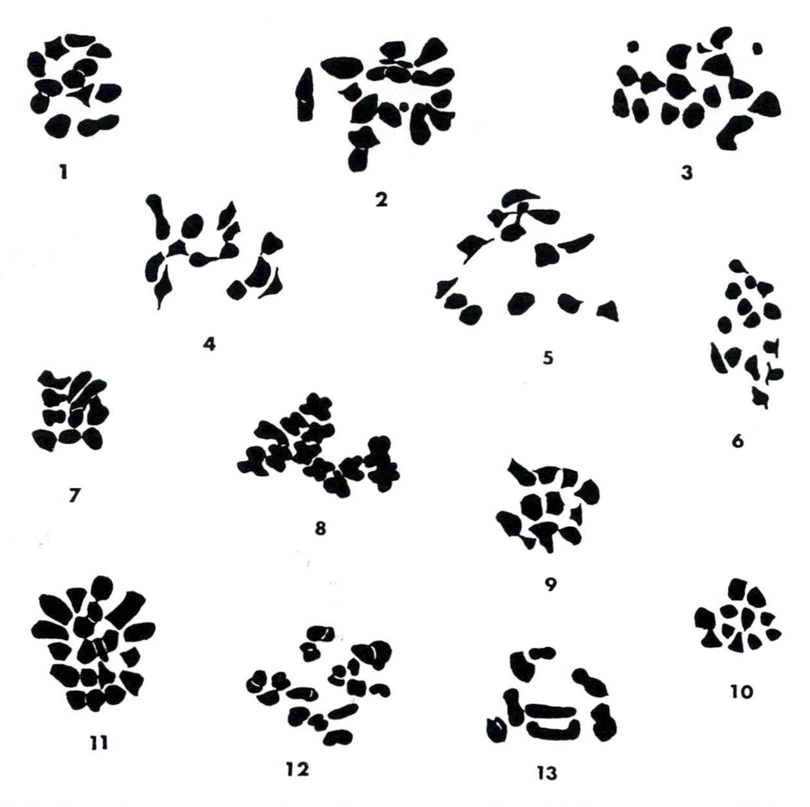

Figure 14–13 Metaphase chromosome configurations in mosses; (8 and 12) are metaphase II figures; all others are metaphase I. (1) *Dicranum sabuletorum* (n = 12); (2) *Barbula unguiculata* (n = 16); (3) *Barbula unguiculata* (n = 16); (4) *Weissia controversa* (n = 13); (5) *Grimmia apocarpa* (n = 14); (6) *Grimmia apocarpa* (n = 14); (7) *Bryum criberrimum* (n = 10); (8) *Mnium cuspidatum* (n = 12); (9) *Leskea gracilescens* (n = 11); (10) *Leskea obscura* (n = 11); (11) *Amblystegium juratzkanum* (n = 20); (12) *Atrichum angustatum* (n = 8); (13) *Atrichum angustatum* (n = 8). (From Messmer, L. W. and Lersten, N. R., *Bryologist*, 71, 348, 1968. With permission.)

Following fertilization, as the sporophyte embryo is developing, a portion of the archegonial tissue forms a cap, or **calyptra,** over the sporophyte. As the sporophyte matures and the capsule develops, the calyptra covers the capsule (Figure 14-1). The capsule itself consists of an **operculum,** or lid, in most mosses, and a spore chamber. The enlarged pedestal attaching the capsule to the seta is called an **apophysis** and often has two, or sometimes more, **stomata,** the function of which is not known. The opening beneath the operculum through which the spores escape is called a **peristome** and is surrounded by **teeth.** In the Andreaeales, the capsule opens by means of four longitudinal slits and there is no operculum.

Most mosses have a peristome (Figures 14-9 to 14-12). The teeth vary considerably from family to family and even among species of the same genus, but is highly constant within species. Some mosses have a single row of small teeth, some a single row of long or of massive teeth, others have a double row of peristome teeth. In the Tetraphidales, there are only four teeth to the peristome, but in most mosses there are 16, 32, or 64, and sometimes more.

Moss tissues are extremely hygroscopic. Most cells hold several times their weight of water. Slight amounts of moisture cause readily visible movement of the peristome teeth in any species of moss with a well-developed peristome. Movement of the teeth results in opening or closing of the peristome and effects the dispersal of the spores.

Few physiological studies have been made of mosses in comparison to the number of such studies in the Tracheophyta or the number of morphological studies that have been made with mosses. It is commonly assumed that the rhizoids are important organs of water and mineral absorption as well as of anchorage and that the lower stems function in conduction of water and minerals. However, numerous studies with several species have indicated that this is not so (Sjös, Uppsala and University of Minnesota class notes). In general, mosses obtain both moisture and minerals from the air and the rhizoids provide anchorage only. Recently, it has been demonstrated that in some species of moss, absorption of water does occur through the rhizoids, though there is still no direct evidence of mineral absorption by rhizoids.

Phylogeny and Classification

Although bryophytes fossilize well, the fossil record is incomplete, perhaps due as much to a shortage of interested paleobotanists as to a shortage of good fossils. The epidermal cells of mosses often have thick walls that are resistant to decay, the spores also have thick, decay-resistant walls (Figure 14-2), and in the acid environments where mosses normally thrive, accumulation of their remains builds up into thick deposits. Few animals feed on mosses to an appreciable extent, and few decomposers break down their tissues. Theoretically, therefore, Bryopsida should be among the best known classes paleobotanically. This, however, is not the case; but as more paleobotanists become better acquainted with the morphology and anatomy of mosses, a better picture of bryophyte phylogeny should emerge.

The oldest known bryophyte fossils date from the Permian period in the late Paleozoic era and resemble *Mnium* and *Bryum* in the Bryaceae, and *Funaria* in the Funariaceae. These two families possess characteristics usually interpreted as primitive, an interpretation supported, therefore, by the fossil record. The Permian genus *Intia,* in the Funariales, is believed by many paleobotanists to be very similar to the ancestral stock from which the Musci evolved. Its growth habit and arrangement of leaves are similar to modern Bryaceae and Funariaceae; however, the cells of the leaf are different, resembling more the cells in Sphagnum leaves than those in any of the Funariales or Bryales.

The Protosphagnales, which also thrived in the Permian and early Mesozoic, but became extinct later in the Mesozoic era, are different from any plants known today, but are believed by many to be the ancestral stock from which the Sphagnales and Andreaeales evolved. Like them, and like the Anthocerotales, the seta was reduced to little more than a foot and capsule. Apparently, the separation of the two main lines of moss evolution (Figure 14-3) occurred very early, *viz.,* in the Permian. Species of Sphagnales have been reported from the Jurassic and are abundant in the Cretaceous.

Although few Paleozoic mosses have been discovered, Mesozoic fossils are relatively abundant in many parts of the world. In general, Mesozoic mosses were similar to modern species. Most of the families and many of the genera may be found in upper Jurassic and Cretaceous formations.

Taking into consideration morphology and anatomy, in addition to the fossil record, we can hypothesize that the first mosses evolved from chlorophycean-hepatic stock, similar to the

Ulotrichales and Anthocerotales, during the Permian period, and were similar to modern Funariaceae, Bryaceae, and possibly Tetraphidaceae. Very early, the Buxbaumaceae, Sphagnaceae, and Andreaeaceae separated from the main moss line. From the Bryaceae have evolved the rest of the Bryales, the Archideales, and the Polytrichales.

The occurrence in the Bryales of characteristics that seem to be primitive—less differentiation between gametophyte and sporophyte generations than there is in the Archideales, Sphagnales, and Polytrichales, for example, including abundant chlorophyll in the sporophytes—suggests that this is an ancient order. The fossil record agrees. On the other hand, the great amount of variation in the Bryales, with genera and families tending to blend into each other, and with many polyploid series, suggests an advanced taxon of recent evolutionary origin. Obviously, much more fossil evidence and much more data from comparative morphology, biochemistry, and physiology are needed before our understanding of Musci phylogeny will be as good as we would like.

Moss taxonomy is characterized by a few taxa with large numbers of species, and many small taxa. Of the eight orders in the class Bryopsida, 7 contain either 1 or 2 families, whereas the Bryales is composed of 19 families (more than 60 in some classification systems!). A logical way to classify the families of Bryophyta would be to follow the outlines in Flowers (1973) along with outlines published by Watson (1971) and equate families in a resulting system with suborders in Engler's Syllabus (Reimers in Melchior and Werdermann 1954). Characteristics of the eight orders are summarized in Table 14-4.

Representative Species and Genera

Fissidens grandifrons is a common and conspicuous species found in the waterfalls and rapids of small streams; other species of *Fissidens* are often abundant on wet soil or rocks or in streams (Figure 14-9). The fronds are flattened, with two rows of dark green leaves, superficially giving the appearance of a liverwort. *F. grandifrons* seldom produces sporophytes, but is readily recognized as a moss by its relatively thick leaves, dark green color, and lack of notched leaves and amphigastra.

Polytrichum juniperinum is a relatively common species well adapted to habitats where dry periods are frequent. Leaf structure in this species is more complex than in most mosses. From the upper surface of the leaves there arise a number of parallel plates of thin-walled photosynthetic cells separated from each other by narrow fissures. The lower part of the leaf is made up of thick, sclerotic cells. During dry periods, the leaves roll up longitudinally, closing the air spaces and protecting the chlorenchyma layer from drying out.

Tortula ruralis is a very common desert moss growing on soil (Figure 14-11). Another desert moss, growing on rock, especially basalt, is the black cushion moss, *Grimmia apocarpa* (Figure 14-6E). In humid areas, species of *Amblystegium* and *Brachythecium* often grow in shaded lawns, around dwellings, and in parks; species of *Funaria* and *Bryum* are frequent along the edges of shrubbery or in footpaths and cracks of sidewalks. Several mosses are illustrated in Figure 14-6.

Andreaea, the black granite moss, grows on granitic rocks in mountainous areas. Partly because of its unique capsules, it is placed in a subclass by itself. *Sphagnum* is a genus of about 350 species common in bogs and on acid soil. It is abundant in arctic and subarctic regions and around alpine lakes. It grows rapidly, and together with various filamentous algae and species of sedge, is responsible for the "floating bogs" and "quaking bogs" of alpine meadows,

TABLE 14-4

DIVERSITY IN MOSSES: GENERAL MORPHOLOGICAL CHARACTERISTICS OF THE ORDERS OF BRYOPSIDA (MUSCI)

Orders (Fam–Gen–Spp)	Peristome Details	General Morphology and Habitat	Representative Genera
Funariales (1-30-433)	16 jointed teeth in 2 series	Small to minute light green annuals or biennials; sporophyte photosynthetic	*Funaria, Physcomitrium, Entosthodon*
Bryales (11-595-12,760)	16 jointed teeth in 1 or 2 rows	Variable, mostly erect with leaves in 4-5 ranks; 2-ranked in *Fissidens*	*Mnium, Pottia, Ceratodon, Grimmia, Bryum, Tortula, Rhacomitrium, Pohlia*
Polytrichales *(2-18-370)*	32-64 (rarely 16) short ligulate, solid teeth	Large, coarse, stiff, dark green, becoming reddish to brown with age; large central cylinder with phloem and other tissues	*Dawsonia, Atrichum, Pogonatum, Polytrichum, Dendoligostrichum*
Archideales (1-1-5)	16 jointed teeth	Small annual, terrestrial mosses of North America and Europe	*Archideum*
Buxbaumilaes (2-3-19)	1-4 rows of linear jointed teeth, 16 per row	Very short seta; common on rotten wood, acidic rock and soil	*Buxbaumia, Physcium, Theriotia*
Tetraphidales (1-2-5)	4 solid, persistent teeth, not hygroscopic	Phloem present but poorly developed; north temperate in distribution	*Tetraphis, Tetrodontium*
Andreaeales (1-2-20)	No peristome	Small black or blackish green cushions growing on alpine granite outcroppings	*Andreaea, Acroschisma*
Sphagnales (1-1-350)	No peristome	Large whitish green mosses, often tinged with red, pink, or brown; stems erect to pendulous, branches whorled, distant below, closer above; plants of bogs	*Sphagnum*

Note: The number of families (Fam), genera (Gen), and species (Spp.) in each order is indicated in the first column.

northern woods, and other areas. Individual stems may become a meter or more long. It is readily recognized by its grayish color often tinged with red, coarse texture, and great water-holding capacity. It is of considerable economic as well as ecological value, being used for packing and as a soil additive to improve the physical properties of soil.

Ecological Significance

In contrast to liverworts, mosses are sometimes adapted to desert conditions. Many species grow in the harsh, cold climates of the far north, the far south, and high mountains, and are often part of the dominant vegetation in such environments.

The most common desert moss in the Great Basin of the U.S. is *Tortula ruralis* (Flowers 1973), which commonly occurs in scattered or concentrated tufts among beds of cacti, around the bases of desert shrubs, or occasionally on open soil. *Crossidium desertorum* is a rare moss limited in distribution to desert ecosystems; other desert mosses occur in a variety of habitats.

While most mosses are best adapted to high moisture and cool temperatures, some tolerate extremely high soil temperatures. The tolerance of terrestrial plants to high soil temperatures has been studied at the Puhimau geothermal area in Hawaii Volcanoes National Park. Three concentric zones of vegetation surround the area: a central zone dominated by *Campylopus* and characterized by much bare soil with a few scattered vascular plants, a peripheral zone dominated by *Andropogon virginicum* (a grass), and a submontane seasonal forest dominated by *Metrosideros collina.* The only species found to possess appreciable heat tolerance was the moss, *Campylopus praemorsus,* which consistently colonizes areas where nighttime soil temperatures are as high as 35°C; it occasionally colonizes areas where the soil temperature is as high as 45°C. This moss is not only the ecological dominant in the central zone but is the second most frequently encountered species in the *Andropogon*-dominated peripheral zone. In the central zone, only four other species of plants occur: two lichens and two angiosperms survive on elevated aggregates of ash, cinder, and humus formed by the *Campylopus.*

Some species of moss grow on slag dumps from smelters where heavy metals, toxic to most plants, are often concentrated. Within species, ecotypes morphologically indistinguishable from each other often vary considerably in their tolerance to such environments. Engelmann and Weaks found that *Ceratodon purpureus* appeared to have the highest potential for the reclamation of strip-mined land in West Virginia of all the species they encountered in their study, 17 mosses and 5 liverworts.

In the forests of Europe, eastern North America, and the mountains of the American West, mosses especially thrive. Some species, like *Funaria hygrometrica,* are essentially cosmopolitan in distribution and occur in a wide variety of habitats; others are very limited in distribution. *Aulocomnium adrogynum* is especially abundant in many lush forests and occurs also in lava tube cave openings of Craters of the Moon National Monument in the central Idaho desert. With increasing moisture, forest mosses become increasingly abundant and increasingly varied. Where soil is acid and boggy, species of *Sphagnum* are abundant.

The sphagnum mosses are the dominant mosses of acid bogs. Frequently, they grow in such abundance as to form thick mats which float on the water surface. As sedges and other flowering plants become rooted in these moss mats, the mat becomes so dense and tight that it will support the weight of human bodies. Often the mats break loose from the shore vegetation and are blown by the wind from one side of the bog to the other. Such "floating bogs" are common, for example, in the North Woods of Minnesota and the high mountains of northeastern Utah. In Denmark and northern Germany, the sphagnum bogs are so acid that fruit, other plants, and the bodies of animals have been preserved for thousand of years. A number of human bodies, well preserved for 2000 years or more, have been recovered from these acid bags. Study of the clothes, hair styles, jewelry, and cause of death, along with analysis of stomach contents, has proved to be of great anthropological value.

Although mosses are not important producers of large quantities of food that can support a great population of animals in any ecosystem, they are important in the pioneer communities of both xerarch and hydrarch primary succession. On xeric sites, clumps of moss catch bits of sand or soil blown by the wind or washed by the rains and provide a place where the plants of the next seral stage can root. In hydric sites, organic soils are gradually built from the vegetative growth of the hydric mosses. In arctic and alpine tundra, the mosses are not only

TABLE 14-5

ANTIBIOTICS IN BRYOPHYTA: NUMBER OF SPECIES OF GRAM-POSITIVE, GRAM-NEGATIVE, AND ACID-FAST BACTERIA INHIBITED BY LIVERWORTS AND MOSSES FROM SELECTED ORDERS

Order		**GR+**	**G−**	**AC F**	**Total**	**Ave. RES**		
Spp.	**No. Bact.**	**4**	**5**	**1**	**10**	**GR+**	**GR−**	**AC F**
HEP								
Anthocerotales	(2)	0	2	1	3	0	4.8	7.5
Marchantiales	(17)	4	5	1	10	4.03	5.25	9.6
Jungermanniales	(1)	0	0	0	0	0	0	0
BRY								
Funariales	(1)	0	0	0	0	0	0	0
Bryales	(14)	3	4	1	8	1.82	2.86	6.71
Polytrichales	(2)	0	0	0	0	0	0	0
Sphagnales	(1)	2	0	0	2	9.50	0	0

Note: A relative effectiveness score (RES) was given each species of bryophyte for each of the 10 species of bacteria (GR+ = Gram positive, GR− = negative, AC F = acid-fast positive.

important pioneers in the ecological succession, but are the chief producers of organic matter in the ecosystem. Mosses increase the water-holding capacity of soil as they become incorporated in it. Some mosses contain antibiotics which inhibit the growth of some species of microorganisms. Little is known, however, of the significance of moss antibiotics, nor is there much known about animal consumption of mosses.

Economic Significance

The only mosses of real economic significance are the sphagnums. They are used primarily as packing material for live plants and as additives for garden and potting soil. Partially decomposed, sphagnum is known as peat moss and is especially valuable in soils having low water-holding capacity. Formerly, peat, consisting of partly decomposed sedge and other plant material as well as partly decomposed moss, was dug from bogs in northern Europe, dried, and used as fuel.

Primitive people relied much more on unprocessed plant products than do civilized people, and to them moss, especially sphagnum moss, was and is a very important product. Because of its water-holding capacity (200 times its dry weight), it has been widely used in all situations requiring such characteristics from babies' diapers to dressings on wounds. During World War I, sphagnum moss from Ireland substituted for cotton gauze to dress wounds. In recent years there has been a revival in the pharmaceutical industry of looking toward algae, mosses, fungi, and lichens for medicines. Table 14-5 summarizes the results of studies involving bryophytes for new antibiotics.

Several species of moss have been widely used, and are still used, to chink log cabins. The general rule is that whatever species having suitable stem characteristics is most abundant near the cabin site will be used. In Alaska, *Hylocomnium splendens, Rhytidiadelphus loreus, Rhacomitrium canescens,* and *Sphagnum* spp. are useful for this purpose. In Antarctica, mosses make up a significant part of the vegetation in fell-field ecosystems (Lewis Smith 1986).

Research Potential

Mosses are widely used in studying the effects of changing humidity on plant metabolism. They have also been used in studying processes of photosynthesis, effects of light quality on pigmentation, protoplasmic streaming, and food translocation. Probably their greatest use in research up to now has been in various phases of ecology, especially ecological succession.

As the role of the nucleus became more and more appreciated by biologists, controversy began to develop between those who believed that all heritable traits were controlled by the nucleus and those who believed that cytoplasmic inheritance existed and was important. For a while, it seemed as though the former had won the argument. The first good evidence of cytoplasmic inheritance was obtained from studies with *Funaria, Physcomitrium,* and *Bryum.* Today there is no controversy; cytoplasmic inheritance is an established fact.

The first studies on the morphological effects of polyploidy were conducted with mosses. These studies also involved *Funaria, Physcomitrium,* and *Bryum.* In 1942, *Bryum corrensii* became one of the first synthetic species to be created in the biological laboratory, thus strengthening the hypothesis that new plant species can evolve from previously existing species providing there is a source of variation, isolation of strains exists or can be accomplished, natural selection will favor some phenotypes over others, and chromosome aberrations capable of preventing interbreeding of the daughter lines has been incorporated into the new population.

Much of the basic work involving descriptive taxonomy and morphology of mosses has been performed by amateurs. Equipment needed to study mosses scientifically includes a compound microscope, glass slides and cover slips, forceps and teasing needles, single-edge razor blades, a metric ruler, and either a good hand lens or a dissecting microscope or both (Flowers 1983).

In the future, mosses should be especially significant in genetic research, especially research on the inheritance of physiological processes, because of their diplohaplontic life cycles with the haploid phase dominant and their variation in photosynthetic and other metabolic processes. Because they are easily cloned and readily reproduce asexually, they should be of special utility in many kinds of genetic studies. Having low chromosome numbers, they would seem to have a potential for various kinds of cytogenetic studies.

SUGGESTED READING

M. S. Conrad's *How to Know the Mosses and Liverworts* is a relatively easy to use guide to North American mosses with keys to many of the common species. Flowers' *Mosses of Utah* is an outstanding guide to western mosses with excellent, easy to use keys.

E. V. Watson's *The Structure and Life of Bryophytes* discusses moss anatomy, life cycles, taxonomy, phylogeny, and paleobotany in a thorough and logical way and is highly recommended to every student. For the student searching for every possible detail of information on liverworts, R. M. Schuster's monumental work, *The Hepaticae and Anthocerotae of North America East of the 100th Meridian* should be examined. Schuster lives up to the statement on the title page, "It is by studying little things that we attain the great knowledge of having as little misery and as much happiness as possible."

Every student should also be acquainted with the *The Bryologist,* a quarterly journal published by the American Society of Bryology and Lichenology, which carries the latest research about mosses and liverworts including a list in each issue of recent publications from all parts of the world in any type of botanical journal.

You may find *The Strange One,* a novel by F. Bodsworth in which he describes some of the uses of mosses by native Americans of the James Bay area, interesting.

STUDENT EXERCISES

1. Find several local mosses and make the following observations and measurements on each (use dissecting microscope):
   (a) phyllotaxy (number of ranks of leaves);
   (b) average leaf length;
   (c) average length of stem;
   (d) average diameter of stem;
   (e) shape of leaf cells;
   (f) shape of alar cells;
   (g) average length of seta (if present); and
   (h) number of peristome teeth (if present).
2. Some botanists believe that the Bryophyta evolved before the Tracheophyta; others believe that the Tracheophyta evolved first. Which do you believe evolved first? Present evidence to support your hypothesis.
3. Which class of bryophytes evolved first? Which order in each class? Present evidence in support of your hypothesis.
4. Examine herbarium specimens of a green *Peltigera* species and of a *Marchantia* species. List several characteristics of each that will enable you to distinguish between them when you encounter them in your field studies.
5. With a sharp razor blade, prepare a thin freehand section of a fresh *Marchantia* thallus. Compare with a similar section of a lichen thallus. (You may need to moisten the lichen thallus before sectioning). What tissues, if any, can be identified in each?
6. How many chromosomes would you expect to find in a leaf cell in a *Jungermania* species? In a stem cell? In a cell of the seta? In a peristome cell? In a calyptra cell? In a spore?
7. How many chromosomes would you expect to find in a leaf cell of *Polytrichum* species? In a stem cell? In a cell of the seta? In a peristome cell? In a calyptra cell? In a spore?
8. Piute Indian women who had a difficult time producing enough milk for their babies, massaged their breasts and nipples with the sap of the milkweed. This is an application of the "doctrine of signatures" which you were informed in this chapter is a worthless superstition. Nevertheless, empirical evidence has demonstrated that milk production does increase in these women. There is even evidence that adopted babies may be breast fed by women who massage their nipples several times a day with milkweed sap. How do you reconcile these two conflicting ideas?

9. If you had funds available to support your research in bryophyte studies, what problem would you like to study? Write a short research proposal outlining hypothesis, objectives, and materials and methods you propose using. Suggest a budget.

10. In Japan, *Grimmia pilifera* is called "giboshi," which is the name of the sphere on the top of a bridge post. American mosses generally have no common names; should they? Why or why not? If you think they should, coin names for a few mosses in the herbarium.

15 THE FERN ALLIES AND ORIGIN OF THE VASCULAR PLANTS

Imagine a world of vast oceans teeming with microscopic life, shark-like fish, and reef-forming seaweeds, but totally void of life on land: a world of low rugged mountains with extensive mudflats covering the lowlands between the mountains and the receding sea. The atmosphere consists mainly of nitrogen, methane, and carbon dioxide gases; there is little oxygen and somewhat less argon than in today's atmosphere. Ultraviolet radiation is intense, for the ozone layer is thin. Except for the mudflats, where organic matter produced by marine life of past millennia has accumulated, there is no true soil, only bare rock and the sand that has resulted from chemical and climatic weathering of the rock. When it rains, rivers become torrents which soon dry up as the storm passes. In such an inhospitable habitat, the first land plants evolved over 400 million years ago.

Green seaweeds, similar to some of the Ulotrichales, such as *Fritschiella tuberosa* (Figure 13-7), and possibly *Chara vulgare* (Figures 13-2a and 13-14), survive on the mudflats of this Silurian landscape, or in the adjacent shallows, and it is probably from one or the other of them that the first vascular plants evolved. Adaptations which made it possible to transport water, from the mud where the plants were anchored to the upper parts of the plant, and food, produced in the aerial parts where light and carbon dioxide were abundant, to the lower parts of the plant, made the invasion onto land successful. These adaptations were the beginnings of a primitive **vascular system.**

The Tracheophyta (*trachea* = the windpipe, hence a tube; *phytum* = a plant), or vascular plants, are structurally the most complex of the Lorobionta and are also the most abundant in number of species of all groups of plants. The name refers to the presence in nearly all species of Tracheophyta of a **stele** consisting of water conducting tissue, the **xylem,** and food conducting tissue, the **phloem.** These tubes extend from the tips of the roots to the apex of each leaf, making possible survival of vascular plants in relatively dry terrestrial habitats.

In terms of total photosynthetic productivity of the entire biosphere, some ecologists have calculated that vascular plants are less important than the Chrysophyta and, possibly, the Pyrrophyta (however, see Table 2-5), but to man they are by far the most important members of the vegetable kingdom. Almost all terrestrial ecosystems are dominated by vascular plants, and it is the Tracheophyta upon which man depends for nearly all of his food, most of his clothing, and much of his shelter; many of our drugs and much of our energy and other needs also come in varying degree, directly or indirectly, from vascular plants.

The division Tracheophyta is large, both in number of species and in number of individual plants. The 230,000 species are classified in 7 classes, 69 orders, 308 families, and 11,250 genera. With the exception of some tundra communities of high altitudes and high latitudes and some rocky ecosystems elsewhere, all terrestrial ecosystems are dominated by two classes of Tracheophyta, the Pinopsida or gymnosperms and the Anthopsida or angiosperms.

The three classes discussed in this chapter are relicts from the Paleozoic era. Well adapted as ecological subordinates in a few specialized habitats, some of the species are widespread. Collectively, they are commonly known as fern allies because they were formerly classified with the ferns in the division Pteridophyta, one of four divisions of plants according to the Schimper-Eichler system.

Of the seven classes of Tracheophyta, the Psilopsida are the most similar to the presumed ancestors of the Division; the Lycopsida are also an ancient group. There is a big gap in the fossil record of the Psilopsida, and it may be that modern psilophytes are not closely related to the fossil species. It is also possible that the three species found at the present time have been in existence continuously since the Paleozoic era, but during most of that time have been rare, as they are now, and have left few fossils. The fossil record of the Lycopsida, on the other hand, is continuous from early Devonian to the present.

The Tracheophyta apparently originated in late Silurian from ancestors similar to either the Ulotrichales or the Charales of the Chlorophyta. A few botanists believe that the first vascular plants evolved from liverworts; however, both the fossil record and ecological considerations suggest that the Tracheophyta are older than the Hepaticopsida. Chemically, the Charophycopsida resemble the Psilopsida and Lycopsida more nearly than any other plants do, including the Ulotrichales (Stewart and Mattox 1975). On the other hand, no trace of them has been found in formations older than early Devonian, even though they fossilize well. By then, the Psilopsida were well established. This leaves the Ulotrichales as the most likely candidates to explain the origin of the vascular plants.

The fossil record clearly indicates that within a short period of time following their origin, perhaps 20 or 30 million years, numerous species had evolved. By mid-Devonian, there was a great diversity of vascular plants representing all three classes treated in this chapter; a few species of Filicopsida and possibly Pinopsida were also apparently present, but the Gnetopsida and Anthopsida evolved much later.

Although the Tracheophyta are rather variable in gross morphological features, they are highly uniform in cell structure and patterns of cell differentiation. The chloroplast pigments, mitochondrial ultrastructure, ribosomes, dictyosomes, flagella, and most other organelles are typical for the subkingdom and similar to those of the Chlorophyta. On the other hand, cell wall structure differs from that of the Chlorophyta in many ways, but is very uniform throughout the division; the same is true of the internal structure of the plastids.

In all seven classes, life cycles are similar: there is always a distinct alternation of generations with the sporophyte generation dominant and the gametophyte, which is often parasitic on the sporophyte, much reduced.

GENERAL CHARACTERISTICS OF THE TRACHEOPHYTA

The plant body of most vascular plants is made up of four organ systems: the food production system consisting of flat leaves, the anchorage system consisting of branched roots, a conduction and support system connecting the two and usually consisting of a more or less hollow cylinder or stem, and a reproductive system consisting of sporogenous and accessory structures. In gross morphology as well as function, these organ systems resemble those of many brown and some red seaweeds, but anatomically they are different.

The Stele

A structure composed largely of vertically elongated cells especially adapted to conduction of water, minerals, and food, is the most distinctive feature of Tracheophyta anatomy. The stele may consist of a single strand of **vascular tissues** or of several separated strands and is usually surrounded by cortical tissues made up mostly of parenchyma cells (Figure 15-1). Often the only stelar tissues are the **xylem,** through which water and minerals flow, and the **phloem,** through which foods are conducted; in the more advanced classes, two meristematic tissues, the **cambium** and the **pericycle,** may also be present. The arrangement of the xylem and phloem relative to each other and to other stelar tissues, and the exact structures of these tissues are major features which enable us to classify members of the Tracheophyta in a meaningful and useful manner. Relationships among the classes of vascular plants are suggested in Figure 15-2.

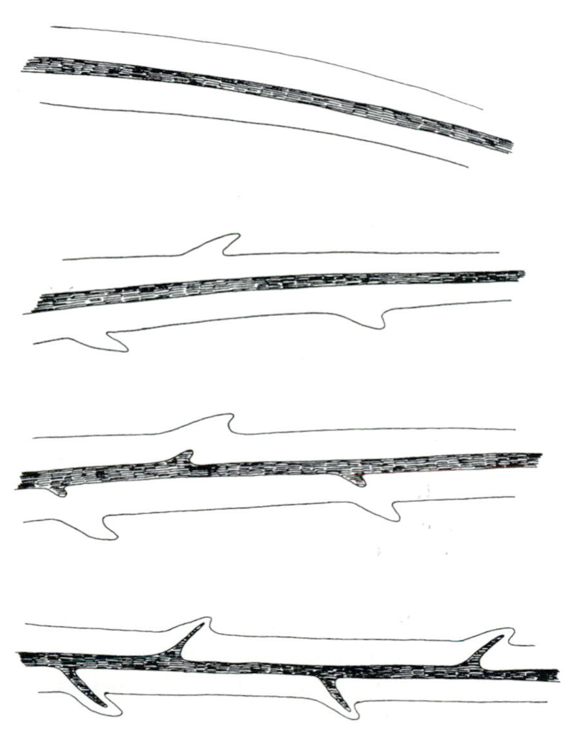

Figure 15–1 Stems of vascular plants showing position of stele relative to leaves. From top to bottom, primitive type with no enations, primitive type with simple enations and no connection to the stem stele, enations with vascular tissue leading into them, microphyll with midrib and vein.

In addition to the stele, the Tracheophyta are characterized by a waxy cuticle, presence of stomates, and spores which are formed in tetrads following meiosis and therefore usually leave a trilete scar. In most vascular plants, the stelar tissues are arranged differently in the different organs, roots and leaves being more uniform in stelar structure than stems. In the Psilopsida, or whisk ferns, there is no clear differentiation of organs and the plant body consists only of a simple vertical axis, often called a stem, plus simple sporangia (Figure 15-3).

Morphologically, there are three basic types of steles: (1) **protosteles** in which the xylem occupies the central portion of the organ and is surrounded by phloem, (2) **siphonosteles** in which a pith, or parenchymatous medulla, occupies the central portion of the organ with xylem and phloem surrounding it, and

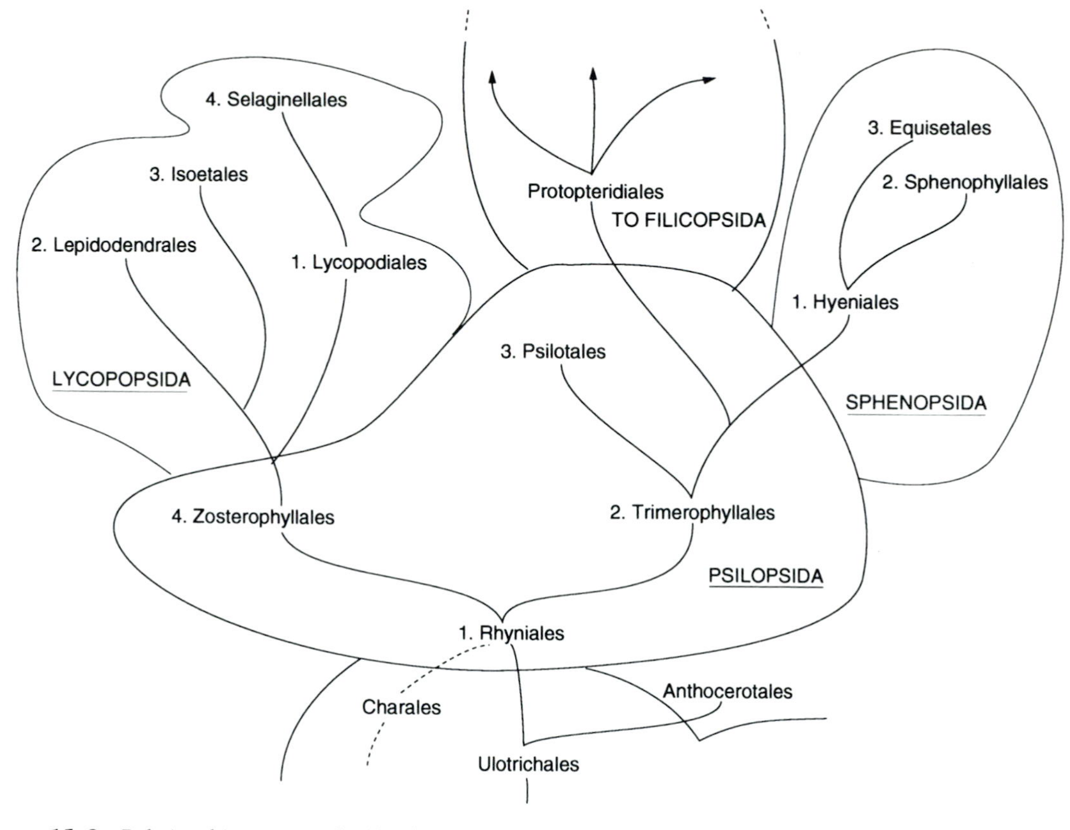

Figure 15–2 Relationships among the Tracheophyta.

(3) **dictyosteles** in which the stele is divided into two or more vascular strands. Each of these types can be further subdivided: star-like protosteles are called **actinosteles;** siphonosteles may be either **amphiphloic** with phloem both inside and outside of the xylem or **ectophloic** with phloem only outside of the xylem; and dictyosteles may consist of bundles which themselves are protosteles (**polysteles**), or they may be modified actinosteles (**plectosteles**), potential siphonosteles (**eusteles**), or **atactosteles.** A summary of stele types is presented in Table 15-1.

The first xylem to mature is called **protoxylem;** later xylem is called **metaxylem.** In young organs, the cell walls of the protoxylem are lignified while the cells in the metaxylem are still parenchymatous and growing; as a consequence, cells of the protoxylem tend to be smaller than those of the metaxylem. In some species, the protoxylem is outside the metaxylem while in others it is inside the metaxylem. If lignification of the cells making up the primary xylem begins nearest the cortex and proceeds toward the center, xylem development is called exarch. If lignification begins at the center and proceeds outward it is called **centrarch** or **endarch** (Figures 15-2, 15-3). In some species the first xylem, or protoxylem, has later xylem both toward the center and toward the outside; such development is called **mesarch.**

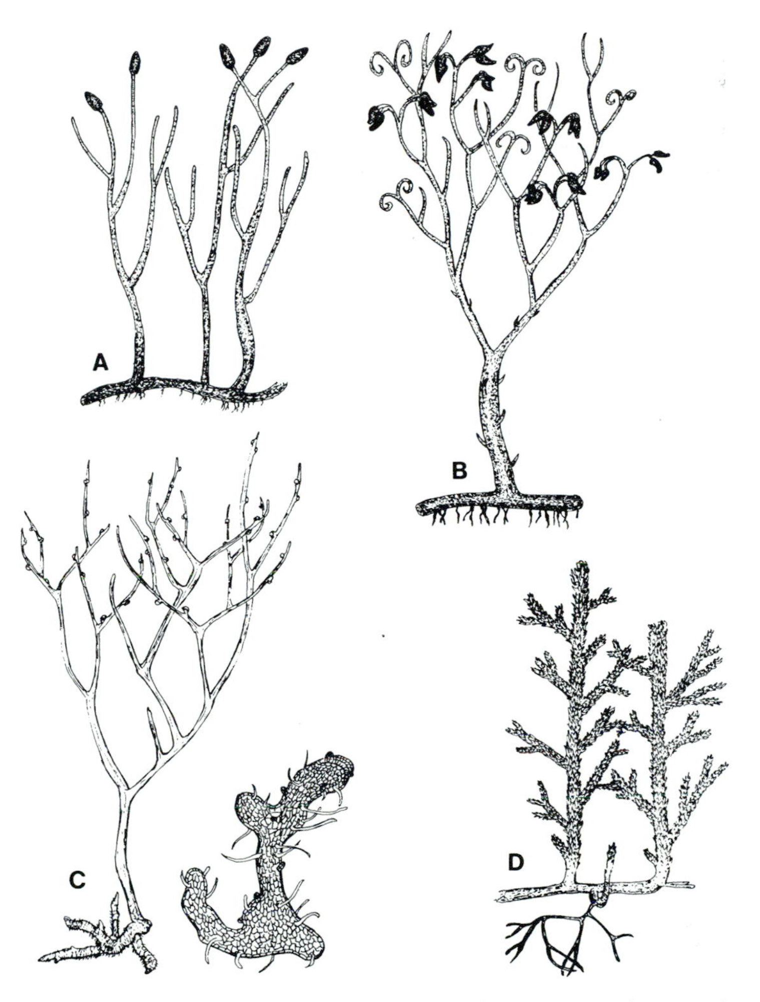

Figure 15–3 Psilopsida: (A) *Rhynia;* (B) *Psilophyton;* (C) *Psilotum,* sporophyte and gametophyte; and (D) *Asteroxylon.*

Nearly all roots have protosteles as do most leaf veins and petioles. The steles of stems, on the other hand, are highly variable. In the most primitive Psilopsida, *Cooksonia* and *Rhynia,* where there is no differentiation of the plant axis into stem and root and there are no leaves, the stele is a protostele consisting of xylem which is made up of only one kind of cell: tracheids with annular (ring-like) thickenings in their walls. Protoxylem is in the center of the axis with metaxylem outside it (endarch development). The axis is dichotomously branched once or twice. *Cooksonia,* which lived in late Silurian time, apparently had only a single vertical axis, but *Rhynia,* which appeared somewhat later during the Gedinnian epoch of the Devonian period, had a subterranean, horizontal rhizome with filamentous rhizoids attached (Figure 15-3). The apex of the axis and each branch is swollen and contains spores.

In *Zosterophyllum,* another early member of the Psilopsida which appeared some 10 or 15 million years after *Cooksonia,* the protostele is also very simple but its development was

TABLE 15-1

MORPHOLOGICAL CLASSIFICATION OF STELES

Categories		Subcategories	Description
Protosteles	Single strand of xylem in center of organ	Haplostele	Single cylindrical strand of xylem surrounded by phloem
		Actinostele	Single star-shaped strand of xylem with phloem around
		Polystele	Several strands of xylem, each one surrounded by a ring of phloem (each one a simple protostele)
		Hypophloic Haplostele	Single strand of xylem with phloem beneath it
Siphonostele	Cylinder of xylem and phloem surrounding a pith	Amphiphloic siphonostele	Ring of xylem with a ring of phloem outside it and another ring inside it but outside the pith
		Ectophloic siphonostele	Ring of xylem surrounding a pith with a ring of phloem surrounding the xylem
Dictyostele	Stele divided into several strands	Polystele	Essentially several protosteles (see above)
		Plectostele	Modified actinostele with rays of the "star" separated by strands of phloem
		Eustele	Modified ectophloic siphonostele; vascular bundles separated by strands of parenchyma
		Atactostele	Random distribution of ectophloic siphonostele

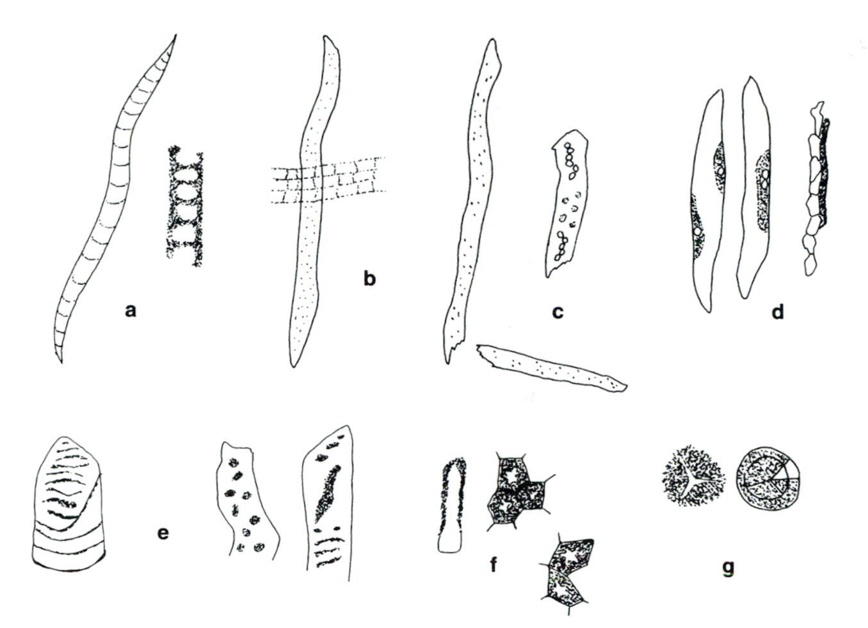

Figure 15–4 Cell Diversity in Tracheophyta: (a) tracheids in *Cooksonia* and *Psilophyton* (Psilopsida) with annular and scalariform secondary wall thickenings, respectively; (b) tracheid (with simple pits) and ray parenchyma (radial view) in *Tetraxylopteris* (progymnosperm); (c) tracheid and vessel element in *Ephedra* (Gnetales) with bordered pits; (d) sieve tube elements and companion cells (stippled) in grape and *Eucalyptus,* respectively; (e) three types of perforation plates found in end walls of vessel elements (scalariform in a fern, foraminate in *Ephedra,* and simple in grape); (f) sclereids (epidermis of pea seed coat, subdermal of same, bean seed coat); (g) spores of Lycopodium and *Selaginella.*

exarch. Exarch protosteles also occur in the stems of Lycopsida and in the roots of most vascular plants. The stems of Sphenopsida are endarch like the axis of *Cooksonia* and *Rhynia.* Xylem development in many of the Filicopsida is mesarch, but in other ferns and in gymnosperms, gnetophytes, and angiosperms, the xylem development in the stems is endarch.

In some Tracheophyta, both primary and secondary xylem occur. The secondary xylem is produced by a special meristematic tissue called the cambium, situated between the xylem and phloem and from which both tissues arise after the primary growth, or growth in length, has ceased. Cambium is found in the stems and roots of many vascular plants, but is less common in the leaves. In most plants having extensive cambium, mitotic activity varies from season to season depending on temperature and available moisture; in many species this variation results in annual growth rings which often make possible quite accurate estimation of age.

Xylem is composed of one or more of five major types of cells: vessel elements, tracheids, fibers, wood parenchyma, and ray parenchyma (Figure 15-4). In most species of five of the classes, the only water-conducting cells are **tracheids:** elongated pointed cells which make contact with each other through various kinds of openings near the ends of the cells. Water is drawn from the soil, through the tracheids, into the leaves, where it is used in synthesizing carbohydrates and other substances and where much of it is lost through an evaporative process called **transpiration.**

Water is taken up by roots or other underground organs by root hairs or, more commonly, by mycorrhizae (Figure 6-12) and is transported to the leaves where it is raw material for

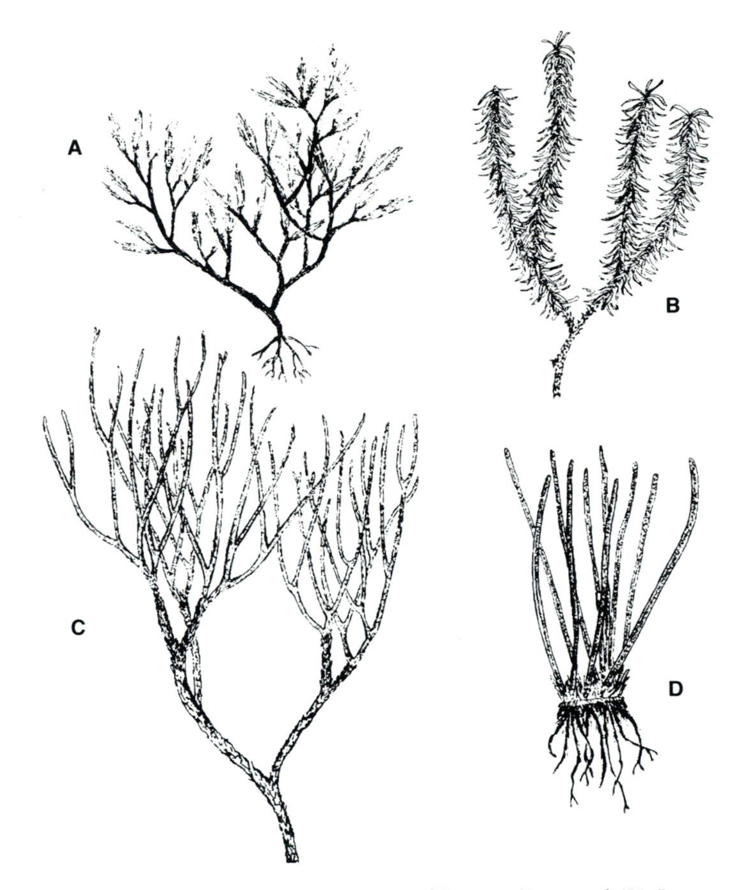

Figure 15–5 Lycopsida: (A) *Selaginella,* (B, C) two species of *Lycopodium,* and (D) *Isoetes.*

making sugars and other food substances by photosynthesis. In the most primitive vascular plants (Figure 15–3A–C), photosynthesis takes place primarily in the epidermis and outer cortex of green stems. Small outgrowths of these tissues, called enations, may aid in the process to a slight extent. In the ferns, gymnosperms, and angiosperms, more complex megaphylls, or "true leaves," carry on photosynthesis in a more efficient manner. **Megaphylls** can be recognized by the presence of a leaf gap just above the **leaf trace** (Figure 15–1). Between the enation and the megaphyll in complexity is the microphyll, having venation but no leaf gaps, characteristic of the Lycopsida and Sphenopsida (Figure 15–5).

The waxy **cuticle** covering the epidermis of leaves aids in reducing transpiration and, along with the presence of a stele, of prime importance in enabling tracheophytes to survive on land. Also of special significance is the type of spore the plant possesses (Figure 15–6). In most Anthopsida, spores (pollen grains) are adapted for insect pollination; in others and also in the Pinopsida, most species have adaptations improving their ability to be spread by wind. In the more primitive Psilopsida and Lycopsida, trilete spores, characterized by a three-pronged scar on the surface of the thick cell wall, enable the spore to germinate regardless of which side is down, where the moisture is, when it lands on a moist surface. Trilete spores are also found in a few species of moss. Types of spores are summarized in Table 15–2.

In many Gnetopsida and most Anthopsida, two kinds of specialized cells replace the tracheids in the xylem: **vessel elements,** which are much larger in diameter than tracheids and are joined end to end into long water-conducting tubes, and **fibers,** which resemble tracheids in form but are smaller in diameter and provide strength and hardness to the tissue but do not conduct water. The other cells, the **wood** and **ray parenchyma** cells, are involved in food storage and metabolism in new xylem, but typically become lignified in older tissues. The patterns in which these different kinds of cells occur in relation to each other are of importance in identifying different kinds of wood.

Phloem also contains several kinds of cells: sieve cells, sieve tube elements, companion cells, parenchyma cells, and fibers. Most vascular plants have **sieve cells** similar to the food-conducting cells of the Bryophyta and Phaeophyta. In the overlapping ends of these cells are sieve areas with numerous small holes or pits. In the angiosperms, more specialized **sieve tube elements** and **companion cells** replace the sieve cells of the other vascular plants (Figure 15–4D). While sieve cells have tapered or pointed ends and contain a distinct nucleus, sieve tube elements have truncated or oblique ends with well-developed, perforated sieve plates, and lack nuclei. Phloem fibers, also called bast fibers, are generally dispersed around the periphery of the tissue, where they form a mechanical sheath.

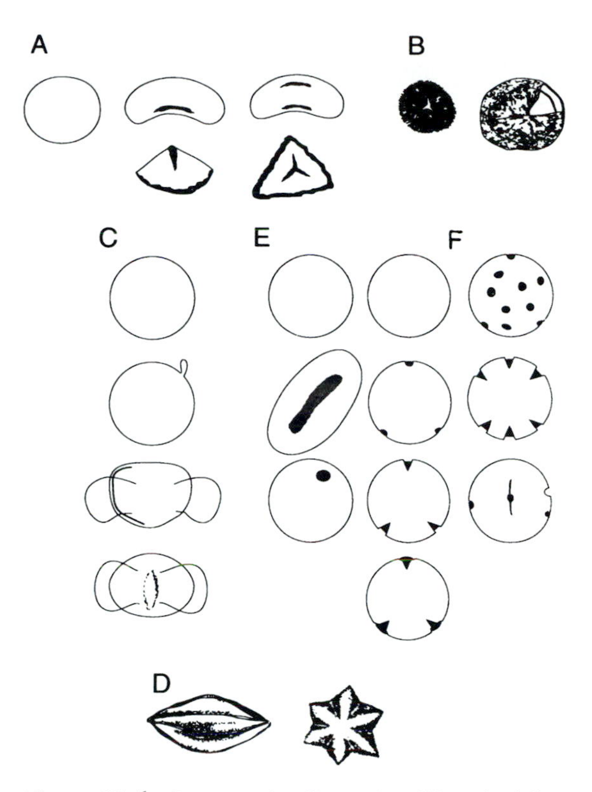

Figure 15–6 Spores and pollen grains. (A) to the left, alete type; to the right, upper, monolete type, top view, left, and side view; lower, trilete type, top view and side view; (B) *Lycopodium* spores; (C) Pinopsida pollen; (D) polyplicate type, Gnetopsida pollen; (E) monocot type pollen; (F) dicot type pollen.

Other tissues of the stele include the **cambium** and the **pericycle.** Both are meristematic. Two kinds of cells are present in cambium, elongated **fusiform initials** and nearly isodiametric **ray initials.** The former give rise to the tracheids, vessel elements, fibers, wood parenchyma, phloem parenchyma, sieve cells, sieve tube elements, and companion cells; the latter give rise to ray parenchyma and horizontal phloem parenchyma cells. The cambium is a thin tissue, typically half a dozen or so cells thick, situated between the xylem and phloem in stems and roots of both gymnosperms and dicots, and also in the leaves of evergreen gymnosperms and many other plants. Differentiation of cambium initials is controlled by hormones and the sugar content of the differentiating cells. At high sugar content, a fusiform initial differentiates into a sieve tube element, at low sugar content into a vessel element or a tracheid. The pericycle is situated just outside the phloem in the roots of many Tracheophyta and in the stems and leaves of some. It is typically one cell layer thick, but may be two or more, especially opposite the protoxylem poles, and is made up of nearly isodiametric, though somewhat elongated, cells. In many plants, secondary and/or adventitious roots arise from the pericycle.

TABLE 15-2

TYPE OF PALYNOMORPHS

	Description	Examples
	Group 1. No Scars, Pores, or Furrows	
Inaperturate	No trilete scar	Basidiomycopsida and many other fungi
	Group 2. Single Scar, Furrow, or Pore	
Trilete	3-Pronged scar	Mosses, lycopods, ferns
Monolete	Kidney-shaped with single scar; outer wall (perine) may be present	Many ferns
Monocolpate	Spherical with one furrow; perine never present	Water lilies, many gymnosperms, some ferns
Monoporate	Single pore	Redwood, other gymnosperms, many club fungi
	Group 3. Ridges or Wings, Adapted to Wind Pollination	
Vesiculate	Wing-like extension	Many gymnosperms (Pinopsida)
Polyplicate	Pronounced longitudinal ridges	Gnetopsida (*Ephedra* and *Welwitschia*); also Bennettitales
	Group 4. One or More Furrows	
Dicolpate	2 Furrows	Figwort family; Araceae
Tricolpate	3 Furrows	Common: buttercups, mustards, geraniums, mints, and other dicots
Stephanocolpate	5–Many furrows medially arranged	Buttercup family, ash, bedstraw
Pericolpate	5–Many furrows scattered over surface of spore	*Portulaca, Dicentra, Opuntia*
Trichotomocolpate	3 Furrows joined at one pole	Coconut palm
Syncolpate[a]	Furrows fused	Primroses, barberries
	Group 5. One or More Pores	
Diporate	2 Pores, usually at opposite poles	Some members of the birch and mulberry families
Triporate	3 Pores, usually on the equator	Birch, elm, walnut, nettle, and related families
Stephanoporate	5–Many pores restricted to the equator	Alder, elm, bluebells, evening primrose, rust fungi
Periporate	5–Many pores regularly or irregularly dispersed but not limited to equator	Redroot, goosefoot, melon families and many other dicot and some monocot genera
	Group 6. Having Both Furrows and Pores	
Tricolporate	3 Furrows associated with pores	Common: Legume, nightshade, maple, and many other dicot families
Stephanocolporate	Apertures centered on equator of pollen grain	*Sarrancenia, Polygala, Urticularia*
Heterocolpate	Some furrows associated with pores; others not associated with pores	*Lythra, Verbena,* some Boraginaceae
	Group 7. Having Pitted Surfaces	
Fenestrate[b]	Lacunae or pits on surface of pollen	Cactus and composite families
	Group 8. Spores or Pollen Grains United in Sets of Two or More	
Dyad	Palynomorphs united in pairs	*Scheuzeria* and many sac and club fungi
Tetrad	Palynomorphs united in sets of four	*Drosera,* cattails, heath family, *Mimosa*
Polyad	Palynomorphs united in large aggregations	Milkweed, many legumes and orchids, some sac fungi

[a] Many dinoflagellate cells, for example *Peridinium* species, resemble syncolpate pollen grains.

[b] Many palynologists distinguish between fenestrate and lophate types of pollen, which has not been done here.

Leaf Morphology

Land plants live in four different kinds of habitats, and can be classified into four corresponding morphological types in reference to the water supply and dryness of the air: xeromorphic, hydromorphic, mesomorphic, and epiphytic. Leaves show some very striking adaptations to these four habitat types. Adaptations also occur in other organs; however, leaves show greater variation in form and structure correlated to the environmental conditions than do roots, stems, or reproductive structures.

Leaves also vary considerably in attachment and basic structure depending on their taxonomic affinities. In the Psilopsida, very simple leaves, more properly called **enations,** commonly occur as superficial outgrowths from the epidermis (Figure 15-3). These lack veins, although a midrib, consisting of one or more series of parenchyma cells longitudinally arranged, may be present. In *Tmesipteris* and in the Lycopsida and Sphenopsida, leaves having a single vein and leaving no leaf gap occur; these are properly called **microphylls** (Figure 15-5). Although *micro* means small and most microphylls are small, microphylls may be larger than **true leaves** or **megaphylls** in some cases. The essential difference is the presence of a **leaf gap,** or a place in the stele of the stem immediately above the leaf trace where the vascular tissues of the stem are replaced by parenchyma cells. True leaves (megaphylls) occur in the Filicopsida, Pinopsida, Gnetopsida, and Anthopsida. In most cases, they have a more or less complex, branching system of veins, often dichotomously branched in the first three classes, but anastomosed to form a closed network of large and small veins in the Anthopsida. It is believed that these large, efficient leaves have evolved from shoot systems and that, basically, the petiole and veins represent modified stems. In the process of evolution and adaptation, however, the steles in the veins and petioles have become quite different from those of stems.

Plants growing in desert, semidesert, alkaline land, and salt marsh environments have available to them a very limited water supply; at the same time, rates of evaporation in such environments is usually very high. The leaves of such plants possess xeromorphic adaptations such as a thick cuticle of wax, thick-walled hypodermic tissue beneath the epidermis, sunken stomates, compact mesophyll, and cells containing water-holding mucilages and gums.

At the other extreme are plants growing in or near freshwater where the relative humidity of the air is high and the water supply to the roots is abundant. Hydromorphic adaptations observed in leaves of such plants include large air spaces in the mesophyll; thin, glabrous epidermis; and small veins with little lignin. Often there are no stomates, no cuticle on the epidermis, and no palisade mesophyll tissue, only spongy mesophyll.

Between the two extremes are mesic habitats in which leaves with mesomorphic adaptations occur. Mesomorphic leaves tend to be larger, less hairy, and less likely to contain gums and mucilages than xeromorphic leaves, but they generally have stomates and a cutinized epidermis in contrast to hydromorphic leaves; for example, their shade leaves typically have a thin palisade layer and little cutin in the epidermis, whereas their sun leaves have a double or triple layer of palisade cells and the epidermis is strongly cutinized.

The fourth type of adaptation occurs in epiphytes. Epiphytes live in the air, attached to other plants but receiving neither food nor water from them, and therefore, their supply of water is very limited. However, they usually grow in habitats where the relative humidity is almost constantly high and where there is abundant rain most of the year; therefore, transpiration losses are not great. Adaptations in epiphytes resemble those of xerophytes in many ways, but specific adaptations for collection of moisture also exist. In all of these types, leaf and other adaptations are superimposed on the basic genetic framework inherited from the taxonomic group. Leaf anatomy is thus able to reveal much concerning both ecological adaptations and taxonomic relationships among plants.

Leaf morphology is of special interest to botanists who study evolutionary trends within the Tracheophyta. Much of our paleobotanical evidence of plant phylogeny has come from study of leaf fossils. For many groups of plants, especially the Anthopsida, leaf fossils are more abundant than fossils of stems, roots, flowers, or fruits. Our knowledge of the origin and evolutionary development of plants in these groups is therefore largely limited by our knowledge of leaf morphology and anatomy.

Most tracheophyte leaf fossils have been named and classified on the basis of gross morphology alone. Leaves which are lanceolate with toothed margins such that they superficially resemble willows have been classified in the genus *Salix* without any attention being paid to the anatomy of the veins or the chemistry of associated chemical residues. This is changing now as paleobotanists use vein and petiole anatomy, venation patterns, and biochemical analysis in addition to gross morphology in classifying fossils; however, existing lists of species of plants of the Cretaceous and Tertiary periods must be interpreted with caution in order to avoid misidentifications.

The petioles and veins of most angiosperms are hypophloic protosteles with endarch xylem and lacking both cambium and pericycle. In some cases, the petioles are quite different from the veins; in other cases, the two have similar anatomy. In conifers, a cambium is often present in the leaves; both primary and secondary xylem therefore occur, and a pericycle may also be present. In most vascular plants, only one or two orders of branching occur in the veins, but in many angiosperms, five or six orders of branching of veins and veinlets are the rule. All of these characteristics are useful not only in classifying living vascular plants, but in both classification and identification of fossil tracheophytes.

Sexual Reproduction and Life Cycles

Life cycles are relatively uniform throughout the division. Meiospores germinate to produce gametophytes that are always much smaller than the sporophytes and may be either autotrophic or heterotrophic, depending on the species. Eggs are produced in relatively simple archegonia, and fertilization takes place there. Sperm may be flagellated with whiplash flagella or, in the Gnetopsida, Anthopsida, and most species of Pinopsida, nonflagellated. The zygote develops directly into an embryo which, in most cases, has distinctive embryonic organs. The embryo is dependent upon the gametophyte for nourishment; as it becomes independent, the plant goes through distinct juvenile and mature stages. Mature sporophytes produce spores by means of meiosis; **heterosporous** species produce two kinds of spores, larger ones which develop into female gametophytes and smaller ones which develop into male gametophytes. **Homosporous** species produce only one kind of spores which develop into gametophytes that produce isogametes.

In all Tracheophyta, there is a pronounced difference between the gametophyte and sporophyte generations, but differentiation is least in the Psilopsida. Gametophytes have never been found for any of the fossil Psilopsida, but in the extant genus, *Psilotum,* which is similar to fossil psilophytes in both gross morphology and anatomy, the gametophytes are subterranean saprophytes which sometimes have a small strand of protostele-like vascular tissue at the center of the fleshy prothallium. In *Cooksonia* and *Rhynia,* the sporophyte consisted of a vertical axis without roots or leaves (Figure 15-3). *Rhynia* had rhizoids attaching it to the muddy substratum. Simple enations served the function of leaves although most of the photosynthesis probably took place in the stems as in *Psilotum;* and a swelling at the apex of each stem or branch housed the meiospores. Branching was dichotomous and the stele was an endarch protostele. Thus, the sporophyte is anatomically simpler in the Psilopsida than in any of the other classes, and the gametophyte is more like the sporophyte than in any of them.

Tracheophyta life cycles vary, of course, in many fine details. In some, the gametophytes are saprophytic, as in *Psilotum,* in others they are autotrophic, as in many ferns, and in still others they are parasitic on the mother sporophyte, as in flowering plants. Likewise, gametophytes vary greatly in size and complexity of tissue, being fleshy, dichotomously branched, and sometimes containing a simple stele in *Psilotum,* and being microscopic and consisting of only three cells in the male gametophytes of most angiosperms. In some vascular plants, the meiospores can remain dormant for long periods of time; in others, the spores are very short-lived. Similarly, the embryos in some species go into a resting or dormant stage before entering into the juvenile stage, whereas in others the embryos develop directly into juveniles.

Spores and Pollen Grains

Spores and pollen grains are collectively called **palynomorphs** (Figure 15-6). Palynomorphs generally fossilize more readily than other plant parts. Occasionally coal beds and other fossiliferous material consist almost exclusively of spore or pollen deposits. Since the 1950s, great progress has been made in **palynology,** the study of palynomorphs. One thing we have learned is that closely related species usually have very similar pollen grains.

Palynomorphs are formed in a variety of ways: (1) by the contents of an ordinary cell rounding up and secreting a thick wall about it; (2) by the formation of a secondary wall inside a cell to form an endospore; (3) by one or both daughter cells following mitosis secreting a wall around it; or (4) by each of the daughter cells following meiosis secreting a spore wall. These last are called meiospores and are part of the sexual reproductive process. Spores resulting from processes not related to sexual reproduction are called asexual spores.

In the angiosperms and other seed plants and in many fungi, the immediate products of meiosis may divide one or more times before the spore wall has completed its development and the mature spore has been formed. Thus the pollen grains of angiosperms and the ascospores of many sac fungi actually consist of two or more cells each containing a haploid nucleus along with adjacent cytoplasm with or without a plasmalemma separating them from each other. Because the formation of a pollen grain and its structure are essentially that of meiospores in nonseed plants, some biologists prefer to call all palynomorphs spores. Other botanists refer to pollen grains, and other spore-like structures having more than one nucleus, as an early stage of the gametophyte generation.

Within each of the 26 palynomorph types, there is considerable variation in ornamentations and markings. These are useful both in spore identification and in studies of paleoecology. Some palynomorphs have reticulate walls, some have little club-like projections on them, others have spines, wrinkles, striations, or slightly rough areas, and others are smooth. Some have elaters, or ribbon-like structures, attached to them, and others have one or more viscin threads attached. Examples of spores and pollen grains are shown in Figure 15-6.

TABLE 15-3

INCREASE IN THE NUMBER OF SPECIES OF FOSSILIZED SPORES DURING THE DEVONIAN PERIOD

	Number of Spore Species in Each Epoch			
Epoch	**Total**	**First Appearance**	**Extinction During**	**Persist to End of Devonian**
Silurian period				
Ludlovian epoch	3	3	1	2
Devonian period				
Gedinnian epoch	8	6	0	7
Siegenian epoch	19	11	0	15
Emsian epoch	41	22	0	31
Eifelian epoch	45	4	3	33
Givetian epoch	56	14	10	42
Frasnian epoch	56	10	8	47
Famennian epoch	51	3	—	51

Note: The three species found in Silurian formations are ***Ambitisporites,*** found only in late Silurian, and ***Punctatisporites*** and ***Lophotriletes,*** which persisted through the Devonian period.

The more complex pollen types, such as dicolpate, tricolpate, syncolpate, fenestrate, and stephanocolporate are found only in the Anthopsida. Polyplicate pollen grains are typical of and limited to the Gnetopsida, while vesiculate pollen is common in the Pinaceae and some other Gymnosperms. Monoporate palynomorphs, very common in the monocots, are also found in many gymnosperms and in some fungi; inapertulate palynomorphs are found in all the major taxonomic groups, including gymnosperms and angiosperms.

The first spores that can be attributed to vascular plants were small, smooth (psilate or aperturate) or with simple ornamentations, simple in structure, and spherical. During the Silurian, three species of vascular plants left spores that have been discovered and described; all three had psilate spores about 18 μm in diameter. In mid-Devonian formations, over 50 different species of spores can be found; their average size is 60 μm with some species exceeding 200 μm (Table 15-3). Complex spores of the monosulcate type, but with exine sculpturing characteristic of angiosperms rather than gymnosperms, first appear in the Aptian epoch of the Cretaceous period about 100 million years ago. Reticulate and crotonoid monosulcate types and tricolpate types appear in the Albian epoch. Obviously, angiosperms were rapidly evolving by that time. Today, the most common angiospermous pollen type is the tricolporate; the most complex and highly specialized pollen grains are the lophate and fenestrate types found in many composites. Also first appearing in the fossil record during the Cretaceous period are polyplicate grains typical of the Gnetopsida.

Summary of Phylogeny

Two main hypotheses attempt to explain the origin of the Tracheophyta. According to one of these, the Tracheophyta evolved from ulotrichalean Chlorophycopsida in the late Silurian Period, probably from algae similar to *Fritschiella,* the same group which likely gave rise to the Bryophyta. According to the other hypothesis, the first vascular plants were descendants of marine Charophycopsida. *Chara intermedia,* with its corticated internodes, resembles a very simple vascular plant having a single "tracheid" (with a large vacuole and thin, nonlignified cell wall) surrounded by a cortex consisting of a single layer of smaller parenchyma cells. Similarities in phragmoplast and microtubular cytoskeleton structure, together with the presence of glycolate oxidase in both Charales and Tracheophyta, add support to this hypothesis (Stewart and Mattox 1975), while the ecological limitation of both extant and fossil species of Charales to freshwater or occasionally brackish water and the lateness of the appearance of charophyte fossils (early Devonian) compared to vascular plant fossils (late Silurian) argue against it (Banks 1970).

There is considerable difference of opinion concerning the origin of the Bryophyta and Tracheophyta, their relationship to each other, and relationships within the two divisions. Some botanists believe that the Lycopsida most nearly resemble the ancestral group from which all of the other vascular plants, including the Psilotales, have descended. Some believe that the Tracheophyta are direct descendants of the Hepaticopsida, specifically *Anthoceros,* and others believe that the two groups evolved independently of each other. One botanist recently suggested that the Psilophytales were ancestral to the Bryophyta. On the basis of comparative morphology together with the fossil record, it is most likely that the Psilopsida represent the ancestral group from which the other classes of vascular plants have descended and that the Bryophyta evolved independently of the Tracheophyta. The suggested relationships among the classes of Tracheophyta are illustrated in Figure 15-2.

CLASS 1. PSILOPSIDA

The Psilopsida (*psilo* = bare or naked, *phytum* = a plant) are morphologically the simplest of the Tracheophyta and are commonly called whisk ferns or simply psilophytes. The three species now in existence are classified in two closely related genera, *Psilotum* with two species and *Tmesipteris* with one. *Psilotum nudum* grows naturally as far north as Florida, South Carolina, Louisiana, and Japan, and as far south as New Zealand and Australia; it is also extensively cultivated in greenhouses. It grows in relatively dense shade on wet soil high in organic matter. *Tmesipteris* is native to Australia, New Zealand, Malaysia, and the Philippines, and also favors highly humid habitats. Both genera, but especially *Tmesipteris,* often grow as epiphytes on tree ferns and tropical shrubs. Unless one lives in the tropics, the only place to find any of the three species is in a well-shaded greenhouse with high humidity. *Psilotum* can be recognized by its leafless, rootless stems which are dichotomously branched, and by the sporangia in the axils of the small, scale-like enations (Figure 15-7); no other plants can be confused with them.

The Psilopsida were once more abundant than they are today. Arising in the Silurian, they were the first land plants and became abundant over a period of 25 million years, then declined and became almost extinct during the next 25 million years. Like *Psilotum* and *Tmesipteris,* they were adapted to wet, marshy areas and warm climate. The environment in which they lived, however, was lower in oxygen and much higher in carbon dioxide than that in which their descendants live. They were not only abundant, but widespread, and their fossils have been found in many parts of the world. While most botanists consider modern Psilopsida to be closely related to *Cooksonia, Rhynia, Zosterophyllum,* and/or other Devonian psilophytes, it disturbs them that a 300 million-year gap in their fossil record occurs. Because of this gap, it has been proposed by some botanists that the three modern species of whisk ferns really belong to the Filicopsida.

Figure 15–7 *Psilotum nudum,* grown in greenhouse.

General Morphological Characteristics

Whisk ferns are herbaceous perennials, lacking roots, and having dichotomously branched stems arising from an underground rhizome (Figure 15-7). In the oldest fossils, *Cooksonia* from late Silurian, and *Rhynia,* from early Devonian, the stem contains a small, simple endarch protostele and there are no roots (Figure 15-3). *Zosterophyllum,* also from early Devonian, has an exarch protostele. The rhizomes of the extant species also have exarch protosteles.

The vertical stem in *Psilotum* is dichotomously branched, commonly 20 to 30 cm tall, but can be up to a meter tall, with scattered scale-like enations and small globose sporangia (Figure 15-7). The stele is exarch. The vertical stem of *Tmesipteris* is unbranched or occasionally dichotomously branched once, 5 to 25 cm tall, and covered with lanceolate leaves about 15 mm long. The stele is mesarch. The leaves have stomata and a single unbranched vein; they are usually interpreted as microphylls (e.g., Scagel et al. 1965) but are considered simple enations by some botanists.

The gametophytes are perennial but lack chlorophyll and live saprophytically on soil high in organic matter in symbiosis with mycorrhizal fungi. They are sometimes dichotomously branched and are covered with rhizoids (Figure 15-3D); to some extent, they resemble the rhizomes of the sporophytes and may even contain a simple stele. The gametophytes produce numerous archegonia and antheridia. Sperm are multicilliate and migrate into the neck of the archegonium in response to chemotactic stimulation. The first mitotic division of the zygote yields an inner cell which develops into the foot, or food-absorbing organ, and an outer cell which develops into the rhizome. By the time the sporophyte has become independent of the gametophyte, as a result of the rhizome splitting free from the foot, it has become infected with a mycorrhizal fungus. Soon after this, some of the branch tips become negatively geotropic and develop into aerial stems.

The apex of the rhizomes and aerial stems consists of meristematic cells which divide mitotically to produce growth in length. These differentiate into three layers of meristematic cells; the protoderm, the ground meristem, and the procambium. The protoderm is a single cell layer thick and gives rise to the epidermis. The ground meristem differentiates into two kinds of cells, the outer ones smaller and containing many chloroplasts, the inner ones larger and lacking chloroplasts. These two groups of cells make up the cortex. The innermost layer of cells in the cortex has slightly thickened cell walls and is sometimes called endodermis. The procambium cells differentiate according to a very exact pattern: at five points, separated from each other by approximately 72°, the outermost cell becomes lignified and differentiates into a tracheid. Lignification proceeds inward from these five points until all of the central part of the stem consists of tracheids. Between the points, cells differentiate into sieve cells beginning with the cells next to the xylem. This continues until the stele consists of a five-pointed star of xylem with pockets of phloem between the points of the star. The cells in the very center of the upright stems of *Psilotum* become fibers instead of tracheids.

The enations are too small to be very significant in the total photosynthesis of *Psilotum,* but the wrinkling or ridging of the stem probably increases its photosynthetic surface. In *Tmesipteris,* the microphylls are large enough and numerous enough that they undoubtedly play a significant role in the energy budgets of these plants.

In the axils of many of the leaves, short branches with swollen tips emerge and become sporangia. The cells in the young sporangia divide meiotically to produce spores that are homosporous, or all alike in appearance and germination behavior. The spores are ovoid, flattened on one face, rather small (12 to 20 fm), colorless, and surrounded by a simple wall made up of one layer of cells. The sporangia are also simple and are subtended by a short stalk. Both sporangia and spores of extant psilophytes are very similar to the sporangia and spores of the extinct Silurian and Devonian species. The gametophytes that develop from the spores are colorless, fleshy, about 20 mm long, and produce both male and female gametes.

Phylogeny and Classification

The oldest known fossils of land plants are stems of *Cooksonia* from the Ludlovian epoch of Silurian formations in southern Germany. *Cooksonia* has also been found in lower Devonian formations in Wales. Also found in upper Silurian formations are three different kinds of wind-borne spores similar to the spores of *Psilotum* in size and morphological characteristics. The number of species of vascular plants increased rapidly from the Silurian through the early epochs of the Devonian period. By the end of the Givetian epoch, about 36 million years after the end of the Silurian, there were at least 56 species according to the palynological record of that time (Table 15-3).

Similar to *Cooksonia,* but better preserved, are fossils of *Rhynia* found in Scotland near Aberdeen, laid down during early Devonian time. Like modern *Psilotum,* the stems of *Rhynia* possessed a single vascular strand with a protostele. The smallest tracheids were in the center of the stem just as in *Cooksonia;* branching was dichotomous and the stems were leafless. It differed from *Cooksonia* in having elongate rather than globose sporangia; it also had a rhizome, something which has not been found in *Cooksonia* yet.

Cooksonia and other early land plants were small, up to 20 cm tall, simple, had protosteles with only one kind of tracheid, and phloem that was hardly distinguishable from the very simple cortex. There was no differentiation into roots, stems, and leaves. Branching was dichotomous.

Soon after the origin of *Cooksonia,* during the 5 million years of the Gedinnian epoch, several other species originated, one of which was *Zosterophyllum.* While similar to *Cooksonia* in many ways, *Zosterophyllum* differed in having lateral, rather than terminal, sporangia, which were reniform (kidney-shaped) and dehisced along the distal, convex edge, and in having lateral branches near the base of the plant which protruded at right angles to the main axis and then forked into an ascending and a descending branch. Also its protostele was exarch. Apparently *Zosterophyllum* grew in mud because there are stomates on the ascending branch but none on the descending branch. It is hypothesized that *Zosterophyllum* evolved very early from *Cooksonia,* or from an unknown ancestral stock to them both, and that from these two genera have evolved all of the modern as well as numerous extinct vascular plants. It appears as though the Lycopsida are descended from the *Zosterophyllum* type and the Sphenopsida, Filicopsida, Pinopsida, Gnetopsida, and Anthopsida from the *Cooksonia-Rhynia* group.

As the fossil record is studied in detail, gradual change can be observed from *Cooksonia* through *Rhynia, Psilophyton,* and the progymnosperms down to modern Cycadales and Coniferales, and although the record is far from complete, it seems safe to assume that such was the line of evolution that led to most of our modern vascular plants. A more puzzling situation exists in the Psilopsida. There are three species of vascular plants extant today that resemble *Cooksonia, Rhynia,* and *Zosterophyllum* in so many ways—morphology of the sporangia and spores, lack of roots, enations or microphylls instead of true leaves, dichotomous branching of rhizome and upright stem, presence of simple protostele having only simple tracheids and simple phloem, sporangia in the axils of the leaves—that it seems only logical to include them in the same class. Yet we find no trace of *Psilotum*-like or *Rhynia*-like fossils among all of the thousands of fossils that have been examined from any of the geological formations younger than mid-Frasnian, or older than recent. Could it be that these plants have continued to live in such small numbers all of these millennia, or in such isolated areas, that the probability of finding them is too low? Have we been searching in the wrong places? Since we cannot answer these questions satisfactorily at the present time, we will apply Occam's razor and accept as a working hypothesis the simplest: that *Psilotum* and *Tmesipteris* are relicts of a once abundant psilophyte flora that have managed to survive in specialized habitats for millions of years; if this hypothesis is correct, we will probably find linking fossils someday. To do so, we will need to learn as much as possible about the ecological requirements of modern Psilopsida.

The three species of extant whisk ferns are closely related to each other and are included in two genera in the same family and order, the Psilotales. The extinct species are generally placed in one order, the Psilophytales; however, Banks (1970) has presented a more logical classification and placed the ancient psilophytes in three orders: the Rhyniales, Zosterophyllales, and Trimerophytales (Figure 15-2). Characteristics of the four orders are shown in Table 15-4.

Ecological, Economic, and Research Significance

Today the Psilopsida are of negligible significance ecologically, economically, or in research, but at one time they were of great ecological significance. Within only a few million years after the origin of the first vascular plants, representatives of all three orders of ancient Psilopsida had vegetated large portions of the earth, as attested to by the abundance of psilophyte fossils in North America, Europe, Asia, Australia, and many other places. Species in existence during the Siegenian epoch differed considerably from the earliest types: some were much branched; some had numerous sporangia, sometimes in clusters; some were leafy. Variation in anatomy kept pace with variation in gross morphology: some had tracheids with spiral thickenings or scalariform thickenings; some had inner cells of the cortex differentiated into an endodermis distinct from the rest of the cortex; some had actinosteles instead of the very simple protostele that *Cooksonia* had. Each of these changes improved the adaptation of these plants to live on land. The original types suffered in the competition that resulted, and by the end of the Devonian, all three of the ancient orders were extinct as far as we can tell from the fossil record.

Where competition is reduced because of some kinds of harsh environmental conditions, the whisk ferns of today seem to have an ecological advantage. Smith (1981) reported a

TABLE 15-4

GENERAL CHARACTERISTICS OF THE ORDERS OF PSILOPSIDA

Order (Fam-Gen-Spp)	Stem	Gametophyte	General Characteristics	Representative Genera
Rhyniales (extinct)	Endarch protostele	Unknown	No differentiation of vertical axis into stem and root; globose, terminal sporangia	*Cooksonia, Rhynia, Hicklingia*
Trimerophyllales (extinct)	Endarch protostele	Unknown	No differentiation of root and stem; alternate branching; larger stele, no phloem	*Psilophyton*
Zosterophyllales (extinct)	Exarch protostele	Unknown	Vertical axis differentiated into rhizome and vertical stem, no phloem; reniform sporangia	*Zosterophylum, Crenaticaulus, Gosslingia*
Psilotales (1-2-3)	Exarch protostele	Large, saprophytic, may contain stele	No roots, rhizomes with rhizoids; enations or microphylls present; lateral, compound sporangia	*Psilotum, Tmesipteris*

frequency of 20% for *Psilotum nudum* along a transect in the peripheral zone of vegetation surrounding the Puhimau geothermal area in Hawaii Volcanoes National Park. This is an area that would ordinarily by covered by tropical montane forest but, because of extremely high soil temperatures, there are no trees, and the zone is dominated by a grass, *Andropogon virginicus,* which is the only vascular plant having a higher frequency than the *Psilotum.* Various mosses and lichens are also present.

CLASS 2. LYCOPSIDA

On the basis of gross morphology, the Lycopsida (*lyco* = a wolf, *pody* = a foot or paw), also known as the Microphyllophyta, consists of two types, a leafy, spreading type commonly called club moss or ground pine, and an erect, grass-like type commonly known as quillwort (Figure 15-5D). There are over 800 species at the present time, but at one time they were much more abundant. Arising in early Devonian, they were the dominant land plants during much of the Paleozoic era, at the end of which most of them became extinct. There are now five genera classified in three families and three orders; other families in these orders along with all of the families of a fourth order became extinct during the Triassic and Cretaceous periods (Figure 15-8). Characteristics of the four orders are summarized in Table 15-5.

Thousands of fossils of primitive vascular plants have been found, examined, identified, and stored away in museums to serve as references for anyone interested in fossils and to assist us in understanding plants of the past and how they have led to our modern day flora. Probably a large majority of vascular plant fossils from the Paleozoic era, and even the Mesozoic, are Lycopsida. However, a paleobotanist doesn't often go out and find a fossil plant with roots, stems, leaves, flowers, and fruit all hooked together. Rather, a leaf is found here, a stem there, and a piece of root somewhere else. For that reason, most fossils have several names (Figure 15-9); occasionally we are able to figure out which belongs to which and get a reasonably good picture of the entire plant.

Living Lycopsida or club mosses, and some fossil species, are readily recognized by the dichotomously branched stems covered with several ranks of microphylls, by the dichotomously branched roots, and by the strobili, or groups of spore-bearing leaves, which are square in cross section because of the pronounced 4-ranked arrangement of the terminal leaves on the fertile branches. The quillworts, on the other hand, could easily be confused with grasses or rushes; they somewhat resemble small clumps of chives but are readily distinguished from angiosperms by the dichotomously branched roots and by the presence of a single vein in the long leaf. The base of the leaf is spoon-shaped and often bears sporangia on its upper surface.

Club mosses grow in humid forests where evaporation is low even though soil moisture is not necessarily high. Some of them are common in the pioneer and settler communities that follow forest fires; others are common in the climax vegetation at or near timberline in high mountains. Quillworts are water plants growing near the edge of ponds or in marshy or soggy places. While not especially common, lycopsids are not rare and can be found in most forested areas.

General Morphological and Physiological Characteristics

The zygote develops within the archegonium into an embryo which obtains its nourishment from the gametophyte tissue in which it is embedded. The first division of the zygote results

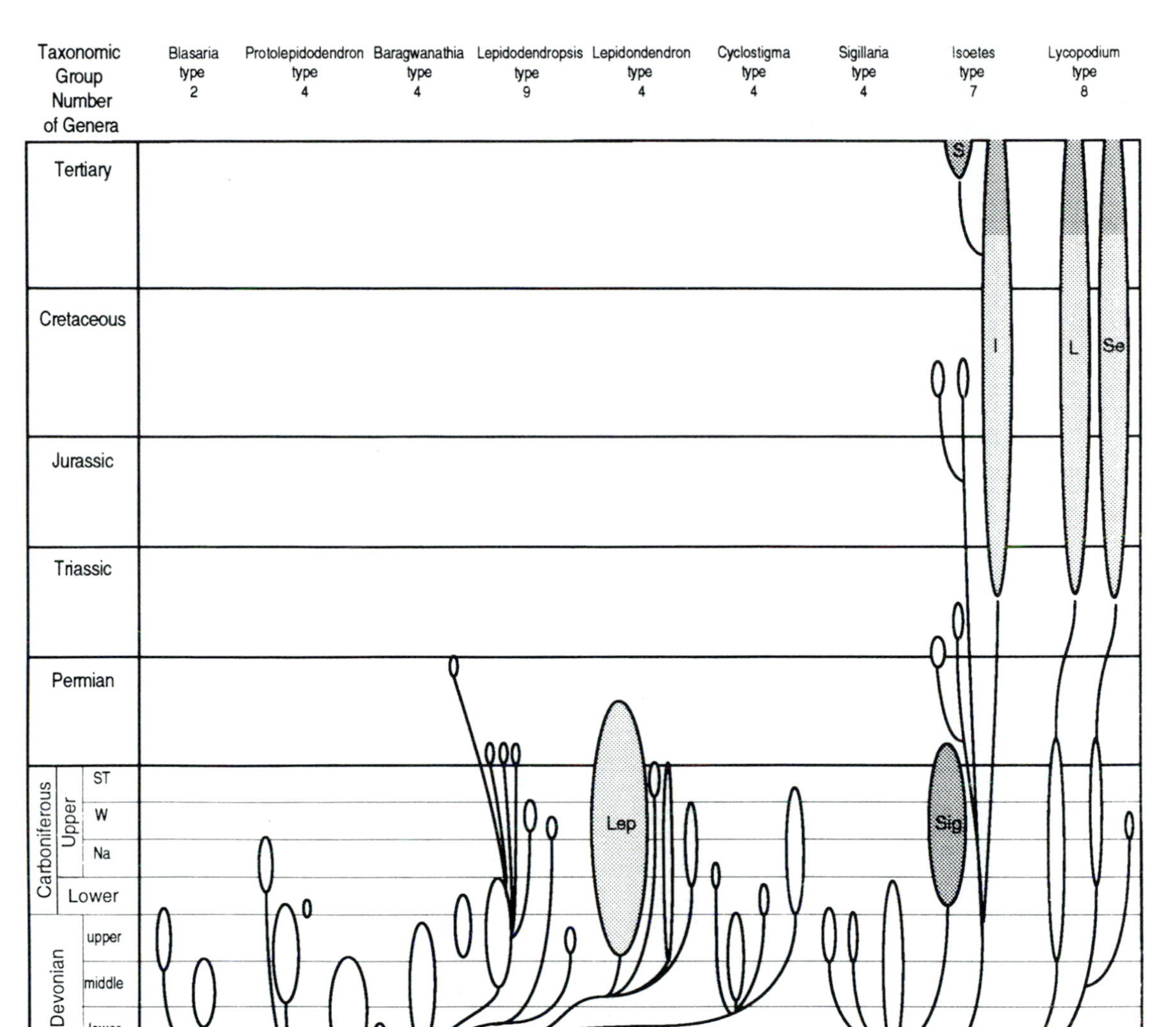

Figure 15–8 Rise and fall of the Lycopsida. (Lep = Lepidodrendron; Sig = Sigillaria; S = Stylites; I = Isoetes; L = Lycopodium; Se = Selaginella.)

in two cells, the upper of which will develop into a multicellular suspensor while the lower one will develop into the embryo consisting of a large foot and root and stem primordia. As the embryo develops, the suspensor forces it slightly deeper into the gametophyte tissue. After the root and stem emerge and become established, they remain attached to the gametophyte for an indefinite period of time.

The embryonic stem is differentiated into three areas. A procambial strand develops near the apex and will differentiate into an exarch protostele. Cortex develops from the ground

TABLE 15–5

MORPHOLOGICAL DIVERSITY OF LYCOPSIDA: CHARACTERISTICS OF FOUR MAJOR ORDERS INCLUDING ALL EXTANT ORDERS

	Lycopodiales	**Lepidodendrales**	**Selaginellales**	**Isoetales**
Growth form	Herbaceous	Woody	Mostly herbaceous, some extinct species woody	Mostly herbaceous, one extinct genus woody
	Prostrate with erect strobils	Erect, tree-like, some trees over 50 m tall	Prostrate, moss-like	Erect, grass-like
Roots	Dichotomously branched, exarch protostele	Dichotomously branched, delicate, exarch protostele	Dichotomously branched, exarch protostele	Dichotomously branched, exarch protostele
Stems	Aerial branches from rhizome, both dichotomously branched with exarch protostele which may become a plectostele	Aerial branches from horizontal rhizophores; erect, monopodial with exarch protostele or ectophloic siphonostele	Aerial branches from rhizomes, both dichotomously branched with exarch protostele or polystele; vessels present	Erect, corm-like, with small exarch protostele; cambium tissue present
Leaves	Eligulate, up to 1 cm long or more	Ligulate, narrow simple, up to 20 cm	Ligulate, 2–3 mm long	Ligulate, up to 60 cm long with spoon-like base
Sporangia	Homosporous; terminal leaves modified to sporophylls	Heterosporous with terminal leaves modified to sporophylls	Heterosporous; terminal leaves modified as microsporophylls, subterminal leaves as megasporophylls	Heterosporous; all leaves function as sporophylls, megasporophylls outside the microsporophylls
Habitat and Ecological Niche	Fairly humid forests as part of understory; common in the tropics	Very humid climate as dominants	Fairly humid forests and alpine meadows as part of understory in forests	Very humid areas in marches and at edges of ponds and lakes
Range in time	Upper Devonian to present	Mostly Mississippian and Pennsylvanian	Mississippian to present	Lower Cretaceous to present with 1 genus in Triassic
Extant genera	*Lycopodium* (180 spp.) *Phylloglossum* (1 sp.)	Became extinct during Permian	*Selaginella* (600 spp.)	*Isoetes* (50 spp.) *Stylites* (2 spp.)

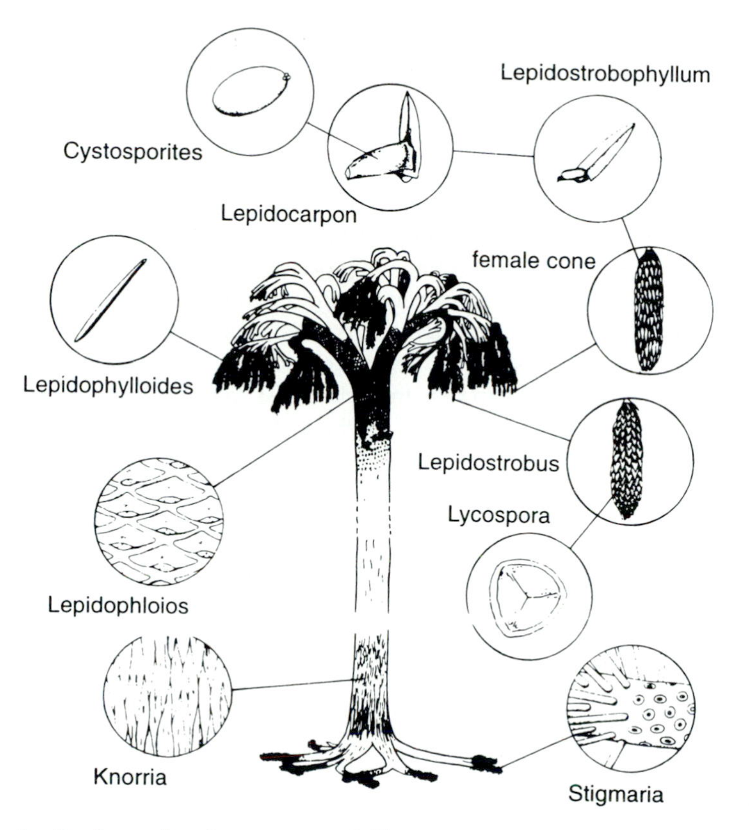

Figure 15–9 How fossils of vascular plants are named. The same species may have several names because all parts of a fossilized plant are seldom found immediately associated with each other. (From Thomas, B. A. and Spicer, R. A., Chapman & Hall, 1987. With permission.)

meristem surrounding the procambial strand. A protoderm differentiates into the epidermis. Differentiation of the embryonic root follows the same pattern.

The outermost cells in the stele apparently remain meristematic, forming a pericycle. The innermost layer of cells in the cortex have thickened walls that hinder the passage of water and gases. This tissue is called endodermis and lies just outside the pericycle. According to the usual interpretation, secondary and adventitious roots develop from the pericycle; however, there is difference of opinion as to whether the pericycle in Lycopsida is really meristematic. In mature plants, all of the roots are adventitious. In the stems of Isoetales, there is a cambium between the xylem and phloem. Cambium was also present in the Lepidodendrales, but not in any of the other Lycopsida. As in most vascular plants, the xylem generally consists entirely of tracheids. In the Selaginellales, however, vessels are also present.

All Lycopodsida have microphylls rather than true leaves. In other words, where a strand of xylem and phloem branch outward from the stem into the leaf, a simple branch is formed and the xylem continues unbroken up the stem. In megaphylls, or true leaves, the entire strand turns into the leaf trace leaving a gap of undifferentiated parenchyma, called a leaf gap, immediately above the leaf trace.

Leaves in Lycopsida are usually in 4 to 16 ranks, occasionally more in *Selaginella,* with 4-ranked terminal strobili. The foliage is usually evergreen, the leaves subulate to oblong, sometimes rounded. Leaf shape and phyllotaxy vary considerably and are useful characteristics for species identification.

A single vein runs the length of the lycopsid leaf. In *Isoetes,* there are air canals on either side of the vein, apparently an adaptation to the aquatic habitat, providing aeration to the roots and stems and buoyancy as well as aeration to the leaves. In *Selaginella* there are stomates on one or both surfaces of the leaves, depending on species, but no stomates on the stems. The mesophyll of lycopsids is made up of very uniform parenchyma cells containing chloroplasts. Surrounding the mesophyll is an epidermis. In some species, the upper mesophyll is made up of palisade cells. The veins consist of a strand of tracheids surrounded by poorly differentiated sieve cells or undifferentiated parenchyma cells. Most lycopsid leaves have a small outgrowth of tissue on the upper surface of the base of the leaf which resembles a tongue and is called a ligule. Its function is unknown.

Apical leaves of many lycopsids are differentiated into sporophylls. In the Lycopodiales, only one kind of sporophyll occurs, but in the other orders, both megasporophylls and microsporophylls occur. The former produce a small number of large spores which develop into female, or archegonia-producing, gametophytes, the latter produce a large number of small spores which develop into male, or antheridia-producing, gametophytes. In the Lycopodiales, the spores are all the same size and the gametophytes are hermaphroditic (or monoecious), producing both archegonia and antheridia. In the Isoetales, nearly all of the leaves are sporophylls, only the outermost and innermost commonly being sterile, but in the other orders, the sporophylls occur in modified groupings called strobili. In the Selaginellales, the male strobilus is situated above and as an extension of the female strobilus (Figure 15–10A).

In the Lycopodiales and Isoetales, the gametophytes may be either autotrophic or saprophytic. In the Selaginellales, the gametophyte is at first parasitic on the parent sporophyte, developing within the strobilus, but continues its development as a free-living, green plant which probably obtains most of its nourishment from the stored foods in the spore wall. In at least two species of *Selaginella, S. apoda* and *S. rupestris,* the female gametophyte develops to maturity within the strobilus and fertilization takes place there.

In the Lycopsida, female gametophytes produce a small number of eggs in the archegonia and male gametophytes in most species produce a large number of biflagellate sperm in antheridia. In *Isoetes,* the sperm are spirally twisted and multiflagellate, resembling the sperm of some ferns and horsetails more than other lycopods. Examples of the Lycopsida are shown in Figure 15–10.

The gametophytes of many Lycopsida are dependent on symbiotic fungi for growth and development. Soon after the spore germinates, the young gametophyte is infected by a fungus which is apparently important in making food available to the lycopsid. These gametophytes are colorless, subterranean, fleshy structures, similar to, though smaller and simpler than, the gametophytes of *Psilotum.* In other species, the gametophytes are only partly subterranean, the portion above ground being green in color and autotrophic. The gametophytes of many lycopsids are perennial.

In all, four morphological types of gametophytes occur in nature (Bruce 1979). Buchmann (1898) described these in some detail and differentiated them on the basis of number of layers of tissue and how they were associated with their fungal symbionts, size, color, number of

Figure 15–10 Lycopsida. (A) *Selaginella;* (B) *Lycopodium;* (C) *Isoetes.*

canal cells in the neck of the archegonium, and orientation of the axis. Some botanists have proposed dividing the genus *Lycopodium* into three or four genera on the basis of these differences.

In *Lycopodium,* the antheridia and archegonia are on the same gametophyte (prothallus), sometimes intermingled and sometimes segregated in different parts of the same gametophyte. The archegonia are sunken with only the neck protruding. The neck canal is composed of one to thirteen cells and is surrounded by three or four rows of neck cells. The ventral canal cell contains a single large egg; at maturity, the neck cells separate, the canal cells dissolve, and sperm are able to swim directly to the egg in the venter. In other orders, the processes are similar, but the archegonia and antheridia are on separate prothalli.

Phylogeny and Classification

Excellent fossil records of the Lycopsida have been analyzed and the pattern of evolution seems fairly well worked out. The earliest lycopods were similar in many ways to the Lycopodiales of today and probably evolved from the Zosterophyllales during early Devonian time. *Baragwanathia longifolia,* discovered in early Devonian formations in Australia, is believed to be the oldest fossil that can definitely be assigned to the Lycopsida. It shows characteristics of both the Psilopsida, especially *Zosterophyllum,* and modern lycopsids. From such plants have evolved the club mosses and quillworts of today as well as the extinct lycopsids, like *Lepidodendron,* which flourished in the carboniferous periods.

Evolution was rapid during the first half of the Devonian period and continued to be rapid during the second half with three important adaptations originating at that time; the origin of the arborescent habit, the origin of cambium, and the origin of megaphylls or true leaves. The first two were of special significance in the phylogeny of lycopsids, and by early Mississippian, about ten distinct types of lycopsids, represented by at least 20 genera, had evolved (Figure 15-8).

The Lycopsida reached their zenith of ecological significance during the Mississippian and Pennsylvanian periods when they were among the chief dominants of terrestrial ecosystems everywhere; but they declined and many species became extinct during the late Paleozoic era (Figure 15-8). Geologists have ascertained that the Permian period was a time of glaciation and increasing aridity. One hypothesis to account for the decline of arborescent lycopods is that the great swamps that covered much of the earth during the Carboniferous periods dried up, and the lycopsids, so well adapted to living under the warm, humid conditions of that time, were no longer able to survive as dominants. Humid microclimates continued to exist, as they do at the present time, in shady nooks near bodies of water, and herbaceous lycopsids survived in these microhabitats after the great forests of arborescent *Sigillaria* and *Lepidodendron* had disappeared. Another hypothesis to account for the extinction of arborescent lycopods during the Permian is that the carbon dioxide level of the atmosphere had decreased to where the lycopsids could not compete with the gymnosperms which were rapidly evolving efficient leaf systems at that time.

The best known of the Carboniferous lycopods, *Sigillaria* and *Lepidodendron* (Figure 15-8), were large trees, often 30 meters tall and a meter in diameter, and were both abundant and widespread at that time. Their success depended in part on the presence of cambium in the stems making it possible to produce strong trunks. Branching near the top of the trunk was dichotomous; the roots were also dichotomously branched. The leaves were microphylls and were apparently deciduous; they were borne on rather unique leaf cushions which make the identification of these plants easy. Lycopods and sphenopsids growing at higher latitudes had growth rings; growth rings were faint during the Devonian and Carboniferous periods, and generally not present in European and North American fossils, both areas being tropical at that time, but good growth rings are found in Asiatic fossils.

Ecological and Economic Significance

Though at one time many lycopsids were widespread dominants, extant species are found primarily in the shade of other vascular plants. Most species are tropical and commonly epiphytic on various trees. Temperate zone lycopsids are all terrestrial except for the aquatic *Isoetes* species. They are found in all parts of the world except in very dry habitats. At one time, some of the species of *Selaginella* were abundant in alpine ecosystems of the Rocky Mountains, but they are fragile and have been destroyed by heavy use of the rangeland by horses and cattle and today survive only in sheltered areas atop the same mesas they once dominated.

Some species of *Selaginella* are grown as ornamentals, especially in home conservatories. Masses of *Lycopodium* are gathered around Christmas time and sold for decorative purposes. *Selaginella lepidophylla* is collected in Mexico and sold by the bale in the U.S. as a novelty, especially around Easter time. The plants roll up into tight balls, appearing to be lifeless, as they dry, but when water is applied, they turn green and take on a normal appearance; hence the common name, "resurrection plant." *Lycopodium* spores have been used for medicinal purposes and as a pill coating and are still available in many drugstores. Commercially, *Lycopodium* is known as princess pine, creeping Jenny, crowfoot, or ground pine.

Economically, the class has been and still is of greatest economic significance for the production of coal (Special Interest Essay 15-1). The arborescent lycopsids that thrived in the Mississippian and Pennsylvanian Periods were responsible for many of our coal deposits. Frequently, petrified bits of *Lepidodendron* or other lycopods are found in the coal beds or in their near vicinity. To "rock hounds" these represent a valuable resource and are frequently cut, polished, and converted into jewelry.

Also of great economic significance is the tremendous cost in both money and health associated with air pollution resulting from the burning of coal and other fossil fuels (Ferry et al. 1973, Huckaby 1993). In addition to toxic substances like sulfur dioxide, carbon monoxide, and particulate matter, all of which cause serious human health problems, the greenhouse gases (GHG), carbon dioxide and methane, are now looked upon as creating problems that will get worse (Gore 1992, Nordhaus 1992, 1994).

Since the 1950s, careful studies have revealed some facts about GHG. (1) Careful atmospheric measurements show a gradual but steady incease, world wide, in GHG. (2) Light rays readily pass through GHG but the heat rays the light converts to as it reaches the earth surface are blocked by GHG, thus tending to warm the lower atmosphere. (3) Any appreciable warming of the lower atmosphere will cause the Arctic ice cap to begin melting resulting in release of water that could raise the oceans' level. (4) Other forces will be brought into play as the GHG increase, some of which tend to mitigate the extent of global warming, others which tend to amplify its effects (Special Interest Essay 15-1). Considering all factors, the consensus of those who have studied the problems most carefully is that something must be done to reduce GHG emissions worldwide, beginning, of course, at home. Nordhaus (1992) considered the effect of carbon taxes, emission control regulations, and other options that might be used to check global warming and concluded that "a modest carbon tax [about $5.00 per ton of coal to begin with] would be an efficient approach to slow global warming, whereas rigid emissions . . . [controls] would impose significant net economic costs."

Research Potential

Because of the abundance of their fossils in all periods of geological time since their origin in early Devonian to the present, the Lycopsida have been the most useful of all land plants in elucidating patterns of evolution and inferring climatic conditions. As more fossils are gathered and existing fossils are studied more critically and in greater detail, the lycopsids should be of even greater value.

Since *Lycopodium* spores were once used in medicine, there may be substances present in the spores or other parts of the plants that have chemotherapeutic value. Many of our most valuable drugs are plants once prescribed by herbalists, later ignored or forgotten, now used again as a source of concentrated drugs.

The relatively large size of lycopsid gametophytes and the fact that some species are autotrophic, others saprophytic and/or symbiotic, and still others dependent on the mother sporophyte suggests that a potential exists in this group for studying the forces of morphogenesis and differentiation. If we can learn how to manipulate the environment in which the gametophytes are grown in such a way as to produce haploid sporophyte plants from them, we will have an advantage in studying genetic traits.

CLASS 3. SPHENOPSIDA

The Sphenopsida (*spheno* = a wedge), also known as the Equisetineae (*equi* = a horse; *seta* = a bristle, hence tail) are commonly called horsetails, joint grass, snake grass, or scouring rushes. They are common in all parts of the world and are found along ditch banks, in seeps, in sloughs and other wet sites, and around springs. There are some 25 species extant today, all classified in one genus, but at one time there were many genera and several families classified in three orders. Characteristics of the three orders are summarized in Table 15-6, and suggested lines of relationship are presented in Figure 15-2.

The stems of some species of Equisetum (*e.g., E. arvense* and *E. sylvaticum*) are branched, whereas other species (*e.g., E. hyemale*) have unbranched stems. When the stems are branched, the long branches typically curve upward so that they are close to the main stem, giving it the appearance, more or less, of a horse's tail. Because the stems easily separate at the nodes into hollow segments, the term "joint grass" is also applied to the group. The name "snake grass" is a reference to the marshy habitat in which many of the species grow and the water snakes which find shelter amidst the plants.

The Sphenopsida are also known as the Arthrophyta and the Articulatae. They differ from all other Tracheophyta in several ways: (1) the leaves and branches are arranged in whorls, (2) they are the only vascular plants to have leaves and branches alternating with each other instead of having the branches arise from buds in the axils of the leaves, (3) the stems are hollow and easily pull apart into internode segments because of extensive meristematic tissue there which weakens the stem at the node, (4) the stems are ribbed and grooved with a silica-impregnated epidermis which makes them so rough that they were formerly widely used for scouring pads, and (5) the sporophylls or cone scales, which are borne on a terminal strobillus, are peltate with the spores produced on the under surface of the scales. Representative specimens are illustrated in Figure 15-11.

Because of these distinctive features, the horsetails are not commonly confused with any other group of plants. Superficially, they resemble stoneworts, but the horsetail stems are much larger in diameter and contain vascular tissue whereas stonewort stems are very delicate, usually only one cell thick.

General Morphological and Physiological Characteristics

The typical horsetail consists of an underground rhizome with vertical stems arising from it. These are usually about 1 m long, but may be 2 m or more, and are of two types, fertile and sterile, or vegetative. In some species, the fertile shoots lack chlorophyll. Spores are produced within a **strobilus,** or cone, at the apex of each fertile shoot. Stems are often branched. In some species, both sterile and fertile stems are branched, in some neither is branched, in others only the sterile, or vegetative, stem is branched. In *Equisetum pratense,* the fertile shoots are flesh colored, solid, unbranched, lack chlorophyll, and develop very early in the spring; the vegetative shoots develop 2 or 3 weeks later, are green, and are profusely branched. Except

TABLE 15-6

GENERAL CHARACTERISTICS OF THE ORDERS OF SPHENOPSIDA

Order	Stem	Gametophyte	General Characteristics	Representative Genera
Hyeniales (extinct)	Exarch (?) siphonostele	Unknown	Large trees and shrubs	*Calamites, Hyenia, Calanophyton, Pseudobornea*
Sphenophyllales (extinct)	Exarch protostele	Unknown	Small to large herbs	*Sphenophyllum*
Equisetales (1-1-25)	Endarch eustele or ectophloic siphonostele	Small, lobed, autotrophic	Erect herbs with ribbed stems and whorls of microphylls at each node	*Equisetum*

Figure 15–11 Sphenopsida: (A) Group of horsetails, *Equisetum hyemale;* (B) vegetative shoot; (C) reproductive shoot.

that both types of shoot arise from the same rhizome, the novice might think he was seeing two different species of *Equisetum.* Both types of shoots die back each fall. In contrast to this, the very common, essentially ubiquitous *E. hyemale* has perennial, dark green, coarse, unbranched vegetative and fertile shoots which are identical to each other except for the sessile cones on the latter. *E. giganteum,* a tropical species growing to about 5 m tall, also has only unbranched stems, is perennial and evergreen, and has fertile and sterile shoots that are identical in appearance except for the presence or absence of strobili at the apex.

Many of the Paleozoic species were woody arborescents with abundant secondary xylem. Modern horsetails, however, have no secondary xylem and all except one are herbaceous. The one woody species is a tropical vine which reaches to a height of 10 to 12 meters. The stems of both extant and extinct horsetails are endarch eusteles or siphonosteles in which the pith typically breaks down, except at the nodes, to become a large central cavity.

Anatomically, the three kinds of stem (rhizomes, fertile shoots, and vegetative shoots) are essentially identical. Three kinds of cavities can be found in each stem: the central cavity, the vallecular canals, and the carinal canals, respectively, in order from largest to smallest.

In most species, the central cavity is very large and the internodes, or "joints," resemble thin-walled pipes. In some species, however, for example *E. palustre,* the central cavity is very small. The carinal canals are situated nearest the central cavity and directly below the ridges; the vallecular canals lie directly beneath the grooves. Between adjacent vallecular canals and directly above the carinal canals are bundles of vascular tissue. Protoxylem is innermost and, as the plant grows, it breaks down to form the small carinal canals. The metaxylem is outside the protoxylem and consists of scalariform tracheids and a few small, simple, reticulate vessels arranged in two lateral horns with phloem in the space between. There is no cambium. Surrounding each vascular bundle is a ring of pericycle tissue.

In the internodes, the stele is a eustele, inasmuch as the bundles of vascular tissue are separated from each other by parenchyma tissue, but at the nodes, the bundles fan out to form a continuous ring, becoming an ectophloic siphonostele. Cortex tissue lies outside the stele and is surrounded by epidermis. The pericycle and epidermis are each one cell layer thick. The outermost layer of cortex is called chlorenchyma inasmuch as the cells have many plastids in them. Beneath the ridges, lignified sclerenchyma cells occur. Stomata are present in the grooves.

The presence of canals through which air can move—the vallecular and carinal canals and the central cavity—is typical of plants that grow in wet habitats, especially where the plants may be partly inundated at times. The sunken stomata, on the other hand, are typical of plants adapted to arid habitats. Horsetails grow in habitats where there is abundant moisture at least part of the time, but some species, for example, *Equisetum hyemale* and *E. arvense,* are well adapted to habitats which dry out later in the season and may remain very dry for long periods of time.

Except for a small primary root, the roots are all adventitious, originating in the pericycle of the rhizomes. Rhizomes and roots are often a meter or more beneath the ground surface, which probably accounts in part for the occurrence of many species of *Equisetum* in relatively arid habitats. The stele in the root is an exarch protostele.

Equisetum leaves are small, wedge-shaped microphylls that occur in whorls. The leaf bases are fused to form a sheath at each node; the length of the unfused portion of the leaves (sheath teeth) compared to the fused portion (sheath) is an important diagnostic characteristic in identifying species of *Equisetum,* as are number of leaves, length of internodes, whether each ridge or longitudinal rib on the stem consists of one or of two rows of tubercles, leaf shape, strobilus size and shape, and size of the central cavity. The number of leaves at each node varies from three to twenty or more; there are usually the same number of ridges on the stem as there are leaves at each node, but in some species there are twice as many.

The leaves of *Equisetum* have a single vein and a leaf gap has been reported in some species. Leaves are small and poorly suited for efficient photosynthesis; the stems are the primary productive organs. In the species included in the section Hippochaete, which consists mostly of tropical species but includes some widespread temperate zone species like *E. hyemale,* the stems are perennial and evergreen. In the section Euequisetum, which is mostly limited to the temperate zone, the stems are generally deciduous and paler in color.

Although there is some variation in spore size, *Equisetum* is generally considered to be homosporous. The spores are all borne in the same strobilus with no distinction between megasporangia and microsporangia as in most vascular plants. Nevertheless, there is a distinction in some species between male and females gametophytes. In *E. brasiliensis,* a bimodal continuum of spore size can be observed. The larger spores produce only female gametophytes or gametophytes that produce exclusively or primarily archegonia, and the smaller spores produce only male gametophytes.

The spores are smooth and spherical, similar to the pollen grains of some gymnosperms, *e.g.,* larch and Douglas fir, except for the presence of four ribbon-like strips called elaters, which are hygroscopic and coil and uncoil in response to moisture change. The spores germinate to produce small, green, circular or irregular ribbon-like gametophytes which, when mature, range in size from pinhead to about 8 mm. In most species, gametophytes produce both antheridia and archegonia on the upper surface, but in some species, primarily one or the other. The sperm are multicilliate and spirally coiled. Fertilization takes place in archegonia similar to those of *Lycopodium* and *Selaginella.* The zygote divides twice to form a quadrant of cells. The upper two develop into the stem and first leaf, the lower two into the foot and root. The embryo lacks a suspensor and the foot is poorly developed. It obtains nourishment from the gametophyte for a while, but the young sporophyte soon becomes independent as the embryonic root grows directly through the prothallus and into the soil and the stem emerges through the neck of the archegonium. The stem soon branches to form several upright stems and a rhizome. Adventitious roots on the rhizome aid in anchoring the plant.

Phylogeny and Classification

The oldest fossils of Equisetineae have been found in the Emsian epoch of the Devonian period (Banks 1970). During the Mississippian and Pennsylvanian periods, arborescent horsetails were very abundant; these along with the arborescent club mosses of that time have produced much of the coal being mined today. Species of *Calamites* were especially important in coal building. *Calamites* often attained heights of 30 m and diameters of 40 cm. With their hollow stems and whorled branches, they resembled huge horsetails.

Hyenia is the oldest genus of fossil plants that can be assigned to the Sphenopsida. These were herbaceous plants approximately 10 cm tall, similar to *Psilophyton* in growth form but possessing dichotomously branched roots and having a protostele with xylem structure similar to that of Equisetum. Closely related to *Hyenia* was *Calamophyton,* also herbaceous but somewhat taller, though seldom more than 60 cm tall. *Calamophyton* stems have narrow transverse markings on them similar to the joints of *Equisetum.* Like *Hyenia,* it originated in mid-Devonian time. Both species were apparently homosporous.

By late Devonian, great forests covered much of the earth. At this time, another herbaceous genus, *Sphenophyllum,* evolved. These species were up to a meter in height and were apparently an important part of the forest understory. The stems were ribbed and plainly jointed, and the leaves, typically about 2 cm long and somewhat wedge-shaped, were in whorls at the nodes. The stems were triangular protosteles with an active cambium outside the xylem. Cambium activity was apparently equal at all times of the year as there are no clear cut growth rings in the secondary xylem.

Calamites evolved during late Devonian. It is so similar to *Equisetum* that it is usually included in the same order, although some botanists place it in an order by itself, the Calamitales. It was extremely abundant for about 60 million years but became extinct during the Permian. Since it was a dominant, it was more susceptible to the macroclimatic changes that occurred at that time than ecological subordinate species would be. *Equisetum,* which originated during the Carboniferous, is a herbaceous subordinate and survived because the microclimates to which it was well adapted continued to exist. Fossil horsetails from the Miocene epoch are morphologically identical to species that are common today.

Including extinct taxa, approximately 100 species of Sphenopsida are known. These are classified into 10 genera, 5 families, and 3 orders (Table 15-6). Relationships among the three orders are suggested in Figure 15-2. At the present time, only 1 genus and about 25 species remain in this once important class.

Ecological and Economic Significance

At the present time, *Equisetum* is neither ecologically important, on a world basis, nor of great economic value. A hundred years ago, the plants were sought after for scouring pots and pans and for scrubbing and polishing floors. Campers, especially those interested in "survival skills," are often aware of their value in cleaning pots. They are of ecological significance as producers in some aquatic ecosystems; however, even along canal banks in eastern Idaho, where they were a conspicuous part of the flora, they contributed only 10% of the total photosynthate (Pearson 1966).

Millions of years ago, on the other hand, the class was ecologically very important, and because of this, their fossils are now economically important. About 70% of the world's known coal deposits were formed during the Carboniferous periods, largely from the remains of *Calamites, Lepidodendron,* and *Sigillaria* which were so abundant at that time (Special Interest Essay 15-1).

Research Potential

Some species of *Equisetum* have an interesting combination of xeric and hydric adaptations and offer a possibility for studying plant needs for water. Inasmuch as *Equisetum* spores appear to be intermediate between homospores and heterospores, this genus should have value in studies of sexual differentiation in plants. It has often been suggested that silicon is an essential element in plant nutrition; however, with the exception of the Chrysophyta, it has not been possible to demonstrate essentiality. *Equisetum* accumulates large amounts of silicon and therefore shows promise as a genus in which to study silicon metabolism. Ridge and Sack (1992) found *Equisetum hyemale* useful in testing the starch-statolith hypothesis proposed by B. Nemec in 1901 to explain gravitropic responses in plant roots. Overlooked for almost a century, Nemec's hypothesis is now widely accepted.

SUGGESTED READING

Bank's *Evolution and Plants of the Past* introduces students to many exciting aspects of evolution and paleobotany. Several plant morphology textbooks go into excellent detail on the structure and anatomy of vascular plants: *Evolutionary Survey of the Plant Kingdom* (Scagel et al. 1965), Haupt's *Plant Morphology,* and Esau's *Plant Anatomy* are especially to be recommended. Kapp's *How to Know Pollen and Spores* can help you gain an appreciation of the types of plants that existed when the "fern allies" were at their peak of dominance.

STUDENT EXERCISES

1. Make a chart in which you show the chief kinds of plants in existence at the beginning of the Devonian, Mississippian, Permian, and Triassic periods.
2. Make thin free-hand sections (transverse and longitudinal) of whisk fern (psilopsid) stems and rhizomes. Stain with phloroglucinol and concentrated hydrochloric acid and examine under a microscope. Sketch or describe the lignified cells.

3. How does the distribution of tracheids in the rhizome of the whisk fern compare with that of the aerial stem?
4. Prepare and stain free-hand sections of a clubmoss (*Lycopodium* or *Selaginella*) and describe or sketch location of tracheids. Compare aerial stems, rhizomes, and roots.
5. Diagram the life cycle of *Equisetum hyemale.*
6. From a flora of the region in which you live, compare three species of *Equisetum* (for example, *E. hyemale, E. silvaticum,* and *E. arvense*) as to habitat, phenology, color, and general growth form.
7. Which classes of vascular plants would you expect to find in sedimentary rocks of Devonian age? Of Silurian age?
8. What kind of meristematic tissue gives rise to adventitious roots? To secondary xylem? To secondary phloem?
9. From an ecological point of view, what are the four kinds of leaves? In what kind of habitat would you find plants with each of these four kinds?
10. Anatomically, what are the four kinds of leaves?
11. The Psilopsida evolved near the end of the Silurian Period; how many years ago was that?

SPECIAL INTEREST ESSAY 15–1

Coal

It was "black gold" that brought about the industrial revolution, beginning in the late 1700s, and "black gold" has continued to be one of our most valuable minerals ever since, although a strong trend toward using petroleum products for heating homes and running many industries began in the late 1930s. Largely because of its greater convenience, but partly because of its apparently relative cleanliness, oil has largely replaced coal in many industries; now, however, a swing back to greater use of coal as an energy source is taking place. Huge electrical generating plants powered by coal have sprung up in coal-producing areas around the world in recent years. Coal is not only a major source of electrical energy today, however; it is also used as a raw material in the production of dyes, textiles, farm fertilizers, and other products.

Our great coal deposits have not always existed. Production of our oldest deposits began about 300 million years ago and depended then, as now, on several factors, including (1) the accumulation of large amounts of organic matter at a rate greater than it can be completely decomposed by bacteria and fungi; (2) exclusion of air; (3) great pressures, and (4) temperatures favorable for partial decomposition. These conditions exist in parts of the world even today, but during some periods in the past, they were especially favorable for coal formation.

In the great Pennsylvanian forests that gave rise to some of our oldest coal deposits, the dominant trees were *Sigillaria, Lepidodendron,* and *Calamites,* and the climate was hot and humid. The atmosphere contained considerably more carbon dioxide than at present, but the oxygen content was lower than now. The mountains were bare of vegetation as were the dry uplands, but the low-lying areas and river bottoms were swampy and covered with thick forest.

In the shade of the dominant species grew many herbaceous plants: *Hyenia, Calomophyton, Cladoxylon, Aneurophyton,* and great seed ferns. Basidiomyceteae and Ascomyceteae were very limited in distribution and were not the efficient wood decomposers they are today, for they were just beginning to evolve at that time.

As the trees died and fell, they partly decomposed in the warm, humid climate, and new plants grew in the peat-like substrate they formed. Because of the abundant water, poor aeration, and lack of efficient decomposer organisms, the organic matter built up into layers hundreds of meters thick. As land forms changed over a period of millions of years, many of these marshy areas sank below the surface of lakes, and under the great pressure that developed from hundreds of meters of water and silt, the organic matter changed into peat and then into coal. Since then, land forms have changed many times, and today some of these coal beds are found in mountainous areas overlain with hundreds of meters of rock and soil. At least 5 to 8 meters of organic matter are needed to produce 1 meter of coal; some coal beds in western Germany are over 100 meters thick. Depending on the thickness of the rock formations deposited above them, and often still there, coal deposits now are being mined either by strip mining or by deep mining methods.

About 75% of the coal we use comes from Mississippian and Pennsylvanian formations; most of the rest comes from Jurassic deposits. The giant horsetails and lycopods that gave rise to Mississippian and Pennsylvanian deposits became extinct during or shortly after the Permian period; the Jurassic and Cretaceous deposits were formed largely from gymnosperms such as *Ginkgo, Bennettia,* and *Cordiates.*

Coal varies greatly in quality, depending on the species of plant from which it was produced, the stage of development of the plants, the environment under which the coal-producing processes were initiated, the amount of pressure under which the processes operated, and the length of time since the processes were initiated. Water content, degree of decomposition, and mineral content are important factors to be considered when judging the quality of a coal deposit. For example, coal produced from plants high in proteins would be expected to have a higher sulfur content than coal produced from plants high in lignin and cellulose, providing the environmental conditions were similar.

The first stage of coal formation yields peat, a material consisting of about 90% water and having very apparent cell structure. In northern Europe, peat was formerly dug during the summer months and cut into large, flat blocks which were stacked in loose stacks for drying. The dried peat was then burned during the winter. Later stages of coal deposit are lignite and brown coal, in which the water content is about 50%, and the fibrous nature of the organic matter, but not much cellular detail, is apparent. Subbituminous coal is a still later stage. It contains about 25% moisture and burns very readily. In the final stage of coal formation, either bituminous or anthracite coal results. Bituminous or "soft coal" contains less than 15% moisture and is the foundation of the steel and other industries of many countries. Usually it contains relatively large amounts of sulfur and when burned gives off particulate matter, or smoke, in large quantity. It produces more heat per ton than any other type of coal. Anthracite or "hard coal" has the lowest moisture content of all types of coal, and usually the lowest sulfur content, and burns with very little smoke; it is therefore the preferred coal for heating homes and for city industries; consequently, it commands the highest price per ton. However, some western bituminous deposits are also much sought after because of their high heat value per ton and sulfur content lower than that of most anthracites.

Air pollution is always associated with the burning of coal. Sulfur dioxide is generally the most harmful of the many pollutants present in coal smoke as far as the health of humans and plants alike is concerned, but particulate matter, consisting of unburned bits of carbon, is the most obvious pollutant, causing the black smoke from which soot settles down on everything near coal burning facilities. Oil produces much less particulate matter, but may produce as much or more sulfur dioxide and other toxic substances.

Both coal and oil produce vast amounts of carbon dioxide as they burn. Since the 1950s, carbon dioxide production has increased far beyond the point at which plants, through increased photosynthesis stimulated by the higher level of CO_2, can maintain a relatively constant CO_2 level in the atmosphere. As the CO_2 level increases, it is expected that surface temperatures on earth will markedly increase, due to the much publicized "greenhouse effect" which high levels of carbon dioxide in the atmosphere create.

Methods of trapping particulate matter and of converting harmful sulfur dioxide gas into useful sulfuric acid have been developed, but are not as widely employed as most conservationists would like to see. Until they are, crops and livestock, as well as humans, will continue to be affected negatively by smoke stack effluents.

16 FERNS

If we were to visit a Devonian valley shortly after the origin of the first fern-like plants, we would find ourselves in an environment alien to everything we know. Breathing would be difficult because the oxygen content of the air was less than half of what it now is while the content of carbon dioxide and methane was much greater. Our watches would be of little use to us for there were not only fewer hours of daylight then, but also fewer hours of night: the Devonian day was only 21.9 hours long because the earth was rotating more rapidly at that time than it is today; the Devonian year was 400 days long.

Devonian hills and mountains were barren of vegetation, but the swamps and marshes in the lowlands supported a great diversity of plants. Forests of giant horsetails and clubmosses were beginning to replace the low-growing psilophytes and lycopods which until then had been the dominant vegetation. In the shady understory of these forests, herbaceous psilophytes, preferns, horsetails, and club mosses grew. During the Mississippian period, true ferns evolved from the early preferns and by the end of the Permian period had entirely replaced them.

Ecologically, the earliest ferns, like their descendents today, were mostly subordinate species growing in the shade of trees. Numerous arthropods fed on the ferns and other Devonian vegetation, sometimes grazing it to near extinction for there were fewer predators then than now. However, a few species of amphibians were also present, preying on the mites, collembolas, and other arthropods, and thus providing some protection to the plants. A variety of fishes, some of them with primitive lungs and leg-like fins, also evolved during the Devonian and Carboniferous periods. From them evolved the higher vertebrates which have dominated the fauna ever since the Mesozoic era.

Recent research in geophysics and paleoecology helps to explain why the environment was so different then from what it is today. Geophysicists inform us that tidal friction is slowing down the earth's rotation at the rate of 0.8 seconds per century. Since there is no friction in outer space, the earth's rate of revolution about the sun is constant. Days are therefore getting longer but years are the same length as they have always been: Christmas day this year will be 28 microseconds longer than it was last year. Studies by paleoecologists confirm these estimates. They also provide estimates of how long it must have taken for the combined photosynthesis of all green plants to raise the level of oxygen in the atmosphere to the point at which an ozone layer thick enough to protect terrestrial life from the harmful effects of ultraviolet radiation could have formed. Paleoecologists have also estimated the lag time between the origin of herbivores, which usually tend to destroy vegetation, and the origin of carnivores, which are the protectors of vegetation; it is typically in the tens of millions of years.

The Filicopsida (*filix* = fern, from *fili* = thread) of today resemble those of the Carboniferous periods, and also the preferns of the Devonian period, both ecologically and morphologically. The class is a relatively large one at the present time, both in number of species and in number of individuals. The preferns apparently originated during the Eifelian epoch of the Devonian period approximately 370 million years ago, the true ferns some 40 million years later during the Mississippian. While some biologists have postulated a close relationship to the Lycopsida, many others disagree. Banks (1970) suggested that they evolved from the *Rhynia-Psilophyton* line referred to in the previous chapter and are therefore more closely related to whisk ferns and horsetails than to club mosses or other members of the *Zosterophyllum-Asteroxylon* line. Their closest relatives are the seed plants as pointed out by Tippo (1942). Anatomically, they are a rather variable group, as expected in a primitive taxon.

CLASS 4. FILICOPSIDA

The ferns make up the fourth of the seven classes of Tracheophyta; they are more abundant than any of the previous three classes with which they were formerly allied as the Division Pteridophyta. They are easily recognized by their rather large, feathery leaves, which are often bipinnately or tripinnately compound, by the presence of sori, or groups of spores, along the veins on the sundersurface of the leaves, by a form of leaf development known as circinate vernation (Figure 16–1A), and, in many species, by the dichotomous venation of their leaves. The ancient Latin name, *filix,* apparently refers to the delicate thread-like petioles and rachises characteristic of most species (Figure 16–1B). The observant student seldom confuses ferns with any other group of plants. But to careless students, paying no attention to venation and leaf vernation, such flowering plants as lousewort and yarrow resemble ferns, and some ferns, such as the pepperwort, resemble clover. To add to the less discriminating students' confusion, florists frequently refer to another flowering plant, a variety of asparagus, as fern.

Ferns are worldwide in distribution, but are especially abundant in wet areas. While most ferns are adapted to moist habitats, others are aquatic or epiphytic; few ferns, however, are truly xeric. Only the sporophyte generation is tolerant of relatively dry conditions; the gametophytes are limited in drought tolerance and thus limit the distribution of the sporophytes as well. There are approximately 9000 species of ferns, the majority of them tropical. These are included in 18 families and 5 orders.

Fern sporophytes, or diploid plants, are large and are differentiated into roots, stems, and leaves, while the gametophytes or haploid plants are small and show little differentiation. Fern gametophytes are thallus-like, usually heart-shaped, and in most species bright green in color (Figure 16–2); they are typically not much larger than the head of a thumb tack, although they may be 2 cm or more in diameter. They are able to grow and reproduce only where there is an abundance of water, either free water or very moist soil. In some species, the gametophytes are ribbon-like or filamentous, and in a few species they are subterranean, saprophytic, and colorless, and form symbiotic relationships with various fungi.

Figure 16–1 (A) Maidenhair fern; note the delicate stems which give ferns the name Filicopsida; (B–C) water clover or pepperwort, *Marsilea;* (D) *Cyathea pinnata* from upper Cretaceous in Utah. (D, courtesy Lee R. Parker, California Polytechnic State University, San Luis Obispo.)

General Morphological Characteristics

Ferns differ from the preceding three classes by having large leaves which show evidence of having originated from flattened stems, by sporangia born on the undersurface of the leaves (Figure 16-2), by leaf gaps in the stems, and by differences in stelar anatomy, gametophyte development, and spore characteristics. They differ from most seed plants—which will be discussed in the next two chapters—by having independent gametophytes, buds which contain single leaves instead of entire shoots, no tap root, usually no cambium, and in other leaf and stem characteristics.

Two main types of growth form (habit), occur in ferns, the prostrate or **rhizomatous** type and the upright or **basket** type. A typical temperate region fern has a horizontal, dichotomously branched underground stem called a **rhizome** from which the leaves, the most conspicuous part of the plant, arise. The undersurface of the rhizomes are covered with fine roots which are most abundant at the nodes opposite the leaves. The rhizomes are usually dark in color and covered with scales which are the remnants of old leaf bases. Modifications of creeping rhizome type of stem are the stems of climbing ferns and epiphytes. In the basket type, the stem is vertical and usually at least partly above ground. These ferns are often characterized by a massive, but usually very short, vertical stem, or crown, occurring in large basket-like or vase-like

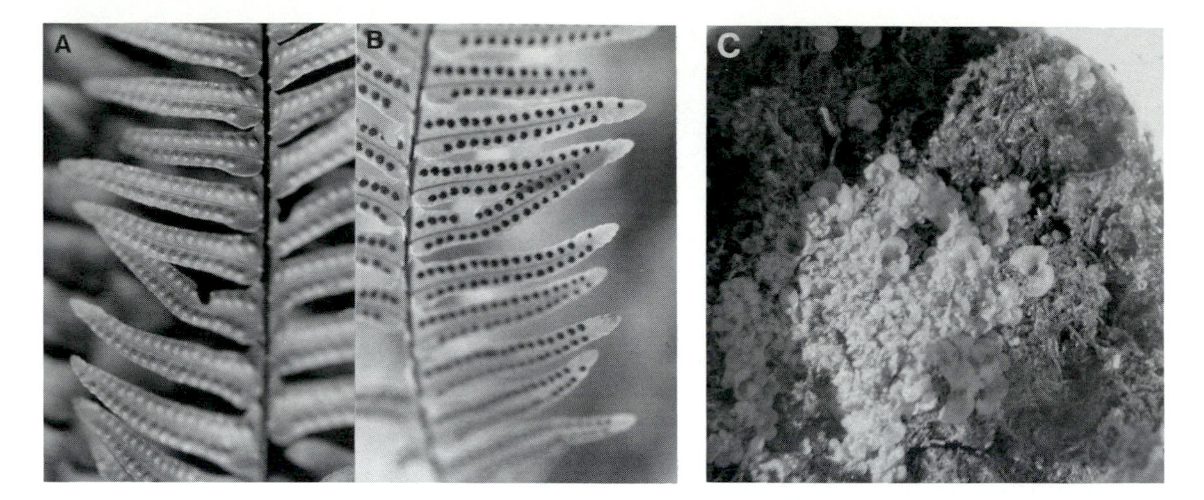

Figure 16–2 (A) Fern frond, upper surface; (B) frond, lower surface, same leaf showing sori with sporangia and spores; (C) gametophytes of the same. Each sorus contains numerous sporangia and thousands of spores.

tufts. An example is the popular house fern, the Boston fern. Tree ferns, which occur in most tropical regions, are modified basket ferns in which the vertical stem may be 3 m long or more.

The leaves generally develop by **circinate vernation** or the uncoiling of a helical bud containing a single leaf (Figure 16–1c). On the undersurface of some or all of the leaves are more or less globular structures called **sori** (singular, sorus) containing **sporangia** and spores. The spores have arisen by the process of meiosis and are hence meiospores; they germinate to produce small green gametophytes which produce the gametes that give rise to the next generation. Each gametophyte is a fleshy, independent plant, anchored to the soil by rhizoids, and consisting of hundreds of cells (Figure 16–2). Under laboratory conditions, it can live for months, producing several crops of gametes, but in nature it lives for only a few weeks.

Each leaf consists of a stalk, or petiole, and a blade which is often divided into many leaflets (Figure 16–3). Between the upper and lower epidermis of the blade, in most species, there is a layer of undifferentiated chlorenchymous tissue called mesophyll. In some species, however, the mesophyll is differentiated into a palisade layer and a spongy layer as in flowering plants. In a few species the leaves are only a single cell layer thick. The petiole contains xylem and phloem tissue and branches out into the veins of the leaves; the veins may be either open or closed and are usually dichotomously forked (Figure 16–1).

In some ferns, e.g., *Phyllitis* and *Pterozonium* species, the leaves are simple with long petioles and kidney-shaped blades. In others, the blades are more or less lance-shaped, as in species of *Botrychium* and *Ophioglossum*. In *Marsilea*, the blades are palmately divided and superficially resemble clover leaves. However, the majority of ferns have pinnately compound leaves (Figure 16–4).

The stele in both petiole and veins is typically a horseshoe-shaped protostele. In most ferns, all of the leaves are very much alike and spores are produced on the lower surface of vegetative leaves. In some ferns, however, there is a differentiation into two kinds of leaves, and spores

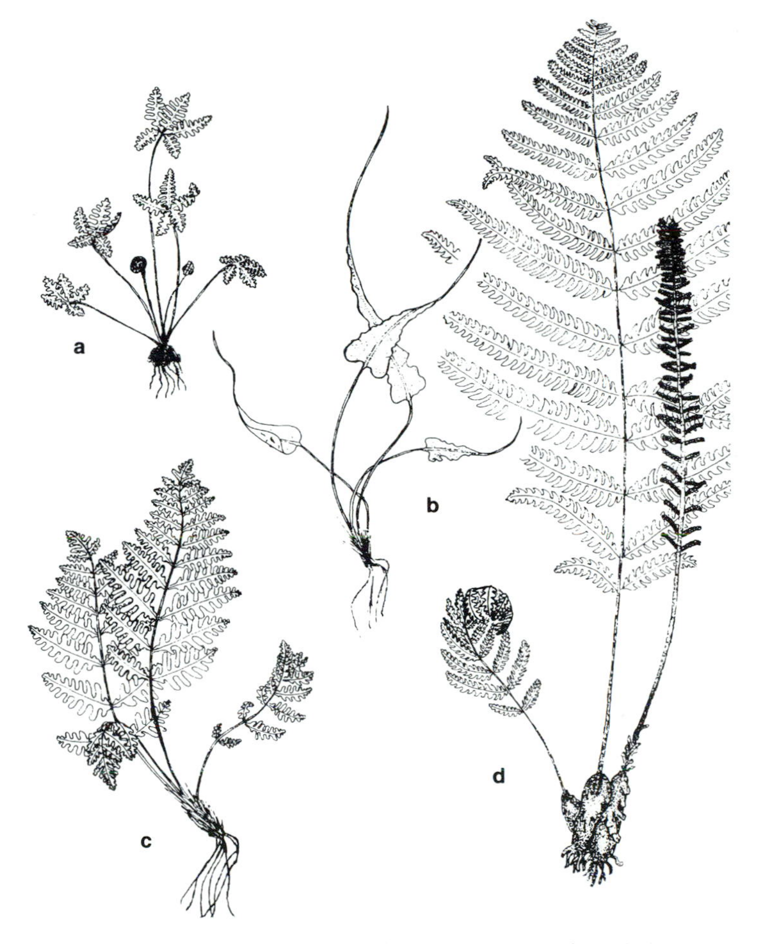

Figure 16–3 Filicopsida: (a) *Notholaena stanleya,* (b) *Camptosorus rhizophyllus,* (c) *Woodsia scopulina,* (d) *Pteretis pennsylvanicus.*

are produced only on specialized **sporophylls.** In the Marsileaceae and Salviniaceae, the spores are born in **sporocarps,** fleshy, parenchymatous structures with spores inside them.

Sporangia are usually contained in sori (Figure 16-2). Surrounding the sorus there is generally a ring of thicker-walled cells called an annulus; dehiscence of the sorus occurs when the annulus ruptures, discharging the spores with some force. Most ferns are homosporous, but *Salvinia, Marsilea* and their relatives are heterosporous. The sporocarps of *Marsilea* are of special interest because of their great longevity; spores kept in laboratories for over 50 years have germinated into vigorous gametophytes within a matter of a few hours. Raghavan (1992) and others are investigating the causes of dormancy and germination in fern species.

Fern gametophytes or **prothalli** are of three types (Figure 16-2C): the green cordate type, the green filamentous type, and the colorless saprophytic type. The saprophytic type is often associated with genera that are anatomically simple, the filamentous type seems to be associated

Figure 16–4 Filicopsida: (a) *Marsilea vestita,* (b) *Asplenium marianum,* (c) *Anoclea sensibilis,* (d) *Mattheucia struthiopteris.*

with genera traditionally regarded as advanced. In the cordate type, which is the most common, archegonia are located near the indentation of the prothallus and consist of a chamber in which eggs are produced and a neck through which sperm can enter. Antheridia are located on the undersurface of the prothallus among the rhizoids which anchor it to the soil. The somewhat coiled, elongate sperm are motile by means of a tuft of flagella at each end.

Following fertilization of the egg by a sperm, the zygote divides twice in parallel planes to form a four-celled embryo. Subsequent mitotic divisions are less regular and the embryo soon shows differentiation into embryonic organs: a foot, which absorbs food from the gametophyte; a hypocotyl and radicle, which soon anchor the young sporeling to the soil, and an epicotyl, which develops into the rhizome or stem. To begin with, the embryo is enclosed within the archegonium and obtains its nourishment there from the mother gametophyte; however, it soon is too large to be contained there longer, and as it develops photosynthetic mechanisms, it is able to live independently. In the laboratory, young sporophytes can be removed from the gametophyte at this stage and can be induced to live independently; when this is done, the gametophyte will continue to function, producing eggs and sperm for several years. In nature, however, the young sporophyte generally remains attached to the gametophyte and gradually destroys it. The life cycle of a typical fern, as it occurs in nature, is illustrated in Figure 16–5).

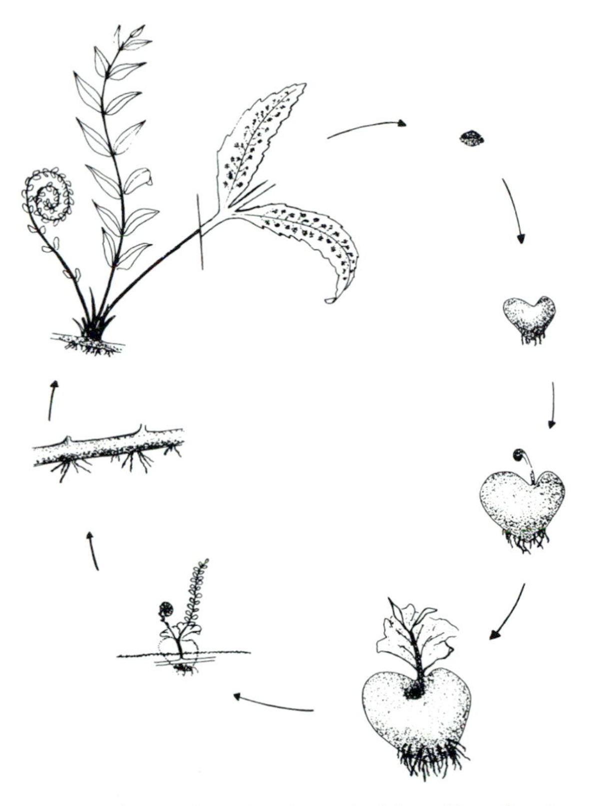

Figure 16–5 Life cycle of a typical fern. Note the circinnate vernation at upper left on the mature plant shedding spores and follow through the stages of gametophyte development, fertilization, and embryo development.

In tree ferns, the verical stem is elongated into a trunk that is frequently 5 m tall and may approach 20 m in length, and up to 50 cm in diameter. Since there is no cambium in ferns, the increase in girth is achieved by development of additional vascular bundles, both inside and outside the original ring of bundles. The trunk of a tree is thus a massive accumulation of sclerenchyma and other tissues; since meristematic activity is greatest in young tissue, the trunk often increases in diameter from the base upward. At the apex, a large bunch of leaves, which in some species are 6 to 7 m long, arise; along the trunk, the scars or previous years' leaves can be seen.

Protosteles, siphonosteles, and dictyosteles all occur in the stems of ferns. The xylem usually consist mainly of scalariform tracheids and develops in a mesarch pattern. The phloem is usually outside the xylem and forms a solid ring around each individual bundle of xylem or it may form a continuous band around all of the bundles together.

In protostelic stems, the size of the stele varies greatly from one species to another; generally, the stele becomes increasingly complex in outline as the size of the stele increases. Outside the phloem in nearly all fern stems, as well as in roots, there is a layer of meristematic tissue, the pericycle, forming the outermost layer of stelar tissues, from which secondary and adventitious roots arise. Outside this, and surrounding the entire stele, is a layer of suberized cells, the endodermis, which prevents lateral movement of gases, water, salts, and sugars, between the stele and the cortex. Cortical cells in ferns are frequently highly lignified and the resulting sclerenchymatous tissue adds strength to the stems of vine-like and tree-like species.

In many ferns, a medullary tissue (pith) forms at the center of the stem, and the vascular tissues form outside this pith either as a continuous ring or as a divided ring of individual bundles. Such a pattern provides greater strength for the same amount of xylem and phloem, and while it is genetically more complex, it certainly must have considerable adaptive advantage.

Anchorage in ferns is accomplished primarily by rhizomes or underground stems with roots occurring at the nodes. Gemma et al. (1992) found mycorrhyzae in 74% of the Hawaiian ferns they examined: 86% in epilithic species, 83% in terricolous species, 55% in epiphytic species, and none in aquatic species. The highest frequency was in the Dicksoniaceae, Dennstaedtiaceae, and Lindsaeaceae; there were none found in the Polypodiaceae.

Roots are more uniform in structure than stems and resemble the roots of angiosperms and gymnosperms except that they lack a cambium and therefore, like the stems, have no secondary xylem or phloem. Lacking secondary tissue, they remain small and fibrous, and they are usually very numerous. The steles are always protosteles and are usually diarch; however, steles vary from monarch to hexarch among the many species of ferns. The cells of the inner cortex are often lignified. In the tree ferns (Dicksoniaceae and Cyathaceae), numerous adventitious roots intertwine to form a massive, fibrous structure that adds strength and supportive power to the anchorage system and often results in the basal portion of the trunk being considerably swollen.

Phylogeny and Classification

There are three groups of plants that can be included in the Class Filicopsida: the preferns, the ferns, and the progymnosperms. The first and last are now extinct. The progymnosperms were transitional between the ferns and the seed plants and could also be classified with the gymnosperms. The preferns and progymnosperms originated in upper Devonian times and became extinct during the Permian; the ferns originated somewhat later and are still a meaningful and important group of plants consisting of 5 orders, 18 families, and almost 9000 species (Figure 16-6).

Characteristics of the nine orders of Filicopsida are summarized in Table 16-1. Their proposed relationships are shown in Figure 16-6. The fossil record suggests that they probably evolved from psilophyte stock during the Devonian period. Protopteridiales resemble both the psilophytes and the ferns and are probably the earliest plants that can be classified as Filicopsida. Of the nine orders included in Table 16-1, four have been extinct since the Permian. The Cycadofilicales or seed ferns, which were once believed to be ferns but are now classified as gymnosperms, were also very abundant during the Carboniferous and Permian periods; they became extinct during the Mesozoic era.

Devonian fossils of ancient preferns, especially Protopteridiales and Coenopteridales, resemble the Psilopsida in many ways; however, the leaf structure is similar to that of modern Filicales and is considerably more complex than any of the Psilopsida. Prefern leaves differ from leaves of modern ferns in two significant ways: (1) the leaflets in the prefern fossils branch out from the rachis in several planes rather than being confined to a single plane as are the leaves of modern ferns and seed plants, and (2) there is little or no anatomical differentiation between stem and leaf of the preferns. In the preferns, stems were branched in all three dimensions and then the apex of each branch was somewhat flattened into a blade-like structure. Spores were produced at the apex of the terminal branches.

The earliest prefern fossil stems are protostelic with small, round steles; slightly later fossils have lobed steles. In many cases, the steles in the petioles and veins are identical in appearance to those in the stem. Late Devonian ferns, however, showed pronounced differentiation between stem steles and leaf (vein and petiole) steles, and some of the ferns of that period, which are now extinct, had active cambium tissue in the stems (Andrews 1963). The sporangia are large in these fossils and are located on the ultimate divisions of the bipinnately or tripinnately compound leaves, there is no annulus present; dehiscence was by a terminal pore or longitudinal slit.

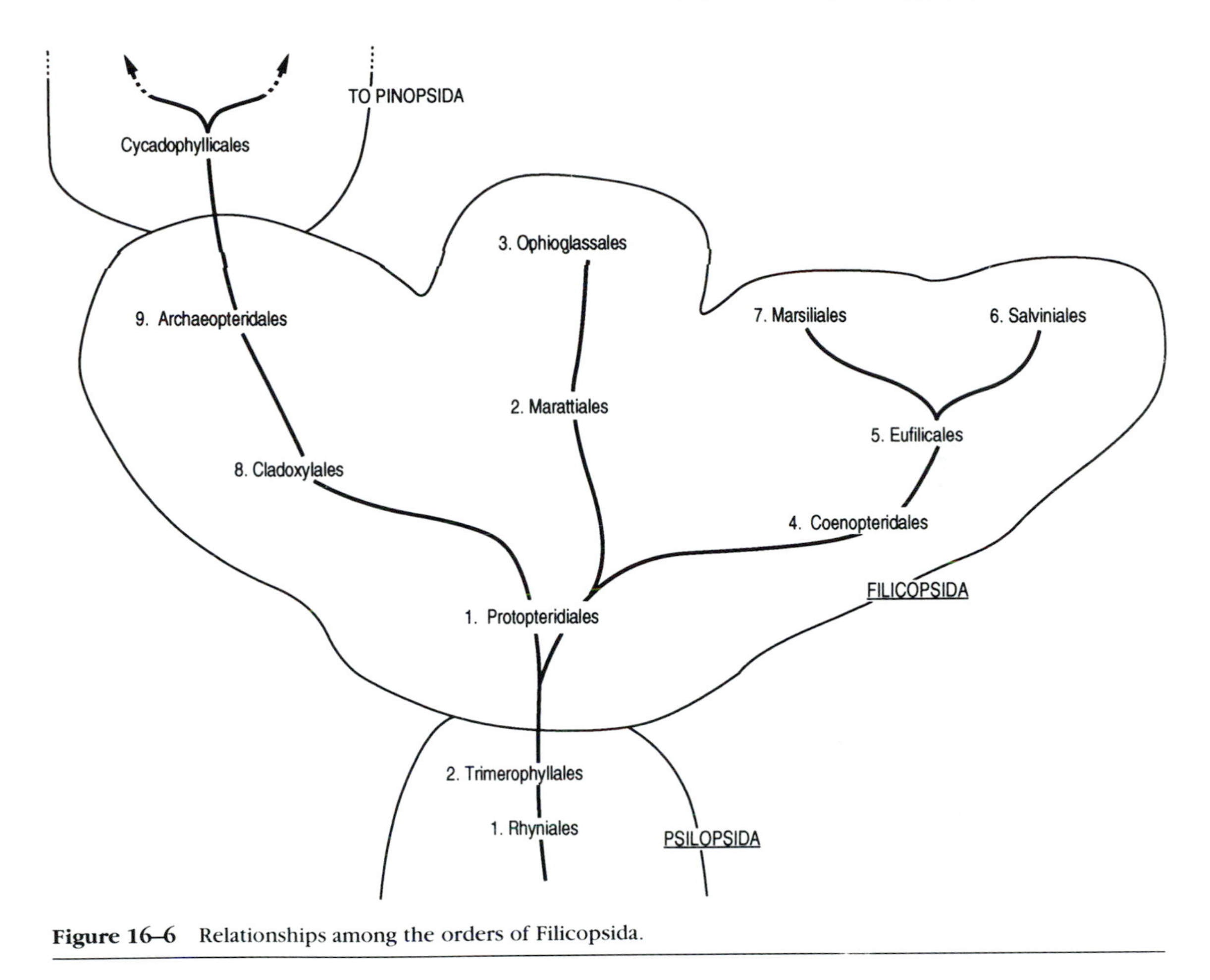

Figure 16–6 Relationships among the orders of Filicopsida.

The fossils of several species of ferns have been found in middle and upper Devonian formations. The oldest of these are classified in the Protopteridiales. Anatomically, and in some other ways, they are so similar to *Psilophyton* that some botanists prefer to include them in the Psilopsida; however, the flattened branches resemble fern fronds and are quite different from the stems or leaves of Psilophyton, suggesting a relationship to the true ferns, and so they are usually classified in the Filicopsida.

From the Protopteridiales, fern evolution appears to have progressed along three lines, one leading to the Cladoxylales and the progymnosperms (Archaeopteridales), one to the Marattiales and Ophioglossales, and one to the Coenopteridales, Filicales, Salviniales, and Marsileales. Fossils of ferns are found in all of the geological periods from upper Devonian to the present time; *Cyathea pinnata* (Figure 16–1D) thrived during late Cretaceous.

The Cladoxylales were evolving about the same time as the Protopteridiales according to the fossil record, and became extinct during the Carboniferous. The arrangement of the vascular tissues is unique, but suggests a relationship to the Protopteridiales, although some

TABLE 16-1

DIVERSITY AMONG FERNS: MORPHOLOGICAL CHARACTERISTICS OF THE ORDERS OF FILICOPSIDA

Orders (Fam–Gen–Spp)	Leaf and Stem Characteristics	Sporangia and Other Characteristics	Representative Genera
Protopteridiales (extinct)	Leaves 3-dimensional, blade dichotomously lobed	Sporangia terminal, no annulus, apical pore present, spores discharged through apical pore	*Svalbardia, Protopteridium, Aneurophyton*
Marattiales (1-7-200)	Vernation circinnate, stipules present, attach to stem by a swollen node	Eusporangiate development, sporangia large, sessile, dehisce from a longitudinal slit	*Danaea, Marattia, Christensenia*
Ophioglossales (1-3-80)	Terrestrial or epiphytic, range in size from 8 cm to over 2 m	Eusporangiate development, large sessile sporangia; xylem endarch or mesarch	*Ophioglossum, Botrychium, Helminthostachys*
Eufilicales (10-197-8450)	Large, dichotomously veined; mostly herbaceous but 2 orders are arborescent	Leptosporangiate development, four major types of sporangia; protosteles, siphonosteles, and dictyosteles all occur in stems of this order	*Camptosorus, Asplenium, Matteucia, Onoclea, Pteris, Notholaena, Woodsia, Schizaea, Cyathea, Osmunda Anemia, Oligocarpia, Matonia*
Salviniales (1-2-18)	Rootless aquatics or having a single root; stems delicate or wanting; vernation is not circinnate	Leptosporangiate, heterosporous, sporocarps represent single sorus, sporocarp wall is a single indusium; male sporocarps large, round, female sporocarps smaller, oval	*Azolla, Salvinia*
Marsileales (1-3-80)	Aquatics in streams or marshes, leaves fanshaped, sometimes clover-shaped	Leptosporangiate, heterosporous, petioles arise from horizontal rhizomes, rhizoids anchor plants to the substrate	*Marsilea, Pilularia, Regnellidium*

Note: The number of families (Fam), genera (Gen), and species (Spp.) in each order is indicated in the first column.

botanists place these plants in a separate class by themselves. The Cladoxylales probably gave rise to the gymnosperms; however, the evidence is not complete.

The Marattiales and Ophioglossales resemble each other in many ways and are considered by nearly all botanists to be closesly related. The Marattiales are an ancient group and their fossils are found in mid-Carboniferous and all later formations. The Ophioglossales, on the other hand, have left a fossil record extending only to mid-Tertiary, but because of some similarities to Psiliopsida in the structure of the sporangia, along with their small leaves and some other simple features, are generally considered to be a primitive group. Bold (1967) has suggested that the apparent simplicity of the Ophioglossales could be the result of reduction rather than an indication of primitiveness. The stele is not simple, compared with *Psilotum* and many other vascular plants, in either the Ophioglossales or the Marattiales. In contrast to the Psilopsida, the Ophioglossales possess cambium tissue and there are thin rays present in the xylem. The subterranean gametophytes are saprophytic, a characteristic that could have evolved from the autotrophic gametophytes of the Marattiales, while the reverse is less likely. Because the Marattiales are a very ancient group, obviously related to the Ophioglossales, and because the Ophioglossales possess some advanced characteristics, including cambium tissue and a rather unique type of xylem having large tracheids with bordered pits instead of scalariform markings as in other ferns, it is suggested that the Ophioglossales have evolved relatively recently from ancestors similar to modern Marattiales. Many will object to this hypothesis and will prefer to regard the Ophiglossales as an ancient group that evolved during the Carboniferous or earlier, without leaving a record of its immediate ancestors, and then survived for hundreds of millions of years in such small numbers that few if any fossils were formed, for none have as yet been found.

The Marattiales are intermediate between the Ophioglossales and Eufilicales in many respects. For example, they have a gametophyte that is autotrophic, as do the Eufilicales, but it has an endotrophic fungus associated with it as do the Ophioglossales. They are also intermediate in leaf and sporangia characteristics. Formerly it was often taught that a simple line of evolution led from the Psilophytales to the Ophioglossales, and then to the Marattiales, and culminated in the Filicales. Few botanists accept this hypothesis today.

Fossils of the third group that evolved from the ancient preferns, the Coenopteridales, have been found in late Devonian formations, and like the Protopteridiales, show a number of distinct similarities to the Psilopsida. In addition to their fern-like habit, which they share with the Protopteridiales, they have sporangia with annuli. They became extinct during the Permian, long after the first Eufilicales (also simply known as Filicales) had evolved.

With 8800 extant species, the Eufilicales is by far the largest of the five orders of modern ferns. Lines of relationship among the 14 families are suggested in Figure 16-7. Fossils almost identical to modern members of the three families at the base of the "family tree" have been found in Carboniferous formations; the other families appear to be of more recent origin. The Marsileales appear to have evolved from Schizacean ancestors during the Jurassic Period and the Salviniales from Hydropteridaceous ancestors duing the Cretaceous.

Representative Genera and Species

Favorite and well-known species of ferns include the maidenhair ferns, *Adiantum pedatum* and *A. capillusveneris* (Figure 16-1B), the moonwort, *Botrychium lunaria,* the grapeferns,

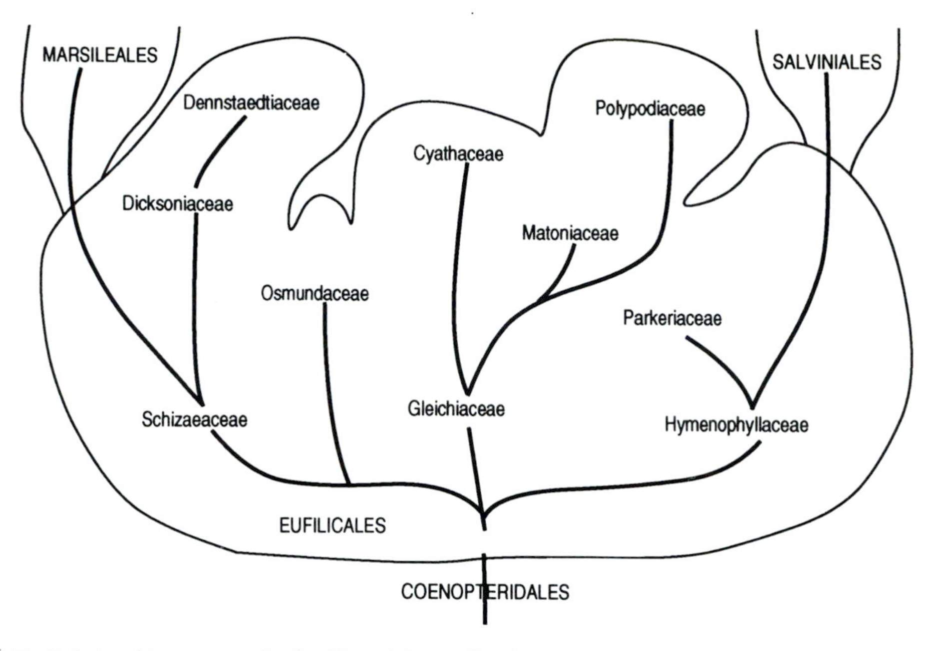

Figure 16–7 Relationships among the families of the Eufilicales.

Botrychium spp., the holly fern, *Polystichum lonchitis,* the male shield fern, *Drypteris filix-mas,* the lady fern, *Athyrium filix-foemina,* the pepperwort, *Marsilea vestita* (Figure 16–4), the bracken fern, *Pteridium aquilinum,* and the Boston fern, *Nephrolepsis exalta* var. *bostonensis.* Some of the characteristics of these and other ferns are given in Table 16–2.

The little grape fern, *Botrychium simplex,* is a small fern up to 15 cm tall, which grows in dry woods and open slopes throughout much of North America. It can be recognized by the two kinds of leaflets borne on a single leaf, the lower leaflet green, dichotomously veined, and photosynthetic, with no spores on it, and the upper leaflet divided into very small, linear segments crowned by large, cup-shaped sporangia. Other species of *Botrychium* grow in rich humus soil in woods and thickets, in wet meadows, and in marshes; some of them are considerably larger than *B. simplex.* All are characterized by the two kinds of leaflets on each petiole, one vegetative and sterile, the other consisting of grape-like clusters of sporangia that give the genus its common and scientific names.

The bracken fern or common brake, *Pteridium aquilinum,* is a widespread species easily recognized by its doubly compound leaves, which are roughly triangular in outline, and its tolerance of open, sunny areas. It is commonly found in crevices of cliffs and ledges and in the talus slopes beneath them, often near springs or seepage areas. Ecologically, it is a **euryhydric** species, tolerating both wet and dry habitats, and because of its widespread and abundant nature, coupled with its large size, is popular with campers for bedding material to place beneath their sleeping bags.

Marsilea vestita, the hairy pepperwort, is a widespread species that occurs in two forms, as an aquatic with long petioles and flat blades divided into three or four segments floating on the surface of ponds and lakes or as an aerial plant rooted in mud and having rather short

TABLE 16–2

CHARACTERISTICS OF THE FAMILIES OF EUFILICALES

Family (Gen-Spp)	Habit; Leaf and Vein Characteristics	Habitat Preference	Spores and Sporangia	Representative Genera
Gleicheniaceae (1–130)	Leaves dimorphic, stems dichotomously branched	Terrestrial, tropical	Sporangia below veins or at vein tips	*Gleichenia*
Cyathaceae (3–700)	Trees and shrubs, venation open	Temperate	Annulus oblique, dehiscense transverse	*Alsophila, Hemitelia, Cyathea*
Matoniaceae (2–3)	Herbaceous, petioles long and forked	Terrestrial, tropics	Sori on either side of the midvein	*Phanerosorus, Matonia*
Polypopodiaceae (170–7000)	Herbaceous, leaves mostly monomorphic with reticulate venation	Cosmopolitan, desert to rain forest	Sori in 2 rows on either side of the midrib	*Dipteris, Polypodium, Clathropteris, Pleopeltis, Platycerium, Woodsia*
Osmundaceae (3–20)	Leaves pinnately compound, dimorphic; terrestrial to subaquatic	Cosmopolitan, southern hemisphere	Large sporangia globose to pyriform	*Todea, Leptopteris, Osmunda*
Schizeaceae (4–160)	Leaves dimorphic, grass-like to delicate vines	Tropical forests	Sori parallel to the margin, in from it	*Schizaea, Lygodium, Mohria, Anemia*
Dicksoniaceae (5–30)	Tree ferns, stem bases covered with dense wool; amphiphloic siphonstele	Mostly tropical; southern hemisphere	Marginal sori at tip of veins	*Cibotium, Dicksonia, Coniopteris, Culcita, Thyrsopteris*
Dennstaedtiaceae (79–2,000)	Highly diverse, simple to compound leaves, herbaceous	North temperate and pantropic	Marginal sori by veins in 2 rows	*Lindsaya, Blechnum, Davallia, Ctenitis*
Hymenophyllaceae (2–40)	Filmy leaves 1 cell layer thick	Very widespread warm, shady, moist habitat	Sori terminal along leaf margins	*Trichomanes, Hydrophyllum*
Parkeriaceae (1–7)	Aquatic annuals	Aquatic, tropical and subtropical	Annulus thick-celled	*Ceratopteris*

Note: the number of genera and species in each family is indicated in column 1.

petioles and small blades. Superficially it resembles a clover, but closer examination reveals the dichotomous venation and circinate vernation that characterize most ferns. Spores are produced in pod-like sporocarps located near the base of the petiole and originates from either the petiole or the rhizome. Morphologically, the sporocarp is a modified pair of leaflets, folded together, with two rows of sori inside it. Two kinds of sporocarps develop on the same plant, one producing megaspores which germinate to produce female gametophytes and the other producing microspores which produce male gametophytes upon germination.

Ecological Significance

The ferns are of greatest significance ecologically in tropical forests; however, numerous species occur in the temperate zone and a few species are found in arctic habitats. Ecologically most ferns are hydric and mesic subordinates in forest communities. A few species favor sunny locations and a small number are inhabitants of very dry areas.

Xerophytic ferns are able to tolerate conditions of prolonged and severe drought by becoming dormant during dry periods and recovering rapidly when moisture returns. Their drought resistance is aided by a thick, waxy epidermis covering the petioles and the very small leaflets. *Lepicystus incanum* is a tropical fern which can lie for weeks without moisture, all dried up under the hot sun, but when moistened, it brightens up and begins growing again. In the deserts of southern Utah, Nevada, and eastern California, *Notholaena jonesii* grows in the crevices of very dry rocks where its ecological niche is apparently identical, or nearly so, to that of the desert mosses. The leaflets are small, few and fleshy, as is typical of xerophytes (Flowers 1944). Some of the lip ferns, such as *Cheilanthes covillei* and *C. feei,* grow in the deserts of California in dry rocky crevices and can be recognized by their bead-like leaf segments. None of these species is abundant. A study of 18 desert area ecosystems, in which a few mesic ferns were the only Filicopsida observed, failed to reveal any that were important to total ecosystem productivity (Pearson 1966).

At the other extreme, many ferns are either aquatic or semiaquatic in their habitat requirements. *Azolla caroliniana,* the mosquito fern, grows with its roots floating in the water, and superficially resembles duckweed. Although relatively rare, it may locally be as abundant as duckweed from which it is readily distinguished by its two lobed leaves of which the upper lobe is inflated with large air cavities. Another aquatic species is the pepperwort or water clover, *Marsilia vestita.*

Epiphytic ferns possess interesting adaptations which enable them to survive the xeric conditions of the habitats in which they live. *Polypodium brunei,* for example, produces pouch-like urns at the base of the stem where it comes in direct contact with the host species. When it rains, water flowing down the stem of the host or falling directly on the pouches is accumulated there. These urns have been described as modified stems. In other species of epiphytic ferns, such as *Asplenium nidis,* the "nest-habit" is useful for water conservation. The leaves are aggregated in dense tufts with the roots immediately below, together they collect litter and humus which provides excellent water-holding capacity. In the arid habitat in which such ferns live, earthworms may be found burrowing in the accumulated humus (Perry 1980).

Many ferns form mycorrhizal relationships with Basidiomycopsida. Mycorrhizae are especially efficient water-obtaining structures consisting of the roots of the fern modified by the mycelium of the club fungus. Some of the mycorrhizal basidiomycetes live in the rhizoids of the gametophytes as well as in the roots of the sporophytes. In the Ophioglossaceae, the underground gametophytes are completely dependent upon basidiomycetes for food as well as water.

Economic Significance

While ferns are not used extensively in industry, they are of economic importance in some rather specialized ways, especially in naturepathic medicine, fad foods, some gourmet foods, and as ornamentals. Some ferns having long hairs on their leaves are dried and used for packing material. In the tropics, the large tree ferns are used locally for lumber.

The licorice-flavored rhizomes of *Polypodium glycerrhiza* are used commercially to a limited extent in flavoring pipe tobacco. The juices extracted from rhizomes of *Ophioglossum vulgatum* and *Botrychium lunaria* were used by early California settlers to curb vomiting, stop nosebleeds, and treat ulcers. The now seldom-used worming drug, aspidium, is obtained from *Dryopteris filix-mas; Athyrium filix-femina* and *Pteridium aquilineum* are also reported to be useful in controlling worms in both livestock and people. The old foliage of all three species, but especially *D. filix-mas,* has been reported to be poisonous when used in large quantity. An extract of *P. aquilinum* is used as an astringent; this versatile fern was also used by western settlers to thatch their cabins. *Adiantum pedatum* was used by the Indians of California in making an ointment for treating inflammation of the skin. Medieval herbalists, following the "doctrine of signatures" and imagining a morphological similarity between *Asplenium* and the human spleen, used extracts from this fern to treat disorders of the spleen such as malaria.

Marsilea vestita has been used in some places as horse fodder; it is also considered a weed in areas where its growth gets out of control. Some species of *Pellaea* are poisonous to livestock; sheep frequently eat the tender young shoots of *Pellaea* on western rangeland, resulting in some loss to ranchers.

A South American native, *Salvinia molesta,* has been introduced into Australia, Africa, India, and islands of the Indian Ocean and has become a serious weed. It has been seriously considered as a source of livestock fodder inasmuch as it is reported to produce over 100 tons of dry matter per hectare annually. However, it is high in lignin and in tannins which may interfere with its value as feed.

Blue-green algae live symbiotically with *Azolla;* Talley and Rains (1980) have reported that China has over 3 million acres of rice fields planted with *Azolla* containing nitrogen-fixing species of *Anabaena* as symbionts.

Several species of fern are cultivated as ornamentals (Figures 16–1 to 16–4); among them may be mentioned the giant chain fern, *Woodwardia fimbrianta,* and several rock garden favorites: the holly fern, *Polystichum lonchitis,* and species of *Adiantum,* and *Woodsia.* The sword fern, *Polystichum minutum,* is widely used in making floral arrangements and wreaths.

Research Potential

Because of their alternation of independent, free-living generations, the ferns have been of special significance in studying the relative influence of genes and environment on both the gametophyte and sporophyte generations. Closely related research has centered on the mechanisms of morphogenesis. Fertilized zygotes, when dissected from the archegonium, apparently do not develop into normal embryos unless a hormone capable of stimulating differentiation is added the medium in which the gametophytes are growing.

The ferns have been considered potential suppliers of drugs, and thousands of species have been tested for drug content; however, few effective drugs have as yet been found in the Filicopsida.

The great amount of variation in arrangement of vascular tissues found in closely related species of ferns provides an opportunity to study the adaptive advantages of different kinds of steles and also provide insight into the evolution of steles. Most ferns are ecological subordinates, growing in the shade of taller plants, and have a potential for providing more information of the ecological niches of subordinates.

SUGGESTED READING

D. L. Jones' *Encyclopedia of Ferns* (Timber Press, 1987) includes sections on fern morphology and life cycles, anatomy, phytogeography, uses in medicine and as food, cultivation, and propagation, along with descriptions of each taxonomic group of ferns. It is highly recommended as a source on ferns.

D. B. Lellinger's (Smithsonian Institution Press, 1985) *Field Manual of Ferns and Fern Allies* has good keys and excellent color photographs.

There are many state and regional fern floras that are good. Among the best is Seville Flowers' "Ferns of Utah" (*Bull. Univ. Utah* 35(7); 1–87), which gives good coverage of the ferns of the Intermountain States and has excellent, easy-to-use keys. Some others are *Ferns of Wisconsin* (Tyron et al. 1953), *Ferns of California* (Grillos 1966), and *Ferns of the Northwest* (Frye 1934).

Clara Hires' *Spores: Ferns* (Mistaire Laboratories, Millburn, NJ, 1965) is of great value for botanists and ecologists working with microfossils.

STUDENT EXERCISES

1. What are the two morphological types of growth form in the Filicopsida? Give an example of a species of each type.
2. What kinds of steles are found in ferns?
3. To which of the 29 classes of plants do the "seed ferns" belong?
4. Compare the roots of ferns and the roots of some common angiosperms. In what ways are they alike? In what ways do they differ from each other?
5. What kinds of cells occur in the phloem of ferns? How does this compare with the phloem of *Selaginella* or other lycopods? How does it compare with the phloem of a flowering plant like the lilac (*Syringa*) or the buttercup (*Ranunculus*)?
6. A root tip squash is prepared, using the Feulgen technique, of a certain species of fern, and it is found to have 18 chromosomes in each metaphase nucleus. How many chromosomes should there be in the nucleus from a cell in a leaf? In the nucleus of a cell in the phloem? In the nucleus of a cell in a leaf bud? In the nucleus of a cell in the

prothallus? In the nucleus of a cell in a rhizome? In a sperm cell? In an egg cell prior to fertilization? In a zygote?

7. What kind of flagella would you expect to find on a sperm cell from a fern?

8. Pennsylvania coal is found largely in Carboniferous formations: how many years ago were these deposits laid down? Would you expect to find fern fossils in them? Would you expect to find angiosperm fossils in them? Would you expect to find diatom fossils in them?

17 PLANTS WITH SEEDS: THE GYMNOSPERMS

In 1892, the discovery of a fern-like Permian fossil with seeds on some of its leaves caused considerable excitement among paleontologists and other botanists. This interesting fossil was soon widely known as a "seed fern." Many biologists speculated that here, at last, was the missing link connecting the seed plants to the ferns.

A diversity of "seed ferns" was soon discovered and neatly classified among the extant families of Filicopsida. Everyone who has visited a dinosaur museum has seen paintings depicting them as a major part of the marshland vegetation of the Jurassic and Cretaceous periods, where they provided food for the huge herbivorous reptiles that thrived through most of the Mesozoic Era (Figure 17–1). But environmental changes and competition from organisms better adapted to the new ecosystems resulted in extinction of both the "seed ferns" and the dinosaurs at about the beginning of the Coenozoic.

Careful examination of the stems sometimes associated with these fossil leaves revealed the presence of a cambium and of tracheids similar to those of modern pines. Cambium tissue is typical of gymnosperms, but not Filicopsida. A gymnosperm-like embryo was embedded in the gametophyte tissue. The so-called "seed ferns" are now recognized to be the earliest of the gymnosperms and are classified in the Pinopsida, not the Filicopsida. The early speculations that the so-called "seed ferns" represent the missing link between the Filicopsida and the Pinopsida is essentially correct, but the classification of these interesting plants in the Filicopsida was not. In the 1960s, a fossil seed found in a Devonian formation dated the origin of the Pinopsida to at least 350 million years ago.

The foliage of the most primitive of these early gymnosperms differs from that of ferns in that some of the leaflets are very narrow with their bases fused to form an **integument** enclosing a single megaspore. In the ferns, many megaspores are produced on each leaflet (Figure 16–5); when the spores are shed, each spore develops on damp soil into an independent gametophyte. The "seed fern" megaspore, on the other hand, developed into a gametophyte while still attached to the leaf. Although two or more archegonia were present on the female gametophyte, only one egg was generally fertilized. The resulting zygote developed into an embryo enclosed within the integument, still attached, therefore, to the leaf. In the foreground,

Figure 17–1 Artist's concept of the late Permian/early Triassic world. Aneurophyton at lower right, behind it a *Ginkgo* and a forerunner of the conifers; *Cordaites* behind it. A member of the Bennettitales, *Williamsonia sewardiana* is prominent near the left margin of the picture.

of Figure 17-1, Aneurophyton has fern-like foliage with the leaflets between the uppermost and lowest ones of some of the leaves modified as seeds.

The name "gymnosperm" means "having naked seeds" and is commonly applied to all seed plants in which the ovules are not enclosed in an ovary. Including species known only as fossils, there are about ten orders that are commonly called gymnosperms; all but four of these became extinct during the Permian and Cretaceous periods (Figure 17-2).

The four extant groups of gymnosperms are the conifers, ginkgoes, cycads, and gnetophytes. The first three evolved from fern-like stock during late Devonian and early Carboniferous time and have many features in common. They make up the class Pinopsida. Ecologically and economically, this is one of the most important of all plant classes. The fourth probably evolved from the Bennettitales (Figure 17-1, left foreground), an extinct order of gymnosperms having fern-like foliage, during the Cretaceous period, and is anatomically sufficiently different from other gymnosperms that it seems best to classify it as a separate class, the Gnetopsida.

Based upon extensive studies of fossils of both living and extinct vascular plants and upon detailed anatomical and embryological analyses of the major groups of gymnosperms, Krassilov

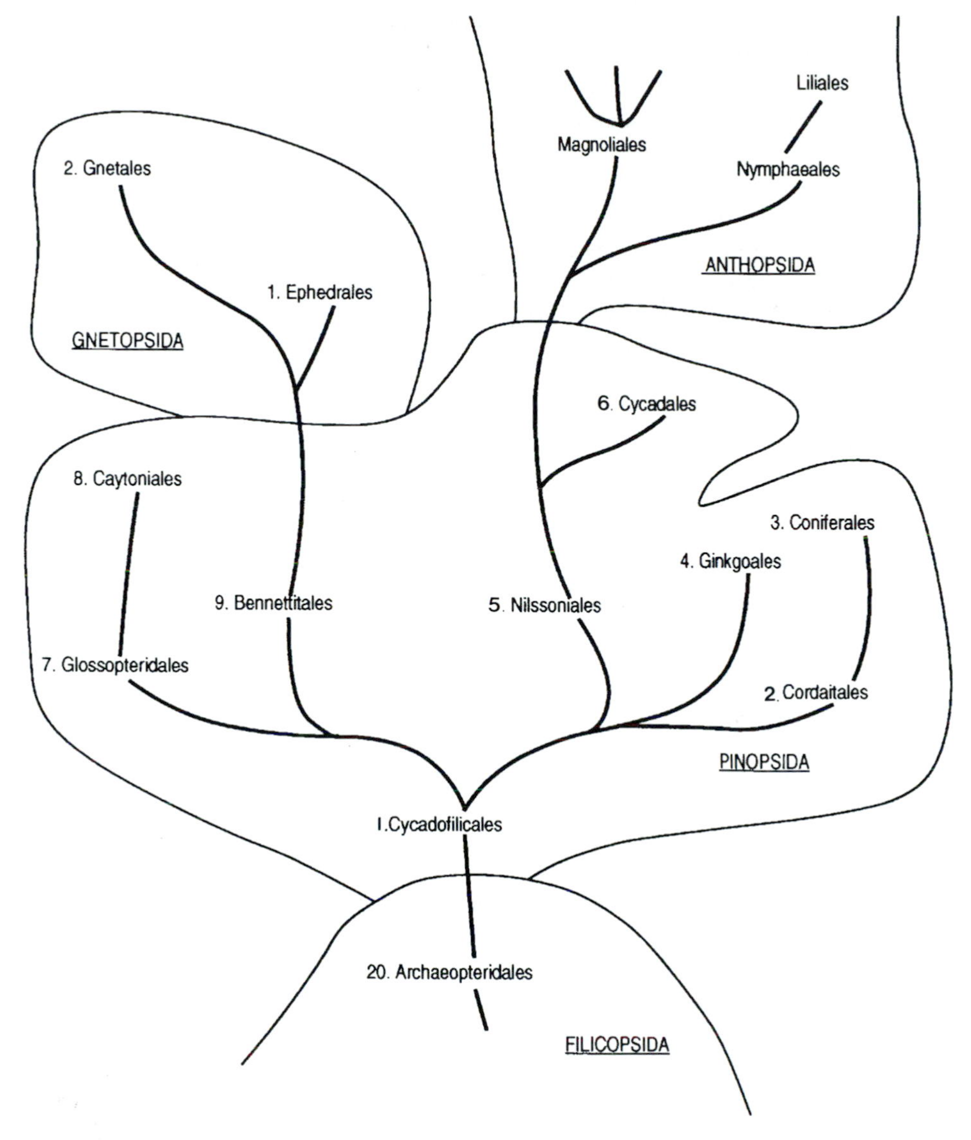

Figure 17–2 Relationships among the Pinopsida and Gnetopsida.

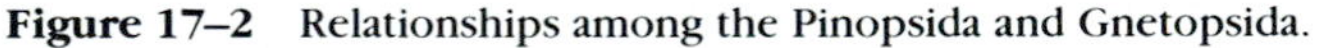

(1977) concluded that evolution of gymnosperms progressed along two main lines, originating with the Cycadofilicales, or "seed ferns," and leading to the Cordaitales and Nilssoniales and their relatives along one line and to the Glossopteridales, Caytoniales, Bennettitales, and their relatives along the other. In the first line, the arborescent habit prevailed, xylem was characterized by mesarch development and the presence of resin canals, and leaves were almost always alternate, often with rather high phyllotactic values. This line led to the Coniferales, Ginkgoales, and Cycadales among the gymnosperms and, by way of the Nilssoniales, to the angiosperms. Plants along the second line were and are mostly shrubs or vines, the xylem is characterized by endarch development, resin canals are never present, the leaves are whorled, and there are vessels in the xylem. This line led by way of the Bennettitales to the gnetophytes.

Opinions differ among botanists as to relationships within, as well as between, the two classes. Bold (1957) divided the gymnosperms into four divisions, each containing a single class, whereas Lawrence (1951) included all of the gymnosperms in a single class with four orders. Krassilov (1977), a "splitter," divided the class into 19 orders, including 15 extinct and 4 extant; Banks (1970), who is more of a "lumper," considered only three extinct orders. When the group is divided into more than two classes, the Cycadales are most logically and frequently separated from the Coniferales and Ginkgoales. The latter orders resemble each other in wood anatomy and other ways more so than either resembles the Cycadales or the Gnetales.

CLASS 5. PINOPSIDA

The Pinopsida (from *pinus,* Latin for the pine tree), also known as the Gymnospermeae (*gymno* = naked or bare, *sperma* = seed), is a large group of plants in number of individuals but not in number of species. It is second only to the Anthopsida among terrestrial plants in ecological importance as producers. Representatives of the class grow in all of the terrestrial biomes and nearly everyone is well acquainted with them.

Modern Pinopsida are included in 3 orders, 9 families, and 61 genera. An additional 5 orders are known only as fossils (Figure 17-2). Students do not ordinarily confuse the Pinopsida with any other group of plants; any woody plant with a simple pattern of leaf venation and seeds produced in cones is either a gymnosperm or a gnetophyte; gnetophyte leaves are produced in whorls, Pinopsida, as a rule, have spirally arranged leaves (Table 17-1). The two also differ in other ways, including pollen morphology (Figure 15-6), and stem and leaf anatomy.

On the basis of exhaustive studies conducted by Rudolf Florin (Banks 1970) over a period of several years, most botanists now believe that the best known and most abundant of the Pinopsida, the Coniferales, evolved from the Cordaitales during the Mississippian or Pennsylvanian periods. The origin of the other orders is less certain. Suggested relationships among the orders is illustrated in Figure 17-2.

General Morphological Characteristics

Pinopsida differ from essentially all ferns and fern allies in a number of respects: (1) possession of cambium and hence secondary xylem and phloem, (2) heterospory, (3) megasporangia enclosed in integuments which are provided with vascular strands and which form a chamber within which the gametophyte can develop and mature, (4) ability of the embryo to enter into a dormant period after reaching a size conducive to becoming an independent plant, (5) buds generally giving rise to shoots consisting of a stem and several leaves, rather than a single leaf, (6) greater uniformity in basic structure of the stele, and (7) greater root development, with the roots performing anchorage and storage functions usually performed by rhizomes in the more primitive vascular plants.

Like other Tracheophyta, Pinopsida produce embryos that develop within an archegonium. Unlike the more primitive vascular plants, the gymnosperm gametophyte develops within integuments and the young embryo, therefore, obtains its nourishment from the "grandmother"

TABLE 17-1

COMPARISON OF MORPHOLOGICAL CHARACTERISTICS OF THE PINOPSIDA, GNETOPSIDA, AND ANTHOPSIDA

Character	Pinopsida	Gnetopsida	Anthopsida
Ovules	Naked on cone scales	Naked on cone scales	Enclosed within an ovary
Female Gametophyte	Very large	Relatively large	Small
Male Gametophyte	Long, often branched	Small, unbranched	3-Nucleate, unbranched
Pollination	Wind; beetles in one genus	Wind	Insects, birds, bats, wind, etc.
Fertilization	Single	Single	Double
Integuments	Single	Single plus bracts	Double
Pollen grains	Vesiculate, and inaperturate	Polyplicate	Tricolpate, polyporate, etc.
Embryo	2–Many cotyledons, often 8–10	2 Cotyledons	1 or 2 Cotyledons
Xylem	Tracheids only	Tracheids and vessels	Vessels and fibers besides tracheids
Phyllotaxy	Spiral, rarely opposite or whorled	Opposite	Opposite, whorled, or spiral
Growth Form	Woody: trees and shrubs	Woody: shrubs and vines	Woody trees, vines and shrubs, herbaceous annuals and perennials
Ecological preference	Cold to warm	Xeric dominants	Xeric, hydric, mesic, both dominants and subordinants

sporophyte rather than from the mother gametophyte alone. When an embryo has developed to the point that it can survive independently, it usually loses moisture, becomes less active metabolically, and enters a period of dormancy. The embryo at that stage is called a seed.

The above characteristics apply equally well to most seed plants. Pinopsida differ from other seed plants, however, in several ways which are summarized in Table 17-1. Especially significant are (1) the presence of vessels and vessel elements in some of the other seed plants but only tracheids in Pinopsida; (2) a double integument around the seeds of Anthopsida and a single integument around the seeds of Pinopsida; (3) flowers and pollen grains adapted to insect pollination in angiosperms, pollen grains adapted to wind pollination in gymnosperms; and (4) leaves having several orders of vein branching in the Anthopsida but simple venation in the Pinopsida.

All Pinopsida are woody perennials, either trees or shrubs. In the Coniferales and Ginkgoales, abundant secondary xylem is produced from an active cambium. In the Cycadales, the cambium is relatively inactive and little secondary xylem is produced. The Ginkgoales and most Coniferales are characterized by stems that taper from a large base to a small apical shoot with whorls of branches revealing where each year's new growth began; hence, it is often easy to tell exactly how old a tree is by counting the whorls of branches on the main trunk. Cycadales, on the other hand, have stems that are almost cylindrical, or even tapered toward the base like tree ferns, and the stems are usually unbranched. A spirally arranged tuft of leaves crowns the stem and below this are old leaves and leaf scars.

Some of the tallest plants as well as some of the oldest are members of the Pinopsida. Many, including species in all three orders, are 25 to 35 m tall and over 1000 years old. Most people are aware of the great size and old age of the giant sequoias or "big trees" and the coast redwoods of California, the tallest of which attain a height of over 110 m, but few are aware of comparable size and age in some other gymnosperms. According to the *Guiness Book of Records,* the tallest tree ever measured was a Douglas fir in British Columbia that was 127 m tall. Some of the mountain sequoias, while not as tall as the coast redwoods and some Douglas fir, have attained the greatest mass of any living things: approximately 2500 tons, dry weight, including 700 tons of marketable saw timber plus roots, branches, and leaves, in a tree the size of the General Sherman tree: 83 m tall and 11.16 m in diameter.

A few giant sequoias are over 3500 years old; so are some Rocky Mountain red cedars (*Juniperus scopulorum,* also known as *J. virginianum* var. *scopulorum*), which are, however, much smaller. A specimen of Mexican baldcypress, *Taxodium mucronatum,* and a close relative of the sequoias, is over 12 m in diameter, 43 m tall, and almost 5000 years old. Some of the bristlecone pines of southern Nevada and adjacent areas have attained an age of over 6000 years.

The ability to live long and attain massive size is due in part to the ability of the trees to protect themselves from insects and fungi. Interspersed among the tracheids of the xylem are resin canals which produce aromatic substances that repel insects and prevent decay. The tracheids in the older xylem, or heartwood, also become filled with resin. The effectiveness of these substances in repelling insects is attested to by the use of some gymnosperm wood, especially the red cedar (*Juniperus virginianum* and *J. scopulorum*), as lining for mothproof clothes closets. In addition, the tough wood, deep root systems, and moderate to excellent drought resistance aid the gymnosperms in becoming both big and ancient.

Gymnosperm wood is commonly referred to as "soft wood." This term is often misleading: the wood of red cedar, for example, is harder than that of cottonwood, balsa, and many other "hardwoods"; but in general, gymnosperm wood is relatively soft, easily worked, and usually quite tough and strong. Because of these characteristics, together with its abundance and availability, the wood of conifers is used more than any other type of wood for construction, cabinet shelving, inexpensive furniture, and so on. The wood of ginkgo (maidenhair tree) has the same desirable characteristics as conifer wood, but is not abundant enough to be of any commercial importance. Cycad wood is used locally in tropical areas for construction. Since the stems often are several hundred years old and over 20 m tall, it provides excellent beams and rafters; however, the stems consist mostly of parenchymatous cortical and medullary tissues with little primary or secondary xylem and are therefore soft, rather weak, and of little commercial value.

The xylem of Pinopsida (Figure 17-3), like that of the four previous classes, is not differentiated into vessels and fibers but consists of undifferentiated tracheids (in addition to ray and wood parenchyma); these usually have circular pit markings rather than the spiral or scalariform thickenings of the ferns and fern allies. The phloem, although more developed than that of the previous classes, is rather simple compared to that of angiosperms, consisting only of sieve cells rather than sieve tube elements and companion cells as in the Anthopsida. The cells in the mature stem are arranged in an ectophloic siphonostele. Primary xylem of stems has endarch development, that of the leaf veins is mesarch, and that of the roots is exarch. This is different from the previous classes in which the steles of all organs tended to be more uniform within species but more variable among species. Gymnosperm leaves are sometimes dichotomously

Figure 17–3 Nonporous wood, typical of Pinopsida, yew above and ginkgo below. Left to right: transverse, radial, and tangential views. Compare with diffuse-porous and ring-porous wood in Chapter 18.

veined, as the leaves of ferns and fern allies usually are, or they may be single veined or parallel veined. In any event, the venation is open, that is, the veins do not form a closed network as in the angiosperms. In Coniferales, the veins are completely surrounded by mesophyll and other tissues and generally cannot be observed except by dissecting the leaves. The veins of 2- and 3-year-old leaves contain secondary xylem and phloem. Gymnosperm mesophyll is usually undifferentiated, consisting of spongy parenchyma alone.

Ginkgo biloba, the only extant species in the Ginkgoales, is a broadleaf deciduous tree often over 35 m tall (Figure 17–4). The leaves superficially resemble angiosperm leaves but are dichotomously veined. There are two kinds of leaves in Ginkgoales: those on the "long shoots," or main stems and main branches, are distinctly bilobed and the mesophyll is differentiated into a palisade layer and a spongy layer, while those on the "spurs" or "short shoots" are entire and the mesophyll is undifferentiated. Long shoots develop from apical buds and have many widely spaced leaves with long internodes between them; spurs develop from lateral buds and have a few closely spaced leaves with very short internodes. As in the conifers, there are two veins in the petiole; each of these forks into two veins which in turn fork into two veins each, and so on, giving the leaf of *Ginkgo biloba* its characteristic fan shape. In some of the extinct species of Ginkgoales, the dichotomy pattern was different, resulting in a diversity of leaf forms. Ginkgo stem anatomy is also similar to that of conifers and the wood has the same desirable working qualities that pine, spruce, and yew possess (Figure 17-3). Although only one species of Ginkgoales has survived the changes in climate and competition that began near the end of the Mesozoic Era, at least 16 genera flourished during the Jurassic and Cretaceous Periods.

In the Cycadales, the leaves are persistent, or evergreen, pinnately compound, and up to 6 m long (Figure 17-5). Venation is open, either dichotomous or parallel, with the veins usually surrounded by mesophyll tissue. In *Cycas,* there is only one vein per leaflet, in the other genera there are two or more. The leaves arise spirally at the apex of the stem by circinate vernation. Stomates are abundant, occur only on the undersurface of the leaves, and are sunken. Beneath the epidermis, which is heavily cutinized, there is a second protective tissue, the **hypodermis.** Beneath the hypodermis, the mesophyll is differentiated into a chlorophyll-rich palisade layer and starch-storing spongy layer. Because the leaves are large, pinnately compound, and appear to be parallel veined (even when dichotomously veined, because the dichotomies are limited to the base of the leaf), cycads are often mistaken by beginning botany students for palm trees. Careful examination of how the veins are arranged at the leaf bases should solve this problem. If flowers, fruit, or cones are present, diagnosis should be immediate. The pith of some cycads is used to prepare a starchy food called sago; cycads are therefore commonly called "sago palms."

Figure 17-6 compares two cycads with two conifers. Note how the conifers taper from base to top with branches and leaves arranged along the main trunk whereas the cycads taper from apex to base with leaves only at the apex. In their natural habitats (Figure 17-7), further diversity within each group is apparent.

The leaf traces in the Cycadales are unique among gymnosperms. Each leaf trace arises at a point some distance from the leaf base and passes through the cortex at an angle almost tangential to the stele so that in passing from its point of origin to the leaf base it partially girdles the stem. While this process of girdling is unknown in other gymnosperms and the lower vascular plants, it is common in many angiosperms, especially in the Monocotyledonae.

Some leaves in all gymnosperms are modified to produce pollen and ovules. This ability contributes significantly to their survival in environments too dry for ferns and fern allies. A seed is "a mature ovule, consisting of an embyronic plant together with a store of food, all surrounded by a protective coat." Because the seed is a complex, multicellular plant having differentiated organs and possessing physiological and protective adaptations which allow it to remain dormant until moisture conditions become favorable for growth, seed bearing plants are able to become established in much drier areas than most other plants can.

Leaves that are specialized to produce spores are called **sporophylls.** According to early interpretations, the cone scales of gymnosperms are **megasporophylls,** or female sporophylls. Actually, each cone scale is anatomically a small branch bearing several sporophylls (the

Figure 17–4 *Ginkgo biloba.* (A) young tree in late April on Ricks College campus; note that leaves are coming out on birch tree behind it; (B) shoots and buds from (A); (C) older tree, University of Utah campus, same time of year; (D) leaves and shoots from Ricks College herbarium specimen.

Figure 17–5 Cycads, South Africa.

integuments). The cone scales are thus only analogous to megasporophylls and are more properly called ovuliferous scales. Megasporangia arise as little blisters on the upper surface of the ovuliferous scales and gradually develop into a mass of cells called an **ovule.** Some of the cells in the ovule become spore mother cells and undergo meiosis. Enclosing each ovule is an **integument,** which is a modified leaf, or the actual megasporophyll, as near as we can tell. At the proximal end of the ovule, where it attaches to the cone scale, there is an opening in the integument called the **micropyle.** Pollen grains gain access to the ovule through the micropyle.

Following meiosis, one of the daughter cells becomes a megaspore; the other three disintegrate. In gymnosperms, the megaspore is not shed, but remains attached to the megasporophyll where it develops by mitosis into a female gametophyte. This is a relatively large, fleshy plant, completely dependent on the mother sporophyte for food and moisture which are obtained through the vascular system of the integument. The integument continues to grow as the gametophyte grows. At maturity, which takes several months to reach, the female gametophyte may be 15 to 20 mm long and 6 to 10 mm in diameter in some pines and *Ginkgo,* and up to 55 mm long in cycads. At maturity, there are two archegonia present, each containing an egg, and fertilization is possible. Before fertilization can take place, however, pollination must occur.

The **microsporophylls** are simpler in structure than the ovuliferous scales: whereas the latter are interpreted by most botanists as modified shoots and each integument as a modified leaf, male cone scales are microsporophylls and are interpreted simply as modified leaves. Following meiosis, the daughter cells become **pollen grains** (Figure 15-6C). Note the little wings on many of the pollen grains; pinopsida pollen is carried by wind and may be carried thousands of miles before being deposited on ovuliferous scales or other surfaces. This process is known as **pollination.** The pollen grains complete their development on the ovuliferous scales following pollination.

Figure 17–6 Two conifers (top) and two cycads (bottom): note how the trunks of the cycads taper from crown to base, just opposite the conifers.

A considerable number of pollen grains will generally be deposited on each cone scale, and some of them inevitably come in contact with a sticky liquid given off by the ovules. As the liquid dries, the pollen grains are drawn through the micropyle into the micropylar chamber where they germinate. The pollen tubes grow slowly, obtaining nourishment from the female cone, and develop into coarse threads which can be seen, in some species, with the unaided eye. These penetrate the sporangial tissue (nucellus) between the micropylar chamber and the female gametophyte. After several weeks or months of gametophyte growth following pollination, fertilization takes place.

Figure 17–7 Pinopsida. (A) A small grove of cycads in South Africa; (B) a cycad in Australia; (C) the Swedish en, *Juniperus communis.*

The details of pollination and fertilization vary somewhat among the three orders and among species within each order. In *Zamia floridana,* a cycad, microsporophylls begin to develop in July. As the sporophylls mature, meiosis takes place and meiospores are formed. A thick, but structurally rather simple, spore wall develops and encloses each of the four daughter cells. Within this wall, development continues and the nucleus divides mitotically to form a multicellular male gametophyte. At the same time, but on different plants, megasporophylls have been developing, undergoing meiosis, and producing megaspores and female gametophytes. Usually by mid-December, sometimes a month or two later, the plants are ready for pollination.

Following pollination, the gametophytes continue development for several months before fertilization occurs. In contrast to ferns and fern allies, free water is not essential for fertilization in *Zamia* and other seed plants. In *Zamia,* the pollen tubes have a very high osmotic concentration which apparently enables them to draw moisture from the tissues of the female plant even when atmospheric humidity is low. Just prior to fertilization, usually in late May in southern Florida, motile sperm are produced. These are very large, the largest sperm in the vegetable kingdom, in fact, and can be seen with the unaided eye. They have several short, spirally arranged, whiplash flagella and swim rapidly in the fluid that makes up the interior of the pollen tubes. As the tip of the pollen tube approaches the archegonium of the female gametophyte, or ovule, the tube bursts and the sperm are discharged. The egg at this time is large and easily visible (3 mm in diameter) and is surrounded by fluids that have arisen from autodigestion of nucellus and archegonial cells. The sperm migrate through this fluid, into the egg cytoplasm, and fertilization takes place. The life cycle of *Zamia* is representative of all cycads and, with slight modification, of other Pinopsida as well.

Fertilization in Coniferales and Ginkgoales follows a similar pattern. However, in the Coniferales, the sperm are not motile. In all gymnosperms, the female gametophyte is quite large, large enough in cycads and some pines to be an important food source in some cultures. Each gametophyte, male and female, typically produces two gametes, only one of which ordinarily functions in fertilization, although polyembryonic seeds are not uncommon in cycads.

Following fertilization, the zygote undergoes a series of free nuclear divisions: usually two or three in conifers, about eight in *Ginkgo,* and ten or eleven in cycads, after which cell walls begin to form and a **proembryo** develops. In the conifers, the proembryo consists of three or four tiers of four cells each; in the cycads, three groups of several cells each. The cells in the middle group or tier elongate and form a **suspensor** which pushes the distal cells deeper and deeper into the gametophyte tissue. In some cycads, suspensors 7 cm long have been reported. As the embryo cells come in contact with the gametophyte cells, the latter disintegrate and are used as food by the developing embryo. In *Ginkgo,* all of the cells at the micropylar end of the proembryo elongate, pushing the embryo into the gametophyte tissue as in the Cycadales and Coniferales, but no definite suspensor is formed.

The cells of the proembryo farthest from the micropyle but nearest the **chalaza** differentiate into the embryo proper. Four embryonic organs develop: The **cotyledons** which function in food absorption, the **epicotyl** which develops into the shoot of the sporophyte, the **radicle** which develops into the root, and the **hypocotyl,** or embryonic stem, which functions in pushing the epicotyl and cotyledons above the ground.

The number of cotyledons depends on the species and is usually constant for each species: in the Cycadales and Ginkgoales, there are two cotyledons, and in the Coniferales, there are usually six to ten. The mature seed consists of three parts: the hard **seed coat** which is derived from the integuments and is therefore diploid tissue from the parent sporophyte, the **gametophyte** (sometimes called "endosperm"), which is fleshy haploid tissue and is the part eaten in a pine nut or pinyon nut, and the **embryo,** which is deeply embedded in the gametophyte. As the seed germinates, usually following a long period of dormancy in the conifers and *Ginkgo* but immediately after reaching maturity in cycads, the radicle develops first, expanding into the soil and developing secondary roots which provide firm anchorage and obtain water and minerals. The hypocotyl develops next and, as it elongates, the cotyledons, still attached to the gametophyte, and the epicotyl are pulled out of the soil. Light striking the upper side of the elongating hypocotyl causes it to straighten. The cotyledons open up, shedding any unconsumed portions of gametophyte remnants, and the epicotyl begins development into leaves and stem. This ends the embryo stage of development and initiates the seedling stage in the life cycle of the gymnosperm.

Phylogeny and Classification

The Pinopsida evolved in very early Mississippian or late Devonian time from fern-like plants similar to *Archaeopteris* or *Callixylon*. *Archaeopteris* is the name applied to leaf fossils and *Callixylon* the name applied to stem fossils of the same species (Beck et al. 1982). These "**progymnosperms**" (Archaeopteridales) are included in the Filicopsida and were discussed in the previous chapter along with the Cladoxylales. From them evolved the earliest gymnosperms, the Cycadofilicales, commonly called "seed ferns." Evolution in the Pinopsida seems to have progressed along two main lines, one leading from the Cycadofilicales to the Cordaitales, conifers, Ginkgoales, cycads, and angiosperms; the other leading to the Glossopteridales (Arberiales), Caytoniales, Bennettitales, and Gnetopsida (Figure 17-2). The suggested classification illustrated in Figure 17-2 is adapted from Krassilov (1977), Chamberlain (1965), Miller (1977), and others.

The progymnosperms (Archaeopteridales) were common during much of the Devonian period (from late Emsian on). Especially abundant was *Archaeopteris*, a genus of tall tree ferns that thrived during late Devonian and Carboniferous. About 25 to 30 species are known. The stems, which in at least one case were 1.5 m in diameter and over 30 m tall, contained cambium and resembled modern gymnosperms in some other ways. *Archaeopteris latifolia* was heterosporous; probably other species were too. Closely related to *Archaeopteris* were *Aneurophyton*, *Tetraxylopteris*, and other genera which also possessed characteristics similar to both the Cladoxylales and the gymnosperms. There is not good agreement as to how these groups should be classified. In this book, both the Cladoxylales and Archeopteridales are included in the Filicopsida, but Andrews (1963) placed them in a class by themselves independent of both the Filicopsida and Pinopsida but probably intermediate between the ancient Psilopsida and the gymnosperms.

Fossil evidence indicates that the first true gymnosperms were fern-like plants which are called Pteridospermales by some botanists, Cycadofilicales by others, and "seed ferns" by both. In Table 17-2 they are called Cycadofilicales. Formerly they were included in the Filicopsida, primarily because the leaves were similar to fern leaves and the seeds seemed to be contained in simple structures developed from the sori on the undersurface of vegetative leaves. More careful study has revealed that they possessed many gymnosperm characteristics, as indicated in the opening paragraphs of this chapter, including the presence of cambium tissue, tracheids having circular pits rather than the scalariform tracheids typical of the ferns, greater anatomical differentiation between roots and stems and between stems and leaves than occurs in the Filicopsida, an exarch eustele, integuments enclosing the seeds, and a much more extensive root system.

TABLE 17-2

DIVERSITY IN GYMNOSPERMS: MORPHOLOGICAL CHARACTERISTICS OF THE THREE EXTANT ORDERS AND TWO EXTINCT ORDERS OF PINOPSIDA AND OF THE TWO EXTANT ORDERS OF GNETOPSIDA

Order (Fam-Gen-Spp)	Habit	Habitat and Geographic Range	Anatomy and Morphogenesis	Gametophyte and Embryo	Representative Genera
Pinopsida					
Cycadofilicales (extinct)	Mostly shrubs, also trees and vines	Moist tropical, Devonian to Permian	Protostele, polystele, ectophloic siphonostele with mesarch development	No strobili (embryo unknown)	*Heterangium, Calymmatotheca, Medullosa, Lyginopteris, Pinus, Picea, Abies*
Coniferales (6-40-500)	Mostly trees, also shrubs; leaves small, alternate, rarely opposite	Mesic to dry; tropical to cold temperate	Endarch ectophloic siphonostele; tracheids with bordered pits	Peltate stamens and vesiculate pollen; gametophyte large, often edible	Araucaria, Podocarpus, Taxus, Torreya, Tsuga, Juniperus
Ginkgoales (1-1-1)	Conifer-like trees 30 m tall; fan-like leaves	Warm to cool temperate, mesic climate, cultivated	Ectophloic siphonostele little pith; dioecious seeds odoriferous; pine-like wood	Epaulet type stamens, large gametophyte, motile sperm	*Ginkgo biloba*
Cycadales (1-9-200)	Stem an underground tuber to 18 m tall trees; dichotomous veins, circinnate vernation	Tropical to warm temperate, mesic to moderately dry habitats	Ectophloic siphonostele with abundant pith, tracheids with bordered pits; strobili large, compact	Large, often edible gametophytes, motile sperm with spiral cilia	*Cycas, Bowenia, Stangeria, Dioon, Zamia, Microcycas, Encephalartos, Macrozamia, Ceratozamia*
Bennettitales (extinct)	Shrubs to medium trees	Probably warm, moist; Mesozoic	Ectophloic siphonostele embryo with two cotyledons; polyplicate pollen	Center leaves were strobili; motile sperm	*Williamsoniella, Cycadoidea, Wielandiella, Williamsonia*
Gnetopsida					
Ephedrales (2-2-35)	Shrubs: almost leafless or with 2 long torn leaves	Hot, dry desert to cold desert; Permian to present	Endarch ectophloic siphonostele; leaves in whorls or opposite; xylem with vessels	Dioecious with woody strobili, embryo with 2 cotyledons	*Ephedra, Welwitschia*
Gnetales (1-1-30)	Shrubs, small trees, vines	Moist tropics; many epiphytes	Endarch eustele; pollen grains spiny, not polyplicate	Embryo with 2 cotyledons	*Gnetum*

Note: The number of extant families (Fam), genera (Gen), and species (Spp) in each order is indicated in the first column.

The Carboniferous period in which the early gymnosperms reached the zenith of their ecological success is often called the "age of ferns" because of the abundance of "seed ferns." It is also often called the "age of dinosaurs." Either appelation is a confusing misnomer inasmuch as "seed ferns" were gymnosperms not ferns, and dinosaurs, like other animals, are subordinate species completely dependent on the dominant vegetation for their survival. The true ferns of that age, like the ferns of today, were ecological subordinates living in the shade of various Cycadofilicales and other trees.

The seeds produced by Cycadofilicales were not as advanced as the seeds of modern gymnosperms. Pollination took place while the ovules were attached to the parent plant, but fertilization apparently occurred after they had been shed. Among the thousands of seeds found attached to leaves of fossil Cycadofilicales, none have contained pollen tubes or embryos. Hence, while these plants possessed a potential for drought tolerance which would allow their descendants to colonize vast areas of land where the humidity and precipitation were too low for any plants of that day to survive, they were not yet entirely freed from their dependence on water. Nevertheless, their ability to nourish both female and male gametophytes while they were attached to the parent sporophyte was an adaptation that gave the gymnosperms the **potential** to become independent of free water for completion of their life cycle.

In *Archaeopteris* and other progymnosperms, the leaves were typically compound with some of the leaflets on some of the leaves being modified as megasporophylls, or leaflets that produce sporangia containing female spores. The number of sporangia per megasporophyll was reduced to one in the course of evolution from progymnosperms to early seed plants. The remaining part of the leaflet, or adjacent leaflets, was apparently modified to form an integument, or nucellus, the protective covering that encloses the sporangium. The first megasporangia to be enclosed in integuments received very little protection from them, but as the millennia sped by, integuments that more closely enclosed the megasporangia evolved. Eventually the integument so completely enclosed the sporangia that it could exist in its own special humid microenvironment, favorable for gametophyte growth, with its only contact with the arid macroclimate outside a small opening at the apex of the structure called the micropyle.

The fossil record has not revealed when seeds were first retained until fertilization had taken place and the embryo progressed to the point that success of independent living was essentially ensured. This could have happened more than once, once the potential had been established.

The earliest known fossil having fernlike foliage and gymnosperm-like seeds was *Lyginopteris oldhamia* from very early Mississippian. Other early seed ferns were **Heterangia** and **Medullosa.** The **Lyginopteris** stem is an ectophloic siphonostele, similar to modern gymnosperms. Steles were more variable in the Cycadofilicales than in extant gymnosperms; *Heterangia* had protosteles and *Medullosa* had a polystelic stem. Seed types were also more variable in this order than in extant orders. The Cycadofilicales thrived during the Pennsylvanian, became almost extinct toward the end of the Permian, increased in numbers during late Triassic and early Jurassic, and were extinct by mid-Jurassic.

The Cordaitales first appeared about 20 million years after the Cycadofilicales in the middle of the Mississippian. They were characterized by strap-shaped leaves, compound strobili, an ectophloic siphonostele containing abundant pith, and dichotomously branched roots. The primary xylem contained both helical and scalariform tracheids. The cambium was active, producing abundant secondary xylem with bordered pits. No growth rings are evident in *Cordaites* fossils, suggesting that they grew in a tropical or subtropical climate lacking distinct seasons. The seeds were small and heart-shaped with the integument expanded into a wing that protruded from the cone scales.

The first conifers made their appearance in mid-Pennsylvanian, having evolved from the Cordaitales. They were members of the Lebachiaceae, a family that became extinct during the Jurassic. The xylem shows weak growth rings, indicating that they were adapted to climates with mild seasonal changes, and the tracheids are similar to those of modern Araucariaceae. All of the conifers of today, with the possible exception of the yews, (Taxaceae), and the Cephalotaxaceae, are descendants of the Lebachiaceae. The evolution of the Coniferales from the Cordaitales has been carefully worked out by Rudolf Florin (Banks 1970). During the Jurassic and the Cretaceous, the conifers reached their greatest dominance in number of species and area covered. Since then, they have been replaced by angiosperms in many of their former habitats, by Gnetales in a few.

Throughout the Permian, the dominant vegetation on the southern continent, Gondwana, were gymnosperms with entire, reticulate-veined leaves now referred to the genus *Glossopteris* in the Cycadofilicales (Pigg and Taylor 1993). On the north continent, Laurasia, species of *Ginkgo* and some of the early Bennettitales were beginning to dominate the landscape. These were tall trees with large, well developed trunks and, in some species at least, deciduous foliage. Fossils of *Glossopteris skaarensis* and *G. schopfii* from the Transantarctic Mountains of Antarctica were studied in detail by Pigg and Taylor (1993). At least nine other species of *Glossopteris* are known from throughout the southern hemisphere (South Africa, Australia, Brazil, India, etc.), mostly from the Permian. Many species of *Ginkgo, Williamsonia* (Bennettitales), and *Nilssonia* were abundant during late Permian and through the Triassic and Jurassic Periods.

Sequoia-like plants dominated the landscape during much of the Cretaceous and Tetiary Periods. Figure 17–8 is *Sequoia cuneata* from a late Cretaceous formation in Wyoming, a species long since extinct. Figure 17–9 shows leaves and twigs of *Metasequoia chinensis* from the Carmen Formation in Idaho, also believed to be extinct until the 1940s when it was found growing in western China. The famous fossil forests of Yellowstone Park are made up primarily of *M. chinensis.*

The origin and evolution of the Ginkgoales is less certain than that of the Coniferales. It is believed they evolved either from Cordaitales or pre-Cordaites stock among the Cycadofilicales during the late Paleozoic era. They became abundant and dominant in many forests at that time, but have since dwindled in ecological significance. At present, one species, cultivated in temple gardens of China and Japan, survives. It has been suggested by some biologists that its survival today is due entirely to man's cultivation of it. This may be true; however, it is reported to grow wild in forests of western China. Furthermore, the rest of the order has been extinct for 100 million years and man has been here only 2 million. Other factors must account for its survival between 100 million years ago and when man began practicing horticulture about 6 or 7 **thousand** years ago.

Superficially, the Cycadales and Bennettitales resemble each other leading some botanists to assume a close relationship between them; however, differences in tracheid structure and careful examination of the fossil record has led others to the conclusion that they evolved along parallel paths from rather different seed fern stock (Krassilov 1977). The Nilssoniales, an order of cycad-like gymnosperms which thrived from the Jurassic to the beginning of the Tertiary,

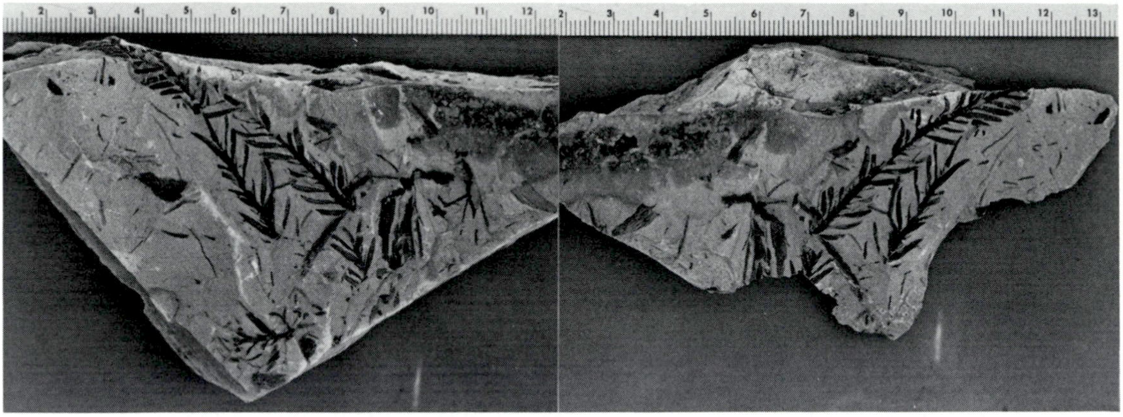

Figure 17–8 Fossil Pinopsida. *Sequoia cuneata* from the upper Cretaceous, Rock Springs Formation, Wyoming, long since extinct. (Courtesy Lee Palmer, California Polytechnic State University, San Luis Obispo.)

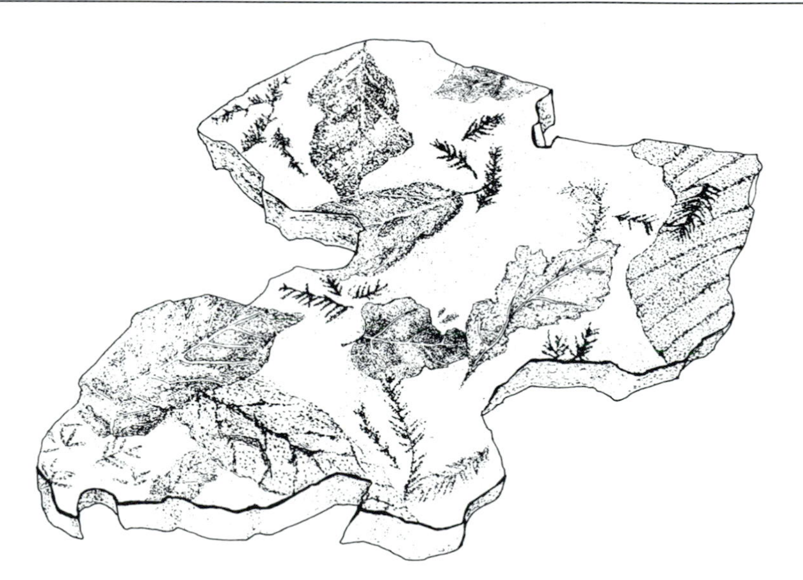

Figure 17–9 Fossils from the Miocene Carmen Formation, Idaho. Both Pinopsida and Anthopsida leaves are abundant here, as well as ferns, horsetails, and others.

appears more likely to form the link between seed ferns and cycads. The Nilssoniales also appear to be the link between the Cycadofilicales and early angiosperms (Figure 17-2).

The presence of "girdling" in leaf trace origin in both the cycads and the angiosperms and the tendency for mesophyll to be differentiated into a palisade and a spongy layer in the cycads as in the flowering plants, but not in other gymnosperms, are among the characteristics that point to the Nilssoniales as the most likely taxon from which angiosperms evolved. Other similarities in anatomy, embryology, and cell characteristics strengthen that hypothesis. Characteristics of the nine orders of gymnosperms are compared in Table 17-2.

Representative Genera and Species

To people living in Europe, the pines, spruces, and junipers are the best known of the gymnosperms. The conifers of temperate North America and northeastern Asia include these and also firs, hemlocks, false hemlocks, sequoias, cypresses, and others.

Pines (*Pinus* spp.) have leaves which are usually divided lengthwise into two to five segments, the complete fascicles being perfectly round in cross section and shaped like sewing needles. They have woody cones, and the tree is typically rugged in appearance and more or less cylindrical.

Firs (*Abies* spp.) have flat leaves and cones with membranous or paper-like scales that stand erect on the upper branches. The yews, hemlocks, and false hemlocks (*Taxus* spp., *Tsuga* spp., and *Pseudotsuga* spp., respectively) also have flat leaves but the cones are membranous and pendant. A distinctive feature of *Pseudotsuga* is the presence of three-toothed bracts on each cone scale. Spruces (*Picea* spp.) have terete leaves attached to the twigs on short pedestals. The leaves are usually very stiff and sharp pointed; the cones have membranous scales and are pendant.

North American cedars, *Juniperus* spp. and *Thuja* spp., have scale-like or awl-like leaves and cones that are modified to berry-like structures in *Juniperus,* or are woody in *Thuja.* Other conifers include the larch, or tamarack, the bald cypress, and dawn redwoods or *Metasequoias,* all of which are deciduous, and the redwoods or *Sequoias.* In the southern hemisphere, *Araucaria, Podocarpus, Callitris,* and *Fitzroya* represent the Coniferales.

Common names always arouse considerable confusion, even in conifers. To many people, "pines" includes not only the genus *Pinus,* but *Abies, Picea,* and *Pseudotsuga* as well. Throughout the American West, there are cities, canyons, and counties named red pine and white pine—in honor of *Pseudotsuga taxifolia* (the false hemlock or Douglas fir) and *Picea engelmannii* (Engelmann spruce), respectively, while "cedar" refers to several cypress-like species in addition to the genus *Cedrus,* the "true cedar" or "cedar of Lebanon." It is a surprise to many Americans to learn that eastern red cedar, the tree from which cedar chests are made, is a juniper while western red cedar, the tree from which cedar shingles are made, is an arborvitae.

For millions of years, a tree having leaves similar to those of the coast redwoods, *Sequoia sempervirens,* flourished in what is now Oregon, Idaho, Montana, Wyoming, and parts of eastern Asia. Known only as a fossil, it was named *Sequoia chinensis* by Chaney; it differed from other *sequoias,* such as *S. sempervirens* and *S. gigantea* from California, and *S. langsdorfii,* a European fossil, in several ways, including possession of stalked cones, obtuse leaf apices, and deciduous branchlets. Therefore, it was transferred by Miki to a new genus, *Metasequoia,* and renamed *Metasequoia chinensis* (Figure 17-9). In 1941, Professor T. Kan of the National Central University in China discovered three tall trees, representing a species new to botany (though well known to local farmers), growing near some rice paddies in an isolated area of western China; upon careful examination, they were found to be essentially identical in every way to the fossil species, *M. chinensis.* Professor Kan collected a number of cones and sent packets of seeds to friends in North America. Since then, a large number of these trees, since named *M. gloptostroboides,* have been found, and many of them are now growing on college campuses throughout the U.S.

The dominant species in the famous fossil forests of Yellowstone National Park is *Metasequoia chinensis.* Hundreds of these trees, fossilized by petrifaction, are found on the north slope of Specimen Ridge which forms the south border of the LaMar Valley near the center of the Park. At the northwest corner of the Park, in the "Montana Strip," another ridge, just north of Specimen Creek, is covered with massive *M. chinensis* petrifactions. Some of these, several meters tall and more than three meters in diameter, are probably the largest fossils of plants standing upright in their natural position anywhere.

Ecological Significance

With the exception of some high altitude, high latitude, and desert ecosystems dominated by lichens and mosses (and low in ecological productivity), all terrestrial ecosystems are dominated by either gymnosperms or angiosperms. According to recent analysis, 8% of the land surface of the earth is covered by coniferous forests which contribute 6% of the total terrestrial productivity and 3% of the total global productivity. Assuming that the dominants in these ecosystems contribute 85% of the total productivity, the Pinopsida are responsible for about 2.5% of the total global production of organic matter.

Ecologically, the Coniferales are by far the most important of the three orders of Pinopsida. Their significance extends beyond their ability to produce food and oxygen; they also provide shelter for birds, mammals, and insects, while various epiphytes, especially lichens and mosses, grow on their bark and branches. Most conifers form mycorrhizal associations with various mushrooms, which in turn provide food for a number of animals, including man. The slowly decomposing litter produced by gymnosperms, and by the subordinate vegetation in the forests they dominate, is important in holding the moisture from rainfall and snow melt so that runoff from hills and mountains remains relatively constant throughout the year instead of coming all at once within a few hours, as in a typical desert flash flood.

Generally tolerant of abundant light, conifers are often the dominants in forest ecosystems; most conifers are also long-lived. The major ecosystems dominated by conifers make up the boreal forest biome, also known as the taiga in its northern reaches, and the montane-coniferous forests of the Rocky Mountains, Cascades, Sierra Nevada, Himalayas, and other mountain ranges. Some Pinopsida, especially pinyon pine and Utah cedar, are well adapted to semiarid habitats where they may be dominant species in pigmy forests.

The climate in the boreal forest is only slightly less severe than that of the tundra; the growing season is cool and short, lasting from mid-June to mid-August, and winters are very cold. Precipitation is moderate, about 50 cm over most of the area, but the evaporation: precipitation ratio is favorable. Over much of the area, the topography is marked by glaciation with numerous lakes and bogs; the soils are generally thin except where deep muck and peat deposits have built up in old lake bottoms.

In North America, the dominant climax species of the boreal forest are *Abies balsamea* and *Picea glauca,* while *Pinus strobus, P. banksiana,* and *Betula papyrifera* are subclimax species. In bogs, and sometimes on the climax sites, *Larix laricina* shares dominance with *Picea* and *Abies.* Across northern Asia and northern Europe, *Picea excelsa* is the climax dominant with *Pinus sylvestris* and *Betula alba* dominant in subclimax forests and often intermingled with *Picea.*

In the boreal-like montane forests of the Rocky Mountains and the Cascades, *Abies, Picea,* and *Pseudotsuga* are climax dominants and *Pinus contorta* and *Populus tremuloides* dominants in the subclimax forests. At lower and drier sites, *Pinus ponderosa* is a climax dominant over a large area extending from the Black Hills of the Dakotas to the east slopes of the Sierras.

At still lower and drier sites, *Juniperus osteospermum* and *Pinus centroides* (*P. edulis*) occupy vast areas, some of it formerly grassland which they have been able to invade because of overgrazing. On the west slopes of the Sierras and along the California and Oregon coast, *Sequoia gigantea* (*S. washingtonia*) and *S. sempervirens* are important. Farther north, *Tsuga mertensiana, Abies amabilis, Abies nobilis,* and *Thuja plicata* are often dominant species.

In contrast to northern Europe, where only six species of Pinopsida occur and only two are dominants in the extensive forests that cover millions of hectares, at least 70 species of conifers, representing 15 genera and 4 of the 7 families, are ecologically important in North America. Another 30 species occur, but in limited numbers. During the Pleistocene epoch, glaciation destroyed most of the forests of Europe; because of the east-west orientation of the mountain ranges, old species could not migrate southward as the glaciers advanced, nor have new species been able to migrate into the area as the climate has warmed. In North America, on the other hand, species that moved further and further south as the glaciers advanced were able to move northward as the glaciers retreated because the mountain ranges have a north-south orientation. East of the Ural Mountains, diversity of trees increases, and eastern Asia has about as many species of conifers as North America. Many of them are found on the north slopes of the Himalayan Mountains and in southeast China (Special Interest Essay 17-1).

In South America, coniferous forests occur from Colombia to southern Chile in the Andes, and are found also in southeastern Brazil. Species of *Fitzroya, Podocarpus,* and *Araucaria* occur in both areas.

On the northern slopes of the Himalayas, *Pseudotsuga, Tsuga,* and *Cephalotaxus* thrive; farther north in western China *Metasequoia* can be found. *Pseudotsuga* and *Tsuga* are well adapted to a wide range of habitats and ecological niches and appear to be expanding their ranges by evolving new ecotypes which may some day become new species; *Metasequoia,* on the other hand, is very limited in distribution and, like modern *Pseudolarix* and ancient *Lebachia,* may be moving toward extinction. Other coniferous forests are found in the mountains of New Guinea and in the Pyrenees, Alps, and Caucasian Mountains of southern Europe. Conifers are also found as subordinate species in angiosperm dominated forests of Australia, South Africa, and elsewhere.

Ecologically, cycads were very important during the Jurassic and Cretaceous periods when they were worldwide in distribution. For example, Hickey and Doyle (1977) reported that almost all of the fossils in the lowest strata of the Cretaceous Potomac formation of the eastern U.S. are ferns and gymnosperms, including many cycad-like species, and only 2% are angiosperms. Now only 9 genera and about 100 species of Cycadales remain, and these are very patchy in their distribution; the cycad-like Nilssoniales, Caytoniales, and Bennettitales are extinct.

Cycads are well adapted to drier sites in the tropical and subtropical forests. Some species of *Macrozamia* are large trees, often more than 15 m tall with reports of trees over 30 m tall. Other cycads are low shrubs not more than 1 m tall. Four genera are found in Mexico and the West Indies, extending northward to Florida and Alabama and southward along the Andes to Chile and across the northern edge of South America. Two genera occur in South Africa. The other three genera are found in Australia and Indonesia with species of *Cycas* extending northward through India and China to Japan and westward to Madagascar. All of the species except those of *Cycas* have very narrow ranges of tolerance for temperature; but some species of *Cycas* are relatively cold hardy and can be grown in gardens as far north as southern California.

TABLE 17-3

U. S. PRODUCTION OF FOREST PRODUCTS (IN MILLIONS OF CUBIC FEET OF SAW TIMBER)

Douglas fir or false hemlock	602,622
Western hemlock	269,935
Ponderosa pine	241,722
Fir (white, noble, and alpine)	234,780[a]
Southern pines	211,925
Oak	163,254
Western spruce	155,404[a]
Western white pine	53,083
Redwood	31,257
Hickory	28,488
Sweetgum	25,879
Hard maple	25,764
Ash, walnut, and cherry	22,923
Tulip or yellow poplar	21,202
Cypress	15,346
Yellow birch	11,594
Other	
Western conifers	177,835
Eastern hardwoods	131,356
Eastern conifers	22,229
Total	**2,536,799**

[a] The totals for fir and western spruce do not include $20,629 \times 10^6$ ft^3 of eastern fir and spruce reported as "other eastern conifers."

Economic Significance

The conifers are of great value in the production of lumber, in the manufacturing of paper, and in other ways (Tables 17-3 to 17-5). In addition, they are the dominants in many of the national forests and national parks and are therefore enjoyed for aesthetic and recreational purposes. Many species that are of no direct commercial value in the production of forest products are of tremendous value in conserving soil and water. To maintain our forests for future generations, much planning and work are required (Table 17-5).

While members of the Cycadales are not of great economic value in world commerce, they are locally important producers of food and building materials. The seeds of some cycads are the size of duck eggs, weighing several ounces, and are used in a variety of ways. Seeds of most, if not all, species contain toxins which must be leached out before the seeds are used as food; in other species, the seeds are used in much the way pine nuts are used (the name *Zamia* is based on the Latin for pine nut). Kaffir bread is made from the seeds of *Encephalartos.* The pith of *Cycas* species yields a form of sago used in bread-making and in other ways.

Thieret (1958) described the uses of cycads in Australia, Africa, and the islands of the southwest Pacific. At the time he wrote, as throughout the first half of the 20th century, sago and cycad seeds were important items of food. Some species of cycads which Thieret reported to be everywhere are now rare because nobody plants or cares for them; other sources of food

TABLE 17-4

WORLD PRODUCTION OF FOREST PRODUCTS

Geographic Area	Sawn Wood ($m^3 \times 10^6$)	Wood-Based Products ($m^3 \times 10^6$)	Round Wood and Fuel ($m^3 \times 10^6$)	Pulp and Paper (tons $\times 10^6$)	Total (tons 10^6)
Far East	59	6.1	452	19.8	330.1
Former U.S.S.R.	111	14.4	140	15.0	174.2
Rest of Europe	87	22.6	98	50.6	175.2
Africa	7	1.1	262	2.5	164.6
North America	108	26.0	45	56.4	163.8
Latin America	25	1.9	232	6.9	162.2
China	20	2.5	130	8.0	99.5
Near East	59	0.6	17	0.6	12.4
Pacific	8	0.6	8	2.2	12.2

Note: The total, in millions of tons, is based on an average density of oven-dried wood of 0.6 grams per cubic centimeter. Density of wood varies greatly, from 0.2 in oven-dried balsa from South America (*Ochroma* spp.) to 1.23 in oven-dried greenheart wood from Africa (*Nectandra rodioei*). In general, the hardwoods (Anthopsida) are heavier than the softwoods (Pinopsida); major exceptions are balsa, basswood (0.40), quaking aspen (0.40), and black willow (0.41) among the "hardwoods," and eastern red cedar (0.49), western larch (0.59), and longleaf pine (0.64) among the "softwoods."

TABLE 17-5

ACRES OF FOREST PLANTED AND PRODUCTION OF LUMBER IN THE U.S. 1976–1990

State	Acres Planted
1. Georgia	343,374
2. Mississippi	198,932
3. Alabama	259,657
4. Florida	239,162
5. Oregon	238,721
6. Washington	220,666
7. South Carolina	174,106
8. Louisiana	163,624
9. Texas	132, 394
10. Arkansas	128,427
11. North Carolina	124,616
12. Virginia	109,548

Year	Millions of Board Feet Produced		
	Softwood	Hardwood	Total
1976	30,571	6,427	36,997
1980	28,239	7,115	35,354
1985	30,479	5,966	36,445
1990	36,602	7,340	43,942

TABLE 17-6

DISTRIBUTION OF THE GENERA OF CYCADALES

Genus	No. of Species	Distribution	Comments
Zamia	35	U.S., Mexico, Central America, Carribean, and South to Venezuela and Chile	
Microcycas	1	Cuba	Largest and among the tallest; leaves are very small
Dioon	3	Mexico	
Ceratozamia	1	Mexico	
Macrozamia	12	Australia	*M. hopei* at 18 m is tallest cycad
Bowenia	2	Australia	
Cycas	4	Australia, New Zealand, and north to Japan	
Stangeria	1	Southeast Africa	Limited distribution
Encephalartos	12	South Africa north to equator, southwest to Port Elizabeth	Widespread

Data from Chamberlain (1965).

are easier to prepare. Lepofsky (1992) discussed the use of cycad seeds as food with a number of women born in the late 1950s and 1960s and was surprised to learn that none of them had ever tasted cycad seeds or knew how to prepare them for eating.

The toxin in cycad seeds is an azoxyglycoside called cycasin. It is not only acutely toxic, capable of causing death (especially in children), if not removed from the seeds, but is now known to be a neurotoxin, even when consumed in relatively small amounts, and has carcinogenic properties as well (Beck 1992). Three methods are used in Australia in detoxifying or processing cycad seeds for food (Beck 1992): rapid leaching, slow leaching, and aging. Leaching with hot or cold water after the seeds have been broken or ground is preferred if the product is to be used quickly before it spoils; aging, usually carried out in shade, is preferred when the product must be kept for a long time.

Why did people go to all the work and risk to process cycad seeds when other edibles were available? Beck (1992) presents three reasons: (1) high yield, (2) dependability, (3) resistance of the trees to drought and fire. Although cycads do not produce a good crop every year, the combination of high yield when they do produce and good storage properties results in dependability.

Commercial uses of cycads include the gathering, drying, baling, and shipping of large quantities of cycad leaves, which are painted and used for decorations, and the use of some 25 species, representing all of the genera except *Stangeria,* as indoor ornamentals or novelties. Distribution of the genera of Cycadales is given in Table 17-6.

The commercial value of *Ginkgo biloba* and of many conifers, especially species of *Juniperus, Thuja, Acauparia, Picea,* and *Pinus,* is as ornamentals. Many Pinopsida are of commercial value in the production of drugs. As early as 1855, arborvitae extracts were being administered orally in Europe and North America for treatment of cancer, with favorable results reported in a few cases. This practice was based on the "Aztec Herbal of 1552" as translated by Jesuit priests in northern Mexico in the 16th Century.

Fitzgerald and associates reported in the *Journal of Cancer Research* in 1953 that they had screened over 100 species of plants, including all five species mentioned in the Aztec Herbal, for control of cancer in mice. Three of the species used by the Aztecs, all three belonging to the *Juniperus virginianum* or red cedar complex, and two other *Juniperus* species, effectively necrotized some tumors and increased the life expectancy of the mice; some of the other species tested also necrotized some of the cancers but the chemotherapeutic index was so low that even minute overdoses killed the mice. Arborvitae was not effective in these experiments. This work of screening and testing has continued, and at the present time, several chemotherapeutic agents are being used in the control of cancer in humans (e.g., Balandrin et al. 1985). All of them have the same problem that Fitzgerald had in his study: a very low chemotherapeutic index. In other words, the amount of chemical that will cause serious problems in the patient, often death, is only slightly more than the amount needed to control the disease.

Research Potential

When most people think of conservation and preservation of the environment, they think of forestry. The fact that the basic principles of ecology and conservation were developed largely from studies of gymnosperms, primarily conifers, and their associates in natural ecosystems is partly responsible for this. One of the greatest conservation developments of all time was the establishment of the world's first national park, Yellowstone National Park in northwestern Wyoming and adjacent portions of Montana and Idaho, in 1873.

The 1871 Washburn-Langford Expedition to explore that part of the Louisiana Purchase known as "Coulter's Hell" had far-reaching influence on conservation philosophy. As the leaders sat around the campfire their last night in the area, discussing the marvels and beauties they had seen during the past 3 weeks, they wove the fabric of a dream that would affect mankind all over the world for years to come. "This area is too valuable to all people of the United States to be turned over to private developers for their monetary gain," they said. This revolutionary idea was accepted by President U. S. Grant and Yellowstone Park came into being.

The purpose of the park was to preserve for future generations several unique geological and biological features in order that these may be enjoyed and studied by generations of Americans and other people. In the past, similar unique features had either been exploited purposefully for economic gain by a few or carelessly destroyed or changed because of lack of protection. From the establishment of this, still one of the largest of the American national parks, other national parks and monuments have been established across the U.S., and the idea has spread to other countries. Basic ecological concepts that have developed from the study of gymnosperm-dominated ecosystems include the concept of succession, the concept of ecosystem structure and of energy flow, the use of dendrochronology in the study of climate, and the concept of annual energy budgets. Conifers and the communities they dominate have been of further value in the development and expansion of other basic ecological concepts, such as Liebig's "law of the minimum," Shelford's "law of tolerance," Allen's "law of size-climate correlations in animals," and the concept of trigger factors.

National Park Service policy in recent years has been one of noninterference of events triggered by natural causes. Except for fires caused by man, for example, forest fires are allowed to burn until stopped by natural causes. Following the Yellowstone Park fires of 1988, caused by a prolonged and very severe drought and by "dry" lightning storms, the general public and many politicians attacked this policy. Few ecologists, on the other hand, have been critical of the policy. The fires, by burning the climax forests, have triggered ecological succession patterns which increase the total diversity, and possibly total productivity, of the area. On the other hand, ecologists have been critical of the Park Service policy of not removing herbivorous animals or allowing any form of big game hunting in the Parks. With many of the natural predators now gone—there are no longer any wolves in Yellowstone Park, for example, although natural reintroduction is occurring and human-directed reintroduction policies are being explored—man is the only species that can fill the secondary consumer niche in many of the national park ecosystems. We have seen in recent years evidence of overgrazing by buffalo and wapiti in Yellowstone Park and, prior to 1988, when management policies were modified, of severe overgrazing by burros in the Grand Canyon.

Gymnosperms have also played an important role in the development of plastics and other modern chemicals, paints and varnishes, industrial alcohol with its many uses, textiles, photographic materials, soaps and disinfectants, and many other products.

As we learn more about the chemistry of the conifers, as a result of the great demand for industrial chemicals, we will be able to study relationships in greater detail. This will also open up ways to study relationships between the conifers and other gymnosperms and between gymnosperms and angiosperms. In the near future, there will undoubtedly be increased research interest in studying relationships among plants employing biochemistry as an important tool.

The Ginkgoales and Cycadales offer excellent possibilities in scientific research. *Ginkgo biloba* is well suited for studies of morphogenesis, for example (Del Tredici 1992). The large egg cells in cycads also present opportunity for this type of study and also for genetic research. The nucleus in the egg cell of cycads increases in size just prior to fertilization until it becomes so large it can be seen with the unaided eye. What causes this increase in size is one question that geneticists will be answering because it bears on the production and function of DNA in a very basic way. Organisms with large nuclei often, though not always, have large chromosomes; chromosome study is thus enhanced in such species. Cycads also have the largest sperm cells known in the Plant Kingdom, and thus present another advantage in genetic and cytological studies.

CLASS 6. GNETOPSIDA

The Gnetopsida (from the Malaysian name for *Gnetum*) is a small class consisting of three genera and about 75 species found mostly in the tropics. Traditionally, the class has been treated as a single order within the class Gymnospermeae. Bold (1957), Scagel et al. (1965), and others have raised it to division status with one class and three orders. In this book, the three genera are classified in one class consisting of two orders: the Ephedrales with two families and the Gnetales with one. Suggested lines of relationship within the class are illustrated in Figure 17-2, and the general characteristics of the two orders are summarized in Table 17-2.

Ephedra is the largest genus with 42 species, 15 of which occur in the deserts of North America. *Ephedra equisetifolia* and *E. nevadensis* (Figure 17-10) are abundant in the Great Basin, the Grand Canyon, and other arid locations, but do not extend northward beyond the southern border of Oregon and Idaho. *Welwitschia mirabilis* is a drought-resistant plant found in a limited area of southwestern Africa, and the only species in the family Welwitschiaceae. *Gnetum,* with about 30 species, is a genus of tropical vines and shrubs.

Of the three genera, *Ephedra* is the easiest for most students to find. It is fairly common in the deserts of North and South America, Asia, and Africa and also occurs in southern Europe. Its unique habit and overall morphology cries out to any student in the area to come closer and examine it in detail. On the other hand, *Gnetum,* which is limited to the tropics of Asia, Africa, South America, and the Pacific islands, hides within the dense undergrowth and escapes notice. *Welwitschia* is so limited in its distribution that even the professional botanist is not likely ever to see one in its native habitat unless he lives in southwestern Africa. Because it is an endangered species, he will not be allowed to collect a specimen if he should see one.

Figure 17–10 *Ephedra equisetifolia.*

Ephedra superficially resembles *Equisetum* and many students have made the mistake of confusing the two. This is almost unforgivable. *Ephedra* has opposite, usually scale-like, leaves or leaves in whorls of 3 to 4, *Equisetum* has whorls of 10 or more scale-like leaves; *Ephedra* has a solid stem, *Equisetum* has a hollow, jointed stem; *Ephedra* grows in extremely dry areas and usually far away from water, *Equisetum* is always found near water or near water courses of some type; and *Ephedra* produces seed-bearing cones in the axils of its leaves while cones of *Equisetum* produce only spores and are terminal on the central stem.

Welwitschia is sufficiently unique in its appearance that once a student has seen a picture or read a description of it he will probably never confuse it with anything else. It has two large leaves growing out of a huge turnip-like crown. Since the leaves are perennial, they are tough and leathery and nearly always torn into longitudinal ribbons by the persistent desert wind.

Considerable evidence points to a fairly close relationship between *Welwitschia* and *Ephedra:* the nature of the vessel elements in the xylem of both, for example, the highly specialized drought resistance that both possess, and the unique polyplicate pollen grains of both (Figure 15-6). The relationship of either or both to *Gnetum* is much less certain.

Gnetum grows either as a climbing or trailing vine, or occasionally as a shrub or tree, in tropical hardwood forests. It resembles an angiosperm in leaf and some other characteristics, the leaves being similar to those of laurel and magnolia and the seeds resembling the hard fruits of many angiosperms; therefore it is easily mistaken for a flowering plant. However, the seeds are borne naked on cone scales, as in other gymnosperms, and the wood anatomy suggests a relationship to *Ephedra* and *Welwitschia.* In the past, some botanists looked to the Gnetales as the link between the Gymnospermeae and Angiospermeae, but most have abandoned this hypothesis for several reasons, including the lack of any fossil evidence in support of it. In this book, *Gnetum* is treated as the only genus in one of the two orders of the Gnetopsida, but the possibility that *Gnetum* is a reduced angiosperm, perhaps related to the Chenopodiaceae, cannot be lightly dismissed.

General Morphological Characteristics

Some of the morphological characteristics of the Gnetopsida are summarized in Table 17-1. Throughout the class, leaves are opposite, there are vessels in the xylem, and habit is woody. In the Ephedrales, pollen grains are unique, being polyplicate (Figure 15-6). Gnetophyte stems are ectophloic siphonosteles; the roots and veins are protosteles. In all three organs, the primary xylem development is endarch. The phloem is essentially identical in structure to that of the Pinopsida and consists of unspecialized sieve cells. Cambium is well developed, even in the leaves. Several layers of cambium are present in *Gnetum,* as in some dicots, the new layers arising from cells in the phloem. The apical meristem is differentiated into a single-layered tunica and multilayered corpus similar to the mother cell zone of the Gymnospermeae, but more complex. In this respect, as in many others, the Gnetopsida resemble Anthopsida more than Pinopsida; in the latter, there is no tunica, only a small number of surface initials from which the central mother cells originate, while in angiosperms there is a tunic which may be one cell layer thick, as in the Gnetopsida, or up to five cell layers thick.

The stems are woody, consisting mostly of secondary xylem, and may live to be a hundred years old or more; in *Welwitschia* they often exceed a meter in diameter in old specimens, but in *Ephedra* and most species of *Gnetum* they are seldom more than 2 or 3 cm thick; however, in *Gnetum gnemon,* a small tree native to Malay, the stems may be up to 25 cm in diameter. None of the species is very tall. *Welwitschia* stems are seldom more than 25 to 30 cm long, including the underground portion, and resemble huge turnips; the stem apex is deeply concave. *Ephedra* and *Gnetum* have much longer stems, occasionally 2 m long in *Ephedra,* though more often 40 to 80 cm, and several meters long in some species of *Gnetum.*

Ephedra and *Welwitschia,* like many other truly xerophytic plants, have very extensive root systems. The long taproot of *Welwitschia* is reported to extend several meters in the soil to the water table. Root systems of *Gnetum* are rather variable, but tend to be shallow with a prominent tap root and short secondary roots. Steles are diarch in most gnetophytes.

The leaves of gnetophytes vary considerably among the three genera. *Ephedra* leaves are small, seldom more than 1 cm long, entire, leathery, and deciduous. Two small vascular bundles are present in each leaf. The pair of guard cells around each stomate originate from a single epidermal cell, as in the Pinopsida; in the rest of the class, their derivation is more complex. The mesophyll is undifferentiated, and most of the photosynthesis takes place in the stems. *Welwitschia* leaves, on the other hand, are very large and leathery, often more than 3 m long, and parallel-veined. Superficially resembling the leaves of *Iris* or *Colchicum* when young, they soon split into longitudinal strips as winds whip them about. Their stomates are sunken and occur on the undersurface of the leaf only. The mesophyll is differentiated into palisade and spongy layers. Photosynthesis is of the crassulacean acid (CAM) type.

Leaf venation in *Gnetum* is different from that of the Ephedrales as well as that of Pinopsida and Anthopsida. Branching is simple, anastomosing only near the margins. The leaves are typically about 15 cm long, lanceolate to narrow.

As in the Pinopsida, the pollen grains are wind-borne; in *Ephedra* and *Welwitschia,* however, they are unique among all vascular plants being elongated egg-shaped or ellipsoid with wing-like ridges extending longitudinally from pole to pole, a type called **polyplicate.** The furrows between the ridges are sometimes branched (*e.g., E. viridis* and *E. nevadensis*), sometimes unbranched with low, gently rounded ridges (*E. trifurca* and *W. mirabilis*), or with high, very angular or undulating ridges (*E. antisyphilitica*). Pollen grains in the Bennettitales were

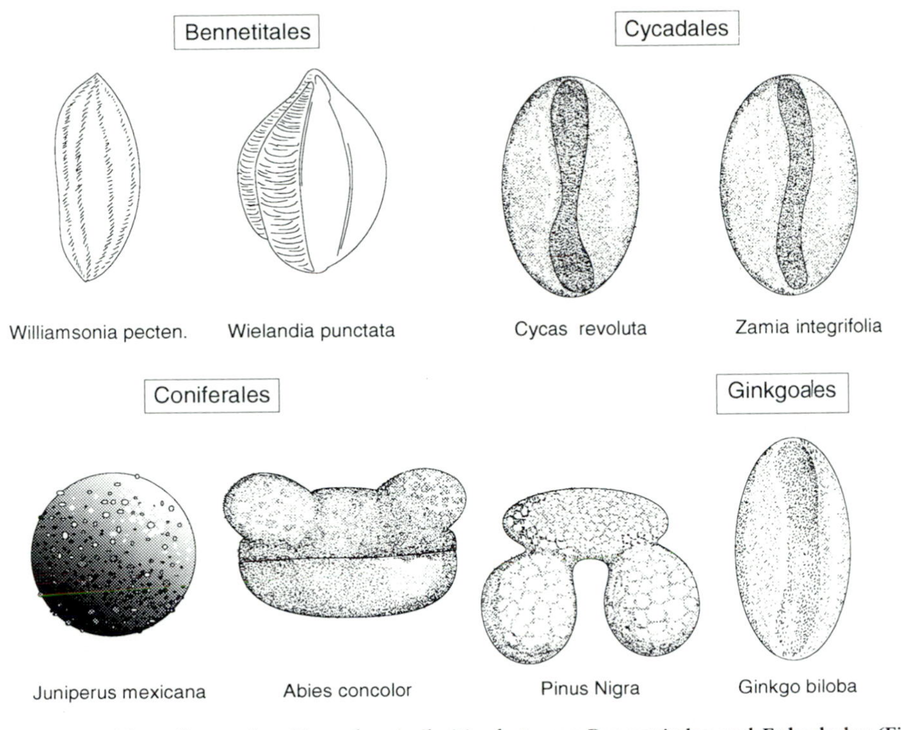

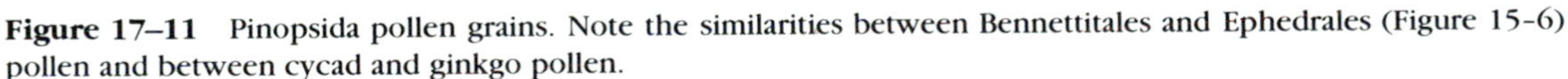

Figure 17–11 Pinopsida pollen grains. Note the similarities between Bennettitales and Ephedrales (Figure 15-6) pollen and between cycad and ginkgo pollen.

also polyplicate (Figure 17-11). In *Gnetum,* the pollen grains have numerous small spines similar to the pollen grains of some of the Malvaceae and Chenopodiaceae except that distinct germination pores are lacking.

Following meiosis in the female *Ephedra* plants, a row of four haploid cells is produced; only the cell at the chalazal end survives. It undergoes a series of about eight, or sometimes ten, free mitotic divisions before cell walls are formed. Two or three archegonia are produced at the micropylar end of the gametophyte; each consists of an egg, a mass of supportive cells, and a neck made up of several tiers of four cells each. As the gametophyte develops, the portion of the nucellus nearest the micropyle undergoes autodigestion and a large pollen chamber or micropylar chamber soon separates the micropyle from the archegonium. A pollination droplet forms at the micropyle and functions in drawing pollen grains into the micropylar chamber. The pollen tube develops rapidly and fertilization may take place within 10 hours following pollination.

Following fertilization, the zygote divides to form two cells which elongate into a tube-like structure in the Ephedrales. The cell nearer the micropyle develops into a suspensor and the chalazal cell develops into a proembryo. Embryonic development is similar to that in the Pinopsida and the mature embryo consists of two cotyledons, an epicotyl, a hypocotyl, and a radicle. Germination does not take place until at least 5 months after the embryos have anatomically and morphologically matured.

In *Welwitschia,* an archegonial tube grows out from the outer layer of gametophyte tissue and carries the egg into the micropylar chamber where it meets the pollination tube and fertilization takes place. In *Gnetum,* cell wall formation never takes place in the female gametophyte and archegonia are not formed. The inner integument (or true integument) in *Gnetum* develops into a beak that protrudes beyond the bracts (or outer integument).

In *Gnetum,* zygotic divisions produce a branching system of primary suspensors from which secondary suspensors sometimes arise and the proembryos develop from the apical cells on these suspensors. There may thus be several embryos in one ovule, but ordinarily only one, usually the one which has grown most rapidly, will become a mature embryo.

Phylogeny and Classification

Formerly, many botanists looked to the Gnetopsida as the probable ancestor of the Angiospermeae; however, the earliest fossils of *Ephedra* and *Welwitschia* occur in upper Cretaceous formations, considerably later than the early angiosperm fossils; and *Gnetum,* the genus which most nearly resembles the angiosperms, has apparently left no fossil record. Reports of Permian pollen grains have not been confirmed. Furthermore, the characteristics in the Gnetopsida which seem to tie the Gymnospermeae to the Angiospermeae, the vessels in the xylem and the double integuments around the seed, show a decidedly different pattern of ontogenesis in the two groups. Consequently, few botanist today regard the Gnetopsida as the link between the angiosperms and the gymnosperms. Similarities between the Chenopodiaceae and the Gnetaceae, however, indicate that more research attention should be paid to the hypothesis that *Gnetum* has evolved from the flowering plants by simplification of some of the structures (Scagel et al. 1965).

The extant species of Gnetopsida can be classified in three families of one genus each, and in two orders, as suggested in Table 17-5 and Figure 17-2. The Ephedrales apparently evolved during the late Cretaceous from the Bennettitales. The primitiveness of *Ephedra* compared to *Welwitschia* as inferred from comparative anatomy is confirmed by the geological record.

During the early Tertiary, the Ephedrales was a larger order than it is today. There were also several species of *Welwitschia* at that time, and some of them were adapted to mesic habitats. Only one species has survived to the present time; it is especially well adapted to desert habitats, growing and surviving in an area where the annual precipitation averages less than 5 cm and where rainless periods of 3 to 5 years have been recorded. *Ephedra* species are also very drought resistant. Presumably, the Gnetales have evolved during the past few million years from plants similar to extant species of *Ephedra.* However, neither comparative anatomy nor paleobotany has provided any clear-cut evidence in support of this hypothesis.

Similarity in pollen grains further supports the hypothesis that the Ephedrales are descended from the Bennettitales (compare Figure 15-6 with Figure 17-11). Pollen morphology also tends to confirm the basic assumption in this chapter of a close relationship between the Cycadales and Ginkgoales (Figure 17-11).

Ecological Significance

The evolution of drought resistance to a greater degree than is found in most other plants has given the Ephedrales a distinct ecological advantage in arid habitats. Both genera have extensive and deep root systems, leathery leaves, sunken stomata, and abundant waxy cuticle, features characteristic of xerophytes. *Welwitschia mirabilis* is found only in the coastal deserts of southwest Africa, less than 150 km from the coast in a 1000 km long strip extending from 14° to 23° south latitude. It is definitely an endangered species. In 1954, it was estimated that there were fewer than 1000 *Welwitschia* plants left; therefore, the law provides a heavy fine and jail sentence for anyone caught collecting without a permit (Fuller and Tippo 1954). Recent visitors to southwest Africa still comment on its rarity.

Ephedra is found in the deserts of southwestern North America, in Mediterranean countries and Asia Minor eastward to India, in western China, and in the mountains of western South America. They grow primarily on sandy or rocky soil. Like *Welwitschia,* the ephedras are true xerophytes. Unlike *Welwitschia,* on the other hand, they are often the dominant species in the ecosystems to which they are adapted. Species of *Gnetum* are mostly vines native to mesic or luxuriant tropical forests. By utilizing the strong stems of other species for support, they can use nearly all of the food they manufacture to produce increase in length rather than girth and thus compete favorably for the light that is available in the forest crown. Other gnetums are trailing vines; if rooted near the top of a precipice or rock wall, these grow down the face of it to where there is abundant light. In less dense forests, shrubby species of *Gnetum* compete with ferns and shrubby angiosperms in the forest understory.

Economic Significance

Gnetophytes are of economic value in the production of ephedrine, a drug used to control asthma and other allergic problems and to increase blood pressure. The commercial source is an Asiatic species, *Ephedra sinica.* Another Asiatic species, *E. vulgaris,* is imported for use in herb teas and naturopathic medicines. Ephedra tea, commonly called "Brigham tea" or "Mormon tea" has a sweet, pleasant, licorice-like flavor. Seeds of several species are ground and added to flour to flavor bread (Kirk 1975).

The leaves and stems of the North American species of *Ephedra* are widely used to brew a "health tea." A common recipe for its preparation is to dump a handful of the leaves and young shoots, fresh or dried, into a cup of boiling water, preferably over a campfire and in an empty bean can or coffee can. Immediately remove the water from the fire and allow to steep for 20 minutes or more, depending on individual taste. Lemon juice, sugar, and jam may be added to bring out the flavor. *Ephedra nevadensis* is supposed to make the best tea, but all species make a pleasant tasting drink containing small amounts of pseudoephedrine, reportedly especially effective in controlling the sniffles that accompany mild head colds and asthma, and other mild allergies.

The American species of *Ephedra,* or Brigham tea, also known as Mexican tea, teamster tea, and joint fir, have a mild licorice flavor that most desert campers seem to enjoy. Indians and early settlers used a very strong solution to treat syphilis and some other diseases. Although it had little, if any, effect on the spirochetes that cause syphilis, it apparently reduced the severity of the symptoms.

As a tourist attraction, *Welwitschia* entices people to travel thousands of miles to see one of the most unique deserts of the world and the bizarre plant that helps to make it so unique.

Research Potential

Unfortunately, *Welwitschia mirabilis* is so rare and in such danger of becoming extinct, that it is not practical to recommend collecting it for scientific study. At times, it produces a fairly abundant crop of seeds, and if methods to grow these in greenhouses successfully can be discovered, it could be of great value in studying the nature of drought resistance in some plants. Any species that can survive 5 years of no precipitation certainly must possess physiologic adaptations of a special nature.

Ephedra is also a true xerophyte which offers potential for research into many aspects of drought tolerance. Detailed studies of the nature of the phloem and other tissues and of the development and nature of the steles, especially the roots, are needed. Research in stem and root anatomy has been neglected in this class of plants. A number of good studies of life cycles have been made but there are still some important details missing. Until more is known about the anatomy and other characteristics of these three genera, it will be impossible to reconstruct accurately their phylogeny and origin as well as their relationships to other vascular plants.

SUGGESTED READING

Two articles in *Botanical Review* are of special interest: "Metasequoia, fossil and living" by E. H. Fulling, 42: 215–316 (1976), and "Mesozoic conifers" by C. N. Miller, 43: 217–280 (1977).

The 1949 Yearbook of Agriculture published by the U.S. Department of Agriculture, entitled *Trees,* is devoted largely to the characteristics of gymnosperms. Some of the 138 articles are: "Pine breeding in the United States," "Cash crops from small forests," "Growing better timber," "Forests of Alaska," "City trees," "Keeping shade trees healthy," and "Key to the identification of woods without the aid of a hand lens or microscope."

R. J. Rodin has published papers on the leaf anatomy of *Welwitschia* in the *American Journal of Botany,* Vol. 45: 90–103 (1958). An old paper published in 1918 by W. F. Thompson in the *Botanical Gazette,* Vol. 65: 83–90, compares the vessel elements in the Gnetales to those in angiosperms.

An article in *Time Magazine,* Nov. 15, 1976, pages 79 and 80, tells about the Jari Forests Products' attempt to exploit the Amazon forest. This is followed up by another article on Sept. 10, 1979, pages 76–78. An article in *Fortune* in April 1981, pages 102–117, and one in *American Scientist,* July–August 1982, vol. 70, pages 394 to 401 discuss this further.

STUDENT EXERCISES

1. List the species of gymnosperms living within a radius of a few miles of your home. Are there any ginkgoes, cycads, or ephedras in your list?

2. Examine several pieces of furniture, observing carefully the patterns of growth rings and grain as seen from the ends of the pieces of wood as well as the surface. With the aid of a key (e.g., the one that begins on page 833 in the 1949 Yearbook of Agriculture) identify the kinds of wood in the furniture.

3. Find a stump of pine, spruce, or other gymnosperm and by examining the growth rings, develop a precipitation record for the past 12 years. Which year was the climate most favorable for the growth of the species you are observing according to this record? Compare your record with local weather bureau records.

4. List five ways in which conifers differ from cycads.
5. In larger cities one can usually find ginkgoes in some of the parks or around college campuses and cycads in greenhouses. List some of the characteristics of each that will enable you or someone else to recognize each of these when you encounter them.
6. What are the kinds of chlorophyll you would expect to find in C_3 plants like yellow pine, *Ginkgo,* or *Cycas revoluta?* In CAM plants like *Welwitschia?* What kinds of flagella would you expect to find on sperm cells of each of the above four species?
7. A fossil leaf attached to a stem to which a seed-bearing cone is also attached is seen to have dichotomous venation. Which of the seven classes of Tracheophyta does this leaf belong to? (Indicate all possible classes.)

SPECIAL INTEREST ESSAY 17–1

Forestry

Forestry is the study, management, and use of ecosystems dominated by trees. In the U.S., where forests cover almost 200 million hectares, it is a multibillion dollar industry. Forests provide (1) timber for lumber, paper, and other uses, (2) livestock grazing, (3) water for irrigation and power, (4) hunting and fishing, and (5) recreation, camping, and research.

For centuries, forests have provided lumber for building homes and fuel for cooking and heating homes, but in recent years, the list of forest products has been rapidly growing. Since 1882, we have been able to make paper from trees; today 90% of our paper comes from wood. The U.S. is the world's largest producer of paper, yet does not produce enough for her own needs; Canada, Finland, and Sweden are important exporters. Most of the paper is made from conifers, or softwoods, but other woods can be used. Species that are high in resins, such as pines, require different techniques in their processing than trees low in resins, like spruce and aspen.

Forest products are also used to manufacture textiles such as rayon, lignin for the manufacturing of plastics and molding resins, food products such as maple syrup and artificial vanilla flavoring, methyl alcohol with its multitude of industrial uses, perfumes, photographic film, solvents, explosives, synthetic rubber, and a host of other products including dyes, medicines, turpentine, cellophane, and artificial leather. Still, the search for new products goes on; and the goal is 100% utilization of the tree. If and when this is accomplished, the problem of what to do with forest wastes will have been solved. A major breakthrough occurred when methods to produce paper from wood chips, and even sawdust, were developed. Today, sawdust is also used to produce filters, glues, and plastics; and even the leaves of some trees are utilized to produce a variety of oils.

A major barrier to full usage of waste products is the leaching of sugars and other substances from the bark of logs being seasoned in sawmill ponds or floated down rivers to mills. Hopefully, even this problem will be solved in the near future. In the meantime, bacterial growth will continue to be stimulated in such bodies of water by the abundance of nutrients, oxygen content will be depleted by bacterial respiration, and the water will turn black from the dissolved organic complexes that form under anaerobic conditions.

While the forests of temperate climates have been utilized for centuries, tropical forests have largely been ignored until quite recently. Now, as our attention has been turned to the tropics, problems have arisen due to our lack of understanding of tropical ecosystems. In Europe and North America, early settlers knew that if the forest consisted of tall, vigorous, dense stands of trees, the soil would be excellent for agriculture after the forest was removed. In the tropics, this rule has not seemed to apply. Decomposition rates are much more rapid in hot tropical rainforests, and minerals are therefore rapidly leached from the soil. But despite the poor soils, lush vigorous forests grow in Central America, the Amazon Basin, and other tropical area because tropical ecosystems function differently from ecosystems in colder climes. Failure to recognize the innate differences in how ecosystems function has led to failure of many forestry projects. Furthermore, these projects may have already had far-reaching effects that are touching each of our lives.

Exploitation of the vast vegetational treasures of the Amazon Basin has tempted many would-be developers. We read in the newspapers of floating saw mills, of vast fields of rice, and of huge farms and cattle ranches cut out of the rain forest. One attempt to gain wealth from the rich resources of the Amazon was begun in the late 1960's by Jari Forest Products under the direction of multibillionaire Daniel K. Ludwig. The Jari Company purchased over a million hectares of land and began clearing the forest. Finding that there were more than 300 species of trees on any given hectare, they soon realized that it would not be economically feasible to harvest the native trees for either lumber or pulp. Some kinds of wood were too hard to use even for furniture. The company decided to plant fast-growing trees of their own choosing. But the heavy equipment brought in to clear the forest was so damaging to the soil, this project had to be abandoned; gmelina, the tree chosen to replace the native forest, would not grow on much of the soil there; Caribbean pine, a fast-growing gymnosperm which was next chosen to replace the native vegetation, grew well for a few years, then began to show signs of mineral malnutrition; rice, planted on the bottom lands, gave yields comparable to those obtained in Arkansas and Louisiana for only 4 or 5 years until its productivity fell to unprofitable levels. Hundreds of millions of dollars were spent on developing the area over a period of 10 or 12 years; but within 15 years of the time the project was started, it was apparent that it, like other large scale projects before it, was failing. Why have such projects generally failed? Perhaps the answer to that question can be found in a statement made by an official of the Jari Company: "Mr. Ludwig doesn't want to waste time with research," he said. "He just wants to begin. Naturally, we make mistakes, but we also get things done a lot faster."

Because tropical soils are generally low in organic matter, they do not hold mineral nutrients well; however, they are often covered by a 15 to 40 cm thick layer of leaves, twigs, and other litter. The native vegetation is well adapted to obtain mineral nutrients from this layer. With the aid of mycorrhizal fungi, the roots wrap around old leaves and twigs and draw nutrients from them. On the living leaves high in the forest canopy are foliicolous lichens which, like all lichens, obtain minerals from the atmosphere; as the leaves fall, the nutrients accumulated by these lichens are also made available to the roots of the trees and shrubs. The crops introduced from other areas lack these adaptations. For a few years following removal of the native forest, the new crops are able to obtain nutrients from the rapid decomposition of the litter layer as they percolate through the soil, but when that layer has been decomposed, the introduced plants stop growing well and begin to show symptoms of mineral deficiency.

The changes that take place in the vast forests of the Amazon Basin are believed to have far-reaching effects. Many ecologists claim that the gradual worldwide increase in atmospheric carbon dioxide, recorded since the 1950s, is due largely to attempted exploitation of tropical forests since that time stemming from (1) elimination of the high rate of photosynthesis by the native vegetation, and (2) the greatly increased rate of decomposition of the natural litter as it is aerated and mixed with the soil.

Recent research suggests that the tropical forests may be utilized for human benefit without destroying the soil. Strip logging, in which narrow bands of forest are harvested, beginning near the river, and then planted to trees (or crops) can allow transfer of nutrients from the unharvested area to the cultivated strips without damage to either. This method will not work, however, if the developers are greedy. "The Amazon ecosystem is like the goose that laid the golden egg," wrote C. F. Jordan (1982). "If the farmer can be satisfied with a little at a time, the system can sustain him well; but if he tries to take everything at once, it will perish."

18 THE FLOWERING PLANTS

Picture yourself living on a platform suspended between giant tropical trees 35 meters above the ground. From your platform, you catch the fragrance of the flowers, and you hear the buzzing of insects attracted to their beautiful blossoms. You see rodents, monkeys, birds, lizards, insects, and spiders that no one on the ground ever sees. A pleasant breeze is blowing and the sluggish mosquitoes that pestered you on the ground are gone, but in their place are other biting insects resembling horseflies and deerflies that you cannot identify, even with the aid of your keys, because they are not included; no one has ever seen them before. The huge boughs are carpeted with vegetation and even small trees are rooted in the layers of organic matter that have accumulated over the years.

One of the last frontiers to be explored by man is the canopy of the tropical rain forests (Figure 18-1). The canopy is that uppermost stratum where high levels of usable energy are available to plants and animals for photosynthesis and other metabolic processes. Ecologically, tropical forests are among the most productive of all ecosystems, and 90% of the food in these forests is produced within the canopy. The diversity of plant life in the canopy is almost unbelievably great; the diversity of the consumer organisms that are adapted to exploit the supply of food that the plants produce is apparently even greater (Perry 1984).

Until the 1970s, little was known about the plants and animals of the rain forest canopy. When a tree, overburdened with layers of epiphytes, came crashing down in a storm, it was possible to examine the plants that came down with it, but the animals of the canopy, and also many of the plants, especially pioneer species in the ecological successions of the canopy communities, remained unknown until methods were developed that made it possible for biologists to live in that mysterious celestial world for days at a time studying first hand its life.

The flora and fauna of the canopy of a tropical rain forest seem odd, or unconventional, to those of us accustomed to spending all our time on the ground. The forest itself may have several hundred species of trees in it, in contrast to forests of temperate zones where more than a dozen is rare. Many of these trees are 30 to 50 meters tall. The lowest branches may be 25 meters above the ground; they reach out in all directions, becoming entangled with branches of other trees to form a dense web. Above those lowest branches, the emerald canopy

Figure 18–1 Studying the canopy of a tropical forest. A new frontier awaits dedicated botanists in the world in the treetops.

reaches upward toward the sunlight, rising another 25 meters, as tall as an eight-story hotel. Lianas, or woody vines, are attached to the main trunk and to the branches; drought-tolerant epiphytes such as cacti, bromeliads, and lichens are common where the light level is high and breezes common (Figure 18-2). At lower levels and in the crotches of the branches where humidity is higher and free moisture available, epiphytic ferns occur and mosses and liverworts are abundant. Far below the canopy, at ground level, plants adapted to living at very low light intensity and constant humidity and temperature, form a thicket of vegetation that hides from view the canopy and its fascinating flora and fauna.

At all levels, the most abundant plants are the Anthopsida. They are characterized by flowers of brilliant colors and pleasing fragrances, designed to attract insects, birds, and other pollinating animals. Many of these animals—possibly 80 to 90% of the insects, for example—had never been seen by man until special projects to study life in the canopy were designed. The same is true of some of the flowering plants, although energetic botanists for decades have scrutinized every giant tree they could get to that has fallen in a tropical storm.

CLASS 7. ANTHOPSIDA

The Anthopsida (*antho* = a flower or a bright color), or flowering plants, are also known as the Angiospermeae (*angio* = a box or case, and *sperma* = semen or seed). Magnoliopsida is another name which is now used by many botanists. With approximately 220,000 known species, they are the largest class of plants; and in terms of primary productivity, they are also the most important class and dominate all of the terrestrial ecosystems of the world, with the exception of the coniferous forests and tundra. A hundred million years ago, they were an insignificant group dominating some disturbed ecosystems and living as subordinates in various gymnosperm-dominated ecosystems. The 220,000 species are included in 11,000 genera, 275 families, and 49 orders.

Angiosperms are most easily recognized by their unique reproductive structures: flowers consisting of pistils and stamens and usually petals and sepals (Figure 18-3). They may also be recognized by their pollen grains uniquely adapted to insect pollination, by their complex net pattern of leaf venation, by the type of buds they produce, and by the fruits which enclose the seeds and give rise to the name angiosperm (Figure 18-4). The flowers and fruit are so distinctive that when they are present, no other plants can be confused with them; in their absence, however, some angiosperms might conceivably be confused with ferns, cycads, or gnetophytes. Careful observation of leaf and other characteristics quickly eliminates these possibilities.

The **pistil,** which is made up of one or more carpels, or megasporophylls, is especially unique in that it consists of a stigma on which the pollen grains are able to germinate while exposed to the drying influence of air. Most pollen grains, in turn, have three or more sutures through which the germination tube can emerge, ensuring that the pollen grain will never land upside-down and thus be in an unfavorable position for germination. The number and arrangement of stamens and petals are also important in the identification of species (and genera and families), both to taxonomists and to the insects which perform the work of pollination.

Figure 18–2 Epiphytes are common in the canopy: a bromeliad on a fig tree (in a greenhouse).

The 49 orders fit into 10 subclasses within 2 natural taxonomic groups, or **series,** the **Dicotyledonae** and **Monocotyledonae,** commonly called **dicots** and **monocots,** respectively. The fossil record indicates that both originated about the same time. Probable lines of relationship among the 49 orders of the class are shown in Figure 18-5. The broad gray line in the chart separates the two taxonomic series from each other. Dashed lines outline the subclasses which are indicated by the first three or four letters of their names. Series characteristics are compared in Table 18-1, and subclass characteristics in Table 18-2.

The Dicotyledonae (*di* = two and *cotyl* = cup or socket, referring to the absorption organ of the embryo) are characterized by net-veined leaves with pinnate or palmate venation, stamens and perianth segments in 4s, 5s, or multiples of 4 or 5, anchorage usually by a taproot, stems with eusteles which develop into ectophloic siphonosteles, stems often woody, leaves often opposite, and embryos having two cotyledons. Oaks, walnuts, buckwheat, willows, roses, peas, pinks, buttercups, potatoes, daisies, cabbages, and waterlilies are examples of dicots.

The Monocotyledonae (*mono* = one, and *cotyl* = cup or socket) are characterized by parallel-veined leaves, stamens and perianth segments in 3s or multiples of 3, root system usually fibrous, stems with atactosteles, seldom woody, leaves never opposite, usually two or three ranked, and embryo having but one cotyledon. Wheat, corn, and other grasses, papyrus, sedges, lilies, bananas, pineapples, palm trees, duckweed, flowering quillwort, orchids, cattails, elodea, and irises are examples of monocots.

Hickey and Doyle (1977) present a convincing argument that all ten subclasses evolved from a common cycad-like ancestor fairly early in the Cretaceous Period. Hughes (1977), Krassilov (1977), and others, however, believe that the angiosperms had a polyphyletic origin; if they are correct, the Anthopsida should be split into at least two, and possibly four, separate classes.

The fossil record seems to favor the **monophyletic origin hypothesis;** Takhtajan (1969), for example, has pointed out that the great uniformity among angiosperms in at least seven characteristics—(1) staminal structure, (2) **carpels** with **style** and **stigma,** (3) constancy of

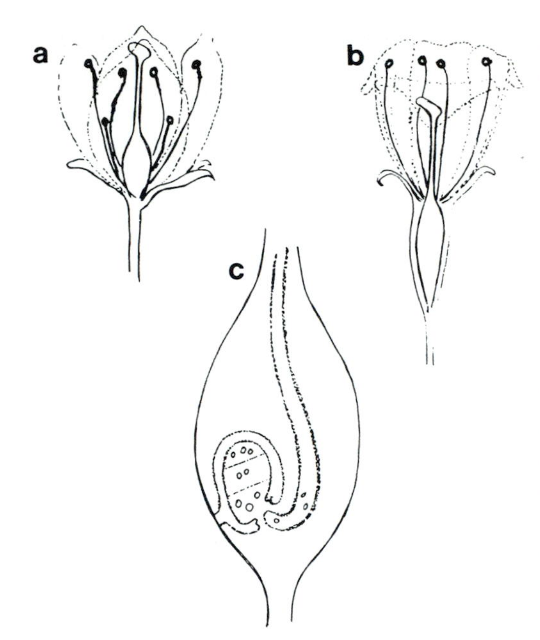

Figure 18–3 Structure of flowers. (a) Mustard family is characterized by flowers with 4 petals, 4 sepals, a lobed stigma, a superior ovary and a seed pod (silique) with 2 chambers, and 6 stamens, 2 shorter than the others; (b) Evening Primrose family is characterized by flowers with 4 petals, 4 sepals, a 4-parted stigma, an inferior ovary, and a seed pod with 2 or 4 chambers; (c) the pollen grains germinate on the stigma and send pollen tubes down through the style; when they reach the ovary, each tube discharges 2 sperm cells, one of which fertilizes the egg; another sperm fertilizes the 2 nuclei in the center of the ovule.

relative position of **gynoecium** and **androecium** in the flower, (4) three-nucleate pollen tubes, (5) characteristic female gametophyte with egg flanked by **synergids** and lacking an archegonium, (6) the **double fertilization** process, and (7) presence of sieve tubes—strongly favors the monophyletic hypothesis. To Takhtajan's list may be added the following unique characteristics of angiosperms: (8) development of the embryo within a closed carpel well protected from the effects of drought, low humidity, and small herbivorous animals; (9) leaves having three or more orders of branching of the veins; (10) pollen grains (microspores) characterized by well-developed, usually reticulate, **exine sculpturing,** the presence of columnar structure, and absence of a laminated exine; (11) well-developed cambium in both stems and roots; and (12) the replacement of tracheids in the xylem by vessels and fibers. Except for the last two characteristics, these are almost completely limited to the angiosperms, and all are characteristically consistent throughout the class.

The earliest flowering plants were probably "weedy" semixerophytic shrubs adapted to stream margins, gravel bars, and similar disturbed habitats (Hickey and Doyle 1977). They differed from the cycad-like gymnosperms from which they evolved in having smaller leaves, a closed carpel, and sculptured pollen grains, and they resembled closely many of the shrubby members of the Magnoliales of today, such as *Drymis, Wintera, Degeneria,* and *Michelia.* Originating during the Barremian epoch of the Lower Cretaceous (Table 18-3), or possibly slightly earlier, they soon spread both north and south from their point of origin, which was apparently near the center of Gondwanaland at about 10° South Latitude. By the beginning of the Cenomanian epoch, some 20 million years later, they were abundant at all latitudes between 75° S and 80° N. It took only a few million years from the origin of the first shrubby angiosperms until species had evolved which were well adapted to aquatic habitats, as understory shrubs and herbs in forests, and as dominant species in deserts and semiarid lands. Slightly later, as geologists count time, angiosperms had evolved which were dominants in climax forest communities. Today, the angiosperms are the most widespread of all the Tracheophyta. Even in forests dominated by gymnosperms, the majority of species are flowering plants.

Ecologically, the Anthopsida are the most important of all plants. They are also the ones people know best. Most people, in fact, when they think of plants, think only of this group with their big leaves, showy flowers, stiff stems, and extensive root systems, and which start out life as seeds. These are the angiosperms, late-comers in the evolution of life, the largest of

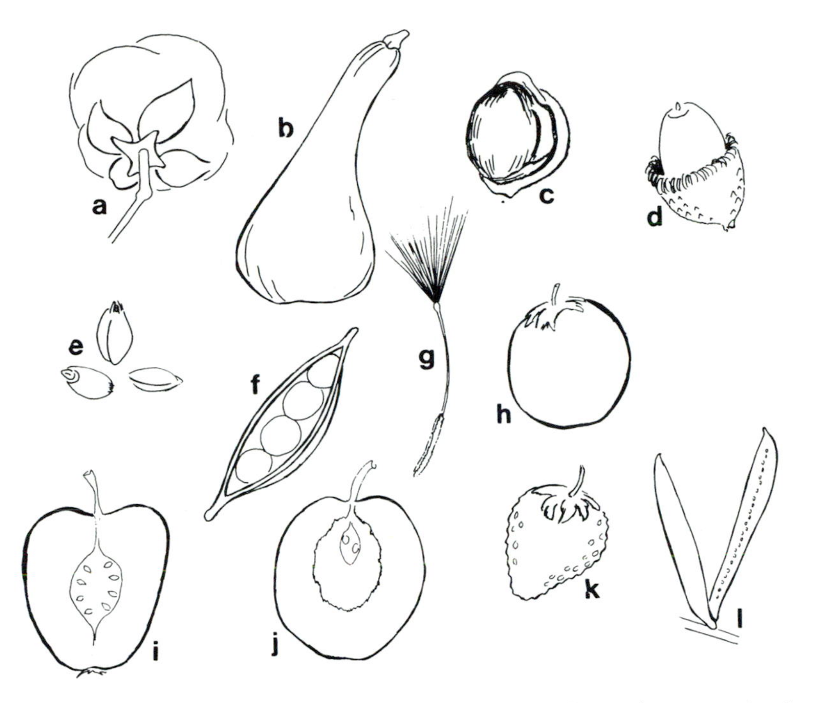

Figure 18–4 Stimulated by hormones secreted by the developing embryo, the ovary develops into a fruit. (a) Capsule (cotton); (b) pepo (gourd); (c) nut; (d) acorn; (e) caryopsis (wheat); (f) legume; (g) achene; (h) berry (tomato); (i) pome; (j) drupe; (k) multiple achenes on a fleshy receptacle (strawberry); (l) silique.

all the classes in numbers of species, and the group of plants that nearly all terrestrial animals, including man, depend on for food and for shelter.

The combination of closed carpel and insect pollination has been hypothesized to have been the most essential feature giving flowering plants their ecological advantage over other vascular plants in terrestrial ecosystems. The tiny reticulations on the walls of the pollen grains of most angiosperms allow them to stick to the bodies of insect vectors, thus promoting cross-pollination. Understory plants that are widely separated in a forest or other ecosystem would have to produce large quantities of pollen, if they were wind-pollinated, in order to ensure cross-fertilization. Angiosperms, however, need to produce only small amounts of pollen which can be carried to other members of the same species by insect or other animal vectors. Colors and odors associated with the perianth are attractive to insects and consequently aid in successful pollination. Even in angiosperms which are wind-pollinated, the structure of the walls of the pollen grains indicates they have evolved from species that were insect-pollinated. With the exception of the cycad genus *Encephalartos,* which is beetle-pollinated, angiosperms are the only known insect-pollinated plants.

General Morphological and Physiological Characteristics

The angiosperm body is clearly differentiated into three distinct organ systems: an anchorage system, consisting of roots, which are also absorption organs for both water and minerals and are frequently storage organs; a production system, consisting primarily of leaves, which are

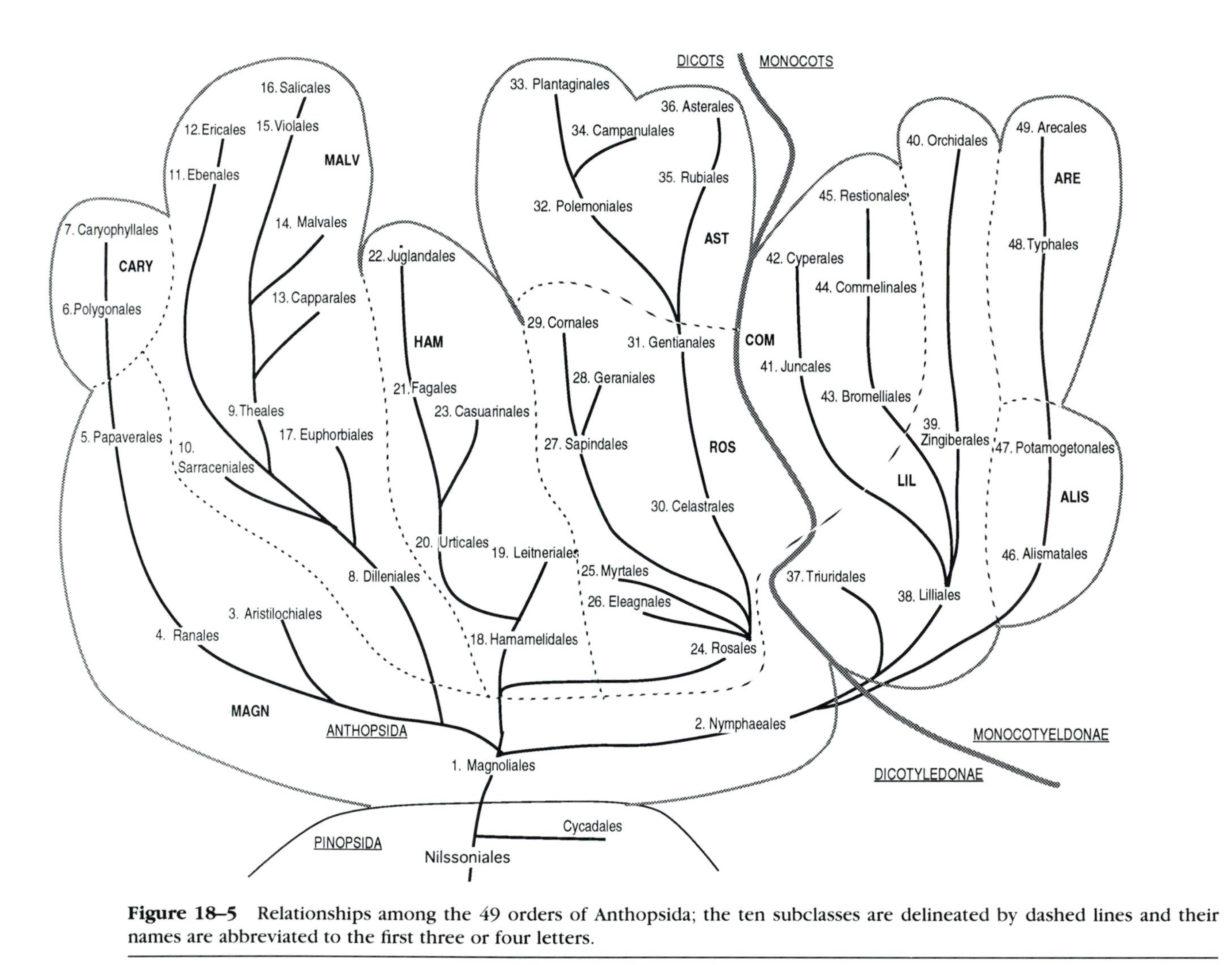

Figure 18–5 Relationships among the 49 orders of Anthopsida; the ten subclasses are delineated by dashed lines and their names are abbreviated to the first three or four letters.

TABLE 18-1

DISTINGUISHING TRAITS OF THE TAXONOMIC SERIES, MONOCOTYLEDONAE AND DICOTYLEDONAE

	Monocots	Dicots
Flower parts	In 3s or multiples of 3	In 4s or 5s or multiples of 4 or 5
Leaf venation	Parallel veins	Net veins, either pinnate or palmate
Pollen	Inaperturate, monocolpate, monoporate types	Polyporate, polycolpate, fenestrate, etc.
Roots	Fibrous roots, primary root usually temporary	Tap root, primary root persisting
Stele	Atactostele in stems; may or may not have a pith in the roots	Eustele in stem developing into an ectophloic siphonostele; protostele in roots
Phyllotaxy	Always alternate	Whorled, opposite, or alternate
Cambium	Absent or very simple	Infrafascicular
Secondary xylem	Seldom woody and then more or less fibrous	Mostly woody with abundant secondary xylem
Embryo	One cotyledon, rarely none	Usually two cotyledons, rarely one, three, or four

well adapted to make efficient use of light in the photosynthetic process; and a display and conduction system, consisting of stems, which aid in displaying the leaves to the most favorable light conditions as well as conducting food from the leaves. In addition, mature plants possess a reproductive system consisting of flowers, each of which possesses either **pistils** or **stamens,** or both, and also **perianth** units which may be differentiated into **petals** and **sepals** (Figure 18-3).

Anatomically, the angiosperms are complex plants. Dicots typically have the eustele-ectophloic siphonostele pattern of vascular tissues in the stems, with exarch protosteles in the roots and leaves. Secondary xylem and phloem are produced by a cambium in the roots, stems, and largest leaf veins. Xylem is made up primarily of vessels and fibers in the roots, stems, petioles, and large veins, but the smaller veins contain only tracheids. Monocots typically have an atactostele pattern of vascular tissues in the stems with an exarch siphonostele in the roots and a modified exarch protostele, similar to that in the dicots, in the leaf veins.

Dispersal of the vessels in stems varies considerably among species. Vessels are relatively long conduits—about 60 cm long in maple and over 3 m in ash—made up of rather long cylindrical or sometimes drum-shaped cells called vessel elements. The number of vessel elements in one vessel varies from species to species, but individual elements range from about 75 to 200 or more micrometers in length. The vessels connect with each other by means of pit pairs through tapered end walls, but within the vessels, the individual elements or cells connect more directly through simple perforations, multiple perforations, or scalariform end walls. In some species, such as maple and poplar (Figure 18-6), the vessel elements very gradually decrease in size from spring through summer, but in other species, such as oak and black locust (Figure 18-7), the summer vessels are much smaller than the spring vessels and the change in size is

TABLE 18–2

MAJOR CHARACTERISTICS OF THE TEN SUBCLASSES OF ANTHOPSIDA

Subclass (Order-Fam)	Leaf Characteristics	Floral Characteristics	Pollen Characteristics	Other Characteristics	Some included Families
Part A. Dicotyledonae					
Magnoliidae (5–23)	Large, mostly simple with entire margins, often thick, narrow, with pinnate venation	Carpels free, separate; flowers polypetalous or with petal-like sepals; many stamens and pistils	Often monosulcate or uniaperturate, always binucleate; mostly beetle-pollinated	Vessels often lacking; endosperm usually copious; woody	Magnoliaceae, Winteraceae, Lauraceae, Degeneriaceae, Nymphaeaceae, Ranunculaceae, Papaveraceae
Caryophyllidae (2–11)	Thin, narrow, leathery leaves; often modified, stipules well-developed	Highly diverse, often with 5 petal-like sepals	Trinucleate, polyporate with thick tectate exine	Betacyanin and betaxanthin instead of anthocyanins, etc	Chenopodiaceae, Nyctaginaceae, Cactaceae, Polygonaceae, Rafflesiaceae
Malvidae (10–64)	Simple leaves, often from resinous buds, often with deep lobes, palmate venation	Syncarpous, parietal placentation; seeds many; stamens usually many	Binucleate, rarely trinucleate; often periporate	Includes some insectivorous species	Paeoniaceae, Theaceae, Tiliaceae, Malvaceae, Droseraceae, Violaceae, Salicaceae
Hamamelidae (6–19)	Mostly simple with deep angular lobes; may be pinnately compound	Perianth poorly developed or lacking; often unisexual; wind-pollinated	Often tricolpate with long furrow or triaperturate	Almost always woody	Platanaceae, Ulmaceae, Moraceae, Urticaceae, Fagaceae, Betulaceae, Juglandaceae
Rosidae (7–93)	Often pinnately or palmately compound with well-developed stipules	Petals 5, separate, or rarely united; frequently many stamens and pistils	Mostly binucleate; tricolpate, also tricolporate	Often shrubs or small trees, also herbaceous	Rosaceae, Leguminosae, Myrtaceae, Cornaceae, Eleagnaceae, Vitaceae, Euphorbiaceae, Onagraceae, Umbelliferae

Asteridae (6–33)	Mostly simple, less often pinnately compound	Mostly 5 united petals, 2 carpels; 5 or less stamens, never opposite corolla lobes	Tricolporate, periporate	Resin or aromatic oils common; latex in some families	Asclepiadaceae, Solanaceae, Labiatae, Scrophulariaceae, Oleaceae, Gentianaceae, Compositae
Part B. Monocotyledonae					
Liliidae (4–18)	Relatively narrow and parallel veined, often broad with net venation	Actinomorphic; highly adapted to insect pollination; also zygomorphic with inferior ovary	Monosulcate, usually binucleate	Vessels usually confined to the roots; food reserves often oils	Liliaceae, Iridaceae, Orchidaceae, Dioscoraceae, Agavaceae, Zingiberaceae, Musaceae
Alismatidae (2–10)	Usually large, often peltate	Carpels usually separate	Always trinucleate, periporate	Endosperm usually lacking, never starchy; chiefly aquatics	Alismataceae, Najadaceae, Zosteraceae, Potamogetonaceae
Aridae (2–7)	Large, often petiolate, pinnately or palmately compound	Flowers often crowded into a spadix; ovary usually superior	Monolete, monocolpate, monoporate, trichotomocolpate	Many aquatics, also trees	Araceae, Lemnaceae, Palmae, Typhaceae, Sparganaceae, Cyclanthaceae
Commelinidae (5–13)	Narrow, parallel veined, ligulate in some families	Petals and sepals well differentiated but petals may be reduced to chaffy bracts	Adapted for wind pollination; monoporate	Endosperm mostly starchy, sometimes lacking	Xiridaceae, Restionaceae, Juncaceae, Bromeliaceae, Gramineae, Cyperaceae

Note: The number of orders and families (Fam) in each subclass is indicated in the first column.

From Takhtajan (1964) and Cronquist (1968).

TABLE 18–3

EPOCHS AND EVENTS OF THE CRETACEOUS PERIOD

Period	Epoch	Years Since Beginning	Events
Tertiary	Oligocene	63 million	First grass-like monocots appear; early pennate diatoms evolve
	Maestrichtian		Ephedrales appear; Bennettitales become extinct
	Campanian		Almost all modern families of Anthopsida present
Upper Cretaceous	Santonian		Nilssoniales become extinct
	Coniacean		
	Turonian		Centric diatoms abundant
	Cenomanian	100 million	
	Albian		Early rosalean fossils appear
	Aptian		
Lower Cretaceous	Barremian		First Angiosperms (Ranales) evolve
	Hauterianian		Cycads and conifers dominant
	Valangian		
	Berriasian	136	Earliest centric diatoms appear
Jurassic		180 million	Ginkgoes decline; modern red seaweeds replace *Solenopora*
Triassic		230 million	Ginkgoes and cycads dominant

rather abrupt. The former species are said to produce **diffuse-porous wood,** the latter produce **ring-porous wood.** This is in sharp contrast to gymnosperms and most other vascular plants, which produce no vessels at all and have, therefore, **nonporous wood.**

Ring-porous angiosperms are more commonly the ecological dominants on drier sites, diffuse-porous species on low wet sites (Guthrie 1989). A hypothesis to explain this is that in the springtime, when soil moisture is abundant even on the dry sites, compared to the rest of the year, plants with large vessels would be able to make better use of that extra moisture. On low wet sites, soil moisture is relatively constant throughout the growing season and diffuse-porous plants would effectively compete with the ring-porous ones.

In addition to vessels, there are tracheids, fibers, wood parenchyma cells, and ray parenchyma cells in the xylem, or wood, of most angiosperms. A knowledge of how these are arranged relative to each other is essential to correct identification of species from observation of wood alone, including petrified wood and leaf compressions (Figure 18–8). The beautiful grain seen in some types of wood and much sought after in furniture is dependent on the arrangement of the different kinds of cells relative to each other in the xylem.

Young stems contain pith, xylem, cambium, phloem, cortex, and epidermis; in older stems, phellogen and cork are present. Phellogen is also called cork cambium. There is usually no pericycle or endodermis in angiosperm stems. The epidermis of young stems is replaced by cork tissue in older stems. The cork arises from the phellogen each year; the phellogen

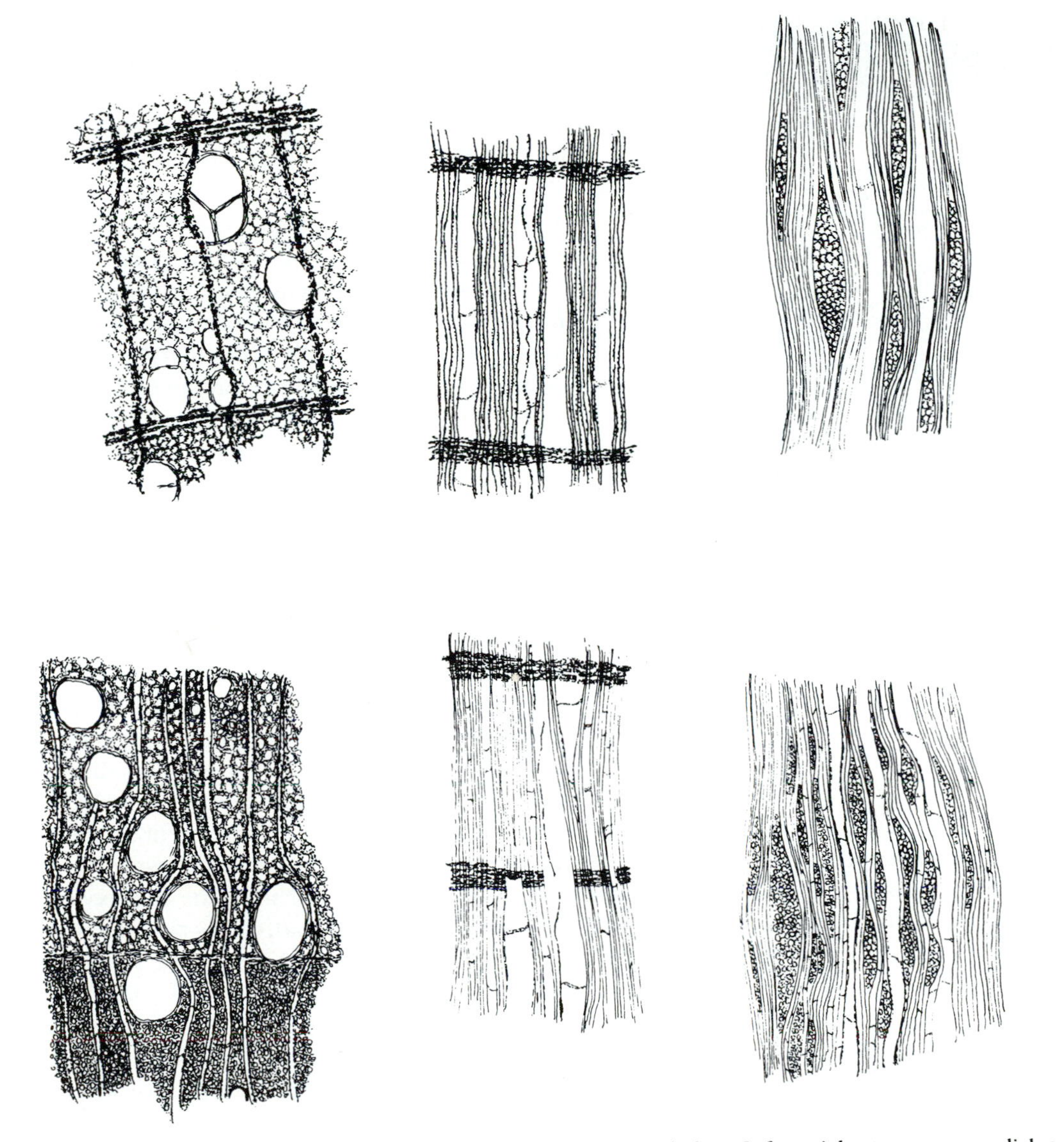

Figure 18–6 Diffuse-porous wood, rock or sugar maple above, poplar below. Left to right: transverse, radial, and tangential views.

differentiates from cortex or phloem early in the season and is active for several weeks. As the stem continues to increase in diameter, as the result of cambial activity in producing secondary xylem and phloem, it also increases in circumference, resulting in the old protective tissue, epidermis or cork, splitting or tearing in some way to make room for the new tissues. Bark, therefore, is fissured or cracked, the exact pattern depending on the species and how the bark tissues, primarily phloem and cork, are laid down.

Figure 18–7 Ring porous-wood, black locust.

Although angiosperm xylem is generally more complex than xylem of other vascular plants, a few of the flowering plants, *Drymis* and *Wintera,* for example, have very simple xylem containing only tracheids. However, most angiosperms have both vessels and fibers in the xylem and both companion cells and sieve tubes in their phloem, in contrast to gymnosperms, ferns, and lower vascular plants which have only tracheids and sieve cells in their xylem and phloem, respectively. Primary xylem and phloem are produced from the apical meristem as the stem elongates in juvenile growth (Figure 18–8); secondary xylem and phloem are produced in woody dicots and mature herbaceous dicots from vascular cambium. In monocots, there is generally no cambium tissue; all vascular tissues (with few exceptions) are primary.

The elongation, or primary growth, of dicot stems comes from a tunica-corpus type of apical meristem. If one were to make a series of cross sections of a stem tip beginning near the growing point and continuing to the mature stem, one would see the following tissues: first would be a section consisting of two layers of undifferentiated meristematic cells, next would be a section in which some of the cells, making up a procambial area, are slightly different from the others, although they are still all meristematic. Next would be a section in which the cells in the center of the procambial region have differentiated into vessels and fibers and are now, therefore, protoxylem. Finally, the remaining sections would show two metaxylem areas, inside and outside the protoxylem, in which the vessels are larger than in the protoxylem. As the xylem differentiates, the other tissues—phloem, cortex, and epidermis—also differentiate, and the mature first year stem contains all of the primary tissues arranged in the form of a eustele which gradually develops into an ectophloic siphonostele as the separated vascular bundles enlarge and grow together.

In monocots, only primary growth of stems and leaves generally occurs, although many species in the Liliidae possess a cambium and produce secondary vascular tissues. In contrast to dicots, the internodes continue to elongate as they thicken. In most monocots, there are both an apical meristem and a peripheral primary thickening meristem in the stems. The leaves of most monocots develop from an apical meristem, maturing from the tip downwards. In the grasses, however, the meristem is intercalary, the base of the leaf blade remaining permanently meristematic (Figure 18–9). Cronquist (1968), in pointing out how this adaptation may account for the great ecological success of the grass family, has commented, "Anyone who has pushed a lawnmower should recognize the significance of the intercalary meristem in permitting a plant to withstand grazing."

Angiosperm roots differ from stems in that the root stele is usually an exarch protostele. Lacking pith, the xylem forms a star-like body in the center of the root. The number of points, or protoxylem poles, on the star depends on the species; the xylem may be diarch, triarch, tetrarch, pentarch, or polyarch (Figure 15-2). Most angiosperm roots contain xylem, phloem, pericycle, endodermis, cortex, epidermis, and often other tissues. Dicot roots also contain cambium, and in mature, perennial dicots phellogen and cork replace the epidermis.

The pericycle is a meristematic tissue forming the outermost layer of cells in the stele. Meristematic activity is usually confined to the cells at the xylem points, and secondary roots arise in rows opposite these points. Thus, underground stems, such as quackgrass rhizomes (often called "roots") and potato tubers, are easily distinguished from roots: branching of stems follows a definite phyllotactic pattern, whereas secondary roots tend to arise in vertical rows; roots never have buds or leaves on them while underground stems do (the "eyes" of a potato, for example, and the scales on quackgrass rhizomes).

Where root and stem meet is called the **collet.** At this point, the xylem of the root branches from a single strand to a cylinder. Primitive man often capitalized on this characteristic in his arrow making; the light shaft of the arrow was made of the pithy stem and the hard point of his arrow of the upper part of the root just below the collet. Interesting furniture pieces are sometimes made by taking advantage of the contrast between root and stem where they meet at the collet.

Figure 18–8 Fossil fern frond from the Upper Cretaceous, Blackhawk Formation, Utah, collected by Lee R. Palmer. (Photo courtesy Lee R. Palmer, California Polytechnic State University, San Luis Obispo.)

The chief functions of roots are anchorage, absorption of water, absorption of minerals, and storage of starch and other food substances. The roots are also primarily responsible for the ability of herbaceous perennials to survive the cold of winter. For example, alfalfa roots which contain abundant starch survive harsh winters very well while those in which the food reserves have been depleted in producing new growth following a late hay harvest winter-kill badly. In order for woody plants to compete well in the continuing contest to display their foliage to the highest possible light intensity, a strong root system that firmly anchors the plant to the soil is a valuable asset.

Absorption of water and minerals is primarily by small, vigorously growing roots. Large quantities of food, mostly carbohydrates, are used by roots in producing thousands of root tips and pushing these through the soil. In their growth characteristics, roots are strongly and positively **hydrotropic,** turning and growing toward areas where moisture is most abundant. Observing this hydrotropic phenomenon, many amateur botanists have assumed that plants

Figure 18–9 Grass bud. (A–E) Successive stages in development.

actively "search" for water and that by withholding water, plants can be forced to grow deeper in their search. Since they have no nervous system and no consciousness, plants do not search for water or anything else; they simply grow fastest where conditions are best. Food, moisture, and oxygen must all be present for cells to divide and grow well. The ideal soil for good root growth is approximately 50% solid particles and 50% pore space. As water is added to soil, it moves through the pores and reaches the roots where it is absorbed, and as it moves, it carries with it dissolved mineral. As water is removed from the pores by the roots, it is replaced by air, which is essential for respiration needed for root growth. When most of the space is filled with water, there will not be enough air, and root growth is poor. When most of the water is gone, it will become limiting and root growth will be poor. As soil moisture is gradually depleted from the optimum level, vigorous root growth brings new rootlets in contact with pores which are still filled with water. To a large extent, plants obtain their moisture by the roots growing to where the water is rather than by water moving to the roots. In some species, the total underground biomass may considerably exceed the aboveground biomass (Pearson 1965).

In nature, many species of plants usually produce a type of root different from the "typical" root seen in the laboratory; this is a thick, dichotomously branching type called **mycorrhizae** in which a symbiotic relationship between a flowering plant and a fungus, usually a basidiomycete, exists (Figure 6–12). Mycorrhizae seem to improve the water-gathering ability of roots and in addition aid in absorbing minerals. In some species, mycorrhizae are able to convert cellulose and other complex molecules into soluble substances that can be utilized by the flowering plant.

Angiosperm leaves contain **epidermis, palisade mesophyll, spongy mesophyll, xylem,** and **phloem** (Figure 18–10). Most of the chlorophyll—over 80% according to measurements made by R. N. Schurhoff in the early 1900s—is located in the palisade layer. The

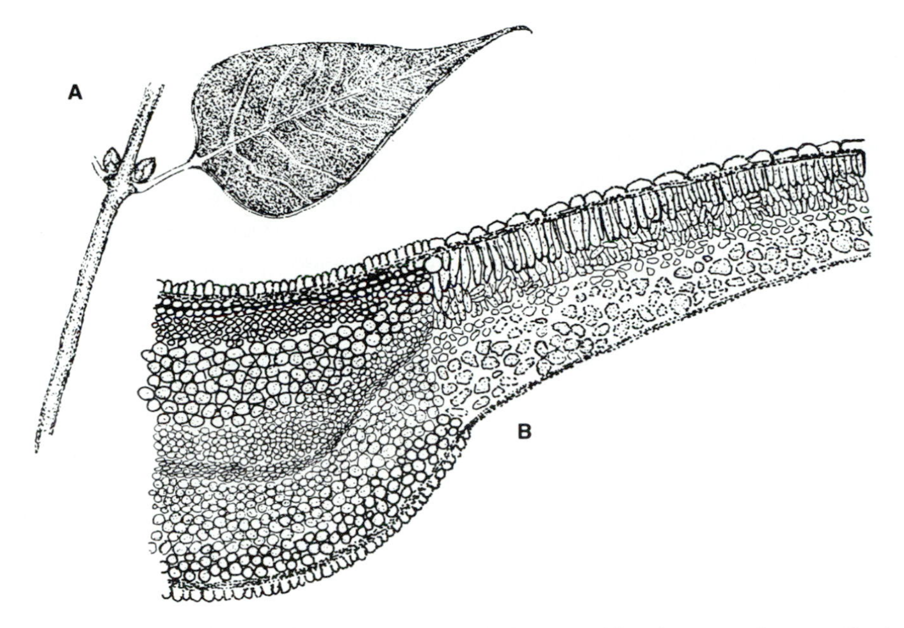

Figure 18–10 Angiosperm leaf (A) and its cross section (B). Upper epidermis at top, then a palisade mesophyll, spongy mesophyll, lower epidermis; vein to the left in center of leaf. Over 80% of the photosynthesis in a plant normally takes place in mesophyll tissue below,

epidermis is covered with a layer of cutin, a waxy substance which prevents evaporation of water. Openings called stomates (Figure 18–11) allow entrance of the gases necessary for photosynthesis and respiration, but also allow water to escape. Under conditions of water stress, the stomates close, thus stopping evaporation. In most angiosperms, the stomates are also genetically programmed to close at night, an adaptation which prevents loss of water at a time when photosynthesis is not taking place.

Whereas the leaves of other vascular plants arise from a single layer of cells, or at most two layers, in the apical meristem, the angiosperm leaf develops from three layers: the petiole and midvein from one, secondary veins from a second, and blade and higher order veins from a third.

Within the plastids of the palisade cells, the thylakoids are stacked into well-defined **grana.** Every quantum, or **photon,** of light energy entering the leaf has a high probability of being trapped by a chlorophyll molecule. Many photons of green and yellow light are reflected from the surface of the grana, but very little energy passes on through the leaf. Within the grana, the energy is used in two ways: to break down water molecules into hydrogen ions, free oxygen, and active electrons, and to attach phosphate radicals to molecules of adenosine diphosphate (ADP) and thus form adenosine triphosphate (ATP).

$$12\ H_2O \rightarrow 24\ H^+ + 24\ e^- + 6\ O_2 \qquad (1)$$

$$n\ Ad(P_i)_2 + n\ P_i \rightarrow n\ Ad(P_i)_3 \qquad (2)$$

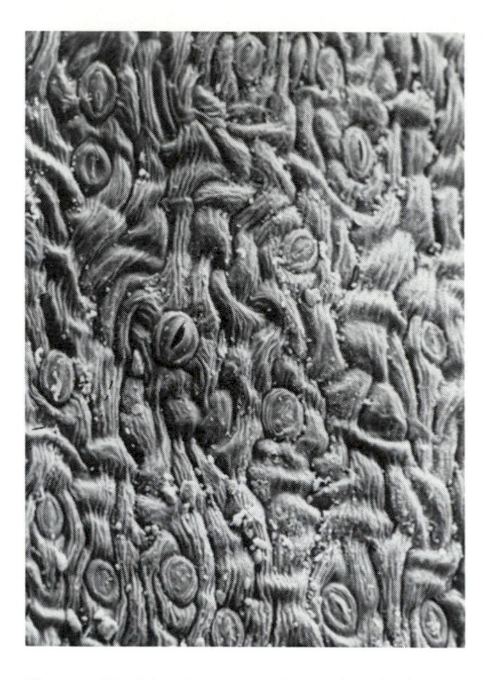

Figure 18–11 Lower surface of leaf of *Chrysothamnus,* showing stomata. (SEM courtesy of W.M. Hess, Brigham Young University Microscopy Lab, Provo, Utah.)

For every molecule of glucose formed and stored in photosynthesis, 24 molecules of water must be reduced and many molecules of ATP must be produced. The hydrogen ions released in Equation 1 combine with a hydrogen receptor, nicotinamide adenine dinucleotide phosphate (NADP) to form $NADPH_2$ which carries the hydrogen ions to the sites where they can be released again in a reverse reaction.

Also in the mesophyll cells of most angiosperms, carbon dioxide, entering through the stomates, combines with a phosphorus-containing 5-carbon sugar, **ribulose diphosphate,** to form phosphoglycerate (PGA):

$$6\ CO_2 + 6\ C_5H_{10}O_5(P_i)_2 \rightarrow 12\ C_3H_5O_3(P_i)_2 + 6\ O^* \qquad (3)$$

The energy to drive this reaction comes from the conversion of ATP to ADP, the reverse reaction of Reaction 2, above. The nascent oxygen (O*) produced combines with half of the hydrogen ions released in the photoreduction of water (Equation 1) which it reoxidizes to water.

$$12\ H^+ + 12\ e^- + 6\ O^* \rightarrow 6\ H_2O\ (4) \qquad (4)$$

The remaining hydrogen ions combine with PGA to produce triose:

$$12\ H^+ + 12\ e^- + 12\ C_3H_5O_3(P_i) \rightarrow 12\ C_3H_6O_3(P_i) \qquad (5)$$

Conversion of ATP to ADP provides the energy needed to convert triose into two more complex sugars: a hexose (glucose), which is temporarily stored in the mesophyll, and a pentose (ribulose diphosphate), which is used to absorb more carbon dioxide and keep the photosynthetic process going. The stored hexose is eventually transported through the phloem to roots, flowers, growing buds, and other areas where it is used in various metabolic processes or else stored, usually as starch. The entire process, consisting of at least 50 individual chemical reactions, in which carbon dioxide reacts with ribulose diphosphate to form PGA, which then reacts with NADPH to form triose molecules that combine with each other to produce glucose and to reconstitute ribulose diphosphate, is known as the **Calvin cycle,** named after Nobel prizewinner Melvin Calvin, who was instrumental in discovering the process.

Within the cells of leaves, there are other organelles, minute structures called **peroxisomes** (Figure 6–5), which are also sensitive to light and which are associated with an increase in the rate of respiration in the presence of light. What adaptive value such **photorespiration** may have is not fully understood at the present time. Peroxisomes are found in all kinds of cells, plant and animal. What is known is that when light intensity increases, the rate of respiration also increases, especially if oxygen is abundant and carbon dioxide is limiting, and that the energy released in the peroxisomes, unlike that released in mitochondria, produces no ATP, only heat. Inasmuch as some of the immediate products of photosynthesis may be toxic to the cell or its organelles, it is believed that the beneficial function of peroxisomes is to degrade these substances before they build up to toxic levels. Under conditions of intense light, or at high temperatures and moderately intense light, as much as 50% of the hexose produced in the plastids is immediately used up in photorespiration in the surrounding peroxisomes.

Some angiosperms are able to avoid the negative effects of photorespiration by utilizing the **Hatch-Slack pathway** of photosynthesis. In this process, carbon dioxide from the atmosphere combines with **phosphoenolpyruvate (PEP)** instead of ribulose diphosphate, to form oxaloacetate in the first "dark reaction" of photosynthesis. The oxaloacetate is then reduced to malate or else converted to aspartate, depending on the species, and the malate or aspartate moves from the mesophyll tissue, where it was formed, to bundle sheath cells surrounding the vascular bundles of the veins. There it is oxidized to yield carbon dioxide and pyruvate. The carbon dioxide reacts with ribulose diphosphate and thus enters the Calvin cycle; the pyruvate moves to the mesophyll tissue where it reacts with ATP to form PEP, and the cycle repeats itself.

Plants which carry on photosynthesis by means of the Calvin cycle only are commonly called C_3 plants because the first detectable product of the cycle is phosphoglycerate, a 3-carbon compound. Plants that utilize the Hatch-Slack pathway are commonly called C_4 plants because the first products in this pathway are the 4-carbon compounds, oxaloacetate and malate. C_4 plants possess two advantages over C_3 plants in environments where there is abundant light: (1) There is less opportunity for photorespiration because carbon dioxide is "pumped" from the area around the stomates, where it is limiting and oxygen is abundant, to the area around the veins where the reverse is true. Thus, photorespiration, which is most rapid in the presence of O_2 and absence of CO_2, is diminished. (2) PEP has a higher affinity for CO_2 than ribulose diphosphate has.

In environments where light and/or temperature are limiting, as in the understory of a forest or in the canopy of forests in cool temperate climates, C_3 plants have an advantage over C_4 plants because the Hatch-Slack pathway increases the complexity of the total process and is of necessity, therefore, less efficient in its use of energy. In C_3 plants, the two processes, photoreduction of water to yield hydrogen ions and active electrons, and incorporation of CO_2 into the Calvin cycle, take place in the same cells in the palisade mesophyll. No malate "pump" is needed to move malate and/or aspartate to the sheathes surrounding the veins. In C_4 plants, these two processes are spatially separated, CO_2 absorption taking place in the mesophyll and photoreduction going on in the bundle sheath cells.

In many succulents, especially desert succulents, a variation of the Hatch-Slack process occurs in which carbon dioxide absorption and photoreduction are separated by time instead of by space. In these plants, carbon dioxide absorption takes place at night, when temperatures are lower and humidity is higher, and transpiration, therefore, is diminished. Photoreduction of water takes place during the daytime when the stomates are closed but light is abundant. Malic acid is stored in large vacuoles in the mesophyll cells of these plants and converted to carbon dioxide and PEP in the presence of light in the same cells. In eastern Idaho, Hui Chea, a Ricks College student, observed that the pH of the fluids in the medulla of the common prickly pear, *Opuntia polyacantha,* decreased from about 5.2 to about 4.1 between sundown and sunup in April as malic acid accumulated, and then increased to about 5.2 again during the daytime as the malic acid reverted to carbon dioxide and PEP (Figure 18–12). The process is

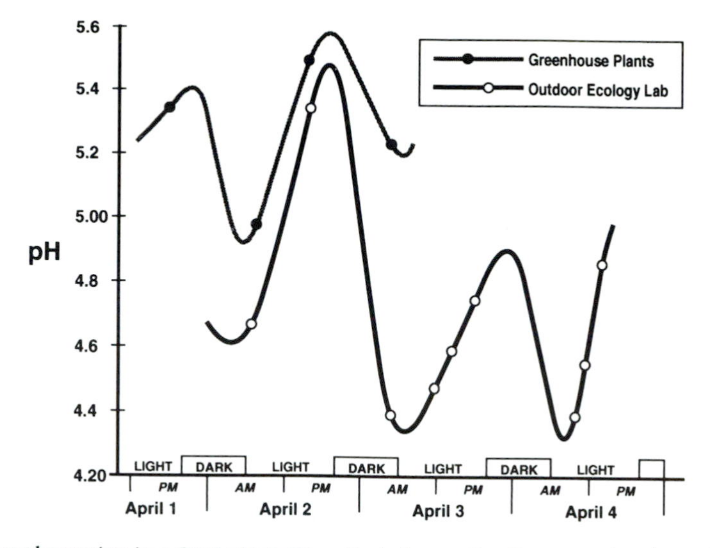

Figure 18–12 Carbon absorption in a CAM plant, *Opuntia polyacantha.* By separating the carboxylation stage from the Calvin cycle of photosynthesis, many desert plants are able to have their stomata open at night when evaporative forces are not as severe as in the daytime.

called crassulacean acid metabolism (CAM) because of the production of crassulacean acid by some of these plants. Many CAM plants belong to the Cactaceae, Crassulaceae, and Bromeliaceae, three families in which many of the species are desert succulents. However, 25 families of angiosperms plus the gnetophyte species *Welwitschia mirabilis* are known to contain CAM species. The process is an obvious adaptation for desert conditions although not all CAM plants are desert plants, and not all desert plants are CAM plants.

Peroxisomes occur in both plants and animals. All contain at least one H_2O_2-producing enzyme. In reference to peroxisomes, there are two kinds of plants: (I) high glycolate and (II) low glycolate. In C_4-I and CAM plants, abundant CO_2 is produced in the vascular bundle sheaths but does not escape from the mesophyll tissue; in C_4-II plants, very little CO_2 is produced in the peroxisomes (Huang et al. 1983).

Because H_2O_2 is a strong oxidizing agent, peroxisomes in animals are able to destroy many toxic substances that sometimes accumulate in cells. Little research has been conducted on plant peroxisomes and what has been conducted has been mostly on spinach leaves. Peroxisome researchers apparently assume that all plants behave like spinach, which, of course, is highly unlikely. In the algae that have been studied, photorespiration rates are much lower than in angiosperms (Huang et al. 1983). In *Trebouxia,* the green alga phycobiont in *Parmelia sulcata,* Pearson and Skye (1965) reported that as light intensity increased from essentially total darkness to very low levels, the rate of carbon dioxide evolution increased; as light intensity continued to increase, the process reversed, carbon dioxide was taken up and oxygen released, suggesting photorespiration at low light intensities in this foliose lichen.

Useful functions postulated for peroxisomes in plants include (1) decomposition of glycolate which at high concentrations is toxic to plants, (2) rescuing part of the glycolate produced in the plastids and converting it into glycerate, (3) production of amino acids (Huang et al. 1983).

The attachment of leaves to stem follows a geometric pattern that is relatively constant for each species of angiosperm. If three or more leaves originate at a single node, the pattern is referred to as whorled; if there are two leaves at each node, they are called opposite. There are many variations of the alternate type of leaf origin, in which only one leaf arises at each node, and these form a **phyllotactic** series. Monocot leaves are always alternate, but all three basic patterns occur in dicots.

The phyllotactic series—known as the Fibonacci series to mathematicians—is such that each number in the series is the sum of the previous two numbers: 1, 1, 2, 3, 5, ... This series may be expressed as fractions: $1/2$, $1/3$, $2/5$, ... Interpreted, this means that in a species having $1/3$ phyllotaxy, each leaf originates $1/3$ of the way around the stem from the leaf just below it, and in a species with $5/13$ phyllotaxy, each leaf arises $5/13$ of the way around the stem from the leaf just below it. Looking at a stem having $1/3$ phyllotaxy from directly above the apex, the leaves can be seen to form three distinct rows or ranks, while the one with $5/13$ has 13 ranks of leaves. The $1/2$, $1/3$, $2/5$, and $3/8$ phyllotactic patterns are the ones most frequently encountered in angiosperms, but higher ranks also occur, especially in dicots and in gymnosperms. Frequently, two or more patterns can be observed on the same plant, the greatest difference being between old and young leaves. On the other hand, phyllotaxy is highly constant in some species; for example, grasses are consistently 2-ranked ($1/2$ phyllotaxy) and sedges consistently 3-ranked ($1/3$ phyllotaxy).

Venation of the angiosperm leaf is unique. The dicot leaf has a basic **reticulate venation** pattern with three or more discrete orders of branching. The monocot leaf has **striate venation,** which is simpler and is often interpreted as a flattened petiole lacking the blade (Cronquist 1968); however, the fossil record has so far failed to confirm this hypothesis. In both dicot and monocot leaves, anastomoses occur freely between veins of the same order and between veins of different orders. Ferns and gymnosperms, in contrast, are limited in leaf architecture, usually have but one order of venation, and frequently have open venation or have anastomoses only at the distal endings of the veins.

In the larger veins of some angiosperms, cambium is present in strips adjacent to the xylem and phloem. The xylem of larger veins contains vessel elements and fibers which provide much of the support needed to keep the leaves flat and relatively rigid. Petioles in some species have the same structure as the midvein, but in other species the petioles are structured more like stems than like veins. Typically, the steles in petioles and veins are **collateral protosteles** with the phloem below the xylem, not surrounding it as in the ectophloic protostele of roots.

The angiosperm leaf is remarkably well adapted for efficient photosynthesis, much more so than the leaves of other vascular plants (Figure 18-10). A complex genetic mechanism causes leaves that are exposed to intense light to develop two or three layers of palisade cells whereas the shaded leaves on the same plant have only one layer. Leaf shape and size are also important to the survival of a species. Diversity in dicot leaves is illustrated in Figures 18-13 and 18-14.

Some students have difficulty in distinguishing between simple leaves and the leaflets of a compound leaf. All of the leaflets in a compound leaf lie in a common plane, which may, however, be folded lengthwise along the rachis. The blades of simple leaves, on the other hand, lie in many planes and are arranged according to the rules of phyllotaxy in two to several ranks. Walnut and hickory have pinnately compound leaves; the others in Figure 18-13 are simple and all have pinnate venation; white oak and sassafras have deeply lobed blades.

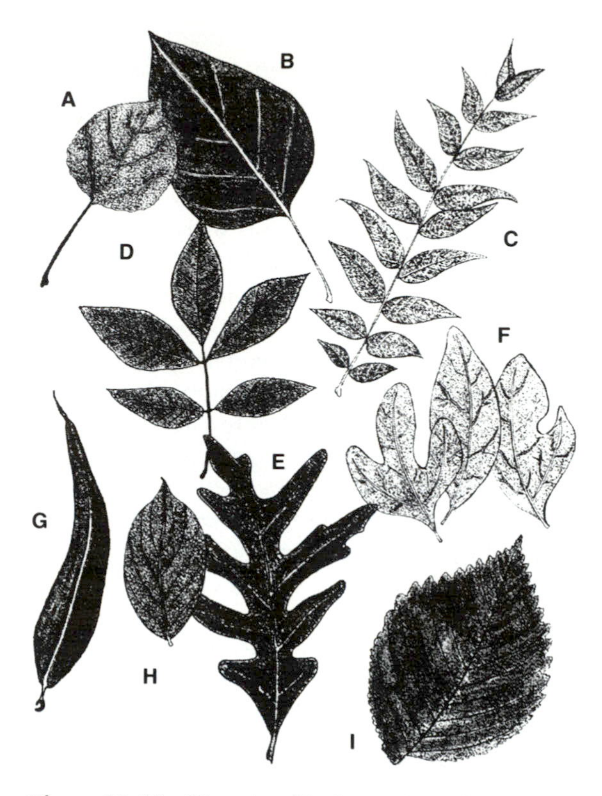

Figure 18–13 Diversity of leaf types in Anthopsida. All are pinnately veined. (A–B) Simple leaves, *Populus tremuloides* and *Syringa vulgare;* (C–D) pinnately compound leaves, *Juglans nigra* and *Garya ovata;* (E–F) deeply lobed simple leaves, *Quercus alba* and *Sassafras albidum;* (G–I) simple leaves, *Salix nigra, Prunus serotina,* and *Ulmus thomasi.*

Nearly all leaves have openings in the epidermis through which the essential gases, CO_2 and O_2, move. These are called stomata, singular stoma (or stomate nowadays), and are able to open and close according to environmental conditions at the moment (Figure 18–11).

Some species of Anthopsida produce leaves with special adaptations. Plants of arid regions often have leaves that are modified into spines or thorns as in the two species of cactus (order Caryophyllales) in Figure 18–14. Where climate is severe and growth slow, spines discourage herbivores from overgrazing the vegetation. Venation is palmate in *Pelargonium domesticum,* a member of the geranium family (order Geraniales), and striate in *Agropyron repens,* a grass (order Restionales).

The buds formed in angiosperms are also different from buds in ferns and lower vascular plants. Each bud contains an apical meristem from which the next season's stem and leaves develop. In many woody angiosperms, buds begin forming soon after the leaves have fully expanded in early summer and by late summer or fall have attained full size. The meristem is protected within the bud during the winter; when spring comes, the cells enlarge and soon an entire shoot has emerged from each bud.

Some buds develop into modified shoots known as flowers. In flowers, the internodes are very short and the leaves are modified to form sepals, petals, stamens, and carpels. In most species, the leaflike characteristics of the carpels and stamens are not immediately apparent; only the sepals are obviously leaflike as a rule. Anatomically and biochemically, too, the sepals are the most leaflike, but the others are also modified leaves.

The sepals collectively make up the **calyx,** the main function of which is to protect the floral bud from the drying influence of wind and sun and from insect predation. Venation of sepals is similar to but usually simpler than that of vegetative leaves, consisting of only two or three orders of branching. The mesophyll is often differentiated into palisade and spongy layers just as the leaves are. The anatomy of petals, on the other hand, is quite different from that of leaves: venation is typically dichotomous, as in cycad, ginkgo, and fern leaves, and the mesophyll is undifferentiated. The chief function of petals is to attract pollinators; however, in some species, this function has been taken over by the sepals, and the petals may be small and inconspicuous or absent. Collectively, the petals make up the **corolla.** In many flowering plants,

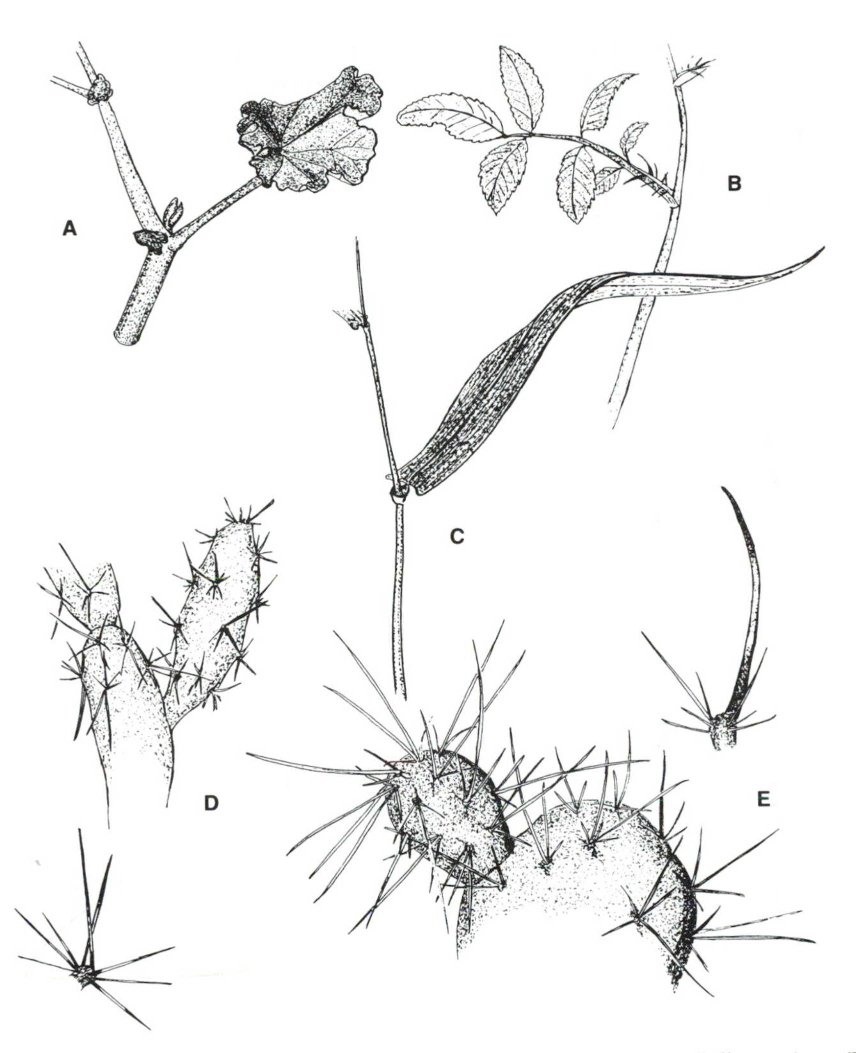

Figure 18–14 Diversity in venation of angiosperm leaves: (A) Palmate venation in *Pellargonium;* (B) compound leaf in a Rosales; (C) striate venation in a grass; (D–E) leaves modified to protective spines in Cactaceae.

especially in the Magnoliidae and Liliidae, there is no clear-cut distinction between sepals and petals. The individual units of such undifferentiated **perianths** are commonly called **tepals.**

The shape of the corolla is important in plants that are insect-pollinated. Flowers that are radially symmetrical are **actinomorphic;** those that are asymmetrical or bilaterally symmetrical only are **zygomorphic.** Bees, hummingbirds, and other specialized pollinators tend to be adapted for pollinating species with zygomorphic flowers. Bats are especially important pollinators of tropical plants with zygomorphic flowers (Fleming 1993). In the morphogenesis of flowers, fewer genes are involved in producing actinomorphic corollas than zygomorphic; the more primitive species, genera, and families of Anthopsida are characterized by actinomorphy.

Bees not only recognize corolla shape, number of petals, and flower color, but they soon learn which plant is offering pollen or nectar at which hour of the day. They pass this information on to other members of the hive (Bünning 1967) who then go foraging at the right time of day. This illustrates a basic biological principle: part of adaptation consists in being **different** from other species in the same ecosystem.

Stamens consist of anthers, in which meiosis takes place and pollen is produced, and filaments, which are slender stalks which support the anthers. In ranalian flowers (Magnoliidae), the anther is not appreciably different in appearance from the filament, but in most angiosperms it is very different. Collectively, the stamens make up the **androecium.**

The function of the carpels is the production of megasporangia or **ovules.** One or more carpels rolled lengthwise with the edges fused to form a single morphological unit comprises a pistil. Each pistil consists of a pollen-receptive surface, the stigma, a more or less elongated mid portion, the style, and a swollen basal portion in which the ovules are produced, the ovary. The pistils collectively make up the **gynoecium.**

In ranalian flowers, stamens and carpels are anatomically very similar to each other, both having three veins in each unit: one midvein and two laterals. In the higher angiosperms, a single vein traverses the filament and ends at the base of the anther. A modified dichotomized venation pattern exists in some stamens with the single vein dividing and one branch leading into each half of the anther. Venation in carpels is uniformly the three vein type observed in ranalian species.

Biochemically, floral leaves also differ from vegetative leaves. The carpels of most angiosperms contain sugar and various fragrant or sweet oils. Similar substances are often present in the stamens or the petals. In many species, some of the stamens are modified to become nectaries, which secrete sufficient quantities of sweetish carbohydrate to attract insects or other pollinating animals. Some monocot and dicot flowers are illustrated in Figure 18–15.

Like other vascular plants, the angiosperms have sporophyte-dominated diplohaplontic life cycles. The female gametophytes are completely dependent on the sporophytes for their existence, but male gametophytes can grow and survive for days in water containing a small amount of sugar and minerals, independent of the parent sporophyte. The male gametophyte consists of but three nuclei and the female of eight in most species. The nuclei in the male gametophytes are not separated by plasma membranes; those in the female gametophyte may or may not be. If they are, the eight nuclei are contained in seven cells, the center cell having two nuclei.

The gametophyte generation has its beginning, in the male, in a pollen grain. Actually, the pollen grain is a three-celled progametophyte (two-celled in some species) enclosed in a sculptured exine (Figure 18–16). Pollination is the transfer of pollen from the anther of a flower to the stigma of the same or another flower where pollen germination and growth may take place. As each pollen grain germinates, it produces a pollen tube containing three nuclei, two of which are functional sperm. One of the sperm nuclei unites with the egg in the female gametophyte to form a zygote which will develop into a sporophyte embryo. The other sperm may unite with two nuclei to produce a triploid nucleus which develops into the endosperm tissue used as food by the developing embryo. Fertilization or syngamy, the uniting of sperm and egg, may take place many hours, or even days, after pollination has occurred.

Co-evolution of pollination patterns and pollinators has resulted in interesting adaptations valuable both to angiosperms and to insects and other animals. When most people think of the

Figure 18–15 Diversity in flowers and fruit of Anthopsida. (A) Six tepals in *Narcissus;* (B) many small flowers and colorful bracts in Arecaceae; (C) four petals and siliques of canola; (D) sepals modified as parachutes in dandelion.

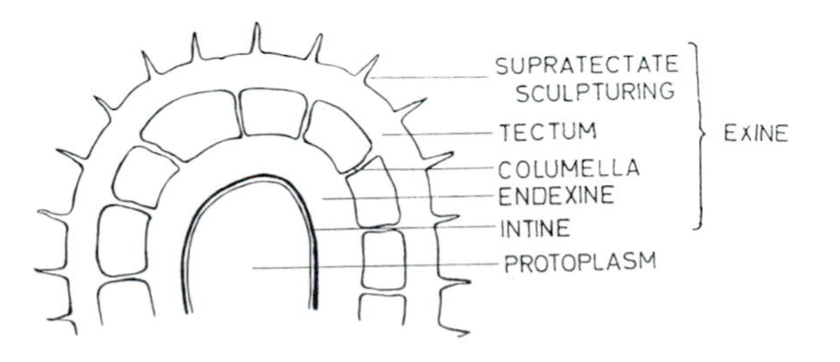

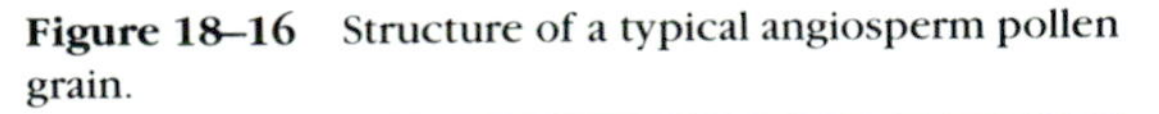
Figure 18–16 Structure of a typical angiosperm pollen grain.

animals that have become specially adapted for pollination of angiosperms, they think of bees, butterflies, and perhaps flies; but there are many other kinds of pollinators including ants, marsupials, mollusks, bats, rodents, and reptiles. The adaptations that entice animal pollinators include color, odor, and taste; geometric patterns are also distinguished by many insects.

The earliest insect pollinators were attracted to flowers by their flavor; beetles ate the pollen and accidentally carried some pollen from flower to flower and thus performed a pollination service. At the present time, flavor still attracts many pollen-eating animals to flowers, but most bees, butterflies, beetles, bats, hummingbirds, and other animals visit flowers to gather nectar which they consume directly or take to their hives or nests for winter use. The position of the nectaries and the shape of the flowers limit the kind of animal that can effectively gather the nectar; these limitations are often reinforced by odors and colors which may be attractive to one kind of animal while offensive to another. Birds and mammals, for example, are much more sensitive to bitter flavors than bees are; flies are attracted by putrid odors that tend to repel bees and butterflies. The taste buds in the feet of the red admiral butterfly make this the most taste-sensitive species known; it can taste sugar at concentrations of 27 mg per liter of water. By way of comparison, humans can taste sugar at concentrations of 4.25 grams per liter and bees at 28.4 g/l.

Odor is important in plants that are pollinated by flies, mosquitoes, and moths. Mosquitoes and moths are night pollinators and favor the same odors that humans find attractive; their attraction for odors is reinforced by white flowers that show up at night. Flies favor strong odors that humans find objectionable, as a rule. Flies favor more humid habitats than bees and butterflies and pollinate plants that grow in dense shade; these plants are characterized by colors that show up under low light intensities, such as whites and reds. Butterflies and bees, on the other hand, are most active during the heat of the day and favor yellows and blue; their preferences for color are reinforced by the geometric patterns that the petals produce. Bees are able to distinguish between flowers with four petals and those with five, for example. Many insects are able to detect colors in the ultraviolet range.

Mimicry also plays a part in animal preference for flowers. The odors of flowers pollinated by moths often mimic the odors given off by the moths themselves and attract members of the opposite sex. Flowers of an African orchid, *Ophrys* are morphologically so similar to female bees of the genus *Gorytes,* including how the stigma and anthers are arranged in mimicry of the female genital plate, that male bees are fooled. Pseudocopulation results in pollination of the orchid.

The adaptive advantage of animal pollination is that a species needs to produce only a fraction as much pollen as one which is wind-pollinated. The "law of parsimony" wears many hats; in this case the advantage persists for as long as pollination is limited to a species of pollinator that visits only one kind of plant during the active pollinating interval of time. Elaborate mechanisms have evolved which ensure a high level of fidelity among the pollinators. The long tubes which accommodate only the long, curved bills of hummingbirds, the odors that imitate those produced by male moths, staminal arches that a beetle or other insect must crawl through to reach the nectaries and which accommodate only one kind of insect, and the pseudocopulatory adaptation in *Ophrys* are examples of these mechanisms. In ecological dominants, such as grasses and forest trees, pollen is readily carried by wind to receptive stigmas and these mechanisms do not occur; but to the understory or subordinate species, insect pollination is an important advantage.

The nature of the exine sculpturing of the pollen grains is equally important to pollination. The pollen of insect-pollinated species has various patterns of reticulation or projections which help the pollen grains stick to each other and to the bodies of insects. In wind-pollinated species, the reticulations are typically much less pronounced and the grains may be almost smooth. Size is less variable in wind-pollinated species since the grains must be small enough to be buoyant yet large enough that they are not deflected by eddy currents around the stigmas. Competition among pollen grains that become attached to the stigma at the same time is probably significant in angiosperm evolution, since genes that provide greater vigor or other advantageous traits to the gametophytes often behave the same way in the sporophytes. Grooves on the surface of the pollen grains aid in absorption of moisture from the stigma and in germination.

The female gametophyte begins as a megaspore that divides to form two nuclei each of which divides twice, resulting in eight nuclei all enclosed in the sporangial wall or nucellus. Migration of nuclei and formation of plasma membranes results in the typical female gametophyte with three cells at each end and a single larger cell containing two nuclei in the center (see Figure 18-3C).

Rate of growth of pollen tubes varies both within and among species. In corn, the tubes grow at a rate of 1 to 2 cm per hour and attain a length of 40 to 50 cm in about 36 hours. Oak pollen tubes attain a total length of only a few millimeters in several weeks. Within a species, the fastest growing tubes possess an obvious advantage. As they grow, they are nurtured by substances in the pistil of the mother sporophyte. When the tip of the pollen tube approaches the female gametophyte, chemical attractants cause it to bend toward the micropyle of the ovule into which it then grows (Figure 18-3C). Upon reaching the female gametophyte, it discharges its two sperm, the first of which moves directly to the egg and fertilizes it. The next sperm to enter the micropyle, which is often, but not necessarily, the other sperm nucleus in the germinating tube that provides the first, bypasses the egg and moves to the two polar nuclei and fertilizes them, thus forming a cell with the triploid number of chromosomes.

Following fertilization, the triploid nucleus divides first and develops into the **endosperm** tissue that will nourish the embryo until it matures into a **seed.** Endosperm is rich in hormones, such as **cytokinin,** the hormone vital for differentiation of the cells of the embryo as they form the embryonic organs. In some species, a similar tissue, **perisperm,** develops from the nucellus and functions in a similar way in providing nourishment for the embryo.

When the typical angiosperm zygote divides, producing two daughter cells, the one farther from the micropyle develops into the embryo and the one adjacent to the micropyle divides to form a **suspensor.** The suspensor is a fast-growing structure, but it does not last long. It is the food-absorbing structure in the early **proembryo** stage. About the time the cotyledons are beginning to differentiate and to take over the function of absorbing food and mineral nutrients, the suspensor has reached maturity and is beginning to degenerate. Suspensor morphology varies greatly from species to species and even within species, especially within the legume family.

The first two division of the zygote result in a four-celled filament. Additional divisions result in a variety of patterns, depending on the species, but typically an elongated embryo becomes a somewhat flattened egg-shaped structure differentiated into **cotyledons, epicotyl, hypocotyl,** and **radicle** (Figure 18-17). The cotyledons absorb food from the endosperm; the other structures develop into shoot, lower stem, and root, respectively. In some species, the endosperm tissue is used up almost as rapidly as it is produced, and by the time the embryo has reached the seed stage, there is none left. In other species, the mature seed contains abundant endosperm, typically a hard, dry tissue in mature seeds. In immature seeds and in mature seeds of some species, like coconut, endosperm is liquid.

Nutrients, in the form of sugars, amino acids, other organic compounds, and mineral salts, enter into the developing ovary through the phloem. The ovule is attached to the **pericarp**

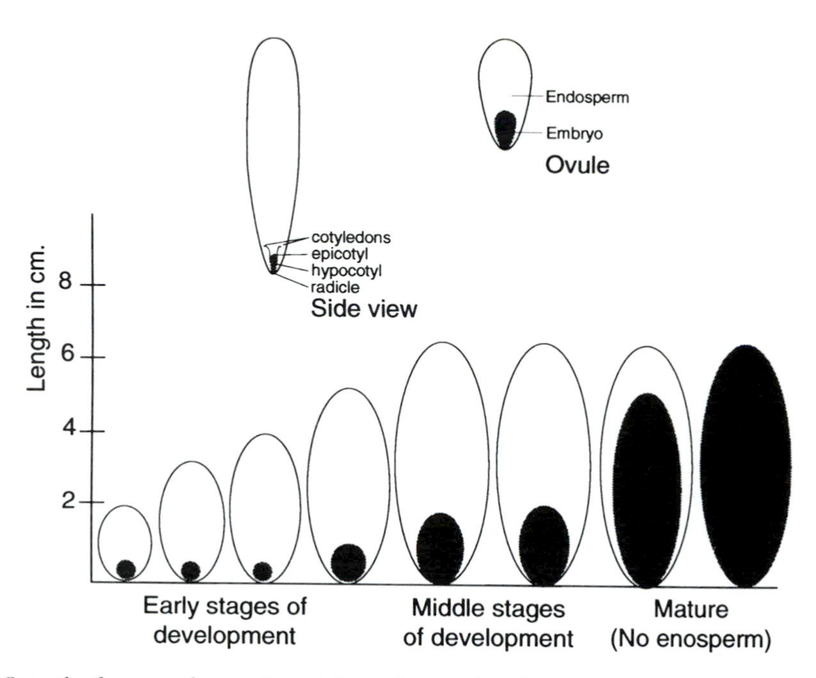

Figure 18–17 Growth of a cucumber ovule and the embryo and endosperm tissue in it. Note how much faster the endosperm grows following fertilization than the embryo.

or wall of the ovary by a delicate placenta, and as the embryo and endosperm increase in mass, the ripening process that will make it possible for the seed to survive adverse conditions begins to take place. Part of a mustard pod, including parts of two seeds, is shown in Figure 18–18). The markings on the seeds are unique for each species and with only a little experience a student can learn to identify hundreds of kinds of seeds by these markings alone. What function, if any, they serve in nature is not known.

At times, it is desirable to grow embryos in test tubes, in order to by-pass the dormancy period which occurs in most seeds, for example, or to enable a hybrid that is genetically incompatible with its own endosperm to survive. Figure 18–17 was prepared for a botany laboratory exercise in embryo culture; more than likely, you, as a student, will receive such instruction as part of your training in botany. With the aid of a dissecting microscope, alcohol and other disinfectants, an alcohol or gas flame, and care, any student can find embryos that are no more than 2 to 4 mm long (depending on species) and transfer them aseptically to test tubes where essentially 100% of them will grow.

Angiosperm seeds are well adapted for dispersal by various agents including mammals, birds, water, and wind. They are also well adapted for survival through periods of unfavorable weather or environmental conditions. Most seeds will not germinate until after an "after-ripening" period of a few weeks or months; however, seeds of the red mangrove germinate while still on the trees after which the young seedlings fall into the shallow water and float around for a few days before becoming established. Aspen seeds must germinate within about 3 weeks after maturation or they die; but wild morning glory seeds are known to have germinated after more than 50 years in storage, mullein seeds after more than 70 years, and lotus seeds after more than a thousand years. However, the embryo is a living organisms and there is a limit to how long metabolism can continue, even at the very reduced rate of a dormant seed. Reports of wheat and corn seeds germinating after thousands of years in storage are undoubtedly fabrications. Under ideal conditions, 10 years of storage results in reduced germination and very reduced vigor, and 32 years is the maximum period of storage on record after which wheat has germinated (Harrington 1972).

Angiosperms go through three phenological stages following germination; the **seedling** stage during which total dry weight gradually decreases as food stored in the cotyledons or in the endosperm is used in growth, the **juvenile** stage in which the plant increases in weight because photosynthesis now exceeds respiration, and the mature or **reproductive** stage when the rate of photosynthesis slows down and the plant produces flowers and fruit. In perennials, the juvenile stage may last for several years; after reproductive maturity has been reached, there may be an alternation of vegetative growth periods, similar to the juvenile or even the seedling stage of annuals, followed by reproductive periods.

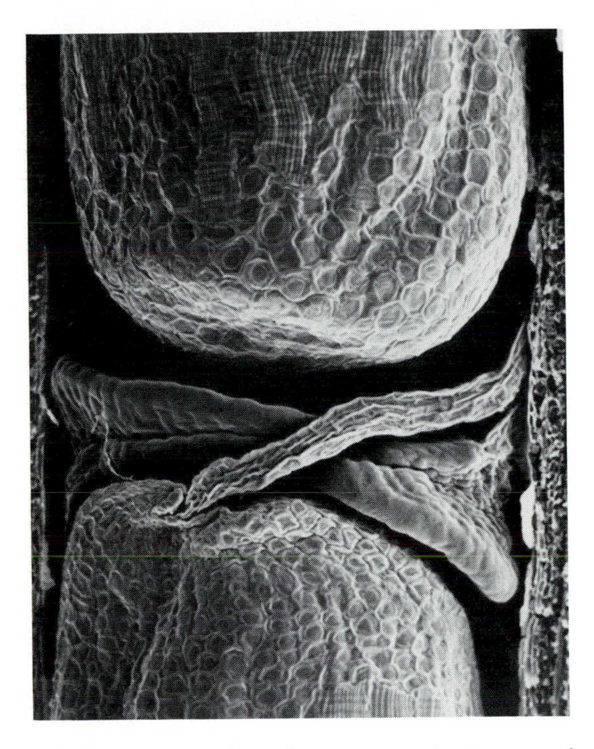

Figure 18–18 As the embryo matures, its tissues, and the surrounding nucellus tissues, harden and become dormant. Note the placenta attaching the embryo, now become a seed, to the inner pericarp wall of ovary. (SEM courtesy of BYU Microscopy Lab.)

The seedling stage begins as soon as the seed breaks dormancy and follows essentially the same pattern in all angiosperms. First, the radicle develops into the primary root; as it forms secondary roots, the young plant becomes firmly anchored to the soil. The hypocotyl develops next in most angiosperms; however, in **hypogeous** species, the hypocotyl does not develop at all. In **epigeous** species, the hypocotyl forms a loop as it pushes up through the soil. As light destroys the hormones on the upper surface of the loop, the cells there stop growing while those on the undersurface continue to elongate, forcing the hypocotyl to straighten out. As it does, the cotyledons are drawn out of the soil. As they spread apart, the epicotyl begins developing into a shoot with stem and leaves. There is some variation here too among species. In *Monophyllea,* for example, there is no development of epicotyl into stem and leaves and the mature plant consists of one very large cotyledon with flowers at the apex of the hypocotyl. *Monophyllea* is a tropical genus in the family Gesneriaceae, order Polemoniales.

In many angiosperms, it takes 5 to 10 days after germination until the seedling has become self-sufficient and photosynthetic rates exceed respiration rates. This is the beginning of the juvenile stage, which continues until **anthesis** or the beginning of flowering. During the juvenile stage, plants are very sensitive to light intensity. At low light, stems become elongated, and leaves are thin with little palisade mesophyll. At higher light, more vigorous but shorter stems are produced and leaves often have two or three layers of palisade cells. Growth is also affected by temperature, moisture, and the abundance and quality of minerals in the soil.

During the mature stage, flowers develop, are pollinated, and fertilization takes place. In angiosperms, successful competition depends to some extent on which pollen grain produces the most rapidly growing pollen tube (Johnston 1993). Hormones produced by the developing embryo stimulate growth in the ovary, which consequently develops into a fruit. In many species, only the ovary is affected and develops into a fruit, but in many others, both the ovary and accessory tissues, from either the calyx or the receptacle or both, become part of the fruit. In apples, pears, and most other fruits that have been studied, fruits fail to mature if fertilization is not complete.

Phylogeny and Classification

Angiosperm fossils are primarily of five types: (1) petrified wood, mostly stems but sometimes roots; (2) petrifactions and occasionally compressions of fruits; (3) compression of leaves; (4) compressions of flowers; and (5) pollen grains. Flowers, which are used most by taxonomists to infer relationships among extant genera, are too rare and too poorly preserved to be of great value to the paleobotanist; and petrifications, for reasons we do not understand, seem to be extremely deficient in Cretaceous formations, although it is possible that as we study more Cretaceous rocks we will find more petrifactions. This means that paleobotanical studies of the origin and evolution of the angiosperms must be based primarily on pollen grains and leaf compressions. Nevertheless, when flowers and petrifactions are found, they can impart a massive amount of information. Herngreen (1974) studied *Chloranthus*-like stamens of fossils from the Turonian epoch (Cretaceous) and reported that the fossilized pollen grains were identical to those of extant *Chloranthus,* thus casting light on the evolution of Chloranthaceae (Magnoliales).

The earliest angiosperm fossils are small, linear leaves having entire margins, and monosulcate pollen grains with a columnellar rather than alveolar inner exine (Hickey and Doyle 1977). The leaves are found only in coarse-grained river channel deposits of Barremian age; the pollen grains are more widespread but are of the same age as the first leaves and are scarcely distinguishable from cycad pollen. Judging from both leaf and pollen morphology and the gravelly nature of the formations in which the leaves are found, we infer that the first angiosperms were evergreen shrubs growing on disturbed sites, that they were pollinated by beetles, and that they resembled both cycads and magnolias in many respects. From these cycad-like earliest angiosperms, all of the flowering plants of today have evolved.

Only 2% of the leaves in the coarse-grained upper Barremian formations are angiosperms, and there are no angiosperm leaves at all in the associated finer-grained rocks. Fern leaves are very abundant, and leaves of various cycad-like gymnosperms are common. By late Aptian, more kinds of flowering plants can be found, and angiosperm leaves not only make up a larger percentage of the total fossils, but can also be found in finer grained deposits. During the Albian and Cenomanian, angiosperms continued to increase in number of species and total abundance and to expand into more and more habitats; however, conifers and Bennettitales continued to be the only plants of swamps and some other habitats until much later (Hickey and Doyle 1977).

Upper Cretaceous fossils, late Albian to Campanian (Table 18-3), are poorly known, but in early Tertiary, angiosperms had become both very abundant and very important. The leaves

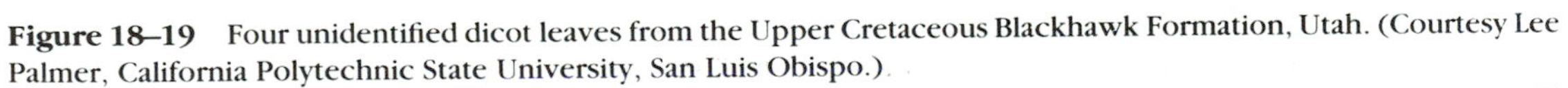

Figure 18–19 Four unidentified dicot leaves from the Upper Cretaceous Blackhawk Formation, Utah. (Courtesy Lee Palmer, California Polytechnic State University, San Luis Obispo.)

in Figure 18-19 are unidentified dicots collected from the Upper Cretaceous Blackhawk Formation of Utah. None of them fit well into present-day genera, and even family is difficult to determine with any high degree of confidence. By mid-Tertiary, angiosperms had become very abundant and occupied about the same position of importance they occupy today with all of our modern families and most genera. For example, angiosperm leaves covered over 50% of the rock surface and made up 88% of the species in a dawn redwood-dominated forest growing near what is now Salmon, Idaho during the Miocene, some 13 million years ago (Bryant and Benson 1973). Elm, apple, hawthorn, birch, and alder were among the most abundant species found in their study, but herbaceous grasses, cattails, horsetails, and ferns were also present (Figures 17-9 and 18-20); examination of the coal-like material near, and sometimes surrounding, the leaves revealed a diversity of palynomorphs (Figure 18-21).

While the immediate ancestors of the angiosperms cannot be ascertained with certainty, the earliest-flowering plant fossils show many similarities to the Cycadales and even more so to the Nilssoniales, a group of gymnosperms which were very abundant during the Jurassic and early Cretaceous, but became extinct during the Tertiary, about 50 million years ago.

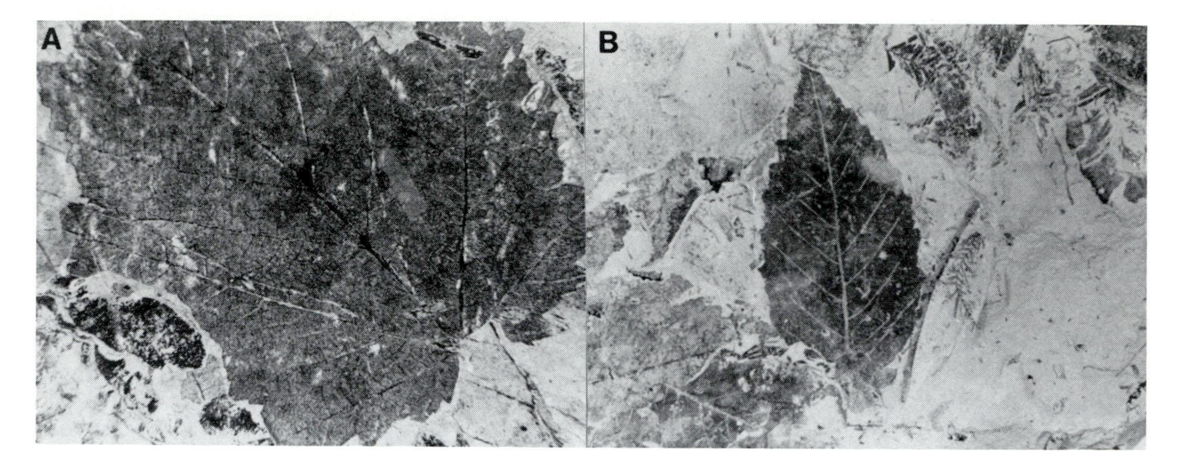

Figure 18–20 Two fossil dicot leaves from the Carmen Formation, Idaho. (A) *Acer osmunti* (Sapindales), with palmately veined leaves; (B) *Betula multinervis* (Fagales), with doubly serrate leaf margins.

Soon after their origin, angiosperms evolved adaptations that gave them distinct ecological advantages over gymnosperms, ferns, and other vascular plants: greater drought resistance and more efficient reproduction. Drought resistance in mid-Cretaceous angiosperms depended on an enclosed fertilization mechanism, trilete pollen grains that could germinate even when the air was very dry, vessels which made water conduction more efficient, improved stomatal structures on the leaves, and the deciduous habit which essentially eliminated transpiration during times unfavorable for photosynthesis. Their greater reproductive efficiency depended from the beginning on insect pollination and later by development of a shortened life cycle and double fertilization. Insect-pollinated plants of the same species can be widely dispersed, yet the plants do not have to produce large quantities of pollen because insects are able to recognize flower color, floral patterns, and flower odors and flavors and thus carry pollen to other members of the same species. Double fertilization ensures that no energy will be spent on producing an embryo unless there is a food supply available for its growth, and no energy will be spent on a food supply unless there is an embryo prepared to make use of it (Figure 18–17).

Knowledge of Cretaceous geography is helpful in understanding the origin and spread of the angiosperms. The world has not always been divided into the continental and island land masses we know today; and even now, some land areas are rising, others are sinking. The area now included in the Grand Canyon National Park, for example, has been submerged below sea level at least seven times since the beginning of the Cretaceous Period, or just prior to the time when the first angiosperms evolved.

During the Barremian epoch of the early Cretaceous period, world climate was relatively humid. Two great continents, Laurasia in the north and Gondwanaland in the south, were separated by a narrow, equatorial seaway, the Tethys (Figure 18–21). Although Laurasia spanned 200 degrees of longitude and Gondwanaland over 100 degrees, climate was uniformly humid and relatively mild because neither continent was mountainous. In the absence of mountain, desert, and ocean barriers, the first flowering plants were able to migrate in all directions quite rapidly. A third continent, Antarctica, and a large island, now India, had recently broken loose from Gondwanaland and were migrating southeastward and northeastward, respectively.

During the Aptian and Albian epochs, mountain building and other factors resulted in increasing aridity of northern Gondwanaland. Southern Laurasia remained relatively humid. Nevertheless, barriers to plant migration were relatively few, and new species still migrated freely from one part of the world to another. Studies of Albian gymnosperm and fern distribution indicate that the floristic provinces were very large at that time and all of southern Laurasia was a uniformly flat, humid, tropical to subtropical floristic region, while northern Gondwanaland, although considerably drier, was also relatively uniform (Brenner 1976). Under such conditions, competition would be expected to be severe, and a plant not as well adapted to an environment as another would gradually become extinct. The late Cretaceous is, in fact, marked by extinction of several major groups of plants including the Bennettitales, Czeckanowskiales, and Caytoniales. The Anthopsida at that time, however, were colonizing niches previously unfilled, while rapidly creating and colonizing new niches. Apparently competition either was not yet severe in most of the habitats to which they were best adapted, or they already possessed a competitive edge over other vascular plants, since only the most primitive angiosperms seem to have become extinct at the end of the Cretaceous.

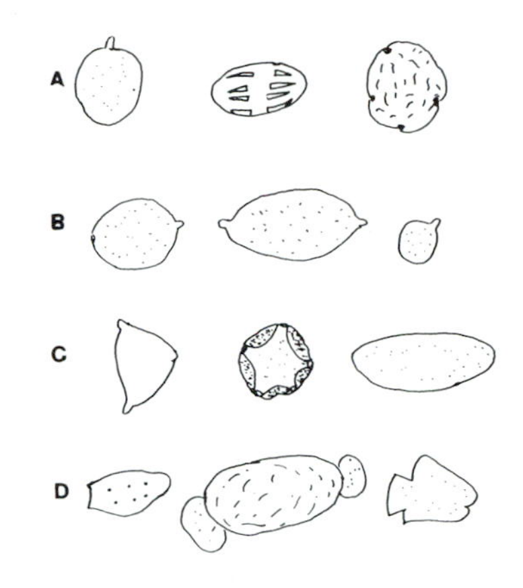

Figure 18–21 Pollen grains from the Carmen Formation: (A) left to right: *Metasequoia, Malus, Ulmus;* (B) an unidentified pollen grain, possibly *Cypress*, unidentified structure (possibly a protozoon), an unidentified pollen grain (probably *Typha*); (C) *Betula, Alnus*, unidentified pollen grain, probably *Cicuta;* (D) unidentified pollen grain (possibly a chenopod), *Picea, Salix*. (From presentation to Idaho Academy of Science by Ruth Baker.)

As we have gradually learned to appreciate the great complexity of both leaves and pollen grains, our knowledge of angiosperm origins has greatly improved. Because pollen is more readily transported by wind, water, and animals, it gives us a quicker and better picture of the total variation in species over a large area at any given time; but the leaves, which generally fall and are fossilized very near their point of origin, give us a much better picture of the ecological requirements of each species. One must keep in mind, of course, that in the absence of flowers attached to leafy stems, it is very difficult to match a given pollen grain with leaves of the same species. One must also keep in mind that in the past, well-intentioned paleobotanists as well as botanically oriented "rock hounds" and others, in their zeal to report new finds, often have been careless in their identification of the generic affinities of their fossils or of the geological strata in which they occurred, or both. Not all deltoid leaves with dentate margins, for example, belong to the genus *Populus*. A wealth of fossil diversity today masquerades in published reports under a few popular names like *Populus, Salix, Sassafras, Fraxinus, Sparganium, Eucalyptus*, and *Aralia*.

From combined studies of leaves and pollen grains, assisted by other kinds of fossils where available, we have concluded that the first angiosperms were "weedy" shrubs of streambanks and gravel bars—leaves were narrow with entire margins, providing drought resistance, pollen grains were either inaperturate or monosulcate, suggesting pollination by pollen-eating insects such as beetles—and that they originated near the center of Gondwanaland at about 10° south latitude in an area now partly in the Congo (Zaire) and partly in eastern Brazil (Figure 18–21).

There is little structural diversity in fossilized leaves from lower Cretaceous formations (Upchurch 1984). Venation is uniformly relatively simple with only two or three orders of branching as compared with up to eight orders in many modern angiosperms and only one order in gymnosperms. Cuticle types conform to a single plan of stomatal construction unusual in its plasticity; on a single leaf will be stomata that would be classified into four or more categories if we were examining different species of modern plants in which all the stomata on a given plant are essentially identical in form (Figure 18–11). Among extant species, stomatal diversity within species occurs in the Chloranthaceae and Illiciales of the Magnoliidae and few other taxa. These characteristics of angiosperm leaves were probably due to chance mutations which gave neither an advantage nor disadvantage to survival, but which conveniently, for us, serve to distinguish between angiosperms and gymnosperms: venation in Anthopsida, for example, is never dichotomous.

Although the earliest angiosperm pollen grains were monosulcate, and in other ways resembled gymnosperm pollen and some fern spores, they possessed minute projections on the exine wall which enabled them to stick to the bodies of the beetles that often fed on them, and they were thus readily carried to other plants of the same species where they could germinate and grow. The pollen was produced in anthers at the apex of an elongated microsporophyll making it easily accessible to insects. The pollen grains lacked the spongy inner wall structure characteristic of gymnosperms and possessed columnellae. This, too, probably had no ecological advantage in the beginning, but has provided a base on which the complex pollen patterns of modern angiosperms, which do have ecological significance, have been built. The carpel was involute (rolled into a cylinder) with the megasporangium, or nucellus, on the inside, providing protection to the gametophytes and young embryo from desiccation and from foraging by insects.

From their point of origin, early angiosperms spread rapidly throughout central and northern Gondwanaland and across the Tethys Sea to southern Laurasia (Figure 18–21). Associated with the very small number of Barremian angiosperm fossils are large numbers of fern and cycad-like fossils. Aptian formations of the eastern Brazil-Congo area contain the first trisulcate pollen grains characteristic of more advanced angiosperms (Figure 15–6). The leaves of Aptian and Albian angiosperms are also more complex than earlier leaves. Additional orders of venation provide rigidity to bigger leaves, which are especially efficient photosynthesizers under favorable moisture conditions. By early Albian, large, lobed leaves, suggesting adaptation to more mesic habitats, had evolved. By mid-Albian, pinnatifid leaves were common in mesic understory sites and peltate leaves in aquatic habitats. By late Albian, pinnately compound leaves had appeared. Somewhat later, specialized leaf and stem adaptations evolved. In the saguaro cactus, the stems (called cladodes) have taken over the function of photosynthesis and the leaves have become spines which help protect the sweet, succulent plants from hungry herbivores (Figure 18–22). Note that the ridges on the saguaro are dichotomously forked.

Figure 18–22 Desert dicot, Saguaro cactus, Arizona; note dichotomous venation of the cladode (stem) ridges.

During the Aptian and Albian epochs, pollen grains also continued to become more and more diversified, specialized, and complex. The three grooves of the trisulcate grains were replaced in some flowering plants by three rows of pores, an adaptation which reduces moisture loss, while still allowing for good germination. By late Albian, angiosperm pollen grains were abundant in many kinds of habitats from 70° south latitude to 80° north latitude, and judging from pollen morphology, all angiosperms were still insect-pollinated.

Angiosperm evolution paralleled insect evolution, and neither could have developed the diversity and ecological success it enjoys today without the other. The first pollinators were pollen-foraging beetles; many primitive angiosperms today are still beetle-pollinated. The beetles are a fairly ancient group of insects with excellent fossils from the Permian. Caddisflies, another ancient order of insects, are also believed to have been important pollen foragers during much of the Cretaceous period. From them, the butterflies and moths presumably evolved early in the Tertiary, at about the time the first flowers with nectaries evolved (Takhtajan 1969). Among nectar-producing angiosperms, moths and butterflies are the most important pollinators of species having actinomorphic flowers, and hive bees and dipterans, both of which evolved during late Cretaceous or early Tertiary, are most important in species having zygomorphic flowers. Birds and sometimes bats are also important in the latter group of flowers.

During early Cenomanian, pollen grains in some lines became smaller, in other lines they lost their reticulation, becoming essentially smooth, and in still other lines, pores replaced the grooves. These characteristics appear in mid-Cenomanian, indicating that angiosperms had finally been able to compete with gymnosperms as dominant or canopy species in some ecosystems so that wind pollination was an advantage. In the grasses, monosulcate angiosperms gave rise to species with a single pore in the pollen grain; in the poplars, oaks, walnuts, and other "amentiferae," three pores replaced the three grooves; and in the chenopods, amaranths, and plantaginous dicots, several pores replaced the grooves (Figure 15-6).

Also during the Cenomanian, the first angiosperms migrated into the colder regions of the two major continents. As late as early Albian, the diversity of angiosperm species was much greater in northern Gondwanaland than in southern Laurasia, although abundant evidence indicates that the climate had become quite severe on the southern continent. Hickey and Doyle (1977) have pointed out that spores of Schizeaceae (ferns) and pollen grains of Taxodiaceae and Cupressaceae all decreased in abundance and diversity from Barremian time, when ferns and gymnosperms remained very abundant in northern Gondwanaland, down through the Aptian and Albian epochs, suggesting increasing aridity. During the same span of time, ferns and gymnosperms remained very abundant in southern Laurasia, suggesting a continuation of the warm, humid climate of the Barremian. Also, slate deposits are frequently found in Gondwanaland formations just above the Barremian, indicating arid conditions. The fact that angiosperm fossils are nevertheless more abundant in Gondwanaland formations during the Aptian than they are in Laurasian formations, and only in Albian times become as abundant in the Laurasian deposits, is further evidence that they originated near the center of Gondwanaland and during much of Cretaceous time were gradually migrating in all directions, including northward, from the point of origin. Part of the lag in migration time to higher latitudes depended on the occurrence of mutations giving cold resistance in several angiosperm lines. Even today, the vast majority of angiosperm species and genera are limited to the tropics; in fact, over 65 of the 275 families of modern angiosperms are entirely limited to tropical and subtropical habitats, while only seven families are limited to cool temperate and polar climates.

TABLE 18-4

MAJOR MORPHOLOGICAL GROUPS OF CRETACEOUS LEAVES ACCORDING TO KRASSILOV (1977), AND MODERN TAXONOMIC ORDERS WHICH THEY MOST NEARLY RESEMBLE

Cretaceous Leaf Fossil	Modern Taxa	
Groups	**Subclass**	**Order**
Laurophylls	Magnoliidae	Magnoliales
Nymphaephylls	Magnoliidae	Nymphaeales
Trochodendroids	Hamamelidae	Hamamelales
Platanophylls	Hamamelidae	Hamamelales
Ficophylls	Rosidae	Ulmales
Proteophylls	Rosidae	Proteales
Rosiphylls	Rosidae	Rosales
Leguminiphylls	Rosidae	Rosales
Palmophylls	Liliidae	Liliales

Evolution from the first "weedy" angiosperms to the complex and varied assemblage of species known today progressed at first along three main lines (Figure 18-5): one leading to the lilies and other monocots; one leading by way of the buttercups and their relatives to the poppies, pinks, buckwheats, and willows, and the third leading by way of peony-like plants to the sycamores and their relatives along one branch and to the roses, walnuts, phloxes, and daisies along a second branch. All three of these lines have further branched out until today we recognize ten subclasses of angiosperms. Four of the ten are recognizable in late Cretaceous formations, according to Krassilov (1977), who placed all Cretaceous leaf fossils in nine morphological categories (Table 18-4).

A visitor to the very early Cretaceous world would find the marine ecosystems much as they are today, but terrestrial ecosystems would look strange to him. Great forests of conifers and ginkgoes covered most of the cool humid areas 135 million years ago, while forests of cycad-like gymnosperms—Bennettitales, Nilssoniales, and Caytoniales—covered the moist tropical and subtropical areas. The forest understory was made up largely of ferns, with lycopods and horsetails often abundant. In the tropical forests, epiphytic ferns grew in the tree tops, though not to the extent that epiphytes are found in the tropics today. The deserts and steppes were more barren than are modern arid and semiarid lands.

If we lived during late Barremian time, about 110 million years ago, and were classifying land plants, we would undoubtedly place *all* angiosperms in one small family within the Nilssoniales or possibly the Cycadales. But if we came back 20 million years later, at the end of the Albian, we would probably recognize the Anthopsida as a distinct class, or at least a subclass, within the Tracheophyta. All of the flowering plants at that time would be accommodated in three orders: the Ranales or Magnoliales, the Rosales, and the Liliales, and about a dozen families.

Angiosperm paleobotany enters into a "dark age" during late Cretaceous. Either the histories written by angiosperm scribes of that day have been lost or destroyed or modern historians have failed to find them. As we emerge from these "dark ages," 40 million years later, we find many earlier angiosperm taxa, such as the once widespread "Normapolles" group, completely extinct. Nevertheless, many of the remaining Cretaceous families and orders were much as they are today, and even species have remained remarkable stable since mid-Tertiary. For example, Walpole and Gallup (oral report, Idaho Academy of Science, 1975) compared venation pattern and protein content of miocene apple and elm leaves from the Carmen Formation of Idaho with 1974 apple and elm leaves, respectively, and found no detectable differences between a fossil species and its extant homolog either in venation or amino acid chromatography, although the differences between the two genera were very pronounced.

Two groups of Cretaceous plants are of special interest because of the light they throw on relationships among modern plants: (1) *Sapindopsis* is a genus of platanoid leaves from the Albian from which both the Hamamelidae and Rosidae have undoubtedly evolved, indicating that these two subclasses are more closely related to each other than most taxonomists formerly supposed; (2) Normapolles are a group of triangular pollen first appearing in mid-Cretaceous among the Rosidae and showing gradual change over a period of a few million years to typical "amentiferae" pollen types, suggesting that the walnuts and their relatives, and probably the elms, birches, and their relatives, have evolved from the roses, rather than the sycamores, as indicated in most taxonomic treatments of the Angiospermeae. If so, Figure 18-5, which is based on Cronquist (1968), should be modified so that the Juglandales, and probably Urticales and Fagales, occupy locations adjacent to the Sapindales within the Rosidae.

In the absence of good paleobotanical evidence, relationships among plants are inferred from morphological, physiological, and biochemical comparisons of extant species. Such taxonomic studies are based upon certain axioms: (1) closely related plants resemble each other more nearly than plants which are not closely related, (2) homologous structures are more valuable than analogous structures in inferring relationships, (3) evolution is mostly, but not exclusively, in the direction of simple to complex, (4) in highly reduced or otherwise greatly specialized plants, the study of vestigial rudiments of organs or of vestigial vascular supply systems to missing organs can often furnish clues of evolutionary trends, and (5) useful phylogenetic information can be gleaned from all parts of a plant and from all stages of growth—in other words, everything is important when trying to ascertain relationships. Fortunately, both paleobotanical and comparative studies are available in the Angiospermeae, and many of the modern systems of angiosperm classification are mostly good, "natural" systems. A natural system of classification is not only a valuable aid in transmitting information to others and helping all to understand evolutionary and other processes, but is also of maximum practical value. For example, in agronomic applications, natural systems result in increased yield of crops because they aid in plant breeding, in weed control, in fertilizer management, in control of diseases, and in other ways.

At the family level, there are many good monographs which almost all taxonomists agree present natural classification of the families and genera included. There is less agreement when it comes to combining the families into natural orders, subclasses, and classes. Systems published by Cronquist in the U.S. and Takhtajan in the former U.S.S.R. (now Russia) are probably the most widely accepted at the present time.

Because of the large number of species of flowering plants, classification is a challenge; nevertheless, numerous attempts to arrange the 220,000 species into natural groupings of closely related species have been attempted since the days of the Greeks, and even before. The most widely accepted modern systems of classification are based largely on ideas expressed by Charles Bessey in the early 1900s. On the other hand, most herbaria and angiosperm floras, until recently, were based on the Englerian system.

Adolph Engler and his co-workers, assuming that simple, non-petalous flowers, wind-pollinated as the gymnosperms are, to be primitive, and that plants with complex and showy flowers, typical of insect-pollinated species, have evolved from the wind-pollinated ones, developed a "phylogenetic" system with the Amentiferae at the base and the composites at the apex. In doing this, Engler ignored the findings of E. C. Jeffrey who, around 1890, had noted the absence of vessels in some Ranales (Magnoliidae), the general similarity between *Magnolia* xylem and gymnosperm xylem, and the resemblance of *Magnolia* floral structure to cycad cones. Nevertheless, Engler's *Syllabus* enjoyed wide utilization from 1895 to the 1960s, largely because of its thoroughness, for it included all groups of plants.

Bessey, a botanist at the University of Nebraska, first published a system of classification of Anthopsida in 1893 which he amplified and modified in 1897 and 1915, and which was well balanced, taking into account stem anatomy, leaf and root characteristics, and habitat preferences in addition to floral characteristics. It suffered by ignoring many of the tropical families and possibly by excessive lumping. Studies of fossils in the past few years have confirmed many of Bessey's hypotheses and, at the same time, modified them.

J. Hutchinson's system, published in two volumes in 1926 and 1934 and revised in 1948 and 1959, has been widely used in Great Britain, but has been criticized far more than the Bessey system. Its greatest strength lay in pointing out characteristics that can be used in classifying but had been largely ignored; its greatest weakness lay in overemphasis of the secondary xylem, and some other characteristics, as *the* "important" taxonomic characteristics.

Carl Skottsberg presented in 1940 a system similar to Bessey's in his treatment of the dicots but differing from all other systems in treatment of the monocots. Published during World War II in a little-known language, Swedish, it is not well known outside of Scandinavia. It is probably the best system available at present in the treatment of the monocots.

Armen Takhtajan and Arthur Cronquist published in 1961 and 1968, respectively, similar systems based primarily on Bessey's 1915 publication, but involving a thorough review of anatomy, embryology, palynology, paleobotany, genetics, and other disciplines. Like Bessey's, both are better in their treatment of the dicots than the monocots. Cronquist's system might be regarded as an integrated modern version of Bessey's, while Takhtajan's might be regarded as a fragmented modernized version of Bessey's dicot classification and Skottsberg's monocot classification.

Takhtajan, professor of botany at the Academy of Sciences in Leningrad (St. Petersburg) and world-renowned plant geographer, is noted for his thoroughness—no flowering plants from any part of the world are ignored—and is somewhat of a "splitter." Cronquist, senior curator at the New York Botanical Gardens, has for many years shown an interest in biochemistry, genetics, and ecology as they relate to taxonomy, and is more of a "lumper." This book follows Cronquist's classification of the dicots very closely, though modified somewhat by R. F. Thorne (unpublished report, X 1BC 1964), Hickey and Doyle (1977), and Lawrence (1951), as well as Takhtajan (1969), but holds closer to Takhtajan's and Skottsberg's classification of the monocots, with considerable "lumping" to bring the size and scope of the families and orders in line with Cronquist's classification and the rest of the book.

The ten subclasses of Anthopsida, six in the Dicotyledonae and four in the Monocotyledonae, are briefly described in Table 18–1. The orders included in each subclass, and how they relate to other orders, are indicated in Figure 18–5.

The fossil record definitely points to the Magnoliidae as the most primitive of the ten subclasses. According to Takhtajan, the most primitive order in the subclass, the Magnoliales, with eight families (Winteraceae, Magnoliaceae, Degeneriaceae, Himantandraceae, Eupomatiaceae, Annonaceae, Canellaceae, and Myristicaceae), can be regarded as an order of "living fossils." Some species have wood lacking vessels; all have primitive flowers with many spirally arranged free carpels and stamens on an elongated axis or receptacle; the stamens are mostly broad and laminar, undifferentiated into anther and filaments, with sporocarps embedded in the sporophyll tissue; the carpels are incompletely closed; and the pollen grains are monosulcate, similar to the pollen of cycads.

Degeneria vitiensis, a species found only on Fiji, is especially interesting to the phylogenist. The only surviving member of a rather unique family, it has a carpel that resembles a folded leaf. The carpel margins are completely free and before anthesis are noticeably distant from each other. The stigmatic surfaces extend the full length of the margins and are hairy. The ovules are laminal and near the midfold of the carpel. The numerous ovules are in two rows. However, there is only one carpel in each flower, in contrast to *Magnolia* and most other members of the Magnoliales in which there are many carpels, just as there are in *Cycas.*

Representative Angiosperms

In number of species, the largest families of angiosperms are the Compositae and the Orchidaceae, each with about 20,000. Four other families have over 5000 species each (Table 18–5). In terms of total abundance, the Gramineae, or grass family, is by far the most important of all angiosperm families.

Closely related to the Gramineae is the Restionaceae, a family of monocots found in the southern hemisphere (Figure 18–23). Also related to the grasses are the palms and other monocots. Figure 18–24 shows three species of palms and one of duckweed, all members of the Arecales. The coconut palm is the largest of the monocots, duckweed the smallest, yet they are close relatives.

Among the best-known grasses are wheat, rice, corn, barley, oats, rye, and other cereal grains; bamboo and sugar cane; crested wheatgrass, smooth brome, orchard grass, blue grama, curly mesquite, Harding grass, and hundreds of other range and pasture species; Kentucky bluegrass, redtop, colonial bent, Bermuda grass, and other lawn and golf course grasses; and a number of weeds like quack grass, Johnson grass, crabgrass, and sandburs. Grasses are found in all of the ecological life zones including arctic and alpine tundra, deserts, coniferous and hardwood forests, ponds and sloughs, and open woodland, but they are most abundant and most important in the steppes or prairies situated between the forested areas and the deserts of each of the continents, making up approximately one fourth of the land area of the world, and on agricultural lands worldwide.

Grasses are easily recognized by their typically narrow, two-ranked, parallel-veined leaves, hollow or pithy stems which are nearly always round in cross section, and unique flowers in which a superior ovary with two feathery stigmas is enclosed in a pair of chaffy bracts, the palea and lemma, and typically having several sessile flowers arranged in spikelets. The fruit is

TABLE 18–5

FAMILIES OF ANGIOSPERMS HAVING 5,000 OR MORE SPECIES[a]

Family	Order	No. Species	Special Characteristics
Compositae	Rubiales	20,000	Inflorescence a head or composite with 2 kinds of flowers; ovary inferior, 2 feathery stigmas
Orchidaceae	Microspermae	20,000	Flowers sympetalous, zygomorphic, inferior ovary; 6 stamens, 6 perianth lobes
Leguminosae	Rosales	13,000	Pea-like flower, pea-like pod, compound leaves, symbiotic bacteria in root nodules
Gramineae	Restionales	8,000	Narrow leaves with parallel venation; 3 stamens, superior ovary with 2 feathery stigmas, fruit a caryopsis
Euphorbiaceae	Euphorbiaceae	7,600	Flowers unisexual; ovary superior, usually trilocular, placentation axile; 5 petals, 5 or 10 stamens; milky sap
Rubiaceae	Rubiales	6,500	Inferior, multicarpellate ovary; opposite leaves; stipules usually absent

[a] Other families with over 2000 species are the Liliaceae (4,850), Cyperaceae (4,000), Melastomaceae (4,000), Asclepiadaceae (4,000), Scrophulariaceae (3,650), Labiatae (3,200), Myrtaceae (3,000), Umbelliferae (3,000), Cruciferae (3,000), Verbenaceae (2,600), Acanthaceae (2,600), Eucaceae (2,500), Azoaceae (2,500), Chenopodiaceae (2,400), Lauraceae (2,200), Ranunculaceae (2,000), Campanulaceae (2,000), Cactaceae (2,000), and Caryophyllaceae (2,000).

Figure 18–23 Monocots from southern hemisphere. (A–B) *Elegia cuspidata,* Restionales; (C) *Aloe* sp., Liliales.

Figure 18–24 Arecales: (A) windmill palm with pinnately veined leaves; (B) *Phoenix roebelenii*, pygmy date palm with palmately veined leaves; (C) *Lemna minor*, the duckweed; (D) coconut palm.

a type of hardened, one-seeded berry called a caryopsis, in which the embryo and the abundant endosperm are fused to each other and to the pericarp.

According to Malme (1928), agrostologist at Lund University, the **palea** represents altered and fused tepals. Stebbins interpreted the lodicules as reduced petals and the palea, with its two veins, as two fused sepals, a reduction in the number of each from three to two occurring in most lines of evolution as the flowers changed from actinomorphic, trimerous ones to zygomorphic ones. That some species of Gramineae have three stigmas, six stamens, three lodicules, and a "split" palea are suggested as part of the evidence favoring this hypothesis. Just inside the ring formed by the lodicules are the three stamens, and inside them the large, egg-shaped ovary with its two large, feathery stigmas. The flower is thus a zygomorphic, sympetalous flower with superior ovary. The feathery stigmas and monoporate pollen grains aid grass species in their adaptation for wind pollination.

The Leguminosae, or pea family, is also a widespread family of tremendous importance to mankind. Some common legumes are the acacias, locusts, and honey locust among the trees; peas, beans, soybeans, peanuts, lentils, cowpeas, and carob among the food crops; the clovers, alfalfa, sweet clover, vetch, kudzu, trefoil, and lespedeza among the cultivated forage crops; the milk vetches, lupines, wild peavines, lotus, prairie clover, and wild clovers among range species; and numerous species of ornamental flowers including wisteria, sweetpea, lupine, redbud, orchid tree, poinciana, wattles, broom, senna, and albizia. Legumes are easily recognized by their "butterfly" (**papilionaceous**) blossoms consisting of a large banner petal, two smaller wing petals, and two keel petals which are sometimes fused at the base and often have interlocking hairs along their common margins; ten stamens, nine of which are fused at the base; a single-chambered, unicarpellate fruit or dehiscent pod; usually compound leaves; and the presence on their roots of nodules containing nitrogen-fixing bacteria (Figure 3–8). In fact, any plant containing three of the above four characteristics (pea-like blossoms, pea-like pods, compound leaves, and root nodules) is certain to be a legume.

There is considerable variation in size among the flowering plants. Some years ago, a eucalyptus tree (family Myrtaceae in the Myrtales) found in Queensland measured 133 meters tall and was 5.5 meters in diameter at breast height. However, the top had obviously been broken off in a storm, or by some other means, and the trunk at the break was almost a meter in diameter. From this it was estimated that the tree before the break had been well over 150 meters tall (1974 *Guinness Book of Records*). At the other extreme is the ubiquitous duckweed, *Lemna minor*, consisting of two leaves, each 2 to 5 mm long, and a single slender root.

Of special interest in the angiosperms is variation in size of flowers. In wind-pollinated species, the flowers are often reduced to individual pistils consisting of a small ovary and feathery stigma or of one or a few small stamens. However, not all small flowers are wind-pollinated. The willows have very small, inconspicuous flowers which produce abundant nectar and are insect-pollinated. Members of the daisy or aster family have numerous small flowers arranged in a large flower-like inflorescence called a head or composite. In the balsamroot, *Balsamorhiza sagitata*, for example, there are two kinds of flowers in each head: ray flowers which are zygomorphic and disk flowers which are actinomorphic. In both kinds of flowers, the ovaries are inferior, the pistil terminates in two stigmas, the sepals are membranous bracts which have no apparent function, and the petals are fused into a sympetalous corolla. In the disk flowers, the anthers are fused in a ring through which the stigmas protrude at maturity. There are no stamens in the ray flowers, only pistils. While the individual flowers are small in balsamroot and its relatives, the inflorescence is often very large. The inflorescence of the sunflowers, *Helianthus annuus*, may be 70 cm across and contain several hundred disk flowers and about 100 ray flowers. In *Helianthus* the ray flowers have neither pistils nor stamens, but are neuter. Palms, cattails, and many other angiosperms have small flowers arranged into large inflorescences. On the other hand, *Rafflesia arnoldii* (Order Rafflesiales) has flowers which are often more than a meter across (Cronquist 1968) and *Selenipedium caudatum* (Orchidales) has petals 45 cm long and an overall diameter of the flower in full bloom of 90 cm.

Seed and fruit characteristics vary greatly among the angiosperms. Achenes occur in nearly all Compositae, but there is still variation: in burdock (*Arctium major*) and cocklebur (*Xanthium pennsylvanicum*), the exocarp, or outer layer of the ovary wall after the achene has matured, is covered with hooked bristles which enable the fruits to attach to the fur or hair of mammals that pass through an area where these weeds are growing (Figure 18–4). In members of the lettuce tribe of the same family, such as dandelion and oyster plant, the sepals are modified into capillary bristles which continue to grow as the seeds and fruits mature and become parachutes which can carry the fruits great distances. In other families, the pericarp is fleshy and sweet or otherwise edible, while the seeds are hard and pass through the digestive tract undamaged. In size, seeds vary from those of the epiphytic orchids, which average about 1.2 million per gram, to those of the double coconut, *Ladoicea seychellarum* (Arecales) which weigh on the average about 18 kg.

Ecological Significance

The Angiospermeae are ecologically the most important of all terrestrial plants, being the dominants in most terrestrial ecosystems and the most abundant of the subordinates in nearly all the rest. They provide from about 30% to over 95% of the food consumed by primary consumers in each of the land biomes. Fifteen percent of total global productivity is by tropical forest angiosperms alone (Table 2–5).

All of the ecosystems of a large geographic province, where macroclimate is sufficiently uniform that the major climax ecosystems are of similar structure, make up a biome. Within any given biome, there may be some ecosystems very different from the major ones because of local microclimate or mesoclimate variation, an oasis in a desert, for example. Figure 2–7 illustrates the relationship of major biomes to each other; it is somewhat oversimplified, since it takes into account only average temperature and average moisture conditions, ignoring the differences between wet winter dry summer and dry winter wet summer climates, for example, as well as wind and other climatic factors. Table 2–5 summarizes the general characteristics of the biomes of the world with the plants that are most typical in each of them. Three major biomes—cold desert (western North America), savannah (southern Africa), and mixed deciduous forest (eastern U.S.)—are illustrated in Figure 18–25. The student in the desert picture is measuring C_3 photosynthesis by an alkali absorption method, now largely replaced by infrared gas analysis.

All of the biomes have special features that make them interesting. In the coniferous forests, for example, pollination patterns are highly diverse among the understory plants. In open areas, wind-pollinated and bee- and butterfly-pollinated angiosperms abound, many of the latter having blue, yellow, or purple flowers. Where the forest is dense and light intensity in the understory low, white flowers have a competitive advantage over blue and purple ones. In the darkest areas, the Diptera are the chief pollinators. In the bogs, which are many in this biome, "carnivorous" plants grow. These are the plants which possess adaptations that enable them to trap insects and other small animals and digest their bodies. Although these plants are autotrophic, there is evidence they obtain some energy from the carbon compounds in the animal bodies; however, the chief advantage of this adaptation is that it provides these plants with a source of minerals, such as iron, which is nearly always in short supply in water-logged, acid soils.

Figure 18–25 Three types of ecosystems. (A) Cold desert dominated by grasses and sagebrush, western North America; (B) measuring photosynthesis in a sagebrush plant; (C) a savannah, South Africa, dominated by *Acacia;* (D) close-up of a branch of *Acacia;* (E) mixed hardwood/conifer forest, eastern U.S.

Tropical rain forests have many interesting plants, but our knowledge of them is limited. Several strata of vegetation—the canopy trees, understory trees, tall shrubs, low shrubs, grasses and forbs, and cryptogams—utilize the available light so completely that the ground is in a constant state of relative darkness except around clearings where trees have fallen during storms (Figure 18-1). The tremendous variety of epiphytes that thrive in the canopy layer is a fascinating area of study largely overlooked until recently (Perry 1980). Pollination by insects, birds, bats, and lizards is much more common in these forests than wind pollination. Total diversity is much greater than in most ecosystems; however, the reports that no species of tree has a density of more than one or two per hectare, made over a century ago (Alfred R. Wallace 1857) and still widely quoted in biology textbooks, is an exaggeration.

The canopy of an evergreen tropical forest consists of the crowns of giant framework trees together with epiphytic herbs, shrubs, and vines growing profusely on major horizontal branches, and other vines which have grown rapidly from the ground into the canopy. In central America, the framework consists of trees like *Anacardium excelsum* and *Lecythis costaricensis* that have their first branches 20 to 30 meters above the ground, but the canopy itself is made up largely of epiphytic shrubs and herbs: mostly bromeliads, orchids, cacti, ferns, and philodendrons. The fig family produces many of the always abundant vines. Some of the epiphytes become quite large, and the weight of all these plants may cause the support trees to come crashing down in a storm (Perry 1980). Often a domino effect results in a fairly large clearing where formerly several forest giants stood; in and near these clearings, light is favorable for the development of an extensive shrub vegetation.

Walking through a tropical rain forest and comparing it with other ecosystems, one is immediately aware of its great productivity. In addition to the lush vegetation, the abundance of animal life is an indicator of high productivity. Because of the great extent of tropical forests, it is obvious that a significant portion of the oxygen released into the atmosphere each year by way of photosynthesis must come from this biome. Conversely, this biome must be a most significant user of carbon dioxide.

Careful observation of the fossils in an area makes possible the analysis of climatic conditions of the past. A recent study of fruits and seeds from the Brandon lignite of Vermont suggested concluded that the Tertiary climate there was subtropical and wet; Bryant and Benson (1973), reported that the plants of the Carmen Formation in central Idaho during the Tertiary indicated a climate also either subtropical or warm temperate and relatively wet. Other studies strongly indicate that warm, subtropical, wet forests covered most of what is now North America during much of the Tertiary period.

Careful measurements of the carbon dioxide in the atmosphere have been made since 1958; the data show that global carbon dioxide content is steadily rising. According to many ecologists, cutting the tropical forests, especially in the Amazon Basin, is the main cause of this steady rising. Cutting has not only decreased the rate of global photosynthesis significantly, and therefore reduced the rate at which carbon dioxide is removed from the atmosphere, but, more importantly, has caused an increase in the rate of conversion of litter and humus to carbon dioxide and methane as the exposed soils have become better aerated. This position has been challenged by Broecker and his colleagues (see Sundquist 1993) who maintain that the great increase in burning fossil fuels since World War II is more important as the cause of the rise in global carbon dioxide level. Regardless of the cause, the rise itself is certain to have far-reaching ecological effects. The productivity of some ecosystems, such as farmland, may increase. Average temperature may also increase, resulting in more rapid melting of the ice caps at the poles and some rise in sea level. However, an increase in ocean temperature could cause a significant increase in precipitation on land. Changes in storm patterns could occur and have far-reaching climatic changes that cannot be fully predicted at this time (Gore, 1992).

The savannah biome occupies an area between the semiarid grasslands and the humid forest biomes. Either groves of trees or single trees are interspersed with broad grassy areas. The codominants, grasses over much of the area and oak or other trees in between, are primarily wind-pollinated, but the subordinates are largely insect-pollinated, and diversity of both plants and animals is rather high. Man-made ecosystems—cities and suburbs with lawn and shade trees, farmland with corn, wheat and other grasses sharing dominance with windbreaks and woodlots—are man-made savannahs. Consequently, the amount of land occupied by this biome is greater today in most parts of the world than it has ever been before. The "true savannah," in which less than 30% of the cover consists of trees, merges into woodland, with 30% to over 90% cover by trees, and scrubland, in which small trees and shrubs predominate. In Africa, single specimens of *Acacia* (Leguminosae) with their picturesque broad crowns are widely spaced, with grasses and other herbaceous plants occupying the land between trees. In Jamaica, the structure of the savannah is similar to that of Africa, but the species are different.

The most fragile of all biomes are the deserts and the tundra. Members of the willow and birch families among dicots and sedges and grasses among monocots share dominance with lichens and mosses in the tundra. Grasses, legumes, composites, cacti, and spurges are among the families dominating deserts. Most desert plants have thorns, contain toxins, or possess other adaptations which discourage excessive herbivory.

Numerous interesting examples of parallel evolution can be observed on deserts. The presence of essentially identical analogous structures can fool the careless botany student. A person well acquainted with the warm deserts of North America with cacti and yucca everywhere feels right at home on the deserts of South Africa until he is told that the cacti he is admiring are either *Asclepias* (milkweed family), *euphorbias* (spurge family), or *senecios* (daisy family), and the yuccas are not yuccas but aloes. Cacti have a well-developed pith which enables them to store water (Mauseth 1993); the spurges and groundsels (*Senecio*) have analogous structure. Growth form, including the general morphology of stems and thorns, of the giant senecios of South Africa is analogous to that of the arborescent cacti of North America (Knox and Kowal 1993).

Economic Significance

Most of the food, fiber, and drugs used by man come from this one class of plants. It is easier to name the plants that are not angiosperms that are used by man for food than to list the angiosperms: there are pine nuts, fiddle head ferns, mushrooms, söl and other seaweeds, and the animals that feed on diatoms, dinoflagellates, and their relatives. Human consumption of fish is important, but in most societies, people eat more beef, pork, and mutton than fish. It is probably safe to say that over 80% of the food we eat comes directly or indirectly from angiosperms.

Table 18–6 lists the major farm crops of the U.S. with the use and farm value of each. Farm crops are used for food, either directly (potatoes, apples, oranges), or following processing (wheat, soybeans, cottonseed); for livestock feed (corn, barley, alfalfa); for fiber (cotton, flax,

TABLE 18-6

ANGIOSPERMS OF SPECIAL ECONOMIC SIGNIFICANCE IN AMERICAN AND WORLD AGRICULTURE

Species	Family	Comments
Restionales		
Zea mays (corn)	Gramineae	Farm value of U.S. crop over $17 billion (b.) per year
Triticum aestivum (wheat)	Gramineae	Annual U.S. farm value almost $6 b.; wheat, rice world's #1 food crops
Oryza sativa (rice)	Gramineae	Alongside wheat, the most valuable food crop in the world
Rosales		
Medicago sativa (alfalfa)	Leguminosae	U.S. farm value c. $7 b. per year; preferred as the top quality hay
Glycine max (soy beans)	Leguminosae	Important to industry; processed to a diversity of products
Arachis hypogaea (peanuts)	Leguminosae	Important oil crop
Pyrus malus (apples)	Rosaceae	
Fragaria chiloensis (strawberries)	Rosaceae	
Malvales		
Gossypium hirsutum (cotton)	Malvaceae	U.S. farm value c. $6 b.
Hibiscus esculenta (okra or gumbo)	Malvaceae	Vegetable crop
Polemoniales		
Nicotiana tabaccum (tobacco)	Solanaceae	
Solanum tuberosum (potato)	Solanaceae	
Lycopersicon esculentum (tomato)	Solanaceae	Combined farm value of U.S. potato and tomato crops is over $1 b. per year
Convolvulus vulgare (wild morning glory)	Convolvulaceae	Noxious weed in every state in U.S.
Ipomoea batatas (sweet potato)	Convolvulaceae	$100 m. per year
Sapindales		
Citrus spp.	Rutaceae	Lemon, lime, orange, grapefruit
Asterales		
Artemisia tridentata (sagebrush)	Compositae	Most widespread shrub in North America
Lactuca sativa (lettuce)	Compositae	U.S. crop worth $1 m.

hemp); for drug or medicinal purposes (tobacco, poppy, quinine); for flavoring of other foods (hops, vanilla, spearmint); and for some miscellaneous uses, such as paper pulp and rubber. Table 18-7 lists important medicinal crops and their dollar value.

In number of species, the Compositae or daisy family is the largest family of flowering plants, but in terms of economic value, the Gramineae or grass family and the Leguminosae or pea family far exceed the Compositae. Other important families from the economic point of

TABLE 18–7

DRUGS USED IN AMERICAN MEDICINE

A. Sources of Drugs (% of total)

Major plant or animal source	1959	1973
Gymnosperms and angiosperms	25.5	25.2
Fungi and bacteria	21.3	13.3
Animals	2.3	2.7
Coal tar derivatives	50.8	58.8

B. Specific Plant Sources: Annual Cost of Drugs to American Public

Species	Family	Cost ($)
Atropa belladonna	Solanaceae	10,418,000
Cephalis ipecacuanha	Rubiaceae	7,047,000
Papaver somniferum	Papaveraceae	6,894,000
Rauvolfia serpentina	Apocynaceae	5,822,000
Rhamnus purshiana	Rhamnaceae	2,451,000
Digitalis purpurea	Scrophulariaceae	2,451,000
Citrus spp.	Rutaceae	1,379,000
Vinatrum viride		1,072,000

C. Prescriptions Filled

Drug	Number
Steroids	225,050,000
Codeine	31,099,000
Atropine	22,980,000
Reserpine	22,214,000
Pseudoephedrine	13,788,000
Ephedrine	11,796,000
Hyoscyamine	11,490,000
Digoxin	11,184,000
Scopolamine	10,111,000
Digitoxin	5,056,000
Pilocarpine	3,983,000
Quinidine	2,758,000
Total prescriptions	1,532,000,000

D. Number of Compounds of Known Structure

Plant Group	Number	Examples
Dicots	2,579	Digitalis, morphine, atropine, quinine
Fungi and bacteria	523	Penicillin, Ergotine, Streptomycin
Monocots	277	Garlic oil
Gymnosperms	221	Ephedrine
Marine organisms	199	
Ferns and fern allies	90	Lycopodium powder
Lichens	44	Usnic acid
Bryophytes	32	

view are the Malvaceae (mallow or hollyhock family), the Solanaceae (nightshade family), the Chenopodiaceae (goosefoot family), the Rosaceae (rose family), the Cruciferae (mustard family), and the Umbelliferae (parsley family).

Plant diseases, insects, and weeds cost farmers billions upon billions of dollars in reduced crop yields every year. The most costly are weeds. Just as all the important crops are angiosperms, nearly all the important weeds are angiosperms. In the U.S., some of the most serious weeds are wild morning glory, *Convolvulus arvense;* white top, *Lepidium draba;* quack grass, *Agropyron repens;* Johnson grass, *Sorghum halepense;* false toadflax, *Linaria vulgaris;* leafy spurge, *Euphorbia esula,* and dogbane, *Apocynum cannabinum.* All of these are perennial herbs difficult to eradicate and easy to spread. All of them have been declared by law to be noxious weeds in several of the states, meaning that it is illegal to sell seed containing the seed of any of them, and it is illegal for a farmer to allow them to remain in his fields without making an earnest attempt to control them. County weed boards are empowered to enter on to premises where these weeds are found but are not being controlled and begin an eradication program, the costs of which are added to the owners' taxes. Weed boards are also required by law to carry on eradication programs along roadways and on other public property. Holloway claims that five species of weeds in the Everglades of Florida cost over 10 million dollars a year to control.

Common annual weeds often cost farmers more in reduced yields than the noxious weeds do. These are species with high light tolerance that are normally pioneers in secondary successions. Most of them have been introduced from other parts of the world: from Asia to North America, for example, or from North America to Europe. They include many representatives from the grass family, the mustard family, and the composite family, and a sprinkling from other families, especially the goosefoot, amaranth, and buckwheat families. Wild buckwheat is an annual but at least as difficult to eradicate as most perennials because its seeds can lie dormant in the ground for 10 years or more (Metzger 1992).

Recent research has indicated that weeds are not always negative. Careful experiments by Giliessman and Altieri in an article in "California Agriculture" (1982) indicate that crop yields may be improved by allowing a limited weed population within the field; however, yields sharply declined in this study when weed populations were moderate to heavy. Other studies have indicated that good weed control is essential early in the growing season, but that a moderate weed control later in the season provides shelter for carnivorous insects which prey on many of the crop pests that would otherwise damage the maturing fruit.

Soil erosion is another major problem limiting food production (Bongaarts 1994). In the U.S., the U.S. Department of Agriculture, Soil Conservation Service, have worked with farmers since about 1933 in coping with problems of wind and water erosion, which take away millions of tons of topsoil every year. Some other nations are beginning to embark on similar programs.

The U.S., Canada, Australia, and Argentina produce much beef and mutton on rangeland. Range species vary greatly from one area to another, but the most important species are members of the grass family with Leguminosae, Rosaceae, Cruciferae, Chenopodiaceae, Compositae, and some other families also important. When rangeland is overgrazed, unpalatable and even poisonous weeds increase in density. Weedy species include members of the Compositae, Cactaceae, Boraginaceae, Liliaceae, Umbelliferae, and Ranunculaceae.

Angiospermeae are also economically important for their forest products. In the great tropical forests which are now being harvested in large quantity, nearly all of the trees are angiosperms with the Meliaceae (mahogany and African mahogany), Verbenaceae (teak), Zygophyllaceae (lignum vitae, the hardest and heaviest wood known), Ebonaceae (ebony), Santalaceae (sandalwood), and Bombacaceae (balsa) being of special importance because the lumber produced by members of these families is so valuable in cabinet making, furniture production, and other uses. A number of other forest products come from tropical forests; members of the Euphorbiaceae produce rubber, castor oil, and tapioca, in addition to a wood called African teak which resembles teak but is not as hard; figs, mulberries, hops, and breadfruit, come from members of the Moraceae or fig family, a predominantly tropical family; chicle, used in chewing gum, comes from a member of the Sapotaceae; and cocoa and chocolate are processed from the cacao seed from the Sterculiaceae.

The **Health Food** industry has grown appreciably in the U.S. and many other countries in recent years. Advocates of "natural" foods and medicines proclaim as an axiom that if it is "natural" it is good, providing it is used as "Nature" originally intended, and if it is "processed," it is suspect and possibly dangerous. Opponents argue that "food is food" regardless of whether it has been milled, preserved, frozen, or otherwise processed, and are often inclined to call the advocates of "natural" foods "health nuts," "food faddists," or "quacks."

Regardless of who is right in this argument, the health food industry continues to grow. There are reasons for this: (1) a growing consciousness on the part of many people of the importance of sound nutrition; (2) an increasing awareness of the legacy left by Native Americans and also American pioneers as to the food value and medicinal properties of wild plants; (3) a fear of being caught in an uninhabited area, perhaps as the result of an automobile or airplane accident, and not knowing what to eat; (4) a desire to live as a frontiersman, pioneer, or mountain man for a few days, living off the land, contemplating the beauties of nature and the expanse of the universe; (5) a realization that not all diseases are curable by modern medicines prescribed by the family physician or obtainable in the local pharmacy.

There are more reasons, but some of the above deserve further comment. In reference to (3): Donald Kirk (1975) reported an incident in which several hunters became lost in the vast wilderness of northern British Columbia.

> "They starved to death in a pine forest composed of trees whose inner bark was not only edible, but rather nourishing. They also had available to them several species of white grubworms in the shallow soil and decaying logs. Admittedly, a fat, juicy grubworm is not easy to eat. For those whose palate rejects such food sources, this book [in which approximately 2,000 species of edible angiosperms are described] will be of great value."

In reference to (2) and (4): every spring for many years, and usually again in the summer and fall, Ricks College in Rexburg, Idaho, has invited townspeople, students, and faculty to two or three all-day botanist-led excursions into the nearby mountains to find mushrooms and wild berries and enjoy a meal of morels or chanterelles or red cap boletes, depending on the season, with french-fried dandelion buds, fresh asparagus, boiled stinging nettle or pigweed shoots, camas bulbs, wild mustard-rose petal-raw stinging nettle salad, dutch oven huckleberry pie, cattail pollen pancakes with elderberry or choke cherry syrup, cattail rhizome biscuits, or anything else available, being always careful not to over-harvest or otherwise harm the environment. Such excursions could not be conducted in many parts of North America or Europe, but where they can be held they are delightful and educational experiences.

And in reference to (5): A recent newspaper article (*Post Register*, Idaho Falls, 1993, Associated Press release) reported that it costs approximately $16,000,000 to test and release one new drug. If the disease it treats is a rare one, or one that is only occasionally severely painful so that patient and doctor are normally not willing to pay enough to enable the company that could produce it to make a profit, the drug company is not going to invest in its development. People suffering from such ailments will gladly purchase *Ephedra* capsules, ground mullein leaves, spearmint and other herb teas, rose hips, *Grindelia* powder, or other products from a health store and hope for relief. Often they get excellent results. Or they will go out and gather dwarf mistletoe, wolf lichens, greasewood and sagebrush leaves, dandelion and stinging nettle greens, or other wild herbs, and brew their own tea or prepare a wild vegetable meal.

Terms like "health food advocate" and "natural" do not have single definitions. Many health food advocates drink teas processed from wild plants or swallow capsules containing concentrated garlic oil or cayenne pepper or *Ephedra sinica* extract. Cost is no hindrance. Others gather "weeds" in order to save money and simultaneously avoid "poisonous substances found in vegetables produced by artificial means." There is a good market for vegetables and fruits that have been produced on organic compost fertilized soil guaranteed to be free of commercial fertilizers and chemical pesticides.

Health food advocates are not alone in wanting to see fewer pesticides and herbicides used. There are many problems in modern agriculture that all conservation-minded people want solved: how to become less dependent on chemical control of pests is among them. Lanini et al. (1989) compared black plastic and living mulches with commercial herbicides and reported that both had significant advantages over the chemicals for managing weeds in vegetables. Thus, progress is being made.

There are advantages and disadvantages in using "Nature's pharmacy" for treatment of ills instead of the corner drugstore. First, the advantages. (1) Availability: if you are out on a camping trip and get a headache, a slice of aspen bark or the bud of a willow or a cottonwood tree is a lot handier than the city drugstore. (2) Your ailment may be one of the rare ones for which no drug has yet been developed. (3) Your body may reject the prepared drug but accept the same drug in the whole plant. Some vitamin pills create problems for some people, who can nevertheless tolerate the same vitamin in a garlic clove or a cabbage or a carrot and raisin salad. (4) The buffering capacity of other substances in the wild herb or vegetable may improve the ability of the body to absorb the needed amount of the vitamin or other nutrient. (5) Cost—for some people, especially in developing countries, but even in the U.S., Japan, and western Europe, the weeds that grow along streambanks and in fields and forests make it possible to have a nutritious and well-balanced meal which they could not otherwise afford to have.

And now the disadvantages. (1) Pesticides and other chemicals may have been sprayed on the vegetation that is gathered; lead from automobile exhaust is often present on roadside vegetation in toxic amounts. (2) Some plants concentrate toxic substances in the soil; selenium is naturally present in some western soils and arsenic is present in the soil of places where apple orchards once stood. (3) Granted that aspirin is the same in willow bark as in aspirin tablets, and should be equally effective, Nature's pharmacists do not test it for concentration; one tree may have extremely high levels and another one very low levels. You are guaranteed the right concentration when you buy from a reliable druggist. (4) People vary in their needs. Brigham Young urged British and Scandinavian immigrants to drink tea made from *Ephedra* shoots and leaves and thus overcome their addiction to tea and coffee (hence the common name, Brigham tea). Since *Ephedra* is not habit forming and is very effective in controlling allergies, this was sound advice for most of the Mormon pioneers, but dangerous for those who had high blood pressure. And you can probably think of other advantages and disadvantages, or consult with your local health food store or family physician for additional ideas.

Research Significance

Flowering plants are the most widely used of all plants in biological research. Our knowledge of the basic principles of genetics, and much of our knowledge of ecology, plant physiology, cytology, and morphogenesis has come from experiments performed with flowering plants. They are ideal for basic botanical research: their importance to man for food and fiber, their large size, their ease of cultivation, and their complex anatomy and physiology all contribute to their value in research.

Much of the research with plants has been and will continue to be in the fields of agronomy and forestry; all of our crop species are flowering plants, and man is almost completely dependent on the Anthopsida for food. We can safely predict that this world's agronomists (the botanists who specialize in cultivated crops) will continue to use flowering plants for most of their research.

The U.S., once the world leader in agricultural research, has fallen behind in recent years (Abelson 1992). Japan devotes 6.5% of its research and development funds to agricultural research, the U.S. only 1.9%. (The U.K. devotes 5.5% of R&D to agricultural research, France 4.6%, Germany 3.2%.) Yields of food grains have fallen in the U.S. and are now lower than in many countries. Budget appropriations for U.S. Department of Agriculture research have not increased, in terms of constant dollars, since 1955. Studies have shown that for every dollar spent on agricultural research, the value of increased productivity increases several dollars. This benefits the farmer whose income increases and the consumer who buys the product for less; yet both farmers and consumers complain about the increased taxes and their fears frequently frustrate good research plans.

The U.S. Department of Agriculture (USDA) has a four-prong mission: to conduct basic agricultural research, to introduce promising plant and animal genetic material from all parts of the world for use by USDA scientists, to publish the results of USDA basic research so that other scientists, U.S. or foreign, can benefit from it and/or build upon it, and to teach farmers good agricultural practices. To further its mission, the Department publishes a *Yearbook of Agriculture* each year. For example, the 1908 USDA *Yearbook of Agriculture* has an article on the processing of vanilla produced in Mexico, Tahiti, and other countries, another on the importance of protecting predaceous birds in order to safeguard crops from rodents and insects, a third on the development of methods plant geneticists can use to produce higher yielding grain, and a fourth on new introductions of leguminous plants and the role of nitrogen fixation in improving crop yields. The latest *Yearbooks* indicate that the USDA is still pursuing its four-prong mission albeit with reduced funding.

Much of the basic research on the role of soil elements in plant nutrition has been conducted by USDA agronomists; the 1938 Yearbook (*Soils and Men*) and the 1957 Yearbook (*Soils*) summarize their findings. These publications have stimulated other botanists and soil scientists to delve deeply into such areas as mineral nutrition. Thus, Epstein (1972) discovered that the principal role of potassium in plants is that of activator of numerous enzymes. This knowledge has now been tested on other plants, including many kinds of algae, and while effective rates vary, the generality holds. Lamont (1982), for example, was able to detect preferential absorption of potassium and phosphorus and has studied mechanisms for enhancing nutrient uptake in plants. And so the work continues, each scientist building on the foundation established by other researchers, often USDA agronomists.

Not only in agronomy and forestry, but in other areas of science have botanists working with the Anthopsida been making significant contributions. Flowering plants have helped us evaluate the plate tectonic theory of geology, for example. The similarity of the Cretaceous angiosperm flora in western Africa and eastern Brazil suggests that those two areas were joined at the time the flora was evolving and dispersing.

The angiosperms are especially valuable in studying climatic changes during the late Mesozoic and through the Cenozoic. In early Cretaceous, for instance, it is possible to infer the ecological requirements of ferns, mosses, mushrooms, mycorrhizal fungi, and other plants associated with angiosperms because the leaf morphology of flowering plant fossils gives us excellent clues to the climatic, edaphic, and other environmental conditions at that time.

Thus, brick by brick, our edifice of scientific knowledge is built, sometimes so rapidly we are amazed at how far we have come, sometimes very slowly. We may be sure that flowering plants will continue to play an important role in basic genetic studies, in studies of hormone effects, in research with organelles and in other areas of cytology, and in analysis of ecosystems and the basic concepts of ecology.

SUGGESTED READING

Every student of angiosperm diversity and evolution should be acquainted with Cronquist's and Takhtajan's books on angiosperm taxonomy, and also with the article on angiosperm paleobotany by Hickey and Doyle in the 1977 *Botanical Review. Taxonomy of Vascular Plants* by Lawrence contains valuable descriptions of most of the families of vascular plants. Most general botany textbooks discuss the morphology and physiology of vascular plants and especially flowering plants in detail. Pearson's *Principles of Agronomy* (1967) discusses the principles of producing agricultural crops.

STUDENT EXERCISES

1. Sketch a stem in cross section, showing relative size of the vessel elements in the primary xylem for each of the following: exarch development, mesarch development, endarch development. Which of these would represent the vast majority of angiosperms?
2. How would a sketch of a cross section of the root of a flowering plant differ from that of the stem?
3. Beginning with $\frac{1}{2}$, $\frac{1}{3}$, $\frac{2}{5}$, calculate the next ten fractions in the Fibonacci series (the phyllotaxy series in vascular plants).
4. Convert the 13 fractions in the Fibonacci series that you obtained in exercise no. 3 into decimals. Comment on any patterns or similarities you may observe.

5. People with ulcers sometimes have trouble taking aspirin for headaches. However, they claim that chewing aspen bark gives the same relief that aspirin gives but does not irritate their ulcers. Give a logical explanation for this observation. What are other advantages and disadvantages of taking aspen or willow bark instead of aspirin for headaches?

6. Name five medicines used in modern pharmacy and find the species of plant from which each is extracted. (See any good book on pharmacognosy.)

7. From time to time suggestions are proposed for domesticating a native species of plant or an introduced weed and developing a new crop. Name four or five species that might conceivably be domesticated and indicate what each would be used for and what its advantages over existing crops would be.

8. Present arguments for and against the hypothesis that global warming will increase worldwide aridity and melting of the polar ice caps. Consider the following facts: (1) CO_2 level is increasing in the atmosphere worldwide; (2) it is a universally accepted assumption that increased CO_2 levels in the atmosphere must lead to global warming including warming of oceans; (3) evaporation increases as water temperature decreases; (4) both low temperatures and high precipitation are necessary to building or maintaining glaciers; (5) over 70% of the Earth's surface is ocean.

GLOSSARY

acervulus (pl. acervuli) — discoid or pillow-shaped sporangium in which asexual spores (conidia) are produced in the Ascomycopsida and Fungi Imperfecti.

acid-fast stain — a staining method used to identify tuberculosis and leprosy bacteria.

adelphoparasite — an organism which parasitizes closely related species, usually within the same family or genus.

adnate — in Eumycophyta, having the gills attached directly to and at right angles to the stem.

adnexed — having the gills of a mushroom attached only to the apex of the stipe so that the gills are almost free from it.

aeciospore — dikaryotic spore produced in an aecium as in the Uredinales where they are produced on the alternate host.

aerobic — requiring oxygen for respiration.

agonomic productivity — rate of production of photosynthates stored in the economically productive part of crop species.

alar cells — cells at the point of attachement of leaves to the stem in mosses, often of distinctive shape.

algal layer — the inner cortex of lichens, the tissue in which the phycobiont or algal partner is found.

allopolyploidy — polyploids in which the parental individual was produced by an interspecific cross after which a doubling of the chromosomes changed the sterile hybrid to a fertile plant.

analogous structures — structures which have the same function though often different ontogeny.

androecium — collective term for all the stamens in a flower.

annulus — a ring on the stipe of a mushroom, a remnant of the partial veil; in the mosses and ferns, a ring of cells surrounding the sporangium and involved in explosive spore discharge.

antheridiophore — in the liverworts, a structure arising from the gametophyte and on which the antheridium is produced.

antheridium — The male gamete-producing structure.

antherozoid — A flagellated, motile cell produced by the antheridium in some algae, especially *Oedogonium;* a sperm.

aplanospore — a thick-walled, often spiny, endospore produced by some Mastigomycopsida, similar in structure and general appearance to the statospores of Xanthophycopsida and other Chrysophyta.

archegoniophore — in the liverworts, a structure arising from the gametophyte and on which the archegonium is produced.

artificial system — in taxonomy, a classification system in which the taxonomic categories are based on one or a very low number of characteristics.

autoecious — in the rust fungi, a species which completes its life cycle on a single host.

autotroph — an organism which requires only raw materials and a source of energy, usually light, in order to produce food.

bacteria — any member of the Schizomycopsida, small organisms lacking nuclei and other organelles; some bacteria have bacteriochlorophyll, but none have chlorophyll *a.*

bacteriophage — a virus which attacks bacteria.

basidiocarp — the sporocarp or spore fruit of the basidiomycetes.
basidiospore — spores produced on a club-shaped structure, usually in fours, typical of the Basidiomycopsida; the meiospores of the Basidiomycopsida.
basidium — a club-shaped structure on the gills or within the gleba or hymenium of fungi (Basidiomycopsida) on which meiosis occurs.
basket fern — a fern having an upright rather than prostrate growth habit.
biliprotein — any protein of ring structure, especially the phycobilin pigments of the Cyanophycopsida and Rhodophycopsida.
binding hyphae — filaments in the medulla of brown seaweeds which run transversely through the prosenchyma tissue.
binomial — the scientific or species name of a plant or animal, consisting of two parts, the genus name and the species epithet and unique for each individual species.
biome — all of the ecosystems of a climatic zone characterized by vegetation of the same general growth form.
blade — the thin, flat part of a leaf, connected by the petiole to the stem.
CAM — crassulacean acid metabolism in which carbon dioxide is absorbed in the dark and converted to carbohydrates in the presence of light; common in desert plants.
Calvin cycle — the series of chemical reactions in which ribulose diphosphate is first carboxylated and then hydrogenated, and after a series of chemical reactions produces a molecule of carbohydrate and one of ribulose diphosphate which again reacts with carbon dioxide and water.
calyptra — a hat-like or cup-like sheath covering the sporangium of mosses at the apex of the sporophyte part of the plant.
calyx — collective term for the sepals of a flower.
cambium — a meristematic tissue in vascular plants which produces the secondary xylem and phloem or ingrease in girth.
capillitium — in some fungi and slime molds, thread-like strands often forming a network interspersed with spores.
carboxysome — minute bodies in cells of lichens and other plants in which the Calvin cycle of photosynthesis takes place.
carpel — the individual megasporophyll in angiosperms on which the ovules are produced; one or more carpels fused together form the pistil.
carpogonium — in red seaweeds, the female gametangium.
carpospore — spores, usually diploid, produced by the carpogonium in the Rhodophyta.
caryogamy (karyogamy) — the fusion of two gametes or the nuclei of two gametes in the process of sexual reproduction.
centriole — an organelle lying next to the nucleus, almost always present in dividing animal cells, where it is probably actively involved in anaphase, but found only in flagellated plant cells and apparently involved in formation of new flagella.
chalaza — basal part of the ovule adjacent to the stalk and opposite the micropylar end.
chemoautotrophic — a plant which produces its own food by utilizing the energy derived from oxidation of inorganic compounds rather than from light.
chromosome aberrations — changes in the number or structure of chromosomes, such as deletions, inversions, interchanges.
chromosome interchange — an aberration in which part of one chromosome is exchanged with another that is not its homolog; also called a translocation, especially by zoologists.
cilium (pl. cilia) — a short flagellum, usually many per cell.

circadian rhythm — a cyclic rhythm of approximately 24 hours' duration from peak to peak (such as jet lag in humans).

cirri — pustules which produce asexual spores in many fungi.

cisternal rings — characteristic concentric rings seen in the cells of lichenized and some other fungi and which are believed to function the same as dictyosomes or Golgi bodies.

cladistics — a method of taxonomic analysis in which relationships among species or other taxa are inferred by numerical analysis of selected characteristics.

clamp connection — connection between protubernces from two adjacent dicaryotic cells typical of many Basidiomycopsida.

cleistothecium — an ascocarp in which the hymenium at meispore maturity completely encloses the asci.

climax forest — according to Frederick Clements and his followers, a stage in vegetational development which is permanent because the dominant species are able to reproduce in their own shade.

coenocyte — a multinucleate cell-like structure.

coenozygote — in the Zygomycopsida, a zygote consisting of many nuclei resulting from the fusion of multinucleate gametes.

coevolution — evolution of two species with adaptations that are usually beneficial to both.

collatoral protostele — a protostele, or stele lacking a pith or medulla, in which the phloem lies below the xylem.

collet — the point at which the stem meets the root in vascular plants.

companion cell — parenchyma cells, in mosses and vascular plants, which are associated with sieve tube elements.

conchocelis stage — the stage in the life cycle of *Porphyra* in the Bangiales (Rhodophycopsida) which was formerly believed to be a different species even classified in a different order (the Nemalionales).

conidium (pl. conidia) — an asexual spore often produced within pycnidia or acervuli in the Fumycophyta.

conjugation — uniting of two cells for sexual reproduction or (in the case of Schizophyta) parasexual reproduction.

connecting band — a cingulum or band connecting the epithecum to the hypothecum at the time of sexual reproduction.

conservation — the wise use of natural resources.

conserved name — a family or order (or other) name which is allowed by the international rules committee to be retained or conserved because of its wide usage, even though it does not have the designated suffix such as -ales or -aceae.

consumer organism — an organism which feeds upon others, usually phagotrophicaly; an animal, either herbivore or carnivore.

contractile vacuole — in the Cryptophycopsida, Euglenophycopsida, and some other groups of unicellular species, a membrane-enclosed structure which rhythmically contracts and expands.

corolla — collectively, all the petals of a flower.

cortex — an outer tissue, though usually not the outermost, made up of parenchyma or pseudoparenchyma, rarely prosenchyma, cells.

corticated — an organ containing cortex tissue.

corticolous — describes lichens, mosses, or other plants attached to the bark of trees.

cortina — in some mushrooms, a remnant of the universal veil or more commonly the partial veil which hangs down from the cap like a spider web, or like the veil worn by women in church in some cultures.

costa (pl. costae) — a ridge as in some diatom cells or the midrib of a moss leaf.

cotyl — in seed plants, the point on the axis of an embryo where the cotyledons arise and where epicotyl meets hypocotyl.

cotyledon — the embryonic organ in seed plants which absorbs food from the gametophyte or endosperm and which becomes the first leaf-like structure of the seedling.

crozier — in sac fungi, a hook-like structure at the apex of a dicaryotic filament which ensures that each daughter cell will receive the proper daughter nuclei following mitosis, the homolog of the clamp connection of the club fungi.

cruciform — in the form of a cross, as in mitosis of the clubroot fungus, where the ring of chromosomes around the elongated nucleus gives the appearance of a crucifix.

crustose — growing like a crust on rock, bark, or other medium, especially as applied to the Rhodophyta and lichens.

cystocarp — in Rhodophycopsida, a structure made up of the diploid carposporophyte and surrounding haploid pericarp.

cytokinin — a plant growth substance, or hormone, involved in the differentiation of embryonic tissues and other developmental processes.

decomposer organism — any organism which gives off carbon dioxide, water, and various minerals as a result of its metabolism, especially very small organisms like flowers.

decurrant — in mushrooms, having gills which are attached to the stipe over a considerable length.

depside — a group of lipoid substances in lichens, often brightly colored.

depsidone — a group of substances in lichens.

Deuteromycetes — the Fungi Imperfecti, treated as a taxonomic category by some biologists.

dictyostele — a divided stele, for example, polystele, eustele, atactostele.

dikaryotic (dicaryotic) — having two haploid nuclei in each cell.

dinoflagellate — unicellular, mostly marine plankters, making up the class Dinophycopsida.

Dinophycopsida — the dinoflagellates, unicellular marine and freshwater organisms which are often bioluminescent.

dioecious plants — plants in which the male and female flowers or gametangia are borne on separate plants.

diplohaplontic — having an alternation of generations; a life cycle in which the gametes are produced by gametophyte individuals and meiospores by sporophyte individuals.

diplontic — having life cycles in which there is no haploid generation but the gametes are produced by meiosis in diploid individuals.

discomycete — Ascomycopsida in which the sporocarp is a flat, dish-like to cup-like apothecium; the Helotialean line.

disjunctive symbiosis — a mutually beneficial association between two species in which the relationship is not so intimate as to develop into a type that is morphologically unique.

diversity — variety; many morphogical and/or physiological types within an area or within a taxon or other group.

dominant — in ecology: a population or species of plants which carries on a major portion of the photosynthesis in the ecosystem, usually because it receives direct light; in genetics: a gene which has the ability to mask the effects of its allele.

donor — in bacterial conjugation, the cell from which the chromosome migrates.

double fertilization — a phenomenon characteristic of the angiosperms in which one sperm fertilizes the egg and a second sperm unites with the two nuclei at the center of the gametophyte to form an endosperm.

ecological homeostasis — having the ability to adjust to widely fluctuating environmental conditions.

ecological niche — the function of a species within an ecosystem.

ecology — the study of the structure and function of ecosystems; the study of how organisms relate to each other and to their physical environment.

ecosystem — all of the plants and all of the animals within a discrete geographical unit which interact with each other and their physical environment.

ecotone — the border or boundary between two ecosystems.

ectophloic siphonostele — a stele in which there is a pith surrounded by xylem and phloem with the phloem external to the xylem.

ectotrophic mycorrhizae — a symbiotic relationship between a fungus, usually a basidiomycete, and the roots of a vascular plant (or rarely the rhizoids of a moss) in which fungal tissue forms a sheath with hyphae intimately associated with cortical cells of the root such that the morphology of the root or rhizoid is appreciably altered.

effigurate — having a margin which is of definite form as in some crustose lichens in which the thallus reaches to the margin without breaks in between.

embryo — early stages in the development of a plant.

enation — a primitive type of leaf which consists of an outgrowth of epidermal tissue without a leaf gap.

endarch — a type of primary xylem development in which the first cells to mature (the protoxylem) are at the center of the stem.

endolithic — describes lichens in which the thallus develops within the substrate, which is usually stone; "fenster flechten."

endosperm — food reserve in the seed of angiosperms resulting from double fertilization; in gymnosperms, gametophyte tissue is often referred to as endosperm.

endotrophic mycorrhizae — a symbiotic relationship between a fungus and the roots of a vascular plant in which the fungus is within the root tissues; therefore the external morphology of the roots is less drastically altered than in the case of an ectotrophic mycorrhiza.

enzymology — the study of enzymes and how they function.

epicotyl — the portion of the embryonic axis above the cotyl or point of attachment of the cotyledons which develops into the shoot.

epidermis — the outermost cell layer of roots, leaves, and stems.

epigeous — development of a seedling by elongation of the hypocotyl resulting in the cotyledons being pulled above the surface layer of soil.

epistatic — genes having the ability to mask the effects of genes which are not their alleles.

epithecium — in ascomycetes, the upper layer or disk of the apothecium.

epithecum — in diatoms and dinoflagellates, in which the cell is made up of two valves or frustules, the upper or outer valve.

epoch — in geology, a subdivision of the period.

era — in geology, one of the five major divisions of time; eras are further divided into periods which are in turn divided into epochs.

etiolated — describes stems with abnormal elongation caused by low light intensity.

eukaryotic — having true nuclei and other organelles.

euryhalic — adapted to a wide range of tolerance to salinity

euryhydric — adapted to a wide range of tolerance of moisture or

eurythermic — adapted to a wide range of temperatures

exine sculpturing — the projections and patterns on the outer layer of tissue of spores and pollen grains.

exotic — an organism introduced from another area, usually from another continent.

exploit, exploitative — to use natural resources in a selfish manner without consideration for conservation or other's needs.

false branching — in Schizophyta and some other filamentous algae, a break in the filament with one or both ends protruding from the sheath.

family — the taxonomic category between genus and order; the highest taxon in which similarities are more pronounced than differences.

fertile — in ferns and "fern allies," the shoots which bear sporangia.

fiber — a specialized cell, similar to but more compact or dense than a tracheid, providing strength and hardness to angiosperm wood.

filament — in algae, a trichome or thread-like row of cells; in angiosperms, the stalk of the stamen on which the anthers are borne.

flagellated fungi — the Mastigomycopsida, fungi in which the cells, or some of them, are propelled by miniature whip-like structures

flagellum (pl. flagella) — whip-like structures on cells which aid in their motility.

foliose — leaf-like; in Rhodophyta, Phaeophyta, liverworts, and lichens, a flat growth form which resembles leaves.

form families — in the Fungi Imperfecti, artificial grouping of genera which have similar conidiophores.

form genus (pl. genera) — genera based upon asexual reproductive structures.

fragmentation — asexual reproduction by means of fragments of a filament, thallus, or other structure breaking free from the parent structure and beginning to grow.

frond — a large leaf or leaf-like structure in seaweeds and ferns; also used for shoots with leaves attached in some angiosperms like asparagus, palms, etc.

frustule — a valve or half of a cell wall in diatoms and other Chrysophyta.

fruticose — shrublike (or sometimes vinelike), especially in red seaweeds and lichens.

fungal sheath — in mycorrhizae, the sheath of fungal tissue which surrounds the root proper.

Fungi Imperfecti — higher fungi, usually ascomycetes, in which sexual reproduction has not yet been documented.

fusiform initials — meristematic cells in the cambium which develop into vessels, fibers, tracheids, and all cells oriented parallel to the long dimension of the stem.

gametangia — structures which produce gametes.

gametophyte — the haploid generation of plants, the generation which produces gametes.

gelatinous lichen — lichens in which the phycobiont is a cyanophyte; often nitrogen-fixing.

gene splicing — process in which chromosomes are cut by specific enzymes and the fragmented portions with their genes introduced into a developing embryo or other tissue.

genetic drift — phenomenon by which a species or other taxon receives certain genotypes purely by chance.

genetic recombination — in sexual reproduction, the diversity of genotypes and phenotypes resulting from meiosis and syngamy.

genus — a category of taxonomy consisting of one or many closely related species; usually gene exchange is possible within members of the same genus but interspecific hybrids are usually sterile.

geologist — one who studies geology.

girdle groove — in diatoms and dinoflagellates, a groove encircling the cell in the area where the two valves meet.

Gnetopsida — a class of cone-bearing plants especially well adapted to desert conditions; *Ephedra, Welwitschia,* and *Gnetum.*

Golgi — an organelle involved in purification of enzymes and other processes.

Gram negative — any bacterium which gives a negative reaction to the Gram staining technique.

granum (pl. grana) — a grouping of the thylakoids or parts of them to form a granular suborganelle within the plastid.

gullet — an opening into the cell of cryptomonads, euglenids, and other unicellular organisms through which food enters the cell.

gynoecium — a collective term for all of the carpels (and pistils) of a flower.

habitat — where an organism lives.

haplontic — a life cycle in which every cell except the zygote is haploid.

haptera — a branched structure at the base of a seaweed stipe which attaches to the substrate; a "holdfast."

hartig net — in ectotrophic mycorrhizae, a net of fungal hyphae just outside the root proper.

haustoria — short hyphae or filaments which penetrate the cells of the host.

herbarium — a library of preserved, usually pressed, plant specimens.

herpokinetic motility — in Schizophyta and some Rhodophyta, a movement of the filaments which resembles the motion of a snake.

heteroecious — describes a parasitic organism, especially a rust fungus, requiring two hosts to complete its sexual life cycle.

heterotrophic — not able to produce its own food; not autotrophic; may be saprophytic, parasitic, or phagotrophic.

holdfast — a rootlike structure at the base of the stipe in red and brown seaweeds which anchors it to the substrate.

homologous — two chromosomes which are capable of pairing or synapsing with each other.

hormogonia — in blue-green algae, portions of a thallus having a weak point at either end, which can break loose and become propagules.

hymenium — the fertile or spore-bearing portion of a spore fruit or sporocarp.

hypha — in fungi, a filament made of end-to-end rows of cells.

hypostatic — in genetics, being masked by another gene not its allele.

hypothecum — the bottom or inner frustule or valve in diatoms and dinoflagellates.

hypothesis — a speculative statement formulated in such a way that it is vulnerable or able to be disproven.

indigenous — native to an area though not necessarily limited to it; not an exotic.

integument — the innermost layer of mother tissues surrounding an ovule.

inversions — chromosome aberrations in which a segment of a chromosome has its genes in reverse order compared to a "norm."

involucre — collective term for all of the bracts subtending a flower.

involucrellum — a usually dark-colored structure surrounding the ostiole or opening to a perithecium in many fungi.

isidia — "coral-like" projections from the thallus of lichens which function in asexual reproduction.

isogametes — gametes that are identical to each other in appearance and function and hence cannot be distinguished as egg or sperm.

isomorphic — having both or all stages of the life cycle morphologically indistinguishable from each other.

juvenile — a seedling or sporeling in its early, immature stages of growth and development.

kelp — any of the large brown seaweeds, especially of the order Laminariales.

lamella — a leaf or leaf-like structure or leaf blade.

late blight — the "potato plague," a plant disease caused by some of the Perinosporales.

leaf gap — a region of parenchyma tissue where a leaf trace departs from the vascular tissue of the stem.

leaf trace — a vascular bundle in a stem extending from the vascular system of the stem into the base of the leaf.

lecanorine — apothecia in lichens which have chlorophyll extending into the rim or exciple, hence the disk and exciple are usually different in color.

lecideine — apothecia having proper exciples, *i.e.,* chlorophyll not extending into the rim of the apothecium; therefore the disk and rim are nearly always the same color.

lemma — the lower or outer of two chaffy bracts which subtend grass flowers.

lichen — a symbiotic association of alga (either Chlorophycopsida or Cyanophycopsida) and fungus (usually Ascomycopsida), having the appearance of a single plant and functioning as one plant unit.

life cycle — the continuous sequence of changes undergone by an organism from one primary form to the development of the same form again.

lignicolous — growing on logs or stumps or any material with a high lignin content.

locule — a compartment, cavity, or chamber; in Ascomycetes, stromatic chambers containing asci; in Anthophyta, a cavity in an ovary containing ovules.

lophotrichous — having proterokonts ("flagella") as a tuft at one or both poles of a bacterial cell.

lorica — in some algae, a surrounding case, often vase-like and siliceous in appearance, external to the cell wall if one is present, and always separate from the protoplast.

lower cortex — in plants with horizontal, or thalloid, growth form, the lowermost tissue, usually parenchymatous or pseudoparenchymatous in structure.

luminescence — the emission of light not caused by incandescence and occurring at a temperature below that of incandescent bodies.

mastigoneme — a hair-like thread or process occurring along the length of some flagella; also known as "flimmer" or tinsel.

medulla — the innermost region of the thallus in lichens, and in some Phaeophyta and Rhodophyta.

megaphyll — leaves which have a leaf gap in addition to a leaf trace where attached to the stem and which usually have several to many orders of branching of their veins.

megasporophyll — a leaf-like appendage bearing megasporangia.

meiospore — a spore produced by meiosis, with a reduction in chromosome number from diploid to haploid (spores usually produced in fours).

meristem — a cell or tissue which can divide by mitosis to form new cells.

meristoderm — the outer meristematic cell layer (epidermis) of some Phaeophyta, analogous to both cambium and epidermis of seed plants.

meromixis — general term for the types of genetic exchange occurring in some Schizophyta; involves a unidirectional transfer of genetic material.

metaboly (metabolic) — capable of changing shape, as in the cells of many Euglenophyta and Chrysophycopsida.
microphyll — a leaf-like structure attached to the stem without a leaf gap.
microsporophyll — a leaf-like appendage bearing microsporangia.
migration — moving from one location to another, generally in a consistent manner.
monoecious — refers to seed plants in which separate pollen-bearing and ovule-bearing strobili (or flowers) are both borne on the same plant.
monophyletic — evolving from a single ancestral stock.
monospore — a single spore produced by metamorphosis of single vegetative cell, the monosporangium; characteristic of Rhodophyta.
monotypic — having only one representative, as a genus with a single species.
monotrichous — having a single flagellum.
multiaxial — a main (central) axis composed of many parallel or almost parallel filaments.
muriform — a spore having septations in two directions giving the impression of a miniature brick wall.
Musci — the mosses; the conserved name for the Bryopsida.
mutation — a heritable change in DNA structure and hence genetic composition.
mycelium (pl. mycelia) — a mass of hyphae; the thallus, or vegetative part, as opposed to reproductive part, of a fungus.
mycobiont — the fungal partner, or component, of a lichen.
mycophagist — one who eats mushrooms.
mycorrhiza (pl. mycorrhizae) — a root-like structure commonly found on naturally growing vascular plants, consisting of vascular plant tissue and fungal hyphae intimately associated with each other and usually mutually beneficial to each other.
naked cell — a cell not enclosed by a cell wall; it may be enclosed only by a plasma membrane or there may be a paraplast or pellicle present interior to the plasma membrane.
nanoplankton — plankton with dimensions less than 70 to 75 μm and usually less than 5 μm.
natural selection — selection of usually better adapted phenotypes by natural, ecological forces.
net plasmodium — a structure resembling the plasmodium of a myxomycete but consisting of nonliving matter laid down by a labyrinthulomycete.
nitrogen fixation — conversion of atmospheric nitrogen or N_2 into nitrates, amino acids, and other metabolically usable substances.
nodules — on the roots of certain plants, an enlargement within which nitrogen-fixing bacteria live.
nucellus — the central cellular mass of the body of the ovule including the integuments and containing the embryo sac; that which becomes the seedcoat.
nucleoids — nuclear material (DNA or chromosomes) not enclosed in a plasma membrane.
obligate anaerobe — an organism which can live only in the absence of free oxygen.
ontogeny — the development of an organism in its various stages from initiation to maturity.
operculum — a lid or cover; in the fungi, part of a cell wall; in the Bryidae, a multicellular tissue in the capsule.
optimum — the conditions (temperature, moisture, etc.) at which an organism grows best.
ostiole — an opening or pore.
ovule — a sporangium surrounded by an integument and maturing into a seed following fertilization.

palisade mesophyll — a tissue in Anthopsida and some other vascular plants, such as Gnetopsida, in which the light-gathering cells are upright like the palisade around a fort.

parallel evolution — evolution of two unrelated species in which very similar adaptations develop.

parasexual — describes the recombination of genes not involving meiosis and syngamy.

parenchyma — a tissue composed of living, thin-walled, randomly arranged cells.

partial veil — a membranous layer covering the developing hymenium in some Basidiomycetes.

pectinate — a flagellum with one file of mastigomenes.

PEP — phosphoenol pyruvate, in C_4 and CAM plants the chemical substance which combines with carbon dioxide to form malic acid.

pericarp — tissue surrounding carposporophyte (sometimes collectively referred to as cystocarp); in Anthophyta, the mature ovary wall.

peroxisome — a cellular organelle which oxidizes unneeded substances and in which photorespiration takes place.

phenology — the study of development of plants and animals in relationship to environmental factors, such as the changing color of leaves in the fall.

phenotype — referring to the external, visible appearance of an organism.

photoautotroph — an autotrophic plant that derives energy from sunlight.

phycobilin — a water-soluble pigment, similar to bile pigment, occurring in Cyanophyta, Rhodophyta, and Cryptophyceae.

phycobilisomes — small organelles located on the lamina of red and bluegreen algae and which contain phycoerythrin and/or phycocyanin.

phycobiont — the algal partner or component of a lichen.

phycocyanin — blue phycobillin pigment occurring in Cyanophyta, Rhodophyta, and Cryptophyceae.

phycoerythrin — red phycobillin pigment occurring in Cyanophyta, Rhodophyta, and Cryptophyceae.

pistil — the ovule-producing part of a flower, consisting of a carpel (simple pistil) or of two or more partly or wholly fused carpels (compound pistil).

plasmids — gene-containing suborganelles which occur in some Cyanophycopsida and Schizomycopsida.

pleiotropic — a gene which controls two or more unrelated, phenotypes.

pneumatocyst — the hollow area of a stipe which helps keep some Phaeophyta afloat.

pollen grains — immature androgametophyte seed plants.

polyphyletic — evolution from more than one ancestral stock.

polystele — having two to several protosteles within a structure.

predict — to state accurately before an event happens that it will happen.

producer organism — an autotrophic organism.

proembryo — earliest stages of embryo before main body and suspensor (if present) are differentiated.

prosenchyma — fungal or algal tissue in which the hyphal elements are recognizable as such.

proterokont — a fine hair-like structure projecting from the plasma membrane of some bacterial cells; they resemble thin flagella, and are often called flagella, from which they differ in basic anatomy, lacking the 9+2 structure of true flagella.

protist — a unicellular or colonial organism, plant or animal.

pseudofilament — a row of individual cells attached end to end like a filament and held together by gelatinous matter.

pseudoparenchyma — a cell or tissue type which originated from filaments in which the filamentous nature is not obvious.

pycnidium — a flask-shaped structure in which conidia are formed in some Ascomycetes and Fungi Imperfecti.

pycnospore — in the rust fungi or Uredinales, a male gamete or spermatium produced in a pycnidium-like structure on the alternate host (e.g., a barberry leaf) and which can also function at times as an asexual spore.

recessive — one of a pair of contrasting characters which is masked, when both are present, by the other, or dominant, character; also refers to the gene determining such characters.

recipient — one who receives; in meromixis the cell that receives the chromosome from the donor.

reference species — a fossil animal, or nowadays sometimes a plant, used to ascertain the age of a geological formation.

retrovirus — a virus in which single-stranded RNA rather than DNA carries the genetic information.

rhizine — (1) a differentiated, multicellular structure, common in lichens and present in some mushrooms, which anchors the plant to the substrate, and superficially resembles the roots of flowering plants; (2) a thick-walled cell in the medulla of some Rhodophycopsida, especially the Gelidiales.

rhizoid — in some fungi, mosses, liverworts, etc., hair-like appendages which penetrate the soil or other substratum, anchoring the plant and absorbing water and other substances.

samara — a dry, indehiscent, one-seeded, winged fruit, such as that of elm or maple.

secondary pits — in the Rhodophyta and Eumycophyta, connections between cells which were not originally connected to each other, as in croziers, clamp connections, and auxiliary cell connections and also between parasite and host.

serological test — a test for analyzing degree of similarity between proteins by inoculating plant substances into the bloodstream of a vertebrate; proteins from the plants produce specific antibodies in the blood serum which if mixed with proteins of the same or a closely related species of plant will cause agglutination.

sheath — a covering external to the cell wall.

species epithet — the second part of a species name or binomial.

species — the basic unit of taxonomy, consisting of all the population of a kind of plant capable of interbreeding to produce fertile offspring. The species name consists of two parts, the genus name and its species epithet.

spermatangium — a structure that produces one or more spermatia in Rhodophyta.

sporophytes — the spore-producing phase in alteration of generations with the diploid chromosome number.

stephanokont — a spore or gamete having a ring of cilia or short flagella near the apex thus suggestive of the hair style of the monks of the order of St. Stephan.

stigma — apex of a pistil receptive to pollen.

stipe — a stem-like structure, commonly observed in large seaweed such as the kelps; hence, the stem of a mushroom.

stromatolite — a club-like or "cabbage head"-like petrified mass of algae.

supporting cell — specialized cell from which the carpogonial branch arises in some Florideophycidae (Rhodophyta).

style — in flowers, a cylindrical structure which rises from the top of an ovary and through which pollen tubes grow.

synergids — micropylar nuclei associated with the egg in Anthophyta; part of the egg apparatus.

syngamy — the process of fusion between two gametes; fertilization.

taxon (pl. taxa) — a general term that can be applied to any taxonomic grouping.

trichogyne — receptive hair-like extension of female gametangium in Rhodophyta and Ascomycetes.

trichome — an outgrowth from the epidermis of plants, as a hair; a microorganism composed of many filamentous cells arranged in strands or chains.

uniaxial — having a main or central axis consisting of a single filamentof usually large cells.

vaccine — a substance composed of weakened or dead bacteria (or other pathogens) or their diluted toxins, injected into the body to induce immunity to the same kind of pathogen or its toxins.

valve — the top or bottom surface of the frustule of diatoms.

vascular system — a plant conductive system composed of xylem and phloem.

vascular tissue — conducting tissue; xylem and phloem.

vernalize — (1) to undergo the phenological changes that occur in the springtime to any plant that is dormant during the winter months; (2) to shorten the growth period before blossoming and fruit or seed bearing by chilling the seed or bulb.

virulent — actively poisonous; intensely noxious.

vitta — a streak or stripe, as of color.

volva — usually a cup-like structure at the base of the stem of a mushroom, which may be reduced to markings of various kinds, formed from remnants of the universal veil.

xerophyte — a plant living in a dry habitat.

xylem — water-conducting tissue of vascular plants; constitutes the major portion of wood.

zoospore — a spore motile by means of one or more flagella; also termed planospore.

zooxanthellae — algal cells (often yellow) living symbiotically in cells of certain invertebrate animals; algae known to be members of Dinophyceae, Cryptophyceae, and Xanthophyceae.

zygomorphic — having bilateral symmetry, i.e., being symmetrical only about a single axis.

zygote — A fertilized egg; a cell, often thick-walled, resulting from fusion or syngamy between egg and sperm or between two isogametes.

BIBLIOGRAPHY

Aaronson, S. and Hutner, S. H., Biochemical markers and microbiological phylogeny, *Q. Rev. Biol.,* 41, 13, 1966.

Abelson, P. H., Editorial: agricultural research, *Science,* 257, 1187, 1992.

Ahmadjian, V., Algal/fungal symbioses, *Progr. Phycological Res.,* 1, 179, 1982.

Ahmadjian, V. and Jacobs, J. B., Algal-fungal relationships in lichens: recognition, synthesis, and development, in Goff, L. J., Ed., *Algal Symbiosis.* Cambridge University Press, London, 1983.

Ahmadjian, V., Russell, L. A., and Hildreth, K. C., Artificial reestablishment of lichens. I. Morphological interactions between the phycobionts of different lichens and the mycobionts *Cladonia cristatella* and *Lecanora chrysoleuca, Mycologia,* 72, 73, 1980.

Ainsworth, G. C., *Ainsworth and Bisby's Dictionary of Fungi,* Commonwealth Microbiological Institute, Kew, U.K., 1961.

Ainsworth, G. C., Sparrow, F. K., and Sussman, A. S., eds., *The Fungi, Vol. IVB,* Academic Press, New York, 1973.

Aldrich, D. V., Ray, S. M., and Wilson, W. B., *Gonyaulax moniliata:* population growth and development of toxicity in cultures, *J. Protozool.,* 14, 636–639, 1967.

Alexopoulos, C. J., and C. W. Mims 1979. *Introductory Mycology.* John Wiley & Sons, New York.

Allen, M. F., Re-establishment of VA mycorrhizae following severe disturbance: comparative patch dynamics of a shrub desert and a subalpine volcano, *Proc. R. Soc. Edinburgh,* 94B, 63, 1988.

Alston, R. E. and Turner, B. L., *Biochemical Systematics,* Prentice-Hall, Englewood Cliffs, NJ, 1963.

Alvarez-Cordero, E. and McKell, C. M., Stem cutting propagation of big sagebrush (*Artemisia tridentata* Nutt.), *J. Range Manage., 32, 141, 1979.*

Ambasht, R. S., Srivastava, A. K., and Sharma, E., Primary productivity and nitrogen cycling in fast growing nitrogen fixing actinorhizal and leguminous trees in Indian tropics and subtropics; *15th Int. Bot. Congr. Abstracts,* 1993, 55.

Andersen, M. C., Diaspore morphology and seed dispersal in several wind-dispersed Asteraceae, *Am. J. Bot.,* 80, 487, 1993.

Andrews, H. N. and Lenz, L. W. A mycorrhizome from the Carboniferous of Illinois, *Bull. Torr. Bot. Club,* 70, 120, 1943.

Andrews, H. N., Early seed plants, *Science,* 142, 925, 1963.

Anfinsen, C. B., *The Molecular Basis of Evolution,* Wiley, New York, 1959.

Arms, K., Camp, P. S., Jenner, J. V., and Zalisko, E. J., *A Journey into Life,* Saunders College Publishing, Orlando, FL, 1994.

Arnold, F. G. C., Zur Lichenflora von München, *Ber. Bayer. Bot. Ges.,* 1–2, 5–8, 1891–1901.

Arora, D., *Mushrooms Demystified,* Ten Speed Press, Berkeley, 1979.

Atkinson, G. F., Phylogeny and relationships in the Ascomycetes, *Ann. Missouri Bot. Gard.* 2, 315–376, 1915.

Baeumer, K., *Allgemeiner Pflanzenbau,* E. Ulmer, Stuttgart, 1971.

Bakerspigel, A., The structure and manner of division of the nuclei in the basidiomycete *Schizophyllum commune, Can. J. Bot.,* 37, 835, 1959.

Balandrin, M. F., Klocke, J. A., Wurtele, E. S., and Bollinger, W. H., Natural plant chemicals, *Science,* 228: 1154, 1985.

Ballard, R. O. and Grassle, J. F., Return to oases of the deep, *Natl. Geogr.,* 156(5), 689, 1979.

Bambach, R. K., Scotese, C. R., and Ziegler, A. M., Before Pangea: the geographies of the Paleozoic world, *Amer. Sci.,* 68, 26–38, 1980.

Banks, H. P., 1970. *Evolution and Plants of the Past,* Wadsworth, Belmont, CA, 1970.

Barghoorn, E. S. and Schopf, J. W., Microorganisms from the late Precambrian of central Australia, *Science,* 150, 337, 1965.

Barghoorn, E. S. and Schopf, J. W., Microorganisms three billion years old from the Precambrian of South Africa, *Science,* 152, 758, 1966.

Barghoorn, E. S. and Tyler, S. A., Microorganisms from the Gunflint Chert, *Science,* 147, 563, 1965.

Bates, M., *The Forest and the Sea,* Random House, New York, 1960.

Bates, M. and Humphrey, P. S., *The Darwin Reader,* Scribner's, New York, 1956.

Bazilevich, N. I., Rodin, L. Y., and Rozov, N. N., Geographical aspects of biological productivity, in *Soviet Geography,* American Geographical Society, New York, 1971.

Bazzazz, F. A. and Fajer, E. D., Plant life in a Co_2-rich world, *Scientific American,* 266(1), 68–77, 1992.

Beadle, G. W. and Tatum, E. L., Genetic control of biochemical reactions in *Neurospora, Proc. Natl. Acad. Sci.,* 27, 499–506, 1941.

Beal, J. L., One man's quest for plant constituents with therapeutic value, *Econ. Bot.,* 44(1), 4–11, 1990.

Beck, C. B., Schmid, R., and Rothwell, G. W., Stelar morphology and the primary vascular system of seed plants, *Bot. Rev.,* 48, 691, 1982.

Beck, W., Aboriginal preparation of *Cycas* seeds in Australia, *Econ. Bot.,* 46, 133, 1992.

Beg, A. R., Present and future prospects of the coniferous forests in West Himalaya, *15th Int. Bot. Congr. Abstr.,* 1993, 51.

Benjamin, R. K., Laboulbeniales on semi-aquatic Hemiptera, *Aliso,* 6(3), 11, 1967.

Berkner, L. V. and Marshall, L. C., On the origin and rise of oxygen concentration in the earth's atmosphere, *J. Atm. Sci.,* 22, 225, 1965.

Berry, A. and Jensen, R. A., Biochemical evidence for phylogenetic branching patterns, *BioScience,* 38, 99, 1988.

Bessey, C. A., The phylogenetic taxonomy of flowering plants, *Ann. Mo. Bot. Gard.,* 2, 109, 1915.

Bessey, E. A., *Morphology and Taxonomy of Fungi,* Blakiston, Philadelphia, 1950.

Billings, W. D., *Plants, Man, and the Ecosystem.* Wadsworth Publishing, Belmont, CA, 1964.

Bischler, H. and Boisselier-Dubayle, M. C., Variation in a polyploid, dioecious liverwort, *Marchantia globosa, Am. J. Bot.,* 80, 953, 1993.

Bogorov, V. G., Ed., *The Pacific Ocean, Biology of the Pacific Ocean,* Vol 1, *Plankton,* Nauka, Moscow, 1967.

Bold, H. C., *Morphology of Plants,* Harper and Brothers, New York, 1957.

Bold, H. C. and Wynne, M. J., *Introduction to the Algae,* Prentice-Hall, Englewood Cliffs, NJ, 1985.

Bold, H. C., Alexopoulos, C. J., and Delevoryas, T., *Morphology of plants and fungi,* Harper & Row, New York, 1987.

Bongaarts, J., Can the growing human population feed itself? *Sci. Am.,* 270, 36, 1944.

Bonham, C. D., *Measurements for Terrestrial Vegetation,* John Wiley & Sons, New York, 1989.

Bonner, J. T., *The Cellular Slime Molds,* Princeton University Press, Princeton, NJ, 1967.

Boucot, A. J., *Evolutionary Paleobiology of Behavior and Coevolution,* Elsevier, Amsterdam, 1990.

Brasier, M. D., *Microfossils,* George Allen and Unwin, London, 1980.

Brasseur, G. and Granier, C., Mt. Pinatubo aerosols, chlorofluorocarbons, and ozone depletion, *Science,* 257, 1239, 1992.

Bray, J. R. and Gorham, E., Litter production in forests of the world, *Adv. Ecol. Res.,* 2, 101, 1964.

Bray, J. R., Lawrence, D. B., and Pearson, L. C., Primary production in some Minnesota terrestrial communities, *Oikos,* 10, 38, 1959.

Breed, R. S., *Bergey's Manual of Determinative Bacteriology,* 7th ed., Williams & Wilkins, Baltimore, 1957.

Bremer, K. and Wanntorp, H. E., The cladistic approach to plant classification, in *Advances in Cladistics,* New York Botanical Garden, New York, 1981, 87.

Brenner, G. J., Middle Cretaceous floral provinces and early migrations of angiosperms, in Beck, C. B., *Origin and Early Evolution of Angiosperms,* Columbia Press, New York, 1976, 23.

Britton, D. M., Chromosome numbers of ferns in Ontario, *Can. J. Bot.,* 42, 1349, 1964.

Brower, J. E. and Zar, J. H., *Field and Laboratory Methods for General Ecology,* Wm. C. Brown, Dubuque, IA, 1977.

Brown, L. H., Human food production as a process in the biosphere, *Sci. Am.,* 223(3), 161, 1970.

Brown, R. W., A bracket fungus from the late Tertiary of southwestern Idaho, *J. Wash. Acad. Sci.,* 30, 422, 1940.

Brown, H. P. and Cox, A., An electron microscope study of protozoan flagella, *Am. Midl. Nat.,* 52, 106, 1954.

Bryant, S. R. and Benson, S. E., Relative abundance of the most common fossil plants in the Carmen Formation, *J. Idaho Acad. Sci.,* 9, 85, 1973.

Buller, A. H. R., *Researchers on Fungi,* Vol. 2, Longmans, Green and Co., London, 1922.

Bünning, E., *The Physiological Clock,* Springer Verlag, Berlin, 1967.

Burns, G. W., *The Science of Genetics,* Macmillan, New York, 1972.

Caporael, L. W., Ergotism: the Satan loosed in Salem? *Science,* 192: 21–26, 1976.

Carmichael, W. W., The toxins of the Cyanobacteria, *Sci. Am.,* 270, 78, 1994.

Challem, J. J., *Spirulina: Green Gold of the Future,* Keats Publishing, New Canaan, CT, 1981.

Chaloner, W. G., Spores and land plant evolution, *Rev. Palaeobotany and Palynology,* 1, 83, 1967.

Chamberlain, C. J., *The Living Cycads,* Hafner, New York, 1965.

Chan, A. T., Comparative physiological study of marine diatoms and dinoflagellates in relation to irradiance and cell size, I. Growth under continuous light, *J. Phycol.,* 14, 396–402, 1978.

Chea, H. and Pearson, L. C., Change in pH in cactus tissue exposed to periods of light and dark, J. Idaho

Chester, K. S., *Nature and Prevention of Plant Diseases,* Blakiston, New York, 1952.

Chesters, K. I. M., Fossil plant taxonomy, in Turrill, W. B., Ed., *Vistas in Botany,* Vol. 4, 1964, 239.

Chopra, R. N., and Bhatla, S. C., Regulation of gametangial formation in bryophytes, *Bot. Rev.,* 49, 29, 1983.

Chuang, T. I. and Constance, L., Seeds and systematics in Hydrophyllaceae: tribe Hydrophylleae, *Am. J. Bot.,* 79, 257, 1992.

Churchill, E. D. and Hanson, H. C., The concept of climax in arctic and alpine vegetation, *Bot. Rev.*, 24, 127, 1958.

Churchill, S. P., A phylogenetic analysis, classification and synopsis of the genera of the Grimmiaceae (Musci), in *Advances in cladistics*, Funk, V. A. and Brooks, D. R., Eds., New York Botanical Garden, New York, 1981, 127.

Chuvaskov, B. I., Ecology of the Foraminifera and algae of the upper Frasnian (in Russian), *Paleont. Zhur., SSSR*, no. 3, 1963, 3.

Clark, R. L. and Rushforth, S. R., Diatom studies of the headwaters of Henry's Fork of the Snake River, Island Park, Idaho, U.S.A. *Bibliotheca Phycologia*, Vol. 33, J. Cramer Veilag.

Clausen, J., Genetics of climatic races of *Potentilla glandulosa*, *Proc. 8th Int. Congr. Genetics*, 162–172, 1949.

Clausen, J., *Stages in the Evolution of Plant Species*, Cornell University Press, Ithaca, 1951.

Clements, F. E., *The Genera of Fungi*, H. W. Wilson, Minneapolis, 1909.

Clements, F. E. and Shear, C. L., *The Genera of fungi*, Hafner, New York, 1931.

Colman, J. S., *The Sea and its Mysteries*, W. W. Norton, New York, 1950.

Conard, H. S., *How to Know the Mosses and Liverworts*, Wm. C. Brown, Dubuque, IA, 1956.

Conrad, W. and Kufferath, H., Recherches sur les eaux saumetres des environs de Lilleo, II., *Inst. R. Soc. Nat. Belg. Mem.*, 127, 1, 1954.

Coon, N., *Using Plants for Healing*, Hearthside Press, 1963.

Corliss, J. B. and Ballard, R. D., Oases of life in the cold abyss, *National Geographic*, 152, 441–453, 1977.

Couch, J. N., The structure and action of the cilia in some aquatic phycomycetes, *Am. J. Bot.*, 28, 704, 1941.

Creber, G. T. and Chaloner, W. G., Influence of environmental factors on the wood structure of living and fossil trees, *Bot. Rev.*, 50, 357, 1984.

Cronquist, A., *The Evolution and Classification of Flowering Plants*, Houghton Mifflin, Boston, 1968.

Culberson, C. F., *Chemical and Botanical Guide to Lichen Products*, University of North Carolina Press, Chapel Hill, NC, 1969.

Culberson, W. L., Culberson, C. F., and Johnson, A., Speciation in lichens of the *Ramalina siliquosa* complex (Ascomycotina, Ramalinaceae): gene flow and reproductive isolation, *Am. J. Bot.*, 80, 1472, 1993.

Culbertson, J. T. and Cowan, M. C., *Living Agents of Disease*, G. P. Putnam's Sons, New York, 1952.

Currie, P. O. and Peterson, G., Using growing season precipitation to predict crested wheatgrass yields, *J. Range Manage.*, 19, 284, 1966.

Darlington, C. D. and Janaki Ammal, E. K., *Chromosome Atlas of Cultivated Plants*, George Allen and Unwin, London, 1945.

Darwin, C. R., *On the Origin of Species by Means of Natural Selection*, Heritage Press, New York, 1859 (1963 reprint).

Davis, R. J., *Flora of Idaho*, Wm. C. Brown, Dubuque, IA, 1952.

Dawson, E. Y., *How to Know the Seaweeds*, Wm. C. Brown, Dubuque, IA, 1956.

Dawson, E. Y., *New Taxa of Benthic Green, Brown and Red Algae Published Since De Toni 1889, 1895, 1924, Respectively as Compiled fromt the Dawson Algal Library*, Beaudette Foundation for Biological Research, Santa Ynez, CA, 1962.

DeLamater, E. D., A new cytological basis for bacterial genetics, *Cold Spring Harbor Symp. Quant. Biol.*, 16, 381, 1951.

Del Tredici, P., Natural regeneration of *Ginkgo biloba* from downward growing cotyledonary buds (basal chichi), *Am. J. Bot.,* 79, 522, 1992.

DeMort, C. L., Lowry, R., Tinslag, I., and Phinney, H. K., The Haptophyceae, *J. Phycol.,* 8, 211, 1972.

Demoulin, V., The origin of ascomycetes and basidiomycetes, the case for a red algal ancestry, *Bot. Rev.,* 40, 315, 1974.

Denison, W. C. and Carroll, G. C., The primitive ascomycete: a new look at an old problem, *Mycologia,* 58, 249, 1966.

Desikachary, T. B., Germination of the heterocysts in two members of the Rivulariaceae, *J. Ind. Bot. Soc.,* 25, 11, 1946.

Desikachary, T. V., *Cyanophyta,* Indian Council of Agricultural Research, New Delhi, 1959.

Dibben, M. J., Whole-lichen culture in a phytotron, *Lichenologist,* 5, 1–10, 1971.

Dickson, J. G., *Diseases of Field Crops,* McGraw-Hill, New York, 1956.

Dilcher, D. L., Approaches to the identification of angiosperm leaf remains, *Bot. Rev,* 40, 1, 1974.

Dineen, C. F., An ecological study of a Minnesota pond, *Am. Midl. Nat.,* 50, 349, 1953.

Dobell, C., *Antony van Leeuwenhoek and his "Little Animals,"* Dover Publications, New York, 1960.

Dodge, J. D., The nucleus and nuclear division in the Dinophyceae. *Arch. Protistenk.* 106, 442–452, 1963.

Doyle, W. T., *The Biology of Higher Cryptogams,* Macmillan, Toronto, 1970.

Edelman, M., Schiff, J. H., and Epstein, H. T., Two types of satellite DNA, *J. Mol. Biol.,* 11, 769, 1965.

Eichler, A. W., *Syllabus der Vorlesungen über specielle und medicinischpharmaceutische Botanik,* Berlin, 1886.

Engelmann, M. D., The role of soil arthropods in the energetics of an old field community, *Ecol. Monogr,* 31, 221, 1961.

Epstein, E., *Mineral Nutrition of Plants: Principles and Perspectives,* Wiley, New York, 1972.

Erwin, D. C., Bartnicki-Garcia, S., and Tsao, P. H., Eds., *Phytophthora, Its Biology, Taxonomy, Ecology, and Pathology,* American Phytopathological Society, St. Paul, Minnesota, 1983.

Evans, J. R., Relationships between photosynthesis rate and leaf nitrogen, *15th Int. Bot. Congr. Abstr.* 1993, 40.

Evans, L. S., Botanical aspects of acidic precipitation, *Bot. Rev.,* 50, 449, 1984.

Fay, P. A., and Knapp, A. K., Photosynthesis and stomatal responses of *Avena sativa* (Poaceae) to a variable light environment, *Am. J. Bot.,* 80, 1369, 1993.

Federici, Fungi as biocontrol of mosquitoes, *California Agriculture,* 34, no. 3, 1980.

Feldman, J. and Feldman, G., Recherches sur les Bonnemaisoniacées et sur alternance de générations, *Ann. Sci. Nat. Bot. Ser.,* 11, 3, 75–175, 1942.

Feldman, J. and Feldman, G., Recherches sur quelques Floridées parasites, *Rev. Gén. Bot.,* 65, 49–128, 1958.

Fenchel, T. and Finlay, B. J., The evolution of life without oxygen, *Am. Sci.,* 82, 22, 1994.

Ferry, B. W., Baddeley, M. S., and Hawksworth, E. L., Eds., *Air Pollution and Lichens,* Athlone Press, London, 1973.

Ferry, B. W. and Coppins, B. J., Lichen transplant experiments and air pollution studies, *Lichenologist,* 11, 63, 1978.

Fitzgerald, J. *Cancer Res.,* 1953.

Fleming, T. H., Plant-visiting bats, *Am. Sci.,* 81, 460, 1993.

Flowers, S., *Ferns of Utah,* Bull. Univ. Utah 35(7), 1-87, 1944.

Flowers, S., Algae of Utah (mimeographed), Biology Department, University of Utah, Salt Lake City, (mimeographed), 1948.

Flowers, S., *Mosses: Utah and the West,* Brigham Young University Press, Provo, UT, 1973.

Flugel, E., *Fossil Algae: Recent Results and Development,* Springer-Verlag, Berlin, 1977.

Fogg, G. E., Nitrogen fixation, in Lewin, R. A., ed., *Physiology and Biochemistry of Algae,* Academic Press, New York, 1962, 161.

Fogg, G. E., Stewart, W. D. P., Fay, P., and Walsby, A. E., *The Blue-Green Algae,* Academic Press, London, 1973.

Forsberg, C., Some remarks on Wood's revision of the Characeae, *Taxon* 12, 141, 1963.

Fraenkel-Conrat, H., *Design and Function at the Threshold of Life: The Viruses,* Academic Press, New York, 1962.

Fritsch, F. E., *Structure and Reproduction of the Algae,* Cambridge University Press, London, 1965.

Frye, T. C., *Ferns of the Northwest,* Metropolitan Press, Portland, OR, 1934.

Fulford, M., Contemporary thought in plant morphology: Hepaticae and Anthocerotae, *Phytomorphology,* 14, 103, 1964.

Fuller, H. J. and Tippo, O., *College Botany,* Henry Holt & Co., New York, 1954.

Fulling, E. H., *Metasequoia,* living and fossil, *Bot. Rev.,* 42, 215, 1976.

Funk, V. A. and Brooks, D. R., Eds., *Advances in Cladistics,* New York Botanical Garden, New York, 1981.

Gantt, E., Ohki, K., and Fujita, Y., *Trichodesmium thiebautii;* structure of a nitrogen-fixing marine blue-green alga (Cyanophyta), *Protoplasma,* 119, 188, 1984.

Gause, G. F., Experimental analysis of Vito Volterra's mathematical theory of the struggle for existence, *Science,* 79, 16, 1934.

Gause, G. F., Experimental demonstration of Volterra's periodic oscillations in the numbers of animals, *J. Exp. Biol.,* 12, 44-48, 1935.

Gebelein, C. D., The effects of the physical, chemical and biological evolution of the earth, in Walter, M. R., Ed., *Stromatolites,* Elsevier Science Pub., Amsterdam, 1976, 499.

Gemma, J. N., Koske, R. E., and Flynn, T., Mycorrhizae in Hawaiian pteridophytes: occurrence and evolutionary significance, *Am. J. Bot.,* 79, 843, 1992.

Gentry, A. H., Correlations between patterns of diversity and dynamics in tropical forests, *15th Int. Bot. Congr. Abstr.,* 1993, 55.

Gezelius, K. and Rànby, B. G., Morphology and fine structure of the slime mold *Dictyostelium discoideum, Exptl. Cell Res.,* 12, 265-289, 1957.

Gibbs, S. P., Plant cells, *J. Ultrastruct. Res.,* 7, 418, 1962.

Ginsburg, R., Rezak, R., and Wray, J. L., *Geology of calcareous algae,* University of Miami, Miami, FL, 1971.

Gliessman, S. R. and Altieri, M. A., Polyculture cropping has advantages, *California Agriculture,* 36(7), 14-16, 1982.

Godward, M. B. E., *The chromosomes of the algae,* St. Martin's, New York, 1966.

Gojdics, M., The genus *Euglena,* University of Wisconsin Press, Madison, 1953.

Golley, F. B., Structure and function of an old-field broomsedge community, *Ecol. Monogr.,* 35, 113, 1965.

Goodwin, T. W., Fungal carotenoids, *Bot. Rev.,* 18, 291, 1952.

Goodwin, T. W., The nature and distribution of carote noids in some blue-green algae, *J. Gen. Microbiol.,* 17, 467, 1957.

Goodwin, T. W., Carotenoids, in *Handbuch der Pflanzenphysiologie,* Vol. 10, Springer-Verlag, Berlin, 1958, 186.

Goodwin, T. W., *Chemistry and Biochemistry of Plant Pigments,* Academic Press, New York, 1965.

Gore, A., *Earth in the Balance,* Houghton Mifflin, Boston, 1992.

Gould, S., Eternal metaphors of paleontology, in Hallam, A., Ed., *Patterns of Evolution,* Elsevier Science Pub., Amsterdam, 1977.

Govindjee, Ed., *Photosynthesis,* Vol. 1. *Energy Conversion by Plants and Bacteria,* Academic Press, New York, 1982.

Grant, B. R. and Borowitzka, M. A., The chloroplasts of giant-celled and coenocytic algae: biochemistry and structure, *Bot. Rev.,* 50, 267, 1984.

Grant, V., *Plant Speciation.* Columbia University Press, New York, 1981.

Gray, W. D. and Alexopoulos, C. J., *Biology of the Myxomycetes,* Ronald Press, New York, 1968.

Green, B., Heilporn, V., Limbausch, S., Boloukhere, M., and Brachet, J., The cytoplasmic DNAs of *Acetabularia mediterranea, Proc. Natl. Acad. Sci.,* 58, 1351, 1967.

Green, B. R., Evidence for the occurrence of meiosis before cyst formation in *Acetabularia mediterranea* (Chlorophyceae, Siphonales), *Phycologia,* 12, 233, 1973.

Green, J. W., Knoll, A. H., Golubic, S., and Swett, K., Paleobiology and distinctive benthic microfossils from the upper Proterozoic limestone-dolomite "series," central east Greenland, *Am. J. Bot.,* 74, 928–940, 1987.

Grillos, S. J., *Ferns and Fern Allies of California,* University of California Press, Berkeley, 1966.

Groves, J. and Bullock-Webster, G. R., *British Charophyta,* Royal Society of London, London, 1924.

Gurin, J., In the beginning, *Science,* 80, 1(5), 44, 1980.

Guthrie, R. L., Xylem structure and ecological dominance in a forest community, *Am. J. Bot.,* 76, 1216, 1989.

Hacskaylo, E., The role of mycorrhizal associations in the evolution of the higher basidiomycetes, in Petersen, T. H., Ed., *Evolution of the Higher Basidiomycetes,* University of Tennessee Press, Knoxville, 1971, 217.

Haldane, J. B. S., The origin of life, *Rationalist Annual,* 1929, 3.

Hale, M. E., Jr., *Lichen Handbook,* Smithsonian Institution, Washington, D.C., 1961.

Hale, M. E., Jr., *How to Know the Lichens,* Wm. C. Brown, Dubuque, IA, 1979.

Hale, M. E., Jr., *The Biology of Lichens,* Third Edition, Edward Arnold, London, 1983.

Halisky, P. M. and Peterson, J. L., Basidiomycetes associated with fairy rings in turf, *Bull. Torr. Bot. Club,* 97, 225, 1970.

Hallam, A., (Ed.), *Patterns of Evolution as Illustrated by the Fossil Record,* Elsevier Publishing Co., Amsterdam, 1977.

Harlan, J. R., *Crops and Man,* American Society of Agronomy, Madison, WI, 1992.

Harley, J. L., Mycorrhiza, in Ainsworth, G. C., and Sussman, A. S., Eds., *The Fungi: an Advanced Treatise,* Vol. 3, 1968, 139.

Harley, J. L. and Smith, S. E., *Mycorrhizal Symbiosis,* Academic Press, New York, 1983.

Harrington, J. F., Seed storage and longevity, in Kozlawski, T. T., Ed., *Seed Biology,* Academic Press, New York, 1972.

Harvey, E. N., *Bioluminescence,* Academic Press, New York, 1952.

Haufler, C. H., Electrophoresis is modifying our concepts of evolution in *Homosporous pteridophytes, Am. J. Bot.,* 74, 953, 966, 1987.

Hauke, R. L., A taxonomic monograph of the genus *Equisetum, Beih. Nova Hedwigia,* 8, 1, 1963.

Haxo, F. F., Phycobilins, *Handbuch der Pflanzenphysiology,* 5(2), 349, 1960.

Haxo, F. F., and Fork, D. C., Phycobilins in cryptomonads, *Nature,* 184, 1051, 1959.

Hayashi, H., Nakamura, S., Hirose, N., Ishiwatari, Y., and Chino, M., Proteins and nucleic acids in the phloem sap of rice plants (*Oryza sativa* L.), *15th Int. Bot. Congr.*, 1993, 41.

Hedgpeth, J. W., *Between Pacific Tides,* Stanford University Press, Stanford, 1962.

Heiser, C. B., Jr., *Nightshades, the Paradoxical Plants,* Freeman, San Francisco, 1969.

Hennig, W., *Phylogenetic Systematics,* University of Illinois Press, Urbana, IL, 1966.

Henriksson, E., Nitrogen fixation by a bacteria-free, symbiotic Nostoc strain isolated from Collema, *Physiologia Plantarum,* 4, 542-545, 1951.

Henriksson, E., Henriksson, L. E., and DaSilva, E. J., A comparison of nitrogen fixation by algae of temperate and tropical soils, in Stewart, W. D. P., Ed., *Nitrogen fixation by Free-Living Micro-Organisms,* International Biological Programme, 6: 199-206.

Henriksson, E. and Pearson, L. C., Carotenoids extracted from mycobionts of *Collema tenax, Baeomyces roseus,* and some other lichens, *Svensk Bot. Tidskr.,* 62, 441, 1968.

Henriksson, E. and Pearson, L. C., Nitrogen fixation rate and chlorophyll content of the lichen *Peltigera canina* exposed to sulfur dioxide, *Am. J. Bot.,* 68, 680, 1981.

Henriksson, L. E. and E. Henriksson, Concerning the biological nitrogen fixation on Surtsey, *Surtsey Research Progress Report 9,* 9-12, The Surtsey Research Society, Reykjavik, 1982.

Henssen, A. and Jahns, H. M., *Lichenes,* Georg Thieme Verlag, Stuttgart, 1974.

Herak, M., Kochansky-Dividé, V., and Gusic, I., The development of the dasyclad algae through the ages, in Flügel, E., Ed., *Fossil Algae,* Springer Verlag, Berlin, 1977, 143.

Herker, R. F., *Introduction to Paleoecology,* American Elsevier, New York, 1965.

Herngreen, G. F. W., Middle Cretaceous palynomorphs from north-eastern Brazil, *Sci. Geol. Bull.,* Strasbourg, 27, 101, 1974.

Hershey, A. D., and Chase, M., Independent functions of viral protein and nucleic acid in growth of bacteriophage, *J. Gen. Physiol.,* 36, 39-56, 1952.

Hesseltine, C. W. and Ellis, J. J., Mucorales, in *Ainsworth and Bisby's Dictionary of Fungi,* in Ainsworth, G. C., Ed., Commonwealth Microbiological Institute, Kew, U.K., 1961, 187.

Hickey, L. J. and Doyle, J. A., Early Cretaceous fossil evidence for angiosperm evolution, *Bot. Rev.,* 43, 3, 1977.

Highfill, J. F. and Pfiester, L. A., The sexual and asexual life cycle of *Glenodiniopsis steinii* (Dinophyceae), *Am. J. Bot.,* 79, 899, 1992.

Hoffman, P., Stromatolite morphogenesis in Shark Bay, Western Australia, *Stromatolites,* in Walter, M. R., Ed., Elsevier Scientific, Amsterdam, 1976, 261.

von Hofsten, A. and Pearson, L. C., Chromatin distribution in Cyanophyceae, *Hereditas,* 53, 212, 1965.

Hole, C. C. and Dearman, J., Sucrose uptake by the phloem parenchyma of carrot storage root, *J. Exp. Bot.,* 45, 7, 1994.

Holton, R. W., Blecker, H. H., and Stevens, T. S., Fatty acids in bluegreen algae: possible relation to phylogenetic position. *Science,* 160, 545-547, 1968.

Hope, A. B. and Walker, N. A., *The physiology of Giant Algal Cells,* Cambridge University Press, London, 1975.

Huang, A. H. C., Trelease, R. N., and Moore, Jr., T. S., *Plant Peroxisomes,* Academic Press, New York, 1983.

Huckaby, L. S., Ed., Lichens as bioindicators of air quality, *U.S. Dept. Agric. For. Serv. Gen. Tech. Rep.,* RM-224, 1993.

Hueber, F. M., *Hepaticites devonicus:* a new fossil liverwort from the Devonian in New York, *Ann. Mo. Bot. Gard.,* 48, 125, 1961.

Hughes, N. F., Paleo-succession of earliest angiosperm evolution, *Bot. Rev.*, 43, 105, 1977.

Hulbert, E. M., *Flagellates from brackish water in the vicinity of Woods Hole*, Massachusetts, *J. Phycol.*, 1, 87, 1965.

Hutchinson, J., *The Families of Flowering Plants*, Clarendon Press, Oxford, 1948.

Hutner, S. H. and Provasoli, L., *Biochemistry and Physiology of Protozoa*, Vol. 2, Academic Press, New York, 1955, 17.

Imbrie, J. and K. P. Imbrie, *Ice Ages: Solving the Mystery*, Enslow, Hillside, NJ, 1979.

Jacobs, W. P., Caulerpa, *Sci. Amer.*, 271(6), 100, 1994.

Jeffrey, E. C., *Anatomy of Woody Plants*, University of Chicago Press, Chicago, 1917.

John, B. S., *The Winters of the World*, David and Charles, London, 1979.

Johnson, J. H., Review of Ordovician algae, *Quarterly of the Colorado School of Mines* 56(2), 1, 1961a.

Johnson, J. H., *Limestone Building Algae and Algal Limestones*, Johnson Publishing, Boulder, CO, 1961b.

Johnston, M. O., Tests of two hypotheses concerning pollen competition in a self-compatible, long-styled species (*Lobelia cardinalis:* Lobeliaceae), *Am. J. Bot.*, 80, 1400, 1993.

Jordan, C. F., Amazon rain forests, *Am. Sci.*, 70, 394, 1982.

Juday, C., The annual energy budget of an inland lake, *Ecology*, 21, 438, 1940.

Kalapos, T., C_4 grasses in the vegetation of Hungary, *15th Int. Bot. Congr. Abstr.*, 1993, 264.

Kanada, T., Biosynthesis of branched chain fatty acids I. Isolation and identification of fatty acids from *Bacillus subtilis* (ATCC 6059), *J. Biol. Chem.*, 238, 1222–1228, 1963.

Kaplan, E. H., *A Field Guide to Coral Reefs*, Houghton-Mifflin, Boston, 1982.

Kapp, R. O., *How to Know Pollen and Spores*, Wm. C. Brown, Dubuque, IA, 1969.

Kaufman, L. and Mallory, K., The last extinction, MIT Press, Cambridge, MA, 1986.

Kauppi, P. E., Mielikäinen, K., and Kuusela, K., Biomass and carbon budget of European forests, 1971 to 1990, *Science*, 256, 70, 1992.

Kawaguchi, H., Productivity and mineral cycling of reforested ecosystem in Thailand, *15th Int. Bot. Congr. Abstr.*, 1993, 55.

Kendrick, B., Ed., *Taxonomy of Fungi Imperfecti*, University of Toronto Press, Toronto, 1971.

Kerr, R. A., The oceans' deserts are blooming, *Science*, 232, 1345, 1986.

Kessin, R. H. and Van Lookeren Campagne, M. M., The development of a social amoeba, *Am. Sci.*, 80, 556, 1992.

Kirk, D. R., *Wild Edible Plants of Western North America*, Naturegraph, Happy Camp, CA, 1975.

Kiss, J. C., Vasconcelos, A. C., and Triener, R. E., Structure of the euglenoid storage carbohydrate paramylon, *Am. J. Bot.*, 74, 877–882, 1987.

Kitamoto, Y. and Gruen, H. E., Distribution of cellular carbohydrates during development of the mycelium and fruit bodies of *Flammulina velutipes*, Plant Physiol., 58, 485, 1976.

Klein, R. M. and Cronquist, A., A consideration of the evolutionary and taxonomic significance of some biochemical, micromorphological, and physiological characters in the thallophytes, *Q. Rev. Biol.*, 42, 105, 1967.

Knox, E. B. and Kowal, R. R., Chromosome numbers of the east African senecios and giant lobelias and their evolutionary significance, *Am. J. Bot.*, 80, 847, 1993.

Kohn, A. J. and Helfrich, P., Primary organic productivity of a Hawaiian coral reef, *Limnol. and Oceanogr.*, 2, 241, 1957.

Kok, B., Photosynthesis: the path of energy, in Bonner, J. and Varner, J. E., Eds., *Plant Biochemistry*, Academic Press, New York, 1965, 903.

Kok, B., Simonis, W., and Urbach, W., Photosynthesis in algae, in *Photosynthesis,* Vol. 1. *Energy Conversion by Plants and Bacteria,* Govindjee, Ed., Academic Press, New York, 1982.

Kole, A. P. and Gielink, A. J., Electron microscope observations on the resting spore germination of *Plasmodiophora brassicae, Proc. K. Ned. Akad. Wetens. Ser. C,* 65, 117, 1962.

Korde, K. B., Formation and systematic position of the conical and cylindrical crusts of the Conophyton type, *Dokl. Akad. Nauk S.S.S.R.,* 89(6), 1091-1094, 1953.

Koske, R. E., Gemma, J. N., and Flynn, T., Mycorrhizae in Hawaiian angiosperms: a survey with implications for the origin of the native flora, *Am. J. Bot.,* 79, 853, 1992.

Krassilov, V. A., Cretaceous fossils, *Bot. Rev.,* 43, 105, 1977.

Krassilov, V. A., New paleobotanical data on origin and early evolution of angiospermy, *Ann. Mo. Bot. Gard.,* 71, 577, 1984.

Kreger-van Rij, N. J. W., Endomycetales, basidiomycetous yeasts and related fungi, in Ainsworth, G. C., Sparrow, F. K., and Sussman, A. S. (eds.), *The Fungi,* Vol. IVA, Academic Press, New York, 1973.

Kroopnick, P. M., Margolis, S. V., and Wong, C. S., *The fate of fossil fuel CO_2 in the oceans,* Plenum, New York, 1977.

Kudo, R. R., *Protozoology,* Charles C Thomas, Springfield, IL, 1966.

Kylin, H., *Die Gattungen der Rhodophyceen,* Gleerups, Lund, 1956.

Lamont, B., Mechanisms for enhancing nutrient uptake in plants, with particular reference to Mediterranean South Africa and western Australia, *Bot. Rev.,* 48, 597, 1982.

Lang, W. H. and Cookson, I. C., On a flora including vascular land plants, associated with *Monograptus,* in rocks of Silurian age from Victoria, Australia, *Phil. Trans. R. London,* 224B, 421, 1935.

Lange, M., *Botanik,* Vol. 2, *Systematisk Botanik,* Nr. 1, *Svampe,* Munksgaard, Copenhagen, 1964.

Lange, O. L., Koch, W., and Schulze, E. D., CO_2-Gaswechsel und Wasserhaushalt von Pflanzen in der Negev-Wüste am Ende der Trochenzeit, *Ber. Deutsch. Bot. Ges.,* 82, 39-61, 1969.

Lanini, W. T., Pittenger, D. R., Graves, W. L., Muñes, F., and Agamalian, H. S., Subclovers as living mulches for managing weeds in vegetables, *Calif. Agric.,* 43(6), 25, 1989.

Larcher, W., *Physiological Plant Ecology,* Springer-Verlag, Heidelberg, 1975.

Lark-Horowitz, K., A permeability test with radioactive indicators, *Nature,* 123, 277, 1929.

Lawrence, D. B., Bray, J. R., and Pearson, L. C., Ecosystem studies at Cedar Creek Natural History Area, *Proc. Minnesota Acad. Sci.,* 25-26: 108-112, 1957-58.

Lawrence, G. H. M., *Taxonomy of Vascular Plants,* Macmillan, New York, 1951.

Lawrey, J. D., *Biology of Lichenized Fungi,* Praeger Scientific, New York, 1984.

Lazaroff, N. and Vishniac, W., The relationship of cellular differentiation to colonial morphogenesis of the blue-green alga *Nostoc muscorum, J. Gen. Microbiol.,* 35, 447, 1964.

LeBlanc, F., Possibilities and methods for mapping air pollution on the basis of lichen sensitivity, *Mitt. Forstl. Bundes Versuchanst. Wien,* 92, 103, 1971.

LeBlanc, F. and DeSloover, J., Relation between industrialization and the distribution and growth of epiphytic lichens and mosses in Montreal, *Can. J. Bot.,* 48, 1485, 1970.

LeBlanc, F., Rao, D. N., and Comeau, G., The epiphytic vegetation of *Populus balsamifera* and its significance as an air pollution indicator in Sudbury, Ontario, *Can. J. Bot.,* 50, 519, 1972.

Leak, L. V., and Wilson, G. B., The distribution of chromatin in a blue green alga *Anabaena variabilis* Kutz., *Can. J. Genet. Cytol.,* 2, 320, 1960.

Lee, N. J. and Nakane, K., Mapping and biomass estimation of forest vegetation at Mountain area, West Japan, based on landsat data, *15th Int. Bot. Congr.,* 1993, 62.

Leonard, L. T., Nitrogen-fixing bacteria and legumes, U.S.D.A. Farmers' Bull., No. 1784, Washington, D.C., 1937.

Lepofsky, D., Arboriculture in the Mussau Islands, Bismarck Archipelego, *Econ. Bot.,* 46, 192–211, 1992.

Levins, R. and nine coauthors, The emergence of new diseases, *Am. Sci.* 82, 52, 1994.

Levinton, J. S., *Marine Ecology,* Prentice-Hall, Engelman Cliffs, New Jersey, 1982.

Lewin, R. A., ed., *Physiology and Biochemistry of Algae,* Academic Press, 1962.

Lewington, A., *Plants for People,* Oxford University Press, New York, 1990.

Lewis Smith, R. I., Plant ecological studies in the fellfield ecosystem near Casey Station, Australian Antarctic Territory 1985–1986, *Bull. Br. Antarct. Surv.,* 72, 81, 1986.

Lieth, H., Analysis of temperate forest ecosystems, *Ecol. Studies,* 1, 29, 46, 1970.

Lincoff, G. H., *The Audubon Society Field Guide to North American Mushrooms,* Alfred A. Knopf, New York, 1981.

Lindegren, C. C., Genetics of the fungi, *Ann. Rev. Microbiology,* 1948.

Lindeman, R. L., The trophic-dynamic apect of ecology, *Ecology,* 23, 284, 1942.

Lindström, K., Selenium as a growth factor for plankton algae in laboratory experiments and in some Swedish lakes, *Hydrobiologia,* 101, 35–48, 1979.

Lindström, K., *Peridinium cinctum* bioassays of selenium in Lake Erken, *Arch. Hydrobiol.,* 89, 110–117, 1980.

Lipps, J. H., Ed., *Fossil Prokaryotes and Protists,* Blackwell Scientific Publications, Oxford, 1993.

Lowe-McConnell, R. H., Ed., *Speciation in Tropical Environments,* Academic Press, New York, 1969.

Lüning, K., *Seaweeds: their Environment, Biogeography, and Ecophysiology,* John Wiley & Sons, Inc., New York, 1990.

Luther, A., Ueber Chlorosaccus, eine neue Gattung der Süsswasseralgen, nebst einigen Bemerkukngen zur Systematik verwandter Algen, *Bih. Kgl. Svensk. Vetensk.-Ak. Handl.,* 24, Afd. 3, No. 13, 1–22, 1899.

MacColl, R. and Guard-Friar, D., *Phycobiliproteins,* CRC Press, Boca Raton, FL, 1987.

Machlis, L., Sex hormones in fungi, in *The Fungi: an Advanced Treatise,* Ainsworth, G. C. and Sussman, A. S., Eds., Academic Press, New York, 1966, 415–433.

MacKey, J., Mutation breeding in Europe, *Brookhaven Symp. Biol.,* 9, 141–156, 1956.

Malme, G., Gräs, *Nordisk Familjebok 8, 1164,* A.B. Familjeboken Förlag, Stockholm, 1928.

Manton, I., *Problems of Cytology and Evolution in the Pteridophyta,* University Press, Cambridge, 1950.

Marden, L., Sea nymphs of Japan, *Nat. Geographic,* 140(7), 122, 1971.

Marden, L., 1978. The continental shelf, *National Geographic,* 153(4), 495–531.

Margalef, R., (Ed.), *Sixth Seaweed Symposium,* Subsecreteria de la Marina Mercante, Direccion General de Pesca Maritima, Madrid, 1968.

Margulis, L., Evolutionary criteria in the thallophytes: a radical alternative, *Science,* 161, 1020, 1968.

Margulis, L., Symbiosis and evolution, *Sci. Am.,* 225(2), 48, 1971.

Martin, G. W., Alexopoulos, C. J., and Farr, M. L., *The Genera of Myxomycetes,* University of Iowa Press, Iowa City, 1983.

Martin, G. W. and Alexopolous, C. J., *The Myxomycetes,* University of Iowa Press, Iowa City, 1969.

Marx, J., New clue to cancer metastasis found, *Science,* 249, 482, 1990.

Matossian, M. K., Ergot and the Salem witchcraft affair, *Amer. Sci.,* 70, 355, 1982.

Mattox, K. R. and Stewart, K. B., Classification of the green algae: a concept based on comparative cytology, in *The systematics of the Green Algae,* Irivine, D. E. G., and Johns, D. M., Eds., Academic Press, London, 1984, 19.

Mauseth, J. D., Medullary bundles and the evolution of cacti, *Am. J. Bot.,* 80, 928, 1993.

Mayr, E., A local flora and the biological species concept, *Am. J. Bot.,* 79, 222, 1992.

McCleary, J. A., Sypherd, P. S., and Walkington, D. L., Mosses as possible sources of antibiotics, *Science,* 131, 108, 1960.

McConnell, W. J., Primary productivity and fish harvest in a small desert impoundment, *Trans. Am. Fish. Soc.,* 92, 1, 1963.

McDonald, K., The ultrastructure of mitosis in the marine red alga *Membranoptera platyphylla, J. Phycol.,* 8, 156, 1972.

McGaughey, W. H. and Whalon, M. E., Managing insect resistance to *Bacillus thuringiensis* toxins, *Science,* 258, 1451, 1992.

McGowan, J. A. and Walker, P. W., Dominance and diversity maintenance in an oceanic ecosystem, *Ecol. Monogr.,* 55, 103, 1985.

McGowan, J. A. and Hayward, T. L., Mixing and oceanic productivity, *Deep-Sea Res.,* 25, 771, 1978.

McMenam, M. A. S. and McMenam, D. L. S., *The Emergence of Animals: the Cambrian Breakthrough,* Columbia University Press, New York, 1990.

Melchior, H. and Werdermann, E. *A. Engler's syllabus der Pflanzenfamilien,* Gebrüder Borntraeger, Berlin, 1954.

Messmer, L. W. and Lersten, N. R., Chromosome studies of ten species of mosses from Iowa, *Bryologist,* 71, 348, 1968.

Metzger, J. D., Physiological basis of achene dormancy in *Polygonum convolvulus* (Polygonaceae), *Am. J. Bot.,* 79, 882, 1992.

Miller, C. N., Jr., Mesozoic conifers, *Bot. Rev.,* 43, 217, 1977.

Miller, O. K., *Mushrooms of North America,* E. P. Dutton, New York, 1981.

Minnis, P., Harrison, E. F., Stowe, L. L., Gibson, G. G., Denn, F. M., Doelling, D. R., and Smith, W. L., Jr., Radioactive climate forcing by the Mount Pinatubo eruption, *Science,* 259, 1411, 1993.

Mirov, N. T. and Hasbrouck, J., *The Story of Pines,* Indiana University Press, Bloomington, 1976.

Mitman, G. G. and van der Meer, J. P., Meiosis, blade development, and sex determination in *Porphyra purpurea* (Rhodophyta), *J. Phycol.,* 30, 143, 1994.

Moldenke, H. N. and Moldenke, A. L., *Plants of the Bible,* Ronald Press, New York, 1965.

Moldowan, J. M., Fago, F. J., Lee, C. Y., Jacobson, S. R., Watt, D. S., Slougi, N.-E., Jeganathan, A., and Young, D. C., Sedimentary 24-*n*-propylcholestanes, molecular fossils diagnostic of marine algae, *Science,* 247, 309, 1990.

Müller, E., Imperfect-Perfect connections in the Ascomycetes, in Kendrick, B., Ed., *Taxonomy of Fungi Imperfecti.* University of Toronto Press, Toronto, 1971, 185.

Nannfeldt, J. A., Studien über die Morphologie und Systematik der nichtlichenisierten inoperculaten Discomyceten, *Nova Acta Regiae Soc. Sci. Upsaliensis,* Ser. IV, 8(2), 1–368, 1932.

Naseath, B., Investigations of food chains in the blind beetle-cave ecosystems of Idaho, *J. Idaho Acad. Sci.,* 16, 77, 1974.

Nealson, K. H., *Bioluminescence: Current Perspectives,* Burgess Pub. Co., Minneapolis, 1981.

Nermut, M. V., Further investigation on the fine structure of influenza virus, *J. Gen. Virol.,* 17, 317, 1972.

Nitsch, J. P. and Nitsch, C., Haploid plants from pollen grains, *Science,* 163, 85, 1969.

Nobel, P. S., Water relations and photosynthesis of a desert CAM plant, *Plant Physiol.,* 58, 576, 1976.

Nordhaus, W. D., An optimal transition path for controlling greenhouse gases, *Science,* 258, 1315, 1992.

Nordhaus, W. D., Expert opinion on climate change, *Am. Sci.,* 82, 45, 1994.

Nylander, W., Les lichens du Jardin du Luxembourg, *Bull. Soc. Bot. Fr.,* 13, 364-372, 1866.

Odum, E. P. and Odum, H. T., *Fundamentals of Ecology,* W. B. Saunders, Philadelphia, 1959.

Oheocha, C., Phycobilins, in Goodwin, T. W., Ed., *Chemistry and Biochemistry of Plant Pigments,* Academic Press, New York, 1965, 175.

Ohki, K. and Gantt, E., Functional phycobilisomes from *Tolypothrix tenuis* (Cyanophyta) grown heterotrophically in the dark, *J. Phycol.,* 19, 359, 1983.

Oltmanns, F., *Morphologie und Biologie der Algen,* Vol. 1, G. Fischer, Jena, 1904.

Oparin, A. I., *The Origin of Life* (English translation by Ann Synge, Academic Press, New York, 1964).

Osborn, T. G. B., The lateral roots of *Amyelon radicans* Will. and their mycorrhiza, *Ann. Bot.,* 23, 603, 1909.

Ovington, J. D., Dry matter production by *Pinus sylvestris, Ann. Bot.,* 21, 287, 1957.

Ovington, J. D., Quantitative ecology and the woodland ecosystem concept, *Adv. Ecol. Res.* 1, 103, 1962.

Ovington, J. D., Organic production, turnover and mineral cycling in woodlands, *Biol. Rev.,* 40, 295, 1965.

Ovington, J. D. and Pearsall, W. H., Production ecology. II. Estimates of average production by trees, *Oikos,* 7, 202, 1957.

Pant, G. and Tewari, S. D., Various human uses of bryophytes in the Kumaun region of northwest Himalaya, *Bryologist,* 92, 120, 1989.

Parker, B. C., Translocation in the giant kelp *Macrocystis,* I. Rates, direction, quantity, C^{14}-labeled products and fluorescein, *J. Phycol.,* 1, 41, 1965.

Parker, J., Photosynthesis of *Picea excelsa* in winter, *Ecology,* 34, 605, 1953.

Parker, P. A., Van Baalen, C., and L. Maurer, Fatty acids in eleven species of blue-green algae: geochemical significance, *Science,* 155, 707-708, 1967.

Pascher, A., *Die Süsswasser-Flora Deutschlands, Österreichs, und der Schweiz,* Vol. 1-12, G. Fischer, Jena, 1913-1932.

Patrick, R. and Reimer, C. W., The Diatoms of the United States. Academy of Natural Sciences of Philadelphia, Monogr. No. 14.

Patterson, G. M. L. and 17 co-authors, Antiviral activity of cultured blue-green algae (Cyanophyceae), *J. Phycol.,* 29, 125, 1993.

Patterson, G. M. L. et al. Antiviral activity of cultured blue-green algae (Cyanophyceae), *J. Phycol.,* 29, 125, 1993.

Pearson, L. C., Genetic segregation in autotetraploid plants, *J. Idaho Acad. Sci.,* 2, 15-25, 1961.

Pearson, L. C., Effect of harvest date on recovery of range grasses and shrubs, *Agron. J.,* 56, 80, 1964.

Pearson, L. C., Primary production in grazed and ungrazed desert communities of eastern Idaho, *Ecology* 46: 1965.

Pearson, L. C., Primary productivity in a northern desert area, *Oikos,* 15, 211, 1966.

Pearson, L. C., *Principles of Agronomy,* Rheinhold, New York, 1967.

Pearson, L. C., Influence of temperature and humidity on distribution of lichens in a Minnesota bog, *Ecology,* 50, 740, 1969.

Pearson, L. C., Varying environmental factors in order to grow intact lichens under laboratory conditions, *Am. J. Bot.,* 57, 659, 1970.

Pearson, L. C., Air pollution and lichen physiology: progress and problems, in B. W. Ferry, M. S. Baddeley, and D. L. Hawksworth (Ed.), *Air Pollution and Lichens,* Athlone Press, London, 1973.

Pearson, L. C., Daily and seasonal patterns of photosynthesis in *Artemisia tridentata, J. Idaho Acad. Sci.,* 11, 11, 1975.

Pearson, L. C., Laboratory growth experiments with lichens based on distribution in nature, *Bryologist,* 80, 318, 1977.

Pearson, L. C., Effects of temperature and moisture on phenology and productivity of Indian ricegrass, *J. Range Manage.,* 32, 127, 1979.

Pearson, L. C., The "three-line hypothesis" and plant classification, *Speculations Sci. Tech.,* 6, 21, 1983.

Pearson, L. C., Air pollution damage to cell membranes in lichens. I. Development of a simple monitoring test, *Atmosph. Environ.,* 19, 209, 1985.

Pearson, L. C., *The Mushroom Manual,* Naturegraph, Happy Camp, CA, 1987.

Pearson, L. C., Evolution annd diversity of plants, *Am. Biol. Teach.,* 50(8), 487, 1988.

Pearson, L. C., Microchemical biomonitoring of air quality with lichens and energy dispersive spectroscopy reveals pattern of atmospheric pollution in a semiarid environment, *XV Intl. Bot. Cong. Abstracts,* Yokohama, 1993, 320.

Pearson, L. C. and Benson, S. E., Laboratory growth experiments with lichens based on distribution in nature, *The Bryologist,* 80, 318–327, 1977.

Pearson, L. C. and Brammer, E., Rate of photosynthesis and respiration in different lichen tissues by the Cartesian diver technique; *Am. J. Bot.,* 65, 276, 1978.

Pearson, L. C. and L. J. Elling 1961. Predicting synthetic varietal performance in alfalfa from clonal cross data. *Crop Science,* 1, 263–266.

Pearson, L. C., Henriksson, E., and Brammer, E., Photosynthesis patterns in lichens following fumigation with sulfur dioxide in laboratory flasks, *XIII Intl. Bot. Cong. Abstracts,* Sydney, Australia, 1981, 82.

Pearson, L. C. and Lawrence, D. B., Photosynthesis in aspen bark during winter months, *Proc. Minn. Acad. Sci.,* 26, 101, 1958.

Pearson, L. C. and Lawrence, D. B. Lichens as microclimate indicators in northwestern Minnesota, *Am. Midl. Nat.,* 74, 257, 1965.

Pearson, L. C. and Rope, S. K., Lichens of the Idaho National Engineering Laboratory, DOE/ID 12110 U. S. Department of Energy, 1987.

Pearson, L. C. and Skye, E., Air pollution affects pattern of photosynthesis in *Parmelia sulcata,* a corticolous lichen, *Science,* 148, 1600, 1965.

Penfound, W. T., Primary production of vascular aquatic plants, *Limnol. and Oceanogr.,* 1, 92, 1956.

Perry, D. R., An arboreal naturalist explores the rain forest's mysterious canopy, *Smithsonian,* 11(3), 43, 1980.

Perry, D. R., The canopy of the tropical rain forest, *Sci. Am.,* 251(5), 138, 1984.

Petersen, R. H., *Evolution in the Higher Basidiomycetes,* University of Tennessee Press, Knoxville, 1971.

Peterson, B., Fry, B., and Deegan, L., The trophic significance of epilithic algal production in a fertilized tundra river ecosystem, *Limnol. Oceanogr.,* 38, 872–878, 1993.

Pia, J., Handbuch der Paläobotanik, M. Hermer, Berlin, 1927.

Pickard, J. and Seppelt, R. D., Phytogeography of Antarctica, *J. Biogeography,* 11, 83, 1984.

Pigg, K. B. and Taylor, T. N., Anatomically preserved *Glossopteris* stems with attached leaves from the central Transantarctic Mountains, Antarctica, *Am. J. Bot.,* 80, 500, 1993.

Playford, R. E. and Cockbain, A. E., Modern algal stromatolites at Hamelin Pool, a hypersaline barred basin in Shark Bay, Western Australia, in *Stromatolites,* Walter, M. R., Ed., Elsevier Scientific Publishing Co., Amsterdam, 1976.

Pitelka, D. R., *Electron Microscope Structure of Protozoa,* Pergamon Press, New York, 1963.

Ponnamperuma, C., Ed., *Chemical Evolution of the Early Precambrian,* Academic Press, New York, 1977.

Pool, R., Illuminating jet lag, *Science,* 244, 1256, 1989.

Pringsheim, E. G., Some aspects of taxonomy in the Cryptophyceae, *New Phytol.,* 43, 143, 1944.

Pringsheim, E. G., Phycology in the field and in the laboratory, *New Phytol.,* 43, 143, 1967.

Puisoux-Deo, S., *Acetabularia and Cell Biology,* Springer Verlag, New York, 1970.

Quay, P. D., Tilbrook, B., and Wong, C. S., Oceanic uptake of fossil fuel CO_2: carbon-13 evidence, *Science,* 256, 74, 1992.

Raff, R. A. and Mahler, H. R., The non symbiotic origin of mitochondria, *Science,* 169, 641, 1972.

Raghavan, V., Germination of fern spores, *Amer. Sci.,* 80, 176, 1992.

Raymont, J. E. G., *Plankton and Productivity of the Sea,* Macmillan, New York, 1963.

Reichle, D. E., *Analysis of Temperate Forest Ecosystems,* Springer-Verlag, Heidelberg, 1970.

Rembert, D. H., Philosophical separation of botany and zoology, *Plant Sci. Bull.,* 18, 22, 1972.

Repetto, R., Deforestation in the tropics, *Sci. Am.,* 262(4), 36, 1990.

Rice, E. R., Allelopathy—an update, *Bot. Rev.,* 45, 15, 1979.

Richerd, S., Destombe, C., Cuguen, J., and Valero, M., Variation of reproductive success in a haplo-diploid red alga, *Gracilaria verrucosa:* effects of parental identities and crossing distance, *Am. J. Bot.,* 80, 1379, 1993.

Ricketts, E. F. and Calvin, J., *Between Pacific Tides,* Stanford University Press, Stanford, CA 1962.

Ridge, R. W. and Sack, F. D., Cortical and cap sedimentation in gravitropic *Equisetum* roots, *Am. J. Bot.,* 79, 328, 1992.

Riley, G. A., The carbon metabolism and photosynthetic efficiency of the earth, *Am. Sci.,* 32, 132, 1944.

Riley, G. A., Phytoplankton of the north central Sargasso Sea, *Limnol. Oceanogr.,* 2, 252, 1957.

Rodin, L. E. and Bazilevich, N. I., World distribution of plant biomass, in *Functioning of terrestrial ecosystems at the primary production level,* Eckhardt, F. E., Ed., Oliver & Boyd, Edinburgh, 1968, 45.

Rodin, R. J., Distribution of *Welwitschia mirabilis, Am. J. Bot.,* 40, 280, 1953.

Rodin, R. J., Leaf anatomy of *Welwitschia.* I. Early development of the leaf; II. a study of mature leaves, *Am. J. Bot.,* 45, 90–103, 1958.

Roe, A., *The Making of a Scientist,* Dodd, Mead, and Co., New York, 1955.

Roe, A., The psychology of the scientist, *Science,* 134, 456, 1961.

Rope, S. K. and Pearson, L. C., Lichens as air pollution biomonitors in a semiarid environment in Idaho, *Bryologist,* 93, 50, 1990.

Rothwell, G. W. and Erwin, D. M., Origin of seed plants: an aneurophyte/seed-fern link elaborated, *Am. J. Bot.,* 740, 970-973, 1987.

Royce, D. J., Allozymes, ribosomal DNA, and breeding in Shiitake (*Lentinula edodes*), *15th Int. Bot. Congr. Abstr.,* 1993, 186.

Round, F. E., *Introduction to the Lower Plants,* Butterworth, London, 1969.

Russell, P. J., *Genetics,* Harper Collins, New York, 1992.

Russell-Hunter, W. D., *Aquatic Productivity,* Macmillan, New York, 1970.

Ryther, J. H., The measurement of primary production, *Limnol. Oceanogr.,* 1, 72, 1956.

Saccardo, P. A., *Sylloge Fungorum Hucusque Cognitorum,* Vol. 14, published by author, Pavia, 1899, 1-1316.

Sachs, J., *Lehrbuch der Botanik: vierte, umgearbeitete Auflage,* Wilhelm Engelmann, Leipzig, 1874.

Safir, G. R., Ed., *Ecophysiology of VA Mycorrhizal Plants,* CRC Press, Boca Raton, FL, 1987.

Sanford, J. H., Japan's laughing mushrooms, *Econ. Bot.,* 26, 174, 1972.

Savile, D. B. O., Fungi as aids in higher plant classification, *Bot. Rev.,* 45, 377, 1979.

Scagel, R. F., Bandoni, R. J., Rouse, G. E., Schofield, W. B., Stein, J. R., and Taylor, T. M. C., *An Evolutionary Survey of the Plant Kingdom,* Wadsworth, Belmont, CA, 1965.

Schmid, R., The terminology and classification of steles: historical perspectives and the outlines of a system, *Bot. Rev.,* 48, 817, 1982.

Schopf, J. W., Microflora of the Bitter Springs Formation, late Precambrian, central Australia, *J. Paleont.,* 42, 651, 1968.

Schopf, J. W., The evolution of the earliest cells, *Sci. Am.,* 239(3), 110, 1978.

Schopf, J. W., Microfossils of the early Archean Apex chert: new evidence of the antiquity of life, *Science,* 260, 640, 1993.

Schopf, J. W. and Barghoorn, E. S., Algal-like fossils from the early Precambrian of South Africa, *Science,* 156, 508, 1967.

Schuster, R. M., *The Hepaticae and Anthocerotae of North America East of the 100th Meridian,* Vol. 1-6, Field Museum of Natural History, Chicago, 1968-1992.

Schürhoff, P. N., Die Plastiden, in Linsbauer, K., Ed., *Handbuch der Pflanzenanatomie,* Vol. 1, 1924, 10.

Schwimmer, M. and Schwimmer D., Medical aspects of phycology, in Jackson, D. F., Ed., *Algae, Man, and the environment,* Syracuse University Press, Syracuse, NY, 1968, 279.

Scott, A. S. and Taylor, T. N., Plant/animal interactions during the upper Carboniferous, *Bot. Rev.,* 49, 259, 1983.

Seaman, W. L., Larson, R. H., and Walker, J. C., Tinsel-type structure in flagella of zoospores associated with cabbage clubroot tissue, *Nature,* 190, 186, 1961.

Setchell, W. A., Limestone reefs, *Proc. Third Pan Pacific Congr.,* 2, 1837, 1928.

Sharkey, T. O., Photosynthesis in intact leaves of C_c plants: physics, physiology, and rate limitations, *Bot. Rev.,* 51, 53, 1985.

Shubert, L. E., Ed., *Algae as Ecological Monitors,* Academic Press, New York, 1984.

Silva, P. C., Classification of algae, in Lewin, R. A., Ed., *Physiology and Biochemistry of Algae,* Academic Press, New York, 1962, 827-833.

Silver, W. S., Biological nitrogen fixation, *Science,* 157, 100, 1967.

Simmonds, N. W., *Evolution of Crop Plants,* Longman Group, London, 1976.

Simonis, W. and Urbach, W., Photophosphorylation in vivo, *Annu. Rev. Plant Physiol.,* 1973.

Singer, R., *The Agaricales (mushrooms) in Modern Taxonomy,* Liloa Revista de Botanic, Tucunan, Argentina, 1949.

Singh, J. H. and Yadara, P. S., Seasonal variation in composition, plant biomass, and net primary productivity of a tropical grassland at kurukshetra, India, *Ecol. Monogr.,* 44, 317, 1974.

Singh, R. N., Role of Blue-green Algae in Nitrogen Economy of Indian Agriculture, *Indian Council Agric. Res.,* 1962.

Skottsberg, C., Västernas Liv, *Nordisk Familjeboks Forlag,* Stockholm, 1940.

Smiley, xx and Huggins, xx, *Pseudofagus idahoensis,* n. gen. et sp. (Fagaceae) from the Miocene Clarkia flora of Idaho, *Am. J. Bot.,* 68, 741, 19–.

Smith, A. H., *The Mushroom Hunter's Field Guide,* University of Michigan Press, Ann Arbor, 1958.

Smith, A. H., *A Field Guide to Western Mushrooms,* University of Michigan Press, Ann Arbor, 1975.

Smith, C. W., Bryophytes and lichens of the Puhimau Geothermal Area, Hawaii Volcanoes National Park, *The Bryologist,* 84, 457, 1981.

Smith, G. M., *Cryptogamic botany.* Vol. 1, *Algae and Fungi,* McGraw-Hill, New York, 1938.

Smith, G. M., *Marine Algae of the Monterey Peninsula of California,* Stanford Press, Stanford, CA, 1944.

Smith, G. M., *Freshwater Algae of the United States,* McGraw Hill, New York, 1950.

Spanos, N. P. and Gottlieb, J., Ergotism and the Salem Village witch trials, *Science,* 194, 1390, 1976.

Stace, C. A., *Plant Taxonomy and Biosystematics,* University Park Press, Baltimore, 1980.

Stackbrandt, E., Phylogenetic considerations of Prochloron, in Lewin, R. A. and Cheng, L., Eds., *Prochloron: a Microbial Enigma,* Chapman & Hall, New York, 1989.

Staley, J. T., Bryant, M. P., and Pfennig, N., Eds., *Bergey's Manual of Systematic Bacteriology, Vol. 3, Bacteria Classification,* Williams and Wilkins, Baltimore, 1984.

Stebbins, G. L., *Variation and Evolution in Plants,* Columbia University Press, New York, 1950.

Stebbins, G. L., *Processes of Organic Evolution,* Prentice-Hall, Englewood Cliffs, NJ, 1966.

Steidinger, K. A. and Haddad, K., Biologic and hydrographic aspects of red tides, *BioScience,* 31, 814–819, 1981.

Stevenson, S. J. and Clark, G. F., Breeding and genetics in potato improvement, *U.S.D.A. Yearbook of Agric., Better Plants and Animals,* II, 405, 1937.

Stewart, K. D. and Mattox, K. R., Comparative cytology, evolution and classification of the green algae with some consideration of the origin of other organisms with chlorophylls *a* and *b, Bot. Rev.,* 41, 104, 1975.

Stockwell, C. H., A time classification of Precambrian rocks and events, *Geol. Survey Can.,* Paper 80–19, Part 1, Ottawa, Canada, 1982.

Stoner, W. A., Miller, P., and Miller, P. C., Seasonal dynamics of standing crops of biomass and nutrients in a subarctic tundra vegetation, *Holarctic Ecology,* 5, 172, 1982.

Stott, P., *Historical Plant Geography,* George Allen and Unwin, London, 1981.

Strain, B. R. and Thomas, R. B., Response of coniferous forests to Co_2 increase and possible global climate change, *15th Int. Bot. Congr. Abstr.,* 1993, 61.

Strain, H. H., *Chloroplast Pigments and Chromatographic Analysis,* Pennsylvania State University (Phi Lambda Upsilon), University Park, PA, 1958.

Summerhays, S., A marine park is born, *Natl. Geogr.,* 159(5), 630, 1981.

Takhtajan, A., *Flowering plants, Origin and Dispersal,* Smithsonian Institution, Washington, 1961 (English translation, 1969).

Talley, S. N. and Rains, D. W., *Azolla filiculoides* Lam. as a fallow-season green manure for rice in a temperate climate, *Agron. Jour.,* 72, 11–18, 1980.

Tamarin, R., *Principles of Genetics,* Wm. C. Brown, Dubuque, IA, 1991.

Taylor, F. J. R., Factors contributing to the increase in number and scope of harmful algal blooms, *15th Int. Bot. Congr. Abstr.,* 1993, 65.

Taylor, R. and Taylor, V. Paradise beneath the sea, *Natl. Geogr.,* 159(5), 636, 1981.

Taylor, W. R., Phaeophycean life-histories in relation to classification, *Bot. Rev.,* 2, 554, 1936.

Taylor, W. R., *Marine algae of the Northeastern Coast of North America,* University of Michigan Press, Ann Arbor, 1957.

Taylor, W. R., *Marine Algae of the Eastern Tropical and Subtropical Coasts of the Americas.* University of Michigan Press, Ann Arbor, 1960.

Thieret, J. W., Economic botany of the cycads, *Econ. Bot.,* 58, 3, 1957.

Thies, P. W., The modification of natural substances in the modern drug synthesis, in Wagner H. and Wolff, P., Eds., *New Natural Products and Plant Drugs with Pharmacological, Biological, or Therapeutic Activity,* Springer-Verlag, Berlin, 1977.

Thinh, L. V., Photosynthetic lamellae of *Prochloron* (Prochlorophyta) associated with the Ascidian *Diplosoma virens* (Hartmeyer) in the vicinity of Townsville, *Austr. J. Bot.,* 26, 617, 1978.

Thom, R. M., Parkwell, T. L., Niyogi, D. K., and Shreffler, D. K., Effects of graveling on the primary productivity, respiration and nutrient flux of two estuarine tidal flats, *Marine Biology,* 118, 329–341, 1994.

Thomas, B. A. and Spicer, R.A., *The Evolution and Palaeo biology of Land Plants,* Dioscorides Press, Portland, OR, 1986.

Thomas, E. A., Ober die Biologie von Flechtenbildnern, *Beitr. Kryptogamenflora der Schweiz,* 9, H.1, 1939.

Thorne, D. W. and Peterson, H. B., *Irrigated Soils,* Blakiston, New York, 1954.

Tippo, O., A modern classification of the plant kingdom, *Chron. Bot.,* 7, 203, 1942.

Trappe, J. M., Phylogenetic and ecologic aspects of mycorrhizae in the angiosperms from an evolutionary standpoint, in *Ecophysiology of VA mycorrhizal plants,* Safir, G. R., Ed., CRC Press, Boca Raton, FL, 1987, 5.

Tucker, S. C., The developmental basis for sexual expression in *Ceratonia siliqua* (Leguminosae: Caesalpinioideae: Cassieae), *Am. J. Bot.,* 79, 318, 1992.

Turesson, G., The genotypical response of the plant species to habitat and climate, *Hereditas* 6, 141, 1925.

Twenhofel, W. H., Ordovician algae, *Bull. Am. Assoc. Petr. Geol.,* 34, 182, 1950.

Tylutki, E. E., *Mushrooms of Idaho and the Pacific Northwest,* Vol. 1–2, University of Idaho Press, Moscow, ID, 1979–1987.

Tyron, R. M., Jr., Fassett, N. C., Dunlop, D. W., and Diemer, M. E., *The Ferns and Fern Allies of Wisconsin,* University of Wisconsin Press, Madison, 1953.

Van Blaricom, G. R. and Estes, J. A., Eds., *The community ecology of sea otters,* Springer-Verlag, New York, 1988.

Vance, B. D., Composition and succession of cyanophycean water blooms, *J. Phycol.,* 1, 81, 1965.

Vasseur, L., Aarssen, L. W., and Bennett, T. Allozymic variation in local apomictic populations of *Lemna minor* (Lemnaceae), *Am. J. Bot.,* 80, 953, 1993.

Vickery, R. K., Jr. and Wullstein, B. M., Comparison of six approaches to the classification of *Mimulus* sect. *Erythranthe* (Scrophulariaceae), *System. Bot.,* 12, 339, 1987.

Vinyard, W. C., *Diatoms of North America,* Mad River Press, Eureka, CA, 1979.

von Arx, J. A., Die Artung der Gattung Colletotrichum Corda, *Phytopath. Zeitschr.,* 29, 413–468, 1957.

Wagner, H. and Wolff, P., Eds., *New Natural Products and Plant Drugs with Pharmacological, Biological, or Therapeutical activity,* Springer-Verlag, Berlin, 1977.

Waksman, S. A., *Soil Microbiology.* Wiley & Sons, New York, 1952.

Walker, C. G., *Origin of the Atmosphere: History of the*

Walker, J. C., *Plant Pathology,* McGraw-Hill, New York, 1950.

Walker, J. C. G., Was the Archaean biosphere upside down? *Nature,* 329, 710–712, 1987.

Walter, H., *Vegetation of the earth,* English Universities Press, London, 1973.

Walter, M. R., Ed., *Stromatolites,* Elsevier Scientific Publishing Co., Amsterdam, 1976.

Walton, J., Improvements in the peel method of preparing sections of fossil plants, *Nature,* 125, 413, 1930.

Washino, Mosquitoes: by-product of rice culture, *California Agriculture,* 34, No. 3, 1980.

Weber, D. J. and Hess, W. M., *The Fungal Spore,* Wiley, New York, 1976.

Weber, J. M., Cai, F., Murali, R., and Burnett, R. M., Sequence and structural analysis of murine adenovirus type-1 hexon, *J. Gen. Virol.,* 75, 141, 1994.

Webster, G. L., At play in the fields of botany (or, Dilemmas of the green cornucopia), *Plant Sci. Bull.,* 39, 8, 1993.

Wheeler, W. A., *Forage and Pasture Crops.* D. Van Nostrand and Company, New York, 1950.

Whittaker, R. H., New concepts of kingdoms of organisms, *Science,* 163, 150, 1969.

Whittaker, R. H., *Communities and Ecosystems,* Macmillan, New York, 1970.

Whittaker, R. H. and Woodwell, G. M., Structure, Production, and diversity the oak-pine forest at Brookhaven, New York, *J. Appl. Ecol.,* 57, 155, 1957.

Wimpenny, R. S., The size of diatoms, *J. Marine Biol. Assoc.,* 21, 29–60, 1936.

Winter, G., Über die Assimilation des Luftstickstoffs durch endophytische Blaualgen, *Beitr. Biol. Pfl.,* 23, 295, 1935.

Wodehouse, R. P., *Pollen Grains,* Hafner, New York, 1959.

Woese, C., Archaebacteria, *Sci. Am.,* 244(6), 98, 1981.

Woese, C. R., *The Genetic Code,* Harper and Row, New York, 1985.

Wolfe, A. D. and Estes, J. R., Pollination and the function of floral parts in *Chamaecrista fasciculata* (Fabaceae), *Am. J. Bot.,* 79, 314, 1992.

Wolken, J. J., *Euglena, An Experimental Organism for Biochemical and Biophysical Studies,* Appleton-Century-Croft, New York, 1967.

Wood, D. M. and Morris, W. F., Ecological constraints to seedling establishment on the Pumice Plains, Mount St. Helens, Washington, *Am. J. Bot.,* 77, 1411, 1990.

Woodcock, C. L. F., Ed., *Progress in Acetabularia Research,* Academic Press, New York, 1977.

Woodwell, G. M., The energy cycle of the biosphere, *Sci. Am.,* 223(3), 64, 1970.

Woodwell, G. M. and Whittaker, R. H., Primary production and the cation budget of the Brookhaven Forest, in Young, H. E., Ed., *Symposium on Primary Productivity and Mineral Cycling in Natural Ecosystems,* University of Maine Press, Orono, 1968, 151.

Xia, B. and Abbott, D., Edible seaweeds and their place in the chinese diet, *Econ. Bot.,* 88(6), 12, 1987.

Yang, H. Y. and Zhou, C., Experimental plant reproductive biology and reproductive cell manipulation in higher plants: now and in the future, *Am. J. Bot.,* 79, 354, 1992.

Yoneda, T. and Ogino, K., Productivity and decomposition of rain forests in west Sumatra, Indonesia, *15th Int. Bot. Congr.,* 1993, 55.

Young, A. T., Seaweed culture and uses, U.S. Dept. Agric. A17.18/4:93–63.

Young, E. L., III, Studies on *Labyrinthula,* the etiologic agent of the wating disease of eelgrass, *Am. J. Bot.,* 30, 586, 1943.

Zhou, J. and Fritz, L., The PAS/accumulation bodies in /itProrocentrum lima and *Prorocentrum maculosum* (Dinophyceae) are dinoflagellate lysosomes, *J. Phycol.*, 30, 39, 1994.

Zingmark, R. G., Ultrastructural studies on two kinds of mesokaryotic nuclei, *Am. J. Bot.*, 57, 586, 1970.

Ziska, L. H., Teramura, A. H., and Sullivan, J. H., Physiological sensitivity of plants along an elevational gradient to UV-B radiation, *Am. J. Bot.*, 79, 863, 1992.

INDEX

A

C

D

F

G

L

M

N

O

P

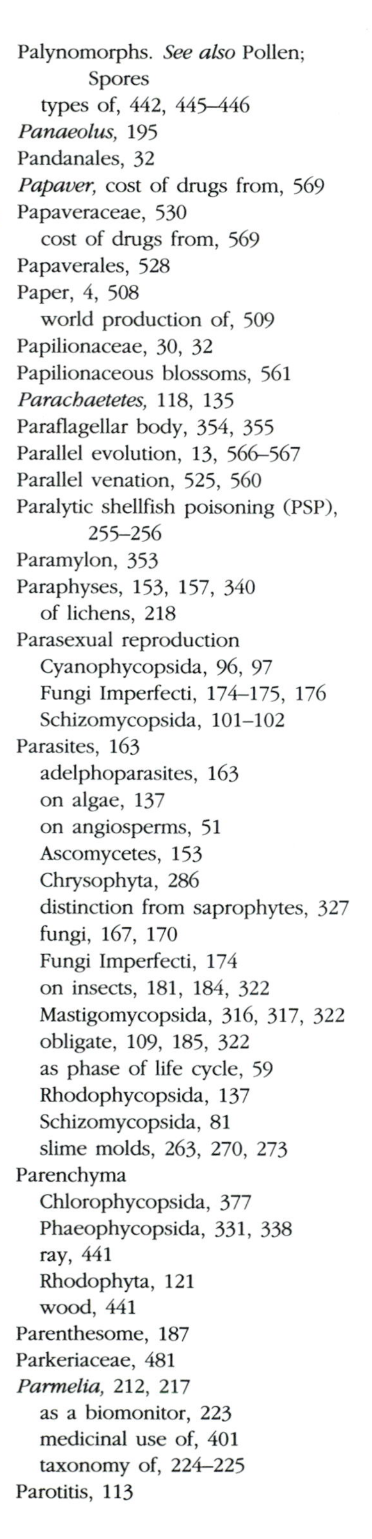

Q

R

S

T

U

V